Signals and Systems
Analysis Using Transform Methods and MATLAB®

Third Edition

Michael J. Roberts
Professor Emeritus, Department of Electrical and Computer Engineering University of Tennessee

SIGNALS AND SYSTEMS: ANALYSIS USING TRANSFORM METHODS AND MATLAB®, THIRD EDITION

Published by McGraw-Hill Education, 2 Penn Plaza, New York, NY 10121. Copyright © 2018 by McGraw-Hill Education. All rights reserved. Printed in the United States of America. Previous editions © 2012, and 2004. No part of this publication may be reproduced or distributed in any form or by any means, or stored in a database or retrieval system, without the prior written consent of McGraw-Hill Education, including, but not limited to, in any network or other electronic storage or transmission, or broadcast for distance learning.

Some ancillaries, including electronic and print components, may not be available to customers outside the United States.

This book is printed on acid-free paper.

1 2 3 4 5 6 QVS 21 20 19 18 17

ISBN 978-0-07-802812-0
MHID 0-07-802812-4

Chief Product Officer, SVP Products & Markets: *G. Scott Virkler*
Vice President, General Manager, Products & Markets: *Marty Lange*
Vice President, Content Design & Delivery: *Betsy Whalen*
Managing Director: *Thomas Timp*
Brand Manager: *Raghothaman Srinivasan/Thomas Scaife, Ph.D.*
Director, Product Development: *Rose Koos*
Product Developer: *Christine Bower*
Marketing Manager: *Shannon O'Donnell*
Director of Digital Content: *Chelsea Haupt, Ph.D.*
Director, Content Design & Delivery: *Linda Avenarius*
Program Manager: *Lora Neyens*
Content Project Managers: *Jeni McAtee; Emily Windelborn; Sandy Schnee*
Buyer: *Jennifer Pickel*
Content Licensing Specialists: *Carrie Burger, photo; Lorraine Buczek, text*
Cover Image: © *Lauree Feldman/Getty Images*
Compositor: *MPS Limited*
Printer: *Quad/Graphics*

All credits appearing on page or at the end of the book are considered to be an extension of the copyright page.

Library of Congress Cataloging-in-Publication Data

Roberts, Michael J., Dr.
  Signals and systems : analysis using transform methods and MATLAB /
  Michael J. Roberts, professor, Department of Electrical and Computer
  Engineering, University of Tennessee.
  Third edition. | New York, NY : McGraw-Hill Education, [2018] |
  Includes bibliographical references (p. 786–787) and index.
  LCCN 2016043890 | ISBN 9780078028120 (alk. paper)
  LCSH: Signal processing. | System analysis. | MATLAB.
  LCC TK5102.9 .R63 2018 | DDC 621.382/2—dc23 LC record
  available at https://lccn.loc.gov/2016043890

The Internet addresses listed in the text were accurate at the time of publication. The inclusion of a website does not indicate an endorsement by the authors or McGraw-Hill Education, and McGraw-Hill Education does not guarantee the accuracy of the information presented at these sites.

mheducation.com/highered

To my wife Barbara for giving me the time and space to complete this effort and to the memory of my parents, Bertie Ellen Pinkerton and Jesse Watts Roberts, for their early emphasis on the importance of education.

CONTENTS

Preface, xii

Chapter 1
Introduction, 1

1.1 Signals and Systems Defined, 1
1.2 Types of Signals, 3
1.3 Examples of Systems, 8
   A Mechanical System, 9
   A Fluid System, 9
   A Discrete-Time System, 11
   Feedback Systems, 12
1.4 A Familiar Signal and System Example, 14
1.5 Use of MATLAB®, 18

Chapter 2
Mathematical Description of Continuous-Time Signals, 19

2.1 Introduction and Goals, 19
2.2 Functional Notation, 20
2.3 Continuous-Time Signal Functions, 20
   Complex Exponentials and Sinusoids, 21
   Functions with Discontinuities, 23
      The Signum Function, 24
      The Unit-Step Function, 24
      The Unit-Ramp Function, 26
      The Unit Impulse, 27
         The Impulse, the Unit Step, and Generalized Derivatives, 29
         The Equivalence Property of the Impulse, 30
         The Sampling Property of the Impulse, 31
         The Scaling Property of the Impulse, 31
      The Unit Periodic Impulse or Impulse Train, 32
   A Coordinated Notation for Singularity Functions, 33
   The Unit-Rectangle Function, 33
2.4 Combinations of Functions, 34
2.5 Shifting and Scaling, 36
   Amplitude Scaling, 36
   Time Shifting, 37
   Time Scaling, 39
   Simultaneous Shifting and Scaling, 43
2.6 Differentiation and Integration, 47
2.7 Even and Odd Signals, 49
   Combinations of Even and Odd Signals, 51
   Derivatives and Integrals of Even and Odd Signals, 53
2.8 Periodic Signals, 53
2.9 Signal Energy and Power, 56
   Signal Energy, 56
   Signal Power, 58
2.10 Summary of Important Points, 60
Exercises, 61
   Exercises with Answers, 61
      Signal Functions, 61
      Shifting and Scaling, 62
      Derivatives and Integrals of Functions, 66
      Generalized Derivative, 67
      Even and Odd Functions, 67
      Periodic Signals, 69
      Signal Energy and Power of Signals, 70
   Exercises without Answers, 71
      Signal Functions, 71
      Scaling and Shifting, 71
      Generalized Derivative, 76
      Derivatives and Integrals of Functions, 76
      Even and Odd Functions, 76
      Periodic Functions, 77
      Signal Energy and Power of Signals, 77

Chapter 3
Discrete-Time Signal Description, 79

3.1 Introduction and Goals, 79
3.2 Sampling and Discrete Time, 80
3.3 Sinusoids and Exponentials, 82
   Sinusoids, 82
   Exponentials, 85
3.4 Singularity Functions, 86
   The Unit-Impulse Function, 86
   The Unit-Sequence Function, 87

The Signum Function, 87
  The Unit-Ramp Function, 88
  The Unit Periodic Impulse Function or Impulse Train, 88
3.5 Shifting and Scaling, 89
  Amplitude Scaling, 89
  Time Shifting, 89
  Time Scaling, 89
    Time Compression, 90
    Time Expansion, 90
3.6 Differencing and Accumulation, 94
3.7 Even and Odd Signals, 98
  Combinations of Even and Odd Signals, 100
  Symmetrical Finite Summation of Even and Odd Signals, 100
3.8 Periodic Signals, 101
3.9 Signal Energy and Power, 102
  Signal Energy, 102
  Signal Power, 103
3.10 Summary of Important Points, 105
Exercises, 105
  Exercises with Answers, 105
    Functions, 105
    Scaling and Shifting Functions, 107
    Differencing and Accumulation, 109
    Even and Odd Functions, 110
    Periodic Functions, 111
    Signal Energy and Power, 112
  Exercises without Answers, 113
    Signal Functions, 113
    Shifting and Scaling Functions, 113
    Differencing and Accumulation, 114
    Even and Odd Functions, 114
    Periodic Signals, 115
    Signal Energy and Power, 116

Chapter 4
Description of Systems, 118

4.1 Introduction and Goals, 118
4.2 Continuous-Time Systems, 119
  System Modeling, 119
    Differential Equations, 120
    Block Diagrams, 124
  System Properties, 127
    Introductory Example, 127
    Homogeneity, 131
    Time Invariance, 132
    Additivity, 133
    Linearity and Superposition, 134
    LTI Systems, 134
    Stability, 138
    Causality, 139
    Memory, 139
    Static Nonlinearity, 140
    Invertibility, 142
  Dynamics of Second-Order Systems, 143
  Complex Sinusoid Excitation, 145
4.3 Discrete-Time Systems, 145
  System Modeling, 145
    Block Diagrams, 145
    Difference Equations, 146
  System Properties, 152
4.4 Summary of Important Points, 155
Exercises, 156
  Exercises with Answers, 156
    System Models, 156
    Block Diagrams, 157
    System Properties, 158
  Exercises without Answers, 160
    System Models, 160
    System Properties, 162

Chapter 5
Time-Domain System Analysis, 164

5.1 Introduction and Goals, 164
5.2 Continuous Time, 164
  Impulse Response, 164
  Continuous-Time Convolution, 169
    Derivation, 169
    Graphical and Analytical Examples of Convolution, 173
    Convolution Properties, 178
    System Connections, 181
    Step Response and Impulse Response, 181
    Stability and Impulse Response, 181
    Complex Exponential Excitation and the Transfer Function, 182
    Frequency Response, 184
5.3 Discrete Time, 186
  Impulse Response, 186
  Discrete-Time Convolution, 189

Derivation, 189
Graphical and Analytical Examples of Convolution, 192
Convolution Properties, 196
Numerical Convolution, 196
  Discrete-Time Numerical Convolution, 196
  Continuous-Time Numerical Convolution, 198
Stability and Impulse Response, 200
System Connections, 200
Unit-Sequence Response and Impulse Response, 201
Complex Exponential Excitation and the Transfer Function, 203
Frequency Response, 204

5.4 Summary of Important Points, 207
Exercises, 207
  Exercises with Answers, 207
    Continuous Time, 207
    Impulse Response, 207
    Convolution, 209
    Stability, 213
    Frequency Response, 214
    Discrete Time, 214
    Impulse Response, 214
    Convolution, 215
    Stability, 219
  Exercises without Answers, 221
    Continuous Time, 221
    Impulse Response, 221
    Convolution, 222
    Stability, 224
    Discrete Time, 225
    Impulse Response, 225
    Convolution, 225
    Stability, 228

Chapter 6
Continuous-Time Fourier Methods, 229

6.1 Introduction and Goals, 229
6.2 The Continuous-Time Fourier Series, 230
  Conceptual Basis, 230
  Orthogonality and the Harmonic Function, 234
  The Compact Trigonometric Fourier Series, 237
  Convergence, 239
    Continuous Signals, 239
    Discontinuous Signals, 240
  Minimum Error of Fourier-Series Partial Sums, 242
  The Fourier Series of Even and Odd Periodic Functions, 243
  Fourier-Series Tables and Properties, 244
  Numerical Computation of the Fourier Series, 248
6.3 The Continuous-Time Fourier Transform, 255
  Extending the Fourier Series to Aperiodic Signals, 255
  The Generalized Fourier Transform, 260
  Fourier Transform Properties, 265
  Numerical Computation of the Fourier Transform, 273
6.4 Summary of Important Points, 281
Exercises, 281
  Exercises with Answers, 281
    Fourier Series, 281
    Orthogonality, 282
    Forward and Inverse Fourier Transforms, 286
    Relation of CTFS to CTFT, 293
    Numerical CTFT, 294
    System Response, 294
  Exercises without Answers, 294
    Fourier Series, 294
    Forward and Inverse Fourier Transforms, 300
    System Response, 305
    Relation of CTFS to CTFT, 306

Chapter 7
Discrete-Time Fourier Methods, 307

7.1 Introduction and Goals, 307
7.2 The Discrete-Time Fourier Series and the Discrete Fourier Transform, 307
  Linearity and Complex-Exponential Excitation, 307
  Orthogonality and the Harmonic Function, 311
  Discrete Fourier Transform Properties, 315
  The Fast Fourier Transform, 321
7.3 The Discrete-Time Fourier Transform, 323
  Extending the Discrete Fourier Transform to Aperiodic Signals, 323
  Derivation and Definition, 324
  The Generalized DTFT, 326
  Convergence of the Discrete-Time Fourier Transform, 327
  DTFT Properties, 327

Numerical Computation of the Discrete-Time Fourier Transform, 334
7.4 Fourier Method Comparisons, 340
7.5 Summary of Important Points, 341
Exercises, 342
   Exercises with Answers, 342
      Orthogonality, 342
      Discrete Fourier Transform, 342
      Discrete-Time Fourier Transform Definition, 344
      Forward and Inverse Discrete-Time Fourier Transforms, 345
   Exercises without Answers, 348
      Discrete Fourier Transform, 348
      Forward and Inverse Discrete-Time Fourier Transforms, 352

Chapter 8
The Laplace Transform, 354

8.1 Introduction and Goals, 354
8.2 Development of the Laplace Transform, 355
   Generalizing the Fourier Transform, 355
   Complex Exponential Excitation and Response, 357
8.3 The Transfer Function, 358
8.4 Cascade-Connected Systems, 358
8.5 Direct Form II Realization, 359
8.6 The Inverse Laplace Transform, 360
8.7 Existence of the Laplace Transform, 360
   Time-Limited Signals, 361
   Right- and Left-Sided Signals, 361
8.8 Laplace-Transform Pairs, 362
8.9 Partial-Fraction Expansion, 367
8.10 Laplace-Transform Properties, 377
8.11 The Unilateral Laplace Transform, 379
   Definition, 379
   Properties Unique to the Unilateral Laplace Transform, 381
   Solution of Differential Equations with Initial Conditions, 383
8.12 Pole-Zero Diagrams and Frequency Response, 385
8.13 MATLAB System Objects, 393
8.14 Summary of Important Points, 395

Exercises, 395
   Exercises with Answers, 395
      Laplace-Transform Definition, 395
      Direct Form II System Realization, 396
      Forward and Inverse Laplace Transforms, 396
      Unilateral Laplace-Transform Integral, 399
      Solving Differential Equations, 399
   Exercises without Answers, 400
      Region of Convergence, 400
      Existence of the Laplace Transform, 400
      Direct Form II System Realization, 400
      Forward and Inverse Laplace Transforms, 401
      Solution of Differential Equations, 403
      Pole-Zero Diagrams and Frequency Response, 403

Chapter 9
The z Transform, 406

9.1 Introduction and Goals, 406
9.2 Generalizing the Discrete-Time Fourier Transform, 407
9.3 Complex Exponential Excitation and Response, 408
9.4 The Transfer Function, 408
9.5 Cascade-Connected Systems, 408
9.6 Direct Form II System Realization, 409
9.7 The Inverse z Transform, 410
9.8 Existence of the z Transform, 410
   Time-Limited Signals, 410
   Right- and Left-Sided Signals, 411
9.9 z-Transform Pairs, 413
9.10 z-Transform Properties, 416
9.11 Inverse z-Transform Methods, 417
   Synthetic Division, 417
   Partial-Fraction Expansion, 418
   Examples of Forward and Inverse z Transforms, 418
9.12 The Unilateral z Transform, 423
   Properties Unique to the Unilateral z Transform, 423
   Solution of Difference Equations, 424
9.13 Pole-Zero Diagrams and Frequency Response, 425
9.14 MATLAB System Objects, 428
   In MATLAB, 429
9.15 Transform Method Comparisons, 430
9.16 Summary of Important Points, 434

Exercises, 435
  Exercises with Answers, 435
    Direct-Form II System Realization, 435
    Existence of the z Transform, 435
    Forward and Inverse z Transforms, 435
    Unilateral z-Transform Properties, 438
    Solution of Difference Equations, 438
    Pole-Zero Diagrams and Frequency Response, 439
  Exercises without Answers, 441
    Direct Form II System Realization, 441
    Existence of the z Transform, 441
    Forward and Inverse z-Transforms, 441
    Pole-Zero Diagrams and Frequency Response, 443

Chapter 10
Sampling and Signal Processing, 446

10.1 Introduction and Goals, 446
10.2 Continuous-Time Sampling, 447
  Sampling Methods, 447
  The Sampling Theorem, 449
    Qualitative Concepts, 449
    Sampling Theorem Derivation, 451
  Aliasing, 454
  Time-limited and Bandlimited Signals, 457
  Interpolation, 458
    Ideal Interpolation, 458
    Practical Interpolation, 459
      Zero-Order Hold, 460
      First-Order Hold, 460
  Sampling Bandpass Signals, 461
  Sampling a Sinusoid, 464
  Bandlimited Periodic Signals, 467
  Signal Processing Using the DFT, 470
    CTFT-DFT Relationship, 470
      CTFT-DTFT Relationship, 471
      Sampling and Periodic-Repetition Relationship, 474
    Computing the CTFS Harmonic Function
      with the DFT, 478
    Approximating the CTFT with the DFT, 478
      Forward CTFT, 478
      Inverse CTFT, 479
    Approximating the DTFT with the DFT, 479
    Approximating Continuous-Time Convolution
      with the DFT, 479
      Aperiodic Convolution, 479
      Periodic Convolution, 479
    Discrete-Time Convolution with the DFT, 479
      Aperiodic Convolution, 479
      Periodic Convolution, 479
    Summary of Signal Processing Using
      the DFT, 480
10.3 Discrete-Time Sampling, 481
  Periodic-Impulse Sampling, 481
  Interpolation, 483
10.4 Summary of Important Points, 486
Exercises, 487
  Exercises with Answers, 487
    Pulse Amplitude Modulation, 487
    Sampling, 487
    Impulse Sampling, 489
    Nyquist Rates, 491
    Time-Limited and Bandlimited Signals, 492
    Interpolation, 493
    Aliasing, 495
    Bandlimited Periodic Signals, 495
    CTFT-CTFS-DFT Relationships, 495
    Windows, 497
    DFT, 497
  Exercises without Answers, 500
    Sampling, 500
    Impulse Sampling, 502
    Nyquist Rates, 504
    Aliasing, 505
    Practical Sampling, 505
    Bandlimited Periodic Signals, 505
    DFT, 506
    Discrete-Time Sampling, 508

Chapter 11
Frequency Response Analysis, 509

11.1 Introduction and Goals, 509
11.2 Frequency Response, 509
11.3 Continuous-Time Filters, 510
  Examples of Filters, 510
  Ideal Filters, 515
    Distortion, 515
    Filter Classifications, 516
    Ideal Filter Frequency Responses, 516
    Impulse Responses and Causality, 517
    The Power Spectrum, 520
    Noise Removal, 520
  Bode Diagrams, 521

The Decibel, 521
   The One-Real-Pole System, 525
   The One-Real-Zero System, 526
   Integrators and Differentiators, 527
   Frequency-Independent Gain, 527
   Complex Pole and Zero Pairs, 530
Practical Filters, 532
   Passive Filters, 532
      The Lowpass Filter, 532
      The Bandpass Filter, 535
   Active Filters, 536
      Operational Amplifiers, 537
      The Integrator, 538
      The Lowpass Filter, 538

11.4 Discrete-Time Filters, 546
  Notation, 546
  Ideal Filters, 547
    Distortion, 547
    Filter Classifications, 548
    Frequency Responses, 548
    Impulse Responses and Causality, 548
    Filtering Images, 549
  Practical Filters, 554
    Comparison with Continuous-Time Filters, 554
    Highpass, Bandpass, and Bandstop Filters, 556
    The Moving Average Filter, 560
    The Almost Ideal Lowpass Filter, 564
    Advantages Compared to Continuous-Time Filters, 566

11.5 Summary of Important Points, 566
Exercises, 567
  Exercises with Answers, 567
    Continuous-Time Frequency Response, 567
    Continuous-Time Ideal Filters, 567
    Continuous-Time Causality, 567
    Logarithmic Graphs, Bode Diagrams, and Decibels, 568
    Continuous-Time Practical Passive Filters, 570
    Continuous-Time Practical Active Filters, 574
    Discrete-Time Frequency Response, 575
    Discrete-Time Ideal Filters, 576
    Discrete-Time Causality, 576
    Discrete-Time Practical Filters, 577
  Exercises without Answers, 579
    Continuous-Time Frequency Response, 579
    Continuous-Time Ideal Filters, 579
    Continuous-Time Causality, 579
    Bode Diagrams, 580
    Continuous-Time Practical Passive Filters, 580
    Continuous-Time Filters, 582
    Continuous-Time Practical Active Filters, 582
    Discrete-Time Causality, 586
    Discrete-Time Filters, 587

Chapter 12
Laplace System Analysis, 592

12.1 Introduction and Goals, 592
12.2 System Representations, 592
12.3 System Stability, 596
12.4 System Connections, 599
  Cascade and Parallel Connections, 599
  The Feedback Connection, 599
    Terminology and Basic Relationships, 599
    Feedback Effects on Stability, 600
    Beneficial Effects of Feedback, 601
    Instability Caused by Feedback, 604
    Stable Oscillation Using Feedback, 608
    The Root-Locus Method, 612
    Tracking Errors in Unity-Gain Feedback Systems, 618

12.5 System Analysis Using MATLAB, 621
12.6 System Responses to Standard Signals, 623
  Unit-Step Response, 624
  Sinusoid Response, 627

12.7 Standard Realizations of Systems, 630
  Cascade Realization, 630
  Parallel Realization, 632

12.8 Summary of Important Points, 632
Exercises, 633
  Exercises with Answers, 633
    Transfer Functions, 633
    Stability, 634
    Parallel, Cascade, and Feedback Connections, 635
    Root Locus, 637
    Tracking Errors in Unity-Gain Feedback Systems, 639
    System Responses to Standard Signals, 640
    System Realization, 641
  Exercises without Answers, 642
    Stability, 642
    Transfer Functions, 642
    Stability, 643

Parallel, Cascade, and Feedback Connections, 643
Root Locus, 646
Tracking Errors in Unity-Gain Feedback Systems, 647
Response to Standard Signals, 647
System Realization, 649

Chapter 13
z-Transform System Analysis, 650

13.1 Introduction and Goals, 650
13.2 System Models, 650
Difference Equations, 650
Block Diagrams, 651
13.3 System Stability, 651
13.4 System Connections, 652
13.5 System Responses to Standard Signals, 654
Unit-Sequence Response, 654
Response to a Causal Sinusoid, 657
13.6 Simulating Continuous-Time Systems with Discrete-Time Systems, 660
z-Transform-Laplace-Transform Relationships, 660
Impulse Invariance, 662
Sampled-Data Systems, 664
13.7 Standard Realizations of Systems, 670
Cascade Realization, 670
Parallel Realization, 670
13.8 Summary of Important Points, 671
Exercises, 672
  Exercises with Answers, 672
    Stability, 672
    Parallel, Cascade, and Feedback Connections, 672
    Response to Standard Signals, 673
    Root Locus, 674
    Laplace-Transform-z-Transform Relationship, 675
    Sampled-Data Systems, 675
    System Realization, 676
  Exercises without Answers, 677
    Stability, 677
    Root Locus, 677
    Parallel, Cascade, and Feedback Connections, 677
    Response to Standard Signals, 677
    Laplace-Transform-z-Transform Relationship, 679
    Sampled-Data Systems, 679
    System Realization, 679
    General, 679

Chapter 14
Filter Analysis and Design, 680

14.1 Introduction and Goals, 680
14.2 Analog Filters, 680
Butterworth Filters, 681
  Normalized Butterworth Filters, 681
  Filter Transformations, 682
  MATLAB Design Tools, 684
Chebyshev, Elliptic, and Bessel Filters, 686
14.3 Digital Filters, 689
Simulation of Analog Filters, 689
Filter Design Techniques, 689
  IIR Filter Design, 689
  Time-Domain Methods, 689
    Impulse-Invariant Design, 689
    Step-Invariant Design, 696
    Finite-Difference Design, 698
    Frequency-Domain Methods, 704
    The Bilinear Method, 706
  FIR Filter Design, 713
    Truncated Ideal Impulse Response, 713
    Optimal FIR Filter Design, 723
    MATLAB Design Tools, 725
14.4 Summary of Important Points, 727
Exercises, 727
  Exercises with Answers, 727
    Continuous-Time Filters, 727
    Finite-Difference Filter Design, 728
    Matched-z Transform and Direct Substitution Filter Design, 729
    Bilinear z-Transform Filter Design, 730
    FIR Filter Design, 730
    Digital Filter Design Method Comparison, 731
  Exercises without Answers, 731
    Analog Filter Design, 731
    Impulse-Invariant and Step-Invariant Filter Design, 732
    Finite-Difference Filter Design, 733
    Matched z-Transform and Direct Substitution Filter Design, 733
    Bilinear z-Transform Filter Design, 733
    FIR Filter Design, 733
    Digital Filter Design Method Comparison, 734

Appendix I   Useful Mathematical Relations, A-1
II   Continuous-Time Fourier Series Pairs, A-4
III   Discrete Fourier Transform Pairs, A-7
IV   Continuous-Time Fourier Transform Pairs, A-10
V   Discrete-Time Fourier Transform Pairs, A-17
VI   Tables of Laplace Transform Pairs, A-22
VII z-Transform Pairs, A-24

Bibliography, B-1

Index, I-1

PREFACE

MOTIVATION

I wrote the first and second editions because I love the mathematical beauty of signal and system analysis. That has not changed. The motivation for the third edition is to further refine the book structure in light of reviewers, comments, correct a few errors from the second edition and significantly rework the exercises.

AUDIENCE

This book is intended to cover a two-semester course sequence in the basics of signal and system analysis during the junior or senior year. It can also be used (as I have used it) as a book for a quick one-semester Master's-level review of transform methods as applied to linear systems.

CHANGES FROM THE SECOND EDITION

1. In response to reviewers, comments, two chapters from the second edition have been omitted: Communication Systems and State-Space Analysis. There seemed to be very little if any coverage of these topics in actual classes.
2. The second edition had 550 end-of-chapter exercises in 16 chapters. The third edition has 710 end-of-chapter exercises in 14 chapters.

OVERVIEW

Except for the omission of two chapters, the third edition structure is very similar to the second edition. The book begins with mathematical methods for describing signals and systems, in both continuous and discrete time. I introduce the idea of a transform with the continuous-time Fourier series, and from that base move to the Fourier transform as an extension of the Fourier series to aperiodic signals. Then I do the same for discrete-time signals. I introduce the Laplace transform both as a generalization of the continuous-time Fourier transform for unbounded signals and unstable systems and as a powerful tool in system analysis because of its very close association with the eigenvalues and eigenfunctions of continuous-time linear systems. I take a similar path for discrete-time systems using the z transform. Then I address sampling, the relation between continuous and discrete time. The rest of the book is devoted to applications in frequency-response analysis, feedback systems, analog and digital filters. Throughout the book I present examples and introduce MATLAB functions and operations to implement the methods presented. A chapter-by-chapter summary follows.

CHAPTER SUMMARIES

CHAPTER 1

Chapter 1 is an introduction to the general concepts involved in signal and system analysis without any mathematical rigor. It is intended to motivate the student by demonstrating the ubiquity of signals and systems in everyday life and the importance of understanding them.

CHAPTER 2

Chapter 2 is an exploration of methods of mathematically describing continuous-time signals of various kinds. It begins with familiar functions, sinusoids and exponentials and then extends the range of signal-describing functions to include continuous-time singularity functions (switching functions). Like most, if not all, signals and systems textbooks, I define the unit-step, the signum, the unit-impulse and the unit-ramp functions. In addition to these I define a unit rectangle and a unit periodic impulse function. The unit periodic impulse function, along with convolution, provides an especially compact way of mathematically describing arbitrary periodic signals.

After introducing the new continuous-time signal functions, I cover the common types of signal transformations, amplitude scaling, time shifting, time scaling, differentiation and integration and apply them to the signal functions. Then I cover some characteristics of signals that make them invariant to certain transformations, evenness, oddness and periodicity, and some of the implications of these signal characteristics in signal analysis. The last section is on signal energy and power.

CHAPTER 3

Chapter 3 follows a path similar to Chapter 2 except applied to discrete-time signals instead of continuous-time signals. I introduce the discrete-time sinusoid and exponential and comment on the problems of determining period of a discrete-time sinusoid. This is the first exposure of the student to some of the implications of sampling. I define some discrete-time signal functions analogous to continuous-time singularity functions. Then I explore amplitude scaling, time shifting, time scaling, differencing and accumulation for discrete-time signal functions pointing out the unique implications and problems that occur, especially when time scaling discrete-time functions. The chapter ends with definitions and discussion of signal energy and power for discrete-time signals.

CHAPTER 4

This chapter addresses the mathematical description of systems. First I cover the most common forms of classification of systems, homogeneity, additivity, linearity, time invariance, causality, memory, static nonlinearity and invertibility. By example I present various types of systems that have, or do not have, these properties and how to prove various properties from the mathematical description of the system.

CHAPTER 5

This chapter introduces the concepts of impulse response and convolution as components in the systematic analysis of the response of linear, time-invariant systems. I present the mathematical properties of continuous-time convolution and a graphical method of understanding what the convolution integral says. I also show how the properties of convolution can be used to combine subsystems that are connected in cascade or parallel into one system and what the impulse response of the overall system must be. Then I introduce the idea of a transfer function by finding the response of an LTI system to complex sinusoidal excitation. This section is followed by an analogous coverage of discrete-time impulse response and convolution.

CHAPTER 6

This is the beginning of the student's exposure to transform methods. I begin by graphically introducing the concept that any continuous-time periodic signal with engineering usefulness can be expressed by a linear combination of continuous-time sinusoids, real or complex. Then I formally derive the Fourier series using the concept of orthogonality to show where the signal description as a function of discrete harmonic number (the harmonic function) comes from. I mention the Dirichlet conditions to let the student know that the continuous-time Fourier series applies to all <u>practical</u> continuous-time signals, but not to all <u>imaginable</u> continuous-time signals.

Then I explore the properties of the Fourier series. I have tried to make the Fourier series notation and properties as similar as possible and analogous to the Fourier transform, which comes later. The harmonic function forms a "Fourier series pair" with the time function. In the first edition I used a notation for harmonic function in which lower-case letters were used for time-domain quantities and upper-case letters for their harmonic functions. This unfortunately caused some confusion because continuous- and discrete-time harmonic functions looked the same. In this edition I have changed the harmonic function notation for continuous-time signals to make it easily distinguishable. I also have a section on the convergence of the Fourier series illustrating the Gibb's phenomenon at function discontinuities. I encourage students to use tables and properties to find harmonic functions and this practice prepares them for a similar process in finding Fourier transforms and later Laplace and z transforms.

The next major section of Chapter 6 extends the Fourier series to the Fourier transform. I introduce the concept by examining what happens to a continuous-time Fourier series as the period of the signal approaches infinity and then define and derive the continuous-time Fourier transform as a generalization of the continuous-time Fourier series. Following that I cover all the important properties of the continuous-time Fourier transform. I have taken an "ecumenical" approach to two different notational conventions that are commonly seen in books on signals and systems, control systems, digital signal processing, communication systems and other applications of Fourier methods such as image processing and Fourier optics: the use of either cyclic frequency, f or radian frequency, ω. I use both and emphasize that the two are simply related through a change of variable. I think this better prepares students for seeing both forms in other books in their college and professional careers.

CHAPTER 7

This chapter introduces the discrete-time Fourier series (DTFS), the discrete Fourier transform (DFT) and the discrete-time Fourier transform (DTFT), deriving and defining them in a manner analogous to Chapter 6. The DTFS and the DFT are almost identical. I concentrate on the DFT because of its very wide use in digital signal processing. I emphasize the important differences caused by the differences between continuous- and discrete-time signals, especially the finite summation range of the DFT as opposed to the (generally) infinite summation range in the CTFS. I also point out the importance of the fact that the DFT relates a finite set of numbers to another finite set of numbers, making it amenable to direct numerical machine computation. I discuss the fast Fourier transform as a very efficient algorithm for computing the DFT. As in Chapter 6, I use both cyclic and radian frequency forms, emphasizing the relationships between them. I use F and Ω for discrete-time frequencies to distinguish them from f and ω, which were used in continuous time. Unfortunately, some authors reverse these symbols. My usage is more consistent with the majority of signals and systems texts. This is another example of the lack of standardization of notation in this area. The last major section is a comparison of the four Fourier methods. I emphasize particularly the duality between sampling in one domain and periodic repetition in the other domain.

CHAPTER 8

This chapter introduces the Laplace transform. I approach the Laplace transform from two points of view, as a generalization of the Fourier transform to a larger class of signals and as result which naturally follows from the excitation of a linear, time-invariant system by a complex exponential signal. I begin by defining the bilateral Laplace transform and discussing significance of the region of convergence. Then I define the unilateral Laplace transform. I derive all the important properties of the Laplace transform. I fully explore the method of partial-fraction expansion for finding inverse transforms and then show examples of solving differential equations with initial conditions using the unilateral form.

CHAPTER 9

This chapter introduces the z transform. The development parallels the development of the Laplace transform except applied to discrete-time signals and systems. I initially define a bilateral transform and discuss the region of convergence. Then I define a unilateral transform. I derive all the important properties and demonstrate the inverse transform using partial-fraction expansion and the solution of difference equations with initial conditions. I also show the relationship between the Laplace and z transforms, an important idea in the approximation of continuous-time systems by discrete-time systems in Chapter 14.

CHAPTER 10

This is the first exploration of the correspondence between a continuous-time signal and a discrete-time signal formed by sampling it. The first section covers how sampling is usually done in real systems using a sample-and-hold and an A/D converter. The second section starts by asking the question of how many samples are enough to describe a continuous-time signal. Then the question is answered by deriving the sampling theorem. Then I discuss interpolation methods, theoretical and practical, the special properties of bandlimited periodic signals. I do a complete development of the relationship between the CTFT of a continuous-time signal and DFT of a finite-length set of samples taken from it. Then I show how the DFT can be used to approximate the CTFT of an energy signal or a periodic signal. The next major section explores the use of the DFT in numerically approximating various common signal-processing operations.

CHAPTER 11

This chapter covers various aspects of the use of the CTFT and DTFT in frequency response analysis. The major topics are ideal filters, Bode diagrams, practical passive and active continuous-time filters and basic discrete-time filters.

CHAPTER 12

This chapter is on the application of the Laplace transform including block diagram representation of systems in the complex frequency domain, system stability, system interconnections, feedback systems including root locus, system responses to standard signals and lastly standard realizations of continuous-time systems.

CHAPTER 13

This chapter is on the application of the z transform including block diagram representation of systems in the complex frequency domain, system stability, system interconnections, feedback systems including root-locus, system responses to standard signals, sampled-data systems and standard realizations of discrete-time systems.

CHAPTER 14

This chapter covers the analysis and design of some of the most common types of practical analog and digital filters. The analog filter types are Butterworth, Chebyshev Types 1 and 2 and Elliptic (Cauer) filters. The section on digital filters covers the most common types of techniques for simulation of analog filters including, impulse- and step-invariant, finite difference, matched z transform, direct substitution, bilinear z transform, truncated impulse response and Parks-McClellan numerical design.

APPENDICES

There are seven appendices on useful mathematical formulae, tables of the four Fourier transforms, Laplace transform tables and z transform tables.

CONTINUITY

The book is structured so as to facilitate skipping some topics without loss of continuity. Continuous-time and discrete-time topics are covered alternately and continuous-time analysis could be covered without reference to discrete time. Also, any or all of the last six chapters could be omitted in a shorter course.

REVIEWS AND EDITING

This book owes a lot to the reviewers, especially those who really took time and criticized and suggested improvements. I am indebted to them. I am also indebted to the many students who have endured my classes over the years. I believe that our relationship is more symbiotic than they realize. That is, they learn signal and system analysis from me and I learn how to teach signal and system analysis from them. I cannot count the number of times I have been asked a very perceptive question by a student that revealed not only that the students were not understanding a concept but that I did not understand it as well as I had previously thought.

WRITING STYLE

Every author thinks he has found a better way to present material so that students can grasp it and I am no different. I have taught this material for many years and through the experience of grading tests have found what students generally do and do not grasp. I have spent countless hours in my office one-on-one with students explaining these concepts to them and, through that experience, I have found out what needs to be said. In my writing I have tried to simply speak directly to the reader in a straightforward conversational way, trying to avoid off-putting formality and, to the extent possible, anticipating the usual misconceptions and revealing the fallacies in them. Transform methods are not an obvious idea and, at first exposure, students can easily get bogged down in a bewildering morass of abstractions and lose sight of the goal, which is to analyze a system's response to signals. I have tried (as every author does) to find the magic combination of accessibility and mathematical rigor because both are important. I think my writing is clear and direct but you, the reader, will be the final judge of whether or not that is true.

EXERCISES

Each chapter has a group of exercises along with answers and a second group of exercises without answers. The first group is intended more or less as a set of "drill" exercises and the second group as a set of more challenging exercises.

CONCLUDING REMARKS

As I indicated in the preface to first and second editions, I welcome any and all criticism, corrections and suggestions. All comments, including ones I disagree with and ones which disagree with others, will have a constructive impact on the next edition because they point out a problem. If something does not seem right to you, it probably will bother others also and it is my task, as an author, to find a way to solve that problem. So I encourage you to be direct and clear in any remarks about what you believe should be changed and not to hesitate to mention any errors you may find, from the most trivial to the most significant.

Michael J. Roberts, Professor
Emeritus Electrical and Computer Engineering
University of Tennessee at Knoxville
mjr@utk.edu

McGraw-Hill Connect®
Learn Without Limits

Connect is a teaching and learning platform that is proven to deliver better results for students and instructors.

Connect empowers students by continually adapting to deliver precisely what they need, when they need it and how they need it, so your class time is more engaging and effective.

73% of instructors who use Connect require it; instructor satisfaction increases by 28% when Connect is required.

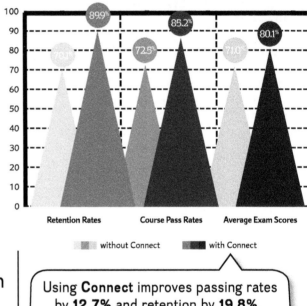

Using **Connect** improves passing rates by **12.7%** and retention by **19.8%**.

Analytics

Connect Insight®

Connect Insight is Connect's new one-of-a-kind visual analytics dashboard that provides at-a-glance information regarding student performance, which is immediately actionable. By presenting assignment, assessment and topical performance results together with a time metric that is easily visible for aggregate or individual results, Connect Insight gives the user the ability to take a just-in-time approach to teaching and learning, which was never before available. Connect Insight presents data that helps instructors improve class performance in a way that is efficient and effective.

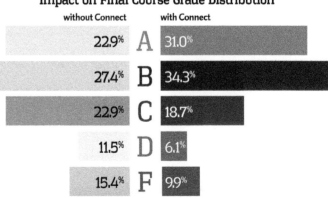

Adaptive

©Getty Images/iStockphoto

THE **ADAPTIVE READING EXPERIENCE** DESIGNED TO TRANSFORM THE WAY STUDENTS READ

More students earn **A's** and **B's** when they use McGraw-Hill Education **Adaptive** products.

SmartBook®

Proven to help students improve grades and study more efficiently, SmartBook contains the same content within the print book, but actively tailors that content to the needs of the individual. SmartBook's adaptive technology provides precise, personalized instruction on what the student should do next, guiding the student to master and remember key concepts, targeting gaps in knowledge and offering customized feedback and driving the student toward comprehension and retention of the subject matter. Available on smartphones and tablets, SmartBook puts learning at the student's fingertips—anywhere, anytime.

Over **5.7 billion questions** have been answered, making McGraw-Hill Education products more intelligent, reliable and precise.

www.mheducation.com

CHAPTER 1

Introduction

1.1 SIGNALS AND SYSTEMS DEFINED

Any time-varying physical phenomenon that is intended to convey information is a **signal**. Examples of signals are the human voice, sign language, Morse code, traffic signals, voltages on telephone wires, electric fields emanating from radio or television transmitters, and variations of light intensity in an optical fiber on a telephone or computer network. **Noise** is like a signal in that it is a time-varying physical phenomenon, but usually it does not carry useful information and is considered undesirable.

Signals are operated on by **systems**. When one or more **excitations** or **input signals** are applied at one or more system **inputs**, the system produces one or more **responses** or **output signals** at its **outputs**. Figure 1.1 is a block diagram of a single-input, single-output system.

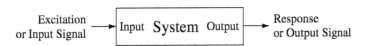

Figure 1.1
Block diagram of a single-input, single-output system

In a communication system, a transmitter produces a signal and a receiver acquires it. A **channel** is the path a signal takes from a transmitter to a receiver. Noise is inevitably introduced into the transmitter, channel and receiver, often at multiple points (Figure 1.2). The transmitter, channel and receiver are all components or subsystems of the overall system. Scientific instruments are systems that measure a physical phenomenon (temperature, pressure, speed, etc.) and convert it to a voltage or current, a signal. Commercial building control systems (Figure 1.3), industrial plant control systems (Figure 1.4), modern farm machinery (Figure 1.5), avionics in airplanes, ignition and fuel pumping controls in automobiles, and so on are all systems that operate on signals.

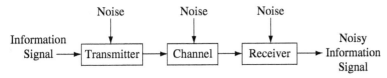

Figure 1.2
A communication system

2 Chapter 1 Introduction

Figure 1.3
Modern office buildings
© Vol. 43 PhotoDisc/Getty

Figure 1.4
Typical industrial plant control room
© Royalty-Free/Punchstock

Figure 1.5
Modern farm tractor with enclosed cab
© Royalty-Free/Corbis

The term *system* even encompasses things such as the stock market, government, weather, the human body and the like. They all respond when excited. Some systems are readily analyzed in detail, some can be analyzed approximately, but some are so complicated or difficult to measure that we hardly know enough to understand them.

1.2 TYPES OF SIGNALS

There are several broad classifications of signals: **continuous-time**, **discrete-time**, **continuous-value**, **discrete-value**, **random** and **nonrandom**. A continuous-time signal is defined at every instant of time over some time interval. Another common name for some continuous-time signals is **analog** signal, in which the variation of the signal with time is *analogous* (proportional) to some physical phenomenon. All analog signals are continuous-time signals but not all continuous-time signals are analog signals (Figure 1.6 through Figure 1.8).

Sampling a signal is acquiring values from a continuous-time signal at discrete points in time. The set of samples forms a discrete-time signal. A discrete-time signal

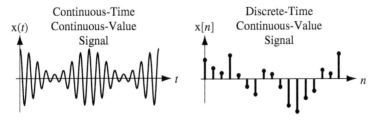

Figure 1.6
Examples of continuous-time and discrete-time signals

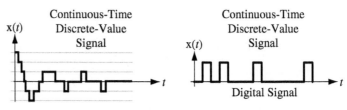

Figure 1.7
Examples of continuous-time, discrete-value signals

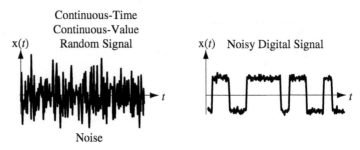

Figure 1.8
Examples of noise and a noisy digital signal

can also be created by an inherently discrete-time system that produces signal values only at discrete times (Figure 1.6).

A continuous-value signal is one that may have any value within a continuum of allowed values. In a continuum any two values can be arbitrarily close together. The real numbers form a continuum with infinite extent. The real numbers between zero and one form a continuum with finite extent. Each is a set with infinitely many members (Figure 1.6 through Figure 1.8).

A discrete-value signal can only have values taken from a discrete set. In a discrete set of values the magnitude of the difference between any two values is greater than some positive number. The set of integers is an example. Discrete-time signals are usually transmitted as **digital** signals, a sequence of values of a discrete-time signal in the form of digits in some encoded form. The term *digital* is also sometimes used loosely to refer to a discrete-value signal that has only two possible values. The digits in this type of digital signal are transmitted by signals that are continuous-time. In this case, the terms *continuous-time* and *analog* are not synonymous. A digital signal of this type is a continuous-time signal but not an analog signal because its variation of value with time is not directly analogous to a physical phenomenon (Figure 1.6 through Figure 1.8).

A random signal cannot be predicted exactly and cannot be described by any mathematical function. A **deterministic** signal can be mathematically described. A common name for a random signal is **noise** (Figure 1.6 through Figure 1.8).

In practical signal processing it is very common to acquire a signal for processing by a computer by sampling, **quantizing** and **encoding** it (Figure 1.9). The original signal is a continuous-value, continuous-time signal. Sampling acquires its values at discrete times and those values constitute a continuous-value, discrete-time signal. Quantization approximates each sample as the nearest member of a finite set of discrete values, producing a discrete-value, discrete-time signal. Each signal value in the set of discrete values at discrete times is converted to a sequence of rectangular pulses that encode it into a binary number, creating a discrete-value, continuous-time signal, commonly called a *digital signal*. The steps illustrated in Figure 1.9 are usually carried out by a single device called an **analog-to-digital converter (ADC)**.

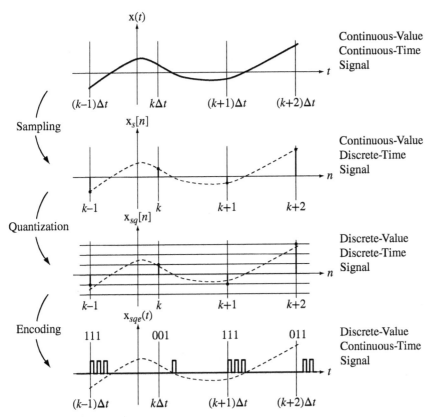

Figure 1.9
Sampling, quantization and encoding of a signal to illustrate various signal types

One common use of binary digital signals is to send text messages using the American Standard Code for Information Interchange (ASCII). The letters of the alphabet, the digits 0–9, some punctuation characters and several nonprinting control characters, for a total of 128 characters, are all encoded into a sequence of 7 binary bits. The 7 bits are sent sequentially, preceded by a **start** bit and followed by 1 or 2 **stop** bits for synchronization purposes. Typically, in direct-wired connections between digital equipment, the bits are represented by a higher voltage (2 to 5 V) for a 1 and a lower voltage level (around 0 V) for a 0. In an asynchronous transmission using one start and one stop bit, sending the message SIGNAL, the voltage versus time would look as illustrated in Figure 1.10.

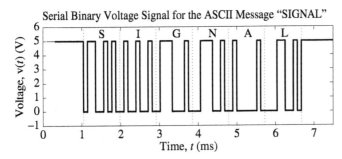

Figure 1.10
Asynchronous serial binary ASCII-encoded voltage signal for the word SIGNAL

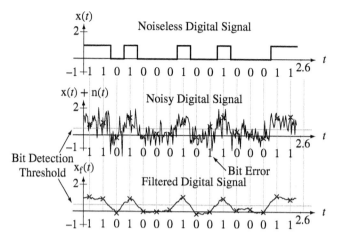

Figure 1.11
Use of a filter to reduce bit error rate in a digital signal

In 1987 ASCII was extended to Unicode. In Unicode the number of bits used to represent a character can be 8, 16, 24 or 32 and more than 120,000 characters are currently encoded in modern and historic language characters and multiple symbol sets.

Digital signals are important in signal analysis because of the spread of digital systems. Digital signals often have better immunity to noise than analog signals. In binary signal communication the bits can be detected very cleanly until the noise gets very large. The detection of bit values in a stream of bits is usually done by comparing the signal value at a predetermined bit time with a threshold. If it is above the threshold it is declared a 1 and if it is below the threshold it is declared a 0. In Figure 1.11, the x's mark the signal value at the detection time, and when this technique is applied to the noisy digital signal, one of the bits is incorrectly detected. But when the signal is processed by a **filter**, all the bits are correctly detected. The filtered digital signal does not look very clean in comparison with the noiseless digital signal, but the bits can still be detected with a very low probability of error. This is the basic reason that digital signals can have better noise immunity than analog signals. An introduction to the analysis and design of filters is presented in Chapters 11 and 15.

In this text we will consider both continuous-time and discrete-time signals, but we will (mostly) ignore the effects of signal quantization and consider all signals to be continuous-value. Also, we will not directly consider the analysis of random signals, although random signals will sometimes be used in illustrations.

The first signals we will study are continuous-time signals. Some continuous-time signals can be described by continuous functions of time. A signal x(t) might be described by a function $x(t) = 50\sin(200\pi t)$ of continuous time t. This is an exact description of the signal at every instant of time. The signal can also be described graphically (Figure 1.12).

Many continuous-time signals are not as easy to describe mathematically. Consider the signal in Figure 1.13. Waveforms like the one in Figure 1.13 occur in various types of instrumentation and communication systems. With the definition of some signal functions and an operation called **convolution,** this signal can be compactly described, analyzed and manipulated mathematically. Continuous-time signals that can be described by mathematical functions can be transformed into another domain called the **frequency domain** through the **continuous-time Fourier transform**. In this context, **transformation** means transformation of a signal to the frequency domain. This is an important tool in signal analysis, which allows certain characteristics of the signal to be more clearly observed

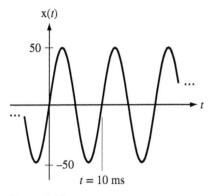

Figure 1.12
A continuous-time signal described by a mathematical function

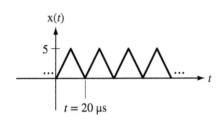

Figure 1.13
A second continuous-time signal

and more easily manipulated than in the time domain. (In the frequency domain, signals are described in terms of the frequencies they contain.) Without frequency-domain analysis, design and analysis of many systems would be considerably more difficult.

Discrete-time signals are only defined at discrete points in time. Figure 1.14 illustrates some discrete-time signals.

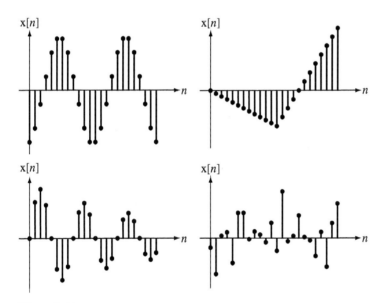

Figure 1.14
Some discrete-time signals

So far all the signals we have considered have been described by functions of time. An important class of "signals" is functions of **space** instead of time: images. Most of the theories of signals, the information they convey and how they are processed by systems in this text will be based on signals that are a variation of a physical phenomenon with time. But the theories and methods so developed also apply, with only minor modifications, to the processing of images. Time signals are described by the variation of a physical phenomenon as a function of a single independent variable, time. Spatial signals, or images, are described by the variation of a physical phenomenon as a

Figure 1.15
An example of image processing to reveal information
(Original X-ray image and processed version provided by the Imaging, Robotics and Intelligent Systems (IRIS) Laboratory of the Department of Electrical and Computer Engineering at the University of Tennessee, Knoxville.)

function of two orthogonal, independent, **spatial** variables, conventionally referred to as x and y. The physical phenomenon is most commonly light or something that affects the transmission or reflection of light, but the techniques of image processing are also applicable to anything that can be mathematically described by a function of two independent variables.

Historically the practical application of image-processing techniques has lagged behind the application of signal-processing techniques because the amount of information that has to be processed to gather the information from an image is typically much larger than the amount of information required to get the information from a time signal. But now image processing is increasingly a practical technique in many situations. Most image processing is done by computers. Some simple image-processing operations can be done directly with optics and those can, of course, be done at very high speeds (at the speed of light!). But direct optical image processing is very limited in its flexibility compared with digital image processing on computers.

Figure 1.15 shows two images. On the left is an unprocessed X-ray image of a carry-on bag at an airport checkpoint. On the right is the same image after being processed by some image-filtering operations to reveal the presence of a weapon. This text will not go into image processing in any depth but will use some examples of image processing to illustrate concepts in signal processing.

An understanding of how signals carry information and how systems process signals is fundamental to multiple areas of engineering. Techniques for the analysis of signals processed by systems are the subject of this text. This material can be considered as an applied mathematics text more than a text covering the building of useful devices, but an understanding of this material is very important for the successful design of useful devices. The material that follows builds from some fundamental definitions and concepts to a full range of analysis techniques for continuous-time and discrete-time signals in systems.

1.3 EXAMPLES OF SYSTEMS

There are many different types of signals and systems. A few examples of systems are discussed next. The discussion is limited to the qualitative aspects of the system with some illustrations of the behavior of the system under certain conditions. These systems will be revisited in Chapter 4 and discussed in a more detailed and quantitative way in the material on system modeling.

A MECHANICAL SYSTEM

A man bungee jumps off a bridge over a river. Will he get wet? The answer depends on several factors:

1. The man's height and weight
2. The height of the bridge above the water
3. The length and springiness of the bungee cord

When the man jumps off the bridge he goes into free fall caused by the force due to gravitational attraction until the bungee cord extends to its full unstretched length. Then the system dynamics change because there is now another force on the man, the bungee cord's resistance to stretching, and he is no longer in free fall. We can write and solve a differential equation of motion and determine how far down the man falls before the bungee cord pulls him back up. The differential equation of motion is a **mathematical model** of this mechanical system. If the man weighs 80 kg and is 1.8 m tall, and if the bridge is 200 m above the water level and the bungee cord is 30 m long (unstretched) with a spring constant of 11 N/m, the bungee cord is fully extended before stretching at $t = 2.47$ s. The equation of motion, after the cord starts stretching, is

$$x(t) = -16.85 \sin(0.3708t) - 95.25 \cos(0.3708t) + 101.3, \quad t > 2.47. \quad (1.1)$$

Figure 1.16 shows his position versus time for the first 15 seconds. From the graph it seems that the man just missed getting wet.

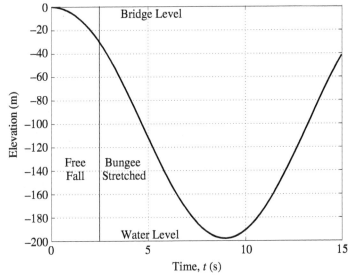

Figure 1.16
Man's vertical position versus time (bridge level is zero)

A FLUID SYSTEM

A fluid system can also be modeled by a differential equation. Consider a cylindrical water tank being fed by an input flow of water, with an orifice at the bottom through which flows the output (Figure 1.17).

The flow out of the orifice depends on the height of the water in the tank. The variation of the height of the water depends on the input flow and the output flow. The rate

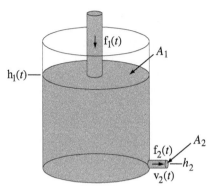

Figure 1.17
Tank with orifice being filled from above

of change of water volume in the tank is the difference between the input volumetric flow and the output volumetric flow and the volume of water is the cross-sectional area of the tank times the height of the water. All these factors can be combined into one differential equation for the water level $h_1(t)$.

$$A_1 \frac{d}{dt}(h_1(t)) + A_2 \sqrt{2g[h_1(t) - h_2]} = f_1(t) \qquad (1.2)$$

The water level in the tank is graphed in Figure 1.18 versus time for four volumetric inflows under the assumption that the tank is initially empty.

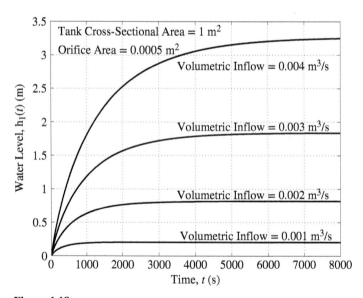

Figure 1.18
Water level versus time for four different volumetric inflows with the tank initially empty

As the water flows in, the water level increases, and that increases the water outflow. The water level rises until the outflow equals the inflow. After that time the water level stays constant. Notice that when the inflow is increased by a factor of two, the final water level is increased by a factor of four. The final water level is proportional to the square of the volumetric inflow. That fact makes the differential equation that models the system nonlinear.

A DISCRETE-TIME SYSTEM

Discrete-time systems can be designed in multiple ways. The most common practical example of a discrete-time system is a computer. A computer is controlled by a clock that determines the timing of all operations. Many things happen in a computer at the integrated circuit level between clock pulses, but a computer user is only interested in what happens at the times of occurrence of clock pulses. From the user's point of view, the computer is a discrete-time system.

We can simulate the action of a discrete-time system with a computer program. For example,

```
yn = 1 ; yn1 = 0 ;
while 1,
   yn2 = yn1 ; yn1 = yn ; yn = 1.97*yn1 - yn2 ;
end
```

This computer program (written in MATLAB) simulates a discrete-time system with an output signal y that is described by the difference equation

$$y[n] = 1.97\,y[n-1] - y[n-2] \tag{1.3}$$

along with initial conditions $y[0] = 1$ and $y[-1] = 0$. The value of y at any time index n is the sum of the previous value of y at discrete time $n-1$ multiplied by 1.97, minus the value of y previous to that at discrete time $n-2$. The operation of this system can be diagrammed as in Figure 1.19.

In Figure 1.19, the two squares containing the letter D are delays of one in discrete time, and the arrowhead next to the number 1.97 represents an amplifier that multiplies the signal entering it by 1.97 to produce the signal leaving it. The circle with the plus sign in it is a **summing junction**. It adds the two signals entering it (one of which is negated first) to produce the signal leaving it. The first 50 values of the signal produced by this system are illustrated in Figure 1.20.

The system in Figure 1.19 could be built with dedicated hardware. Discrete-time delay can be implemented with a shift register. Multiplication by a constant can be done with an amplifier or with a digital hardware multiplier. Summation can also be done with an operational amplifier or with a digital hardware adder.

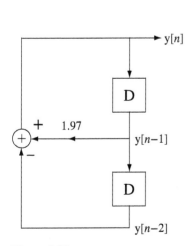

Figure 1.19
Discrete-time system example

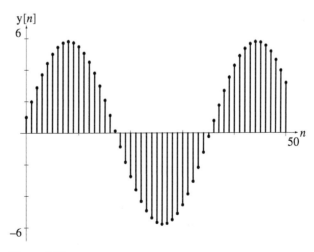

Figure 1.20
Signal produced by the discrete-time system in Figure 1.19

FEEDBACK SYSTEMS

Another important aspect of systems is the use of **feedback** to improve system performance. In a feedback system, something in the system observes its response and may modify the input signal to the system to improve the response. A familiar example is a thermostat in a house that controls when the air conditioner turns on and off. The thermostat has a temperature sensor. When the temperature inside the thermostat exceeds the level set by the homeowner, a switch inside the thermostat closes and turns on the home air conditioner. When the temperature inside the thermostat drops a small amount below the level set by the homeowner, the switch opens, turning off the air conditioner. Part of the system (a temperature sensor) is sensing the thing the system is trying to control (the air temperature) and feeds back a signal to the device that actually does the controlling (the air conditioner). In this example, the feedback signal is simply the closing or opening of a switch.

Feedback is a very useful and important concept and feedback systems are everywhere. Take something everyone is familiar with, the float valve in an ordinary flush toilet. It senses the water level in the tank and, when the desired water level is reached, it stops the flow of water into the tank. The floating ball is the sensor and the valve to which it is connected is the feedback mechanism that controls the water level.

If all the water valves in all flush toilets were exactly the same and did not change with time, and if the water pressure upstream of the valve were known and constant, and if the valve were always used in exactly the same kind of water tank, it should be possible to replace the float valve with a timer that shuts off the water flow when the water reaches the desired level, because the water would always reach the desired level at exactly the same elapsed time. But water valves do change with time and water pressure does fluctuate and different toilets have different tank sizes and shapes. Therefore, to operate properly under these varying conditions the tank-filling system must adapt by sensing the water level and shutting off the valve when the water reaches the desired level. The ability to adapt to changing conditions is the great advantage of feedback methods.

There are countless examples of the use of feedback.

1. Pouring a glass of lemonade involves feedback. The person pouring watches the lemonade level in the glass and stops pouring when the desired level is reached.

2. Professors give tests to students to report to the students their performance levels. This is feedback to let the student know how well she is doing in the class so she can adjust her study habits to achieve her desired grade. It is also feedback to the professor to let him know how well his students are learning.

3. Driving a car involves feedback. The driver senses the speed and direction of the car, the proximity of other cars and the lane markings on the road and constantly applies corrective actions with the accelerator, brake and steering wheel to maintain a safe speed and position.

4. Without feedback, the F-117 stealth fighter would crash because it is aerodynamically unstable. Redundant computers sense the velocity, altitude, roll, pitch and yaw of the aircraft and constantly adjust the control surfaces to maintain the desired flight path (Figure 1.21).

Feedback is used in both continuous-time systems and discrete-time systems. The system in Figure 1.22 is a discrete-time feedback system. The response of the system $y[n]$ is "fed back" to the upper summing junction after being delayed twice and multiplied by some constants.

Figure 1.21
The F-117A Nighthawk stealth fighter
© Vol. 87/Corbis

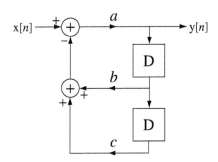

Figure 1.22
A discrete-time feedback system

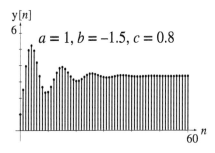

Figure 1.23
Discrete-time system response with
$b = -1.5$ and $c = 0.8$

Let this system be initially at rest, meaning that all signals throughout the system are zero before time index $n = 0$. To illustrate the effects of feedback let $a = 1$, let $b = -1.5$, let $c = 0.8$ and let the input signal x[n] change from 0 to 1 at $n = 0$ and stay at 1 for all time, $n \geq 0$. We can see the response y [n] in Figure 1.23.

Now let $c = 0.6$ and leave a and b the same. Then we get the response in Figure 1.24.
Now let $c = 0.5$ and leave a and b the same. Then we get the response in Figure 1.25.

The response in Figure 1.25 increases forever. This last system is unstable because a bounded input signal produces an unbounded response. So feedback can make a system unstable.

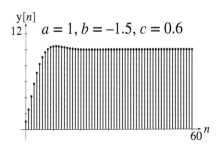

Figure 1.24
Discrete-time system response with
$b = -1.5$ and $c = 0.6$

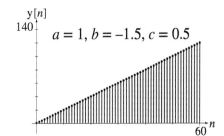

Figure 1.25
Discrete-time system response with
$b = -1.5$ and $c = 0.5$

The system illustrated in Figure 1.26 is an example of a continuous-time feedback system. It is described by the differential equation $y''(t) + ay(t) = x(t)$. The homogeneous solution can be written in the form

$$y_h(t) = K_{h1} \sin(\sqrt{a}t) + K_{h2} \cos(\sqrt{a}t). \tag{1.4}$$

If the excitation x(t) is zero and the initial value y(t_0) is nonzero or the initial derivative of y(t) is nonzero and the system is allowed to operate in this form after $t = t_0$, y(t)

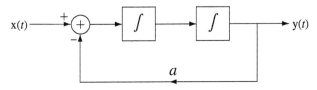

Figure 1.26
Continuous-time feedback system

will oscillate sinusoidally forever. This system is an oscillator with a stable amplitude. So feedback can cause a system to oscillate.

1.4 A FAMILIAR SIGNAL AND SYSTEM EXAMPLE

As an example of signals and systems, let's look at a signal and system that everyone is familiar with, sound, and a system that produces and/or measures sound. Sound is what the ear senses. The human ear is sensitive to acoustic pressure waves typically between about 15 Hz and about 20 kHz with some sensitivity variation in that range. Below are some graphs of air-pressure variations that produce some common sounds. These sounds were recorded by a system consisting of a microphone that converts air-pressure variation into a continuous-time voltage signal, electronic circuitry that processes the continuous-time voltage signal, and an ADC that changes the continuous-time voltage signal to a digital signal in the form of a sequence of binary numbers that are then stored in computer memory (Figure 1.27).

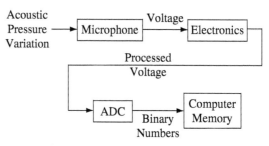

Figure 1.27
A sound recording system

Consider the pressure variation graphed in Figure 1.28. It is the continuous-time pressure signal that produces the sound of the word "signal" spoken by an adult male (the author).

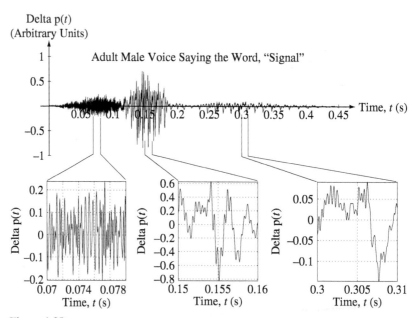

Figure 1.28
The word "signal" spoken by an adult male voice

Analysis of sounds is a large subject, but some things about the relationship between this graph of air-pressure variation and what a human hears as the word "signal" can be seen by looking at the graph. There are three identifiable "bursts" of signal, #1 from 0 to about 0.12 seconds, #2 from about 0.12 to about 0.19 seconds and #3 from about 0.22 to about 0.4 seconds. Burst #1 is the *s* in the word "signal." Burst #2 is the *i* sound. The region between bursts #2 and #3 is the double consonant *gn* of the word "signal." Burst #3 is the *a* sound terminated by the *l* consonant stop. An *l* is not quite as abrupt a stop as some other consonants, so the sound tends to "trail off" rather than stopping quickly. The variation of air pressure is generally faster for the *s* than for the *i* or the *a*. In signal analysis we would say that it has more "high-frequency content." In the blowup of the *s* sound the air-pressure variation looks almost random. The *i* and *a* sounds are different in that they vary more slowly and are more "regular" or "predictable" (although not *exactly* predictable). The *i* and *a* are formed by vibrations of the vocal cords and therefore exhibit an approximately oscillatory behavior. This is described by saying that the *i* and *a* are **tonal** or **voiced** and the *s* is not. Tonal means having the basic quality of a single **tone** or **pitch** or **frequency**. This description is not mathematically precise but is useful qualitatively.

Another way of looking at a signal is in the frequency domain, mentioned above, by examining the frequencies, or pitches, that are present in the signal. A common way of illustrating the variation of signal power with frequency is its **power spectral density**, a graph of the power density in the signal versus frequency. Figure 1.29 shows the three bursts (*s*, *i* and *a*) from the word "signal" and their associated power spectral densities (the G(f) functions).

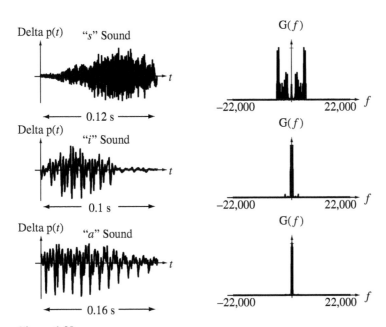

Figure 1.29
Three sounds in the word "signal" and their associated power spectral densities

Power spectral density is just another mathematical tool for analyzing a signal. It does not contain any new information, but sometimes it can reveal things that are difficult to see otherwise. In this case, the power spectral density of the *s* sound is widely distributed in frequency, whereas the power spectral densities of the *i* and *a* sounds are narrowly distributed in the lowest frequencies. There is more power in the *s* sound at

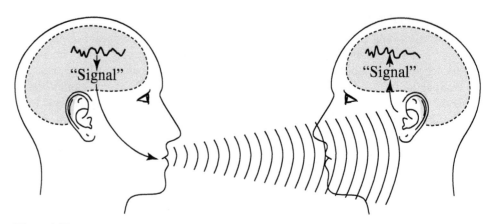

Figure 1.30
Communication between two people involving signals and signal processing by systems

higher frequencies than in the *i* and *a* sounds. The *s* sound has an "edge" or "hissing" quality caused by the high frequencies in the *s* sound.

The signal in Figure 1.28 carries **information**. Consider what happens in conversation when one person says the word "signal" and another hears it (Figure 1.30). The speaker thinks first of the concept of a signal. His brain quickly converts the concept to the word "signal." Then his brain sends nerve impulses to his vocal cords and diaphragm to create the air movement and vibration and tongue and lip movements to produce the sound of the word "signal." This sound then propagates through the air between the speaker and the listener. The sound strikes the listener's eardrum and the vibrations are converted to nerve impulses, which the listener's brain converts first to the sound, then the word, then the concept *signal*. Conversation is accomplished by a system of considerable sophistication.

How does the listener's brain know that the complicated pattern in Figure 1.28 is the word "signal"? The listener is not aware of the detailed air-pressure variations but instead "hears sounds" that are caused by the air-pressure variation. The eardrum and brain convert the complicated air-pressure pattern into a few simple features. That conversion is similar to what we will do when we convert signals into the frequency domain. The process of recognizing a sound by reducing it to a small set of features reduces the amount of information the brain has to process. Signal processing and analysis in the technical sense do the same thing but in a more mathematically precise way.

Two very common problems in signal and system analysis are noise and **interference**. Noise is an undesirable random signal. Interference is an undesirable nonrandom signal. Noise and interference both tend to obscure the information in a signal. Figure 1.31 shows examples of the signal from Figure 1.28 with different levels of noise added.

As the noise power increases there is a gradual degradation in the intelligibility of the signal, and at some level of noise the signal becomes unintelligible. A measure of the quality of a received signal corrupted by noise is the ratio of the signal power to the noise power, commonly called **signal-to-noise ratio** and often abbreviated SNR. In each of the examples of Figure 1.31 the SNR is specified.

Sounds are not the only signals, of course. Any physical phenomenon that is measured or observed is a signal. Also, although the majority of signals we will consider in this text will be functions of time, a signal can be a function of some other independent

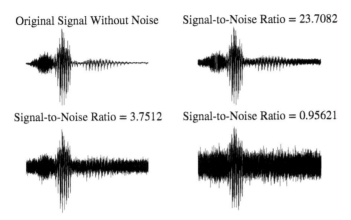

Figure 1.31
Sound of the word "signal" with different levels of noise added

variable, like frequency, wavelength, distance and so on. Figure 1.32 and Figure 1.33 illustrate some other kinds of signals.

Just as sounds are not the only signals, conversation between two people is not the only system. Examples of other systems include the following:

1. An automobile suspension for which the road surface excites the automobile and the position of the chassis relative to the road is the response.
2. A chemical mixing vat for which streams of chemicals are the input signals and the mixture of chemicals is the output signal.
3. A building environmental control system for which the exterior temperature is the input signal and the interior temperature is the response.

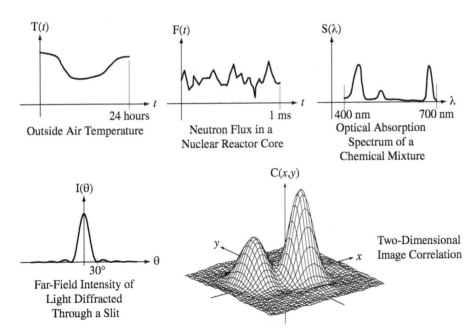

Figure 1.32
Examples of signals that are functions of one or more continuous independent variables

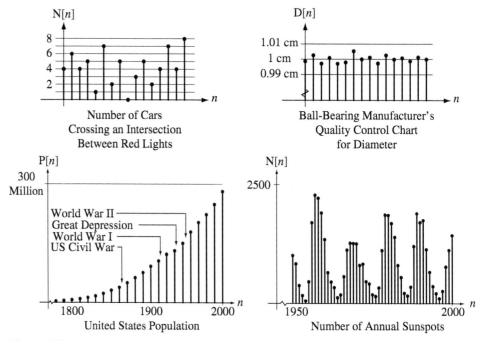

Figure 1.33
Examples of signals that are functions of a discrete independent variable

4. A chemical spectroscopy system in which white light excites the specimen and the spectrum of transmitted light is the response.
5. A telephone network for which voices and data are the input signals and reproductions of those voices and data at a distant location are the output signals.
6. Earth's atmosphere, which is excited by energy from the sun and for which the responses are ocean temperature, wind, clouds, humidity and so on. In other words, the weather is the response.
7. A thermocouple excited by the temperature gradient along its length for which the voltage developed between its ends is the response.
8. A trumpet excited by the vibration of the player's lips and the positions of the valves for which the response is the tone emanating from the bell.

The list is endless. Any physical entity can be thought of as a system, because if we excite it with physical energy, it has a physical response.

1.5 USE OF MATLAB®

Throughout the text, examples will be presented showing how signal and system analysis can be done using MATLAB. MATLAB is a high-level mathematical tool available on many types of computers. It is very useful for signal processing and system analysis. There is an introduction to MATLAB in Web Appendix A.

CHAPTER 2

Mathematical Description of Continuous-Time Signals

2.1 INTRODUCTION AND GOALS

Over the years, signal and system analysts have observed many signals and have realized that signals can be classified into groups with similar behavior. Figure 2.1 shows some examples of signals.

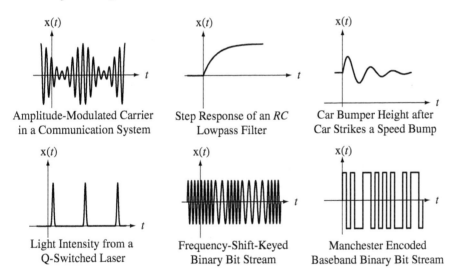

Figure 2.1
Examples of signals

In signal and system analysis, signals are described by mathematical functions. Some of the functions that describe real signals should already be familiar, exponentials and sinusoids. These occur frequently in signal and system analysis. One set of functions has been defined to describe the effects on signals of switching operations that often occur in systems. Some other functions arise in the development of certain system analysis techniques, which will be introduced in later chapters. These functions are all carefully chosen to be simply related to each other and to be easily changed by a well-chosen set of shifting and/or scaling operations. They are prototype functions, which have simple definitions and are easily remembered. The types of symmetries and patterns that most frequently occur in real signals will be defined and their effects on signal analysis explored.

CHAPTER GOALS

1. To define some mathematical functions that can be used to describe signals
2. To develop methods of shifting, scaling and combining those functions to represent real signals
3. To recognize certain symmetries and patterns to simplify signal and system analysis

2.2 FUNCTIONAL NOTATION

A function is a correspondence between the **argument** of the function, which lies in its **domain**, and the **value** returned by the function, which lies in its **range**. The most familiar functions are of the form g(x) where the argument x is a real number and the value returned g is also a real number. But the domain and/or range of a function can be complex numbers or integers or a variety of other choices of allowed values.

In this text five types of functions will appear,

1. Domain—Real numbers, Range—Real numbers
2. Domain—Integers, Range—Real numbers
3. Domain—Integers, Range—Complex numbers
4. Domain—Real numbers, Range—Complex numbers
5. Domain—Complex numbers, Range—Complex numbers

For functions whose domain is either real numbers or complex numbers the argument will be enclosed in parentheses (·). For functions whose domain is integers the argument will be enclosed in brackets [·]. These types of functions will be discussed in more detail as they are introduced.

2.3 CONTINUOUS-TIME SIGNAL FUNCTIONS

If the independent variable of a function is time t and the domain of the function is the real numbers, and if the function g(t) has a defined value at every value of t, the function is called a **continuous-time** function. Figure 2.2 illustrates some continuous-time functions.

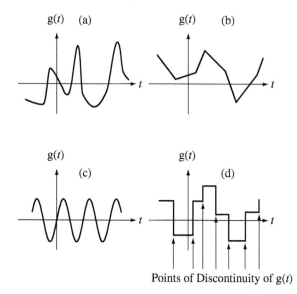

Figure 2.2
Examples of continuous-time functions

Figure 2.2(d) illustrates a discontinuous function for which the limit of the function value as we approach the discontinuity from above is not the same as when we approach it from below. If $t = t_0$ is a point of discontinuity of a function g(t) then

$$\lim_{\varepsilon \to 0} g(t_0 + \varepsilon) \neq \lim_{\varepsilon \to 0} g(t_0 - \varepsilon).$$

All four functions, (a)–(d), are continuous-time functions because their values are defined for all real values of t. Therefore the terms *continuous* and *continuous-time* mean slightly different things. All continuous functions of time are continuous-time functions, but not all continuous-time functions are continuous functions of time.

COMPLEX EXPONENTIALS AND SINUSOIDS

Real-valued sinusoids and exponential functions should already be familiar. In

$$g(t) = A\cos(2\pi t/T_0 + \theta) = A\cos(2\pi f_0 t + \theta) = A\cos(\omega_0 t + \theta)$$

and

$$g(t) = Ae^{(\sigma_0 + j\omega_0)t} = Ae^{\sigma_0 t}[\cos(\omega_0 t) + j\sin(\omega_0 t)]$$

A is the amplitude, T_0 is the fundamental period, f_0 is the fundamental cyclic frequency and ω_0 is the fundamental radian frequency of the sinusoid, t is time and σ_0 is the decay rate of the exponential (which is the reciprocal of its time constant, τ) (Figure 2.3 and Figure 2.4). All these parameters can be any real number.

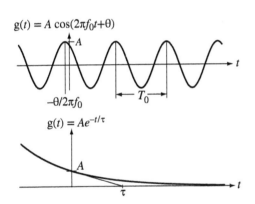

Figure 2.3
A real sinusoid and a real exponential with parameters indicated graphically

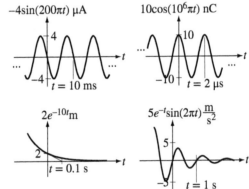

Figure 2.4
Examples of signals described by real sines, cosines and exponentials

In Figure 2.4 the units indicate what kind of physical signal is being described. Very often in system analysis, when only one kind of signal is being followed through a system, the units are omitted for the sake of brevity.

Exponentials (**exp**) and sinusoids (**sin** and **cos**) are intrinsic functions in MATLAB. The arguments of the **sin** and **cos** functions are interpreted by MATLAB as radians, not degrees.

```
>> [exp(1),sin(pi/2),cos(pi)]
ans =
    2.7183  1.0000  -1.0000
```
(pi is the MATLAB symbol for π.)

Sinusoids and exponentials are very common in signal and system analysis because most continuous-time systems can be described, at least approximately, by linear, constant-coefficient, ordinary differential equations whose eigenfunctions are **complex exponentials**, complex powers of e, the base of the natural logarithms. Eigenfunction means "characteristic function" and the eigenfunctions have a particularly important relation to the differential equation. If the exponent of e is real, complex exponentials are the same as real exponentials. Through Euler's identity $e^{jx} = \cos(x) + j\sin(x)$ and the relations $\cos(x) = (1/2)(e^{jx} + e^{-jx})$ and $\sin(x) = (1/j2)(e^{jx} - e^{-jx})$, complex exponentials and real-valued sinusoids are closely related. If, in a function of the form e^{jx}, x is a real-valued independent variable, this special form of the complex exponential is called a **complex sinusoid** (Figure 2.5).

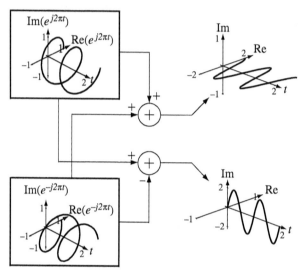

Figure 2.5
The relation between real and complex sinusoids

In signal and system analysis, sinusoids are expressed in either the cyclic frequency f form $A\cos(2\pi f_0 t + \theta)$ or the radian frequency ω form $A\cos(\omega_0 t + \theta)$. The advantages of the f form are the following:

1. The fundamental period T_0 and the fundamental cyclic frequency f_0 are simply reciprocals of each other.
2. In communication system analysis, a spectrum analyzer is often used and its display scale is usually calibrated in Hz. Therefore f is the directly observed variable.
3. The definition of the Fourier transform (Chapter 6) and some transforms and transform relationships are simpler in the f form than in the ω form.

The advantages of the ω form are the following:

1. Resonant frequencies of real systems, expressed directly in terms of physical parameters, are more simply expressed in the ω form than in the f form. The resonant frequency of an LC oscillator is $\omega_0^2 = 1/LC = (2\pi f_0)^2$ and the half-power corner frequency of an RC lowpass filter is $\omega_c = 1/RC = 2\pi f_c$.
2. The Laplace transform (Chapter 8) is defined in a form that is more simply related to the ω form than to the f form.

3. Some Fourier transforms are simpler in the ω form.
4. Use of ω in some expressions makes them more compact. For example, $A\cos(\omega_0 t + \theta)$ is a little more compact than $A\cos(2\pi f_0 t + \theta)$.

Sinusoids and exponentials are important in signal and systems analysis because they arise naturally in the solutions of the differential equations that often describe system dynamics. As we will see in the study of the Fourier series and Fourier transform, even if signals are not sinusoids, most of them can be expressed as linear combinations of sinusoids.

FUNCTIONS WITH DISCONTINUITIES

Continuous-time sines, cosines and exponentials are all continuous and differentiable at every point in time. But many other types of important signals that occur in practical systems are not continuous or differentiable everywhere. A common operation in systems is to switch a signal on or off at some time (Figure 2.6).

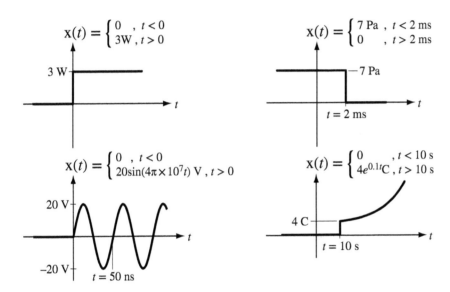

Figure 2.6
Examples of signals that are switched on or off at some time

The functional descriptions of the signals in Figure 2.6 are complete and accurate but are in a cumbersome form. Signals of this type can be better described mathematically by multiplying a function that is continuous and differentiable for all time by another function that switches from zero to one or one to zero at some finite time.

In signal and system analysis **singularity functions**, which are related to each other through integrals and derivatives, can be used to mathematically describe signals that have discontinuities or discontinuous derivatives. These functions, and functions that are closely related to them through some common system operations, are the subject of this section. In the consideration of singularity functions we will extend, modify and/or generalize some basic mathematical concepts and operations to allow us to efficiently analyze real signals and systems. We will extend the concept of what a derivative is, and we will also learn how to use an important mathematical entity, the impulse, which is a lot like a function but is not a function in the usual sense.

The Signum Function

For nonzero arguments, the value of the signum function has a magnitude of one and a sign that is the same as the sign of its argument:

$$\text{sgn}(t) = \begin{cases} 1, & t > 0 \\ 0, & t = 0 \\ -1, & t < 0 \end{cases} \quad (2.1)$$

(See Figure 2.7.)

Figure 2.7
The signum function

The graph on the left in Figure 2.7 is of the exact mathematical definition. The graph on the right is a more common way of representing the function for engineering purposes. No practical signal can change discontinuously, so if an approximation of the signum function were generated by a signal generator and viewed on an oscilloscope, it would look like the graph on the right. The signum function is intrinsic in MATLAB (and called the `sign` function).

The Unit-Step Function

The unit-step function is defined by

$$u(t) = \begin{cases} 1, & t > 0 \\ 1/2, & t = 0 \\ 0, & t < 0 \end{cases} \quad (2.2)$$

(See Figure 2.8.) It is called the *unit* step because the step is one unit high in the system of units used to describe the signal.[1]

Figure 2.8
The unit-step function

---

[1] Some authors define the unit step by

$$u(t) = \begin{cases} 1, & t \geq 0 \\ 0, & t < 0 \end{cases} \quad \text{or} \quad u(t) = \begin{cases} 1, & t > 0 \\ 0, & t < 0 \end{cases} \quad \text{or} \quad u(t) = \begin{cases} 1, & t > 0 \\ 0, & t \leq 0 \end{cases}$$

In the middle definition the value at $t = 0$ is undefined but finite. The unit steps defined by these definitions all have an identical effect on any real physical system.

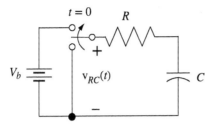

Figure 2.9
Circuit with a switch whose effect can be represented by a unit step

The unit step can mathematically represent a common action in real physical systems, fast switching from one state to another. In the circuit of Figure 2.9 the switch moves from one position to the other at time $t = 0$. The voltage applied to the RC network is $v_{RC}(t) = V_b u(t)$. The current flowing clockwise through the resistor and capacitor is

$$i(t) = (V_b/R)e^{-t/RC}u(t)$$

and the voltage across the capacitor is $v(t) = V_b(1 - e^{-t/RC})u(t)$.

There is an intrinsic function in MATLAB, called heaviside[2] which returns a one for positive arguments, a zero for negative arguments and an NaN for zero arguments. The MATLAB constant NaN is "not a number" and indicates an undefined value. There are practical problems using this function in numerical computations because the return of an undefined value can cause some programs to prematurely terminate or return useless results.

We can create our own functions in MATLAB, which become functions we can call upon just like the intrinsic functions cos, sin, exp, etc. MATLAB functions are defined by creating an m file, a file whose name has the extension ".m". We could create a function that finds the length of the hypotenuse of a right triangle given the lengths of the other two sides.

```
%   Function to compute the length of the hypotenuse of a
%   right triangle given the lengths of the other two sides
%
%   a - The length of one side
%   b - The length of the other side
%   c - The length of the hypotenuse
%
%   function c = hyp(a,b)
%
function c = hyp(a,b)
    c = sqrt(a^2 + b^2) ;
```

The first nine lines in this example, which are preceded by %, are **comment** lines that are not executed but serve to document how the function is used. The first executable line must begin with the keyword function. The rest of the first line is in the form

$$\text{result = name(arg1, arg2,...)}$$

---

[2] Oliver Heaviside was a self-taught English electrical engineer who adapted complex numbers to the study of electrical circuits, invented mathematical techniques for the solution of differential equations and reformulated and simplified Maxwell's field equations. Although at odds with the scientific establishment for most of his life, Heaviside changed the face of mathematics and science for years to come. It has been reported that a man once complained to Heaviside that his writings were very difficult to read. Heaviside's response was that they were even more difficult to write!

where **result** will contain the returned value, which can be a scalar, a vector or a matrix (or even a cell array or a structure, which are beyond the scope of this text), **name** is the function name, and **arg1, arg2,...** are the parameters or **arguments** passed to the function. The arguments can also be scalars, vectors or matrices (or cell arrays or structures). The name of the file containing the function definition must be *name*.m.

Below is a listing of a MATLAB function to implement the unit-step function in numerical computations.

```
%   Unit-step function defined as 0 for input argument values
%   less than zero, 1/2 for input argument values equal to zero,
%   and 1 for input argument values greater than zero. This
%   function uses the sign function to implement the unit-step
%   function. Therefore value at t = 0 is defined. This avoids
%   having undefined values during the execution of a program
%   that uses it.
%
%   function y = us(x)
%
function y = us(x)
    y = (sign(x) + 1)/2 ;
```

This function should be saved in a file named "us.m".

The Unit-Ramp Function

Another type of signal that occurs in systems is one that is switched on at some time and changes linearly after that time or changes linearly before some time and is switched off at that time (Figure 2.10). Signals of this kind can be described with the use of the **ramp** function. The unit-ramp function (Figure 2.11) is the integral of the unit-step function. It is called the *unit*-ramp function because, for positive t, its slope is one amplitude unit per time unit.

$$\text{ramp}(t) = \begin{cases} t, & t > 0 \\ 0, & t \leq 0 \end{cases} = \int_{-\infty}^{t} u(\lambda)d\lambda = tu(t) \qquad (2.3)$$

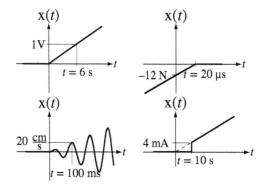

Figure 2.10
Functions that change linearly before or after some time, or are multiplied by functions that change linearly before or after some time

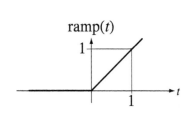

Figure 2.11
The unit-ramp function

The ramp is defined by ramp$(t) = \int_{-\infty}^{t} u(\tau)d\tau$. In this equation, the symbol τ is the independent variable of the unit-step function and the variable of integration. But t is the independent variable of the ramp function. The equation says, "to find the value of the ramp function at any value of t, start with τ at negative infinity and move in τ up to $\tau = t$, while accumulating the area under the unit-step function." The total area accumulated from $\tau = -\infty$ to $\tau = t$ is the value of the ramp function at time t (Figure 2.12). For t less than zero, no area is accumulated. For t greater than zero, the area accumulated equals t because it is the area of a rectangle with width t and height one.

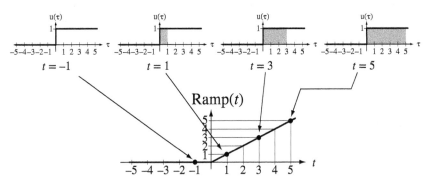

Figure 2.12
Integral relationship between the unit step and the unit ramp

Some authors prefer to use the expression $tu(t)$ instead of ramp(t). Since they are equal, the use of either one is correct and just as legitimate as the other one. Below is a MATLAB m file for the ramp function.

```
%   Function to compute the ramp function defined as 0 for
%   values of the argument less than or equal to zero and
%   the value of the argument for arguments greater than zero.
%   Uses the unit-step function us(x).
%
%   function y = ramp(x)
%
function y = ramp(x)
    y = x.*us(x) ;
```

The Unit Impulse
Before we define the unit impulse we will first explore an important idea. Consider a unit-area, rectangular pulse defined by

$$\Delta(t) = \begin{cases} 1/a, & |t| \leq a/2 \\ 0, & |t| > a/2 \end{cases}$$

(See Figure 2.13.) Let this function multiply a function $g(t)$ that is finite and continuous at $t = 0$ and find the area A under the product of the two functions $A = \int_{-\infty}^{\infty} \Delta(t)g(t)dt$ (Figure 2.14).

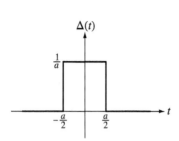

Figure 2.13
A unit-area rectangular pulse of width a

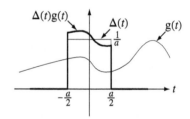

Figure 2.14
Product of a unit-area rectangular pulse centered at $t = 0$ and a function $g(t)$ that is continuous and finite at $t = 0$

Using the definition of $\Delta(t)$ we can rewrite the integral as

$$A = \frac{1}{a} \int_{-a/2}^{a/2} g(t)dt.$$

The function $g(t)$ is continuous at $t = 0$. Therefore it can be expressed as a McLaurin series of the form

$$g(t) = \sum_{m=0}^{\infty} \frac{g^{(m)}(0)}{m!} t^m = g(0) + g'(0)t + \frac{g''(0)}{2!} t^2 + \cdots + \frac{g^{(m)}(0)}{m!} t^m + \cdots$$

Then the integral becomes

$$A = \frac{1}{a} \int_{-a/2}^{a/2} \left[g(0) + g'(0)t + \frac{g''(0)}{2!} t^2 + \cdots + \frac{g^{(m)}(0)}{m!} t^m + \cdots \right] dt$$

All the odd powers of t contribute nothing to the integral because it is taken over symmetrical limits about $t = 0$. Carrying out the integral,

$$A = \frac{1}{a} \left[ag(0) + \left(\frac{a^3}{12}\right) \frac{g''(0)}{2!} + \left(\frac{a^5}{80}\right) \frac{g^{(4)}(0)}{4!} + \cdots \right]$$

Take the limit of this integral as a approaches zero.

$$\lim_{a \to 0} A = g(0).$$

In the limit as a approaches zero, the function $\Delta(t)$ extracts the value of any continuous finite function $g(t)$ at time $t = 0$, when the product of $\Delta(t)$ and $g(t)$ is integrated over any range of time that includes time $t = 0$.

Now try a different definition of the function $\Delta(t)$. Define it now as

$$\Delta(t) = \begin{cases} (1/a)(1 - |t|/a), & |t| \le a \\ 0, & |t| > a \end{cases}$$

(See Figure 2.15.)

If we make the same argument as before we get the area

$$A = \int_{-\infty}^{\infty} \Delta(t)g(t)dt = \frac{1}{a} \int_{-a}^{a} \left(1 - \frac{|t|}{a}\right) g(t)dt.$$

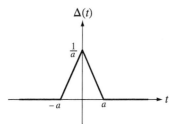

Figure 2.15
A unit-area triangular pulse of base half-width a

Taking the limit as a approaches zero, we again get g(0), exactly the same result we got with the previous definition of $\Delta(t)$. The two definitions of $\Delta(t)$ have the same effect in the limit as a approaches zero (but not before). The *shape* of the function is not what is important in the limit, but its *area is* important. In either case $\Delta(t)$ is a function with an area of one, independent of the value of a. (As a approaches zero these functions do not have a "shape" in the ordinary sense because there is no time in which to develop one.) There are many other definitions of $\Delta(t)$ that could be used with exactly the same effect in the limit.

The unit impulse $\delta(t)$ can now be implicitly defined by the property that when it is multiplied by any function g(t) that is finite and continuous at $t = 0$ and the product is integrated over a time range that includes $t = 0$, the result is g(0):

$$g(0) = \int_\alpha^\beta \delta(t)g(t)dt, \ \alpha < 0 < \beta.$$

In other words,

$$\int_{-\infty}^{\infty} \delta(t)g(t)dt = \lim_{a \to 0} \int_{-\infty}^{\infty} \Delta(t)g(t)dt \qquad (2.4)$$

where $\Delta(t)$ is any of many functions that have the characteristics described above. The notation $\delta(t)$ is a convenient shorthand notation that avoids having to constantly take a limit when using impulses.

The Impulse, the Unit Step and Generalized Derivatives One way of introducing the unit impulse is to define it as the derivative of the unit-step function. Strictly speaking, the derivative of the unit step u(t) is undefined at $t = 0$. But consider a function g(t) of time and its time derivative g'(t) in Figure 2.16.

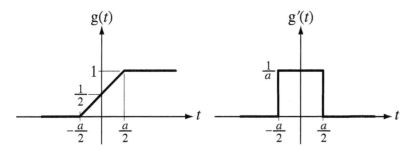

Figure 2.16
Functions that approach the unit step and unit impulse

The derivative of g(t) exists for all t except at $t = -a/2$ and at $t = +a/2$. As a approaches zero, the function g(t) approaches the unit step. In that same limit the nonzero part of the function g'(t) approaches zero while its area remains the same, one. So g'(t) is a short-duration pulse whose area is always one, the same as the initial definition of $\Delta(t)$ above, with the same implications. The limit as a approaches zero of g'(t) is called the **generalized derivative** of u(t). Therefore the unit impulse is the generalized derivative of the unit step.

The generalized derivative of any function g(t) with a discontinuity at $t = t_0$ is

$$\frac{d}{dt}(g(t)) = \frac{d}{dt}(g(t))_{t \neq t_0} + \underbrace{\lim_{\varepsilon \to 0}[g(t+\varepsilon) - g(t-\varepsilon)]}_{\text{Size of the discontinuity}} \delta(t - t_0), \quad \varepsilon > 0.$$

The unit step is the integral of the unit impulse

$$u(t) = \int_{-\infty}^{t} \delta(\lambda) d\lambda.$$

The derivative of the unit step u(t) is zero everywhere except at $t = 0$, so the unit impulse is zero everywhere except at $t = 0$. Since the unit step is the integral of the unit impulse, a definite integral of the unit impulse whose integration range includes $t = 0$ must have the value, one. These two facts are often used to define the unit impulse.

$$\delta(t) = 0, \quad t \neq 0 \quad \text{and} \quad \int_{t_1}^{t_2} \delta(t) dt = \begin{cases} 1, & t_1 < 0 < t_2 \\ 0, & \text{otherwise} \end{cases} \quad (2.5)$$

The area under an impulse is called its **strength** or sometimes its **weight**. An impulse with a strength of one is called a *unit impulse*. The exact definition and characteristics of the impulse require a plunge into generalized function theory. It will suffice here to consider a unit impulse simply to be a pulse of unit area whose duration is so small that making it any smaller would not significantly change any signals in the system to which it is applied.

The impulse cannot be graphed in the same way as other functions because its value is undefined when its argument is zero. The usual convention for graphing an impulse is to use a vertical arrow. Sometimes the strength of the impulse is written beside it in parentheses, and sometimes the height of the arrow indicates the strength of the impulse. Figure 2.17 illustrates some ways of representing impulses graphically.

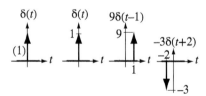

Figure 2.17
Graphical representations of impulses

The Equivalence Property of the Impulse A common mathematical operation in signal and system analysis is the product of an impulse with another function, $g(t)A\delta(t - t_0)$. Consider that the impulse $A\delta(t - t_0)$ is the limit of a pulse with area A centered at $t = t_0$, with width a, as a approaches zero (Figure 2.18). The product is a pulse whose height

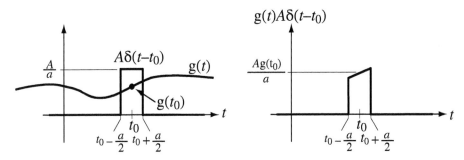

Figure 2.18
Product of a function g(t) and a rectangular function that becomes an impulse as its width approaches zero

at the mid-point is $Ag(t_0)/a$ and whose width is a. As a approaches zero, the pulse becomes an impulse and the strength of that impulse is $Ag(t_0)$. Therefore

$$\boxed{g(t)A\delta(t - t_0) = g(t_0)A\delta(t - t_0)}. \tag{2.6}$$

This is sometimes called the **equivalence** property of the impulse.

The Sampling Property of the Impulse Another important property of the unit impulse that follows from the equivalence property is its **sampling property**.

$$\boxed{\int_{-\infty}^{\infty} g(t)\delta(t - t_0)\,dt = g(t_0)} \tag{2.7}$$

According to the equivalence property, the product $g(t)\delta(t - t_0)$ is equal to $g(t_0)\delta(t - t_0)$. Since t_0 is one particular value of t, it is a constant and $g(t_0)$ is also a constant and

$$\int_{-\infty}^{\infty} g(t)\delta(t - t_0)\,dt = g(t_0)\underbrace{\int_{-\infty}^{\infty} \delta(t - t_0)\,dt}_{=1} = g(t_0).$$

Equation (2.7) is called the *sampling property of the impulse* because in an integral of this type it samples the value of the function $g(t)$ at time $t = t_0$. (An older name is **sifting property**. The impulse "sifts out" the value of $g(t)$, at time $t = t_0$.)

The Scaling Property of the Impulse Another important property of the impulse is its **scaling property**

$$\boxed{\delta(a(t - t_0)) = \frac{1}{|a|}\delta(t - t_0)} \tag{2.8}$$

which can be proven through a change of variable in the integral definition and separate consideration of positive and negative values for a (see Exercise 35). Figure 2.19 illustrates some effects of the scaling property of the impulse.

There is a function in MATLAB called `dirac` that implements the unit impulse in a limited sense. It returns zero for nonzero arguments and it returns `inf` for zero arguments. This is not often useful for numerical computations but it is useful for symbolic analysis. The continuous-time impulse is not an ordinary function. It is sometimes

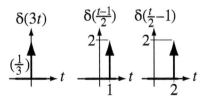

Figure 2.19
Examples of the effect of the scaling property of impulses

possible to write a MATLAB function that can, in certain types of computations, be used to simulate the impulse and obtain useful numerical results. But this must be done with great care, based on a complete understanding of impulse properties. No MATLAB function will be presented here for the continuous-time impulse because of these complications.

The Unit Periodic Impulse or Impulse Train

Another useful generalized function is the **periodic impulse** or **impulse train** (Figure 2.20), a uniformly spaced infinite sequence of unit impulses.

$$\delta_T(t) = \sum_{n=-\infty}^{\infty} \delta(t - nT) \tag{2.9}$$

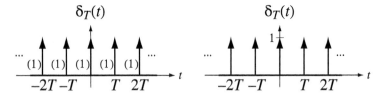

Figure 2.20
The periodic impulse

We can derive a scaling property for the periodic impulse. From the definition

$$\delta_T(a(t - t_0)) = \sum_{k=-\infty}^{\infty} \delta(a(t - t_0) - kT).$$

Using the scaling property of the impulse

$$\delta_T(a(t - t_0)) = (1/|a|) \sum_{k=-\infty}^{\infty} \delta(t - t_0 - kT/a)$$

and the summation can be recognized as a periodic impulse of period T/a

$$\delta_T(a(t - t_0)) = (1/|a|)\delta_{T/a}(t - t_0).$$

The impulse and periodic impulse may seem very abstract and unrealistic. The impulse will appear later in a fundamental operation of linear system analysis, the convolution integral. Although, as a practical matter, a true impulse is impossible to generate, the mathematical impulse and the periodic impulse are very useful in signal and system analysis. Using them and the convolution operation we can mathematically represent, in a compact notation, many useful signals that would be more cumbersome to represent in another way.[3]

A Coordinated Notation for Singularity Functions

The unit step, unit impulse and unit ramp are the most important members of the singularity functions. In some signal and system literature these functions are indicated by the coordinated notation $u_k(t)$ in which the value of k determines the function. For example, $u_0(t) = \delta(t)$, $u_{-1}(t) = u(t)$ and $u_{-2}(t) = \text{ramp}(t)$. In this notation, the subscript indicates how many times an impulse is differentiated to obtain the function in question and a negative subscript indicates that integration is done instead of differentiation. The **unit doublet** $u_1(t)$ is defined as the generalized derivative of the unit impulse, the **unit triplet** $u_2(t)$ is defined as the generalized derivative of the unit doublet, and so on. Even though the unit doublet and triplet and higher generalized derivatives are even less practical than the unit impulse, they are sometimes useful in signal and system theory.

The Unit-Rectangle Function

A very common type of signal occurring in systems is one that is switched on at some time and then off at a later time. It is convenient to define the unit rectangle function (Figure 2.21) for use in describing this type of signal.

$$\text{rect}(t) = \begin{cases} 1, & |t| < 1/2 \\ 1/2, & |t| = 1/2 \\ 0, & |t| > 1/2 \end{cases} = u(t + 1/2) - u(t - 1/2) \qquad (2.10)$$

Figure 2.21
The unit-rectangle function

It is a *unit*-rectangle function because its width, height and area are all one. Use of the rectangle function shortens the notation when describing some signals. The unit-rectangle function can be thought of as a "gate" function. When it multiplies another function, the product is zero outside its nonzero range and is equal to the other function inside its nonzero range. The rectangle "opens a gate," allowing the other function through and then "closes the gate" again. Table 2.1 summarizes the functions and the impulse and periodic impulse described above.

```
%   Unit rectangle function. Uses the unit-step function us(x).
%
%   function y = rect(x)
%
function y = rect(x)
    y = us(x+0.5) - us(x-0.5) ;
```

---

[3] Some authors prefer to always refer to the periodic impulse as a summation of impulses $\sum_{n=-\infty}^{\infty} \delta(t - nT)$. This notation is less compact than $\delta_T(t)$ but may be considered easier than remembering how to use the new function name. Other authors may use different names.

Table 2.1 Summary of continuous-time signal functions, the impulse and the periodic impulse

Sine	$\sin(2\pi f_0 t)$ or $\sin(\omega_0 t)$
Cosine	$\cos(2\pi f_0 t)$ or $\cos(\omega_0 t)$
Exponential	e^{st}
Unit Step	$u(t)$
Signum	$\text{sgn}(t)$
Unit Ramp	$\text{ramp}(t) = t\, u(t)$
Unit Impulse	$\delta(t)$
Periodic Impulse	$\delta_T(t) = \sum_{n=-\infty}^{\infty} \delta(t - nT)$
Unit Rectangle	$\text{rect}(t) = u(t + 1/2) - u(t - 1/2)$

2.4 COMBINATIONS OF FUNCTIONS

Standard functional notation for a continuous-time function is $g(t)$ in which g is the function name and everything inside the parentheses (·) is called the *argument of the function*. The argument is written in terms of the **independent variable**. In the case of $g(t)$, t is the independent variable and the expression is the simplest possible expression in terms of t, t itself. A function $g(t)$ returns a value g for every value of t it accepts. In the function $g(t) = 2 + 4t^2$, for any value of t there is a corresponding value of g. If t is 1, then g is 6 and that is indicated by the notation $g(1) = 6$.

The argument of a function need not be simply the independent variable. If $g(t) = 5e^{-2t}$, what is $g(t + 3)$? We replace t by $t + 3$ everywhere on both sides of $g(t) = 5e^{-2t}$ get $g(t + 3) = 5e^{-2(t+3)}$. *Observe that we do not get $5e^{-2t+3}$*. Since t was multiplied by minus two in the exponent of e, the entire expression $t + 3$ must also be multiplied by minus two in the new exponent of e. Whatever was done with t in the function $g(t)$ must be done with the entire expression involving t in any other function g(expression). If $g(t) = 3 + t^2 - 2t^3$ then $g(2t) = 3 + (2t)^2 - 2(2t)^3 = 3 + 4t^2 - 16t^3$ and $g(1 - t) = 3 + (1 - t)^2 - 2(1 - t)^3 = 2 + 4t - 5t^2 + 2t^3$. If $g(t) = 10\cos(20\pi t)$ then $g(t/4) = 10\cos(20\pi t/4) = 10\cos(5\pi t)$ and $g(e^t) = 10\cos(20\pi e^t)$. If $g(t) = 5e^{-10t}$, then $g(2x) = 5e^{-20x}$ and $g(z - 1) = 5e^{10}e^{-10z}$.

In MATLAB, when a function is invoked by passing an argument to it, MATLAB evaluates the argument, then computes the function value. For most functions, if the argument is a vector or matrix, a value is returned for each element of the vector or matrix. Therefore MATLAB functions do exactly what is described here for arguments that are functions of the independent variable: They accept numbers and return other numbers.

```
>> exp(1:5)
ans =
    2.7183    7.3891   20.0855   54.5982  148.4132
>> us(-1:0.5:1)
ans =
         0         0    0.5000    1.0000    1.0000
>> rect([-0.8:0.4:0.8]')
ans =
    0
    1
    1
    1
    0
```

2.4 Combinations of Functions

In some cases a single mathematical function may completely describe a signal. But often one function is not enough for an accurate description. An operation that allows versatility in the mathematical representation of arbitrary signals is combining two or more functions. The combinations can be sums, differences, products and/or quotients of functions. Figure 2.22 shows some examples of sums, products and quotients of functions. (The sinc function will be defined in Chapter 6.)

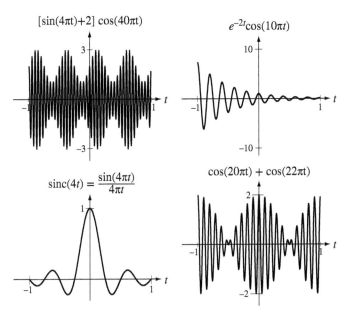

Figure 2.22
Examples of sums, products and quotients of functions

EXAMPLE 2.1

Graphing function combinations with MATLAB

Using MATLAB, graph the function combinations,

$$x_1(t) = e^{-t}\sin(20\pi t) + e^{-t/2}\sin(19\pi t)$$

$$x_2(t) = \text{rect}(t)\cos(20\pi t).$$

```
% Program to graph some demonstrations of continuous-time
% function combinations

t = 0:1/240:6 ;        % Vector of time points for graphing x1

% Generate values of x1 for graphing
x1 = exp(-t).*sin(20*pi*t) + exp(-t/2).*sin(19*pi*t) ;

subplot(2,1,1) ;       % Graph in the top half of the figure window
p = plot(t,x1,'k') ;   % Display the graph with black lines
set(p,'LineWidth',2) ; % Set the line width to 2
% Label the abscissa and ordinate
xlabel('\itt','FontName','Times','FontSize',24) ;
```

```
ylabel('x_1({\itt})','FontName','Times','FontSize',24) ;
set(gca,'FontName','Times','FontSize',18) ; grid on ;

t = -2:1/240:2 ;        % Vector of time points for graphing x2

% Generate values of x2 for graphing
x2 = rect(t).*cos(20*pi*t) ;

subplot(2,1,2);         % Graph in the bottom half of the figure window
p = plot(t,x2,'k');     % Display the graph with black lines
set(p,'LineWidth',2);   % Set the line width to 2
% Label the abscissa and ordinate
xlabel('\itt','FontName','Times','FontSize',24) ;
ylabel('x_2({\itt})','FontName','Times','FontSize',24) ;
set(gca,'FontName','Times','FontSize',18) ; grid on ;
```

The graphs that result are shown in Figure 2.23.

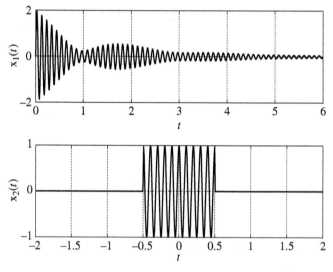

Figure 2.23
MATLAB graphical result

2.5 SHIFTING AND SCALING

It is important to be able to describe signals both analytically and graphically and to be able to relate the two different kinds of descriptions to each other. Let g(t) be defined by Figure 2.24 with some selected values in the table to the right of the figure. To complete the function description let g(t) = 0, $|t| > 5$.

AMPLITUDE SCALING

Consider multiplying a function by a constant. This can be indicated by the notation g(t) → Ag(t). Thus g(t) → Ag(t) multiplies g(t) at every value of t by A. This is called **amplitude scaling**. Figure 2.25 shows two examples of amplitude scaling the function g(t) defined in Figure 2.24.

2.5 Shifting and Scaling

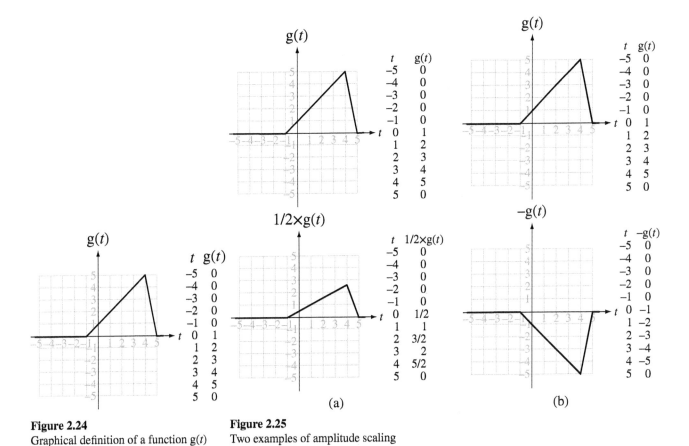

Figure 2.24
Graphical definition of a function g(t)

Figure 2.25
Two examples of amplitude scaling

A negative amplitude-scaling factor flips the function vertically. If the scaling factor is −1 as in this example, flipping is the only action. If the scaling factor is some other factor A and A is negative, amplitude scaling can be thought of as two successive operations g(t) → −g(t) → |A|(−g(t)), a flip followed by a positive amplitude scaling. Amplitude scaling directly affects the *dependent* variable g. The following two sections introduce the effects of changing the *independent* variable t.

TIME SHIFTING

If the graph in Figure 2.24 defines g(t), what does g(t − 1) look like? We can understand the effect by graphing the value of g(t − 1) at multiple points as in Figure 2.26. It should be apparent after examining the graphs and tables that replacing t by t − 1 shifts the function one unit to the right (Figure 2.26). The change t → t − 1 can be described by saying "for every value of t, look back one unit in time, get the value of g at that time and use it as the value for g(t − 1) at time t." This is called **time shifting** or **time translation**.

We can summarize time shifting by saying that the change of independent variable t → t − t_0 where t_0 is any constant, has the effect of shifting g(t) to the right by t_0 units. (Consistent with the accepted interpretation of negative numbers, if t_0 is negative, the shift is to the *left* by |t_0| units.)

Figure 2.27 shows some time-shifted and amplitude-scaled unit-step functions. The rectangle function is the difference between two unit-step functions time-shifted in opposite directions rect(t) = u(t + 1/2) − u(t − 1/2).

38   **Chapter 2**   Mathematical Description of Continuous-Time Signals

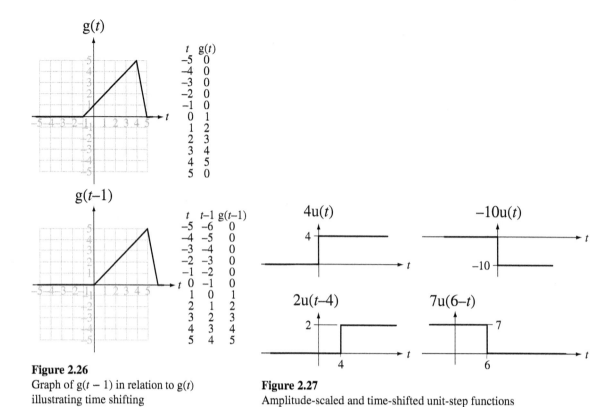

Figure 2.26
Graph of g(t − 1) in relation to g(t) illustrating time shifting

Figure 2.27
Amplitude-scaled and time-shifted unit-step functions

    Time shifting is accomplished by a change of the independent variable. This type of change can be done on any independent variable; it need not be time. Our examples here are using time, but the independent variable could be a spatial dimension. In that case we could call it space shifting. Later, in the chapters on transforms, we will have functions of an independent variable frequency, and this change will be called *frequency shifting*. The mathematical significance is the same regardless of the name used for the independent variable.

    Amplitude scaling and time shifting occur in many practical physical systems. In ordinary conversation there is a propagation delay, the time required for a sound wave to propagate from one person's mouth to the other person's ear. If that distance is 2 m and sound travels at about 330 meters per second, the propagation delay is about 6 ms, a delay that is not noticeable. But consider an observer watching a pile driver drive a pile from 100 m away. First the observer senses the image of the driver striking the pile. There is a slight delay due to the speed of light from the pile driver to the eye but it is less than a microsecond. The sound of the driver striking the pile arrives about 0.3 seconds later, a noticeable delay. This is an example of a time shift. The sound of the driver striking the pile is much louder near the driver than at a distance of 100 m, an example of amplitude scaling. Another familiar example is the delay between seeing a lightning strike and hearing the thunder it produces.

    As a more technological example, consider a satellite communication system (Figure 2.28). A ground station sends a strong electromagnetic signal to a satellite. When the signal reaches the satellite the electromagnetic field is much weaker than when it left the ground station, and it arrives later because of the propagation delay. If the satellite is geosynchronous it is about 36,000 km above the earth, so if the ground station is directly below the satellite the propagation delay on the uplink is about 120 ms. For ground

Figure 2.28
Communication satellite in orbit
© Vol. 4 PhotoDisc/Getty

stations not directly below the satellite the delay is a little more. If the transmitted signal is $A\,\mathrm{x}(t)$, the received signal is $B\,\mathrm{x}(t - t_p)$ where B is typically much smaller than A and t_p is the propagation time. In communication links between locations on earth that are very far apart, more than one up and down link may be required to communicate. If that communication is voice communication between a television anchor in New York and a reporter in Calcutta, the total delay can easily be one second, a noticeable delay that can cause significant awkwardness in conversation. Imagine the problem of communicating with the first astronauts on Mars. The minimum one-way delay when Earth and Mars are in their closest proximity is more than 4 minutes!

In the case of long-range, two-way communication, time delay is a problem. In other situations it can be quite useful, as in radar and sonar. In this case the time delay between when a pulse is sent out and when a reflection returns indicates the distance to the object from which the pulse reflected, for example, an airplane or a submarine.

TIME SCALING

Consider next the change of independent variable indicated by $t \to t/a$. This expands the function $\mathrm{g}(t)$ horizontally by the factor a in $\mathrm{g}(t/a)$. This is called **time scaling**. As an example, let's compute and graph selected values of $\mathrm{g}(t/2)$ (Figure 2.29).

Consider next the change $t \to -t/2$. This is identical to the last example except the scaling factor is now -2 instead of 2 (Figure 2.30). Time scaling $t \to t/a$ expands the function horizontally by a factor of $|a|$ and, if $a < 0$, the function is also **time reversed**. Time reversal means flipping the curve horizontally. The case of a negative a can be conceived as $t \to -t$ followed by $t \to t/|a|$. The first step $t \to -t$ time-reverses the function without changing its horizontal scale. The second step $t \to t/|a|$ time-scales the already-time-reversed function by the scaling factor $|a|$.

Time scaling can also be indicated by $t \to bt$. This is not really new because it is the same as $t \to t/a$ with $b = 1/a$. So all the rules for time scaling still apply with that relation between the two scaling constants a and b.

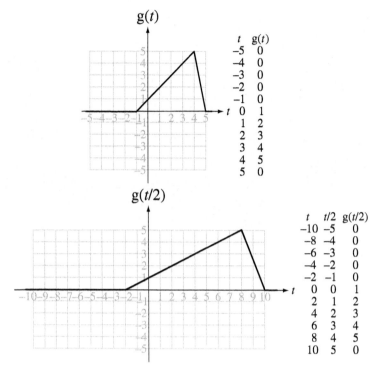

Figure 2.29
Graph of g(t/2) in relation to g(t) illustrating time scaling

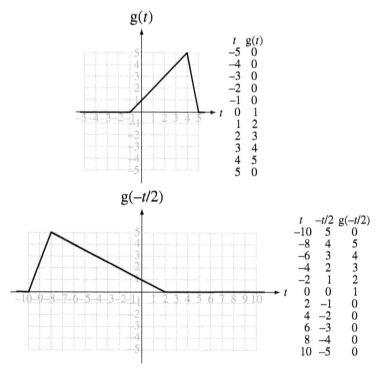

Figure 2.30
Graph of g(−t/2) in relation to g(t) illustrating time scaling for a negative scaling factor

2.5 Shifting and Scaling

Figure 2.31
Firefighters on a fire truck
© Vol. 94 Corbis

A common experience that illustrates time scaling is the Doppler effect. If we stand by the side of a road and a fire truck approaches while sounding its horn, as the fire truck passes, both the volume and the pitch of the horn seem to change (Figure 2.31). The volume changes because of the proximity of the horn; the closer it is to us, the louder it is. But why does the pitch change? The horn is doing exactly the same thing all the time, so it is not the pitch of the sound produced by the horn that changes, but rather the pitch of the sound that arrives at our ears. As the fire truck approaches, each successive compression of air caused by the horn occurs a little closer to us than the last one, so it arrives at our ears in a shorter time than the previous compression and that makes the frequency of the sound wave at our ear higher than the frequency emitted by the horn. As the fire truck passes, the opposite effect occurs, and the sound of the horn arriving at our ears shifts to a lower frequency. While we are hearing a pitch change, the firefighters on the truck hear a constant horn pitch.

Let the sound heard by the firefighters be described by g(t). As the fire truck approaches, the sound we hear is A(t)g(at) where A(t) is an increasing function of time, which accounts for the volume change, and a is a number slightly greater than one. The change in amplitude as a function of time is called **amplitude modulation** in communication systems. After the fire truck passes, the sound we hear shifts to B(t)g(bt) where B(t) is a decreasing function of time and b is slightly less than one (Figure 2.32). (In Figure 2.32 modulated sinusoids are used to represent the horn sound. This is not precise but it serves to illustrate the important points.)

The Doppler shift also occurs with light waves. The red shift of optical spectra from distant stars is what first indicated that the universe was expanding. When a star is receding from the earth, the light we receive on earth experiences a Doppler shift that reduces the frequency of all the light waves emitted by the star (Figure 2.33). Since the color red has the lowest frequency detectable by the human eye, a reduction in frequency is called a *red shift* because the visible spectral characteristics all seem to

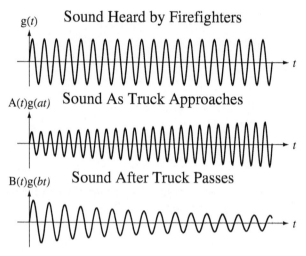

Figure 2.32
Illustration of the Doppler effect

Figure 2.33
The Lagoon nebula
© Vol. 34 PhotoDisc/Getty

move toward the red end of the spectrum. The light from a star has many characteristic variations with frequency because of the composition of the star and the path from the star to the observer. The amount of shift can be determined by comparing the spectral patterns of the light from the star with known spectral patterns measured on Earth in a laboratory.

Time scaling is a change of the independent variable. As was true of time shifting, this type of change can be done on any independent variable; it need not be time. In later chapters we will do some frequency scaling.

SIMULTANEOUS SHIFTING AND SCALING

All three function changes, amplitude scaling, time scaling and time shifting, can be applied simultaneously:

$$g(t) \to Ag\left(\frac{t - t_0}{a}\right). \qquad (2.11)$$

To understand the overall effect, it is usually best to break down a multiple change like (2.11) into successive simple changes:

$$g(t) \xrightarrow{\text{amplitude scaling, } A} Ag(t) \xrightarrow{t \to t/a} Ag(t/a) \xrightarrow{t \to t - t_0} Ag\left(\frac{t - t_0}{a}\right). \qquad (2.12)$$

Observe here that the order of the changes is important. If we exchange the order of the time-scaling and time-shifting operations in (2.12) we get

$$g(t) \xrightarrow{\text{amplitude scaling, } A} Ag(t) \xrightarrow{t \to t - t_0} Ag(t - t_0) \xrightarrow{t \to t/a} Ag(t/a - t_0) \neq Ag\left(\frac{t - t_0}{a}\right).$$

This result is different from the preceding result (unless $a = 1$ or $t_0 = 0$). For a different kind of multiple change, a different sequence may be better, for example, $Ag(bt - t_0)$. In this case the sequence of amplitude scaling, time shifting and then time scaling is the simplest path to a correct result.

$$g(t) \xrightarrow{\text{amplitude scaling, } A} Ag(t) \xrightarrow{t \to t - t_0} Ag(t - t_0) \xrightarrow{t \to bt} Ag(bt - t_0).$$

Figure 2.34 and Figure 2.35 illustrate some steps graphically for two functions. In these figures certain points are labeled with letters, beginning with "a" and proceeding

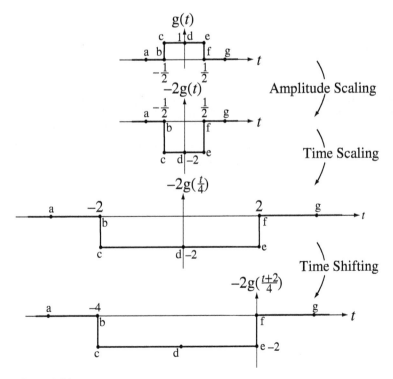

Figure 2.34
A sequence of amplitude scaling, time scaling and time shifting a function

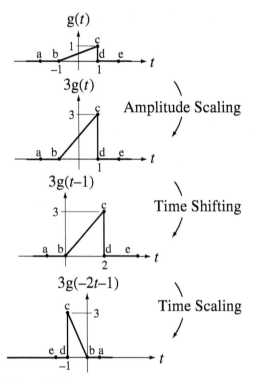

Figure 2.35
A sequence of amplitude scaling, time shifting and time scaling a function

alphabetically. As each functional change is made, corresponding points have the same letter designation.

The functions previously introduced, along with function scaling and shifting, allow us to describe a wide variety of signals. A signal that has a decaying exponential shape after some time $t = t_0$ and is zero before that can be represented in the compact mathematical form $x(t) = Ae^{-t/\tau}u(t - t_0)$ (Figure 2.36).

A signal that has the shape of a negative sine function before time $t = 0$ and a positive sine function after time $t = 0$ can be represented by $x(t) = A \sin(2\pi f_0 t) \, \text{sgn}(t)$ (Figure 2.37).

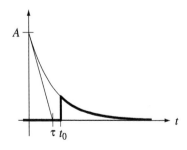

Figure 2.36
A decaying exponential "switched" on at time $t = t_0$

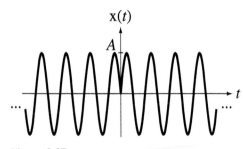

Figure 2.37
Product of a sine and a signum function

A signal that is a burst of a sinusoid between times $t = 1$ and $t = 5$ and zero elsewhere can be represented by $x(t) = A \cos(2\pi f_0 t + \theta) \, \text{rect}((t - 3)/4)$ (Figure 2.38).

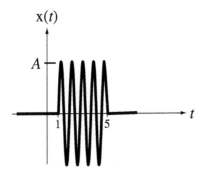

Figure 2.38
A sinusoidal "burst"

EXAMPLE 2.2

Graphing function scaling and shifting with MATLAB

Using MATLAB, graph the function defined by

$$g(t) = \begin{cases} 0, & t < -2 \\ -4 - 2t, & -2 < t < 0 \\ -4 - 3t, & 0 < t < 4 \\ 16 - 2t, & 4 < t < 8 \\ 0, & t > 8 \end{cases}.$$

Then graph the functions $3g(t + 1)$, $(1/2)g(3t)$, $-2g((t - 1)/2)$.

We must first choose a range of t over which to graph the function and the space between points in t to yield a curve that closely approximates the actual function. Let's choose a range of $-5 < t < 20$ and a space between points of 0.1. Also, let's use the function feature of MATLAB that allows us to define the function $g(t)$ as a separate MATLAB program, an m file. Then we can refer to it when graphing the transformed functions and not have to retype the function description. The g.m file contains the following code.

```
function y = g(t)
   % Calculate the functional variation for each range of time, t
   y1 = -4 - 2*t ; y2 = -4 + 3*t ; y3 = 16 - 2*t ;
   % Splice together the different functional variations in
   % their respective ranges of validity
   y = y1.*(-2<t & t<=0) + y2.*(0<t & t<=4) + y3.*(4<t & t<=8) ;
```

The MATLAB program contains the following code.

```
%   Program to graph the function, g(t) = t^2 + 2*t - 1 and then to
%   graph 3*g(t+1), g(3*t)/2 and -2*g((t-1)/2).
tmin = -4 ; tmax = 20 ;      % Set the time range for the graph
dt = 0.1 ;                    % Set the time between points
t = tmin:dt:tmax ;            % Set the vector of times for the graph
g0 = g(t) ;                   % Compute the original "g(t)"
```

```
g1 = 3*g(t+1) ;                 % Compute the first change
g2 = g(3*t)/2 ;                 % Compute the second change
g3 = -2*g((t-1)/2) ;            % Compute the third change

% Find the maximum and minimum g values in all the scaled or shifted
% functions and use them to scale all graphs the same

gmax = max([max(g0), max(g1), max(g2), max(g3)]) ;
gmin = min([min(g0), min(g1), min(g2), min(g3)]) ;

% Graph all four functions in a 2 by 2 arrangement
% Graph them all on equal scales using the axis command
% Draw grid lines, using the grid command, to aid in reading values

subplot(2,2,1) ; p = plot(t,g0,'k') ; set(p,'LineWidth',2) ;
xlabel('t') ; ylabel('g(t)') ; title('Original Function, g(t)') ;
axis([tmin,tmax,gmin,gmax]) ; grid ;
subplot(2,2,2) ; p = plot(t,g1,'k') ; set(p,'LineWidth',2) ;
xlabel('t') ; ylabel('3g(t+1)') ; title('First Change) ;
axis([tmin,tmax,gmin,gmax]) ; grid ;
subplot(2,2,3) ; p = plot(t,g2,'k') ; set(p,'LineWidth',2) ;
xlabel('t') ; ylabel('g(3t)/2') ; title('Second Change) ;
axis([tmin,tmax,gmin,gmax]) ; grid ;
subplot(2,2,4) ; p = plot(t,g3,'k') ; set(p,'LineWidth',2) ;
xlabel('t') ; ylabel('-2g((t-1)/2)') ; title('Third Change) ;
axis([tmin,tmax,gmin,gmax]) ; grid ;
```

The graphical results are displayed in Figure 2.39.

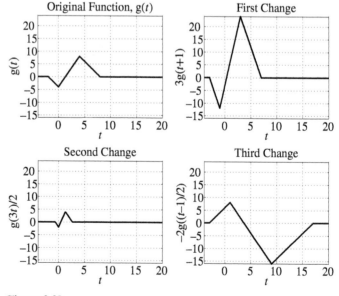

Figure 2.39
MATLAB graphs of scaled and/or shifted functions

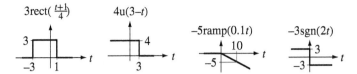

Figure 2.40
More examples of amplitude-scaled, time-shifted and time-scaled functions

Figure 2.40 shows more examples of amplitude-scaled, time-shifted and time-scaled versions of the functions just introduced.

2.6 DIFFERENTIATION AND INTEGRATION

Integration and differentiation are common signal processing operations in practical systems. The derivative of a function at any time t is its slope *at* that time and the integral of a function at any time t is the accumulated area under the function *up to* that time. Figure 2.41 illustrates some functions and their derivatives. The zero crossings of all the derivatives have been indicated by light vertical lines that lead to the maxima and minima of the corresponding function.

There is a function `diff` in MATLAB that does symbolic differentiation.

```
>> x = sym('x') ;
>> diff(sin(x^2))
ans =
2*cos(x^2)*x
```

This function can also be used numerically to find the differences between adjacent values in a vector. These finite differences can then be divided by the increment of the independent variable to approximate some derivatives of the function that produced the vector.

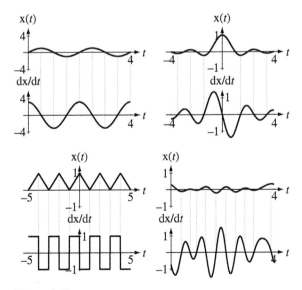

Figure 2.41
Some functions and their derivatives

```
>> dx = 0.1 ; x = 0.3:dx:0.8 ; exp(x)
ans =
    1.3499    1.4918    1.6487    1.8221    2.0138    2.2255
>> diff(exp(x))/dx
ans =
    1.4197    1.5690    1.7340    1.9163    2.1179
```

Integration is a little more complicated than differentiation. Given a function, its derivative is unambiguously determinable (if it exists). However, its integral is not unambiguously determinable without some more information. This is inherent in one of the first principles learned in integral calculus. If a function g(x) has a derivative g'(x), then the function g(x) + K (K a constant) has the same derivative g'(x) regardless of the value of the constant K. Since integration is the opposite of differentiation, what is the integral of g'(x)? It could be g(x) but it could also be g(x) + K.

The term integral has different meanings in different contexts. Generally speaking, integration and differentiation are inverse operations. An **antiderivative** of a function of time g(t) is any function of time that, when differentiated with respect to time, yields g(t). An antiderivative is indicated by an integral sign without limits. For example,

$$\frac{\sin(2\pi t)}{2\pi} = \int \cos(2\pi t)dt.$$

In words, $\sin(2\pi t)/2\pi$ is an antiderivative of $\cos(2\pi t)$. An **indefinite integral** is an antiderivative plus a constant. For example, $h(t) = \int g(t)dt + C$. A **definite integral** is an integral taken between two limits. For example, $A = \int_\alpha^\beta g(t)dt$. If α and β are constants, then A is also a constant, the area under g(t) between α and β. In signal and system analysis, a particular form of definite integral $h(t) = \int_{-\infty}^{t} g(\tau)d\tau$ is often used. The variable of integration is τ, so during the integration process, the upper integration limit t is treated like a constant. But after the integration is finished t is the independent variable in h(t). This type of integral is sometimes called a **running integral** or a **cumulative integral**. It is the accumulated area under a function for all time before t and that depends on what t is.

Often, in practice, we know that a function of time is zero before $t = t_0$. Then we know that $\int_{-\infty}^{t_0} g(t)dt$ is zero. Then the integral of that function from any time $t_1 < t_0$ to any time $t > t_0$ is unambiguous. It can only be the area under the function from time $t = t_0$ to time t:

$$\int_{t_1}^{t} g(\tau)d\tau = \underbrace{\int_{t_1}^{t_0} g(\tau)d\tau}_{=0} + \int_{t_0}^{t} g(\tau)d\tau = \int_{t_0}^{t} g(\tau)d\tau.$$

Figure 2.42 illustrates some functions and their integrals.

In Figure 2.42 the two functions on the right are zero before time $t = 0$ and the integrals illustrated assume a lower limit on the integral less than zero, thereby producing a single, unambiguous result. The two on the left are illustrated with multiple possible integrals, differing from each other only by constants. They all have the same derivative and are all equally valid candidates for the integral in the absence of extra information.

There is a function `int` in MATLAB that can do symbolic integration.

```
>> sym('x') ;
>> int(1/(1+x^2))
ans =
atan(x)
```

2.7 Even and Odd Signals

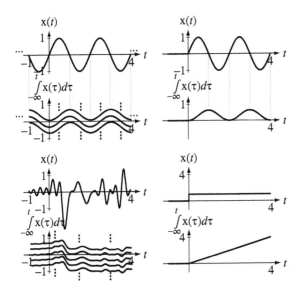

Figure 2.42
Some functions and their integrals

This function cannot be used to do numerical integration. Another function cumsum can be used to do numerical integration.

```
>> cumsum(1:5)
ans =
     1     3     6    10    15
>> dx = pi/16 ; x = 0:dx:pi/4 ; y = sin(x)
y =
     0    0.1951    0.3827    0.5556    0.7071
>> cumsum(y)*dx
ans =
     0    0.0383    0.1134    0.2225    0.3614
```

There are also other more sophisticated numerical integration functions in MATLAB, for example, trapz, which uses a trapezoidal approximation, and quad, which uses adaptive Simpson quadrature.

2.7 EVEN AND ODD SIGNALS

Some functions have the property that, when they undergo certain types of shifting and/or scaling, the function values do not change. They are **invariant** under that shifting and/or scaling. An **even** function of t is invariant under time reversal $t \to -t$ and an **odd** function of t is invariant under the amplitude scaling and time reversal $g(t) \to -g(-t)$.

> An even function $g(t)$ is one for which $g(t) = g(-t)$ and an odd function is one for which $g(t) = -g(-t)$.

A simple way of visualizing even and odd functions is to imagine that the ordinate axis (the $g(t)$ axis) is a mirror. For even functions, the part of $g(t)$ for $t > 0$ and the part of $g(t)$ for $t < 0$ are mirror images of each other. For an odd function, the same two parts of the function are *negative* mirror images of each other (Figure 2.43 and Figure 2.44).

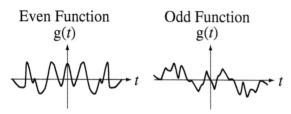

Figure 2.43
Examples of even and odd functions

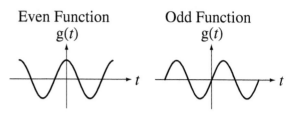

Figure 2.44
Two very common and useful functions, one even and one odd

Some functions are even, some are odd, and some are neither even nor odd. But any function $g(t)$ is the sum of its even and odd *parts*, $g(t) = g_e(t) + g_o(t)$. The even and odd parts of a function $g(t)$ are

$$g_e(t) = \frac{g(t) + g(-t)}{2}, \quad g_o(t) = \frac{g(t) - g(-t)}{2}. \tag{2.13}$$

If the odd part of a function is zero, the function is even, and if the even part of a function is zero, the function is odd.

EXAMPLE 2.3

Even and odd parts of a function

What are the even and odd parts of the function $g(t) = t(t^2 + 3)$?
They are

$$g_e(t) = \frac{g(t) + g(-t)}{2} = \frac{t(t^2+3) + (-t)[(-t)^2+3]}{2} = 0$$

$$g_o(t) = \frac{t(t^2+3) - (-t)[(-t)^2+3]}{2} = t(t^2+3)$$

Therefore $g(t)$ is an odd function.

```
%   Program to graph the even and odd parts of a function

function GraphEvenAndOdd
    t = -5:0.1:5 ;                  % Set up a time vector for the graph
    ge = (g(t) + g(-t))/2 ;         % Compute the even-part values
    go = (g(t) - g(-t))/2 ;         % Compute the odd-part values
    %   Graph the even and odd parts
    subplot(2,1,1) ;
    ptr = plot(t,ge,'k') ; set(ptr,'LineWidth',2) ; grid on ;
    xlabel('\itt','FontName','Times','FontSize',24) ;
    ylabel('g_e({\itt})','FontName','Times','FontSize',24) ;
    subplot(2,1,2) ;
    ptr = plot(t,go,'k') ; set(ptr,'LineWidth',2) ; grid on ;
    xlabel('\itt','FontName','Times','FontSize',24) ;
    ylabel('g_o({\itt})','FontName','Times','FontSize',24) ;

function y = g(x)       % Function definition for g(x)
    y = x.*(x.^2+3) ;
```

Figure 2.45 illustrates the graphical output of the MATLAB program.

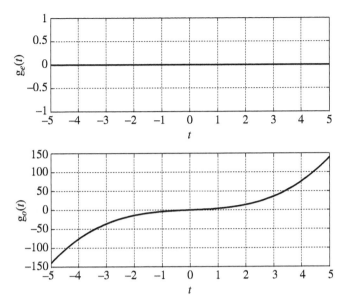

Figure 2.45
Graphical output of the MATLAB program

This MATLAB code example begins with the keyword function. A MATLAB program file that does not begin with function is called a **script file**. One that does begin with function defines a function. This code example contains two function definitions. The second function is called a **subfunction**. It is used only by the main function (in this case GraphEvenAndOdd) and is not accessible by any functions or scripts exterior to this function definition. A function may have any number of subfunctions. A script file cannot use subfunctions.

■

COMBINATIONS OF EVEN AND ODD SIGNALS

Let $g_1(t)$ and $g_2(t)$ both be even functions. Then $g_1(t) = g_1(-t)$ and $g_2(t) = g_2(-t)$. Let $g(t) = g_1(t) + g_2(t)$. Then $g(-t) = g_1(-t) + g_2(-t)$ and, using the evenness of $g_1(t)$ and $g_2(t)$, $g(-t) = g_1(t) + g_2(t) = g(t)$, proving that the sum of two even functions is also even. Now let $g(t) = g_1(t) g_2(t)$. Then $g(-t) = g_1(-t)g_2(-t) = g_1(t)g_2(t) = g(t)$, proving that the product of two even functions is also even.

Now let $g_1(t)$ and $g_2(t)$ both be odd. Then $g(-t) = g_1(-t) + g_2(-t) = -g_1(t) - g_2(t) = -g(t)$, proving that the sum of two odd functions is odd. Then let $g(-t) = g_1(-t)g_2(-t) = [-g_1(t)][-g_2(t)] = g_1(t) g_2(t) = g(t)$, proving that the *product* of two *odd* functions is *even*.

By similar reasoning we can show that if two functions are even, their sum, difference, product and quotient are even too. If two functions are odd, their sum and difference are odd but their product and quotient are even. If one function is even and the other is odd, their product and quotient are odd (Figure 2.46).

Function Types	Sum	Difference	Product	Quotient
Both Even	Even	Even	Even	Even
Both Odd	Odd	Odd	Even	Even
One Even, One Odd	Neither	Neither	Odd	Odd

Figure 2.46
Combinations of even and odd functions

The most important even and odd functions in signal and system analysis are cosines and sines. Cosines are even and sines are odd. Figure 2.47 through Figure 2.49 give some examples of products of even and odd functions.

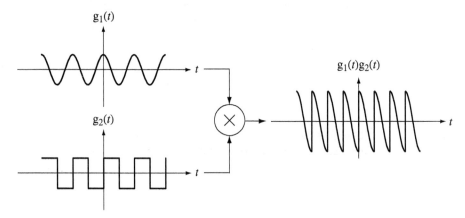

Figure 2.47
Product of even and odd functions

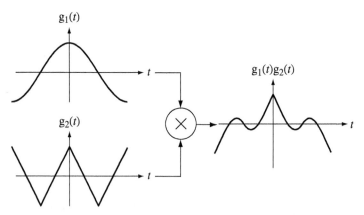

Figure 2.48
Product of two even functions

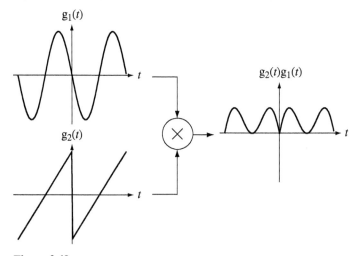

Figure 2.49
Product of two odd functions

Let g(t) be an even function. Then g(t) = g(−t). Using the chain rule of differentiation, the derivative of g(t) is g'(t) = −g'(−t), an odd function. So the derivative of any even function is an odd function. Similarly, the derivative of any odd function is an even function. We can turn the arguments around to say that the integral of any even function is an odd function *plus a constant of integration*, and the integral of any odd function is an even function plus a constant of integration (and therefore still even because a constant is an even function) (Figure 2.50).

Function Type	Derivative	Integral
Even	Odd	Odd+Constant
Odd	Even	Even

Figure 2.50
Function types and the types of their derivatives and integrals

DERIVATIVES AND INTEGRALS OF EVEN AND ODD SIGNALS

The definite integrals of even and odd functions can be simplified in certain common cases. If g(t) is an even function and a is a real constant,

$$\int_{-a}^{a} g(t)dt = \int_{-a}^{0} g(t)dt + \int_{0}^{a} g(t)dt = -\int_{0}^{-a} g(t)dt + \int_{0}^{a} g(t)dt.$$

Making the change of variable $\tau = -t$ in the first integral on the right and using $g(\tau) = g(-\tau)$, $\int_{-a}^{a} g(t)dt = 2\int_{0}^{a} g(t)dt$, which should be geometrically obvious by looking at the graph of the function (Figure 2.51(a)). By similar reasoning, if g(t) is an odd function, then $\int_{-a}^{a} g(t)dt = 0$, which should also be geometrically obvious (Figure 2.51(b)).

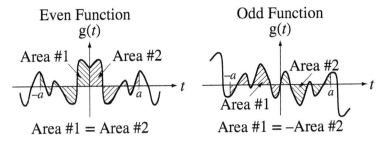

Figure 2.51
Integrals of (a) even functions and (b) odd functions over symmetrical limits

2.8 PERIODIC SIGNALS

A periodic signal is one that has been repeating a pattern for a semi-infinite time and will continue to repeat that pattern for a semi-infinite time.

> A periodic function g(t) is one for which $g(t) = g(t + nT)$ for any integer value of n, where T is a **period** of the function.

Another way of saying that a function of t is periodic is to say that it is invariant under the time shift $t \rightarrow t + nT$. The function repeats every T seconds. Of course it also repeats every $2T$, $3T$ or nT seconds (n is an integer). Therefore $2T$ or $3T$ or nT are all periods of the function. The *minimum positive* interval over which a function repeats is called its **fundamental period** T_0. The **fundamental cyclic frequency** f_0 is the reciprocal of the fundamental period $f_0 = 1/T_0$ and the **fundamental radian frequency** is $\omega_0 = 2\pi f_0 = 2\pi/T_0$.

Some common examples of periodic functions are real or complex sinusoids and combinations of real and/or complex sinusoids. We will see later that other, more complicated types of periodic functions with different periodically repeating shapes can be generated and mathematically described. Figure 2.52 gives some examples of periodic functions. A function that is not periodic is called an **aperiodic** function. (Because of the similarity of the phrase "aperiodic function" and the phrase "a periodic function," it is probably better when speaking to use the term "nonperiodic" or "not periodic" to avoid confusion.)

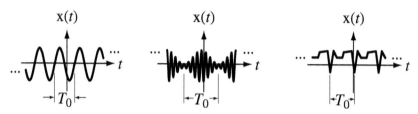

Figure 2.52
Examples of periodic functions with fundamental period T_0

In practical systems, a signal is never actually periodic because it did not exist until it was created at some finite time in the past, and it will stop at some finite time in the future. However, often a signal has been repeating for a very long time before the time we want to analyze the signal and will repeat for a very long time after that. In many cases, approximating the signal by a periodic function introduces negligible error. Examples of signals that would be properly approximated by periodic functions would be rectified sinusoids in an AC to DC converter, horizontal sync signals in a television, the angular shaft position of a generator in a power plant, the firing pattern of spark plugs in an automobile traveling at constant speed, the vibration of a quartz crystal in a wristwatch, the angular position of a pendulum on a grandfather clock and so on. Many natural phenomena are, for all practical purposes, periodic; most planet, satellite and comet orbital positions, the phases of the moon, the electric field emitted by a Cesium atom at resonance, the migration patterns of birds, the caribou mating season and so forth. Periodic phenomena play a large part both in the natural world and in the realm of artificial systems.

A common situation in signal and system analysis is to have a signal that is the sum of two periodic signals. Let $x_1(t)$ be a periodic signal with fundamental period T_{01}, and let $x_2(t)$ be a periodic signal with fundamental period T_{02}, and let $x(t) = x_1(t) + x_2(t)$. Whether or not $x(t)$ is periodic depends on the relationship between the two periods T_{01} and T_{02}. If a time T can be found that is an integer multiple of T_{01} and also an integer multiple of T_{02}, then T is a period of both $x_1(t)$ and $x_2(t)$ and

$$x_1(t) = x_1(t + T) \text{ and } x_2(t) = x_2(t + T). \tag{2.14}$$

Time shifting $x(t) = x_1(t) + x_2(t)$ with $t \to t + T$,

$$x(t + T) = x_1(t + T) + x_2(t + T). \tag{2.15}$$

Then, combining (2.15) with (2.14),

$$x(t + T) = x_1(t) + x_2(t) = x(t)$$

proving that $x(t)$ is periodic with period T. The smallest positive value of T that is an integer multiple of both T_{01} and T_{02} is the fundamental period T_0 of $x(t)$. This smallest value of T is called the **least common multiple (LCM)** of T_{01} and T_{02}. If T_{01}/T_{02} is a rational number (a ratio of integers), the LCM is finite and x(t) is periodic. If T_{01}/T_{02} is an irrational number, x(t) is aperiodic.

Sometimes an alternate method for finding the period of the sum of two periodic functions is easier than finding the LCM of the two periods. If the fundamental period of the sum is the LCM of the two fundamental periods of the two functions, then the fundamental frequency of the sum is the **greatest common divisor (GCD)** of the two fundamental frequencies and is therefore the reciprocal of the LCM of the two fundamental periods.

EXAMPLE 2.4

Fundamental period of a signal

Which of these functions are periodic and, if one is, what is its fundamental period?

(a) $g(t) = 7 \sin(400\pi t)$

The sine function repeats when its total argument is increased or decreased by any integer multiple of 2π radians. Therefore

$$\sin(400\pi t \pm 2n\pi) = \sin[400\pi(t \pm nT_0)]$$

Setting the arguments equal,

$$400\pi t \pm 2n\pi = 400\pi(t \pm nT_0)$$

or

$$\pm 2n\pi = \pm 400\pi n T_0$$

or

$$T_0 = 1/200.$$

An alternate way of finding the fundamental period is to realize that $7 \sin(400\pi t)$ is in the form $A \sin(2\pi f_0 t)$ or $A \sin(\omega_0 t)$, where f_0 is the fundamental cyclic frequency and ω_0 is the fundamental radian frequency. In this case, $f_0 = 200$ and $\omega_0 = 400\pi$. Since the fundamental period is the reciprocal of the fundamental cyclic frequency, $T_0 = 1/200$.

(b) $g(t) = 3 + t^2$

This is a second-degree polynomial. As t increases or decreases from zero, the function value increases monotonically (always in the same direction). No function that increases monotonically can be periodic because if a fixed amount is added to the argument t, the function must be larger or smaller than for the current t. This function is not periodic.

(c) $g(t) = e^{-j60\pi t}$

This is a complex sinusoid. That is easily seen by expressing it as the sum of a cosine and a sine through Euler's identity,

$$g(t) = \cos(60\pi t) - j \sin(60\pi t).$$

The function g(t) is a linear combination of two periodic signals that have the same fundamental cyclic frequency $60\pi/2\pi = 30$. Therefore the fundamental frequency of g(t) is 30 Hz and the fundamental period is 1/30 s.

(d) $g(t) = 10 \sin(12\pi t) + 4 \cos(18\pi t)$

This is the sum of two functions that are both periodic. Their fundamental periods are 1/6 second and 1/9 second. The LCM is 1/3 second. (See Web Appendix B for a systematic method for finding least common multiples.) There are two fundamental periods of the first function and three fundamental periods of the second function in that time. Therefore the fundamental period of the overall function is 1/3 second (Figure 2.53). The two fundamental frequencies are 6 Hz and 9 Hz. Their GCD is 3 Hz, which is the reciprocal of 1/3 second, the LCM of the two fundamental periods.

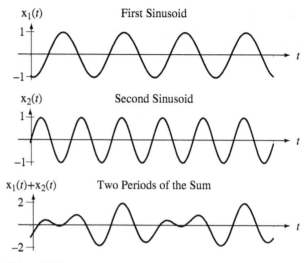

Figure 2.53
Signals with frequencies of 6 Hz and 9 Hz and their sum

(e) $g(t) = 10 \sin(12\pi t) + 4 \cos(18t)$

This function is exactly like the function in (d) except that a π is missing in the second argument. The fundamental periods are now 1/6 second and $\pi/9$ seconds, and the ratio of the two fundamental periods is either $2\pi/3$ or $3/2\pi$, both of which are irrational. Therefore $g(t)$ is aperiodic. This function, although made up of the sum of two periodic functions, is not periodic because it does not repeat exactly in a finite time. (It is sometimes referred to as "almost periodic" because, in looking at a graph of the function, it seems to repeat in a finite time. But, strictly speaking, it is aperiodic.)

There is a function lcm in MATLAB for finding least common multiples. It is somewhat limited because it accepts only two arguments, which can be scalar integers or arrays of integers. There is also a function gcd, which finds the greatest common divisor of two integers or two arrays of integers.

```
>> lcm(32,47)
ans =
        1504
>> gcd([93,77],[15,22])
ans =
     3    11
```

2.9 SIGNAL ENERGY AND POWER

SIGNAL ENERGY

All physical activity is mediated by a transfer of energy. Real physical systems respond to the energy of an excitation. It is important at this point to establish some terminology describing the energy and power of signals. In the study of signals in systems, the

2.9 Signal Energy and Power

signals are often treated as mathematical abstractions. Often the physical significance of the signal is ignored for the sake of simplicity of analysis. Typical signals in electrical systems are voltages or currents, but they could be charge or electric field or some other physical quantity. In other types of systems a signal could be a force, a temperature, a chemical concentration, a neutron flux and so on. Because of the many different kinds of physical signals that can be operated on by systems, the term **signal energy** has been defined. Signal energy (as opposed to just energy) of a signal is defined as the area under the square of the magnitude of the signal. The signal energy of a signal x(t) is

$$E_x = \int_{-\infty}^{\infty} |x(t)|^2 dt \quad . \qquad (2.16)$$

The units of signal energy depend on the units of the signal. If the signal unit is the volt (V), the signal energy is expressed in $V^2 \cdot s$. Signal energy is *proportional to* the actual physical energy delivered by a signal but *not necessarily equal to* that physical energy. In the case of a current signal i(t) through a resistor R the actual energy delivered to the resistor would be

$$\text{Energy} = \int_{-\infty}^{\infty} |i(t)|^2 R \, dt = R \int_{-\infty}^{\infty} |i(t)|^2 dt = RE_i.$$

Signal energy is proportional to actual energy and the proportionality constant, in this case, is R. For a different kind of signal, the proportionality constant would be different. In many kinds of system analysis the use of signal energy is more convenient than the use of actual physical energy.

EXAMPLE 2.5

Signal energy of a signal

Find the signal energy of $x(t) = \begin{cases} 3(1 - |t/4|), & |t| < 4 \\ 0, & \text{otherwise} \end{cases}$.

From the definition of signal energy

$$E_x = \int_{-\infty}^{\infty} |x(t)|^2 dt = \int_{-4}^{4} |3(1 - |t/4|)|^2 dt$$

Taking advantage of the fact that x(t) is an even function

$$E_x = 2 \times 3^2 \int_0^4 (1 - t/4)^2 dt = 18 \int_0^4 \left(1 - \frac{t}{2} + \frac{t^2}{16}\right) dt = 18 \left[t - \frac{t^2}{4} + \frac{t^3}{48}\right]_0^4 = 24$$

```
dt = 0.01 ;         %   Time increment
t = -8:dt:8 ;       %   Time vector for computing samples of x(t)
%   Compute samples of x(t)
x = 3*(1-abs(t/4)).*(abs(t)<4) ;
%   Compute energy of x(t) using trapezoidal rule approximation to the
%   integral
Ex = trapz(x.^2)*dt ;
disp(['Signal Energy = ',num2str(Ex)]) ;

Signal Energy = 24.0001
```

SIGNAL POWER

For many signals, the integral $E_x = \int_{-\infty}^{\infty} |x(t)|^2 dt$ does not converge because the signal energy is infinite. This usually occurs because the signal is not **time limited**. (The term *time limited* means that the signal is nonzero over only a finite time.) An example of a signal with infinite energy is the sinusoidal signal $x(t) = A\cos(2\pi f_0 t)$, $A \neq 0$. Over an infinite time interval, the area under the square of this signal is infinite. For signals of this type, it is more convenient to deal with the average signal power instead of the signal energy. Average signal power of a signal $x(t)$ is defined by

$$P_x = \lim_{T \to \infty} \frac{1}{T} \int_{-T/2}^{T/2} |x(t)|^2 dt . \qquad (2.17)$$

The integral is the signal energy of the signal over a time T and it is then divided by T, yielding the average signal power over time T. Then, as T approaches infinity, this average signal power becomes the average signal power over all time.

For periodic signals, the average signal power calculation may be simpler. The average value of any periodic function is the average over any period. Therefore, since the square of a periodic function is also periodic, for periodic signals,

$$P_x = \frac{1}{T} \int_{t_0}^{t_0+T} |x(t)|^2 dt = \frac{1}{T} \int_T |x(t)|^2 dt$$

where the notation $\int_T$ means the same thing as $\int_{t_0}^{t_0+T}$ for any arbitrary choice of t_0, where T can be any period of $|x(t)|^2$.

EXAMPLE 2.6

Signal power of a sinusoidal signal

Find the average signal power of $x(t) = A\cos(2\pi f_0 t + \theta)$.
From the definition of average signal power for a periodic signal,

$$P_x = \frac{1}{T} \int_T |A\cos(2\pi f_0 t + \theta)|^2 dt = \frac{A^2}{T_0} \int_{-T_0/2}^{T_0/2} \cos^2(2\pi t/T_0 + \theta) dt.$$

Using the trigonometric identity

$$\cos(x)\cos(y) = (1/2)[\cos(x-y) + \cos(x+y)]$$

we get

$$P_x = \frac{A^2}{2T_0} \int_{-T_0/2}^{T_0/2} [1 + \cos(4\pi t/T_0 + 2\theta)] dt = \frac{A^2}{2T_0} \int_{-T_0/2}^{T_0/2} dt + \underbrace{\frac{A^2}{2T_0} \int_{-T_0/2}^{T_0/2} \cos(4\pi t/T_0 + 2\theta) dt}_{=0} = \frac{A^2}{2}$$

The second integral on the right is zero because it is the integral of a sinusoid over two fundamental periods. The signal power is $P_x = A^2/2$. This result is independent of the phase θ and the frequency f_0. It depends only on the amplitude A.

```
A = 1 ;             % Amplitude of x(t)
th = 0 ;            % Phase shift of x(t)
f0 = 1 ;            % Fundamental frequency
T0 = 1/f0 ;         % Fundamental period
dt = T0/100 ;       % Time increment for sampling x(t)
```

```
t = 0:dt:T0 ;        %   Time vector for computing samples of x(t)
%   Compute samples of x(t) over one fundamental period
x = A*cos(2*pi*f0*t + th) ;
%   Compute signal power using trapezoidal approximation to integral
Px = trapz(x.^2)*dt/T0 ;
disp(['Signal Power = ',num2str(Px)]) ;

Signal Power = 0.5
```

■

Signals that have finite signal energy are referred to as **energy signals** and signals that have infinite signal energy but finite average signal power are referred to as **power signals**. No real physical signal can actually have infinite energy or infinite average power because there is not enough energy or power in the universe available. But we often analyze signals that, according to their strict mathematical definition, have infinite energy, a sinusoid, for example. How relevant can an analysis be if it is done with signals that cannot physically exist? Very relevant! The reason mathematical sinusoids have infinite signal energy is that they have always existed and will always exist. Of course practical signals never have that quality. They all had to begin at some finite time and they will all end at some later finite time. They are actually time limited and have finite signal energy. But in much system analysis the analysis is steady-state analysis of a system in which all signals are treated as periodic. The analysis is still relevant and useful because it is a good approximation to reality, it is often much simpler than an exact analysis, and it yields useful results. All periodic signals are power signals (except for the trivial signal $x(t) = 0$) because they all endure for an infinite time.

EXAMPLE 2.7

Finding signal energy and power of signals using MATLAB

Using MATLAB, find the signal energy or power of the signals

(a) $x(t) = 4e^{-t/10}\text{rect}\left(\frac{t-4}{3}\right)$,

(b) A periodic signal of fundamental period 10 described over one period by
$x(t) = -3t, \quad -5 < t < 5$.

Then compare the results with analytical calculations.

```
%   Program to compute the signal energy or power of some example signals
%   (a)

dt = 0.1 ; t = -7:dt:13 ;    %   Set up a vector of times at which to
                             %   compute the function. Time interval
                             %   is 0.1

%   Compute the function values and their squares
x = 4*exp(-t/10).*rect((t-4)/3) ;
xsq = x.^2 ;

Ex = trapz(t,xsq) ;          %   Use trapezoidal-rule numerical
                             %   integration to find the area under
```

```
                                    %   the function squared and display the
                                    %   result
disp(['(a) Ex = ',num2str(Ex)]) ;

%   (b)
T0 = 10 ;                           %   The fundamental period is 10.
dt = 0.1 ; t = -5:dt:5 ;            %   Set up a vector of times at which to
                                    %   compute the function. Time interval
                                    %   is 0.1.
x = -3*t ; xsq = x.^2 ;             %   Compute the function values and
                                    %   their squares over one fundamental
                                    %   period
Px = trapz(t,xsq)/T0 ;              %   Use trapezoidal-rule numerical
                                    %   integration to find the area under
                                    %   the function squared, divide the
                                    %   period and display the result
disp(['(b) Px = ',num2str(Px)]) ;
```

The output of this program is

(a) Ex = 21.5177
(b) Px = 75.015

Analytical computations:

(a) $E_x = \int_{-\infty}^{\infty} |x(t)|^2 dt = \int_{2.5}^{5.5} |4e^{-t/10}|^2 dt = 16 \int_{2.5}^{5.5} e^{-t/5} d\tau = -5 \times 16 [e^{-t/5}]_{2.5}^{5.5} = 21.888$

(The small difference in results is probably due to the error inherent in trapezoidal-rule integration. It could be reduced by using time points spaced more closely together.)

(b) $P_x = \frac{1}{10} \int_{-5}^{5} (-3t)^2 dt = \frac{1}{5} \int_{0}^{5} 9t^2 dt = \frac{1}{5}(3t^3)_0^5 = \frac{375}{5} = 75$ Check.

2.10 SUMMARY OF IMPORTANT POINTS

1. The term *continuous* and the term *continuous-time* mean different things.
2. A continuous-time impulse, although very useful in signal and system analysis, is not a function in the ordinary sense.
3. Many practical signals can be described by combinations of shifted and/or scaled standard functions, and the order in which scaling and shifting are done is significant.
4. Signal energy is, in general, not the same thing as the actual physical energy delivered by a signal.
5. A signal with finite signal energy is called an *energy signal* and a signal with infinite signal energy and finite average power is called a *power signal*.

EXERCISES WITH ANSWERS

(Answers to each exercise are in random order.)

Signal Functions

1. If $g(t) = 7e^{-2t-3}$ write out and simplify

   (a) $g(3)$ (b) $g(2-t)$ (c) $g(t/10 + 4)$

   (d) $g(jt)$ (e) $\dfrac{g(jt) + g(-jt)}{2}$

   (f) $\dfrac{g\left(\dfrac{jt-3}{2}\right) + g\left(\dfrac{-jt-3}{2}\right)}{2}$

   Answers: $0.3485 \cos(2t)$, $7\cos(t)$, 8.6387×10^{-4}, $7e^{-j2t-3}$, $7e^{-t/5-11}$, $7e^{-7-2t}$

2. If $g(x) = x^2 - 4x + 4$ write out and simplify

   (a) $g(z)$ (b) $g(u+v)$ (c) $g(e^{jt})$

   (d) $g(g(t))$ (e) $g(2)$

   Answers: $u^2 + v^2 + 2uv - 4u - 4v + 4$, $t^4 - 8t^3 + 20t^2 - 16t + 4$, 0, $(e^{jt} - 2)^2$, $z^2 - 4z + 4$

3. Find the magnitudes and phases of these complex quantities.

   (a) $e^{-(3+j2.3)}$ (b) e^{2-j6} (c) $\dfrac{100}{8 + j13}$

   Answers: -1.0191 radians, 6.5512, 0.0498, -2.3 radians, 7.3891, -6 radians

4. Let $G(f) = \dfrac{j4f}{2 + j7f/11}$.

   (a) What value does the magnitude of this function approach as f approaches positive infinity?
   (b) What value (in radians) does the phase of this function approach as f approaches zero from the positive side?

   Answer: $\pi/2$ radians, 6.285

5. Let $X(f) = \dfrac{jf}{jf + 10}$

   (a) Find the magnitude $|X(4)|$ and the angle $\angle X(4)$ in radians.
   (b) What value (in radians) does $\angle X(f)$ approach as f approaches zero from the positive side?

   Answers: $\pi/2$ radians, 1.19 radians, 0.3714

Shifting and Scaling

6. For each function g(t) graph g(−t), −g(t), g(t − 1) and g(2t).

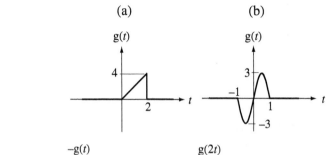

Answers:

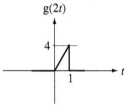 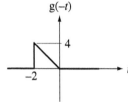

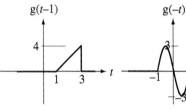 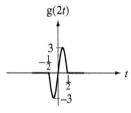

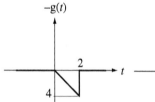 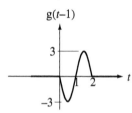

7. Find the values of the following signals at the indicated times.

   (a) $x(t) = 2\,\text{rect}(t/4)$, $x(-1)$
   (b) $x(t) = 5\,\text{rect}(t/2)\,\text{sgn}(2t)$, $x(0.5)$
   (c) $x(t) = 9\,\text{rect}(t/10)\,\text{sgn}(3(t-2))$, $x(1)$

   Answers: 2, −9, 5

8. For each pair of functions in Figure E.8 provide the values of the constants A, t_0 and w in the functional transformation $g_2(t) = A g_1((t - t_0)/w)$.

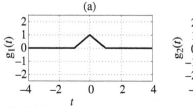 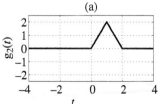

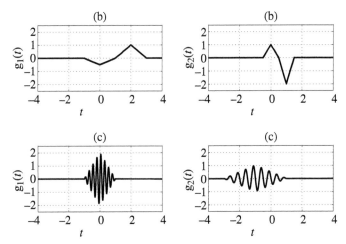

Figure E.8

Answers: $A = -1/2, t_0 = -1, w = 2$, $A = -2, t_0 = 0, w = 1/2$,
$A = 2, t_0 = 1, w = 1$

9. For each pair of functions in Figure E.9 provide the values of the constants A, t_0 and a in the functional transformation $g_2(t) = Ag_1(w(t - t_0))$.

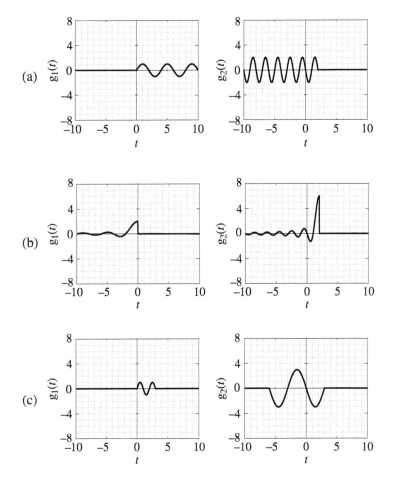

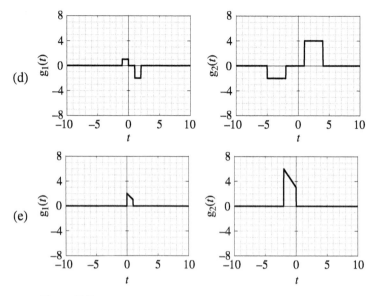

(d) ...
(e) ...

Figure E.9

Answers: $A = 3$, $w = 2$, $t_0 = 2$, $A = -2$, $w = 1/3$, $t_0 = -2$,
$A = 2$, $w = -2$, $t_0 = 2$, $A = 3$, $w = 1/2$, $t_0 = -2$
$A = -3$, $w = 1/3$ or $-1/3$, $t_0 = -6$ or 3

10. In Figure E.10 is plotted a function $g_1(t)$ that is zero for all time outside the range plotted. Let some other functions be defined by

$$g_2(t) = 3g_1(2-t), \quad g_3(t) = -2g_1(t/4), \quad g_4(t) = g_1\left(\frac{t-3}{2}\right)$$

Find these values.

(a) $g_2(1)$

(b) $g_3(-1)$

(c) $[g_4(t)g_3(t)]_{t=2}$

(d) $\int_{-3}^{-1} g_4(t)dt$

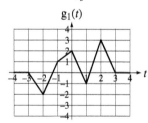

Figure E.10

Answers: $-3, -2, -3/2, -3.5$

11. A function $G(f)$ is defined by

$$G(f) = e^{-j2\pi f}\text{rect}(f/2).$$

Graph the magnitude and phase of $G(f-10) + G(f+10)$ over the range, $-20 < f < 20$.

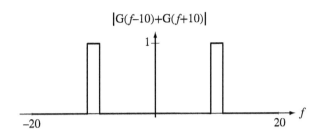

Answer:

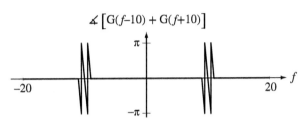

12. Let $x_1(t) = 3\operatorname{rect}((t-1)/6)$ and $x_2(t) = \operatorname{ramp}(t)[u(t) - u(t-4)]$
    (a) Graph them in the time range $-10 < t < 10$.
    (b) Graph $x(t) = x_1(2t) - x_2(t/2)$ in the time range $-10 < t < 10$.

    Answers:

    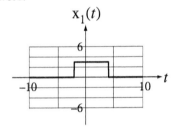
    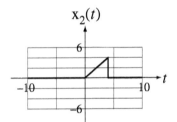

    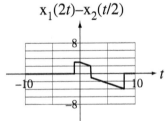

13. Write an expression consisting of a summation of unit-step functions to represent a signal that consists of rectangular pulses of width 6 ms and height 3 that occur at a uniform rate of 100 pulses per second with the leading edge of the first pulse occurring at time $t = 0$.

    Answer: $x(t) = 3 \sum_{n=0}^{\infty} [u(t - 0.01n) - u(t - 0.01n - 0.006)]$

14. Find the strengths of the following impulses.
    (a) $-3\delta(-4t)$ (b) $5\delta(3(t-1))$

    Answers: $-3/4$, $5/3$

15. Find the strengths and spacing between the impulses in the periodic impulse $-9\delta_{11}(5t)$.

    Answers: $-9/5$, $11/5$

Derivatives and Integrals of Functions

16. Graph the derivative of $x(t) = (1 - e^{-t})u(t)$.

    Answer:

17. Find the numerical value of each integral.

    (a) $\int_{-2}^{11} u(4 - t)\,dt$

    (b) $\int_{-1}^{8} [\delta(t + 3) - 2\delta(4t)]\,dt$

    (c) $\int_{1/2}^{5/2} \delta_2(3t)\,dt$

    (d) $\int_{-\infty}^{\infty} \delta(t + 4)\,\mathrm{ramp}(-2t)\,dt$

    (e) $\int_{-3}^{10} \mathrm{ramp}(2t - 4)\,dt$

    (f) $\int_{11}^{82} 3\sin(200t)\,\delta(t - 7)\,dt$

    (g) $\int_{-5}^{5} \sin(\pi t/20)\,dt$

    (h) $\int_{-2}^{10} 39 t^2 \delta_4(t - 1)\,dt$

    (i) $\int_{-\infty}^{\infty} e^{-18t} u(t)\,\delta(10t - 2)\,dt$

    (j) $\int_{2}^{9} 9\delta((t - 4)/5)\,dt$

    (k) $\int_{-6}^{3} 5\delta(3(t - 4))\,dt$

    (l) $\int_{-\infty}^{\infty} \mathrm{ramp}(3t)\,\delta(t - 4)\,dt$

    (m) $\int_{1}^{17} \delta_3(t)\cos(2\pi t/3)\,dt$

    Answers: 45, 0, 8, 0.002732, 4173, −1/2, 6, 0, 5, 0, 1, 12, 64

18. Graph the integral from negative infinity to time t of the functions in Figure E.18 which are zero for all time $t < 0$.

    Figure E.18

    Answers: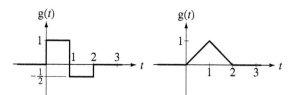

19. If $4u(t - 5) = \dfrac{d}{dt}(x(t))$, what is the function $x(t)$?

    Answer: $x(t) = 4\,\mathrm{ramp}(t - 5)$

Generalized Derivative

20. The generalized derivative of $18\,\text{rect}\left(\frac{t-2}{3}\right)$ consists of two impulses. Find their numerical locations and strengths.

    Answers: 3.5 and -18, 0.5 and 18

Even and Odd Functions

21. Classify the following functions as even, odd, or neither.

    (a) $\cos(2\pi t)\,\text{tri}(t-1)$ (b) $\sin(2\pi t)\,\text{rect}(t/5)$

    Answers: Odd, Neither

22. An even function $g(t)$ is described over the time range $0 < t < 10$ by

    $$g(t) = \begin{cases} 2t, & 0 < t < 3 \\ 15 - 3t, & 3 < t < 7 \\ -2, & 7 < t < 10 \end{cases}.$$

    (a) What is the value of $g(t)$ at time $t = -5$?
    (b) What is the value of the first derivative of $g(t)$ at time $t = -6$?

    Answers: 0, 3

23. Find the even and odd parts of these functions.

    (a) $g(t) = 2t^2 - 3t + 6$ (b) $g(t) = 20\cos(40\pi t - \pi/4)$

    (c) $g(t) = \dfrac{2t^2 - 3t + 6}{1 + t}$ (d) $g(t) = t(2 - t^2)(1 + 4t^2)$

    (e) $g(t) = t(2 - t)(1 + 4t)$ (f) $g(t) = \dfrac{20 - 4t^2 + 7t}{1 + |t|}$

    Answers: $\dfrac{20 - 4t^2}{1 + |t|}$ and $\dfrac{7t}{1 + |t|}$, $(20/\sqrt{2})\cos(40\pi t)$ and $(20/\sqrt{2})\sin(40\pi t)$,
    $7t^2$ and $t(2 - 4t^2)$, $2t^2 + 6$ and $-3t$, 0 and $t(2 - t^2)(1 + 4t^2)$,
    $\dfrac{6 + 5t^2}{1 - t^2}$ and $-t\dfrac{2t^2 + 9}{1 - t^2}$

24. Graph the even and odd parts of the functions in Figure E.24.

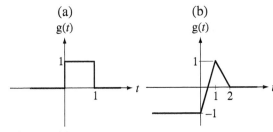

Figure E.24

Answers:

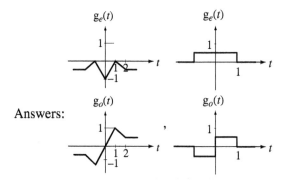

25. Graph the indicated product or quotient g(t) of the functions in Figure E.25.

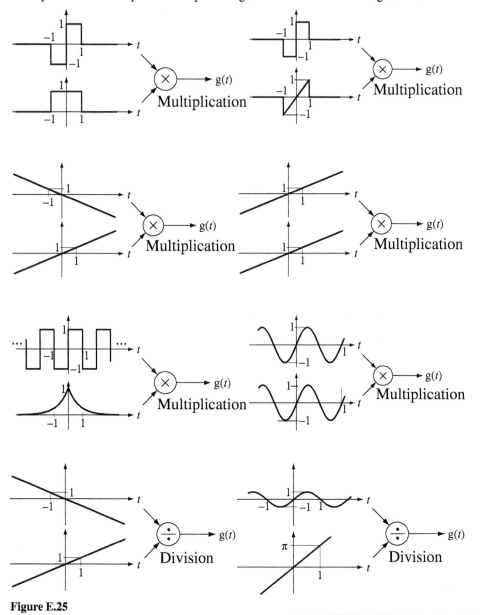

Figure E.25

Answers:

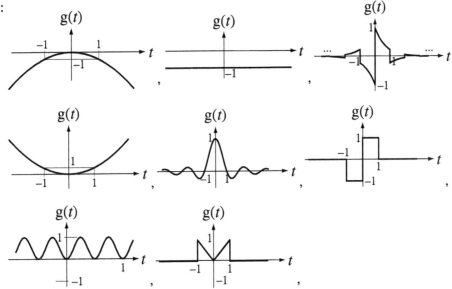

26. Use the properties of integrals of even and odd functions to evaluate these integrals in the quickest way.

    (a) $\int_{-1}^{1} (2 + t) dt$

    (b) $\int_{-1/20}^{1/20} [4 \cos(10\pi t) + 8 \sin(5\pi t)] dt$

    (c) $\int_{-1/20}^{1/20} 4t \cos(10\pi t) dt$

    (d) $\int_{-1/10}^{1/10} t \sin(10\pi t) dt$

    (e) $\int_{-1}^{1} e^{-|t|} dt$

    (f) $\int_{-1}^{1} t e^{-|t|} dt$

    Answers: 0, 1/50π, 4, 8/10π, 0, 1.264

Periodic Signals

27. Find the fundamental period and fundamental frequency of each of these functions.

    (a) $g(t) = 10 \cos(50\pi t)$
    (b) $g(t) = 10 \cos(50\pi t + \pi/4)$
    (c) $g(t) = \cos(50\pi t) + \sin(15\pi t)$
    (d) $g(t) = \cos(2\pi t) + \sin(3\pi t) + \cos(5\pi t - 3\pi/4)$
    (e) $g(t) = 3 \sin(20t) + 8 \cos(4t)$
    (f) $g(t) = 10 \sin(20t) + 7 \cos(10\pi t)$
    (g) $g(t) = 3 \cos(2000\pi t) - 8 \sin(2500\pi t)$
    (h) $g(t) = g_1(t) + g_2(t)$, $g_1(t)$ is periodic with fundamental period $T_{01} = 15\mu s$
    $g_2(t)$ is periodic with fundamental period $T_{02} = 40\mu s$

    Answers: 120 μs and $8333\tfrac{1}{3}$ Hz, 1/25 s and 25 Hz, π/2 s and 2/π Hz, 2 s and 1/2 Hz, 1/25 s and 25 Hz, Not Periodic, 0.4 s and 2.5 Hz, 4 ms and 250 Hz

28. Find a function of continuous time t for which the two successive transformations $t \to -t$ and $t \to t - 1$ leave the function unchanged.

    Answer: Any even periodic function with a period of one.

29. One period of a periodic signal x(t) with period T_0 is graphed in Figure E.29. Assuming x(t) has a period T_0, what is the value of x(t) at time, $t = 220$ ms?

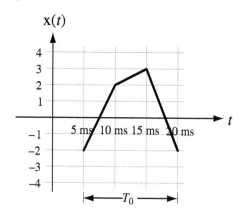

Figure E.29

Answer: 2

Signal Energy and Power of Signals

30. Find the signal energy of each of these signals.

    (a) $x(t) = 2\,\text{rect}(t)$
    (b) $x(t) = A(u(t) - u(t - 10))$
    (c) $x(t) = u(t) - u(10 - t)$
    (d) $x(t) = \text{rect}(t)\cos(2\pi t)$
    (e) $x(t) = \text{rect}(t)\cos(4\pi t)$
    (f) $x(t) = \text{rect}(t)\sin(2\pi t)$
    (g) $x(t) = \begin{cases} |t| - 1, & |t| < 1 \\ 0, & \text{otherwise} \end{cases}$

    Answers: 1/2, 1/2, 1/2, $10A^2$, 4, ∞, 2/3

31. A signal is described by $x(t) = A\,\text{rect}(t) + B\,\text{rect}(t - 0.5)$. What is its signal energy?

    Answer: $A^2 + B^2 + AB$

32. Find the average signal power of the periodic signal x(t) in Figure E.32.

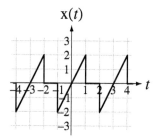

Figure E.32

Answer: 8/9

33. Find the average signal power of each of these signals.

    (a) $x(t) = A$
    (b) $x(t) = u(t)$
    (c) $x(t) = A\cos(2\pi f_0 t + \theta)$

(d) $x(t)$ is periodic with fundamental period four and one fundamental period is described by $x(t) = t(1 - t)$, $1 < t < 5$.

(e) $x(t)$ is periodic with fundamental period six. This signal is described over one fundamental period by

$$x(t) = \text{rect}\left(\frac{t-2}{3}\right) - 4\,\text{rect}\left(\frac{t-4}{2}\right), \; 0 < t < 6.$$

Answers: A^2, 5.167, $A^2/2$, 88.5333, 1/2

EXERCISES WITHOUT ANSWERS

Signal Functions

34. Let the unit impulse function be represented by the limit,

$$\delta(x) = \lim_{a \to 0} (1/a)\,\text{rect}(x/a), \; a > 0.$$

The function $(1/a)\text{rect}(x/a)$ has an area of one regardless of the value of a.

(a) What is the area of the function $\delta(4x) = \lim_{a \to 0}(1/a)\text{rect}(4x/a)$?

(b) What is the area of the function $\delta(-6x) = \lim_{a \to 0}(1/a)\text{rect}(-6x/a)$?

(c) What is the area of the function $\delta(bx) = \lim_{a \to 0}(1/a)\text{rect}(bx/a)$ for b positive and for b negative?

35. Using a change of variable and the definition of the unit impulse, prove that

$$\delta(a(t - t_0)) = (1/|a|)\delta(t - t_0).$$

36. Using the results of Exercise 35,

(a) Show that $\delta_1(ax) = (1/|a|) \sum_{n=-\infty}^{\infty} \delta(x - n/a)$

(b) Show that the average value of $\delta_1(ax)$ is one, independent of the value of a

(c) Show that even though $\delta(at) = (1/|a|)\delta(t)$, $\delta_1(ax) \neq (1/|a|)\delta_1(x)$

Scaling and Shifting

37. A signal is zero for all time before $t = -2$, rises linearly from 0 to 3 between $t = -2$ and $t = 4$ and is zero for all time after that. This signal can be expressed in the form $x(t) = A\,\text{rect}\left(\frac{t - t_{01}}{w_1}\right)\text{tri}\left(\frac{t - t_{02}}{w_2}\right)$. Find the numerical values of the constants.

38. Let $x(t) = -3e^{-t/4}u(t - 1)$ and let $y(t) = -4x(5t)$.

(a) What is the smallest value of t for which $y(t)$ is not zero?
(b) What is the maximum value of $y(t)$ over all time?
(c) What is the minimum value of $y(t)$ over all time?
(d) What is the value of $y(1)$?
(e) What is the largest value of t for which $y(t) > 2$?

39. Graph these singularity and related functions.

(a) $g(t) = 2u(4 - t)$ (b) $g(t) = u(2t)$
(c) $g(t) = 5\,\text{sgn}(t - 4)$ (d) $g(t) = 1 + \text{sgn}(4 - t)$
(e) $g(t) = 5\,\text{ramp}(t + 1)$ (f) $g(t) = -3\,\text{ramp}(2t)$

(g) $g(t) = 2\delta(t+3)$ (h) $g(t) = 6\delta(3t+9)$
(i) $g(t) = -4\delta(2(t-1))$ (j) $g(t) = 2\delta_1(t-1/2)$
(k) $g(t) = 8\delta_1(4t)$ (l) $g(t) = -6\delta_2(t+1)$
(m) $g(t) = 2\operatorname{rect}(t/3)$ (n) $g(t) = 4\operatorname{rect}((t+1)/2)$
(o) $g(t) = -3\operatorname{rect}(t-2)$ (p) $g(t) = 0.1\operatorname{rect}((t-3)/4)$

40. Graph these functions.

    (a) $g(t) = u(t) - u(t-1)$ (b) $g(t) = \operatorname{rect}(t - 1/2)$
    (c) $g(t) = -4\operatorname{ramp}(t)\, u(t-2)$ (d) $g(t) = \operatorname{sgn}(t)\sin(2\pi t)$
    (e) $g(t) = 5e^{-t/4}u(t)$ (f) $g(t) = \operatorname{rect}(t)\cos(2\pi t)$
    (g) $g(t) = -6\operatorname{rect}(t)\cos(3\pi t)$ (h) $g(t) = u(t+1/2)\operatorname{ramp}(1/2 - t)$
    (i) $g(t) = \operatorname{rect}(t + 1/2) - \operatorname{rect}(t - 1/2)$
    (j) $g(t) = \left[\int_{-\infty}^{t} \delta(\lambda+1) - 2\delta(\lambda) + \delta(\lambda-1)\right]d\lambda$
    (k) $g(t) = 2\operatorname{ramp}(t)\operatorname{rect}((t-1)/2)$
    (l) $g(t) = 3\operatorname{rect}(t/4) - 6\operatorname{rect}(t/2)$

41. A continuous-time signal x(t) is defined by the graph below. Let $y(t) = -4x(t+3)$ and let $z(t) = 8x(t/4)$. Find the numerical values.

    (a) y(3)   (b) z(−4)   (c) $\left.\dfrac{d}{dt}(z(t))\right|_{t=10}$

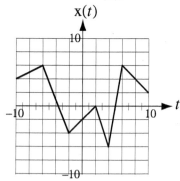

42. Find the numerical values of

    (a) $\operatorname{ramp}(-3(-2)) \times \operatorname{rect}(-2/10)$ (b) $\int_{-\infty}^{\infty} 3\delta(t-4)\cos(\pi t/10)\,dt$
    (c) $\left[\dfrac{d}{dt}(2\operatorname{sgn}(t/5)\operatorname{ramp}(t-8))\right]_{t=13}$

43. Let a function be defined by $g(t) = \operatorname{tri}(t)$. Below are four other functions based on this function. All of them are zero for large negative values of t.

    $g_1(t) = -5g\left(\dfrac{2-t}{6}\right)$ $g_2(t) = 7g(3t) - 4g(t-4)$
    $g_3(t) = g(t+2) - 4g\left(\dfrac{t+4}{3}\right)$ $g_4(t) = -5g(t)g\left(t - \dfrac{1}{2}\right)$

    (a) Which of these transformed functions is the first to become nonzero (becomes nonzero at the earliest time)?
    (b) Which of these transformed functions is the last to go back to zero and stay there?
    (c) Which of these transformed functions has a maximum value that is greater than all the other maximum values of all the other transformed functions?

("Greater than" in the strict mathematical sense of "more positive than." For example, $2 > -5$.)

(d) Which of these transformed functions has a minimum value that is less than all the other minimum values of all the other transformed functions?

44. (a) Write a functional description of the time-domain energy signal in Figure E.44 as the product of two functions of t.

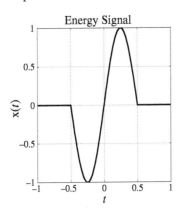
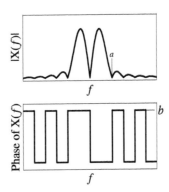

Figure E.44

(b) Write a functional description of the frequency-domain signal as the sum of two functions of f.

(c) Find the numerical values of a and b.

45. A function $g(t)$ has the following description. It is zero for $t < -5$. It has a slope of -2 in the range $-5 < t < -2$. It has the shape of a sine wave of unit amplitude and with a frequency of 1/4 Hz plus a constant in the range $-2 < t < 2$. For $t > 2$ it decays exponentially toward zero with a time constant of 2 seconds. It is continuous everywhere.

(a) Write an exact mathematical description of this function.
(b) Graph $g(t)$ in the range $-10 < t < 10$.
(c) Graph $g(2t)$ in the range $-10 < t < 10$.
(d) Graph $2g(3 - t)$ in the range $-10 < t < 10$.
(e) Graph $-2g((t + 1)/2)$ in the range $-10 < t < 10$.

46. A signal occurring in a television set is illustrated in Figure E.46. Write a mathematical description of it.

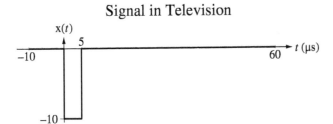

Figure E.46
Signal occurring in a television set

47. The signal illustrated in Figure E.47 is part of a binary-phase-shift-keyed (BPSK) binary data transmission. Write a mathematical description of it.

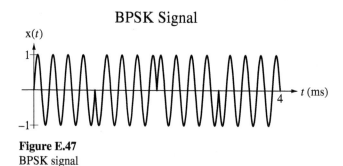

Figure E.47
BPSK signal

48. The signal illustrated in Figure E.48 is the response of an *RC* lowpass filter to a sudden change in excitation. Write a mathematical description of it.

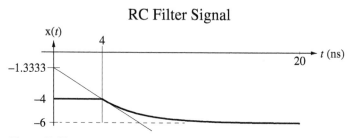

Figure E.48
Transient response of an RC filter

49. Describe the signal in Figure E.49 as a ramp function minus a summation of step functions.

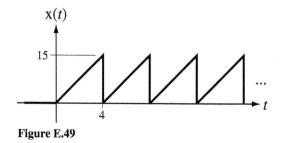

Figure E.49

50. Mathematically describe the signal in Figure E.50.

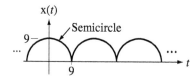

Figure E.50

51. Let two signals be defined by

$$x_1(t) = \begin{cases} 1, & \cos(2\pi t) \geq 0 \\ 0, & \cos(2\pi t) < 0 \end{cases} \quad \text{and} \quad x_2(t) = \sin(2\pi t/10),$$

Graph these products over the time range, $-5 < t < 5$.

(a) $x_1(2t)x_2(-t)$
(b) $x_1(t/5)x_2(20t)$
(c) $x_1(t/5)x_2(20(t+1))$
(d) $x_1((t-2)/5)x_2(20t)$

52. Given the graphical definitions of functions in Figure E.52, graph the indicated transformations.

(a)

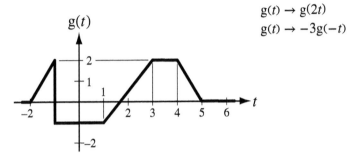

$g(t) = 0, t > 6$ or $t < -2$

$g(t) \to g(2t)$
$g(t) \to -3g(-t)$

(b)

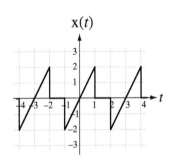

$t \to t + 4$
$g(t) \to -2g((t-1)/2)$

$g(t)$ is periodic with fundamental period, 4

Figure E.52

53. For each pair of functions graphed in Figure E.53 determine what transformation has been done and write a correct functional expression for the transformed function.

(a)

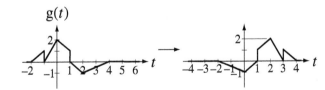

(b)

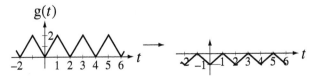

Figure E.53

54. Graph the magnitude and phase of each function versus f.

   (a) $G(f) = \dfrac{jf}{1 + jf/10}$

   (b) $G(f) = \left[\text{rect}\left(\dfrac{f - 1000}{100}\right) + \text{rect}\left(\dfrac{f + 1000}{100}\right)\right] e^{-j\pi f/500}$

   (c) $G(f) = \dfrac{1}{250 - f^2 + j3f}$

Generalized Derivative

55. Graph the generalized derivative of $g(t) = 3 \sin(\pi t/2) \text{rect}(t)$.

56. Find the generalized derivative of the function described by $x(t) = \begin{cases} 4, & t < 3 \\ 7t, & t \geq 3 \end{cases}$.

Derivatives and Integrals of Functions

57. What is the numerical value of each of the following integrals?

   (a) $\displaystyle\int_{-\infty}^{\infty} \delta(t) \cos(48\pi t) dt$,

   (b) $\displaystyle\int_{-\infty}^{\infty} \delta(t - 5) \cos(\pi t) dt$

   (c) $\displaystyle\int_{0}^{20} \delta(t - 8) \text{rect}(t/16) dt$

   (d) $\displaystyle\int_{-8}^{22} 8e^{4t} \delta(t - 2) dt$

   (e) $\displaystyle\int_{11}^{82} 3 \sin(200t) \delta(t - 7) dt$

   (f) $\displaystyle\int_{-2}^{10} 39t^2 \delta_4(t - 1) dt$

58. Graph the time derivatives of these functions.

   (a) $g(t) = \sin(2\pi t) \, \text{sgn}(t)$

   (b) $g(t) = |\cos(2\pi t)|$

Even and Odd Functions

59. Find the even and odd parts of each of these functions.

   (a) $g(t) = 10 \sin(20\pi t)$

   (b) $g(t) = 20t^3$

   (c) $x(t) = 8 + 7t^2$

   (d) $x(t) = 1 + t$

   (e) $x(t) = 6t$

   (f) $g(t) = 4t \cos(10\pi t)$

   (g) $g(t) = \dfrac{\cos(\pi t)}{\pi t}$

   (h) $g(t) = 12 + \dfrac{\sin(4\pi t)}{4\pi t}$

   (i) $g(t) = (8 + 7t)\cos(32\pi t)$

   (j) $g(t) = (8 + 7t^2)\sin(32\pi t)$

60. Is there a function that is both even and odd simultaneously? Discuss.

61. Find and graph the even and odd parts of the function x(t) in Figure E.61

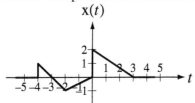

Figure E.61

Periodic Functions

62. For each of the following signals decide whether it is periodic and, if it is, find the fundamental period.

   (a) $g(t) = 28 \sin(400\pi t)$
   (b) $g(t) = 14 + 40 \cos(60\pi t)$
   (c) $g(t) = 5t - 2 \cos(5000\pi t)$
   (d) $g(t) = 28 \sin(400\pi t) + 12 \cos(500\pi t)$
   (e) $g(t) = 10 \sin(5t) - 4 \cos(7t)$
   (f) $g(t) = 4 \sin(3t) + 3 \sin(\sqrt{3}t)$

63. Is a constant a periodic signal? Explain why it is or is not periodic and, if it is periodic what is its fundamental period?

Signal Energy and Power of Signals

64. Find the signal energy of each of these signals.

   (a) $2 \operatorname{rect}(-t)$
   (b) $\operatorname{rect}(8t)$
   (c) $3 \operatorname{rect}\left(\frac{t}{4}\right)$
   (d) $2 \sin(200\pi t)$
   (e) $\delta(t)$ (Hint: First find the signal energy of a signal which approaches an impulse some limit, then take the limit.)
   (f) $x(t) = \frac{d}{dt}(\operatorname{rect}(t))$
   (g) $x(t) = \int_{-\infty}^{t} \operatorname{rect}(\lambda) d\lambda$
   (h) $x(t) = e^{(-1-j8\pi)t} u(t)$
   (i) $x(t) = 2 \operatorname{rect}(t/4) - 3 \operatorname{rect}\left(\frac{t-1}{4}\right)$
   (j) $x(t) = 3 \operatorname{rect}\left(\frac{t-1}{6}\right) \operatorname{tri}\left(\frac{t-4}{6}\right)$
   (k) $x(t) = 5e^{-4t} u(t)$
   (l) A signal x(t) has the following description:

   1. It is zero for all time $t < -4$.
   2. It is a straight line from the point $t = -4, x = 0$ to the point $t = -4, x = 4$.
   3. It is a straight line from the point $t = -4, x = 4$ to the point $t = 3, x = 0$.
   4. It is zero for all time $t > 3$.

65. An even continuous-time energy signal x(t) is described in positive time by

$$x(t) = \begin{cases} 3u(t) + 5u(t-4) - 11u(t-7), & 0 \leq t < 10 \\ 0, & t \geq 10 \end{cases}.$$

Another continuous-time energy signal y(t) is described by $y(t) = -3x(2t - 2)$.

   (a) Find the signal energy E_x of x(t).
   (b) Find the signal energy E_y of y(t).

66. Find the average signal power of each of these signals.
    (a) $x(t) = 2\sin(200\pi t)$
    (b) $x(t) = \delta_1(t)$
    (c) $x(t) = e^{j100\pi t}$
    (d) A periodic continuous-time signal with fundamental period 12 described over one fundamental period by
    $$x(t) = 3\operatorname{rect}\left(\frac{t+3}{4}\right) - 4\operatorname{rect}(t/2 + 1),\ -6 < t < 6$$
    (e) $x(t) = -3\operatorname{sgn}(2(t-4))$

67. A signal x is periodic with fundamental period $T_0 = 6$. This signal is described over the time period $0 < t < 6$ by
    $$\operatorname{rect}((t-2)/3) - 4\operatorname{rect}((t-4)/2).$$
    What is the signal power of this signal?

CHAPTER 3

Discrete-Time Signal Description

3.1 INTRODUCTION AND GOALS

In the 20th century digital computing machinery developed from its infancy to its position today as a ubiquitous and indispensable part of our society and economy. The effect of digital computation on signals and systems is equally broad. Every day operations that were once done by continuous-time systems are being replaced by discrete-time systems. There are systems that are inherently discrete time but most of the application of discrete-time signal processing is on signals that are created by sampling continuous-time signals. Figure 3.1 shows some examples of discrete-time signals.

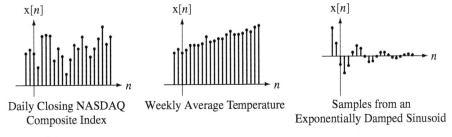

Figure 3.1
Examples of discrete-time signals

Most of the functions and methods developed for describing continuous-time signals have very similar counterparts in the description of discrete-time signals. But some operations on discrete-time signals are fundamentally different, causing phenomena that do not occur in continuous-time signal analysis. The fundamental difference between continuous-time and discrete-time signals is that the signal values occurring as time passes in a discrete-time signal are countable and in a continuous-time signal they are uncountable.

CHAPTER GOALS

1. To define mathematical functions that can be used to describe discrete-time signals
2. To develop methods of shifting, scaling and combining those functions to represent practical signals and to appreciate why these operations are different in discrete-time than in continuous-time

3. To recognize certain symmetries and patterns to simplify analysis of discrete-time signals

3.2 SAMPLING AND DISCRETE TIME

Of increasing importance in signal and system analysis are **discrete-time** functions that describe discrete-time signals. The most common examples of discrete-time signals are those obtained by sampling continuous-time signals. Sampling means acquiring the values of a signal at discrete points in time. One way to visualize sampling is through the example of a voltage signal and a switch used as an ideal sampler (Figure 3.2(a)).

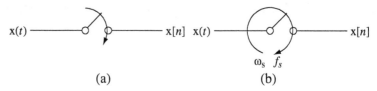

Figure 3.2
(a) An ideal sampler, (b) an ideal sampler sampling uniformly

The switch closes for an infinitesimal time at discrete points in time. Only the values of the continuous-time signal x(t) at those discrete times are assigned to the discrete-time signal x[n]. If there is a fixed time T_s between samples, the sampling is called **uniform sampling** in which the sampling times are integer multiples of a **sampling period or sampling interval** T_s. The time of a sample $n\,T_s$ can be replaced by the integer n, which indexes the sample (Figure 3.3).

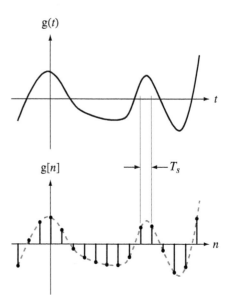

Figure 3.3
Creating a discrete-time signal by sampling a continuous-time signal

This type of operation can be envisioned by imagining that the switch simply rotates at a constant cyclic velocity f_s cycles per second as in Figure 3.2(b) with the time between samples being $T_s = 1/f_s = 2\pi/\omega_s$. We will use a simplified notation for a discrete-time function g[n] formed by sampling which, at every point of continuity of g(t), is the same as $g(nT_s)$, and in which n is an integer. The square brackets [·] enclosing the argument indicate a discrete-time function, as contrasted with the parentheses (·) that indicate a continuous-time function. The independent variable n is called discrete time because it indexes discrete points in time, even though it is dimensionless, not having units of seconds as t and T_s do. Since discrete-time functions are only defined for integer values of n, the values of expressions like g[2.7] or g[3/4] are undefined.

Functions that are defined for continuous arguments can also be given discrete time as an argument, for example, $\sin(2\pi f_0 n T_s)$. We can form a discrete-time function from a continuous-time function by sampling, for example, $g[n] = \sin(2\pi f_0 n T_s)$. Then, although the sine is defined for any real argument value, the function g[n] is only defined for integer values of n. That is, g[7.8] is undefined even though $\sin(2\pi f_0 (7.8) T_s)$ is defined.[1]

In engineering practice the most important examples of discrete-time systems are **sequential-state machines**, the most common example being a computer. Computers are driven by a **clock**, a fixed-frequency oscillator. The clock generates pulses at regular intervals in time, and at the end of each clock cycle the computer has executed an instruction and changed from one logical state to the next. The computer has become a very important tool in all phases of the modern economy, so understanding how discrete-time signals are processed by sequential-state machines is very important, especially to engineers. Figure 3.4 illustrates some discrete-time functions that could describe discrete-time signals.

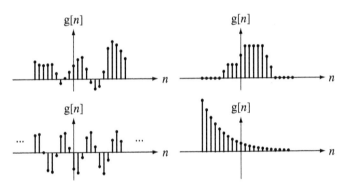

Figure 3.4
Examples of discrete-time functions

The type of graph used in Figure 3.4 is called a **stem plot** in which a dot indicates the functional value and the stems always connect the dot to the discrete time n axis. This is a widely used method of graphing discrete-time functions. MATLAB has a command **stem** that generates stem plots.

The use of MATLAB to draw graphs is an example of sampling. MATLAB can only deal with finite-length vectors, so to draw a graph of a continuous-time function we must decide how many points to put in the time vector so that when MATLAB draws straight lines between the function values at those times, the graph looks like a continuous-time function. Sampling will be considered in much more depth in Chapter 10.

---

[1] If we were to define a function as $g(n) = 5\sin(2\pi f_0 n T_s)$, the parentheses in g(n) would indicate that any real value of n would be acceptable, integer or otherwise. Although this is mathematically legal, it is not a good idea because we are using the symbol t for continuous time and the symbol n for discrete time, and the notation g(n), although mathematically defined, would be confusing.

3.3 SINUSOIDS AND EXPONENTIALS

Exponentials and sinusoids are as important in discrete-time signal and system analysis as in continuous-time signal and system analysis. Most discrete-time systems can be described, at least approximately, by difference equations. The eigenfunction of linear, constant-coefficient, ordinary difference equations is the complex exponential, and the real exponential is a special case of the complex exponential. Any sinusoid is a linear combination of complex exponentials. The solution of difference equations will be considered in more detail in Chapter 4.

Discrete-time exponentials and sinusoids can be defined in a manner analogous to their continuous-time counterparts as

$$g[n] = Ae^{\beta n} \text{ or } g[n] = Az^n, \text{ where } z = e^{\beta},$$

and

$$g[n] = A\cos(2\pi F_0 n + \theta) \quad \text{or} \quad g[n] = A\cos(\Omega_0 n + \theta)$$

where z and β are complex constants, A is a real constant, θ is a real phase shift in radians, F_0 is a real number, $\Omega_0 = 2\pi F_0$, and n is discrete time.

SINUSOIDS

There are some important differences between continuous-time and discrete-time sinusoids. One difference is that if we create a discrete-time sinusoid by sampling a continuous-time sinusoid, the period of the discrete-time sinusoid may not be readily apparent and, in fact, *the discrete-time sinusoid may not even be periodic*. Let a discrete-time sinusoid $g[n] = A\cos(2\pi F_0 n + \theta)$ be related to a continuous-time sinusoid $g(t) = A\cos(2\pi f_0 t + \theta)$ through $g[n] = g(nT_s)$. Then $F_0 = f_0 T_s = f_0/f_s$ where $f_s = 1/T_s$ is the sampling rate. The requirement on a discrete-time sinusoid that it be periodic is that, for some discrete time n and some integer m, $2\pi F_0 n = 2\pi m$. Solving for F_0, $F_0 = m/n$ indicating that F_0 must be a rational number (a ratio of integers). Since sampling forces the relationship $F_0 = f_0/f_s$, this also means that, for a discrete-time sinusoid to be periodic, the ratio of the fundamental frequency of the continuous-time sinusoid to the sampling rate must be rational. What is the fundamental period of the sinusoid

$$g[n] = 4\cos\left(\frac{72\pi n}{19}\right) = 4\cos(2\pi(36/19)n) \ ?$$

F_0 is 36/19 and the smallest positive discrete time n that solves $F_0 n = m$, m an integer, is $n = 19$. So the fundamental period is 19. If F_0 is a rational number and is expressed as a ratio of integers $F_0 = q/N_0$, and if the fraction has been reduced to its simplest form by canceling common factors in the numerator and denominator, then the fundamental period of the sinusoid is N_0, not $(1/F_0) = N_0/q$ unless $q = 1$. Compare this result with the fundamental period of the continuous-time sinusoid $g(t) = 4\cos(72\pi t/19)$, whose fundamental period T_0 is 19/36, not 19. Figure 3.5 illustrates some discrete-time sinusoids.

When F_0 is not the reciprocal of an integer, a discrete-time sinusoid may not be immediately recognizable from its graph as a sinusoid. This is the case for Figure 3.5(c) and (d). The sinusoid in Figure 3.5(d) is aperiodic.

A source of confusion for students when first encountering a discrete-time sinusoid of the form $A\cos(2\pi F_0 n)$ or $A\cos(\Omega_0 n)$ is the question, "What are F_0 and Ω_0?" In the continuous-time sinusoids $A\cos(2\pi f_0 t)$ and $A\cos(\omega_0 t)$ f_0 is the cyclic frequency in Hz or cycles/second and ω_0 is the radian frequency in radians/second. The argument of the cosine must be dimensionless and the products $2\pi f_0 t$ and $\omega_0 t$ are dimensionless, because the cycle and radian are ratios of lengths and the second in t and the (second)$^{-1}$

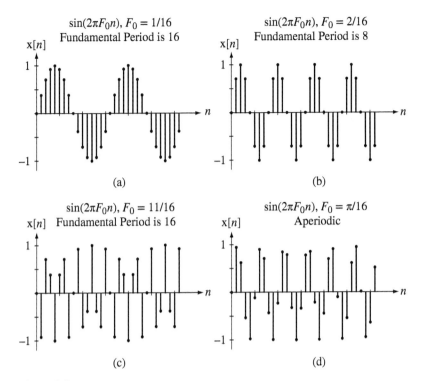

Figure 3.5
Four discrete-time sinusoids

in f_0 or ω_0 cancel. Likewise, the arguments $2\pi F_0 n$ and $\Omega_0 n$ must also be dimensionless. Remember n does not have units of seconds. Even though we call it discrete time, it is really a time index, not time itself. If we think of n as indexing the samples then, for example, $n = 3$ indicates the third sample taken after the initial discrete time $n = 0$. So we can think of n as having units of samples. Therefore F_0 should have units of cycles/sample to make $2\pi F_0 n$ dimensionless and Ω_0 should have units of radians/sample to make $\Omega_0 n$ dimensionless. If we sample a continuous-time sinusoid $A\cos(2\pi f_0 t)$ with fundamental frequency f_0 cycles/second at a rate of f_s samples/second, we form the discrete-time sinusoid

$$A\cos(2\pi f_0 n T_s) = A\cos(2\pi n f_0/f_s) = A\cos(2\pi F_0 n),$$

$F_0 = f_0/f_s$ and the units are consistent.

$$F_0 \text{ in cycles/sample} = \frac{f_0 \text{ in cycles/second}}{f_s \text{ in samples/second}}$$

So F_0 is a cyclic frequency normalized to the sampling rate. Similarly $\Omega_0 = \omega_0/f_s$ is a normalized radian frequency in radians/sample

$$\Omega_0 \text{ in radians/sample} = \frac{\omega_0 \text{ in radians/second}}{f_s \text{ in samples/second}}$$

One other aspect of discrete-time sinusoids that will be very important in Chapter 10 in the consideration of sampling is that two discrete-time sinusoids $g_1[n] = A\cos(2\pi F_1 n + \theta)$ and $g_2[n] = A\cos(2\pi F_2 n + \theta)$ can be identical, even if F_1 and F_2 are different. For example,

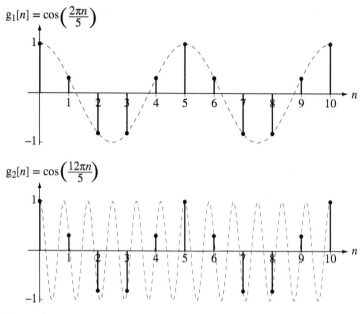

Figure 3.6
Two cosines with different F's but the same functional behavior

the two sinusoids, $g_1[n] = \cos(2\pi n/5)$ and $g_2[n] = \cos(12\pi n/5)$ are described by different-looking expressions, but when we graph them versus discrete time n we see that they are identical (Figure 3.6).

The dashed lines in Figure 3.6 are the continuous-time functions $g_1(t) = \cos(2\pi t/5)$ and $g_2(t) = \cos(12\pi t/5)$, where n and t are related by $t = nT_s$. The continuous-time functions are obviously different, but the discrete-time functions are not. The reason the two discrete-time functions are identical can be seen by rewriting $g_2[n]$ in the form

$$g_2[n] = \cos\left(\frac{2\pi}{5}n + \frac{10\pi}{5}n\right) = \cos\left(\frac{2\pi}{5}n + 2\pi n\right).$$

Then, using the principle that if any integer multiple of 2π is added to the angle of a sinusoid the value is not changed,

$$g_2[n] = \cos\left(\frac{2\pi}{5}n + 2\pi n\right) = \cos\left(\frac{2\pi}{5}n\right) = g_1[n]$$

because discrete-time n is an integer. Since the two discrete-time cyclic frequencies in this example are $F_1 = 1/5$ and $F_2 = 6/5$, that must mean that they are equivalent as frequencies in a discrete-time sinusoid. That can be seen by realizing that at a frequency of 1/5 cycles/sample the angular change per sample is $2\pi/5$, and at a frequency of 6/5 cycles/sample the angular change per sample is $12\pi/5$. As shown above, those two angles yield exactly the same values as arguments of a sinusoid. So, in a discrete-time sinusoid of the form $\cos(2\pi F_0 n + \theta)$, if we change F_0 by adding any integer, the sinusoid is unchanged. Similarly, in a discrete-time sinusoid of the form $\cos(\Omega_0 n + \theta)$, if we change Ω_0 by adding any integer multiple of 2π, the sinusoid is unchanged. One can then imagine an experiment in which we generate a sinusoid $\sin(2\pi F n)$ and let F be a variable. As F changes in steps of 0.25 from 0 to 1.75, we get a sequence of

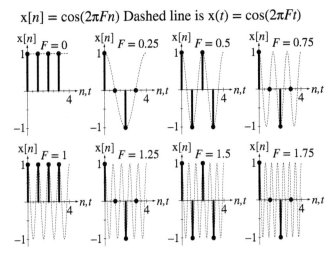

Figure 3.7
Illustration that a discrete-time sinusoid with frequency F repeats every time F changes by one

discrete-time sinusoids (Figure 3.7). Any two discrete-time sinusoids whose F values differ by an integer are identical.

EXPONENTIALS

The most common way of writing a discrete-time exponential is in the form $g[n] = Az^n$. This does not look like a continuous-time exponential, which has the form $g(t) = Ae^{\beta t}$, because there is no "e," but it is still an exponential, because $g[n] = Az^n$ could have been written as $g[n] = Ae^{\beta n}$ where $z = e^{\beta}$. The form Az^n is a little simpler and is generally preferred.

Discrete-time exponentials can have a variety of functional behaviors depending on the value of z in $g[n] = Az^n$. Figure 3.8 and Figure 3.9 summarize the functional form of an exponential for different values of z.

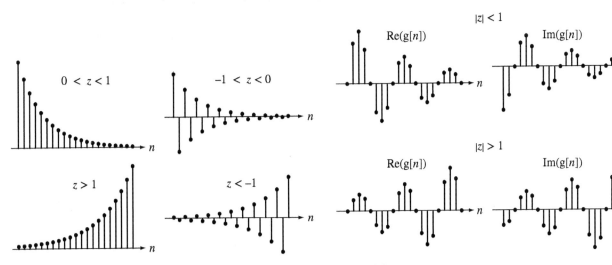

Figure 3.8
Behavior of $g[n] = Az^n$ for different real z's

Figure 3.9
Behavior of $g[n] = Az^n$ for some complex z's

3.4 SINGULARITY FUNCTIONS

There is a set of discrete-time functions that are analogous to continuous-time singularity functions and have similar uses.

THE UNIT-IMPULSE FUNCTION

The unit-impulse function (sometimes called the unit-sample function) (Figure 3.10) is defined by

$$\delta[n] = \begin{cases} 1, & n = 0 \\ 0, & n \neq 0 \end{cases}. \tag{3.1}$$

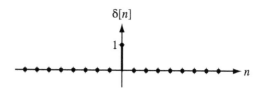

Figure 3.10
The unit-impulse function

The discrete-time unit-impulse function suffers from none of the mathematical peculiarities that the continuous-time unit impulse has. The discrete-time unit impulse does not have a property corresponding to the scaling property of the continuous-time unit impulse. Therefore $\delta[n] = \delta[an]$ for any nonzero, finite, integer value of a. But the discrete-time impulse does have a sampling property. It is

$$\sum_{n=-\infty}^{\infty} A\delta[n - n_0]x[n] = Ax[n_0]. \tag{3.2}$$

Since the impulse is only nonzero where its argument is zero, the summation over all n is a summation of terms that are all zero except at $n = n_0$. When $n = n_0$, $x[n] = x[n_0]$ and that result is simply multiplied by the scale factor A.

We did not have a MATLAB function for continuous-time impulses, but we can make one for discrete-time impulses.

```
%      Function to generate the discrete-time impulse
%      function defined as one for input integer arguments
%      equal to zero and zero otherwise. Returns "NaN" for
%      non-integer arguments.
%
%      function y = impD(n)
%
function y = impD(n)
    y = double(n == 0);           % Impulse is one where argument
                                  % is zero and zero otherwise
    I = find(round(n) ~= n);      % Index non-integer values of n
    y(I) = NaN;                   % Set those return values to NaN
```

This MATLAB function implements the functional behavior of $\delta[n]$ including returning undefined values (NaN) for arguments that are not integers. The "D" at the

3.4 Singularity Functions

end of the function name indicates that it is a discrete-time function. We cannot use the convention of square brackets [·] enclosing the argument in MATLAB to indicate a discrete-time function. Square brackets in MATLAB have a different meaning.

THE UNIT-SEQUENCE FUNCTION

The discrete-time function that corresponds to the continuous-time unit step is the **unit-sequence** function (Figure 3.11).

$$\boxed{\mathrm{u}[n] = \begin{cases} 1, & n \geq 0 \\ 0, & n < 0 \end{cases}} \tag{3.3}$$

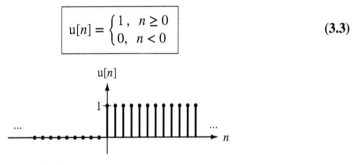

Figure 3.11
The unit-sequence function

For this function there is no disagreement or ambiguity about its value at $n = 0$. It is one, and every author agrees.

```
%     Unit sequence function defined as 0 for input integer
%     argument values less than zero, and 1 for input integer
%     argument values equal to or greater than zero. Returns
%     "NaN" for non-integer arguments.
%
%     function y = usD(n)
%
function y = usD(n)
   y = double(n >= 0);         % Set output to one for non-
                               % negative arguments
   I = find(round(n) ~= n);    % Index non-integer values of n
   y(I) = NaN ;                % Set those return values to NaN
```

THE SIGNUM FUNCTION

The discrete-time function corresponding to the continuous-time signum function is defined in Figure 3.12.

$$\boxed{\mathrm{sgn}[n] = \begin{cases} 1, & n > 0 \\ 0, & n = 0 \\ -1, & n < 0 \end{cases}} \tag{3.4}$$

```
%     Signum function defined as -1 for input integer argument
%     values less than zero, +1 for input integer argument
%     values greater than zero and zero for input argument values
%     equal to zero. Returns "NaN" for non-integer arguments.
%
%     function y = signD(n)
```

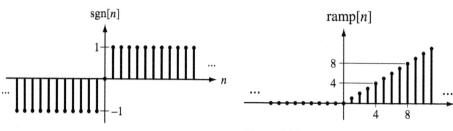

Figure 3.12
The signum function

Figure 3.13
The unit-ramp function

```
function y = signD(n)
    y = sign(n);              % Use the MATLAB sign function
    I = find(round(n) ~= n);  % Index non-integer values of n
    y(I) = NaN;               % Set those return values to NaN
```

THE UNIT-RAMP FUNCTION

The discrete-time function corresponding to the continuous-time unit ramp is defined in Figure 3.13.

$$\text{ramp}[n] = \begin{cases} n, & n \geq 0 \\ 0, & n < 0 \end{cases} = n\,\text{u}[n] \tag{3.5}$$

```
%    Unit discrete-time ramp function defined as 0 for input
%    integer argument values equal to or less than zero, and
%    "n" for input integer argument values greater than zero.
%    Returns "NaN" for non-integer arguments.
%
%    function y = rampD(n)
function y = rampD(n)
    y = ramp(n);              % Use the continuous-time ramp
    I = find(round(n) ~= n);  % Index non-integer values of n
    y(I) = NaN;               % Set those return values to NaN
```

THE UNIT PERIODIC IMPULSE FUNCTION OR IMPULSE TRAIN

The unit discrete-time periodic impulse or impulse train (Figure 3.14) is defined by

$$\delta_N[n] = \sum_{m=-\infty}^{\infty} \delta[n - mN]. \tag{3.6}$$

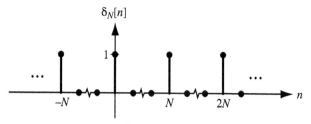

Figure 3.14
The unit periodic impulse function

```
%   Discrete-time periodic impulse function defined as 1 for
%   input integer argument values equal to integer multiples
%   of "N" and 0 otherwise. "N" must be a positive integer.
%   Returns "NaN" for non-positive integer values.
%
%   function y = impND(N,n)
function y = impND(N,n)
  if N == round(N) & N > 0,
    y = double(n/N == round(n/N));   % Set return values to one
                                     % at all values of n that are
                                     % integer multiples of N
    I = find(round(n) ~= n);         % Index non-integer values of n
    y(I) = NaN;                      % Set those return values to NaN
  else
    y = NaN*n;                       % Return a vector of NaN's
    disp('In impND, the period parameter N is not a positive integer');
  end
```

The new discrete-time signal functions are summarized in Table 3.1.

Table 3.1 Summary of discrete-time signal functions

Sine	$\sin(2\pi F_0 n)$	Sampled Continuous-Time
Cosine	$\cos(2\pi F_0 n)$	Sampled Continuous-Time
Exponential	z^n	Sampled Continuous-Time
Unit Sequence	$u[n]$	Inherently Discrete-Time
Signum	$\text{sgn}[n]$	Inherently Discrete-Time
Ramp	$\text{ramp}[n]$	Inherently Discrete-Time
Impulse	$\delta[n]$	Inherently Discrete-Time
Periodic Impulse	$\delta_N[n]$	Inherently Discrete-Time

3.5 SHIFTING AND SCALING

The general principles that govern scaling and shifting of continuous-time functions also apply to discrete-time functions, but with some interesting differences caused by the fundamental differences between continuous time and discrete time. Just as a continuous-time function does, a discrete-time function accepts a number and returns another number. The general principle that the *expression* in g[*expression*] is treated in exactly the same way that *n* is treated in the definition g[*n*] still holds.

AMPLITUDE SCALING

Amplitude scaling for discrete-time functions is exactly the same as it is for continuous-time functions.

TIME SHIFTING

Let a function g[*n*] be defined by the graph and table in Figure 3.15. Now let $n \rightarrow n + 3$. Time shifting is essentially the same for discrete-time and for continuous-time functions, except that the shift must be an integer, otherwise the shifted function would have undefined values (Figure 3.16).

TIME SCALING

Amplitude scaling and time shifting for discrete-time and continuous-time functions are very similar. That is not so true when we examine time scaling for discrete-time functions. There are two cases to examine, time compression and time expansion.

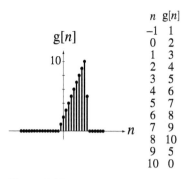

Figure 3.15
Graphical definition of a function
g[n], g[n] = 0, |n| ≥ 15

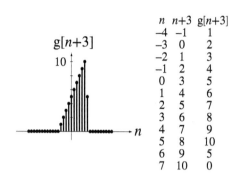

Figure 3.16
Graph of g[n + 3] illustrating time shifting

Time Compression
Time compression is accomplished by a scaling of the form $n \rightarrow Kn$, where $|K| > 1$ and K is an integer. Time compression for discrete-time functions is similar to time compression for continuous-time functions, in that the function seems to occur faster in time. But in the case of discrete-time functions there is another effect called **decimation**. Consider the time scaling $n \rightarrow 2n$ illustrated in Figure 3.17.

For each integer n in g[2n], the value 2n must be an even integer. Therefore, for this scaling by a factor of two, the odd-integer-indexed values of g[n] are never needed to find values for g[2n]. The function has been decimated by a factor of two because the graph of g[2n] only uses every other value of g[n]. For larger scaling constants, the decimation factor is obviously higher. Decimation does not happen in scaling continuous-time functions because, in using a scaling $t \rightarrow Kt$, all real t values map into real Kt values without any missing values. The fundamental difference between continuous-time and discrete-time functions is that the domain of a continuous-time function is all real numbers, an *uncountable* infinity of times, and the domain of discrete-time functions is all integers, a *countable* infinity of discrete times.

Time Expansion
The other time-scaling case, time expansion, is even stranger than time compression. If we want to graph, for example, g[n/2], for each integer value of n we must assign a value to g[n/2] by finding the corresponding value in the original function definition. But when n is one, n/2 is one-half and g[1/2] is not defined. The value of the time-scaled function g[n/K] is undefined unless n/K is an integer. We could simply leave those values undefined or we could **interpolate** between them using the values of g[n/K] at the next higher and next lower values of n at which n/K is an integer. (Interpolation is a process of computing functional values between two known values according to some formula.) Since interpolation begs the question of what interpolation formula to use, we will simply leave g[n/K] undefined if n/K is not an integer.

Even though time expansion, as described above, seems to be totally useless, there is a type of time expansion that is actually often useful. Suppose we have an original function x[n] and we form a new function

$$y[n] = \begin{cases} x[n/K], & n/K \text{ an integer} \\ 0, & \text{otherwise} \end{cases}$$

as in Figure 3.18 where $K = 2$.

3.5 Shifting and Scaling

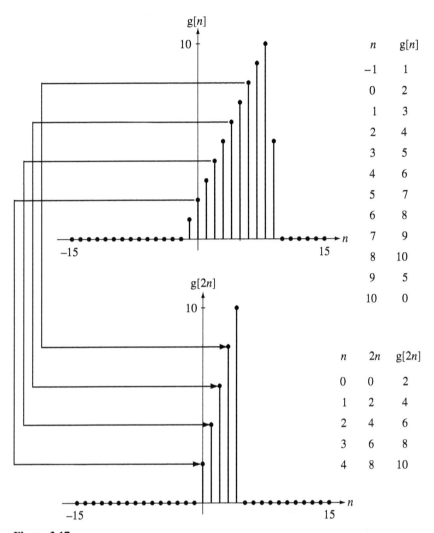

Figure 3.17
Time compression for a discrete-time function

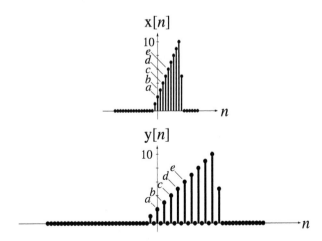

Figure 3.18
Alternate form of time expansion

All the values of the expanded function are defined, and all the values of x that occur at discrete time *n* occur in y at discrete time *Kn*. All that has really been done is to replace all the undefined values from the former time expansion with zeros. If we were to time-compress y by a factor *K* we would get all the x values back in their original positions, and all the values that would be removed by decimating y would be zeros.

EXAMPLE 3.1

Graphing shifting and scaling of discrete-time functions

Using MATLAB, graph the function $g[n] = 10(0.8)^n \sin(3\pi n/16)u[n]$. Then graph the functions $g[2n]$ and $g[n/3]$.

Discrete-time functions are easier to program in MATLAB than continuous-time functions because MATLAB is inherently oriented toward calculation of functional values at discrete values of the independent variable. For discrete-time functions there is no need to decide how close together to make the time points to make the graph look continuous because the function is *not* continuous. A good way to handle graphing the function and the time-scaled functions is to define the original function as an m file. But we need to ensure that the function definition includes its discrete-time behavior; for noninteger values of discrete time the function is undefined. MATLAB handles undefined results by assigning to them the special value NaN. The only other programming problem is how to handle the two different functional descriptions in the two different ranges of *n*. We can do that with logical and relational operators as demonstrated below in g.m.

```
function y = g(n),
    % Compute the function
    y = 10*(0.8).^n.*sin(3*pi*n/16).*usD(n);
    I = find(round(n) ~= n);        % Find all non-integer "n's"
    y(I) = NaN;                      % Set those return values to "NaN"
```

We still must decide over what range of discrete times to graph the function. Since it is zero for negative times, we should represent that time range with at least a few points to show that it suddenly turns on at time zero. Then, for positive times it has the shape of an exponentially decaying sinusoid. If we graph a few time constants of the exponential decay, the function will be practically zero after that time. So the time range should be something like $-5 < n < 16$ to draw a reasonable representation of the original function. But the time-expanded function $g[n/3]$ will be wider in discrete time and require more discrete time to see the functional behavior. Therefore, to really see all of the functions on the same scale for comparison, let's set the range of discrete times to $-5 < n < 48$.

```
%    Graphing a discrete-time function and compressed and expanded
%    transformations of it

%    Compute values of the original function and the time-scaled
%    versions in this section
```

3.5 Shifting and Scaling

```
n = -5:48 ;                            % Set the discrete times for
                                       % function computation
g0 = g(n) ;                            % Compute the original function
                                       % values
g1 = g(2*n) ;                          % Compute the compressed function
                                       % values
g2 = g(n/3) ;                          % Compute the expanded function
                                       % values
%      Display the original and time-scaled functions graphically
%      in this section
%
%      Graph the original function
%
subplot(3,1,1) ;                       % Graph first of three graphs
                                       % stacked vertically
p = stem(n,g0,'k','filled');           % "Stem plot" the original function
set(p,'LineWidth',2,'MarkerSize',4);   % Set the line weight and dot
                                       % size
ylabel('g[n]');                        % Label the original function axis
%
%      Graph the time-compressed function
%
subplot(3,1,2);                        % Graph second of three plots
                                       % stacked vertically
p = stem(n,g1,'k','filled');           % "Stem plot" the compressed
                                       % function
set(p,'LineWidth',2,'MarkerSize',4);   % Set the line weight and dot
                                       % size
ylabel('g[2n]');                       % Label the compressed function
                                       % axis
%
%      Graph the time-expanded function
%
subplot(3,1,3);                        % Graph third of three graphs
                                       % stacked vertically
p = stem(n,g2,'k','filled') ;          % "Stem plot" the expanded
                                       % function
set(p,'LineWidth',2,'MarkerSize',4);   % Set the line weight and dot
                                       % size
xlabel('Discrete time, n');            % Label the expanded function axis
ylabel('g[n/3]');                      % Label the discrete-time axis
```

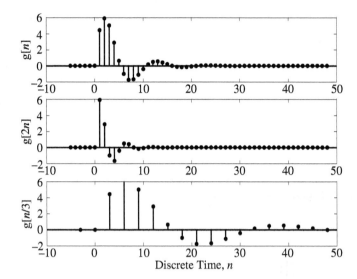

Figure 3.19
Graphs of g[n], g[2n] and g[n/3].

Figure 3.19 illustrates the output of the MATLAB program. ∎

3.6 DIFFERENCING AND ACCUMULATION

Just as differentiation and integration are important for continuous-time functions, the analogous operations, differencing and accumulation are important for discrete-time functions. The first derivative of a continuous-time function g(t) is usually defined by

$$\frac{d}{dt}(g(t)) = \lim_{\Delta t \to 0} \frac{g(t + \Delta t) - g(t)}{\Delta t}.$$

But it can also be defined by

$$\frac{d}{dt}(g(t)) = \lim_{\Delta t \to 0} \frac{g(t) - g(t - \Delta t)}{\Delta t}$$

or

$$\frac{d}{dt}(g(t)) = \lim_{\Delta t \to 0} \frac{g(t + \Delta t) - g(t - \Delta t)}{2\Delta t}.$$

In the limit, all these definitions yield the same derivative (if it exists). But if Δt remains finite, these expressions are not identical. The operation on a discrete-time signal that is analogous to the derivative is the **difference**. The first **forward difference** of a discrete-time function g[n] is g[n + 1] − g[n]. (See Web Appendix D for more on differencing and difference equations.) The first **backward difference** of a discrete-time function is g[n] − g[n − 1], which is the first forward difference of g[n − 1]. Figure 3.20 illustrates some discrete-time functions and their first forward or backward differences.

The differencing operation applied to samples from a continuous-time function yields a result that looks a lot like (but not *exactly* like) samples of the derivative of that continuous-time function (to within a scale factor).

The discrete-time counterpart of integration is **accumulation** (or summation). The accumulation of g[n] is defined by $\sum_{m=-\infty}^{n} g[m]$. The ambiguity problem that

3.6 Differencing and Accumulation

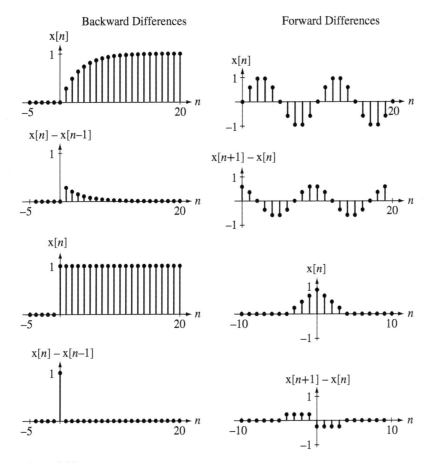

Figure 3.20
Some functions and their backward or forward differences

occurs in the integration of a continuous-time function exists for accumulation of discrete-time functions. The accumulation of a function is not unique. Multiple functions can have the same first forward or backward difference but, just as in integration, these functions only differ from each other by an additive constant.

Let $h[n] = g[n] - g[n-1]$, the first backward difference of $g[n]$. Then accumulate both sides,

$$\sum_{m=-\infty}^{n} h[m] = \sum_{m=-\infty}^{n} (g[m] - g[m-1])$$

or

$$\sum_{m=-\infty}^{n} h[m] = \cdots + (g[-1] - g[-2]) + (g[0] - g[-1]) + \cdots + (g[n] - g[n-1]).$$

Gathering values of $g[n]$ occurring at the same time,

$$\sum_{m=-\infty}^{n} h[m] = \cdots + \underbrace{(g[-1] - g[-1])}_{=0} + \underbrace{(g[0] - g[0])}_{=0} + \cdots + \underbrace{(g[n-1] - g[n-1])}_{=0} + g[n]$$

and

$$\sum_{m=-\infty}^{n} \mathrm{h}[m] = \mathrm{g}[n].$$

This result proves that accumulation and first-backward-difference are inverse operations. The first backward difference of the accumulation of any function g[n] is g[n]. Figure 3.21 illustrates two functions h[n] and their accumulations g[n]. In each of the graphs of Figure 3.21 the accumulation was done based on the assumption that all function values of h[n] before the time range graphed are zero.

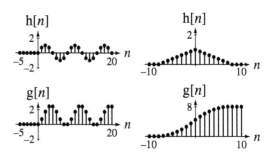

Figure 3.21
Two functions h[n] and their accumulations g[n]

In a manner analogous to the integral-derivative relationship between the continuous-time unit step and the continuous-time unit impulse, the unit sequence is the accumulation of the unit impulse $\mathrm{u}[n] = \sum_{m=-\infty}^{n} \delta[m]$, and the unit impulse is the first backward difference of the unit sequence $\delta[n] = \mathrm{u}[n] - \mathrm{u}[n-1]$. Also, the discrete-time unit ramp is defined as the accumulation of a unit sequence function *delayed by one in discrete time*,

$$\mathrm{ramp}[n] = \sum_{m=-\infty}^{n} \mathrm{u}[m-1]$$

and the unit sequence is the first *forward* difference of the unit ramp $\mathrm{u}[n] = \mathrm{ramp}[n+1] - \mathrm{ramp}[n]$ and the first backward difference of $\mathrm{ramp}[n+1]$.

MATLAB can compute differences of discrete-time functions using the `diff` function. The `diff` function accepts a vector of length N as its argument and returns a vector of forward differences of length $N - 1$. MATLAB can also compute the accumulation of a function using the `cumsum` (cumulative summation) function. The `cumsum` function accepts a vector as its argument and returns a vector of equal length that is the accumulation of the elements in the argument vector. For example,

```
»a = 1:10
a =
   1   2   3   4   5   6   7   8   9   10
»diff(a)
ans =
   1   1   1   1   1   1   1   1   1
»cumsum(a)
ans =
   1   3   6   10   15   21   28   36   45   55
»b = randn(1,5)
```

```
b =
  1.1909   1.1892  -0.0376   0.3273   0.1746
»diff(b)
ans =
  -0.0018  -1.2268   0.3649  -0.1527
»cumsum(b)
ans =
  1.1909   2.3801   2.3424   2.6697   2.8444
```

It is apparent from these examples that cumsum assumes that the value of the accumulation is zero before the first element in the vector.

EXAMPLE 3.2

Graphing the accumulation of a function using MATLAB

Using MATLAB, graph the accumulation of the function $x[n] = \cos(2\pi n/36)$ from $n = 0$ to $n = 36$ under the assumption that the accumulation before time $n = 0$ is zero.

```
%       Program to demonstrate accumulation of a function over a finite
%       time using the cumsum function.
n = 0:36 ;                          % Discrete-time vector
x = cos(2*pi*n/36);                 % Values of x[n]
%       Graph the accumulation of the function x[n]
p = stem(n,cumsum(x),'k','filled');
set(p,'LineWidth',2,'MarkerSize',4);
xlabel('\itn','FontName','Times','FontSize',24);
ylabel('x[{\itn}]','FontName','Times','FontSize',24);
```

Figure 3.22 illustrates the output of the MATLAB program.

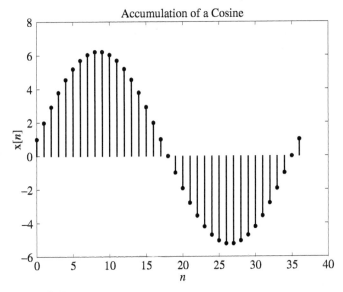

Figure 3.22
Accumulation of a cosine

Notice that this cosine accumulation looks a lot like (but not exactly like) a sine function. That occurs because the accumulation process is analogous to the integration process for continuous-time functions and the integral of a cosine is a sine.

3.7 EVEN AND ODD SIGNALS

Like continuous-time functions, discrete-time functions can also be classified by the properties of evenness and oddness. The defining relationships are completely analogous to those for continuous-time functions. If $g[n] = g[-n]$, then $g[n]$ is even, and if $g[n] = -g[-n]$, $g[n]$ is odd. Figure 3.23 shows some examples of even and odd functions.

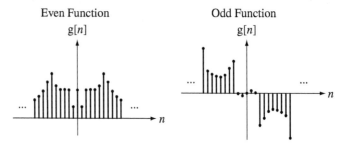

Figure 3.23
Examples of even and odd functions

The even and odd parts of a function $g[n]$ are found exactly the same way as for continuous-time functions.

$$g_e[n] = \frac{g[n] + g[-n]}{2} \text{ and } g_o[n] = \frac{g[n] - g[-n]}{2} \quad (3.7)$$

An even function has an odd part that is zero and an odd function has an even part that is zero.

EXAMPLE 3.3

Even and odd parts of a function

Find the even and odd parts of the function, $g[n] = \sin(2\pi n/7)(1 + n^2)$.

$$g_e[n] = \frac{\sin(2\pi n/7)(1 + n^2) + \sin(-2\pi n/7)(1 + (-n)^2)}{2}$$

$$g_e[n] = \frac{\sin(2\pi n/7)(1 + n^2) - \sin(2\pi n/7)(1 + n^2)}{2} = 0$$

$$g_o[n] = \frac{\sin(2\pi n/7)(1 + n^2) - \sin(-2\pi n/7)(1 + (-n)^2)}{2} = \sin\left(\frac{2\pi n}{7}\right)(1 + n^2)$$

The function $g[n]$ is odd.

```
function EvenOdd
    n = -14:14 ;        %  Discrete-time vector for graphing ge[n] and
                        %  go[n]
%   Compute the even part of g[n]
```

```
ge = (g(n)+g(-n))/2 ;
%   Compute the odd part of g[n]
go = (g(n)-g(-n))/2 ;
close all ; figure('Position',[20,20,1200,800]) ;
subplot(2,1,1) ;
ptr = stem(n,ge,'k','filled') ; grid on ;
set(ptr,'LineWidth',2,'MarkerSize',4) ;
xlabel('\itn','FontName','Times','FontSize',24) ;
ylabel('g_{\ite}[{\itn}]','FontName','Times','FontSize',24) ;
title('Even Part of g[n]','FontName','Times','FontSize',24) ;
set(gca,'FontName','Times','FontSize',18) ;
subplot(2,1,2) ;
ptr = stem(n,go,'k','filled') ; grid on ;
set(ptr,'LineWidth',2,'MarkerSize',4) ;
xlabel('\itn','FontName','Times','FontSize',24) ;
ylabel('g_{\ito}[{\itn}]','FontName','Times','FontSize',24) ;
title('Odd Part of g[n]','FontName','Times','FontSize',24) ;
set(gca,'FontName','Times','FontSize',18) ;

function y = g(n),
    y = sin(2*pi*n/7)./(1+n.^2) ;
```

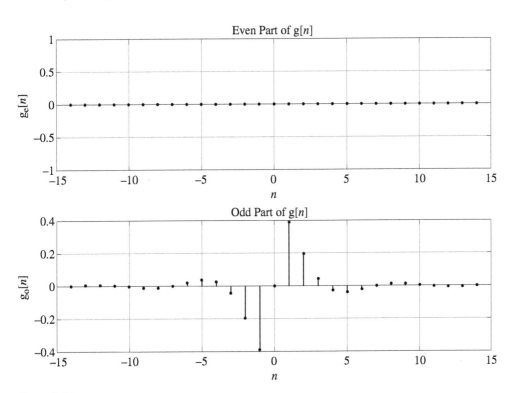

Figure 3.24

COMBINATIONS OF EVEN AND ODD SIGNALS

All the properties of combinations of functions that apply to continuous-time functions also apply to discrete-time functions. If two functions are even, their sum, difference, product and quotient are even too. If two functions are odd, their sum and difference are odd but their product and quotient are even. If one function is even and the other is odd, their product and quotient are odd.

In Figure 3.25 through Figure 3.27 are some examples of products of even and odd functions.

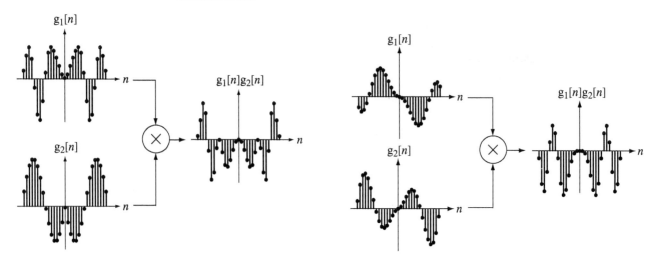

Figure 3.25
Product of two even functions

Figure 3.26
Product of two odd functions

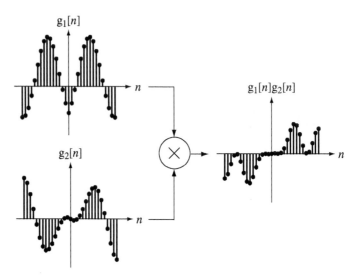

Figure 3.27
Product of an even and an odd function

SYMMETRICAL FINITE SUMMATION OF EVEN AND ODD SIGNALS

The definite integral of continuous-time functions over symmetrical limits is analogous to summation of discrete-time functions over symmetrical limits. Properties hold for summations of discrete-time functions that are similar to (but not identical to) those for integrals of continuous-time functions. If g[n] is an even function and N is a positive integer,

$$\sum_{n=-N}^{N} g[n] = g[0] + 2\sum_{n=1}^{N} g[n]$$

and, if g[n] is an odd function,

$$\sum_{n=-N}^{N} g[n] = 0$$

(See Figure 3.28.)

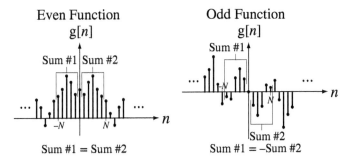

Figure 3.28
Summations of even and odd discrete-time functions

3.8 PERIODIC SIGNALS

A periodic function is one that is invariant under the time shift $n \to n + mN$, where N is a period of the function and m is any integer. The fundamental period N_0 is the *minimum positive* discrete time in which the function repeats. Figure 3.29 shows some examples of periodic functions.

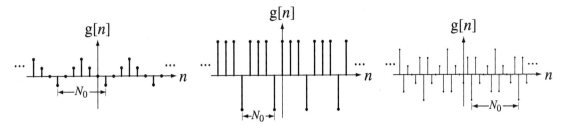

Figure 3.29
Examples of periodic functions with fundamental period N_0

The fundamental frequency is $F_0 = 1/N_0$ in cycles/sample or $\Omega_0 = 2\pi/N_0$ in radians/sample. Remember that the units of discrete-time frequency are not Hz or radians/second because the unit of discrete-time is not the second.

EXAMPLE 3.4

Fundamental period of a function

Graph the function $g[n] = 2\cos(9\pi n/4) - 3\sin(6\pi n/5)$ over the range $-50 \leq n \leq 50$. From the graph determine the fundamental period.

Figure 3.30 shows the function g[n].

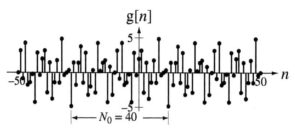

Figure 3.30
The function, $g[n] = 2\cos(9\pi n/4) - 3\sin(6\pi n/5)$

As a check on this graphically determined answer, the function can also be written in the form $g[n] = 2\cos(2\pi(9/8)n) - 3\sin(2\pi(3/5)n)$. The two fundamental periods of the two individual sinusoids are then 8 and 5 and their LCM is 40, which is the fundamental period of g[n].

3.9 SIGNAL ENERGY AND POWER

SIGNAL ENERGY

Signal energy is defined by

$$E_x = \sum_{n=-\infty}^{\infty} |x[n]|^2 , \qquad (3.8)$$

and its units are simply the square of the units of the signal itself.

EXAMPLE 3.5

Signal energy of a signal

Find the signal energy of $x[n] = (1/2)^n u[n]$. From the definition of signal energy,

$$E_x = \sum_{n=-\infty}^{\infty} |x[n]|^2 = \sum_{n=-\infty}^{\infty} \left|\left(\frac{1}{2}\right)^n u[n]\right|^2 = \sum_{n=0}^{\infty} \left|\left(\frac{1}{2}\right)^n\right|^2 = \sum_{n=0}^{\infty} \left(\frac{1}{2}\right)^{2n} = 1 + \frac{1}{2^2} + \frac{1}{2^4} + \cdots.$$

This infinite series can be rewritten as

$$E_x = 1 + \frac{1}{4} + \frac{1}{4^2} + \cdots.$$

We can use the formula for the summation of an infinite geometric series

$$\sum_{n=0}^{\infty} r^n = \frac{1}{1-r}, \quad |r| < 1$$

to get

$$E_x = \frac{1}{1 - 1/4} = \frac{4}{3}.$$

SIGNAL POWER

For many signals encountered in signal and system analysis, the summation

$$E_x = \sum_{n=-\infty}^{\infty} |\mathrm{x}[n]|^2$$

does not converge because the signal energy is infinite, and this usually occurs because the signal is not time limited. The unit sequence is an example of a signal with infinite energy. For signals of this type, it is more convenient to deal with the average signal power of the signal instead of the signal energy. The definition of average signal power is

$$P_x = \lim_{N \to \infty} \frac{1}{2N} \sum_{n=-N}^{N-1} |\mathrm{x}[n]|^2 , \qquad (3.9)$$

and this is the average signal power over all time. (Why is the upper summation limit $N-1$ instead of N?)

For periodic signals, the average signal power calculation may be simpler. The average value of any periodic function is the average over any period and

$$P_x = \frac{1}{N} \sum_{n=n_0}^{n_0+N-1} |\mathrm{x}[n]|^2 = \frac{1}{N} \sum_{n=\langle N \rangle} |\mathrm{x}[n]|^2, \quad n_0 \text{ any integer} \qquad (3.10)$$

where the notation $\sum_{n=\langle N \rangle}$ means the summation over any range of consecutive n's that is N in length, where N can be any period of $|\mathrm{x}[n]|^2$.

EXAMPLE 3.6

Finding signal energy and power of signals using MATLAB

Using MATLAB find the signal energy or power of the signals,

(a) $\mathrm{x}[n] = (0.9)^{|n|} \sin(2\pi n/4)$ and (b) $\mathrm{x}[n] = 4\delta_5[n] - 7\delta_7[n]$.

Then compare the results with analytical calculations.

```
%   Program to compute the signal energy or power of some example signals
%   (a)
n = -100:100 ;                  % Set up a vector of discrete times at
                                % which to compute the value of the
                                % function
%    Compute the value of the function and its square
x = (0.9).^abs(n).*sin(2*pi*n/4) ; xsq = x.^2 ;
Ex = sum(xsq) ;                 % Use the sum function in MATLAB to
                                % find the total energy and display
                                % the result.
disp(['(b) Ex = ',num2str(Ex)]);
%    (b)
```

```
NO = 35;                              % The fundamental period is 35
n = 0:NO-1;                           % Set up a vector of discrete times
                                      % over one period at which to compute
                                      % the value of the function
%    Compute the value of the function and its square
x = 4*impND(5,n) - 7*impND(7,n) ; xsq = x.^2 ;
Px = sum(xsq)/NO;                     % Use the sum function in MATLAB to
                                      % find the average power and display
                                      % the result.
disp(['(d) Px = ',num2str(Px)]);
```

The output of this program is

(a) Ex = 4.7107

(b) Px = 8.6

Analytical computations:

(a) $E_x = \sum_{n=-\infty}^{\infty} |x[n]|^2 = \sum_{n=-\infty}^{\infty} |(0.9)^{|n|}\sin(2\pi n/4)|^2$

$E_x = \sum_{n=0}^{\infty} |(0.9)^n \sin(2\pi n/4)|^2 + \sum_{n=-\infty}^{0} |(0.9)^{-n}\sin(2\pi n/4)|^2 - \underbrace{|x[0]|^2}_{=0}$

$E_x = \sum_{n=0}^{\infty} (0.9)^{2n}\sin^2(2\pi n/4) + \sum_{n=-\infty}^{0} (0.9)^{-2n}\sin^2(2\pi n/4)$

$E_x = \frac{1}{2}\sum_{n=0}^{\infty} (0.9)^{2n}(1 - \cos(\pi n)) + \frac{1}{2}\sum_{n=-\infty}^{0} (0.9)^{-2n}(1 - \cos(\pi n))$

Using the even symmetry of the cosine function, and letting $n \to -n$ in the second summation,

$E_x = \sum_{n=0}^{\infty} (0.9)^{2n}(1 - \cos(\pi n))$

$E_x = \sum_{n=0}^{\infty} \left((0.9)^{2n} - (0.9)^{2n}\frac{e^{j\pi n} + e^{-j\pi n}}{2}\right) = \sum_{n=0}^{\infty} (0.81)^n - \frac{1}{2}\left[\sum_{n=0}^{\infty}(0.81e^{j\pi})^n + \sum_{n=0}^{\infty}(0.81e^{-j\pi})^n\right]$

Using the formula for the sum of an infinite geometric series,

$\sum_{n=0}^{\infty} r^n = \frac{1}{1-r}, \quad |r| < 1$

$E_x = \frac{1}{1 - 0.81} - \frac{1}{2}\left[\frac{1}{1 - 0.81e^{j\pi}} + \frac{1}{1 - 0.81e^{-j\pi}}\right]$

$E_x = \frac{1}{1 - 0.81} - \frac{1}{2}\left[\frac{1}{1 + 0.81} + \frac{1}{1 + 0.81}\right] = \frac{1}{1 - 0.81} - \frac{1}{1 + 0.81} = 4.7107$ Check.

(b) $P_x = \frac{1}{N_0}\sum_{n=\langle N_0\rangle} |x[n]|^2 = \frac{1}{N_0}\sum_{n=0}^{N_0-1} |x[n]|^2 = \frac{1}{35}\sum_{n=0}^{34} |4\delta_5[n] - 7\delta_7[n]|^2$

The impulses in the two impulse train functions only coincide at integer multiples of 35. Therefore in this summation range they coincide only at $n = 0$. The net impulse strength at

$n = 0$ is therefore -3. All the other impulses occur alone and the sum of the squares is the same as the square of the sum. Therefore

$$P_x = \frac{1}{35}\left(\underbrace{(-3)^2}_{n=0} + \underbrace{4^2}_{n=5} + \underbrace{(-7)^2}_{n=7} + \underbrace{4^2}_{n=10} + \underbrace{(-7)^2}_{n=14} + \underbrace{4^2}_{n=15} + \underbrace{4^2}_{n=20} + \underbrace{(-7)^2}_{n=21} + \underbrace{4^2}_{n=25} + \underbrace{(-7)^2}_{n=28} + \underbrace{4^2}_{n=30}\right)$$

$$P_x = \frac{9 + 6 \times 4^2 + 4 \times (-7)^2}{35} = \frac{9 + 96 + 196}{35} = 8.6 \quad \text{Check.}$$

3.10 SUMMARY OF IMPORTANT POINTS

1. A discrete-time signal can be formed from a continuous-time signal by sampling.
2. A discrete-time signal is not defined at noninteger values of discrete time.
3. Discrete-time signals formed by sampling periodic continuous-time signals may be aperiodic.
4. Two different-looking analytical descriptions can produce identical discrete-time functions.
5. A time-shifted version of a discrete-time function is only defined for integer shifts in discrete time.
6. Time scaling a discrete-time function can produce decimation or undefined values, phenomena that do not occur when time scaling continuous-time functions.

EXERCISES WITH ANSWERS

(Answers to each exercise are in random order.)

Functions

1. In Figure E.1 is a circuit in which a voltage $x(t) = A\sin(2\pi f_0 t)$ is connected periodically to a resistor by a switch. The switch rotates at a frequency f_s of 500 rpm. The switch is closed at time $t = 0$ and each time the switch closes it stays closed for 10 ms.

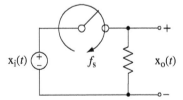

Figure E.1

(a) If $A = 5$ and $f_0 = 1$, graph the excitation voltage $x_i(t)$ and the response voltage $x_o(t)$ for $0 < t < 2$.
(b) If $A = 5$ and $f_0 = 10$, graph the excitation voltage $x_i(t)$ and the response voltage $x_o(t)$ for $0 < t < 1$.
(c) This is an approximation of an ideal sampler. If the sampling process were ideal what discrete-time signal $x[n]$ would be produced in parts (a) and (b)? Graph them versus discrete time n.

Answers:

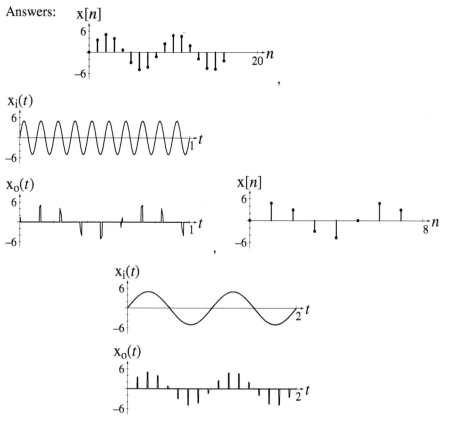

2. Let $x_1[n] = 5\cos(2\pi n/8)$ and $x_2[n] = -8e^{-(n/6)^2}$. Graph the following combinations of those two signals over the discrete-time range, $-20 \leq n < 20$. If a signal has some defined and some undefined values, just plot the defined values.

   (a) $x[n] = x_1[n]x_2[n]$
   (b) $x[n] = 4x_1[n] + 2x_2[n]$
   (c) $x[n] = x_1[2n]x_2[3n]$
   (d) $x[n] = \dfrac{x_1[2n]}{x_2[-n]}$
   (e) $x[n] = 2x_1[n/2] + 4x_2[n/3]$

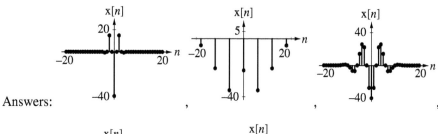

Answers:

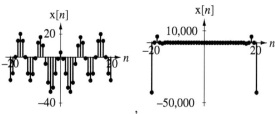

3. Find the numerical values of

   (a) $\sum_{n=-18}^{33} 38n^2\delta[n+6]$ (b) $\sum_{n=-4}^{7} -12(0.4)^n u[n]\delta_3[n]$

   Answers: -12.8172, 1368

Scaling and Shifting Functions

4. A discrete-time function is defined by $g[n] = 3\delta[n-4]\text{ramp}[n+1]$. What is its only nonzero value?

   Answer: 15

5. A discrete-time signal has the following values:

n	−5	−4	−3	−2	−1	0	1	2	3	4	5	6	7	8	9	10	11	12	13
x[n]	4	−2	5	−4	−10	−6	−9	−9	1	9	6	2	−2	2	0	−2	−9	−5	3

   For all other n, $x[n]$ is zero. Let $y[n] = x[2n-1]$. Find the numerical values $y[-2]$, $y[-1]$, $y[4]$, $y[7]$, $y[12]$.

   Answers: 3, 0, −2, 4, 5

6. Find the numerical values of these functions.

   (a) $\text{ramp}[6] - u[-2]$

   (b) $\sum_{n=-\infty}^{7} (u[n+9] - u[n-10])$

   (c) $g[4]$ where $g[n] = \sin(2\pi(n-3)/8) + \delta[n-3]$

   Answers: 6, 0.707, 17

7. For each pair of functions in Figure E.7 provide the values of the constants in $g_2[n] = Ag_1[a(n - n_0)]$.

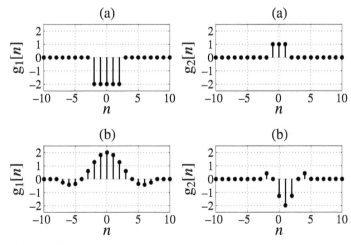

Figure E.7

Answers: $A = -1/2$, $n_0 = 0$, $a = 2$ or -2, $A = -1$, $n_0 = 1$, $a = 2$ or -2

Chapter 3 Discrete-Time Signal Description

8. A function g[n] is defined by

$$g[n] = \begin{cases} -2 & , \ n < -4 \\ n & , \ -4 \leq n < 1 \\ 4/n & , \ 1 \leq n \end{cases}$$

Graph g[−n], g[2 − n], g[2n] and g[n/2].

Answers:

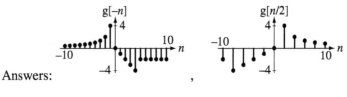

,

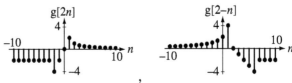

,

9. Find the maximum value over all discrete time of $x[n] = \text{ramp}[n+4]\text{ramp}[5-n]$.

Answer: 20

10. A discrete-time signal has the following values for times $n = -8$ to $n = 8$ and is zero for all other times.

n	−8	−7	−6	−5	−4	−3	−2	−1	0	1	2	3	4	5	6	7	8
x[n]	9	4	9	9	4	9	9	4	9	9	4	9	9	4	9	9	4

This signal can be expressed in the form $x[n] = (A - B\delta_{N_1}[n - n_0])(u[n + N_2] - u[n - N_2 - 1])$. Find the numerical values of the constants.

$$x[n] = (9 - 5\delta_3[n+1])(u[n+8] - u[n-9])$$

or

$$x[n] = (9 - 5\delta_3[n-2])(u[n+8] - u[n-9])$$

Answers: 3, 5, 8, 9, −1 or 3, 2, 8, 5, 9

11 A signal x[n] is nonzero only in the range $1 \leq n < 14$. If y[n] = x[3n]. If y[n] = x[3n] how many of the nonzero values of x appear in y?

Answer: 4

12 A discrete-time function, $g_1[n]$ is illustrated in Figure E.12. It is zero for all time outside the range graphed below. Let some other functions be defined by

$$g_2[n] = -g_1[2n], \quad g_3[n] = 2g_1[n-2], \quad g_4[n] = 3g_1\left[\frac{n}{3}\right]$$

Find the following numerical values.

(a) $g_4[2]$

(b) $[g_4(t)g_3(t)]_{t=2}$

(c) $\left(\dfrac{g_2[n]}{g_3[n]}\right)_{n=-1}$

(d) $\displaystyle\sum_{n=-1}^{1} g_2[n]$

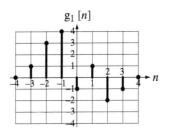

Figure E.12

Answers: Undefined, −3/2, 0, −3/2

Differencing and Accumulation

13. Graph the backward differences of the discrete-time functions in Figure E.13.

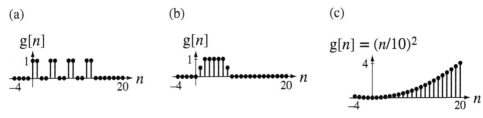

Figure E.13

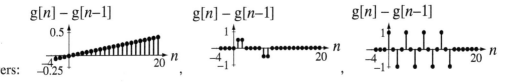

Answers:

14. The signal x[n] is defined in Figure E.14. Let y[n] be the first backward difference of x[n] and let z[n] be the accumulation of x[n]. (Assume that x[n] is zero for all $n < 0$).

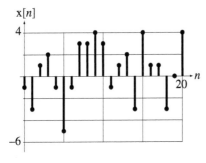

Figure E.14

(a) What is the value of y[4]?   (b) What is the value of z[6]?

Answers: −8, −3

15. Graph the accumulation g[n] of each of these discrete-time functions h[n] which are zero for all times $n < -16$.

    (a) $h[n] = \delta[n]$
    (b) $h[n] = u[n]$
    (c) $h[n] = \cos(2\pi n/16)u[n]$
    (d) $h[n] = \cos(2\pi n/8)u[n]$
    (e) $h[n] = \cos(2\pi n/16)u[n+8]$

Answers: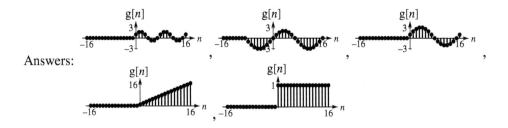

Even and Odd Functions

16. Find and graph the even and odd parts of these functions.

    (a) $g[n] = u[n] - u[n-4]$
    (b) $g[n] = e^{-n/4}u[n]$
    (c) $g[n] = \cos(2\pi n/4)$
    (d) $g[n] = \sin(2\pi n/4)u[n]$

Answers: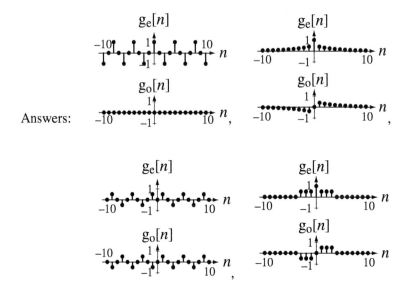

17. Graph g[n] for the signals in Figure E.17.

(a)

(b)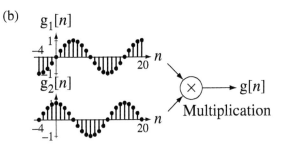

(c)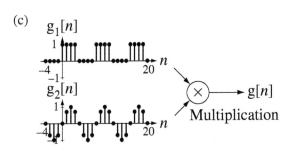

(d)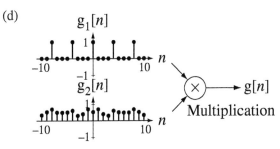

Figure E.17

Answers: 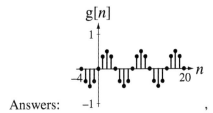 , 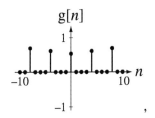 , 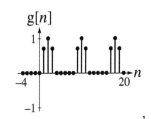 ,

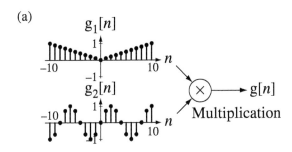

Periodic Functions

18. Find the fundamental period of each these functions.
    (a) $g[n] = \cos(2\pi n/10)$
    (b) $g[n] = \cos(\pi n/10) = \cos(2\pi n/20)$
    (c) $g[n] = \cos(2\pi n/5) + \cos(2\pi n/7)$

(d) $g[n] = e^{j2\pi n/20} + e^{-j2\pi n/20}$
(e) $g[n] = e^{-j2\pi n/3} + e^{-j2\pi n/4}$
(f) $g[n] = \sin(13\pi n/8) - \cos(9\pi n/6) = \sin(2 \times 13\pi n/16) - \cos(2 \times 3\pi n/4)$
(g) $g[n] = e^{-j6\pi n/21} + \cos(22\pi n/36) - \sin(11\pi n/33)$

Answers: 20, 10, 20, 12, 252, 35, 16

19. If $g[n] = 15\cos(-2\pi n/12)$ and $h[n] = 15\cos(2\pi Kn)$ what are the two smallest positive values of K for which $g[n] = h[n]$ for all n?

Answers: 11/12 and 1/12

Signal Energy and Power

20. Identify each of these signals as either an energy signal or a power signal and find the signal energy of energy signals and the average signal power of power signals.

(a) $x[n] = u[n]$ Power Signal

$$P_x = \lim_{N\to\infty} \frac{1}{2N+1} \sum_{n=-N}^{N} |u[n]|^2 = \lim_{N\to\infty} \frac{1}{2N+1} \sum_{n=0}^{N} 1^2 = \lim_{N\to\infty} \frac{N+1}{2N+1} = \frac{1}{2}$$

(b) $x[n] = u[n] - u[n-10]$ Energy Signal

$$E_x = \sum_{n=-\infty}^{\infty} |u[n] - u[n-10]|^2 = \sum_{n=0}^{9} 1^2 = 10$$

(c) $x[n] = u[n] - 2u[n-4] + u[n-10]$ Energy Signal

$$E_x = \sum_{n=-\infty}^{\infty} |u[n] - 2u[n-4] + u[n-10]|^2 = \sum_{n=0}^{3} 1^2 + \sum_{n=4}^{9} (-1)^2 = 4 + 6 = 10$$

(d) $x[n] = -4\cos(\pi n)$ Power Signal

$$P_x = \frac{1}{N} \sum_{n=\langle N\rangle} |-4\cos(\pi n)|^2 = \frac{16}{N} \sum_{n=\langle N\rangle} |\cos^2(\pi n)|$$

$$= \frac{16}{N} \sum_{n=\langle N\rangle} [(-1)^n]^2 = \frac{16}{N} \sum_{n=\langle N\rangle} 1^2 = 16$$

(e) $x[n] = 5(u[n+3] - u[n-5]) - 8(u[n-1] - u[n-7])$ Energy Signal

$$E_x = \sum_{n=-\infty}^{\infty} |5(u[n+3] - u[n-5]) - 8(u[n-1] - u[n-7])|^2$$

$$E_x = \sum_{n=-3}^{6} [5(u[n+3] - u[n-5]) - 8(u[n-1] - u[n-7])]^2$$

$$E_x = \sum_{n=-3}^{0} 5^2 + \sum_{1}^{4} (-3)^2 + \sum_{5}^{6} (-8)^2 = 25 \times 4 + 9 \times 4 + 64 \times 2 = 264$$

Answers: 10, 16, 10, 1/2, 264

21. Find the signal energy of each of these signals.

(a) $x[n] = A\delta[n]$ (b) $x[n] = \delta_{N_0}[n]$
(c) $x[n] = \text{ramp}[n]$ (d) $x[n] = \text{ramp}[n] - 2\text{ramp}[n-4] + \text{ramp}[n-8]$
(e) $x[n] = \text{ramp}[n+3]u[-(n-4)]$

(f) $x[n] = (\delta_3[n] - 3\delta_6[n])(u[n+1] - u[n-12])$.

Answers: $140, \infty, 44, A^2, \infty, 10$

22. A signal $x[n]$ is periodic with period $N_0 = 6$. Some selected values of $x[n]$ are $x[0] = 3$, $x[-1] = 1$, $x[-4] = -2$, $x[-8] = -2$, $x[3] = 5$, $x[7] = -1$, $x[10] = -2$ and $x[-3] = 5$. What is its average signal power?

Answer: 7.333

23. Find the average signal power of a periodic signal described over one period by $x[n] = 2n, -2 \leq n < 2$.

Answer: 6

24. Find the average signal power of these signals.

   (a) $x[n] = A$
   (b) $x[n] = u[n]$
   (c) $x[n] = \delta_{N_0}[n]$
   (d) $x[n] = \text{ramp}[n]$

Answers: $\infty, 1/N_0, A^2, 1/2$

EXERCISES WITHOUT ANSWERS

Signal Functions

25. Graph these discrete-time exponential and trigonometric functions.

   (a) $g[n] = -4\cos(2\pi n/10)$
   (b) $g[n] = -4\cos(2.2\pi n)$
   (c) $g[n] = -4\cos(1.8\pi n)$
   (d) $g[n] = 2\cos(2\pi n/6) - 3\sin(2\pi n/6)$
   (e) $g[n] = (3/4)^n$
   (f) $g[n] = 2(0.9)^n \sin(2\pi n/4)$

Shifting and Scaling Functions

26. Graph these functions.

   (a) $g[n] = 2u[n+2]$
   (b) $g[n] = u[5n]$
   (c) $g[n] = -2\text{ramp}[-n]$
   (d) $g[n] = 10\text{ramp}[n/2]$
   (e) $g[n] = 7\delta[n-1]$
   (f) $g[n] = 7\delta[2(n-1)]$
   (g) $g[n] = -4\delta[2n/3]$
   (h) $g[n] = -4\delta[2n/3 - 1]$
   (i) $g[n] = 8\delta_4[n]$
   (j) $g[n] = 8\delta_4[2n]$

27. Graph these functions.

   (a) $g[n] = u[n] + u[-n]$
   (b) $g[n] = u[n] - u[-n]$
   (c) $g[n] = \cos(2\pi n/12)\delta_3[n]$
   (d) $g[n] = \cos(2\pi n/12)\delta_3[n/2]$
   (e) $g[n] = \cos\left(\dfrac{2\pi(n+1)}{12}\right)u[n+1] - \cos\left(\dfrac{2\pi n}{12}\right)u[n]$
   (f) $g[n] = \displaystyle\sum_{m=0}^{n} \cos\left(\dfrac{2\pi m}{12}\right)u[m]$
   (g) $g[n] = \displaystyle\sum_{m=0}^{n} (\delta_4[m] - \delta_4[m-2])$

(h) $g[n] = \sum_{m=-\infty}^{n} (\delta_4[m] + \delta_3[m])(u[m+4] - u[m-5])$

(i) $g[n] = \delta_2[n+1] - \delta_2[n]$ (j) $g[n] = \sum_{m=-\infty}^{n+1} \delta[m] - \sum_{m=-\infty}^{n} \delta[m]$

28. Graph the magnitude and phase of each function versus k.

    (a) $G[k] = 20 \sin(2\pi k/8) e^{-j\pi k/4}$

    (b) $G[k] = (\delta[k+8] - 2\delta[k+4] + \delta[k] - 2\delta[k-4] + \delta[k-8])e^{j\pi k/8}$

29. Using MATLAB, for each function below graph the original function and the shifted and/or scaled function.

    (a) $g[n] = \begin{cases} 5, & n \le 0 \\ 5 - 3n, & 0 < n \le 4 \\ -23 + n^2, & 4 < n \le 8 \\ 41, & n > 8 \end{cases}$ $\quad g[3n]$ vs. n

    (b) $g[n] = 10 \cos(2\pi n/20)\cos(2\pi n/4)$ $\quad 4g[2(n+1)]$ vs. n

    (c) $g[n] = |8e^{j2\pi n/16} u[n]|$ $\quad g[n/2]$ vs. n

30. Graph versus k, in the range $-10 < k < 10$ the magnitude and phase of

    (a) $X[k] = \dfrac{1}{1 + jk/2}$ (b) $X[k] = \dfrac{jk}{1 + jk/2}$

    (c) $X[k] = \delta_2[k] e^{-j2\pi k/4}$

Differencing and Accumulation

31. Graph the accumulation of each of these discrete-time functions.

    (a) $g[n] = \cos(2\pi n)u[n]$ (b) $g[n] = \cos(4\pi n)u[n]$

32. A discrete-time function is defined by $g[n] = (3 + (-1)^n)u[n]$. Another discrete-time function h[n] is defined as the accumulation of g[n] from $-\infty$ to n. Find the numerical value of h[7].

33. In the equation $\sum_{m=-\infty}^{n} u[m] = g[(n - n_0)/N_w]$

    (a) What is name of the function, g? (b) Find the values of n_0 and N_w.

34. What is the numerical value of each of the following accumulations?

    (a) $\sum_{n=0}^{10} \text{ramp}[n]$ (b) $\sum_{n=0}^{6} \dfrac{1}{2^n}$ (c) $\sum_{n=-\infty}^{\infty} u[n]/2^n$

    (d) $\sum_{n=-10}^{10} \delta_3[n] = 7$ (e) $\sum_{n=-10}^{10} \delta_3[2n] = 7$

Even and Odd Functions

35. Find and graph the magnitude and phase of the even and odd parts of the "discrete-k" function, $G[k] = \dfrac{10}{1 - j4k}$.

36. The even part of g[n] is $g_e[n] = \text{tri}(n/16)$ and the odd part of g[n] is $g_o[n] = u[n + 3] - u[n] - u[n - 1] + u[n - 4]$. Find the value of g[3].

37. Find and graph the even and odd parts of the function in Figure E.37.

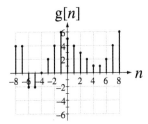

Figure E.37

38. Graph the even and odd parts of these signals.

   (a) $x[n] = \delta_3[n - 1]$ (b) $x[n] = 15 \cos(2\pi n/9 + \pi/4)$

Periodic Signals

39. Find the fundamental periods (if they exist) of

   (a) $x[n] = \delta_{14}[n] - 6\delta_8[n]$

   (b) $x[n] = -2 \cos(3\pi n/12) + 11 \cos(14\pi n/10)$

   (c) $x[n] = \cos(n/3)$

   (d) $x[n] = \cos(7\pi n/3) - 4 \sin(11\pi n/5)$

   (e) $x[n] = 9 \cos(7\pi n/15) + 14 \sin(8\pi n/11)$

   (f) $x[n] = 22 \sin(13n/12) - 13 \cos(18n/23)$

40. What is the smallest positive value of n_0 that makes the signal

$$[4 \cos(6\pi n/7) - 2 \sin(15\pi n/12)] * (u[n] - u[n - n_0])$$

   zero for all n?

41. Using MATLAB, graph each of these discrete-time functions. If a function is periodic, find the period analytically and verify the period from the plot.

   (a) $g[n] = \sin(3\pi n/2) = \sin(6\pi n/4)$

   (b) $g[n] = \sin(2\pi n/3) + \cos(10\pi n/3)$

   (c) $g[n] = \underbrace{5\cos(2\pi n/8)}_{\text{Period of 8}} + \underbrace{3\sin(2n/5)}_{\text{Period of 5}}$

   (d) $g[n] = 10 \cos(n/4)$

   (e) $g[n] = -3 \cos(2\pi n/7)\sin(2\pi n/6)$
   (A trigonometric identity will be useful here.)

42. If $x[n] = \sin(2\pi n/15)$ and $y[n] = \sin(2A\pi n)$, what is the smallest positive numerical value of A greater than 1/15 that makes x and y equal for all n?

Signal Energy and Power

43. Find the signal energy of each of these signals.
    (a) $x[n] = 2\delta[n] + 5\delta[n-3]$
    (b) $x[n] = u[n]/n$
    (c) $x[n] = (-1/3)^n u[n]$
    (d) $x[n] = \cos(\pi n/3)(u[n] - u[n-6])$
    (e) $x[n] = n(-1.3)^n (u[n] - u[n-6])$
    (f) $x[n] = 2(u[n+2] - u[n-3]) - 3\text{tri}(n/3)$

44. Find the average signal power of each of these signals.
    (a) $x[n] = (-1)^n$
    (b) $x[n] = A\cos(2\pi F_0 n + \theta)$
    (c) $x[n] = \begin{cases} A, & n = \cdots, 0, 1, 2, 3, 8, 9, 10, 11, 16, 17, 18, 19, \cdots \\ 0, & n = \cdots, 4, 5, 6, 7, 12, 13, 14, 15, 20, 21, 22, 23, \cdots \end{cases}$
    (d) $x[n] = e^{-j\pi n/2}$
    (e) $x[n] = \delta_3[n] - 3\delta_9[n] - 1$
    (f) $x_1[n] = -2\cos(2\pi n/6)$

45. Find the average signal power of $x[n] = 11\delta_4[2n] - 15\delta_5[3n]$.

46. A periodic discrete-time signal $x_1[n]$ is described over its fundamental period by
$\begin{array}{c|ccc} n & 0 & 1 & 2 \\ \hline x_1[n] & 8 & -3 & 5 \end{array}$. A second periodic discrete-time signal $x_2[n]$ is described over its fundamental period by $\begin{array}{c|cccc} n & 0 & 1 & 2 & 3 \\ \hline x_2[n] & -5 & 2 & 11 & -4 \end{array}$.

    (a) Find the average signal power of $x_1[n]$.
    (b) Find the average signal power of $x_2[n]$.
    (c) If $y[n] = x_1[n] - x_2[n]$, find the average signal power of $y[n]$.

47. A discrete-time function $x[n]$ is defined by Figure E.47.

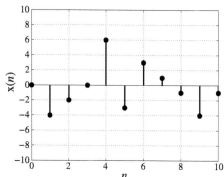

Figure E.47

A second discrete-time function is defined by $y[n] = x[n] - x[n-1]$, a third discrete-time function is defined by $z[n] = \sum_{m=-\infty}^{n} x[m]$, and a fourth discrete-time function is defined by $w[n] = x[2n]$. Find the following.

(a) The signal energy of $x[n]$.
(b) The signal energy of $w[n]$.
(c) The signal energy of $y[n]$.
(d) The average signal power of $z[n]$.

48. A periodic discrete-time signal $x[n]$ is described over exactly one fundamental period by $x[n] = 5 - \text{ramp}[n]$, $3 \le n < 7$. Find the average signal power of $x[n]$.

49. Let two discrete-time signals be defined by

$$g_1[n] = \begin{cases} 0, & n < -3 \\ 2n, & -3 \le n < 3 \\ 0, & n \ge 3 \end{cases} \quad \text{and} \quad g_2[n] = -3\sin(2\pi n/4)$$

(a) Find the signal energy of $g_1[n]$.
(b) Find the average signal power of $g_2[n]$.
(c) Perform the transformation $n \to 2n$ on $g_2[n]$ to form $g_3[n]$ and then perform the transformation $n \to n - 1$ on $g_3[n]$ to form $g_4[n]$. What is the average signal power of $g_4[n]$?
(d) Reverse the order of the two transformations of part (c). What is the new signal power of the new $g_4[n]$?

CHAPTER 4

Description of Systems

4.1 INTRODUCTION AND GOALS

The words *signal* and *system* were defined very generally in Chapter 1. Analysis of systems is a discipline that has been developed by engineers. Engineers use mathematical theories and tools and apply them to knowledge of the physical world to design things that do something useful for society. The things an engineer designs are systems, but, as indicated in Chapter 1, the definition of a system is broader than that. The term *system* is so broad it is difficult to define. A system can be almost anything.

One way to define a system is as anything that performs a function. Another way to define a system is as anything that responds when stimulated or excited. A system can be an electrical system, a mechanical system, a biological system, a computer system, an economic system, a political system and so on. Systems designed by engineers are artificial systems; systems that have developed organically over a period of time through evolution and the rise of civilization are natural systems. Some systems can be analyzed completely with mathematics. Some systems are so complicated that mathematical analysis is extremely difficult. Some systems are just not well understood because of the difficulty in measuring their characteristics. In engineering the term system usually refers to an artificial system that is excited by certain signals and responds with other signals.

Many systems were developed in earlier times by artisans who designed and improved their systems based on their experiences and observations, apparently with the use of only the simplest mathematics. One of the most important distinctions between engineers and artisans is in the engineer's use of higher mathematics, especially calculus, to describe and analyze systems.

CHAPTER GOALS

1. To introduce nomenclature that describes important system properties
2. To illustrate the modeling of systems with differential and difference equations and block diagrams
3. To develop techniques for classifying systems according to their properties

4.2 CONTINUOUS-TIME SYSTEMS

SYSTEM MODELING

One of the most important processes in signal and system analysis is the modeling of systems: describing them mathematically or logically or graphically. A good model is one that includes all the significant effects of a system without being so complicated that it is difficult to use.

Common terminology in system analysis is that if a system is **excited** by **input signals** applied at one or more inputs, **responses** or **output signals** appear at one or more outputs. To excite a system means to apply energy that causes it to respond. One example of a system would be a boat propelled by a motor and steered by a rudder. The thrust developed by the propeller, the rudder position and the current of the water excite the system, and the heading and speed of the boat are responses (Figure 4.1).

Notice the statement above says that the heading and speed of the boat are responses, but it does not say that they are *the* responses, which might imply that there are not any others. Practically every system has multiple responses, some significant and some insignificant. In the case of the boat, the heading and speed of the boat are significant, but the vibration of the boat structure, the sounds created by the water splashing on the sides, the wake created behind the boat, the rocking and/or tipping of the boat and a myriad of other physical phenomena are not significant, and would probably be ignored in a practical analysis of this system.

An automobile suspension is excited by the surface of the road as the car travels over it, and the position of the chassis relative to the road is a significant response (Figure 4.2). When we set a thermostat in a room, the setting and the room temperature are input signals to the heating and cooling system, and a response of the system is the introduction of warm or cool air to move the temperature inside the room closer to the thermostat setting.

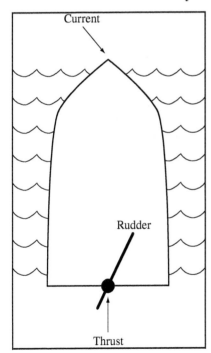

Figure 4.1
A simplified diagram of a boat

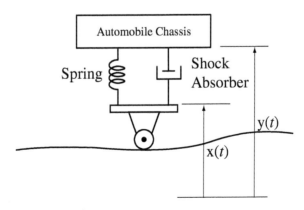

Figure 4.2
Simplified model of an automobile suspension system

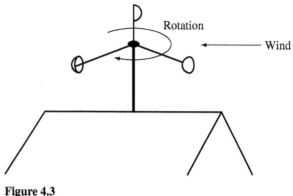

Figure 4.3
Cup anemometer

A whole class of systems, measurement instruments, are single-input, single-output systems. They are excited by the physical phenomenon being measured, and the response is the instrument's indication of the value of that physical phenomenon. A good example is a cup anemometer. The wind excites the anemometer and the angular velocity of the anemometer is the significant response (Figure 4.3).

Something that is not ordinarily thought of as a system is a highway bridge. There is no obvious or deliberate input signal that produces a desired response. The ideal bridge would not respond at all when excited. A bridge is excited by the traffic that rolls across it, the wind that blows onto it and the water currents that push on its support structure, and it *does move*. A very dramatic example showing that bridges respond when excited was the failure of the Tacoma Narrows suspension bridge in Washington state. On one very windy day the bridge responded to the wind by oscillating wildly and was eventually torn apart by the forces on it. This is a very dramatic example of why good analysis is important. The conditions under which the bridge would respond so strongly should have been discovered in the design process so that the design could have been changed to avoid this disaster.

A single biological cell in a plant or animal is a system of astonishing complexity, especially considering its size. The human body is a system comprising a huge number of cells and is, therefore, an almost unimaginably complicated system. But it can be modeled as a much simpler system in some cases to calculate an isolated effect. In pharmacokinetics the human body is often modeled as a single compartment, a volume containing liquid. Taking a drug is an excitation and the concentration of drug in the body is the significant response. Rates of infusion and excretion of the drug determine the variation of the drug concentration with time.

Differential Equations
Below are some examples of the thinking involved in modeling systems using differential equations. These examples were first presented in Chapter 1.

EXAMPLE 4.1

Modeling a mechanical system

A man 1.8 m tall and weighing 80 kg bungee jumps off a bridge over a river. The bridge is 200 m above the water surface and the unstretched bungee cord is 30 m long. The spring constant of the bungee cord is $K_s = 11$ N/m, meaning that, when the cord is stretched, it resists the stretching with a force of 11 newtons per meter of stretch. Make a mathematical model of the dynamic position of the man as a function of time and graph the man's position versus time for the first 15 seconds.

When the man jumps off the bridge he goes into free fall until the bungee cord is extended to its full unstretched length. This occurs when the man's feet are at 30 m below the bridge. His initial velocity and position are zero (using the bridge as the position reference). His acceleration is 9.8 m/s^2 until he reaches 30 m below the bridge. His position is the integral of his velocity and his velocity is the integral of his acceleration. So, during the initial free-fall time, his velocity is $9.8t$ m/s, where t is time in seconds and his position is $4.9t^2$ m below the bridge. Solving for the time of full unstretched bungee-cord extension we get 2.47 s. At that time his velocity is 24.25 meters per second, straight down. At this point the analysis changes because the bungee cord starts having an effect. There are two forces on the man:

1. The downward pull of gravity mg where m is the man's mass and g is the acceleration caused by the earth's gravity
2. The upward pull of the bungee cord $K_s(y(t) - 30)$ where $y(t)$ is the vertical position of the man below the bridge as a function of time

Then, using the principle that force equals mass times acceleration and the fact that acceleration is the second derivative of position, we can write

$$mg - K_s(y(t) - 30) = my''(t)$$

or

$$my''(t) + K_s y(t) = mg + 30 K_s.$$

This is a second-order, linear, constant-coefficient, inhomogeneous, ordinary differential equation. Its total solution is the sum of its homogeneous solution and its particular solution.

The homogeneous solution is a linear combination of the equation's eigenfunctions. The eigenfunctions are the functional forms that can satisfy *this form* of equation. There is one eigenfunction for each eigenvalue. The eigenvalues are the parameters in the eigenfunctions that make them satisfy *this particular* equation. The eigenvalues are the solutions of the characteristic equation, which is $m\lambda^2 + K_s = 0$. The solutions are $\lambda = \pm j\sqrt{K_s/m}$. (See Web Appendix D for more on the solution of differential equations.) Since the eigenvalues are complex numbers, it is somewhat more convenient to express the solution as a linear combination of a real sine and a real cosine instead of two complex exponentials. So the homogeneous solution can be expressed in the form

$$y_h(t) = K_{h1} \sin\left(\sqrt{K_s/m}\ t\right) + K_{h2} \cos\left(\sqrt{K_s/m}\ t\right).$$

The particular solution is in the form of a linear combination of the forcing function and all its unique derivatives. In this case the forcing function is a constant and all its derivatives are zero. Therefore the particular solution is of the form $y_p(t) = K_p$, a constant. Substituting in the form of the particular solution and solving, $y_p(t) = mg/K_s + 30$. The total solution is the sum of the homogeneous and particular solutions

$$y(t) = y_h(t) + y_p(t) = K_{h1} \sin\left(\sqrt{K_s/m}\ t\right) + K_{h2} \cos\left(\sqrt{K_s/m}\ t\right) + \underbrace{mg/K_s + 30}_{K_p}.$$

The boundary conditions are $y(2.47) = 30$ and $y'(t)_{t=2.47} = 24.25$. Putting in numerical values for parameters, applying boundary conditions and solving, we get

$$y(t) = -16.85 \sin(0.3708t) - 95.25 \cos(0.3708t) + 101.3, \quad t > 2.47.$$

The initial variation of the man's vertical position versus time is parabolic. Then at 2.47 s the solution becomes a sinusoid chosen to make the two solutions and the derivatives of the two solutions continuous at 2.47 s, as is apparent in Figure 4.4.

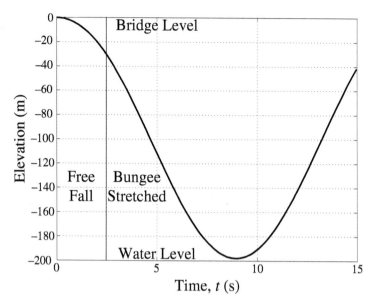

Figure 4.4
Man's vertical position versus time (bridge level is zero)

In Example 4.1 the differential equation

$$my''(t) + K_s y(t) = mg + 30K_s$$

describes the system. This is a linear, constant-coefficient, inhomogeneous ordinary differential equation. The right side of the equation is called its **forcing function**. If the forcing function is zero, we have a homogeneous differential equation and the solution of that equation is the homogeneous solution. In signal and system analysis this solution is called the **zero-input response**. It is nonzero only if the initial conditions of the system are nonzero, meaning the system has stored energy. If the system has no stored energy and the forcing function is not zero, the response is called the **zero-state response**.

Many physical processes were ignored in the mathematical model used in Example 4.1, for example,

1. Air resistance
2. Energy dissipation in the bungee cord
3. Horizontal components of the man's velocity
4. Rotation of the man during the fall
5. Variation of the acceleration due to gravity as a function of position
6. Variation of the water level in the river

Omitting these factors kept the model mathematically simpler than it would otherwise be. System modeling is always a compromise between the accuracy and the simplicity of the model.

Example 4.2

Modeling a fluid-mechanical system

A cylindrical water tank has cross-sectional area A_1 and water level $h_1(t)$ and is fed by an input volumetric flow of water $f_1(t)$ with an orifice at height h_2 whose effective cross-sectional area is

A_2, through which flows the output volumetric flow $f_2(t)$ (Figure 4.5). Write a differential equation for the water level as a function of time and graph the water level versus time for a tank that is initially empty, under different assumptions of inflow.

Under certain simplifying assumptions, the velocity of the water flowing out of the orifice is given by Toricelli's equation,

$$v_2(t) = \sqrt{2g[h_1(t) - h_2]}$$

where g is the acceleration due to earth's gravity (9.8 m/s^2). The rate of change of the volume $A_1 h_1(t)$ of water in the tank is the volumetric inflow rate minus the volumetric outflow rate

$$\frac{d}{dt}(A_1 h_1(t)) = f_1(t) - f_2(t)$$

and the volumetric outflow rate is the product of the effective area A_2 of the orifice and the output flow velocity $f_2(t) = A_2 v_2(t)$. Combining equations we can write one differential equation for the water level

$$A_1 \frac{d}{dt}(h_1(t)) + A_2 \sqrt{2g[h_1(t) - h_2]} = f_1(t). \tag{4.1}$$

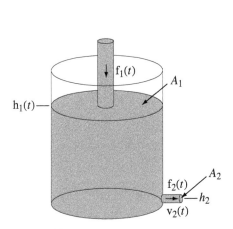

Figure 4.5
Tank with orifice being filled from above

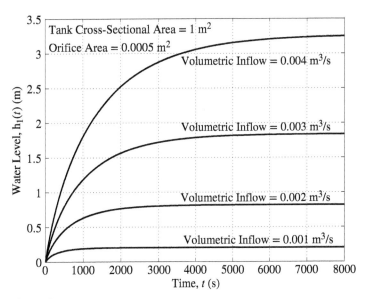

Figure 4.6
Water level versus time for four different volumetric inflows with the tank initially empty

The water level in the tank is graphed in Figure 4.6 versus time for four constant volumetric inflows under the assumption that the tank is initially empty. As the water flows in, the water level increases and the increase of water level increases the water outflow. The water level rises until the outflow equals the inflow and after that time the water level stays constant. As first stated in Chapter 1, when the inflow is increased by a factor of two, the final water level is increased by a factor of four, and this fact makes the differential equation (4.1) nonlinear. A method of finding the solution to this differential equation will be presented later in this chapter.

Block Diagrams

In system analysis it is very useful to represent systems by block diagrams. A system with one input and one output would be represented as in Figure 4.7. The signal at the input x(t) is operated on by the operator $\mathcal{H}$ to produce the signal at the output y(t). The operator $\mathcal{H}$ could perform just about any operation imaginable.

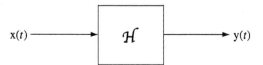

Figure 4.7
A single-input, single-output system

A system is often described and analyzed as an assembly of **components**. A component is a smaller, simpler system, usually one that is standard in some sense and whose properties are already known. Just what is considered a component as opposed to a system depends on the situation. To a circuit designer, components are resistors, capacitors, inductors, operational amplifiers and so on, and systems are power amplifiers, A/D converters, modulators, filters and so forth. To a communication system designer components are amplifiers, modulators, filters, antennas and systems are microwave links, fiber-optic trunk lines, telephone central offices. To an automobile designer components are wheels, engines, bumpers, lights, seats and the system is the automobile. In large, complicated systems like commercial airliners, telephone networks, supertankers or power plants, there are many levels of hierarchy of components and systems.

By knowing how to mathematically describe and characterize all the components in a system and how the components interact with each other, an engineer can predict, using mathematics, how a system will work, without actually building it and testing it. A system made up of components is diagrammed in Figure 4.8.

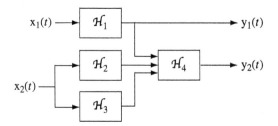

Figure 4.8
A two-input, two-output system composed of four interconnected components

In block diagrams each input signal may go to any number of blocks, and each output signal from a block may go to any number of other blocks. These signals are not affected by being connected to any number of blocks. There is no loading effect as there is in circuit analysis. In an electrical analogy, it would be as though the signals were voltages and the blocks all have infinite input impedance and zero output impedance.

In drawing block diagrams of systems, some types of operations appear so often they have been assigned their own block-diagram graphical symbols. They are the **amplifier**, the **summing junction** and the **integrator**.

The amplifier multiplies its input signal by a constant (its gain) to produce its response. Different symbols for amplification are used in different applications of system

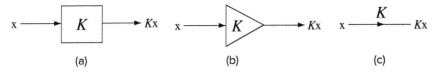

Figure 4.9
Three different graphical representations of an amplifier in a system block diagram

analysis and by different authors. The most common forms are shown in Figure 4.9. We will use Figure 4.9(c) in this text to represent an amplifier.

A summing junction accepts multiple input signals and responds with the sum of those signals. Some of the signals may be negated before being summed, so this component can also produce the difference between two signals. Typical graphical symbols used to represent a summing junction are illustrated in Figure 4.10.

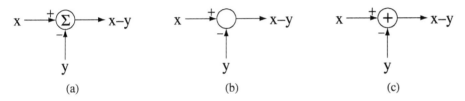

Figure 4.10
Three different graphical representations of a summing junction in a system block diagram

We will use Figure 4.10(c) in this text to represent a summing junction. If there is no plus or minus sign next to a summing junction input, a plus sign is assumed.

An integrator, when excited by any signal, responds with the integral of that signal (Figure 4.11).

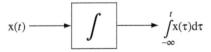

Figure 4.11
The graphical block-diagram symbol for an integrator

There are also symbols for other types of components that do special signal-processing operations. Each engineering discipline has its own preferred set of symbols for operations that are common in that discipline. A hydraulic system diagram might have dedicated symbols for a valve, a venturi, a pump and a nozzle. An optical system diagram might have symbols for a laser, a beamsplitter, a polarizer, a lens and a mirror.

In signals and systems there are common references to two general types of systems, open-loop and closed-loop. An open-loop system is one that simply responds directly to an input signal. A closed-loop system is one that responds to an input signal but also senses the output signal and "feeds it back" to add to or subtract from the input signal to better satisfy system requirements. Any measuring instrument is an open-loop system. The response simply indicates what the excitation is without altering it. A human driving a car is a good example of a closed-loop feedback system. The driver signals the car to move at a certain speed and in a certain direction by pressing the accelerator or brake and by turning the steering wheel. As the car moves down a road, the driver is constantly sensing the speed and position of the car relative to the road and the other cars. Based on what the driver senses she modifies the input signals (steering wheel, accelerator and/or brakes) to maintain the desired direction of the car and to keep it at a safe speed and position on the road.

EXAMPLE 4.3

Modeling a continuous-time feedback system

For the system illustrated in Figure 4.12,

(a) Find its zero-input response, the response with x(t) = 0, if the initial value of y(t) is y(0) = 1, the initial rate of change of y(t) is $y'(t)|_{t=0} = 0$, $a = 1$, $b = 0$ and $c = 4$.
(b) Let $b = 5$ and find the zero-input response for the same initial conditions as in part (a).
(c) Let the system be initially at rest and let the input signal x(t) be a unit step. Find the zero-state response for $a = 1$, $c = 4$ and $b = -1, 1, 5$.

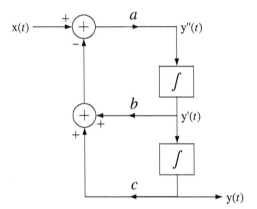

Figure 4.12
Continuous-time feedback system

(a) From the diagram we can write the differential equation for this system by realizing that the output signal from the summing junction is $y''(t)$ and it must equal the sum of its input signals (because $a = 1$)

$$y''(t) = x(t) - [by'(t) + cy(t)]$$

With $b = 0$ and $c = 4$, the response is described by the differential equation $y''(t) + 4y(t) = x(t)$. The eigenfunction is the complex exponential e^{st} and the eigenvalues are the solutions of the characteristic equation $s^2 + 4 = 0 \Rightarrow s_{1,2} = \pm j2$. The homogeneous solution is $y(t) = K_{h1}e^{j2t} + K_{h2}e^{-j2t}$. Since there is no excitation, this is also the total solution. Applying the initial conditions, $y(0) = K_{h1} + K_{h2} = 1$ and $y'(t)|_{t=0} = j2K_{h1} - j2K_{h2} = 0$ and solving, $K_{h1} = K_{h2} = 0.5$. The total solution is $y(t) = 0.5(e^{j2t} + e^{-j2t}) = \cos(2t)$, $t \geq 0$. So, with $b = 0$, the zero-input response is a sinusoid.

(b) Now $b = 5$. The differential equation is $y''(t) + 5y'(t) + 4y(t) = x(t)$, the eigenvalues are $s_{1,2} = -1, -4$ and the solution is $y(t) = K_{h1}e^{-t} + K_{h2}e^{-4t}$. Applying initial conditions, $y(0) = K_{h1} + K_{h2} = 1$ and $y'(t)|_{t=0} = -K_{h1} - 4K_{h2} = 0$. Solving for the constants, $K_{h1} = 4/3$, $K_{h2} = -1/3$ and $y(t) = (4/3)e^{-t} - (1/3)e^{-4t}$, $t \geq 0$. This zero-input response approaches zero for times, $t > 0$.

(c) In this case x(t) is not zero and the total solution of the differential equation includes the particular solution. After $t = 0$ the input signal is a constant, so the particular solution is also a constant K_p. The differential equation is $y''(t) + by'(t) + 4y(t) = x(t)$. Solving for K_p we get $K_p = 0.25$ and the total solution is $y(t) = K_{h1}e^{s_1 t} + K_{h2}e^{s_2 t} + 0.25$ where $s_{1,2} = (-b \pm \sqrt{b^2 - 16})/2$. The response and its first derivative are both zero at $t = 0$. Applying initial conditions and solving for the remaining two constants,

b	s_1	s_2	K_{h1}	K_{h2}
-1	$0.5 + j1.9365$	$0.5 - j1.9365$	$-0.125 - j0.0323$	$-0.125 + j0.0323$
1	$-0.5 + j1.9365$	$-0.5 - j1.9365$	$-0.125 + j0.0323$	$-0.125 - j0.0323$
5	-4	-1	0.0833	-0.3333

The solutions are

b	$y(t)$
-1	$0.25 - e^{0.5t}[0.25 \cos(1.9365t) - 0.0646 \sin(1.9365t)]$
1	$0.25 - e^{-0.5t}[0.25 \cos(1.9365t) + 0.0646 \sin(1.9365t)]$
5	$0.08333e^{-4t} - 0.3333e^{-t} + 0.25$

These zero-state responses are graphed in Figure 4.13.

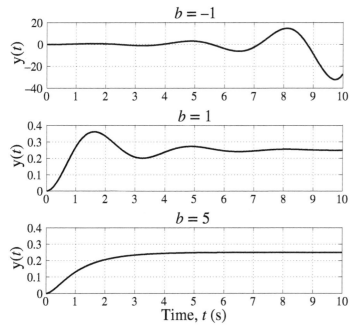

Figure 4.13
System responses for $b = -1$, 1 and 5

Obviously, when $b = -1$ the zero-state response grows without bound and this feedback system is unstable. System dynamics are strongly influenced by feedback.

∎

SYSTEM PROPERTIES

Introductory Example
To build an understanding of large, generalized systems, let us begin with examples of some simple systems that will illustrate some important system properties. Circuits are

familiar to electrical engineers. Circuits are electrical systems. A very common circuit is the *RC* lowpass filter, a single-input, single-output system, illustrated in Figure 4.14.

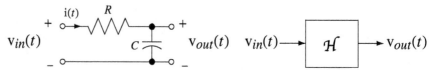

Figure 4.14
An *RC* lowpass filter, a single-input, single-output system

This circuit is called a *lowpass filter* because if the excitation is a constant-amplitude sinusoid, the response will be larger at low frequencies than at high frequencies. So the system tends to "pass" low frequencies through while "stopping" or "blocking" high frequencies. Other common filter types are highpass, bandpass and bandstop. Highpass filters pass high-frequency sinusoids and stop or block low-frequency sinusoids. Bandpass filters pass mid-range frequencies and block both low and high frequencies. Bandstop filters pass low and high frequencies while blocking mid-range frequencies. Filters will be explored in much more detail in Chapters 11 and 14.

The voltage at the input of the *RC* lowpass filter $v_{in}(t)$ excites the system and the voltage at the output $v_{out}(t)$ is the response of the system. The input voltage signal is applied to the left-hand pair of terminals, and the output voltage signal appears at the right-hand pair of terminals. This system consists of two components familiar to electrical engineers, a resistor and a capacitor. The mathematical voltage-current relations for resistors and capacitors are well known and are illustrated in Figure 4.15.

Using Kirchhoff's voltage law, we can write the differential equation

$$\underbrace{RCv'_{out}(t)}_{=i(t)} + v_{out}(t) = v_{in}(t).$$

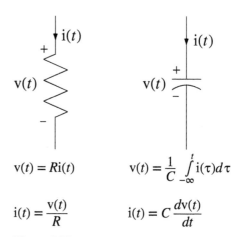

Figure 4.15
Mathematical voltage-current relationships for a resistor and a capacitor

The solution of this differential equation is the sum of the homogeneous and particular solutions. (See Web Appendix D for more on the solution of differential equations.) The homogeneous solution is $v_{out,h}(t) = K_h e^{-t/RC}$ where K_h is, as yet, unknown. The particular solution depends on the functional form of $v_{in}(t)$. Let the input voltage signal $v_{in}(t)$ be a constant A volts. Then, since the input voltage signal is constant, the particular solution is $v_{out,p}(t) = K_p$, also a constant. Substituting that into the differential equation and solving, we get $K_p = A$ and the total solution is $v_{out}(t) = v_{out,h}(t) + v_{out,p}(t) = K_h e^{-t/RC} + A$. The constant K_h can be found by knowing the output voltage at any particular time. Suppose we know the voltage across the capacitor at $t = 0$, which is $v_{out}(0)$. Then

$$v_{out}(0) = K_h + A \Rightarrow K_h = v_{out}(0) - A$$

and the output voltage signal can be written as

$$v_{out}(t) = v_{out}(0)e^{-t/RC} + A(1 - e^{-t/RC}), \qquad (4.2)$$

and it is illustrated in Figure 4.16.

This solution is written and illustrated as though it applies for all time t. In practice that is impossible because, if the solution were to apply for all time, it would be unbounded as time approaches negative infinity, and unbounded signals do not occur in real physical systems. It is more likely in practice that the circuit's initial voltage was placed on the capacitor by some means and held there until $t = 0$. Then at $t = 0$ the A-volt excitation was applied to the circuit and the system analysis is concerned with what happens after $t = 0$. This solution would then apply only for that range of time and is bounded in that range of time. That is $v_{out}(t) = v_{out}(0)e^{-t/RC} + A(1 - e^{-t/RC}), t \geq 0$ as illustrated in Figure 4.17.

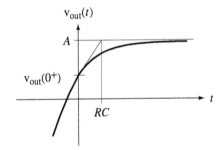

Figure 4.16
RC lowpass filter response to a constant excitation

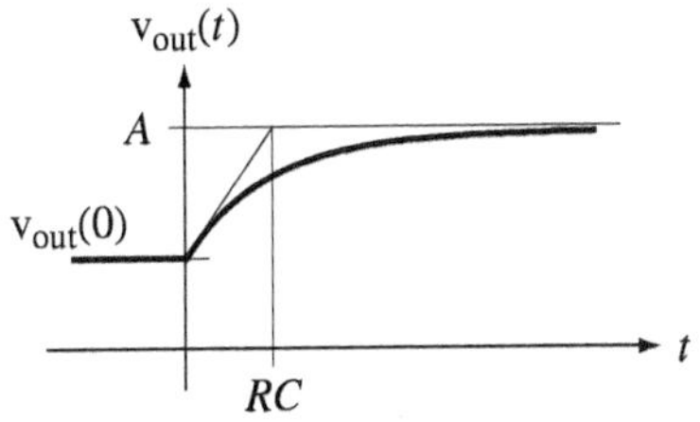

Figure 4.17
RC circuit response to an initial voltage and a constant excitation applied at time $t = 0$

There are four determinants of the voltage response of this circuit for times $t \geq 0$, the resistance R, the capacitance C, the initial capacitor voltage $v_{out}(0)$ and the applied voltage $v_{in}(t)$. The resistance and capacitance values determine the interrelationships among the voltages and currents in the system. From (4.2) we see that if the applied voltage A is zero, the response is

$$v_{out}(t) = v_{out}(0)e^{-t/RC}, \quad t > 0 \qquad (4.3)$$

and if the initial capacitor voltage $v_{out}(0)$ is zero, the response is

$$v_{out}(t) = A(1 - e^{-t/RC}), \quad t > 0 \qquad (4.4)$$

So the response (4.3) is the zero-input response and the response (4.4) is the zero-state response. Zero-state means no stored energy in the system and in the case of the *RC* lowpass filter, zero state would mean the capacitor voltage is zero. For this system the total response is the sum of the zero-input and zero-state responses.

If the excitation is zero for all negative time, then we can express it as a step of voltage $v_{in}(t) = Au(t)$. If we assume that the circuit has been connected with this excitation between the input terminals for an infinite time (since $t = -\infty$), the initial capacitor voltage at time $t = 0$ would have to be zero (Figure 4.18(a)). The system would initially be in its zero state and the response would be the zero-state response. Sometimes an expression like $v_{in}(t) = Au(t)$ for the input signal is intended to represent the situation illustrated in Figure 4.18(b). In this case we are not just applying a voltage to the system, we are actually changing the system by closing a switch. If the initial capacitor voltage is zero in both circuits of Figure 4.18, the responses for times, $t \geq 0$, are the same.

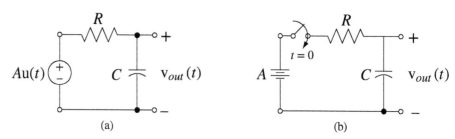

Figure 4.18
Two ways of applying *A* volts to the *RC* lowpass filter at time $t = 0$

It is possible to include the effects of initial energy storage in a system by injecting signal energy into the system when it is in its zero state at time $t = 0$ with a second system excitation, an impulse. For example, in the *RC* lowpass filter we could put the initial voltage on the capacitor with an impulse of current from a current source in parallel with the capacitor (Figure 4.19).

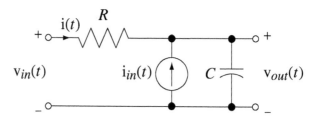

Figure 4.19
RC lowpass filter with a current impulse to inject charge onto the capacitor and establish the initial capacitor voltage

When the impulse of current occurs, all of its charge flows into the capacitor during the time of application of the impulse (which has zero duration). If the strength of the impulse is Q, then the change in capacitor voltage due to the charge injected into it by the current impulse is

$$\Delta v_{out} = \frac{1}{C}\int_{0^-}^{0^+} i_{in}(t)\, dt = \frac{1}{C}\int_{0^-}^{0^+} Q\delta(t)\, dt = \frac{Q}{C}.$$

So choosing $Q = Cv_{out}(0)$ establishes the initial capacitor voltage as $v_{out}(0)$. Then the analysis of the circuit continues as though we were finding the zero-state response to $v_{in}(t)$ and $i_{in}(t)$ instead of the zero-state response to $v_{in}(t)$ and the zero-input response to $v_{out}(0)$. The total response for times $t > 0$ is the same either way.

Most continuous-time systems in practice can be modeled (at least approximately) by differential equations in much the same way as the *RC* lowpass filter above was modeled. This is true of electrical, mechanical, chemical, optical and many other kinds of systems. So the study of signals and systems is important in a very broad array of disciplines.

Homogeneity

If we were to double the input voltage signal of the *RC* lowpass filter to $v_{in}(t) = 2Au(t)$, the factor $2A$ would carry through the analysis and the zero-state response would double to $v_{out}(t) = 2A(1 - e^{-t/RC})u(t)$. Also, if we were to double the initial capacitor voltage, the zero-input response would double. In fact, if we multiply the input voltage signal by any constant, the zero-state response is also multiplied by the same constant. The quality of this system that makes these statements true is called **homogeneity**.

> In a **homogeneous** system, multiplying the input signal by any constant (including *complex* constants) multiplies the zero-state response by the same constant.

Figure 4.20 illustrates, in a block-diagram sense, what homogeneity means.

A very simple example of a system that is not homogeneous is a system characterized by the relationship $y(t) - 1 = x(t)$. If x is 1, y is 2 and if x is 2, y is 3. The input signal was doubled, but the output signal was not doubled. What makes this system inhomogeneous is the presence of the constant -1 on the left side of the equation. This system has a nonzero, zero-input response. Notice that if we were to add $+1$ to both sides of the equation and redefine the input signal to be $x_{new}(t) = x(t) + 1$ instead of just $x(t)$, we would have $y(t) = x_{new}(t)$, and doubling $x_{new}(t)$ would double $y(t)$. The system would then be homogeneous under this new definition of the input signal.

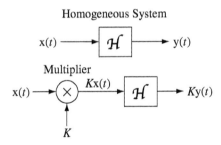

Figure 4.20
Block diagram illustrating the concept of homogeneity for a system initially in its zero state (K is any complex constant)

EXAMPLE 4.4

Determining whether a system is homogeneous

Test, for homogeneity, the system whose input-output relationship is

$$y(t) = \exp(x(t))$$

Let $x_1(t) = g(t)$. Then $y_1(t) = \exp(g(t))$. Let $x_2(t) = Kg(t)$. Then $y_2(t) = \exp(Kg(t)) = [\exp(g(t))]^K \neq Ky_1(t)$. Therefore this system is not homogeneous.

∎

The analysis in Example 4.4 may seem like an unnecessarily formal proof for such a simple function. But it is very easy to get confused in evaluating some systems, even simple-looking ones, unless one uses this kind of structured proof.

Time Invariance

Suppose the system of Figure 4.14 were initially in its zero-state and the excitation were delayed by t_0 changing the input signal to $x(t) = Au(t - t_0)$. What would happen to the response? Going through the solution process again we would find that the zero-state response is $v_{out}(t) = A(1 - e^{-(t-t_0)/RC})u(t - t_0)$, which is exactly the original zero-state response except with t replaced by $t - t_0$. Delaying the excitation delayed the zero-state response by the same amount without changing its functional form. The quality that makes this happen is called **time invariance**.

> If a system is initially in its zero state and an arbitrary input signal $x(t)$ causes a response $y(t)$ and an input signal $x(t - t_0)$ causes a response $y(t - t_0)$ for any arbitrary t_0, the system is said to be **time invariant**.

Figure 4.21 illustrates the concept of time invariance.

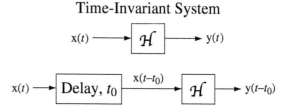

Figure 4.21
Block diagram illustrating the concept of time invariance for a system initially in its zero state

EXAMPLE 4.5

Determining whether a system is time invariant

Test for time invariance the system whose input-output relationship is $y(t) = \exp(x(t))$.

Let $x_1(t) = g(t)$. Then $y_1(t) = \exp(g(t))$. Let $x_2(t) = g(t - t_0)$. Then $y_2(t) = \exp(g(t - t_0)) = y_1(t - t_0)$. Therefore this system is time invariant.

EXAMPLE 4.6

Determining whether a system is time invariant

Test for time invariance the system whose input-output relationship is $y(t) = x(t/2)$.

Let $x_1(t) = g(t)$. Then $y_1(t) = g(t/2)$. Let $x_2(t) = g(t - t_0)$. Then $y_2(t) = g(t/2 - t_0) \neq y_1(t - t_0) = g\left(\frac{t - t_0}{2}\right)$. Therefore this system is not time invariant; it is time *variant*.

Additivity

Let the input voltage signal to the RC lowpass filter be the sum of two voltages $v_{in}(t) = v_{in1}(t) + v_{in2}(t)$. For a moment let $v_{in2}(t) = 0$ and let the zero-state response for $v_{in1}(t)$ acting alone be $v_{out1}(t)$. The differential equation for that situation is

$$RCv'_{out1}(t) + v_{out1}(t) = v_{in1}(t) \tag{4.5}$$

where, since we are finding the zero-state response, $v_{out1}(0) = 0$. Equation (4.5) and the initial condition $v_{out1}(0) = 0$ uniquely determine the solution $v_{out1}(t)$. Similarly, if $v_{in2}(t)$ acts alone, its zero-state response obeys

$$RCv'_{out2}(t) + v_{out2}(t) = v_{in2}(t), \tag{4.6}$$

and $v_{out2}(t)$ is similarly uniquely determined. Adding (4.5) and (4.6),

$$RC[v'_{out1}(t) + v'_{out2}(t)] + v_{out1}(t) + v_{out2}(t) = v_{in1}(t) + v_{in2}(t) \tag{4.7}$$

The sum $v_{in1}(t) + v_{in2}(t)$ occupies the same position in (4.7) as $v_{in1}(t)$ does in (4.5) and $v_{out1}(t) + v_{out2}(t)$ and $v'_{out1}(t) + v'_{out2}(t)$ occupy the same positions in (4.7) that $v_{out1}(t)$ and $v'_{out1}(t)$ do in (4.5). Also, for the zero-state response, $v_{in1}(0) + v_{in2}(0) = 0$. Therefore, if $v_{in1}(t)$ produces $v_{out1}(t)$, then $v_{in1}(t) + v_{in2}(t)$ must produce $v_{out1}(t) + v_{out2}(t)$ because both responses are uniquely determined by the same differential equation and the same initial condition. This result depends on the fact that the derivative of a sum of two functions equals the sum of the derivatives of those two functions. If the excitation is the sum of two excitations, the solution of *this* differential equation, *but not necessarily other differential equations*, is the sum of the responses to those excitations acting alone. A system in which added excitations produce added zero-state responses is called **additive** (Figure 4.22).

> If a system when excited by an arbitrary x_1 produces a zero-state response y_1 and when excited by an arbitrary x_2 produces a zero-state response y_2 and $x_1 + x_2$ always produces the zero-state response $y_1 + y_2$, the system is **additive**.

A very common example of a nonadditive system is a simple diode circuit (Figure 4.23). Let the input voltage signal of the circuit V be the series connection of two constant-voltage sources V_1 and V_2, making the overall input voltage signal the sum of the two individual input voltage signals. Let the overall response be the

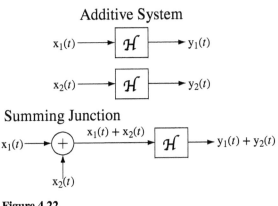

Figure 4.22
Block diagram illustrating the concept of additivity for a system initially in its zero state

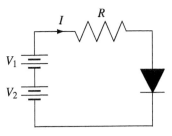

Figure 4.23
A DC diode circuit

current I and let the individual current responses to the individual voltage sources acting alone be I_1 and I_2. To make the result obvious, let $V_1 > 0$ and let $V_2 = -V_1$. The response to V_1 acting alone is a positive current I_1. The response to V_2 acting alone is an extremely small (ideally zero) negative current I_2. The response I to the combined input signal $V_1 + V_2$ is zero, but the sum of the individual responses $I_1 + I_2$ is approximately I_1, not zero. So this is not an additive system.

Linearity and Superposition

Any system that is both homogeneous and additive is called a **linear** system.

> If a linear system when excited by $x_1(t)$ produces a zero-state response $y_1(t)$, and when excited by $x_2(t)$ produces a zero-state response $y_2(t)$, then $x(t) = \alpha x_1(t) + \beta x_2(t)$ will produce the zero-state response $y(t) = \alpha y_1(t) + \beta y_2(t)$.

This property of linear systems leads to an important concept called **superposition**. The term *superposition* comes from the verb superpose. The "pose" part of superpose means to put something into a certain position and the "super" part means "on top of." Together, superpose means to place something on top of something else. That is what is done when we add one input signal to another and, in a linear system, the overall response is one of the responses "on top of" (added to) the other.

The fact that superposition applies to linear systems may seem trivial and obvious, but it has far-reaching implications in system analysis. It means that the zero-state response to any arbitrary input signal can be found by breaking the input signal down into simple pieces that add up to the original input signal, finding the response to each simple piece, and then adding all those responses to find the overall response to the overall input signal. It also means that we can find the zero-state response and then, in an independent calculation, find the zero-input response, and then add them to find the total response. This is a "divide-and-conquer" approach to solving linear-system problems and its importance cannot be overstated. Instead of solving one large, complicated problem, we solve multiple small, simple problems. And, after we have solved one of the small, simple problems, the others are usually very easy to solve because the process is similar. Linearity and superposition are the basis for a large and powerful set of techniques for system analysis. Analysis of nonlinear systems is much more difficult than analysis of linear systems because the divide-and-conquer strategy usually does not work on nonlinear systems. Often the only practical way to analyze a nonlinear system is with numerical, as opposed to analytical, methods.

Superposition and linearity also apply to multiple-input, multiple-output linear systems. If a linear system has two inputs and we apply $x_1(t)$ at the first input and $x_2(t)$ at the second input and get a response $y(t)$, we would get the same $y(t)$ if we added the response to the first input signal acting alone $y_1(t)$ and the response to the second input signal acting alone $y_2(t)$.

LTI Systems

By far the most common type of system analyzed in practical system design and analysis is the **linear, time-invariant** system. If a system is both linear and time invariant, it is called an **LTI** system. Analysis of LTI systems forms the overwhelming majority of the material in this text.

One implication of linearity that will be important later can now be proven. Let an LTI system be excited by a signal $x_1(t)$ and produce a zero-state response $y_1(t)$. Also, let $x_2(t)$ produce a zero-state response $y_2(t)$. Then, invoking linearity, $\alpha x_1(t) + \beta x_2(t)$ will produce the zero-state response $\alpha y_1(t) + \beta y_2(t)$. The constants α and β can be any numbers, including complex numbers. Let $\alpha = 1$ and $\beta = j$. Then $x_1(t) + jx_2(t)$ produces the response $y_1(t) + jy_2(t)$. We already know that $x_1(t)$ produces $y_1(t)$ and that $x_2(t)$ produces $y_2(t)$. So we can now state the general principle.

> When a complex excitation produces a response in an LTI system, the real part of the excitation produces the real part of the response and the imaginary part of the excitation produces the imaginary part of the response.

This means that instead of applying a real excitation to a system to find its real response, we can apply a complex excitation whose real part is the actual physical excitation, find the complex response and then take its real part as the actual physical response to the actual physical excitation. This is a roundabout way of solving system problems but, because the eigenfunctions of real systems are complex exponentials and because of the compact notation that results when applying them in system analysis, this is often a more efficient method of analysis than the direct approach. This basic idea is one of the principles underlying transform methods and their applications to be presented in Chapters 6 through 9.

EXAMPLE 4.7

Response of an *RC* lowpass filter to a square wave using superposition

Use the principle of superposition to find the response of an *RC* lowpass filter to a square wave that is turned on at time $t = 0$. Let the *RC* time constant be 1 ms, let the time from one rising edge of the square wave to the next be 2 ms, and let the amplitude of the square wave be 1 V (Figure 4.24).

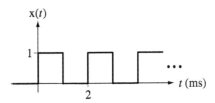

Figure 4.24
Square wave that excites an *RC* lowpass filter

We have no formula for the response of the *RC* lowpass filter to a square wave, but we do know how it responds to a unit step. A square wave can be represented by the sum of some positive and negative time-shifted unit steps. So $x(t)$ can be expressed analytically as

$$x(t) = x_0(t) + x_1(t) + x_2(t) + x_3(t) + \cdots$$

$$x(t) = u(t) - u(t - 0.001) + u(t - 0.002) - u(t - 0.003) + \cdots$$

The *RC* lowpass filter is a linear, time-invariant system. Therefore, the response of the filter is the sum of the responses to the individual unit steps. The response to one unshifted positive unit step is $y_0(t) = (1 - e^{-1000t})u(t)$. Invoking time invariance,

$$y_1(t) = -(1 - e^{-1000(t-0.001)})u(t - 0.001)$$

$$y_2(t) = (1 - e^{-1000(t-0.002)})u(t - 0.002)$$

$$y_3(t) = -(1 - e^{-1000(t-0.003)})u(t - 0.003)$$

$$\vdots \qquad \vdots \qquad \vdots$$

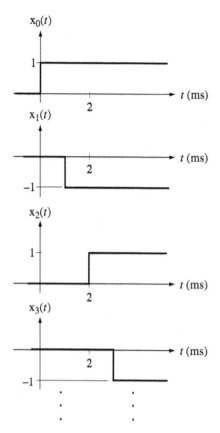

Figure 4.25
Unit steps that can be added to form a square wave

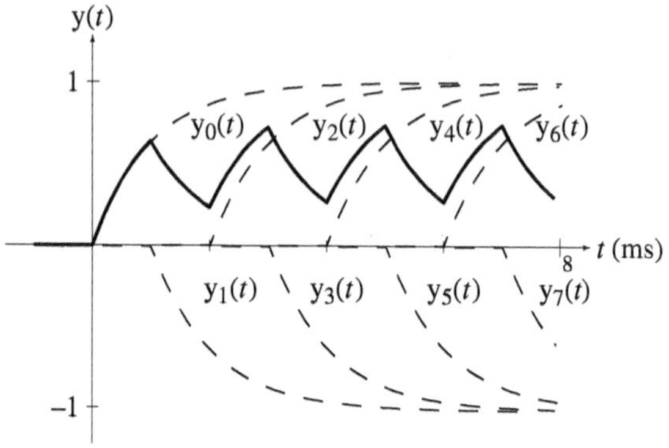

Figure 4.26
Response to the square wave

Then, invoking linearity and superposition,

$$y(t) = y_0(t) + y_1(t) + y_2(t) + y_3(t) + \cdots$$

$$y(t) = (1 - e^{-1000t})u(t) - (1 - e^{-1000(t-0.001)})u(t - 0.001)$$
$$+ (1 - e^{-1000(t-0.002)})u(t - 0.002) - (1 - e^{-1000(t-0.003)})u(t - 0.003) \cdots$$

(see Figure 4.26).

∎

Superposition is the basis of a powerful technique for finding the response of a linear system. The salient characteristic of equations that describe linear systems is that the dependent variable and its integrals and derivatives appear only to the first power. To illustrate this rule, consider a system in which the excitation and response are related by the differential equation $ay''(t) + by^2(t) = x(t)$, where $x(t)$ is the excitation and $y(t)$ is the response. If $x(t)$ were changed to $x_{new}(t) = x_1(t) + x_2(t)$, the differential equation would be $ay''_{new}(t) + by^2_{new}(t) = x_{new}(t)$. The differential equations for $x_1(t)$ and $x_2(t)$ acting alone would be

$$ay''_1(t) + by^2_1(t) = x_1(t) \quad \text{and} \quad ay''_2(t) + by^2_2(t) = x_2(t).$$

The sum of these two equations is

$$a[y_1''(t) + y_2''(t)] + b[y_1^2(t) + y_2^2(t)] = x_1(t) + x_2(t) = x_{new}(t),$$

which is (in general) not equal to

$$a[y_1(t) + y_2(t)]'' + b[y_1(t) + y_2(t)]^2 = x_1(t) + x_2(t) = x_{new}(t).$$

The difference is caused by the $y^2(t)$ term that is not consistent with a differential equation that describes a linear system. Therefore, in this system, superposition does not apply.

A very common analysis technique in signal and system analysis is to use the methods of linear systems to analyze nonlinear systems. This process is called **linearizing** the system. Of course, the analysis is not exact because the system is not actually linear and the linearization process does not make it linear. Rather, linearization replaces the exact nonlinear equations of the system by approximate linear equations. Many nonlinear systems can be usefully analyzed by linear-system methods if the input and output signals are small enough. As an example consider a pendulum (Figure 4.27). Assume that the mass is supported by a massless rigid rod of length L. If a force $x(t)$ is applied to the mass m, it responds by moving. The vector sum of the forces acting on the mass tangential to the direction of motion is equal to the product of the mass and the acceleration in that same direction. That is, $x(t) - mg \sin(\theta(t)) = mL\theta''(t)$ or

$$mL\theta''(t) + mg \sin(\theta(t)) = x(t) \tag{4.8}$$

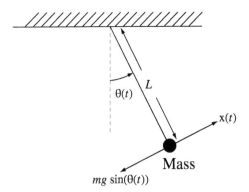

Figure 4.27
A pendulum

where m is the mass at the end of the pendulum, $x(t)$ is a force applied to the mass tangential to the direction of motion, L is the length of the pendulum, g is the acceleration due to gravity and $\theta(t)$ is the angular position of the pendulum. This system is excited by $x(t)$ and responds with $\theta(t)$. Equation (4.8) is nonlinear. But if $\theta(t)$ is small enough, $\sin(\theta(t))$ can be closely approximated by $\theta(t)$. In that approximation,

$$mL\theta''(t) + mg\theta(t) \cong x(t) \tag{4.9}$$

and this is a linear equation. So, for small perturbations from the rest position, this system can be usefully analyzed by using (4.9).

Stability

In the *RC*-lowpass-filter example, the input signal, a step of voltage, was bounded, meaning its absolute value is less than some finite upper bound B for all time, $|x(t)| < B$, for all t. The response of the *RC* lowpass filter to this bounded input signal was a bounded output signal.

> Any system for which the zero-state response to any arbitrary bounded excitation is also bounded is called a **bounded-input–bounded-output (BIBO) stable system**.[1]

The most common type of system studied in signals and systems is a system whose input-output relationship is determined by a linear, constant-coefficient, ordinary differential equation. The eigenfunction for differential equations of this type is the complex exponential. So the homogeneous solution is in the form of a linear combination of complex exponentials. The behavior of each of those complex exponentials is determined by its associated eigenvalue. The form of each complex exponential is $e^{st} = e^{\sigma t}e^{j\omega t}$ where $s = \sigma + j\omega$ is the eigenvalue, σ is its real part and ω is its imaginary part. The factor $e^{j\omega t}$ has a magnitude of one for all t. The factor $e^{\sigma t}$ has a magnitude that gets smaller as time proceeds in the positive direction if σ is negative and gets larger if σ is positive. If σ is zero, the factor $e^{\sigma t}$ is simply the constant one. If the exponential is growing as time passes, the system is unstable because a finite upper bound cannot be placed on the response. If $\sigma = 0$, it is possible to find a bounded input signal that makes the output signal increase without bound. An input signal that is of the same functional form as the homogeneous solution of the differential equation (which is bounded if the real part of the eigenvalue is zero) will produce an unbounded response (see Example 4.8).

> For a continuous-time LTI system described by a differential equation, if the real part of *any* of the eigenvalues is greater than or equal to zero (non-negative), the system is BIBO unstable.

EXAMPLE 4.8

Finding a bounded excitation that produces an unbounded response

Consider an integrator for which $y(t) = \int_{-\infty}^{t} x(\tau)d\tau$. Find the eigenvalues of the solution of this equation and find a bounded excitation that will produce an unbounded response.

By applying Leibniz's formula for the derivative of an integral of this type, we can differentiate both sides and form the differential equation $y'(t) = x(t)$. This is a very simple differential equation with one eigenvalue and the homogeneous solution is a constant because the eigenvalue is zero. Therefore this system should be BIBO unstable. A bounded excitation that has the same functional form as the homogeneous solution produces an unbounded response. In this case, a constant excitation produces an unbounded response. Since the response

---

[1] The discussion of BIBO stability brings up an interesting point. Is any practical system ever actually unstable by the BIBO criterion? Since no practical system can ever produce an unbounded response, strictly speaking, all practical systems are stable. The ordinary operational meaning of BIBO instability is a system described approximately by linear equations that would develop an unbounded response to a bounded excitation *if the system remained linear*. Any practical system will become nonlinear when its response reaches some large magnitude and can never produce a truly unbounded response. So a nuclear weapon is a BIBO-unstable system in the ordinary sense but a BIBO-stable system in the strict sense. Its energy release is not unbounded even though it is extremely large compared to most other artificial systems on earth.

is the integral of the excitation, it should be clear that as time t passes, the magnitude of the response to a constant excitation grows linearly without a finite upper bound. ∎

Causality

In the analysis of the systems we have considered so far, we observe that each system responds only during or after the time it is excited. This should seem obvious and natural. How could a system respond before it is excited? It seems obvious because we live in a physical world in which real physical systems always respond while or after they are excited. But, as we shall later discover in considering ideal filters (in Chapter 11), some system design approaches may lead to a system that responds before it is excited. Such a system cannot actually be built.

The fact that a real system response occurs only while, or after, it is excited is a result of the commonsense idea of cause and effect. An effect has a cause, and the effect occurs during or after the application of the cause.

> Any system for which the zero-state response occurs only during or after the time in which it is excited is called a **causal** system.

All physical systems are causal because they are unable to look into the future and respond before being excited.

The term *causal* is also commonly (albeit somewhat inappropriately) applied to signals. A causal signal is one that is zero before time $t = 0$. This terminology comes from the fact that if an input signal that is zero before time $t = 0$ is applied to a causal system, the response is also zero before time $t = 0$. By this definition, the response would be a causal signal because it is the response of a causal system to a causal excitation. The term **anticausal** is sometimes used to describe signals that are zero *after* time $t = 0$.

In signal and system analysis we often find what is commonly referred to as the **forced response** of a system. A very common case is one in which the input signal is periodic. A periodic signal has no identifiable starting point because, if a signal $x(t)$ is periodic, that means that $x(t) = x(t + nT)$, where T is a period and n is any integer. No matter how far back in time we look, the signal repeats periodically. So the relationship between a periodic input signal and the forced response of an LTI system (which is also periodic, with the same period) cannot be used to determine whether a system is causal. Therefore, in analyzing a system for causality, the system should be excited by a test signal that has an identifiable time before which it has always been zero. A simple signal to test an LTI system for causality would be the unit impulse $\delta(t)$. It is zero before $t = 0$ and is zero after $t = 0$. If the zero-state response of the system to a unit impulse occurring at $t = 0$ is not zero before $t = 0$, the system is not causal. Chapter 5 introduces methods of determining how LTI systems respond to impulses.

Memory

The responses of the systems we have considered so far depend on the present and past excitations. In the RC lowpass filter, the charge on the capacitor is determined by the current that has flowed through it in the past. By this mechanism it, in a sense, remembers something about its past. The present response of this system depends on its past excitations, and that memory, along with its present excitation, determines its present response.

> If any system's zero-state response at any arbitrary time depends on its excitation at any other time, the system has **memory** and is a **dynamic** system.

There are systems for which the present value of the response depends only on the present value of the excitation. A resistive voltage divider is a good example (Figure 4.28).

$$v_o(t) = \frac{R_2}{R_1 + R_2} v_i(t)$$

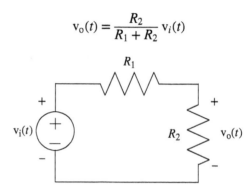

Figure 4.28
A resistive voltage divider

> If any system's response at an arbitrary time depends only on the excitation at that same time, the system has no memory and is a **static** system.

The concepts of causality and memory are related. All static systems are causal. Also, the testing for memory can be done with the same kind of test signal used to test for causality, the unit impulse. If the response of an LTI system to the unit impulse $\delta(t)$ is nonzero at any time other than $t = 0$, the system has memory.

Static Nonlinearity

We have already seen one example of a nonlinear system, one with a nonzero, zero-input response. It is nonlinear because it is not homogeneous. The nonlinearity is not an intrinsic result of nonlinearity of the components themselves, but rather a result of the fact that the zero-input response of the system is not zero.

The more common meaning of the term *nonlinear system* in practice is a system in which, even with a zero-input response of zero, the output signal is still a nonlinear function of the input signal. This is often the result of components in the system that have **static nonlinearities**. A statically nonlinear system is one without memory and for which the input-output relationship is a nonlinear function. Examples of statically nonlinear components include diodes, transistors and square-law detectors. These components are nonlinear because if the input signal is changed by some factor, the output signal can change by a different factor.

The difference between linear and nonlinear components of this type can be illustrated by graphing the relationship between the input and output signals. For a linear resistor, which is a static system, the relation is determined by Ohm's law,

$$v(t) = Ri(t).$$

A graph of voltage versus current is linear (Figure 4.29).

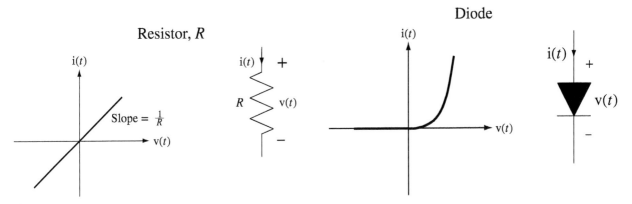

Figure 4.29
Voltage-current relationship for a resistor

Figure 4.30
Voltage-current relationship for a diode at a fixed temperature

A diode is a good example of a statically nonlinear component. Its voltage-current relationship is $i(t) = I_s(e^{qv(t)/kT} - 1)$, where I_s is the reverse saturation current, q is the charge on an electron, k is Boltzmann's constant and T is the absolute temperature, as illustrated in Figure 4.30.

Another example of a statically nonlinear component is an analog multiplier used as a squarer. An analog multiplier has two inputs and one output, and the output signal is the product of the signals applied at the two inputs. It is memoryless, or static, because the present output signal depends only on the present input signals (Figure 4.31).

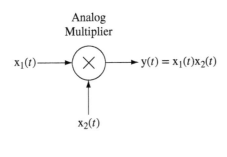

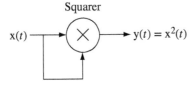

Figure 4.31
An analog multiplier and a squarer

The output signal $y(t)$ is the product of the input signals $x_1(t)$ and $x_2(t)$. If $x_1(t)$ and $x_2(t)$ are the same signal $x(t)$, then $y(t) = x^2(t)$. This is a statically nonlinear relationship because if the excitation is multiplied by some factor A, the response is multiplied by the factor A^2, making the system inhomogeneous.

A very common example of a static nonlinearity is the phenomenon of saturation in real, as opposed to ideal, operational amplifiers. An operational amplifier has two inputs, the inverting input and the noninverting input, and one output. When input voltage signals are applied to the inputs, the output voltage signal of the operational amplifier is a fixed multiple of the difference between the two input voltage signals, *up to a point*. For small differences, the relationship is $v_{out}(t) = A[v_{in+}(t) - v_{in-}(t)]$. But the output voltage signal is constrained by the power supply voltages and can only approach those voltages, not exceed them. Therefore, if the difference between the input voltage signals is large enough that the output voltage signal calculated from $v_{out}(t) = A[v_{in+}(t) - v_{in-}(t)]$ would cause it to be outside the range $-V_{ps}$ to $+V_{ps}$ (where *ps* means power supply), the operational amplifier will saturate. The output voltage signal will go that far and no farther. When the operational amplifier is saturated, the relationship between the input and output signals becomes statically nonlinear. That is illustrated in Figure 4.32.

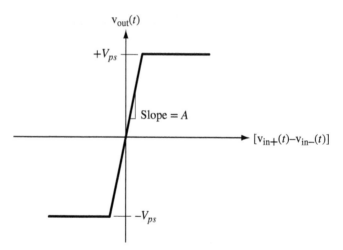

Figure 4.32
Input-output signal relationship for a saturating operational amplifier

Even if a system is statically nonlinear, linear system analysis techniques may still be useful in analyzing it. See Web Appendix C for an example of using linear system analysis to approximately analyze a nonlinear system.

Invertibility

In the analysis of systems we usually find the zero-state response of the system, given an excitation. But we can often find the excitation, given the zero-state response, if the system is **invertible**.

> A system is said to be invertible if unique excitations produce unique zero-state responses.

If unique excitations produce unique zero-state responses then it is possible, in principle at least, given the zero-state response, to associate it with the excitation that produced it. Many practical systems are invertible.

Another way of describing an invertible system is to say that if a system is invertible there exists an inverse system which, when excited by the response of the first system, responds with the excitation of the first system (Figure 4.33).

An example of an invertible system is any system described by a linear, time-invariant, constant-coefficient, differential equation of the form

$$a_k y^{(k)}(t) + a_{k-1} y^{(k-1)}(t) + \cdots + a_1 y'(t) + a_0 y(t) = x(t).$$

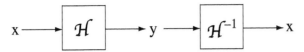

Figure 4.33
A system followed by its inverse

If the response y(t) is known, then so are all its derivatives. The equation indicates exactly how to calculate the excitation as a linear combination of y(t) and its derivatives.

An example of a system that is not invertible is a static system whose input-output functional relationship is

$$y(t) = \sin(x(t)). \quad (4.10)$$

For any x(t) it is possible to determine the zero-state response y(t). Knowledge of the excitation uniquely determines the zero-state response. However, if we attempt to find the excitation, given the response, by rearranging the functional relationship (4.10) into $x(t) = \sin^{-1}(y(t))$, we encounter a problem. The inverse sine function is multiple-valued. Therefore, knowledge of the zero-state response does not uniquely determine the excitation. This system violates the principle of invertibility because different excitations can produce the same zero-state response. If, at $t = t_0$, $x(t_0) = \pi/4$, then $y(t_0) = \sqrt{2}/2$. But if, at $t = t_0$, $x(t_0) = 3\pi/4$, then $y(t_0)$ would have the same value $\sqrt{2}/2$. Therefore, by observing only the zero-state response we would have no idea which excitation value caused it.

Another example of a system that is not invertible is one that is very familiar to electronic circuit designers, the full-wave rectifier (Figure 4.34). Assume that the transformer is an ideal, 1:2-turns-ratio transformer and that the diodes are ideal, so that there is no voltage drop across them in forward bias and no current through them in reverse bias. Then the output voltage signal $v_o(t)$ and input voltage signal $v_i(t)$ are related by $v_o(t) = |v_i(t)|$. Suppose that at some particular time the output voltage signal is +1 V. The input voltage signal at that time could be +1 V or −1 V. There is no way of knowing which of these two input voltage signals is the excitation just by observing the output voltage signal. Therefore we could not be assured of correctly reconstructing the excitation from the response. This system is not invertible.

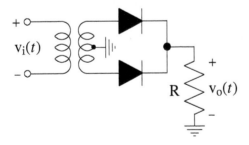

Figure 4.34
A full-wave rectifier

DYNAMICS OF SECOND-ORDER SYSTEMS

First-order and second-order systems are the most common types of systems encountered in system design and analysis. First-order systems are described by first-order differential equations and second-order systems are described by second-order differential equations. We have seen examples of first-order systems. As an example of a second-order system consider the *RLC* circuit excited by a step in Figure 4.35.

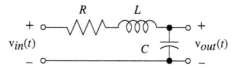

Figure 4.35
An RLC circuit

The sum of voltages around the loop yields

$$LCv''_{out}(t) + RCv'_{out}(t) + v_{out}(t) = Au(t) \tag{4.11}$$

and the solution for the output voltage signal is

$$v_{out}(t) = K_1 e^{\left(-R/2L + \sqrt{(R/2L)^2 - 1/LC}\right)t} + K_2 e^{\left(-R/2L - \sqrt{(R/2L)^2 - 1/LC}\right)t} + A$$

and K_1 and K_2 are arbitrary constants.

This solution is more complicated than the solution for the RC lowpass filter was. There are two exponential terms, each of which has a much more complicated exponent. The exponent involves a square root of a quantity that could be negative. Therefore, the exponent could be complex-valued. For this reason, the eigenfunction e^{st} is called a **complex exponential**. The solutions of ordinary linear differential equations with constant coefficients are always linear combinations of complex exponentials.

In the RLC circuit, if the exponents are real, the response is simply the sum of two real exponentials. The more interesting case is complex exponents. The exponents are complex if

$$(R/2L)^2 - 1/LC < 0. \tag{4.12}$$

In this case the solution can be written in terms of two standard parameters of second-order systems, the **natural radian frequency** ω_n and the **damping factor** α as

$$v_{out}(t) = K_1 e^{\left(-\alpha + \sqrt{\alpha^2 - \omega_n^2}\right)t} + K_2 e^{\left(-\alpha - \sqrt{\alpha^2 - \omega_n^2}\right)t} + A \tag{4.13}$$

where

$$\omega_n^2 = 1/LC \quad \text{and} \quad \alpha = R/2L.$$

There are two other widely used parameters of second-order systems, which are related to ω_n and α, the **critical radian frequency** ω_c and the **damping ratio** ζ. They are defined by $\zeta = \alpha/\omega_n$ and $\omega_c = \omega_n\sqrt{1 - \zeta^2}$. Then we can write as

$$v_{out}(t) = K_1 e^{\left(-\alpha + \omega_n \sqrt{\zeta^2 - 1}\right)t} + K_2 e^{\left(-\alpha - \omega_n \sqrt{\zeta^2 - 1}\right)t} + A$$

When condition (4.12) is satisfied, the system is said to be underdamped and the response can be written as

$$v_{out}(t) = K_1 e^{(-\alpha + j\omega_c)t} + K_2 e^{(-\alpha - j\omega_c)t} + A.$$

The exponents are complex conjugates of each other as they must be for $v_{out}(t)$ to be a real-valued function.

Assuming the circuit is initially in its zero state and applying initial conditions, the output voltage signal is

$$v_{out}(t) = A\left[\frac{1}{2}\left(-1 + j\frac{\alpha}{\omega_c}\right)e^{(-\alpha + j\omega_c)t} + \frac{1}{2}\left(-1 - j\frac{\alpha}{\omega_c}\right)e^{(-\alpha - j\omega_c)t} + 1\right].$$

This response appears to be a complex response of a real system with real excitation. But, even though the coefficients and exponents are complex, the overall solution is real because, using trigonometric identities, the output voltage signal can be reduced to

$$v_{out}(t) = A\{1 - e^{-\alpha t}[(\alpha/\omega_c)\sin(\omega_c t) + \cos(\omega_c t)]\}.$$

This solution is in the form of a damped sinusoid, a sinusoid multiplied by a decaying exponential. The natural frequency $f_n = \omega_n/2\pi$ is the frequency at which the response voltage would oscillate if the damping factor were zero. The rate at which the sinusoid is damped is determined by the damping factor α. Any system described by a second-order linear differential equation could be analyzed by an analogous procedure.

COMPLEX SINUSOID EXCITATION

An important special case of linear system analysis is an LTI system excited by a complex sinusoid. Let the input voltage signal of the *RLC* circuit be $v_{in}(t) = Ae^{j2\pi f_0 t}$. It is important to realize that $v_{in}(t)$ is described exactly for all time. Not only is it going to be a complex sinusoid from now on, it *has always been* a complex sinusoid. Since it began an infinite time in the past, any transients that may have occurred have long since died away (if the system is stable, as this *RLC* circuit is). Thus the only solution that is left at this time is the forced response. The forced response is the particular solution of the describing differential equation. Since all the derivatives of the complex sinusoid are also complex sinusoids, the particular solution of $v_{in}(t) = Ae^{j2\pi f_0 t}$ is simply $v_{out,p}(t) = Be^{j2\pi f_0 t}$ where B is yet to be determined. So if this LTI system is excited by a complex sinusoid, the response is also a complex sinusoid, at the same frequency, but with a different multiplying constant (in general). Any LTI system excited by a complex exponential responds with a complex exponential of the same functional form except multiplied by a complex constant.

The forced solution can be found by the method of undetermined coefficients. Substituting the form of the solution into the differential equation (4.11),

$$(j2\pi f_0)^2 LCBe^{j2\pi f_0 t} + j2\pi f_0 RCBe^{j2\pi f_0 t} + Be^{j2\pi f_0 t} = Ae^{j2\pi f_0 t}$$

and solving,

$$B = \frac{A}{(j2\pi f_0)^2 LC + j2\pi f_0 RC + 1}.$$

Using the principle of superposition for LTI systems, if the input signal is an arbitrary function that is a linear combination of complex sinusoids of various frequencies, then the output signal is also a linear combination of complex sinusoids at those same frequencies. This idea is the basis for the methods of Fourier series and Fourier transform analysis that will be introduced in Chapters 6 and 7, which express arbitrary signals as linear combinations of complex sinusoids.

4.3 DISCRETE-TIME SYSTEMS

SYSTEM MODELING

Block Diagrams

Just as in continuous-time systems, in drawing block diagrams of discrete-time systems there are some operations that appear so often they have been assigned their own block-diagram graphical symbols. The three essential components in a discrete-time system are the **amplifier**, the **summing junction** and the **delay**. The amplifier and summing junction serve the same purposes in discrete-time systems as in continuous-time systems. A delay is excited by a discrete-time signal and responds with that same signal, except delayed by one unit in discrete time (see Figure 4.36). This is the most commonly used symbol but sometimes the D is replaced by an S (for shift).

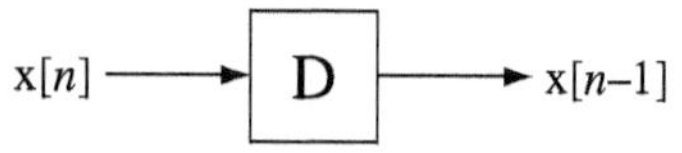

$$x[n] \longrightarrow \boxed{D} \longrightarrow x[n-1]$$

Figure 4.36
The graphical block-diagram symbol for a discrete-time delay

Difference Equations

Below are some examples of the thinking involved in modeling discrete-time systems. Three of these examples were first presented in Chapter 1.

EXAMPLE 4.9

Approximate modeling of a continuous-time system using a discrete-time system

One use of discrete-time systems is in the approximate modeling of nonlinear continuous-time systems like the fluid-mechanical system in Figure 4.37. The fact that its differential equation

$$A_1 \frac{d}{dt}(h_1(t)) + A_2\sqrt{2g[h_1(t) - h_2]} = f_1(t)$$

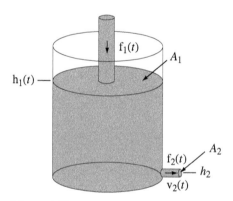

Figure 4.37
Tank with orifice being filled from above

(Toricelli's equation) is nonlinear makes it harder to solve than linear differential equations.

One approach to finding a solution is to use a numerical method. We can approximate the derivative by a finite difference

$$\frac{d}{dt}(h_1(t)) \cong \frac{h_1((n+1)T_s) - h_1(nT_s)}{T_s}$$

where T_s is a finite time duration between values of h_1 at uniformly separated points in time and n indexes those points. Then Toricelli's equation can be approximated at those points in time by

$$A_1 \frac{h_1((n+1)T_s) - h_1(nT_s)}{T_s} + A_2\sqrt{2g[h_1(nT_s) - h_2]} \cong f_1(nT_s),$$

which can be rearranged into

$$h_1((n+1)T_s) \cong \frac{1}{A_1}\{T_s f_1(nT_s) + A_1 h_1(nT_s) - A_2 T_s \sqrt{2g[h_1(nT_s) - h_2]}\} \quad (4.14)$$

which expresses the value of h_1 at the next time index $n+1$ in terms of the values of f_1 at the present time index n and h_1, also at the present time index. We could write (4.14) in the simplified discrete-time notation as

$$h_1[n+1] \cong \frac{1}{A_1}\{T_s f_1[n] + A_1 h_1[n] - A_2 T_s \sqrt{2g(h_1[n] - h_2)}\}$$

or, replacing n by $n-1$,

$$h_1[n] \cong \frac{1}{A_1}\{T_s f_1[n-1] + A_1 h_1[n-1] - A_2 T_s \sqrt{2g(h_1[n-1] - h_2)}\} \quad (4.15)$$

In (4.15), knowing the value of h_1 at any n we can (approximately) find its value at any other n. The approximation is made better by making T_s smaller. This is an example of solving a continuous-time problem using discrete-time methods. Because (4.15) is a difference equation, it defines a discrete-time system (Figure 4.38).

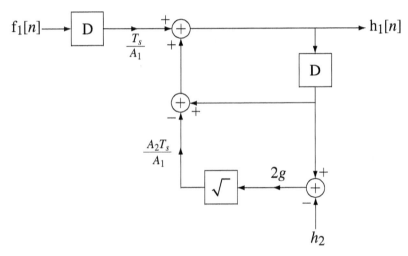

Figure 4.38
A system that approximately solves numerically the differential equation of fluid flow

Figure 4.39 shows examples of the numerical solution of Toricelli's equation using the discrete-time system of Figure 4.38 for three different sampling times 100 s, 500 s and 1000 s. The result for $T_s = 100$ is quite accurate. The result for $T_s = 500$ has the right general behavior and approaches the right final value, but arrives at the final value too early. The result for $T_s = 1000$ has a completely wrong shape, although it does approach the correct final value. The choice of a sampling time that is too large makes the solution inaccurate and, in some cases, can actually make a numerical algorithm unstable.

Below is the MATLAB code that simulates the system in Figure 4.38 used to solve the differential equation describing the tank with orifice.

```
g = 9.8 ;               % Acceleration due to gravity m/s^2
A1 = 1 ;                % Area of free surface of water in tank, m^2
A2 = 0.0005 ;           % Effective area of orifice, m^2
h1 = 0 ;                % Height of free surface of water in tank, m^2
```

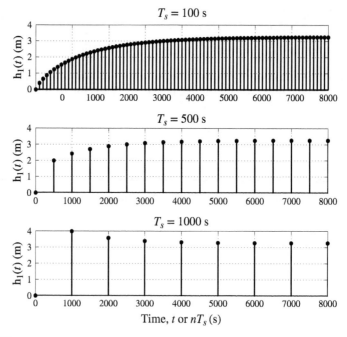

Figure 4.39
Numerical solution of Toricelli's equation using the discrete-time system of Figure 4.38 for a volumetric inflow rate of 0.004 m³/s

```
h2 = 0 ;                     % Height of orifice, m^2
f1 = 0.004 ;                 % Water volumetric inflow, m^3/s

Ts = [100,500,1000] ;        % Vector of time increments, s
N = round(8000./Ts) ;        % Vector of numbers of time steps

for m = 1:length(Ts),        % Go through the time increments
  h1 = 0 ;                   % Initialize h1 to zero
  h = h1 ;                   % First entry in water-height vector
%  Go through the number of time increments computing the
%  water height using the discrete-time system approximation to the
%  actual continuous-time system
  for n = 1:N(m),
%      Compute next free-surface water height
    h1 = (Ts(m)*f1 + A1*h1 - A2*Ts(m)*sqrt(2*g*h1-h2))/A1 ;
    h = [h ; h1] ;    %    Append to water-height vector
  end
%  Graph the free-surface water height versus time and
%  annotate graph
  subplot(length(Ts),1,m) ;
  p = stem(Ts(m)*[0:N(m)]',h,'k','filled') ;
  set(p,'LineWidth',2,'MarkerSize',4) ; grid on ;
  if m == length(Ts),
```

```
        p = xlabel('Time, t or {\itnT_s} (s)',...
                'FontName','Times','FontSize',18) ;
    end
    p = ylabel('h_1(t) (m)','FontName','Times','FontSize',18) ;
    p = title(['{\itT_s} = ',num2str(Ts(m)),...
                    ' s'],'FontName','Times','FontSize',18) ;
end
```

EXAMPLE 4.10

Modeling a feedback system without excitation

Find the output signal generated by the system illustrated in Figure 4.40 for times $n \geq 0$. Assume the initial conditions are y[0] = 1 and y[−1] = 0.

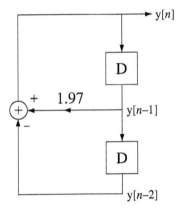

Figure 4.40
A discrete-time system

The system in Figure 4.40 is described by the difference equation

$$y[n] = 1.97y[n-1] - y[n-2] \qquad (4.16)$$

This equation, along with initial conditions y[0] = 1 and y[−1] = 0, completely determines the response y[n], which is the zero-input response. The zero-input response can be found by iterating on (4.16). This yields a correct solution, but it is in the form of an infinite sequence of values of the response. The zero-input response can be found in closed form by solving the difference equation (see Web Appendix D). Since there is no input signal exciting the system, the equation is homogeneous. The functional form of the homogeneous solution is the complex exponential Kz^n. Substituting that into the difference equation we get $Kz^n = 1.97Kz^{n-1} - Kz^{n-2}$. Dividing through by Kz^{n-2} we get the characteristic equation and solving it for z, we get

$$z = \frac{1.97 \pm \sqrt{1.97^2 - 4}}{2} = 0.985 \pm j0.1726 = e^{\pm j0.1734}$$

The fact that there are two eigenvalues means that the homogeneous solution is in the form

$$y[n] = K_{h1}z_1^n + K_{h2}z_2^n. \qquad (4.17)$$

We have initial conditions y[0] = 1 and y[−1] = 0 and we know from (4.17) that $y[0] = K_{h1} + K_{h2}$ and $y[−1] = K_{h1}z_1^{-1} + K_{h2}z_2^{-1}$. Therefore

$$\begin{bmatrix} 1 & 1 \\ e^{-j0.1734} & e^{+j0.1734} \end{bmatrix} \begin{bmatrix} K_{h1} \\ K_{h2} \end{bmatrix} = \begin{bmatrix} 1 \\ 0 \end{bmatrix}$$

Solving for the two constants, $K_{h1} = 0.5 − j2.853$ and $K_{h2} = 0.5 + j2.853$. So the complete solution is

$$y[n] = (0.5 − j2.853)(0.985 + j0.1726)^n + (0.5 + j2.853)(0.985 − j0.1726)^n$$

This is a correct solution but it is not in a very convenient form. We can rewrite it in the form

$$y[n] = (0.5 − j2.853)e^{j0.1734n} + (0.5 + j2.853)e^{-j0.1734n}$$

or

$$y[n] = 0.5\underbrace{(e^{j0.1734n} + e^{-j0.1734n})}_{=2\cos(0.1734n)} − j2.853\underbrace{(e^{j0.1734n} − e^{-j0.1734n})}_{=j2\sin(0.1734n)}$$

or

$$y[n] = \cos(0.1734n) + 5.706\sin(0.1734n).$$

The first 50 values of the signal produced by this system are illustrated in Figure 4.41.

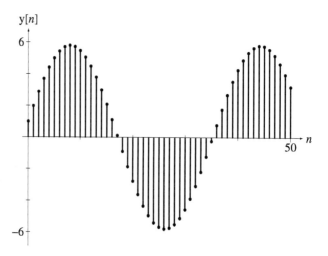

Figure 4.41
Signal produced by the discrete-time system in Figure 4.40

EXAMPLE 4.11

Modeling a simple feedback system with excitation

Find the response of the system in Figure 4.42 if $a = 1$, $b = −1.5$, $x[n] = \delta[n]$, and the system is initially at rest.

The difference equation for this system is

$$y[n] = a(x[n] − by[n − 1]) = x[n] + 1.5y[n − 1].$$

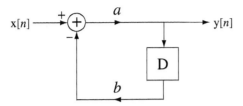

Figure 4.42
A simple discrete-time feedback system with a nonzero excitation

The solution for times $n \geq 0$ is the homogeneous solution of the form $K_h z^n$. Substituting and solving for z, we get $z = 1.5$. Therefore $y[n] = K_h(1.5)^n$, $n \geq 0$. The constant can be found by knowing the initial value of the response, which, from the system diagram, must be 1. Therefore

$$y[0] = 1 = K_h(1.5)^0 \Rightarrow K_h = 1$$

and

$$y[n] = (1.5)^n, \quad n \geq 0.$$

This solution obviously grows without bound so the system is unstable. If we chose b with a magnitude less than one, the system would be stable because the solution is of the form $y[n] = b^n, n \geq 0$.

EXAMPLE 4.12

Modeling a more complicated feedback system with excitation

Find the zero-state response of the system in Figure 4.43, for times $n \geq 0$ to $x[n] = 1$ applied at time $n = 0$, by assuming all the signals in the system are zero before time $n = 0$ for $a = 1$, $b = -1.5$ and three different values of c, 0.8, 0.6 and 0.5.

The difference equation for this system is

$$y[n] = a(x[n] - by[n-1] - cy[n-2]) = x[n] + 1.5y[n-1] - cy[n-2] \quad (4.18)$$

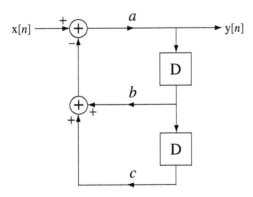

Figure 4.43
A system with more complicated feedback

The response is the total solution of the difference equation with initial conditions. We can find a closed-form solution by finding the total solution of the difference equation. The homogeneous solution is $y_h[n] = K_{h1} z_1^n + K_{h2} z_2^n$ where $z_{1,2} = 0.75 \pm \sqrt{0.5625 - c}$. The particular solution is in the form of a linear combination of the input signal and all its unique differences. The input signal is a constant. So all its differences are zero. Therefore the particular solution is simply a constant K_p. Substituting into the difference equation,

$$K_p - 1.5K_p + cK_p = 1 \Rightarrow K_p = \frac{1}{c - 0.5}.$$

Using (4.18) we can find the initial two values of y[n] needed to solve for the remaining two unknown constants K_{h1} and K_{h2}. They are y[0] = 1 and y[1] = 2.5.

In Chapter 1 three responses were illustrated for $a = 1$, $b = -1.5$ and $c = 0.8$, 0.6 and 0.5. Those responses are replicated in Figure 4.44.

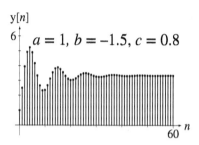
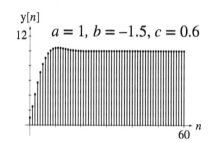
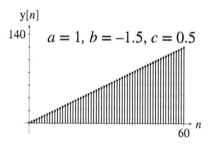

Figure 4.44
System zero-state responses for three different feedback configurations

The results of Example 4.12 demonstrate the importance of feedback in determining the response of a system. In the first two cases the output signal is bounded. But in the third case the output signal is unbounded, even though the input signal is bounded. Just as for continuous-time systems, any time a discrete-time system can exhibit an unbounded zero-state response to a bounded excitation of any kind, it is classified as a BIBO unstable system. So the stability of feedback systems depends on the nature of the feedback.

SYSTEM PROPERTIES

The properties of discrete-time systems are almost identical, qualitatively, to the properties of continuous-time systems. In this section we explore examples illustrating some of the properties in discrete-time systems.

Consider the system in Figure 4.45. The input and output signals of this system are related by the difference equation y[n] = x[n] + (4/5)y[n − 1]. The homogeneous solution is $y_h[n] = K_h(4/5)^n$. Let x[n] be the unit sequence. Then the particular solution is $y_p[n] = 5$ and the total solution is $y[n] = K_h(4/5)^n + 5$. (See Web Appendix D for methods of solving difference equations.) If the system is in its zero state before time $n = 0$ the total solution is

$$y[n] = \begin{cases} 5 - 4(4/5)^n, & n \geq 0 \\ 0, & n < 0 \end{cases}$$

or

$$y[n] = [5 - 4 4/5^n]u[n]$$

(see Figure 4.46).

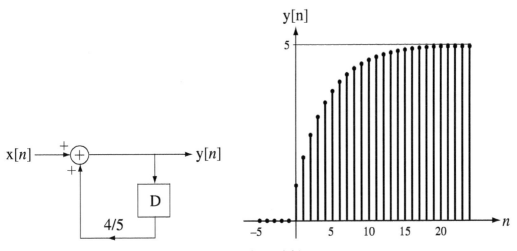

Figure 4.45
A system

Figure 4.46
System zero-state response to a unit-sequence excitation

The similarity between the shape of the *RC* lowpass filter's response to a unit-step excitation and the envelope of this system's response to a unit-sequence is not an accident. This system is a digital lowpass filter (more on digital filters in Chapters 11 and 14).

If we multiply the excitation of this system by any constant, the response is multiplied by the same constant, so this system is homogeneous. If we delay the excitation of this system by any time n_0, we delay the response by that same time. Therefore, this system is also time invariant. If we add any two signals to form the excitation of the system, the response is the sum of the responses that would have occurred by applying the two signals separately. Therefore, this system is an LTI discrete-time system. This system also has a bounded response for any bounded excitation. Therefore it is also stable.

A simple example of a system that is not time invariant would be one described by $y[n] = x[2n]$. Let $x_1[n] = g[n]$ and let $x_2[n] = g[n-1]$, where $g[n]$, is the signal illustrated in Figure 4.47, and let the response to $x_1[n]$ be $y_1[n]$, and let the response to $x_2[n]$ be $y_2[n]$. These signals are illustrated in Figure 4.48.

Since $x_2[n]$ is the same as $x_1[n]$ except delayed by one discrete time unit, for the system to be time invariant $y_2[n]$ must be the same as $y_1[n]$ except delayed by one discrete-time unit, but it is not. Therefore this system is time *variant*.

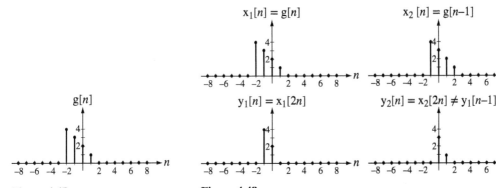

Figure 4.47
An excitation signal

Figure 4.48
Responses of the system described by $y[n] = x[2n]$ to two different excitations

A good example of a system that is not BIBO stable is the financial system of accruing compound interest. If a principle amount P of money is deposited in a fixed-income investment at an interest rate r per annum compounded annually, the amount A$[n]$, which is the value of the investment n years later, is A$[n] = P(1 + r)^n$. The amount A$[n]$ grows without bound as discrete-time n passes. Does that mean our banking system is unstable? The amount does grow without bound and, in an infinite time, would approach infinity. But, since no one who is alive today (or at any time in the future) will live long enough to see this happen, the fact that the system is unstable according to our definition is really of no great concern. When we also consider the effects of the inevitable withdrawals from the account and monetary inflation, we see that this theoretical instability is not significant.

The most common type of discrete-time system studied in signals and systems is a system whose input-output relationship is determined by a linear, constant-coefficient, ordinary difference equation. The eigenfunction is the complex exponential and the homogeneous solution is in the form of a linear combination of complex exponentials. The form of each complex exponential is $z^n = |z|^n e^{j(\angle z)n}$ where z is the eigenvalue. If the magnitude of z is less than one, the solution form z^n gets smaller in magnitude as discrete time passes, and if the magnitude of z is greater than one the solution form gets larger in magnitude. If the magnitude of z is exactly one, it is possible to find a bounded excitation that will produce an unbounded response. As was true for continuous-time systems, an excitation that is of the same functional form as the homogeneous solution of the differential equation will produce an unbounded response.

> For a discrete-time system, if the magnitude of *any* of the eigenvalues is greater than or equal to one, the system is BIBO unstable.

EXAMPLE 4.13

Finding a bounded excitation that produces an unbounded response

Consider an accumulator for which y$[n] = \sum_{m=-\infty}^{n}x[m]$. Find the eigenvalues of the solution of this equation and find a bounded excitation that will produce an unbounded response.

We can take the first backward difference of both sides of the difference equation yielding y$[n]$ − y$[n − 1]$ = x$[n]$. This is a very simple difference equation with one eigenvalue, and the homogeneous solution is a constant because the eigenvalue is one. Therefore this system should be BIBO unstable. The bounded excitation that produces an unbounded response has the same functional form as the homogeneous solution. In this case a constant excitation produces an unbounded response. Since the response is the accumulation of the excitation, it should be clear that as discrete time n passes, the magnitude of the response to a constant excitation grows linearly without an upper bound.

■

The concepts of memory, causality, static nonlinearity and invertibility are the same for discrete-time systems as for continuous-time systems. Figure 4.49 is an example of a static system.

One example of a statically nonlinear system would be a two-input OR gate in a digital logic system. Suppose the logic levels are 0 V for a logical 0 and 5 V for a logical 1. If we apply 5 V to either of the two inputs, with 0 V on the other, the response is 5 V. If we then apply 5 V to both inputs simultaneously, the response is still 5 V. If the

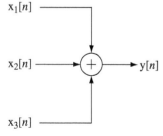

Figure 4.49
A static system

system were linear, the response to 5 V on both inputs simultaneously would be 10 V. This is also a noninvertible system. If the output signal is 5 V, we do not know which of three possible input-signal combinations caused it, and therefore knowledge of the output signal is insufficient to determine the input signals.

Even though all real physical systems must be causal in the strict sense that they cannot respond before being excited, there are real signal-processing systems that are sometimes described, in a superficial sense, as noncausal. These are data-processing systems in which signals are recorded and then processed "off-line" at a later time to produce a computed response. Since the whole history of the input signals has been recorded, the computed response at some designated time in the data stream can be based on values of the already-recorded input signals that occurred later in time (Figure 4.50). But, since the whole data processing operation occurs after the input signals have been recorded, this kind of system is still causal in the strict sense.

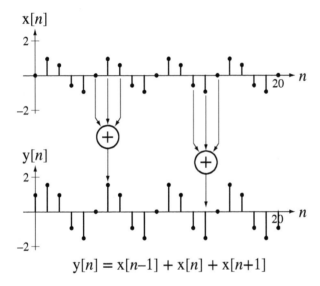

Figure 4.50
A so-called noncausal filter calculating responses from a prerecorded record of excitations

4.4 SUMMARY OF IMPORTANT POINTS

1. A system that is both homogeneous and additive is linear.
2. A system that is both linear and time invariant is called an LTI system.
3. The total response of any LTI system is the sum of its zero-input and zero-state responses.
4. Often nonlinear systems can be analyzed with linear system techniques through an approximation called linearization.
5. A system is said to be BIBO stable if arbitrary bounded input signals always produce bounded output signals.
6. A continuous-time LTI system is stable if all its eigenvalues have negative real parts.
7. All real physical systems are causal, although some may be conveniently and superficially described as noncausal.

8. Continuous-time systems are usually modeled by differential equations and discrete-time systems are usually modeled by difference equations.
9. The solution methods for difference equations are very similar to the solution methods for differential equations.
10. One common use for difference equations is to approximate differential equations.
11. A discrete-time LTI system is stable if all its eigenvalues are less than one in magnitude.

EXERCISES WITH ANSWERS

(Answers to each exercise are in random order.)

System Models

1. Write the differential equation for the voltage $v_C(t)$ in the circuit in Figure E.1 for time $t > 0$, then find an expression for the current $i(t)$ for time $t > 0$.

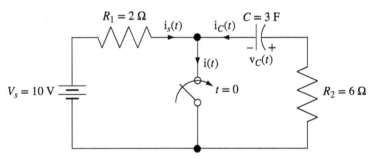

Figure E.1

Answer: $i(t) = 5 + \dfrac{5}{3} e^{-\frac{t}{18}}$, $t > 0$

2. The water tank in Figure E.2 is filled by an inflow $x(t)$ and is emptied by an outflow $y(t)$. The outflow is controlled by a valve which offers resistance R to the flow of water out of the tank. The water depth in the tank is $d(t)$ and the surface area of the water is A, independent of depth (cylindrical tank). The outflow is related to the water depth (head) by

$$y(t) = \frac{d(t)}{R}.$$

The tank is 1.5 m high with a 1-m diameter and the valve resistance is $10 \frac{s}{m^2}$.

(a) Write the differential equation for the water depth in terms of the tank dimensions and valve resistance.
(b) If the inflow is 0.05 m³/s, at what water depth will the inflow and outflow rates be equal, making the water depth constant?
(c) Find an expression for the depth of water versus time after 1 m³ of water is dumped into an empty tank.
(d) If the tank is initially empty at time $t = 0$ and the inflow is a constant $0.2 \frac{m^3}{s}$ after time $t = 0$, at what time will the tank start to overflow?

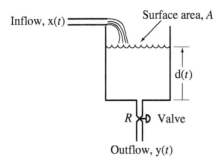

Figure E.2

Answers: $Ad'(t) + \frac{d(t)}{R} = x(t)$, $d(t) = 0.5$ m, $d(t) = (4/\pi)e^{-4t/10\pi}$,

$t_{of} = 1.386 \times 10\pi/4 = 10.88$ s

3. As derived in the text, a simple pendulum is approximately described for small angles θ by the differential equation,

$$mL\theta''(t) + mg\theta(t) \cong x(t)$$

where m is the mass of the pendulum, L is the length of the massless rigid rod supporting the mass and θ is the angular deviation of the pendulum from vertical. If the mass is 2 kg and the rod length is 0.5 m, at what cyclic frequency will the pendulum oscillate?

Answer: 0.704 Hz

4. A block of aluminum is heated to a temperature of 100°C. It is then dropped into a flowing stream of water which is held at a constant temperature of 10°C. After 10 seconds the temperature of the block is 60°C. (Aluminum is such a good heat conductor that its temperature is essentially uniform throughout its volume during the cooling process.) The rate of cooling is proportional to the temperature difference between the block and the water.

   (a) Write a differential equation for this system with the temperature of the water as the excitation and the temperature of the block as the response.
   (b) Compute the time constant of the system.
   (c) If the same block is cooled to 0°C and dropped into a flowing stream of water at 80°C, at time $t = 0$ at what time will the temperature of the block reach 75°C?

Answers: $T_a(t) = 90e^{-\lambda t} + 10$, $\tau = 17$, $t_{75} = -17 \ln(0.0625) = 47.153$

Block Diagrams

5. The systems represented by these block diagrams can each be described by a differential equation of the form,

$$a_N \frac{d^N}{dt^N}(y(t)) + a_{N-1}\frac{d^{N-1}}{dt^{N-1}}y(t) + \cdots + a_2\frac{d^2}{dt^2}y(t) + a_1\frac{d}{dt}y(t) + a_0 y(t) = x(t).$$

For each system what is the value of N? For each system what are the a coefficients, starting with a_N and going down to a_0? In system (b), what range of values of A will make the system stable?

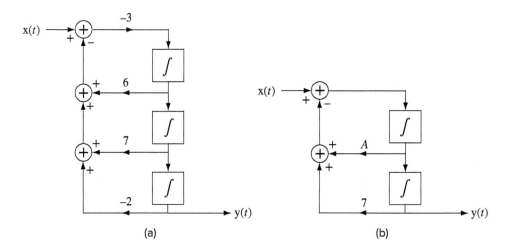

Answers: $N = 3$, $-(1/3)y'''(t) + 6y''(t) + 7y'(t) - 2y(t) = x(t)$, $N = 2$,
$y''(t) + Ay'(t) + 7y(t) = x(t)$. For $A > 0$, the system is stable.

System Properties

6. Show that a system with excitation x(t) and response y(t) described by

$$y(t) = u(x(t))$$

is nonlinear, time invariant, BIBO stable and noninvertible.

7. Show that a system with excitation x(t) and response y(t) described by

$$y(t) = x(t/2)$$

is linear, time variant and noncausal.

8. Show that the system in Figure E.8 is linear, time invariant, BIBO unstable and dynamic.

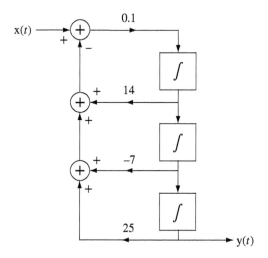

Figure E.8

9. Show that a system with excitation x[n] and response y[n] described by

$$y[n] = nx[n],$$

is linear, time variant and static.

10. Show that the system described by $y[n] = x[-n]$ is time variant, dynamic non-causal and BIBO stable.

11. Show that the system described by $y(t) = \sin(x(t))$ is nonlinear, time invariant, BIBO stable, static and causal.

12. Show that the system described by $y(t) = x(\sin(t))$ is linear, time invariant, BIBO stable and noncausal.

13. Show that the system described by the equation $y[n] = e^{x[n]}$ is nonlinear, time invariant, BIBO stable, static, causal and invertible.

14. Show that a system described by $y(t) = t^2 x(t-1)$ is time variant, BIBO unstable, causal, dynamic and noninvertible.

15. Show that a system described by $y(t) = \dfrac{dx(t)}{dt}$ is not invertible and that a system described by $y(t) = \int\limits_{-\infty}^{t} x(\lambda) d\lambda$ is invertible.

16. Show that a system described by $y[n] = \begin{cases} x[n/2], & n \text{ even} \\ 0, & n \text{ odd} \end{cases}$ is invertible and that a system described by $y[n] = x[2n]$ is not invertible.

17. A continuous-time system is described by the equation $y(t) = tx(t)$. Show that it is linear, BIBO unstable and time variant.

18. Show that the system of Figure E.18 is nonlinear, time invariant, static, and invertible.

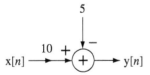

Figure E.18

$$y[n] = 10x[n] - 5,$$

19. Show that the system described by

$$y(t) = \begin{cases} 10, & x(t) > 2 \\ 5x(t), & -2 < x(t) \le 2 \\ -10, & x(t) \le -2 \end{cases}$$

is nonlinear, static, BIBO stable, noninvertible and time invariant.

20. Show that a system described by the equation $y[n] = |x[n+1]|$ is nonlinear, BIBO stable, time invariant, noncausal and noninvertible.

21. Show that the system described by $y[n] = \dfrac{x[n]x[n-2]}{x[n-1]}$ is homogeneous but not additive.

22. Show that the system of Figure E.22 is time invariant, BIBO stable and causal.

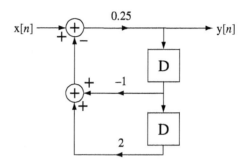

Figure E.22

EXERCISES WITHOUT ANSWERS

System Models

23. Pharmacokinetics is the study of how drugs are absorbed into, distributed through, metabolized by and excreted from the human body. Some drug processes can be approximately modeled by a "one compartment" model of the body in which V is the volume of the compartment, $C(t)$ is the drug concentration in that compartment, k_e is a rate constant for excretion of the drug from the compartment and k_0 is the infusion rate at which the drug enters the compartment.

   (a) Write a differential equation in which the infusion rate is the excitation and the drug concentration is the response.
   (b) Let the parameter values be $k_e = 0.4\,\text{hr}^{-1}$, $V = 20$ l and $k_0 = 200$ mg/hr (where "l" is the symbol for "liter"). If the initial drug concentration is $C(0) = 10\,\text{mg/l}$, plot the drug concentration as a function of time (in hours) for the first 10 hours of infusion. Find the solution as the sum of the zero-excitation response and the zero-state response.

24. A well-stirred vat has been fed for a long time by two streams of liquid, fresh water at 0.2 cubic meters per second and concentrated blue dye at 0.1 cubic meters per second. The vat contains 10 cubic meters of this mixture and the mixture is being drawn from the vat at a rate of 0.3 cubic meters per second to maintain a constant volume. The blue dye is suddenly changed to red dye at the same flow rate. At what time after the switch does the mixture drawn from the vat contain a ratio of red to blue dye of 99:1?

25. A car rolling on a hill can be modeled as shown in Figure E.25. The excitation is the force f(t) for which a positive value represents accelerating the car forward with the motor and a negative value represents slowing the car by braking action. As it rolls, the car experiences drag due to various frictional phenomena that can be approximately modeled by a coefficient k_f which multiplies the car's velocity to produce a force which tends to slow the car when it moves in either direction. The mass of the car is m and gravity acts on it at all times tending to make it roll down the hill in the absence of other forces. Let the mass m of the car be 1000 kg, let the friction coefficient k_f be $5\,\dfrac{\text{N}\cdot\text{s}}{\text{m}}$, and let the angle θ be $\dfrac{\pi}{12}$.

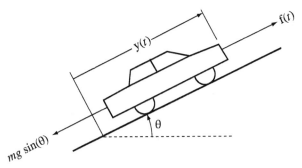

Figure E.25

(a) Write a differential equation for this system with the force f(t) as the excitation and the position of the car y(t) as the response.
(b) If the nose of the car is initially at position y(0) = 0 with an initial velocity $[y'(t)]_{t=0} = 10 \frac{m}{s}$ and no applied acceleration or braking force, graph the velocity of the car y'(t) for positive time.
(c) If a constant force f(t) of 200 N is applied to the car what is its terminal velocity?

26. At the beginning of the year 2000, the country Freedonia had a population p of 100 million people. The birth rate is 4% per annum and the death rate is 2% per annum, compounded daily. That is, the births and deaths occur every day at a uniform fraction of the current population and the next day the number of births and deaths changes because the population changed the previous day. For example, every day the number of people who die is the fraction 0.02/365 of the total population at the end of the previous day (neglect leap-year effects). Every day 275 immigrants enter Freedonia.

    (a) Write a difference equation for the population at the beginning of the nth day after January 1, 2000 with the immigration rate as the excitation of the system.
    (b) By finding the zero-excitation and zero-state responses of the system determine the population of Freedonia at the beginning of the year 2050.

27. In Figure E.27 is a MATLAB program simulating a discrete-time system.

    (a) Without actually running the program, find the value of x when n = 10 by solving the difference equation for the system in closed form.
    (b) Run the program and check the answer in part (a).

```
x = 1 ; y = 3 ; z = 0 ; n = 0 ;
while n <= 10,
        z = y ;
        y = x ;
        x = 2*n + 0.9*y - 0.6*z ;
        n = n + 1 ;
end
```

Figure E.27

System Properties

28. A system is described by the block diagram in Figure E.28.

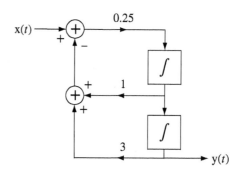

Figure E.28

Classify the system as to homogeneity, additivity, linearity, time invariance, BIBO stability, causality, memory and invertibility.

29. A system is described by the differential equation,

$$ty'(t) - 8y(t) = x(t).$$

Classify the system as to linearity, time invariance and BIBO stability.

30. System #1 is described by $y(t) = \text{ramp}(x(t))$ and system #2 is described by $y(t) = x(t)\,\text{ramp}(t)$. Classify both systems as to BIBO stability, linearity, invertibility and time invariance.

31. A system is described by $y(t) = \begin{cases} 0 & , x(t) < 0 \\ x(t) + x(t-2) & , x(t) \geq 0 \end{cases}$.

Classify the system as to linearity, time invariance, memory, causality and stability.

32. A system is described by $y(t) = \begin{cases} 0 & , t < 0 \\ x(t) + x(t-2) & , t \geq 0 \end{cases} = [x(t) + x(t-2)]u(t)$.

Classify the system as to linearity, time invariance, memory, causality and stability.

33. A system is described by $y[n] = \sum_{m=n-n_0}^{n+n_0} x[m]$. Classify the system as to linearity, time invariance and stability.

34. A system is described by $y[n] = (n^2 + 1)(x[n] - x[n-1])$. Is it invertible?

35. A system is described by $y[n] = \begin{cases} 4 & , x[n] > 3 \\ x[n] & , -1 \leq x[n] < 4 \\ -1 & , x[n] < -1 \end{cases}$. Classify the system as to linearity, time invariance, BIBO stability and invertibility.

36. A system is described by the equation $y[n] = 9\,\text{tri}(n/4)x[n-3]$. Classify the system as to linearity, time invariance, BIBO stability, memory, causality and invertibility.

37. A system is described by the equation $y(t) = \int_{-\infty}^{t/3} x(\lambda)d\lambda$. Classify the system as to time invariance, BIBO stability and invertibility.

38. A system is described by $y'(t+1) + x(t) = 6$. Classify the system as to linearity, time invariance, BIBO stability, memory and invertibility.

39. A system is described by the equation $y(t) = \int_{-\infty}^{t+3} x(\lambda)d\lambda$. Classify the system as to linearity, causality and invertibility.

40. A system is described by $y[n] + 1.8y[n-1] + 1.62y[n-2] = x[n]$. Classify this system as to linearity, time invariance, BIBO stability, memory, causality and invertibility.

41. A system is described by $y[n] = \sum_{m=-\infty}^{n+1} x[m]$. Classify this system as to time invariance, BIBO stability and invertibility.

42. A system is described by $y(t) = \frac{1}{x(t)}\left(\frac{dx(t)}{dt}\right)^2$. Classify this system as to homogeneity and additivity.

43. A system is described by $y[n] = x[n]x[n-1]$. Classify this system as to invertibility.

44. In Figure E.44 are some system descriptions in which x is the excitation and y is the response.

System A.	$y(t) = \int_{-\infty}^{t} x(\tau)d\tau$
System B.	$y[n] - 2y[n-1] = x[n+1]$
System C.	$y''(t) + y(t) = x(t)$
System D.	$y[n] = \begin{cases} 8 & , x[n] > 2 \\ 4x[n] & , -2 < x[n] < 2 \\ -8 & , x[n] < -2 \end{cases}$
System E.	$y(t) = tx(t)$

Figure E.44

(a) Which systems are linear?
(b) Which systems are time invariant?
(c) Which systems are BIBO stable?
(d) Which systems are dynamic?
(e) Which systems are causal?
(f) Which systems are invertible?

5

CHAPTER

Time-Domain System Analysis

5.1 INTRODUCTION AND GOALS

The essential goal in designing systems is that they respond in the right way. Therefore, we must be able to calculate the response of a system to any arbitrary input signal. As we will see throughout this text, there are multiple approaches to doing that. We have already seen how to find the response of a system described by differential or difference equations by finding the total solution of the equations with boundary conditions. In this chapter we will develop another technique called *convolution*. We will show that, for an LTI system, if we know its response to a unit impulse occurring at $t = 0$ or $n = 0$, that response completely characterizes the system and allows us to find the response to any input signal.

CHAPTER GOALS

1. To develop techniques for finding the response of an LTI system to a unit impulse occurring at $t = 0$ or $n = 0$

2. To understand and apply convolution, a technique for finding the response of LTI systems to arbitrary input signals for both continuous-time and discrete-time systems

5.2 CONTINUOUS TIME

IMPULSE RESPONSE

We have seen techniques for finding the solutions to differential equations that describe systems. The total solution is the sum of the homogeneous and particular solutions. The homogeneous solution is a linear combination of eigenfunctions. The particular solution depends on the form of the forcing function. Although these methods work, there is a more systematic way of finding how systems respond to input signals and it lends insight into important system properties. It is called **convolution**.

The convolution technique for finding the response of a continuous-time LTI system is based on a simple idea. If we can find a way of expressing a signal as a linear combination of simple functions, we can, using the principles of linearity and time invariance, find the response to that signal as a linear combination of the responses to those simple functions. If we can find the response of an LTI system to a unit-impulse occurring at $t = 0$ and if we can express the input signal as a linear combination of impulses, we can find the response to it. Therefore, use of the convolution technique begins with the assumption that the response to a unit impulse occurring at $t = 0$ has already been found. We will call that response h(t) the **impulse response**. So the first requirement in using the convolution technique to find a system response is to find the impulse response by applying a unit impulse $\delta(t)$ occurring at $t = 0$. The impulse injects signal energy into the system and then goes away. After the energy is injected into the system, it responds with a signal determined by its dynamic character.

We could, in principle, find the impulse response experimentally by actually applying an impulse at the system input. But, since a true impulse cannot actually be generated, this would only be an approximation. Also, in practice an approximation to an impulse would be a very tall pulse lasting for a very short time. In reality, for a real physical system a very tall pulse might actually drive it into a nonlinear mode of response and the experimentally measured impulse response would not be accurate. There are other less direct but more practical ways of experimentally determining an impulse response.

If we have a mathematical description of the system we may be able to find the impulse response analytically. The following example illustrates some methods for finding the impulse response of a system described by a differential equation.

EXAMPLE 5.1

Impulse response of continuous-time system 1

Find the impulse response h(t) of a continuous-time system characterized by the differential equation

$$y'(t) + ay(t) = x(t) \tag{5.1}$$

where x(t) excites the system and y(t) is the response.

We can rewrite (5.1) for the special case of an impulse exciting the system as

$$h'(t) + ah(t) = \delta(t) \tag{5.2}$$

METHOD #1:

Since the only excitation is the unit impulse at $t = 0$ and the system is causal, we know that the impulse response before $t = 0$ is zero. That is, h(t) = 0, $t < 0$. The homogeneous solution for $t > 0$ is of the form Ke^{-at}. This is the form of the impulse response for $t > 0$ because in that time range the system is not being excited. We now know the form of the impulse response before $t = 0$ and after $t = 0$. All that is left is to find out what happens *at* $t = 0$. The differential equation (5.1) must be satisfied at all times. That is, h'(t) + ah(t) must be a unit impulse occurring at time $t = 0$. We can determine what happens *at* $t = 0$ by integrating both sides of (5.2) from $t = 0^-$ to $t = 0^+$, infinitesimal times just before and just after zero.

The integral of h'(t) is simply h(t). We know that at time $t = 0^-$ it is zero and at time $t = 0^+$ it is K.

$$\underbrace{h(0^+)}_{=K} - \underbrace{h(0^-)}_{=0} + a \int_{0^-}^{0^+} h(t)\,dt = \int_{0^-}^{0^+} \delta(t)\,dt = 1 \tag{5.3}$$

The homogeneous solution applies for all $t > 0$ but *at* $t = 0$ we must also consider the particular solution because the impulse is driving the system at that time. The general rule for the form of the particular solution of a differential equation is a linear combination of the forcing function and all its unique derivatives. The forcing function is an impulse and an impulse has infinitely many unique derivatives, the doublet, the triplet and so on, and all of them occur exactly at $t = 0$. Therefore, until we can show a reason why an impulse and/or all its derivatives cannot be in the solution, we have to consider them as possibilities. If h(t) does not have an impulse or higher-order singularity at $t = 0$, then $\int_{0^-}^{0^+} h(t)\,dt = K\int_{0}^{0^+} e^{-at}dt = (-K/a)\underbrace{(e^{-0^+} - e^{-0^-})}_{=0} = 0$. If $h(t)$ does have an impulse or higher-order singularity at $t = 0$, then the integral may not be zero.

If $h(t)$ has an impulse or higher-order singularity at $t = 0$, then $h'(t)$, which appears on the left side of (5.2), must contain a doublet or higher-order singularity. Since there is no doublet or higher-order singularity on the right side of (5.2), the equation cannot be satisfied. Therefore, in this example, we know that there is no impulse or higher-order singularity in $h(t)$ at $t = 0$ and, therefore, $\int_{0^-}^{0^+} h(t)\,dt = 0$, the form of the impulse response is $Ke^{-at}\,u(t)$, and from (5.3), $h(0^+) = Ke^{-a(0^+)} = K = 1$. This is the needed initial condition to find a numerical form of the homogeneous solution that applies after $t = 0$. The total solution is then $h(t) = e^{-at}\,u(t)$. Let's verify this solution by substituting it into the differential equation,

$$h'(t) + ah(t) = e^{-at}\delta(t) - ae^{-at}u(t) + ae^{-at}u(t) = \delta(t)$$

or, using the equivalence property of the impulse,

$$e^{-at}\delta(t) = e^{0}\delta(t) = \delta(t) \quad \text{Check.}$$

METHOD #2:

Another way to find the impulse response is to find the response of the system to a rectangular pulse of width w and height $1/w$ beginning at $t = 0$ and, after finding the solution, to let w approach zero. As w approaches zero, the rectangular pulse approaches a unit impulse at $t = 0$ and the response approaches the impulse response.

Using the principle of linearity, the response to the pulse is the sum of the responses to a step of height $1/w$ at $t = 0$, and the response to a step of height $-1/w$, at $t = w$. The equation for $x(t) = u(t)$ is

$$h'_{-1}(t) + ah_{-1}(t) = u(t). \tag{5.4}$$

The notation $h_{-1}(t)$ for step response follows the same logic as the coordinated notation for singularity functions. The subscript indicates the number of differentiations of the impulse response. In this case there is -1 differentiation or one integration in going from the unit-impulse response to the unit-step response. The total response for $t > 0$ to a unit-step is $h_{-1}(t) = Ke^{-at} + 1/a$. If $h_{-1}(t)$ has a discontinuity at $t = 0$, then $h'_{-1}(t)$ must contain an impulse at $t = 0$. Therefore, since $x(t)$ is the unit step, which does not contain an impulse, $h_{-1}(t)$ must be continuous at $t = 0$, otherwise (5.4) could not be correct. Also, since $h_{-1}(t)$ has been zero for all negative time and it is continuous at $t = 0$, it must also be zero at $t = 0^+$. Then

$$h_{-1}(0^+) = 0 = Ke^0 + 1/a \Rightarrow K = -1/a$$

and $h_{-1}(t) = (1/a)(1 - e^{-at})$, $t > 0$. Combining this with the fact that $h_{-1}(t) = 0$ for $t < 0$, we get the solution for all time

$$h_{-1}(t) = \frac{1 - e^{-at}}{a}u(t).$$

Using linearity and time invariance, the response to a unit step occurring at $t = w$ would be

$$h_{-1}(t - w) = \frac{1 - e^{-a(t-w)}}{a}u(t - w).$$

Therefore the response to the rectangular pulse described above is

$$h_p(t) = \frac{(1 - e^{-at})u(t) - (1 - e^{-a(t-w)})u(t - w)}{aw}$$

Then, letting w approach zero,

$$h(t) = \lim_{w \to 0} h_p(t) = \lim_{w \to 0} \frac{(1 - e^{-at})u(t) - (1 - e^{-a(t-w)})u(t-w)}{aw}.$$

This is an indeterminate form, so we must use L'Hôpital's rule to evaluate it.

$$\lim_{w \to 0} h_p(t) = \lim_{w \to 0} \frac{\frac{d}{dw}((1 - e^{-at})u(t) - (1 - e^{-a(t-w)})u(t-w))}{\frac{d}{dw}(aw)}$$

$$\lim_{w \to 0} h_p(t) = \lim_{w \to 0} \frac{-\frac{d}{dw}((1 - e^{-a(t-w)})u(t-w))}{a}$$

$$\lim_{w \to 0} h_p(t) = -\lim_{w \to 0} \frac{(1 - e^{-a(t-w)})(-\delta(t-w)) - ae^{-a(t-w)}u(t-w)}{a}$$

$$\lim_{w \to 0} h_p(t) = \frac{(1 - e^{-at})(-\delta(t)) - ae^{-at}u(t)}{a} = -\frac{-ae^{-at}u(t)}{a} = e^{-at}u(t)$$

The impulse response is $h(t) = e^{-at} u(t)$ as before.

∎

The principles used in Example 5.1 can be generalized to apply to finding the impulse response of a system described by a differential equation of the form

$$a_N y^{(N)}(t) + a_{N-1} y^{(N-1)}(t) + \cdots + a_1 y'(t) + a_0 y(t)$$
$$= b_M x^{(M)}(t) + b_{M-1} x^{(M-1)}(t) + \cdots + b_1 x'(t) + b_0 x(t) \quad (5.5)$$

or

$$\sum_{k=0}^{N} a_k y^{(k)}(t) = \sum_{k=0}^{M} b_k x^{(k)}(t)$$

The response $h(t)$ to a unit impulse must have a functional form such that

1. When it is differentiated multiple times, up to the Nth derivative, all those derivatives must match a corresponding derivative of the impulse up to the Mth derivative at $t = 0$, and
2. The linear combination of all the derivatives of $h(t)$ must add to zero for any $t \neq 0$.

Requirement 2 is met by a solution of the form $y_h(t) u(t)$ where $y_h(t)$ is the homogeneous solution of (5.5). To meet requirement 1 we may need to add another function or functions to $y_h(t) u(t)$. Consider three cases:

Case 1 $M < N$

The derivatives of $y_h(t)u(t)$ provide all the singularity functions necessary to match the impulse and derivatives of the impulse on the right side and no other terms need to be added.

Case 2 $M = N$

We only need to add an impulse term $K_\delta \delta(t)$.

Case 3 $M > N$

The Nth derivative of the function we add to $y_h(t)\,u(t)$ must have a term that matches the Mth derivative of the unit impulse. So the function we add must be of the form $K_{M-N}u_{M-N}(t) + K_{M-N-1}u_{M-N-1}(t) + \cdots + \underbrace{K_0 u_0(t)}_{=\delta(t)}$. All the other derivatives of the impulse will be accounted for by differentiating the solution form $y_h(t)\,u(t)$ multiple times.

Case 1 is the most common case in practice and Case 3 is rare in practice.

EXAMPLE 5.2

Impulse response of continuous-time system 2

Find the impulse response of a system described by $y'(t) + ay(t) = x'(t)$.
The impulse response must satisfy

$$h'(t) + ah(t) = \delta'(t) \tag{5.6}$$

The highest derivative is the same for the excitation and response. The form of the impulse response is $h(t) = Ke^{-at}u(t) + K_\delta \delta(t)$ and its first derivative is

$$h'(t) = Ke^{-at}\delta(t) - aKe^{-at}u(t) + K_\delta \delta'(t).$$

Using the equivalence property of the impulse,

$$h'(t) = K\delta(t) - aKe^{-at}u(t) + K_\delta \delta'(t)$$

Integrating (5.6) from $t = 0^-$ to $t = 0^+$

$$\underbrace{h(0^+)}_{=K} - \underbrace{h(0^-)}_{=0} + a\int_{0^-}^{0^+}[Ke^{-at}u(t) + K_\delta \delta(t)]dt = \underbrace{\delta(0^+)}_{=0} - \underbrace{\delta(0^-)}_{=0}$$

$$K + aK\int_0^{0^+} e^{-at}dt + aK_\delta \underbrace{\int_{0^-}^{0^+}\delta(t)dt}_{=1} = 0$$

$$K + aK\left[\frac{e^{-at}}{-a}\right]_0^{0^+} + aK_\delta = K - K\underbrace{[e^{0^+} - e^0]}_{=0} + aK_\delta = 0$$

or $K + aK_\delta = 0$. Integrating (5.6) from $-\infty$ to t and then from $t = 0^-$ to $t = 0^+$ we get

$$\int_{0^-}^{0^+}dt \int_{-\infty}^{t}[K\delta(\lambda) - aKe^{-a\lambda}u(\lambda) + K_\delta \delta'(\lambda)]d\lambda$$

$$+\int_{0^-}^{0^+}dt \int_{-\infty}^{t}[Ke^{-a\lambda}u(\lambda) + K_\delta \delta(\lambda)]d\lambda = \int_{0^-}^{0^+}dt \int_{-\infty}^{t}\delta'(\lambda)d\lambda$$

$$\int_{0^-}^{0^+}[Ku(t) + K(e^{-at} - 1)u(t) + K_\delta \delta(t)]dt + \underbrace{\frac{K}{a}\int_{0^-}^{0^+}(1 - e^{-at})u(t)dt}_{=0}$$

$$+ K_\delta \underbrace{\int_{0^-}^{0^+} u(t)dt}_{=0} = \int_{0^-}^{0^+}dt \int_{-\infty}^{t}\delta'(\lambda)d\lambda$$

$$\int_{0^-}^{0^+} [Ke^{-at}u(t) + K_\delta \delta(t)]\,dt = \underbrace{\int_{0^-}^{0^+} \delta(t)\,dt}_{=u(0^+)-u(0^-)}$$

$$\frac{K}{a}\underbrace{[1 - e^{-at}]_0^{0^+}}_{=0} + K_\delta \Big[\underbrace{u(0^+)}_{=1} - \underbrace{u(0^-)}_{=0}\Big] = 1 \Rightarrow K_\delta = 1 \Rightarrow K = -a.$$

Therefore the impulse response is $h(t) = \delta(t) - ae^{-at}u(t)$. Checking the solution, by substituting it into (5.6),

$$\delta'(t) - a\underbrace{e^{-at}\delta(t)}_{=e^0\delta(t)\,=\delta(t)} + a^2 e^{-at}u(t) + a[\delta(t) - ae^{-at}u(t)] = \delta'(t)$$

or $\delta'(t) = \delta'(t)$. Check.

CONTINUOUS-TIME CONVOLUTION

Derivation
Once the impulse response of a system is known, we can develop a method for finding its response to a general input signal. Let a system be excited by an arbitrary input signal $x(t)$ (Figure 5.1). How could we find the response? We could find an approximate response by approximating this signal as a sequence of contiguous rectangular pulses, all of the same width T_p (Figure 5.2).

Figure 5.1
An arbitrary signal

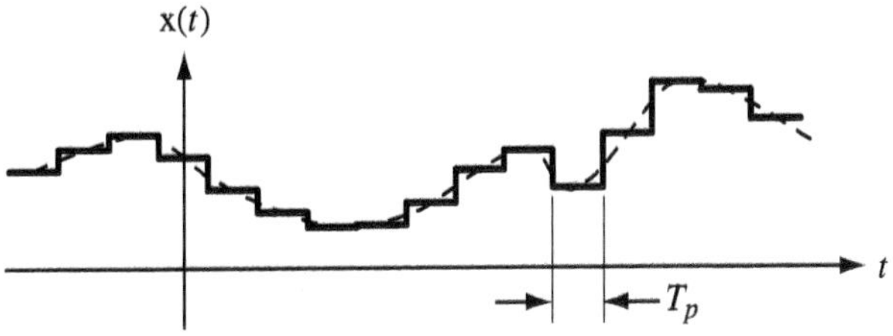

Figure 5.2
Contiguous-pulse approximation to an arbitrary signal

Now we can (approximately) find the response to the original signal as the sum of the responses to all those pulses, acting individually. Since all the pulses are rectangular and the same width, the only differences between pulses are when they occur and how tall they are. So the pulse responses all have the same form except delayed by some amount, to account for time of occurrence, and multiplied by a weighting constant, to account for the height. We can make the approximation as good as necessary by using more pulses, each of shorter duration. In summary, the problem of finding the response of an LTI system to an arbitrary signal becomes the problem of adding responses of a known functional form, but weighted and delayed appropriately.

Using the rectangle function, the description of the approximation to the arbitrary signal can now be written analytically. The height of a pulse is the value of the signal at the time the center of the pulse occurs. Then the approximation can be written as

$$x(t) \cong \cdots + x(-T_p)\,\text{rect}\left(\frac{t+T_p}{T_p}\right) + x(0)\,\text{rect}\left(\frac{t}{T_p}\right) + x(T_p)\,\text{rect}\left(\frac{t-T_p}{T_p}\right) + \cdots$$

or

$$x(t) \cong \sum_{n=-\infty}^{\infty} x(nT_p) \, \text{rect}\left(\frac{t - nT_p}{T_p}\right). \quad (5.7)$$

Let the response to a single pulse of width T_p and unit area centered at $t = 0$ be a function $h_p(t)$ called the *unit-pulse response*. The unit pulse is $(1/T_p) \, \text{rect}(1/T_p)$. Therefore (5.7) could be written in terms of shifted unit pulses as

$$x(t) \cong \sum_{n=-\infty}^{\infty} T_p x(nT_p) \underbrace{\frac{1}{T_p} \text{rect}\left(\frac{t - nT_p}{T_p}\right)}_{\text{shifted unit pulse}}. \quad (5.8)$$

Invoking linearity and time invariance, the response to each of these pulses must be the unit pulse response $h_p(t)$, amplitude scaled by the factor $T_p x(nT_p)$ and time shifted from the time origin the same amount as the pulse. Then the approximation to the response is

$$y(t) \cong \sum_{n=-\infty}^{\infty} T_p x(nT_p) h_p(t - nT_p). \quad (5.9)$$

As an illustration, let the unit pulse response $h_p(t)$ be that of the *RC* lowpass filter introduced above (Figure 5.3). Let the input signal $x(t)$ be the smooth waveform in Figure 5.4, which is approximated by a sequence of pulses as illustrated.

In Figure 5.5 the pulses are separated and then added to form the approximation to $x(t)$.

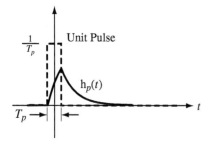

Figure 5.3
Unit pulse response of an *RC* lowpass filter

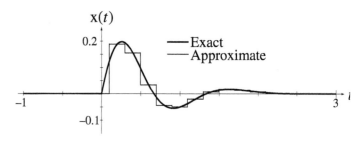

Figure 5.4
Exact and approximate $x(t)$

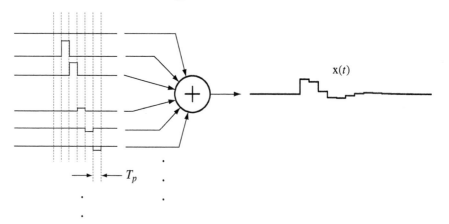

Figure 5.5
Approximation of $x(t)$ as a sum of individual pulses

Since the sum of the individual pulses is the approximation of x(t), the approximate response can be found by applying the approximation of x(t) to the system. But, because the system is LTI, we can alternately use the principle of superposition and apply the pulses one at a time to the system. Then those responses can be added to form the approximate system response (Figure 5.6).

The system exact and approximate input signals, the unit-impulse response, the unit-pulse response and the exact and approximate system responses are illustrated in Figure 5.7, based on a pulse width of 0.2 seconds. As the pulse duration is reduced, the approximation becomes better (Figure 5.8). With a pulse width of 0.1, the exact and approximate responses are indistinguishable as graphed on this scale.

Recall from the concept of rectangular-rule integration in basic calculus that a real integral of a real variable can be defined as the limit of a summation

$$\int_a^b g(x)\,dx = \lim_{\Delta x \to 0} \sum_{n=a/\Delta x}^{b/\Delta x} g(n\Delta x)\,\Delta x. \qquad (5.10)$$

We will apply (5.10) to the summations of pulses and pulse responses (5.8) and (5.9) in the limit as the pulse width approaches zero. As the pulse width T_p becomes smaller, the excitation and response approximations become better. In the limit as T_p approaches zero, the summation becomes an integral and the approximations become exact. In that same limit,

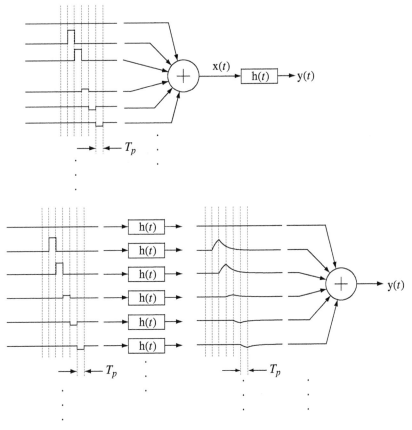

Figure 5.6
Application of linearity and superposition to find the approximate system response

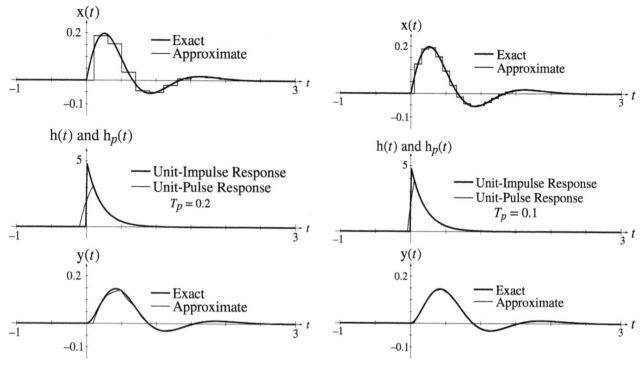

Figure 5.7
Exact and approximate excitation, unit-impulse response, unit-pulse response and exact and approximate system response with $T_p = 0.2$

Figure 5.8
Exact and approximate excitation, unit-impulse response, unit-pulse response and exact and approximate system response with $T_p = 0.1$

the unit pulse $(1/T_p)\,\text{rect}(1/T_p)$ approaches a unit impulse. As T_p approaches zero, the points in time nT_p become closer and closer together. In the limit, the discrete time shifts nT_p merge into a continuum of time-shifts. It is convenient (and conventional) to call that new continuous time shift τ. Changing the name of the time shift amount nT_p to τ, and taking the limit as T_p approaches zero, the width of the pulse T_p approaches a differential $d\tau$ and

$$x(t) = \sum_{n=-\infty}^{\infty} \underbrace{T_p}_{\int d\tau} \underbrace{x(nT_p)}_{(\tau)} \underbrace{\frac{1}{T_p}\text{rect}\left(\frac{t-nT_p}{T_p}\right)}_{\delta(t-\tau)}$$

and

$$y(t) = \sum_{n=-\infty}^{\infty} \underbrace{T_p}_{\int d\tau} \underbrace{x(nT_p)}_{(\tau)} \underbrace{h_p(t-nT_p)}_{h(t-\tau)}.$$

Therefore, in the limit, these summations become integrals of the forms

$$x(t) = \int_{-\infty}^{\infty} x(\tau)\delta(t-\tau)d\tau \qquad (5.11)$$

and

$$y(t) = \int_{-\infty}^{\infty} x(\tau)h(t-\tau)d\tau \qquad (5.12)$$

where the unit-pulse response $h_p(t)$ approaches the unit-impulse response $h(t)$ (more commonly called just the **impulse response**) of the system. The integral in (5.11) is easily verified by application of the sampling property of the impulse. The integral in (5.12) is called the **convolution integral**. The convolution of two functions is conventionally indicated by the operator $*$,[1]

$$y(t) = x(t) * h(t) = \int_{-\infty}^{\infty} x(\tau)h(t-\tau)d\tau. \qquad (5.13)$$

Another way of developing the convolution integral is to start with (5.11), which follows directly from the sampling property of the impulse. The integrand of (5.11) is an impulse at $t = \tau$ of strength $x(\tau)$. Since, by definition, $h(t)$ is the response to an impulse $\delta(t)$ and the system is homogeneous and time invariant, the response to $x(\tau)\delta(t-\tau)$ must be $x(\tau)h(t-\tau)$. Then, invoking additivity, if $x(t) = \int_{-\infty}^{\infty} x(\tau)\delta(t-\tau)d\tau$, an integral (the limit of a summation) of x values, then $y(t) = \int_{-\infty}^{\infty} x(\tau)h(t-\tau)d\tau$, an integral of the y's that respond to those x's. This derivation is more abstract and sophisticated and much shorter than the derivation above and is an elegant application of the properties of LTI systems and the sampling property of the impulse.

The impulse response of an LTI system is a very important descriptor of the way it responds because, once it is determined, the response to any arbitrary input signal can be found. The effect of convolution can be depicted by a block diagram (Figure 5.9).

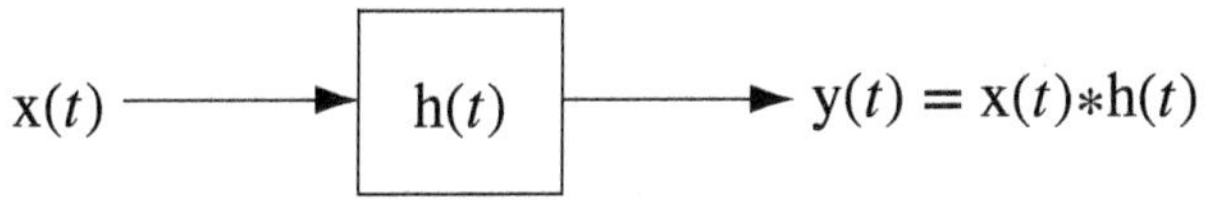

$x(t) \longrightarrow \boxed{h(t)} \longrightarrow y(t) = x(t) * h(t)$

Figure 5.9
Block diagram depiction of convolution

Graphical and Analytical Examples of Convolution
The general mathematical form of the convolution integral is

$$x(t) * h(t) = \int_{-\infty}^{\infty} x(\tau)h(t-\tau)\,d\tau.$$

A graphical example of the operations implied by the convolution integral is very helpful in a conceptual understanding of convolution. Let $h(t)$ and $x(t)$ be the functions in Figure 5.10.

---

[1] Do not confuse the convolution operator $*$ with the indicator for the complex conjugate of a complex number or function $*$. For example, $x[n] * h[n]$ is $x[n]$ convolved with $h[n]$, but $x[n]^* h[n]$, is the product of the complex conjugate of $x[n]$ and $h[n]$. Usually the difference is clear in context.

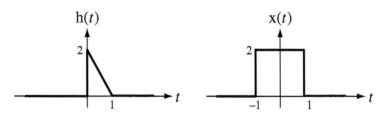

Figure 5.10
Two functions to be convolved

This impulse response h(*t*) is not typical of a practical linear system but will serve to demonstrate the process of convolution. The integrand in the convolution integral is x(τ)h(*t* − τ). What is h(*t* − τ)? It is a function of the two variables *t* and τ. Since the variable of integration in the convolution integral is τ, we should consider h(*t* − τ) to be a function of τ in order to see how to do the integral. We can start by graphing h(τ) and then h(−τ) versus τ (Figure 5.11).

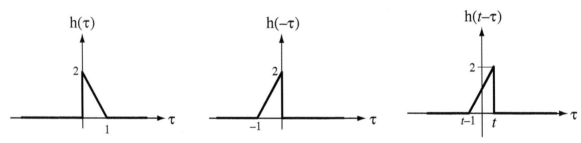

Figure 5.11
h(τ) and h(−τ) graphed versus τ

Figure 5.12
h(*t* − τ) graphed versus τ

The addition of the *t* in h(*t* − τ) shifts the function *t* units to the right (Figure 5.12). The transformation from h(τ) to h(*t* − τ) can be described as two successive shifting/scaling operations

$$h(\tau) \xrightarrow{\tau \to -\tau} h(-\tau) \xrightarrow{\tau \to \tau - t} h(-(\tau - t)) = h(t - \tau).$$

If we substitute *t* for τ in h(*t* − τ), we have h(0). From the first definition of the function h(*t*) we see that that is the point of discontinuity where h(*t*) goes from 0 to 1. That is the same point on h(*t* − τ). Do the same for τ = *t* − 1 and see if it works.

One common confusion is to look at the integral and not understand what the process of integrating from τ = −∞ to τ = +∞ means. Since *t* is not the variable of integration, it is like a constant during the integration process. But it is the variable in the final function that results from the convolution. Think of convolution as two general procedures. First pick a value for *t*, do the integration and get a result. Then pick another value of *t* and repeat the process. Each integration yields one point on the curve describing the final function. Each point on the y(*t*) curve will be found by finding the total area under the product x(τ)h(*t* − τ).

Visualize the product x(τ)h(*t* − τ). The product depends on what *t* is. For most values of *t*, the nonzero portions of the two functions do not overlap and the product is zero. (This is not typical of real impulse responses because they usually are not time limited. Real impulse responses of stable systems usually begin at some time and approach zero as *t* approaches infinity.) But for some times *t* their nonzero portions do overlap and there is nonzero area under their product curve. Consider *t* = 5 and *t* = 0.

When $t = 5$, the nonzero portions of $x(\tau)$ and $h(5 - \tau)$ do not overlap and the product is zero everywhere (Figure 5.13).

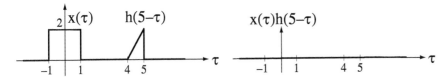

Figure 5.13
Impulse response, input signal and their product when $t = 5$

When $t = 0$, the nonzero portions of $x(\tau)$ and $h(5 - \tau)$ do overlap and the product is not zero everywhere (Figure 5.14).

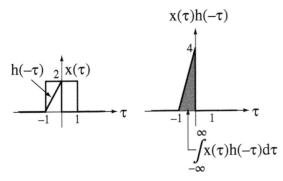

Figure 5.14
Impulse response, input signal and their product when $t = 0$

For $-1 < t < 0$ the convolution of the two functions is twice the area of the h function (which is 1) minus the area of a triangle of width $|t|$ and height $4|t|$ (Figure 5.15).

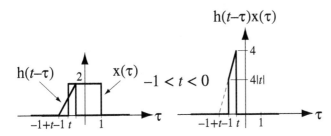

Figure 5.15
Product of $h(t - \tau)$ and $x(\tau)$ for $-1 < t < 0$

Therefore the convolution function value over this range of t is

$$y(t) = 2 - (1/2)(-t)(-4t) = 2(1 - t^2), -1 < t < 0.$$

For $0 < t < 1$ the convolution of the two functions is the constant 2. For $1 < t < 2$, the convolution of the two functions is the area of a triangle whose base width is $(2 - t)$ and whose height is $(8 - 4t)$ or $y(t) = (1/2)(2 - t)(8 - 4t) = 2(2 - t)^2$. The final function $y(t)$ is illustrated in Figure 5.16.

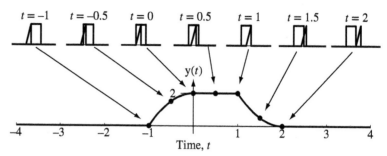

Figure 5.16
Convolution of x(t) with h(t)

As a more practical exercise let us now find the unit-step response of an *RC* lowpass filter using convolution. We already know from prior analysis that the answer is $v_{out}(t) = (1 - e^{-t/RC})\,u(t)$. First we need to find the impulse response. The differential equation is

$$RC v'_{out}(t) + v_{out}(t) = v_{in}(t) \Rightarrow RC\,h'(t) + h(t) = \delta(t)$$

The form of the impulse response is $h(t) = K e^{-t/RC}\,u(t)$. Integrating once from 0^- to 0^+

$$RC\Big[h(0^+) - \underbrace{h(0^-)}_{=0}\Big] + \underbrace{\int_{0^-}^{0^+} h(t)\,dt}_{=0} = \underbrace{u(0^+)}_{=1} - \underbrace{u(0^-)}_{=0} \Rightarrow h(0^+) = 1/RC$$

Then $1/RC = K$ and $h(t) = (1/RC)e^{-t/RC}\,u(t)$ (Figure 5.17).

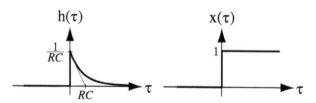

Figure 5.17
The impulse response and excitation of the *RC* lowpass filter

Then the response $v_{out}(t)$ to a unit step $v_{in}(t)$ is $v_{out}(t) = v_{in}(t) * h(t)$ or

$$v_{out}(t) = \int_{-\infty}^{\infty} v_{in}(\tau) h(t-\tau)\,d\tau = \int_{-\infty}^{\infty} u(\tau) \frac{e^{-(t-\tau)/RC}}{RC} u(t-\tau)\,d\tau.$$

We can immediately simplify the integral some by observing that the first unit step $u(\tau)$ makes the integrand zero for all negative τ. Therefore

$$v_{out}(t) = \int_{0}^{\infty} \frac{e^{-(t-\tau)/RC}}{RC} u(t-\tau)\,d\tau.$$

Consider the effect of the other unit step $u(t-\tau)$. Since we are integrating over a range of τ from zero to infinity, if t is negative, for any τ in that range this unit step has a value

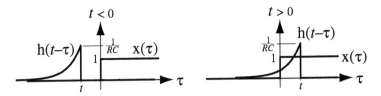

Figure 5.18
The relation between the two functions that form the product in the convolution integrand for t negative and t positive

of zero (Figure 5.18). Therefore for negative t, $v_{out}(t) = 0$. For positive t, the unit step $u(t - \tau)$ will be one for $\tau < t$ and zero for $\tau > t$. Therefore for positive t,

$$v_{out}(t) = \int_0^t \frac{e^{-(t-\tau)/RC}}{RC} d\tau = \left[e^{-(t-\tau)/RC}\right]_0^t = 1 - e^{-t/RC}, \quad t > 0.$$

Combining the results for negative and positive ranges of t, $v_{out}(t) = (1 - e^{-t/RC})u(t)$.

Figure 5.19 and Figure 5.20 illustrate two more examples of convolution. In each case the top row presents two functions $x_1(t)$ and $x_2(t)$ to be convolved and the "flipped" version of the second function $x_2(-\tau)$, which is $x(t - \tau)$ with $t = 0$, the flipped but not-yet-shifted version. On the second row are the two functions in the convolution integral $x_1(\tau)$ and $x_2(t - \tau)$ graphed versus τ for five choices of t, illustrating the shifting of the second function $x_2(t - \tau)$ as t is changed. On the third row are the products of

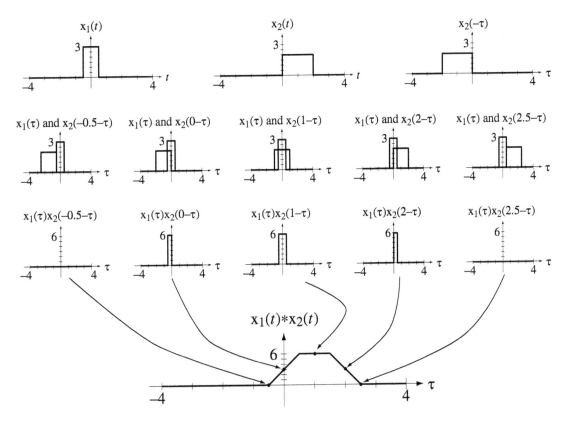

Figure 5.19
Convolution of two rectangular pulses

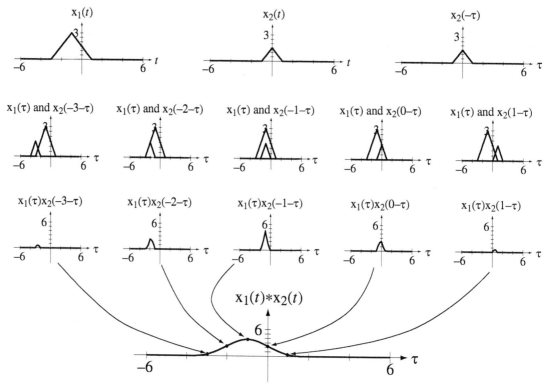

Figure 5.20
Convolution of two triangular pulses

the two functions $x_1(\tau)x_2(t - \tau)$ in the convolution integral at those same times. And at the bottom is a graph of the convolution of the two original functions with small dots indicating the convolution values at the five times t, which are the same as the areas $\int_{-\infty}^{\infty} x_1(\tau)x_2(t - \tau)d\tau$ under the products at those times.

Convolution Properties

An operation that appears frequently in signal and system analysis is the convolution of a signal with an impulse

$$x(t) * A\delta(t - t_0) = \int_{-\infty}^{\infty} x(\tau)A\delta(t - \tau - t_0)d\tau.$$

We can use the sampling property of the impulse to evaluate the integral. The variable of integration is τ. The impulse occurs in τ where $t - \tau - t_0 = 0$ or $\tau = t - t_0$. Therefore

$$x(t) * A\delta(t - t_0) = Ax(t - t_0). \quad (5.14)$$

This is a very important result and will show up many times in the exercises and later material (Figure 5.21).

If we define a function $g(t) = g_0(t) * \delta(t)$, then a time-shifted version $g(t - t_0)$ can be expressed in either of the two alternate forms

$$g(t - t_0) = g_0(t - t_0) * \delta(t) \quad \text{or} \quad g(t - t_0) = g_0(t) * \delta(t - t_0)$$

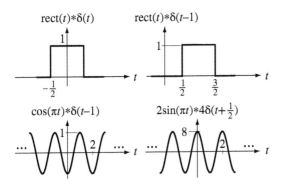

Figure 5.21
Examples of convolution with impulses

but not in the form $g_0(t - t_0) * \delta(t - t_0)$. Instead, $g_0(t - t_0) * \delta(t - t_0) = g(t - 2t_0)$. This property is true not only when convolving with impulses, but with any functions. A shift of either of the functions being convolved (but not both) shifts the convolution by the same amount.

The commutativity, associativity, distributivity, differentiation, area and scaling properties of the convolution integral are proven in Web Appendix E and are summarized here.

Commutativity	$x(t) * y(t) = y(t) * x(t)$
Associativity	$(x(t) * y(t)) * z(t) = x(t) * (y(t) * z(t))$
Distributivity	$(x(t) + y(t)) * z(t) = x(t) * z(t) + y(t) * z(t)$
If $y(t) = x(t) * h(t)$, then	
Differentiation Property	$y'(t) = x'(t) * h(t) = x(t) * h'(t)$
Area Property	Area of y = (Area of x) × (Area of h)
Scaling Property	$y(at) = |a| x(at) * h(at)$

Let the convolution of $x(t)$ with $h(t)$ be $y(t) = \int_{-\infty}^{\infty} x(t - \tau)h(\tau)d\tau$. Let $x(t)$ be bounded. Then $|x(t - \tau)| < B$, for all τ where B is a finite upper bound. The magnitude of the convolution integral is

$$|y(t)| = \left| \int_{-\infty}^{\infty} x(t - \tau)h(\tau)d\tau \right|.$$

Using the principles that the magnitude of an integral of a function is less than or equal to the integral of the magnitude of the function

$$\left| \int_{\alpha}^{\beta} g(x) \, dx \right| \leq \int_{\alpha}^{\beta} |g(x)| dx$$

and that the magnitude of a product of two functions is equal to the product of their magnitudes, $|g(x)h(x)| = |g(x)||h(x)|$, we can conclude that

$$|y(t)| \leq \int_{-\infty}^{\infty} |x(t - \tau)||h(\tau)|d\tau.$$

Since x($t - \tau$) is less than B in magnitude for any τ

$$|y(t)| \le \int_{-\infty}^{\infty} |x(t-\tau)||h(\tau)|\,d\tau < \int_{-\infty}^{\infty} B|h(\tau)|\,d\tau$$

or

$$|y(t)| < B \int_{-\infty}^{\infty} |h(\tau)|\,d\tau.$$

Therefore, the convolution integral converges if $\int_{-\infty}^{\infty} |h(t)|\,dt$ is bounded or, in other words, if h(t) is *absolutely integrable*. Since convolution is commutative, we can also say that, if h(t) is bounded, the condition for convergence is that x(t) be absolutely integrable.

> For a convolution integral to converge, the signals being convolved must both be bounded and at least one of them must be absolutely integrable.

EXAMPLE 5.3

Convolution of two unit rectangles

Find the convolution y(t) of two unit rectangles x(t) = rect (t) and h(t) = rect (t).

This convolution can be done in a direct way using the convolution integral, analytically or graphically. But we can exploit the differentiation property to avoid explicit integration altogether.

$$y(t) = x(t) * h(t) \Rightarrow y''(t) = x'(t) * h'(t)$$

$$y''(t) = [\delta(t+1/2) - \delta(t-1/2)] * [\delta(t+1/2) - \delta(t-1/2)]$$

$$y''(t) = \delta(t+1) - 2\delta(t) + \delta(t-1)$$

$$y'(t) = u(t+1) - 2u(t) + u(t-1)$$

$$y(t) = \text{ramp}\,(t+1) - 2\,\text{ramp}\,(t) + \text{ramp}\,(t-1)$$

(See Figure 5.22.)

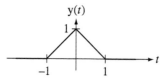

Figure 5.22
Convolution of two unit rectangles

The result of convolving two unit rectangles (Example 5.3) is important enough to be given a name for future reference. It is called the **unit triangle** function (see Figure 5.23).

$$\text{tri}\,(t) = \begin{cases} 1 - |t|, & |t| < 1 \\ 0, & \text{otherwise} \end{cases}$$

It is called a unit triangle because its peak height and its area are both one.

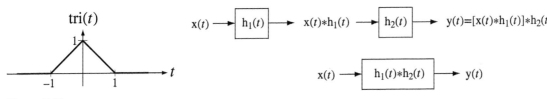

Figure 5.23
The unit triangle function

Figure 5.24
Cascade connection of two systems

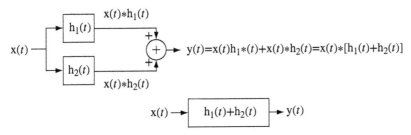

Figure 5.25
Parallel connection of two systems

System Connections

Two very common connections of systems are the cascade connection and the parallel connection (Figure 5.24 and Figure 5.25).

Using the associativity property of convolution we can show that the cascade connection of two systems can be considered as one system whose impulse response is the convolution of the impulse responses of the two systems. Using the distributivity property of convolution we can show that the parallel connection of two systems can be considered as one system whose impulse response is the sum of the impulse responses of the two systems.

Step Response and Impulse Response

In actual system testing, a system is often tested using some standard signals that are easy to generate and do not drive the system into a nonlinearity. One of the most common signals of this type is the step function. The response of an LTI system to a unit step is

$$h_{-1}(t) = h(t) * u(t) = \int_{-\infty}^{\infty} h(\tau)u(t-\tau)d\tau = \int_{-\infty}^{t} h(\tau)d\tau.$$

This proves that the response of an LTI system excited by a unit step is the integral of the impulse response. Therefore we can say that just as the unit step is the integral of the impulse, the unit-step *response* is the integral of the unit-impulse *response*. In fact, this relationship holds not just for impulse and step excitations, but for any excitation. If any excitation is changed to its integral, the response also changes to its integral. We can also turn these relationships around and say that since the first derivative is the inverse of integration, if the excitation is changed to its first derivative, the response is also changed to its first derivative (Figure 5.26).

Stability and Impulse Response

Stability was generally defined in Chapter 4 by saying that a stable system has a bounded output signal in response to any bounded input signal. We can now find a way to determine whether a system is stable by examining its impulse response. We proved

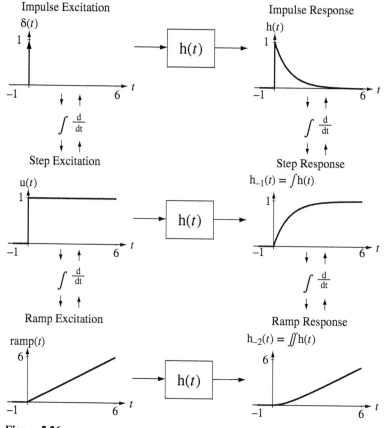

Figure 5.26
Relations between integrals and derivatives of excitations and responses for an LTI system

above that the convolution of two signals converges if both of them are bounded and at least one is absolutely integrable. The response $y(t)$ of a system to $x(t)$ is $y(t) = x(t) * h(t)$. Then, if $x(t)$ is bounded we can say that $y(t)$ is bounded if $h(t)$ is absolutely integrable. That is, if $\int_{-\infty}^{\infty} |h(t)|\,dt$ is bounded.

> A continuous-time system is BIBO stable if its impulse response is absolutely integrable.

Complex Exponential Excitation and the Transfer Function

Let a stable LTI system be described by a differential equation of the form

$$\sum_{k=0}^{N} a_k y^{(k)}(t) = \sum_{k=0}^{M} b_k x^{(k)}(t) \tag{5.15}$$

where

1. The a's and b's are constants, and
2. The notation $x^{(k)}(t)$ means the kth derivative of $x(t)$ with respect to time and, if k is negative, that indicates integration instead of differentiation.

Then let the excitation be in the form of a complex exponential $x(t) = Xe^{st}$ where X and s are, in general, complex valued. This description of the excitation is valid for all time. Therefore, not only is the excitation a complex exponential now and in the future but it *always has been* a complex exponential. The solution of the differential equation is the sum of the homogeneous and particular solutions. The system is stable, so the eigenvalues have negative real parts and the homogeneous solution approaches zero as time passes. The system has been operating with this excitation for a semi-infinite time, so the homogeneous solution has decayed to zero and the total solution now is the particular solution.

The functional form of the particular solution consists of a linear combination of the excitation's functional form and all its unique derivatives. Since the derivative of an exponential is another exponential of the same form, the response $y(t)$ must have the form $y(t) = Ye^{st}$ where Y is a complex constant. Then, in the differential equation, the kth derivatives take the forms $x^{(k)}(t) = s^k X e^{st}$ and $y^{(k)}(t) = s^k Y e^{st}$, and (5.15) can be written in the form

$$\sum_{k=0}^{N} a_k s^k Y e^{st} = \sum_{k=0}^{M} b_k s^k X e^{st}.$$

The equation is no longer a differential equation with real coefficients. It is now an algebraic equation with complex coefficients. The factors Xe^{st} and Ye^{st} can be factored out, leading to

$$Y e^{st} \sum_{k=0}^{N} a_k s^k = X e^{st} \sum_{k=0}^{M} b_k s^k.$$

The ratio of the response to the excitation is then

$$\frac{Ye^{st}}{Xe^{st}} = \frac{Y}{X} = \frac{\sum_{k=0}^{M} b_k s^k}{\sum_{k=0}^{N} a_k s^k}$$

a ratio of two polynomials in s called a **rational function**. This is the system's **transfer function**

$$\boxed{H(s) = \frac{\sum_{k=0}^{M} b_k s^k}{\sum_{k=0}^{N} a_k s^k} = \frac{b_M s^M + b_{M-1} s^{M-1} + \cdots + b_2 s^2 + b_1 s + b_0}{a_N s^N + a_{N-1} s^{N-1} + \cdots + a_2 s^2 + a_1 s + a_0}} \quad (5.16)$$

and the response is therefore $Ye^{st} = H(s)Xe^{st}$ or $y(t) = H(s)x(t)$.

For systems of this type, the transfer function can be written directly from the differential equation. If the differential equation describes the system, so does the transfer function. The transfer function is a fundamental concept in signals and systems and we will be using it many times in the material to follow. We can also find the response to a complex exponential excitation using convolution. The response $y(t)$ of an LTI system with impulse response $h(t)$ to $x(t) = Xe^{st}$ is

$$y(t) = h(t) * Xe^{st} = X \int_{-\infty}^{\infty} h(\tau) e^{s(t-\tau)} d\tau = \underbrace{Xe^{st}}_{x(t)} \int_{-\infty}^{\infty} h(\tau) e^{-st} d\tau$$

Equating the two forms of $y(t)$ we get

$$H(s)Xe^{st} = Xe^{st} \int_{-\infty}^{\infty} h(\tau) e^{-st} d\tau \Rightarrow H(s) = \int_{-\infty}^{\infty} h(\tau) e^{-st} d\tau,$$

which shows how the transfer function and the impulse response are related. Since both the impulse response and the transfer function completely characterize an LTI system, they had to be uniquely related. The integral $\int_{-\infty}^{\infty} h(\tau) e^{-s\tau} d\tau$ will be revisited in Chapter 8 and will be identified as the Laplace transform of h(t).

Frequency Response

The variable s in the complex exponential e^{st} is, in general, complex valued. Let it be of the form $s = \sigma + j\omega$ where σ is the real part and ω is the imaginary part. For the special case $\sigma = 0$, $s = j\omega$, the complex exponential e^{st} becomes the **complex sinusoid** $e^{j\omega t}$ and the transfer function of the system H(s) becomes the **frequency response** of the system H($j\omega$). The function $e^{j\omega t}$ is called a complex sinusoid because, by Euler's identity, $e^{j\omega t} = \cos(\omega t) + j\sin(\omega t)$, the sum of a real cosine and an imaginary sine, both of radian frequency ω. From $Ye^{st} = H(s)Xe^{st}$, letting $s = j\omega$

$$Ye^{j\omega t} = |Y|e^{j\angle Y}e^{j\omega t} = H(j\omega)Xe^{j\omega t} = |H(j\omega)|e^{j\angle H(j\omega)}|X|e^{j\angle X}e^{j\omega t}$$

or, dividing through by $e^{j\omega t}$,

$$|Y|e^{j\angle Y} = |H(j\omega)||X|e^{j(\angle H(j\omega) + \angle X)}.$$

Equating magnitudes we get $|Y| = |H(j\omega)||X|$ and equating phases we get $\angle Y = \angle H(j\omega) + \angle X$. The function H($j\omega$) is called the *frequency response* of the system because, at any radian frequency ω, if we know the magnitude and phase of the excitation and the magnitude and phase of the frequency response, we can find the magnitude and phase of the response.

In Chapter 4 we showed, using principles of linearity and superposition, that if a complex excitation x(t) is applied to a system and causes a response y(t), that the real part of x(t) causes the real part of y(t) and the imaginary part of x(t) causes the imaginary part of y(t). Therefore, if the actual excitation of a system is $x(t) = A_x \cos(\omega t + \theta_x)$, we can find the response of the system to an excitation

$$x_C(t) = A_x \cos(\omega t + \theta_x) + jA_x \sin(\omega t + \theta_x) = A_x e^{j(\omega t + \theta_x)}$$

in the form

$$y_C(t) = A_y \cos(\omega t + \theta_y) + jA_y \sin(\omega t + \theta_y) = A_y e^{j(\omega t + \theta_y)}$$

and we can take the real part $y(t) = A_y \cos(\omega t + \theta_y)$ as the response to the real excitation $x(t) = A_x \cos(\omega t + \theta_x)$. Using $|Y| = |H(j\omega)||X|$ and $\angle Y = \angle H(j\omega) + \angle X$ we get

$$A_y = |H(j\omega)| A_x \quad \text{and} \quad \theta_y = \angle H(j\omega) + \theta_x.$$

EXAMPLE 5.4

Transfer Function and Frequency Response

An LTI system is described by the differential equation

$$y''(t) + 3000y'(t) + 2 \times 10^6 y(t) = 2 \times 10^6 x(t).$$

(a) Find its transfer function.

For this differential equation of the form $\sum_{k=0}^{N} a_k y^{(k)}(t) = \sum_{k=0}^{M} b_k x^{(k)}(t)$,

$N = 2$, $M = 0$, $a_0 = 2 \times 10^6$, $a_1 = 3000$, $a_2 = 1$ and $b_0 = 2 \times 10^6$. Therefore the transfer function is

$$H(s) = \frac{2 \times 10^6}{s^2 + 3000s + 2 \times 10^6}.$$

(b) If $x(t) = Xe^{j400\pi t}$ and $y(t) = Ye^{j400\pi t}$ and $X = 3e^{j\pi/2}$, find the magnitude and phase of Y.

The frequency response is

$$H(j\omega) = \frac{2 \times 10^6}{(j\omega)^2 + 3000(j\omega) + 2 \times 10^6} = \frac{2 \times 10^6}{2 \times 10^6 - \omega^2 + j3000\omega}.$$

The radian frequency is $\omega = 400\pi$. Therefore,

$$H(j400\pi) = \frac{2 \times 10^6}{2 \times 10^6 - (400\pi)^2 + j3000 \times 400\pi} = 0.5272 e^{-j1.46}.$$

$$|Y| = |H(j400\pi)| \times 3 = 0.5272 \times 3 = 1.582$$

$$\angle Y = \angle H(j400\pi) + \pi/2 = 0.1112 \text{ radians}$$

(c) If $x(t) = 8\cos(200\pi t)$ and $y(t) = A_y \cos(200\pi t + \theta_y)$, find A_y and θ_y

Using

$$A_y = |H(j200\pi)| A_x \text{ and } \theta_y = \angle H(j200\pi) + \theta_x,$$

$A_y = 0.8078 \times 8 = 6.4625$ and $\theta_y = -0.8654 + 0 = -0.8654$ radians.

EXAMPLE 5.5

Frequency response of a continuous-time system

A continuous-time system is described by the differential equation

$$y''(t) + 5y'(t) + 2y(t) = 3x''(t).$$

Find and graph the magnitude and phase of its frequency response.
  The differential equation is in the general form

$$\sum_{k=0}^{N} a_k y^{(k)}(t) = \sum_{k=0}^{M} b_k x^{(k)}(t)$$

where $N = M = 2$, $a_2 = 1$, $a_1 = 5$, $a_0 = 2$, $b_2 = 3$, $b_1 = 0$ and $b_0 = 0$. The transfer function is

$$H(s) = \frac{b_2 s^2 + b_1 s + b_0}{a_2 s^2 + a_1 s + a_0} = \frac{3s^2}{s^2 + 5s + 2}.$$

The frequency response is (replacing s by $j\omega$)

$$H(j\omega) = \frac{3(j\omega)^2}{(j\omega)^2 + j5\omega + 2}.$$

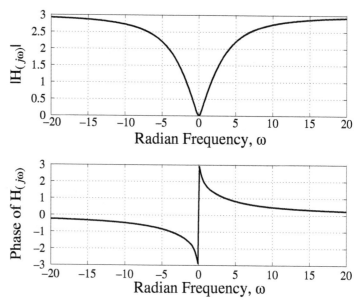

Figure 5.27
Magnitude and phase of frequency response

(See Figure 5.27.)

These graphs were generated by the following MATLAB code:

```
wmax = 20;                  % Maximum radian frequency magnitude for graph
dw = 0.1;                   % Spacing between frequencies in graph
w = [-wmax:dw:wmax]';       % Vector of radian frequencies for graph
%    Compute the frequency response
H = 3*(j*w).^2./((j*w).^2 + j*5*w + 2);
%    Graph and annotate the frequency response
subplot(2,1,1); p = plot(w, abs(H),'k'); set(p,'LineWidth',2);
grid on ;
xlabel('Radian frequency, {\omega}','FontSize',18,'FontName','Times');
ylabel('|H({\itj}{\omega})|','FontSize',18,'FontName','Times');
subplot(2,1,2); p = plot(w,angle(H),'k'); set(p,'LineWidth',2);
grid on;
xlabel('Radian frequency, {\omega}','FontSize',18,'FontName','Times');
ylabel('Phase of H({\itj}{\omega})','FontSize',18,'FontName','Times');
```

5.3 DISCRETE TIME

IMPULSE RESPONSE

Just as was true for continuous-time systems, there is a convolution method for discrete-time systems and it works in an analogous way. It is based on knowing the impulse response of the system, treating the input signal as a linear combination of impulses and then adding the responses to all the individual impulses.

No matter how complicated a discrete-time signal is, it is a sequence of impulses. If we can find the response of an LTI system to a unit impulse occurring at $n = 0$, we can find the response to any other signal. Therefore, use of the convolution technique begins with the assumption that the response to a unit impulse occurring at $n = 0$ has already been found. We will call the impulse response h$[n]$.

In finding the impulse response of a system, we apply a unit impulse $\delta[n]$ occurring at $n = 0$, and that is the only excitation of the system. The impulse puts signal energy into the system and then goes away. After the impulse energy is injected into the system, it responds with a signal determined by its dynamic character.

In the case of continuous-time systems the actual application of an impulse to determine impulse response experimentally is problematical for practical reasons. But in the case of discrete-time systems this technique is much more reasonable because the discrete-time impulse is a true discrete-time function, and a simple one at that.

If we have a mathematical description of the system, we may be able to find the impulse response analytically. Consider first a system described by a difference equation of the form

$$a_0 y[n] + a_1 y[n-1] + \cdots + a_N y[n-N] = x[n]. \qquad (5.17)$$

This is not the most general form of difference equation describing a discrete-time LTI system but it is a good place to start because, from the analysis of this system, we can extend to finding the impulse responses of more general systems. This system is causal and LTI. To find the impulse response, we let the excitation x$[n]$ be a unit impulse at $n = 0$. Then we can rewrite (5.17) for this special case as

$$a_0 h[n] + a_1 h[n-1] + \cdots + a_N h[n-N] = \delta[n].$$

The system has never been excited by anything before $n = 0$, the response h$[n]$ has been zero for all negative time h$[n] = 0$, $n < 0$ and the system is in its zero state before $n = 0$. For all times after $n = 0$, x$[n]$ is also zero and the solution of the difference equation is the homogeneous solution. All we need to find the homogeneous solution after $n = 0$ are N initial conditions we can use to evaluate the N arbitrary constants in the homogeneous solution. We need an initial condition for each order of the difference equation. We can always find these initial conditions by recursion. The difference equation of a causal system can always be put into a recursion form in which the present response is a linear combination of the present excitation and previous responses

$$h[n] = \frac{\delta[n] - a_1 h[n-1] - \cdots - a_N h[n-N]}{a_0}.$$

Then we can find an exact homogeneous solution that is valid for $n \geq 0$. That solution, together with the fact that h$[n] = 0$, $n < 0$, forms the total solution, the impulse response. The application of an impulse to a system simply establishes some initial conditions and the system relaxes back to its former equilibrium after that (if it is stable).

Now consider a more general system described by a difference equation of the form

$$a_0 y[n] + a_1 y[n-1] + \cdots + a_N y[n-N] = b_0 x[n] + b_1 x[n-1] + \cdots + b_M x[n-M].$$

or

$$\sum_{k=0}^{N} a_k y[n-k] = \sum_{k=0}^{M} b_k x[n-k].$$

Since the system is LTI we can find the impulse response by first finding the impulse responses to systems described by the difference equations,

$$\begin{aligned} a_0 y[n] + a_1 y[n-1] + \cdots + a_N y[n-N] &= b_0 x[n] \\ a_0 y[n] + a_1 y[n-1] + \cdots + a_N y[n-N] &= b_1 x[n-1] \\ a_0 y[n] + a_1 y[n-1] + \cdots + a_N y[n-N] &= \vdots \\ a_0 y[n] + a_1 y[n-1] + \cdots + a_N y[n-N] &= b_M x[n-M] \end{aligned} \quad (5.18)$$

and then adding all those impulse responses. Since all the equations are the same except for the strength and time of occurrence of the impulse, the overall impulse response is simply the sum of a set of impulse responses of the systems in (5.18), weighted and delayed appropriately. The impulse response of the general system must be

$$h[n] = b_0 h_1[n] + b_1 h_1[n-1] + \cdots + b_M h_1[n-M]$$

where $h_1[n]$ is the impulse response found earlier.

EXAMPLE 5.6

Impulse response of a system

Find the impulse response $h[n]$ of the system described by the difference equation

$$8y[n] + 6y[n-1] = x[n]. \quad (5.19)$$

If the excitation is an impulse

$$8h[n] + 6h[n-1] = \delta[n].$$

This equation describes a causal system so $h[n] = 0$, $n < 0$. We can find the first response to a unit impulse at $n = 0$ from (5.19)

n	$x[n]$	$h[n-1]$	$h[n]$
0	1	0	1/8

For $n \geq 0$ the solution is the homogeneous solution of the form $K_h(-3/4)^n$. Therefore, $h[n] = K_h(-3/4)^n u[n]$. Applying initial conditions, $h[0] = 1/8 = K_h$. Then the impulse response of the system is $h[n] = (1/8)(-3/4)^n u[n]$.

EXAMPLE 5.7

Impulse response of a system

Find the impulse response $h[n]$ of the system described by the difference equation

$$5y[n] + 2y[n-1] - 3y[n-2] = x[n]. \quad (5.20)$$

This equation describes a causal system so $h[n] = 0$, $n < 0$. We can find the first two responses to a unit impulse at $n = 0$ from (5.20)

n	$x[n]$	$h[n-1]$	$h[n]$
0	1	0	1/5
1	0	1/5	$-2/25$

The eigenvalues are -1 and 0.6. So the impulse response is $h[n] = (K_1(-1)^n + K_2(0.6)^n)u[n]$. Evaluating the constants

$$\left\{\begin{array}{l} h[0] = K_1 + K_2 = 1/5 \\ h[1] = -K_1 + 0.6 K_2 = -2/25 \end{array}\right\} \Rightarrow K_1 = 0.125, K_2 = 0.075$$

and the impulse response is

$$h[n] = (0.125(-1)^n + 0.075(0.6)^n)u[n].$$

∎

DISCRETE-TIME CONVOLUTION

Derivation

To demonstrate discrete-time convolution, suppose that an LTI system is excited by a signal $x[n] = \delta[n] + \delta[n-1]$ and that its impulse response is $h[n] = (0.7788)^n u[n]$ (Figure 5.28).

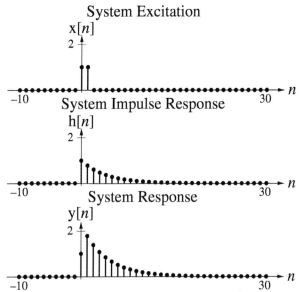

Figure 5.28
System excitation x[n], system impulse response h[n] and system response y[n]

The excitation for any discrete-time system is made up of a sequence of impulses with different strengths, occurring at different times. Therefore, invoking linearity and time invariance, the response of an LTI system will be the sum of all the individual responses to the individual impulses. Since we know the response of the system to a single unit impulse occurring at $n = 0$, we can find the responses to the individual impulses by shifting and scaling the unit-impulse response appropriately.

In the example, the first nonzero impulse in the excitation occurs at $n = 0$ and its strength is one. Therefore the system will respond to this with exactly its impulse response. The second nonzero impulse occurs at $n = 1$ and its strength is also one. The response of the system to this single impulse is the impulse response, except delayed by one in discrete time. So, by the additivity and time-invariance properties of LTI systems, the overall system response to $x[n] = \delta[n] + \delta[n-1]$ is

$$y[n] = (0.7788)^n u[n] + (0.7788)^{n-1} u[n-1]$$

(see Figure 5.28).

Let the excitation now be $x[n] = 2\delta[n]$. Then, since the system is LTI and the excitation is an impulse of strength two occurring at $n = 0$ by the homogeneity property of LTI systems the system response is twice the impulse response or $y[n] = 2(0.7788)^n u[n]$.

Now let the excitation be the one illustrated in Figure 5.29 while the impulse response remains the same. The responses to the four impulses beginning at $n = -5$ are graphed in Figure 5.30.

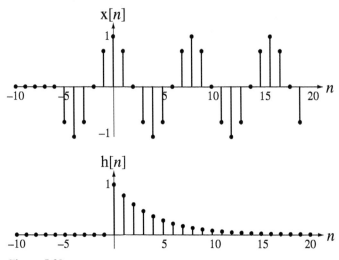

Figure 5.29
A sinusoid applied at time $n = -5$ and the system impulse response

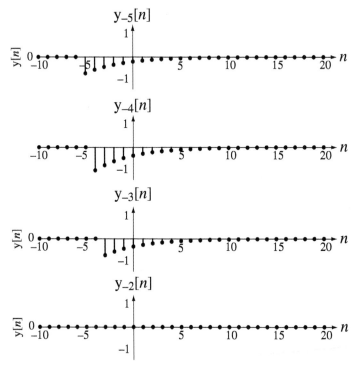

Figure 5.30
System responses to the impulses $x[-5]$, $x[-4]$, $x[-3]$ and $x[-2]$

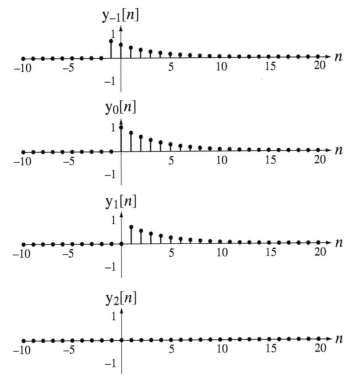

Figure 5.31
System responses to the impulses x[−1], x[0], x[1] and x[2]

Figure 5.31 illustrates the next four impulse responses.

When we add all the responses to all the impulses, we get the total system response to the total system excitation (Figure 5.32).

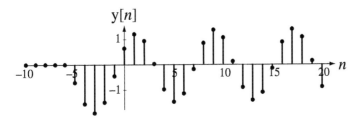

Figure 5.32
The total system response

Notice that there is an initial transient response, but the response settles down to a sinusoid after a few discrete-time units. The forced response of any stable LTI system excited by a sinusoid is another sinusoid of the same frequency but generally with a different amplitude and phase.

We have seen graphically what happens. Now it is time to see analytically what happens. The total system response can be written as

$$y[n] = \cdots x[-1]h[n+1] + x[0]h[n] + x[1]h[n-1] + \cdots$$

or

$$y[n] = \sum_{m=-\infty}^{\infty} x[m]\, h[n-m]\,.\tag{5.21}$$

The result (5.21) is called the **convolution sum** expression for the system response. In words, it says that the value of the response y at any discrete time n can be found by summing all the products of the excitation x at discrete times m with the impulse response h at discrete times $n - m$ for m ranging from negative to positive infinity. To find a system response we only need to know the system's impulse response and we can find its response to any arbitrary excitation. For an LTI system, the impulse response of the system is a complete description of how it responds to any signal. So we can imagine first testing a system by applying an impulse to it and recording the response. Once we have it, we can compute the response to any signal. This is a powerful technique. In system analysis we only have to solve the difference equation for the system once for the simplest possible nonzero input signal, a unit impulse and then, for any general signal, we can find the response by convolution.

Compare the convolution integral for continuous-time signals with the convolution sum for discrete-time signals,

$$y(t) = \int_{-\infty}^{\infty} x(\tau) h(t-\tau) d\tau \quad \text{and} \quad y[n] = \sum_{m=-\infty}^{\infty} x[m]\, h[n-m].$$

In each case one of the two signals is time-reversed and shifted and then multiplied by the other. Then, for continuous-time signals, the product is integrated to find the total area under the product. For discrete-time signals the product is summed to find the total value of the product.

Graphical and Analytical Examples of Convolution

Although the convolution operation is completely defined by (5.21), it is helpful to explore some graphical concepts that aid in actually performing convolution. The two functions that are multiplied and then summed over $-\infty < m < \infty$ are $x[m]$ and $h[n-m]$. To illustrate the idea of graphical convolution let the two functions $x[n]$ and $h[n]$ be the simple functions illustrated in Figure 5.33.

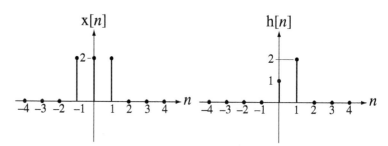

Figure 5.33
Two functions

Since the summation index in (5.21) is m, the function $h[n-m]$ should be considered a function of m for purposes of performing the summation in (5.21). With that point of view, we can imagine that $h[n-m]$ is created by two transformations, $m \to -m$, which changes $h[m]$ to $h[-m]$ and then $m \to m - n$, which changes $h[-m]$ to $h[-(m-n)] = h[n-m]$. The first transformation $m \to -m$ forms the discrete-time reverse of $h[m]$,

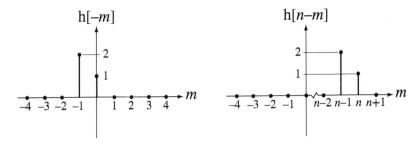

Figure 5.34
h[−m] and h[n−m] versus m

and the second transformation $m \to m - n$ shifts the already-time-reversed function n units to the right (Figure 5.34).

Now, realizing that the convolution result is $y[n] = \sum_{m=-\infty}^{\infty} x[m]h[n - m]$, the process of graphing the convolution result $y[n]$ versus n is to pick a value of n and do the operation $\sum_{m=-\infty}^{\infty} x[m]h[n - m]$ for that n, plot that single numerical result for $y[n]$ at that n and then repeat the whole process for each n. Every time a new n is chosen, the function $h[n - m]$ shifts to a new position, $x[m]$ stays right where it is because there is no n in $x[m]$ and the summation $\sum_{m=-\infty}^{\infty} x[m]h[n - m]$ is simply the sum of the products of the values of $x[m]$ and $h[n - m]$ for that choice of n. Figure 5.35 is an illustration of this process.

For all values of n not represented in Figure 5.35, $y[n] = 0$, so we can now graph $y[n]$ as illustrated in Figure 5.36.

It is very common in engineering practice for both signals being convolved to be zero before some finite time. Let x be zero before $n = n_x$ and let h be zero before $n = n_h$. The convolution sum is

$$x[n] * h[n] = \sum_{m=-\infty}^{\infty} x[m]h[n - m].$$

Since x is zero before $n = n_x$ all the terms in the summation for $m < n_x$ are zero and

$$x[n] * h[n] = \sum_{m=n_x}^{\infty} x[m]h[n - m].$$

Also, when $n - m < n_h$, the h terms are zero. That puts an upper limit on m of $n - n_h$ and

$$x[n] * h[n] = \sum_{m=n_x}^{n-n_h} x[m]h[n - m].$$

For those n's for which $n - n_h < n_x$, the lower summation limit is greater than the upper summation limit and the convolution result is zero. So it would be more complete and accurate to say that the convolution result is

$$x[n] * h[n] = \begin{cases} \sum_{m=n_x}^{n-n_h} x[m]h[n - m], & n - n_h \geq n_x \\ 0, & n - n_h < n_x. \end{cases}$$

194 Chapter 5 Time-Domain System Analysis

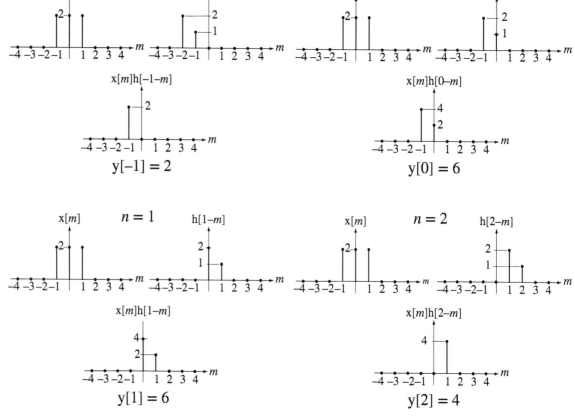

Figure 5.35
y[n] for n = −1, 0, 1 and 2

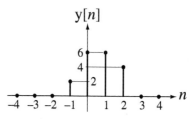

Figure 5.36
Graph of y[n]

EXAMPLE 5.8

Response of a moving-average digital filter

A moving-average digital filter has an impulse response of the form

$$h[n] = (u[n] - u[n - N])/N.$$

Find the response of a moving-average filter with $N = 8$ to $x[n] = \cos(2\pi n/16)$. Then change the excitation to $x[n] = \cos(2\pi n/8)$ and find the new response.

Using convolution, the response is

$$y[n] = x[n] * h[n] = \cos(2\pi n/16) * (u[n] - u[n - 8])/8.$$

Applying the definition of the convolution sum,

$$y[n] = \frac{1}{8} \sum_{m=-\infty}^{\infty} \cos(2\pi m/16)(u[n - m] - u[n - m - 8]).$$

The effect of the two unit sequence functions is to limit the summation range.

$$y[n] = \frac{1}{8} \sum_{m=n-7}^{n} \cos(2\pi m/16)$$

Using the trigonometric identity $\cos(x) = \dfrac{e^{jx} + e^{-jx}}{2}$

$$y[n] = \frac{1}{16} \sum_{m=n-7}^{n} (e^{j2\pi m/16} + e^{-j2\pi m/16})$$

Let $q = m - n + 7$. Then

$$y[n] = \frac{1}{16} \sum_{q=0}^{7} (e^{j2\pi(q+n-7)/16} + e^{-j2\pi(q+n-7)/16})$$

$$y[n] = \frac{1}{16}\left(e^{j2\pi(n-7)/16} \sum_{q=0}^{7} e^{j2\pi q/16} + e^{-j2\pi(n-7)/16} \sum_{q=0}^{7} e^{-j2\pi q/16}\right)$$

The summation of a geometric series with N terms is

$$\sum_{n=0}^{N-1} r^n = \begin{cases} N, & r = 1 \\ \dfrac{1 - r^N}{1 - r}, & r \neq 1 \end{cases}$$

This formula works for any complex value of r. Therefore, summing these geometric series each of length 8,

$$y[n] = \frac{1}{16}\left(e^{j\pi(n-7)/8} \underbrace{\frac{1 - e^{j\pi}}{1 - e^{j\pi/8}}}_{=2} + e^{-j\pi(n-7)/8} \underbrace{\frac{1 - e^{-j\pi}}{1 - e^{-j\pi/8}}}_{=2}\right)$$

Finding a common denominator and simplifying,

$$y[n] = \frac{1}{8}\left(\frac{e^{j\pi(n-7)/8}}{1 - e^{j\pi/8}} + \frac{e^{-j\pi(n-7)/8}}{1 - e^{-j\pi/8}}\right) = \frac{1}{8} \frac{\cos(\pi(n-7)/8) - \cos(\pi(n-8)/8)}{1 - \cos(\pi/8)}$$

Then, using the periodicity of the cosine, $\cos(\pi(n - 8)/8) = \cos(\pi n/8)$ and

$$y[n] = 1.6421[\cos(\pi(n - 7)/8) + \cos(\pi n/8)]$$

Now letting $x[n] = \cos(2\pi n/8)$, the process is essentially the same except for the period of the cosine. The results are

$$y[n] = x[n] * h[n] = \cos(2\pi n/8) * (u[n] - u[n - 8])/8.$$

$$y[n] = \frac{1}{16} \sum_{m=n-7}^{n} (e^{j2\pi m/8} + e^{-j2\pi m/8})$$

$$y[n] = \frac{1}{16}\left(e^{j2\pi(n-7)/8}\sum_{q=0}^{7}e^{j2\pi q/8} + e^{-j2\pi(n-7)/8}\sum_{q=0}^{7}e^{-j2\pi q/8}\right)$$

$$y[n] = \frac{1}{16}\left(e^{j\pi(n-7)/4}\frac{1-e^{j2\pi}}{1-e^{j2\pi/8}} + e^{-j\pi(n-7)/4}\frac{1-e^{-j2\pi}}{1-e^{-j2\pi/8}}\right) = 0, \text{ (because } e^{j2\pi} = e^{-j2\pi} = 1\text{)}.$$

If the averaging time of the moving-average filter is exactly an integer number of periods of the sinusoid, the response is zero because the average value of any sinusoid over any integer number of periods is zero. Otherwise, the response is nonzero.

■

Convolution Properties
Convolution in discrete time, just as in continuous time, is indicated by the operator $*$.

$$y[n] = x[n] * h[n] = \sum_{m=-\infty}^{\infty} x[m]h[n-m]. \tag{5.22}$$

The properties of convolution in discrete time are similar to the properties in continuous time.

$$x[n] * A\delta[n - n_0] = Ax[n - n_0] \tag{5.23}$$

$$y[n - n_0] = x[n] * h[n - n_0] = x[n - n_0] * h[n]. \tag{5.24}$$

The commutativity, associativity, distributivity, differencing and sum properties of the convolution sum are proven in Web Appendix E and are summarized below.

Commutativity Property $\quad x[n] * y[n] = y[n] * x[n]$

Associativity Property	$(x[n] * y[n]) * z[n] = x[n] * (y[n] * z[n])$
Distributivity Property	$(x[n] + y[n]) * z[n] = x[n] * z[n] + y[n] * z[n]$
If $y[n] = x[n] * h[n]$, then	
Differencing Property	$y[n] - y[n-1] = x[n] * (h[n] - h[n-1])$
Sum Property	Sum of y = (Sum of x) × (Sum of h)

For a convolution sum to converge, both signals being convolved must be bounded and at least one of them must be absolutely summable.

Numerical Convolution
Discrete-Time Numerical Convolution MATLAB has a command conv that computes a convolution sum. The syntax is y = conv(x,h) where x and h are vectors of values of discrete-time signals and y is the vector containing the values of the convolution of x with h. Of course, MATLAB cannot actually compute an infinite sum as indicated by (5.22). MATLAB can only convolve time-limited signals, and the vectors x and h should contain all the nonzero values of the signals they represent. (They can also contain extra zero values if desired.) If the time of the first element in x is n_{x0} and the time of the first element of h is n_{h0}, the time of the first element of y is $n_{x0} + n_{h0}$.

If the time of the last element in x is n_{x1} and the time of the last element in h is n_{h1}, the time of the last element in y is $n_{x1} + n_{h1}$. The length of x is $n_{x1} - n_{x0} + 1$ and the length of h is $n_{h1} - n_{h0} + 1$. So the extent of y is in the range $n_{x0} + n_{h0} \leq n < n_{x1} + n_{h1}$ and its length is

$$n_{x1} + n_{h1} - (n_{x0} + n_{h0}) + 1 = \underbrace{n_{x1} - n_{x0} + 1}_{\text{length of x}} + \underbrace{n_{h1} - n_{h0} + 1}_{\text{length of h}} - 1.$$

So the length of y is one less than the sum of the lengths of x and h.

EXAMPLE 5.9

Computing a convolution sum with MATLAB

Let $x[n] = u[n-1] - u[n-6]$ and $h[n] = \text{tri}((n-6)/4)$. Find the convolution sum $x[n] * h[n]$ using the MATLAB conv function.

$x[n]$ is time limited to the range $1 \leq n \leq 5$ and $h[n]$ is time limited to the range $3 \leq n \leq 9$. Therefore any vector describing $x[n]$ should be at least five elements long and any vector describing $h[n]$ should be at least seven elements long. Let's put in some extra zeros, compute the convolution and graph the two signals and their convolutions using the following MATLAB code whose output is illustrated in Figure 5.37.

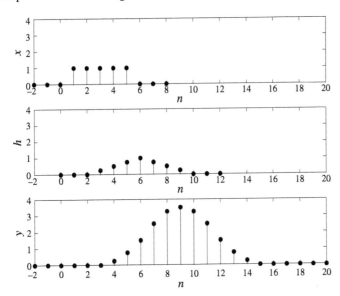

Figure 5.37
Excitation, impulse response and response of a system found using the MATLAB conv command

```
nx = -2:8 ; nh = 0:12;            % Set time vectors for x and h
x = usD(n-1) - usD(n-6);          % Compute values of x
h = tri((nh-6)/4);                % Compute values of h
y = conv(x,h);                    % Compute the convolution of x with h
%
% Generate a discrete-time vector for y
%
ny = (nx(1) + nh(1)) + (0:(length(nx) + length(nh) - 2)) ;
```

```
%
%   Graph the results
%
subplot(3,1,1) ; stem(nx,x,'k','filled');
xlabel('n') ; ylabel('x'); axis([-2,20,0,4]);
subplot(3,1,2) ; stem(nh,h,'k','filled') ;
xlabel('n') ; ylabel('h'); axis([-2,20,0,4]);
subplot(3,1,3) ; stem(ny,y,'k','filled');
xlabel('n') ; ylabel('y'); axis([-2,20,0,4]);
```

Continuous-Time Numerical Convolution At this point a natural question arises. Since there is no built-in MATLAB function for doing a convolution integral, can we do a convolution integral using the **conv** function? The short answer is no. But if we can accept a reasonable approximation (and engineers usually can), the longer answer is yes, approximately. We can start with the convolution integral

$$y(t) = x(t) * h(t) = \int_{-\infty}^{\infty} x(\tau) h(t-\tau) d\tau.$$

Approximate $x(t)$ and $h(t)$ each as a sequence of rectangles of width T_s.

$$x(t) \cong \sum_{n=-\infty}^{\infty} x(nT_s) \text{rect}\left(\frac{t - nT_s - T_s/2}{T_s}\right) \quad \text{and} \quad h(t) \cong \sum_{n=-\infty}^{\infty} h(nT_s) \text{rect}\left(\frac{t - nT_s - T_s/2}{T_s}\right)$$

The integral can be approximated at discrete points in time as

$$y(nT_s) \cong \sum_{m=-\infty}^{\infty} x(mT_s) h((n-m)T_s) T_s.$$

This can be expressed in terms of a convolution sum as

$$y(nT_s) \cong T_s \sum_{m=-\infty}^{\infty} x[m] h[n-m] = T_s x[n] * h[n] \qquad (5.25)$$

where $x[n] = x(nT_s)$ and $h[n] = h(nT_s)$, and the convolution integral can be approximated as a convolution sum under the same criteria as in the use of the **conv** function to do convolution sums. For the convolution integral to converge, $x(t)$ or $h(t)$ (or both) must be an energy signal. Let $x(t)$ be nonzero only in the time interval $n_{x0} T_s \leq t < n_{x1} T_s$ and let $h(t)$ be nonzero only in the time interval $n_{h0} T_s \leq t < n_{h1} T_s$. Then $y(t)$ is nonzero only in the time interval $(n_{x0} + n_{h0}) T_s \leq n < (n_{x1} + n_{h1}) T_s$ and the values of $T_s x[n] * h[n]$ found using the **conv** function cover that range. To get a reasonably good approximate convolution result, T_s should be chosen such that the functions $x(t)$ and $h(t)$ don't change much during that time interval.

EXAMPLE 5.10

Graphing the convolution of two continuous-time signals using the MATLAB conv function

Graph the convolution $y(t) = \text{tri}(t) * \text{tri}(t)$.

5.3 Discrete Time

Although this convolution can be done analytically, it is rather tedious, so this is a good candidate for an approximate convolution using numerical methods, namely the conv function in MATLAB. The slopes of these two functions are both either plus or minus one. To make a reasonably accurate approximation, choose the time between samples to be 0.01 seconds, which means that the functions change value by no more than 0.01 between any two adjacent samples. Then, from (5.25)

$$y(0.01n) \cong 0.01 \sum_{m=-\infty}^{\infty} \text{tri}(0.01m)\,\text{tri}(0.01(n-m)).$$

The limits on the nonzero parts of the functions are $-1 \le t < 1$, which translate into limits on the corresponding discrete-time signals of $-100 \le n < 100$. A MATLAB program that accomplishes this approximation is below.

```
%       Program to do a discrete-time approximation of the
%       convolution of two unit triangles
%       Convolution computations

Ts = 0.01;                              % Time between samples
nx = [-100:99]' ; nh = nx ;             % Discrete time vectors for
                                        % x and h
x = tri(nx*Ts) ; h = tri(nh*Ts) ;       % Generate x and h
ny = [nx(1)+nh(1):nx(end)+nh(end)]';    % Discrete time vector for y
y = Ts*conv(x,h) ;                      % Form y by convolving x and h

% Graphing and annotation

p = plot(ny*Ts,y,'k') ; set(p,'LineWidth',2); grid on ;
xlabel('Time, {\itt} (s)','FontName','Times','FontSize',18) ;
ylabel('y({\itt})','FontName','Times','FontSize',18) ;
title('Convolution of Two Unshifted Unit Triangle Functions',...
    'FontName','Times','FontSize',18);
set(gca,'FontName','Times','FontSize',14);
```

The graph produced is Figure 5.38.

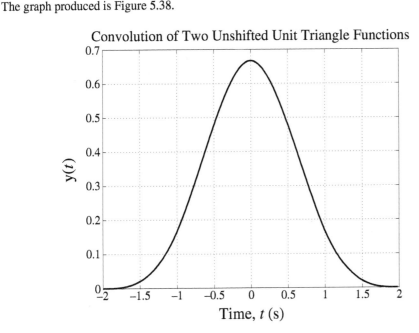

Figure 5.38
Continuous-time convolution approximation using numerical methods

This graphical result agrees very closely with the analytical solution

$$y(t) = (1/6)\begin{bmatrix}(t+2)^3u(t+2) - 4(t+1)^3u(t+1) + 6t^3u(t)\\ -4(t-1)^3u(t-1) + (t-2)^3u(t-2)\end{bmatrix}.$$

Stability and Impulse Response

Stability was generally defined in Chapter 4 by saying that a stable system has a bounded output signal when excited by any bounded input signal. We can now find a way to determine whether a system is stable by examining its impulse response. The convolution of two signals converges if both of them are bounded and at least one of them is absolutely summable. The response $y[n]$ of a system $x[n]$ is $y[n] = x[n] * h[n]$. If $x[n]$ is bounded $y[n]$ is bounded if $h[n]$ is absolutely summable (and therefore also bounded). That is, if $\sum_{n=-\infty}^{\infty}|h[n]|$ is bounded.

A system is BIBO stable if its impulse response is absolutely summable.

System Connections

Two very common interconnections of systems are the cascade connection and the parallel connection (Figure 5.39 and Figure 5.40).

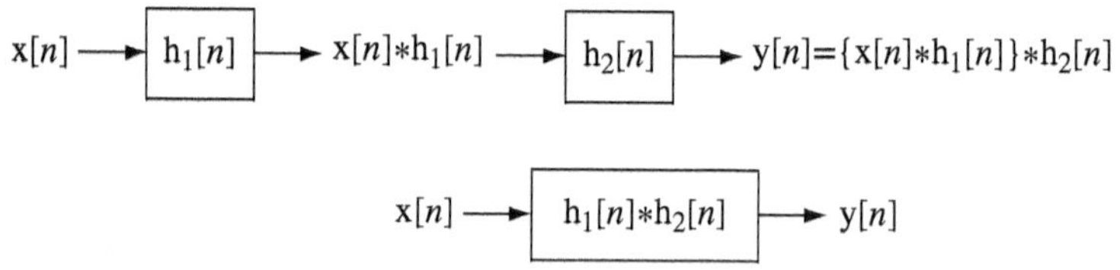

x[n]
h1[n]
x[n]*h1[n]
h2[n]
y[n]={x[n]*h1[n]}*h2[n]
x[n]
h1[n]*h2[n]
y[n]

Figure 5.39
Cascade connection of two systems

x[n]*h1[n]
h1[n]
x[n]
+
y[n]=x[n]*h1[n]+x[n]*h2[n]=x[n]*{h1[n]+h2[n]}
h2[n]
x[n]*h2[n]
x[n]
h1[n]+h2[n]
y[n]

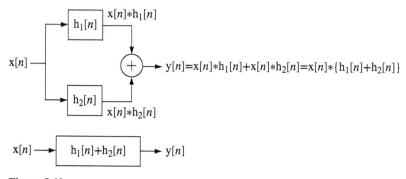

Figure 5.40
Parallel connection of two systems

Using the associativity property of convolution we can show that the cascade connection of two systems can be considered as one system whose impulse response is the convolution of the two individual impulse responses of the two systems. Using the

Unit-Sequence Response and Impulse Response

The response of any LTI system is the convolution of the excitation with the impulse response

$$y[n] = x[n] * h[n] = \sum_{m=-\infty}^{\infty} x[m]h[n-m].$$

Let the excitation be a unit sequence and let the response to a unit sequence be designated $h_{-1}[n]$. Then

$$h_{-1}[n] = u[n] * h[n] = \sum_{m=-\infty}^{\infty} u[m]h[n-m] = \sum_{m=0}^{\infty} h[n-m]$$

Let $q = n - m$. Then

$$h_{-1}[n] = \sum_{q=n}^{-\infty} h[q] = \sum_{q=-\infty}^{n} h[q].$$

So the response of a discrete-time LTI system excited by a unit sequence is the accumulation of the impulse response. Just as the unit sequence is the accumulation of the impulse response, the unit-sequence *response* is the accumulation of the unit-impulse *response*. The subscript on $h_{-1}[n]$ indicates the number of differences. In this case there is -1 difference or one accumulation in going from the impulse response to the unit-sequence response. This relationship holds for any excitation. If any excitation is changed to its accumulation, the response also changes to its accumulation, and if the excitation is changed to its first backward difference, the response is also changed to its first backward difference.

EXAMPLE 5.11

Finding the response of a system using convolution

Find the response of the system in Figure 5.41 to the excitation in Figure 5.42.

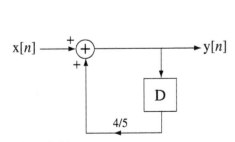

Figure 5.41
A system

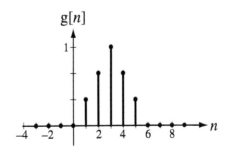

Figure 5.42
Excitation of the system

Chapter 5 Time-Domain System Analysis

First we need the impulse response of the system. We could find it directly using the methods presented earlier but, in this case, since we have already found its unit-sequence response $h_{-1}[n] = [5 - 4(4/5)^n]u[n]$ (see Chapter 4, section on discrete-time system properties, pages 152 and 153), we can find the impulse response as the first backward difference of the unit-sequence response $h[n] = h_{-1}[n] - h_{-1}[n - 1]$. Combining equations,

$$h[n] = [5 - 4(4/5)^n]u[n] - [5 - 4(4/5)^{n-1}]u[n - 1].$$

$$h[n] = 5\underbrace{(u[n] - u[n - 1])}_{=\delta[n]} - 4(4/5)^{n-1}[(4/5)u[n] - u[n - 1]]$$

$$h[n] = 5\delta[n] - 4(4/5)^n\underbrace{\delta[n]}_{=\delta[n]} + (4/5)^n u[n - 1]$$

$$h[n] = (4/5)^n u[n]$$

All that remains is to perform the convolution. We can do that using the MATLAB program below.

```
%   Program to demonstrate discrete-time convolution
nx = -5:15 ;            % Set a discrete-time vector for the excitation
x = tri((n-3)/3;        % Generate the excitation vector
nh = 0:20 ;             % Set a discrete-time vector for the impulse
                        % response
%   Generate the impulse response vector
h = ((4/5).^nh).*usD(nh);
%   Compute the beginning and ending discrete times for the system
%   response vector from the discrete-time vectors for the
    excitation and the impulse response
nymin = nx(1) + nh(1); nymax = nx(length(nx)) + length(nh);
ny = nymin:nymax-1;
%   Generate the system response vector by convolving the
    excitation with the impulse response
y = conv(x,h);
%   Graph the excitation, impulse response and system response, all
%   on the same time scale for comparison
%   Graph the excitation
subplot(3,1,1); p = stem(nx,x,'k','filled');
set(p,'LineWidth',2,'MarkerSize',4);
axis([nymin,nymax,0,3]);
xlabel('n'); ylabel('x[n]');
%   Graph the impulse response
subplot(3,1,2); p = stem(nh,h,'k','filled');
set(p,'LineWidth',2,'MarkerSize',4);
axis([nymin,nymax,0,3]);
xlabel('n'); ylabel('h[n]');
%   Graph the system response
subplot(3,1,3); p = stem(ny,y,'k','filled');
```

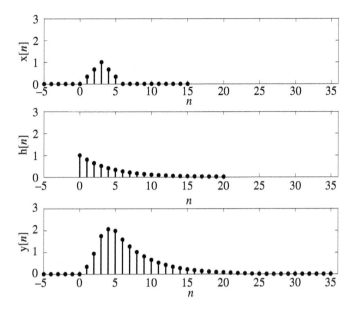

Figure 5.43
Excitation, impulse response and system response

```
set(p,'LineWidth',2,'MarkerSize',4);
axis([nymin,nymax,0,3]);
xlabel('n'); ylabel('y[n]');
```

The three signals as graphed by MATLAB are illustrated in Figure 5.43.

∎

Complex Exponential Excitation and the Transfer Function

In engineering practice the most common form of description of a discrete-time system is a difference equation or a system of difference equations. Consider the general form of a difference equation for a discrete-time system

$$\sum_{k=0}^{N} a_k y[n-k] = \sum_{k=0}^{M} b_k x[n-k]. \quad (5.26)$$

A complex exponential excitation causes a complex exponential response in continuous-time systems and the same is true for discrete-time systems. Therefore, if $x[n] = Xz^n$, $y[n]$ has the form $y[n] = Yz^n$ where X and Y are complex constants. Then, in the difference equation,

$$x[n-k] = Xz^{n-k} = z^{-k}Xz^n \quad \text{and} \quad y[n-k] = z^{-k}Yz^n$$

and (5.26) can be written in the form

$$\sum_{k=0}^{N} a_k z^{-k} Y z^n = \sum_{k=0}^{M} b_k z^{-k} X z^n.$$

Chapter 5 Time-Domain System Analysis

The Xz^n and Yz^n can be factored out, leading to

$$Yz^n \sum_{k=0}^{N} a_k z^{-k} = Xz^n \sum_{k=0}^{M} b_k z^{-k} \Rightarrow \frac{Yz^n}{Xz^n} = \frac{Y}{X} = \frac{\sum_{k=0}^{M} b_k z^{-k}}{\sum_{k=0}^{N} a_k z^{-k}}.$$

This ratio Y/X is a ratio of polynomials in z. This is the transfer function for discrete-time systems symbolized by H(z). That is,

$$\boxed{H(z) = \frac{\sum_{k=0}^{M} b_k z^{-k}}{\sum_{k=0}^{N} a_k z^{-k}} = \frac{b_0 + b_1 z^{-1} + b_2 z^{-2} + \cdots + b_M z^{-M}}{a_0 + a_1 z^{-1} + a_2 z^{-2} + \cdots + a_N z^{-N}}} \qquad (5.27)$$

and $y[n] = Yz^n = H(z)Xz^n = H(z)x[n]$. The transfer function can then be written directly from the difference equation and, if the difference equation describes the system, so does the transfer function. By multiplying numerator and denominator of (5.27) by z^N we can express it in the alternate form

$$\boxed{H(z) = \frac{\sum_{k=0}^{M} b_k z^{-k}}{\sum_{k=0}^{N} a_k z^{-k}} = z^{N-M} \frac{b_0 z^M + b_1 z^{M-1} + \cdots + b_{M-1} z + b_M}{a_0 z^N + a_1 z^{N-1} + \cdots + a_{N-1} z + a_N}} \qquad (5.28)$$

The two forms are equivalent but either form may be more convenient in certain situations.

We can also find the system response using convolution. The response $y[n]$ of an LTI system with impulse response $h[n]$ to a complex exponential excitation $x[n] = Xz^n$ is

$$y[n] = h[n] * Xz^n = X \sum_{m=-\infty}^{\infty} h[m] z^{n-m} = \underbrace{Xz^n}_{=x[n]} \sum_{m=-\infty}^{\infty} h[m] z^{-m}.$$

Equating the two forms of the response

$$H(z)Xz^n = Xz^n \sum_{m=-\infty}^{\infty} h[m]z^{-m} \Rightarrow H(z) = \sum_{m=-\infty}^{\infty} h[m]z^{-m}$$

which shows the relationship between the transfer function and the impulse response of discrete-time LTI systems. The summation $\sum_{m=-\infty}^{\infty} h[m]z^{-m}$ will be identified in Chapter 9 as the z transform of $h[n]$.

Frequency Response
The variable z in the complex exponential z^n is, in general, complex valued. Consider the special case in which z is confined to the unit circle in the complex plane such that $|z| = 1$. Then z can be expressed as $z = e^{j\Omega}$ where Ω is the real variable representing radian frequency in discrete time, z^n becomes $e^{j\Omega n}$, a discrete-time complex sinusoid $e^{j\Omega} = \cos(\Omega) + j\sin(\Omega)$ and the transfer function of the system H(z) becomes the **frequency response** of the system H($e^{j\Omega}$). From $Yz^n = H(z)Xz^n$, letting $z = e^{j\Omega}$

$$Y e^{j\Omega n} = |Y|e^{j\angle Y} e^{j\Omega n} = H(e^{j\Omega}) X e^{j\Omega n} = |H(e^{j\Omega})| e^{j\angle H(e^{j\Omega})} e^{j\Omega n} |X| e^{j\angle X} e^{j\Omega n}$$

or, dividing through by $e^{j\Omega n}$,

$$|Y|e^{j\angle Y} = |H(e^{j\Omega})||X|e^{j\angle(H(e^{j\Omega})+\angle X)}$$

Equating magnitudes we get $|Y| = |H(e^{j\Omega})||X|$ and equating phases we get $\angle Y = \angle H(e^{j\Omega}) + \angle X$. The function $H(e^{j\Omega})$ is called the *frequency response* of the system because, at any radian frequency Ω, if we know the magnitude and phase of the excitation and the magnitude and phase of the frequency response, we can find the magnitude and phase of the response.

As was true for continuous-time systems, if a complex excitation $x[n]$ is applied to a system and causes a response $y[n]$, then the real part of $x[n]$ causes the real part of $y[n]$ and the imaginary part of $x[n]$ causes the imaginary part of $y[n]$. Therefore, if the actual excitation of a system is $x[n] = A_x \cos(\Omega n + \theta_x)$, we can find the response of the system to an excitation

$$x_C[n] = A_x \cos(\Omega n + \theta_x) + jA_x \sin(\Omega n + \theta_x) = A_x e^{j(\Omega n + \theta_x)}$$

in the form

$$y_C[n] = A_y \cos(\Omega n + \theta_y) + jA_y \sin(\Omega n + \theta_y) = A_y e^{j(\Omega n + \theta_y)}$$

and we can take the real part $y[n] = A_y \cos(\Omega n + \theta_y)$ as the response to the real excitation $x[n] = A_x \cos(\Omega n + \theta_x)$. Using $|Y| = |H(j\omega)||X|$ and $\angle Y = \angle H(j\omega) + \angle X$, we get

$$A_y = |H(e^{j\Omega})|A_x \quad \text{and} \quad \theta_y = \angle H(e^{j\Omega}) + \theta_x.$$

EXAMPLE 5.12

Transfer Function and Frequency Response

An LTI system is described by the difference equation

$$y[n] - 0.75\,y[n-1] + 0.25\,y[n-2] = x[n].$$

(a) Find its transfer function.

For this difference equation of the form $\sum_{k=0}^{N} a_k y[n-k] = \sum_{k=0}^{M} b_k x[n-k]$, $N = 2$, $M = 0$,

$a_0 = 0.25$, $a_1 = -0.75$, $a_2 = 1$ and $b_0 = 1$.

Therefore the transfer function is

$$H(z) = \frac{1}{z^2 - 0.75z + 0.25}.$$

(b) If $x[n] = Xe^{j0.5n}$, $y(t) = Ye^{j0.5n}$ and $X = 12e^{-j\pi/4}$, find the magnitude and phase of Y.
The frequency response is

$$H(e^{j\Omega}) = \frac{1}{(e^{j\Omega})^2 - 0.75(e^{j\Omega}) + 0.25} = \frac{1}{e^{j2\Omega} - 0.75e^{j\Omega} + 0.25} \quad \text{(see Figure 5.44)}.$$

The radian frequency is $\Omega = 0.5$. Therefore,

$$H(e^{j\Omega}) = \frac{1}{e^j - 0.75e^{j/2} + 0.25} = 2.001 e^{-j1.303}$$

$$|Y| = |H(e^{j0.5})| \times 12 = 2.001 \times 12 = 24.012$$

$$\measuredangle Y = \measuredangle H(e^{j0.5}) - \pi/4 = -1.3032 - \pi/4 = -2.0886 \text{ radians}$$

(c) If $x[n] = 25\cos(2\pi n/5)$ and $y[n] = A_y \cos(2n/5 + \theta_y)$, find A_y and θ_y.

$$A_y = |H(e^{j\pi/9})| A_x = 1.2489 \times 25 = 31.2225$$

and

$$\theta_y = \measuredangle H(e^{j2\pi/5}) + \theta_x = 2.9842 + 0 = 2.9842 \text{ radians}$$

```
dW = 2*pi/100 ;            %   Increment in discrete-time radian frequency
                           %   for sampling the frequency response
W = -2*pi:dW:2*pi ;        %   Discrete-time radian frequency vector for
                           %   graphing the frequency response
%   Compute the frequency response
H = 1./(exp(j*2*W) - 0.75*exp(j*W) + 0.25) ;
close all ; figure('Position',[20,20,1200,800]) ;
subplot(2,1,1) ;
ptr = plot(W,abs(H),'k') ; % Graph the magnitude of the frequency response
grid on ;
set(ptr,'LineWidth',2) ;
xlabel('\Omega','FontName','Times','FontSize',36) ;
ylabel('|H({\ite^{{\itj}\Omega}})|','FontName','Times','FontSize',36) ;
title('Frequency Response Magnitude','FontName','Times','FontSize',36) ;
set(gca,'FontName','Times','FontSize',24) ;
axis([-2*pi,2*pi,0,2.5]) ;
subplot(2,1,2) ;
ptr = plot(W,angle(H),'k') ; % Graph the phase of the frequency response
grid on ;
set(ptr,'LineWidth',2) ;
xlabel('\Omega','FontName','Times','FontSize',36) ;
ylabel('Phase of H({\ite^{{\itj}\Omega}})','FontName','Times','FontSize',36) ;
title('Frequency Response Phase','FontName','Times','FontSize',36) ;
set(gca,'FontName','Times','FontSize',24) ;
axis([-2*pi,2*pi,-pi,pi]) ;
```

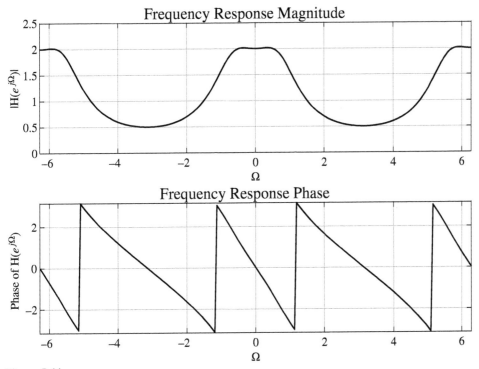

Figure 5.44
Frequency response magnitude and phase

5.4 SUMMARY OF IMPORTANT POINTS

1. Every LTI system is completely characterized by its impulse response.
2. The response of any LTI system to an arbitrary input signal can be found by convolving the input signal with its impulse response.
3. The impulse response of a cascade connection of LTI systems is the convolution of the individual impulse responses.
4. The impulse response of a parallel connection of LTI systems is the sum of the individual impulse responses.
5. A continuous-time LTI system is BIBO stable if its impulse response is absolutely integrable.
6. A discrete-time LTI system is BIBO stable if its impulse response is absolutely summable.

EXERCISES WITH ANSWERS

(Answers to each exercise are in random order.)

Continuous Time

Impulse Response

1. A continuous-time system is described by the differential equation $y'(t) + 6y(t) = x'(t)$.

    (a) Write the differential equation for the special case of impulse excitation and impulse response.

(b) The impulse response is $h(t) = -6e^{-6t}u(t) + \delta(t)$. What is the value of the integral from $t = 0^-$ to $t = 0^+$ of $h(t)$?

(c) What is the value of the integral from $t = 0^-$ to $t = 0^+$ of $h'(t)$?

Answers: -6, $h'(t) + 6h(t) = \delta'(t)$, 1

2. Find the values of these integrals.

(a) $\int_{0^-}^{0^+} 4u(t)dt$

(b) $\int_{0^-}^{0^+} \left[-2e^{-3t}u(t) + 7\delta(t)\right]dt$

(c) $\int_{0^-}^{0^+}\left[\int_{-\infty}^{t} -3\delta(\lambda)d\lambda\right]dt$

Answers: 0, 0, 7

3. Find the impulse responses of the systems described by these equations.

(a) $y'(t) + 5y(t) = x(t)$

(b) $y''(t) + 6y'(t) + 4y(t) = x(t)$

(c) $2y'(t) + 3y(t) = x'(t)$

(d) $4y'(t) + 9y(t) = 2x(t) + x'(t)$

Answers: $h(t) = -(1/16)e^{-9t/4}u(t) + (1/4)\delta(t)$,

$h(t) = -(3/4)e^{-3t/2}u(t) + (1/2)\delta(t)$, $h(t) = e^{-5t}u(t)$,

$h(t) = 0.2237(e^{-0.76t} - e^{-5.23t})u(t)$

4. In the system of Figure E.4 $a = 7$ and $b = 3$.

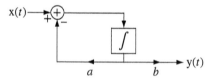

Figure E.4

(a) Write the differential equation describing it.
(b) The impulse response can be written in the form $h(t) = Ke^{\lambda t}u(t)$. Find the numerical values of K and λ.

Answers: $(1/3)y'(t) = x(t) - (7/3)y(t)$, -7, 3

5. A continuous-time system is described by the differential equation,

$$y''(t) + 6y'(t) + 3y(t) = x(t)$$

where x is the excitation and y is the response. Can the impulse response of this system contain

(a) An impulse?
(b) A discontinuity at $t = 0$?
(c) A discontinuous first derivative at $t = 0$?

Answers: Yes, No, No

6. In Figure E.6 is an RC lowpass filter with excitation $v_{in}(t)$ and response $v_{out}(t)$. Let $R = 10\,\Omega$ and $C = 10\,\mu\text{F}$.

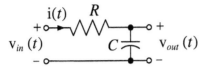

Figure E.6

(a) Write the differential equation for this circuit in terms of $v_{in}(t)$, $v_{out}(t)$, R and C.
(b) Find the impulse response of this system $h(t)$.
(c) Find the numerical value $h(200\,\mu s)$.

Answers: 1353.35, $h(t) = \dfrac{e^{-t/RC}}{RC} u(t)$, $v'_{out}(t) + \dfrac{v_{out}(t)}{RC} = \dfrac{v_{in}(t)}{RC}$

Convolution

7. If $x(t) = 2\,\text{tri}(t/4) * \delta(t-2)$, find the values of

   (a) $x(1)$ (b) $x(-1)$

   Answers: 1/2, 3/2

8. If $y(t) = -3\,\text{rect}(t/2) * \text{rect}((t-3)/2)$, what are the maximum and minimum values of y for all time?

   Answers: 0, 6

9. An LTI system has an impulse response $h(t) = 2e^{-3t}u(t)$.

   (a) Write an expression for $h(t) * u(t)$.
   (b) Let the excitation of the system be $x(t) = u(t) - u(t - 1/3)$. Write an expression for the response $y(t)$.
   (c) Find the numerical value of $y(t)$ at $t = 1/2$.

   Answers: 0.2556, $y(t) = (2/3)[(1 - e^{-3t})u(t) - (1 - e^{-3(t-1/3)})u(t - 1/3)]$,
   $h(t) * u(t) = (2/3)(1 - e^{-3t})u(t)$

10. Let $y(t) = x(t) * h(t)$ and let $x(t) = \text{rect}(t + 4) - \text{rect}(t - 1)$ and let $h(t) = \text{tri}(t - 2) + \text{tri}(t - 6)$. Find the range of times over which $y(t)$ is not zero.

    Answer: $-3.5 < t < 8.5$

11. Find the following values.

    (a) $g(3)$ if $g(t) = 3e^{-2t}u(t) * 4\delta(t - 1)$
    (b) $g(3)$ if $g(t) = e^{-t}u(t) * [\delta(t) - 2\delta(t - 1)]$
    (c) $g(-1)$ if $g(t) = 4\sin\left(\dfrac{\pi t}{8}\right) * \delta(t - 4)$
    (d) $g(1)$ if $g(t) = -5\,\text{rect}\left(\dfrac{t+4}{2}\right) * \delta(3t)$
    (e) If $y(t) = x(t) * h(t)$ and $x(t) = 4\,\text{rect}(t - 1)$ and $h(t) = 3\,\text{rect}(t)$, find $y(1/2)$

    Answers: $-0.2209, -3.696, 0.2198, 5/3, 6$

12. If $x(t) = \text{rect}(t/10) * 3\,\text{rect}((t-1)/8)$, find the following numerical values.

    (a) x(1)   (b) x(5)

    Answers: 24, 15

13. If $x(t) = -5\,\text{rect}(t/2) * [\delta(t+1) + \delta(t)]$, find the values of

    (a) x(1/2)   (b) x(-1/2)   (c) x(-5/2)

    Answers: -10, -5, 0

14. Let $x(t) = 0.5\,\text{rect}\left(\dfrac{t-2}{4}\right)$ and $h(t) = 3\delta(2t) - 5\delta(t-1)$ and let $y(t) = h(t) * x(t)$. Graph y(t).

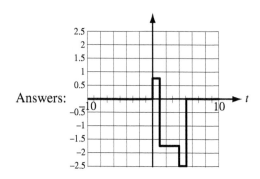

15. Graph g(t).

    (a) $g(t) = \text{rect}(t) * \text{rect}(t/2)$ (b) $g(t) = \text{rect}(t-1) * \text{rect}(t/2)$
    (c) $g(t) = [\text{rect}(t-5) + \text{rect}(t+5)] * [\text{rect}(t-4) + \text{rect}(t+4)]$

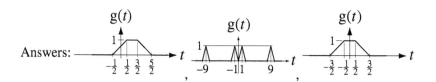

16. Graph these functions.

    (a) $g(t) = \text{rect}(4t)$ (b) $g(t) = \text{rect}(4t) * 4\delta(t)$
    (c) $g(t) = \text{rect}(4t) * 4\delta(t-2)$ (d) $g(t) = \text{rect}(4t) * 4\delta(2t)$
    (e) $g(t) = \text{rect}(4t) * \delta_1(t)$ (f) $g(t) = \text{rect}(4t) * \delta_1(t-1)$
    (g) $g(t) = (1/2)\text{rect}(4t) * \delta_{1/2}(t)$ (h) $g(t) = (1/2)\text{rect}(t) * \delta_{1/2}(t)$

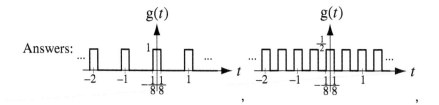

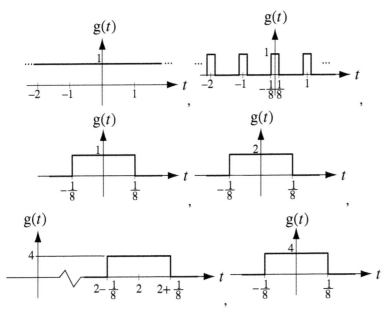

17. Graph these functions.
    (a) $g(t) = \text{rect}(t/2) * [\delta(t+2) - \delta(t+1)]$
    (b) $g(t) = \text{rect}(t) * \text{tri}(t)$
    (c) $g(t) = e^{-t}u(t) * e^{-t}u(t)$
    (d) $g(t) = \left[\text{tri}\left(2\left(t+\frac{1}{2}\right)\right) - \text{tri}\left(2\left(t-\frac{1}{2}\right)\right)\right] * \delta_2(t)$
    (e) $g(t) = [\text{tri}(2(t+1/2)) - \text{tri}(2(t-1/2))] * \delta_1(t)$

Answers:

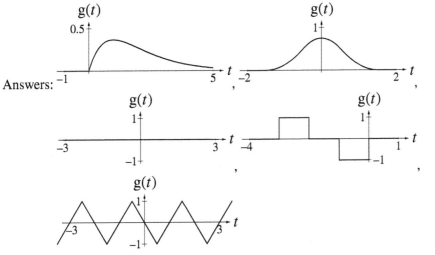

18. A system has an impulse response $h(t) = 4e^{-4t}u(t)$. Find and plot the response of the system to the excitation $x(t) = \text{rect}(2(t-1/4))$.

Answer:

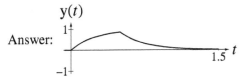

19. Change the system impulse response in Exercise 18 to $h(t) = \delta(t) - 4e^{-4t}u(t)$ and find and plot the response to the same excitation $x(t) = \text{rect}(2(t - 1/4))$.

Answer: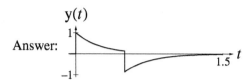

20. Two systems have impulse responses, $h_1(t) = u(t) - u(t - a)$ and $h_2(t) = \text{rect}\left(\frac{t - a/2}{a}\right)$. If these two systems are connected in cascade, find the response $y(t)$ of the overall system to the excitation $x(t) = \delta(t)$.

Answer: $h(t) = 4\text{tri}\left(\frac{t - a}{a}\right)$

21. In the circuit of Figure E.21 the input signal voltage is $v_i(t)$ and the output signal voltage is $v_o(t)$.

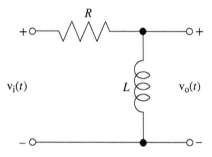

Figure E.21

(a) Find the impulse response in terms of R and L.
(b) If $R = 10\ \text{k}\Omega$ and $L = 100\ \mu\text{H}$ graph the unit-step response.

Answers: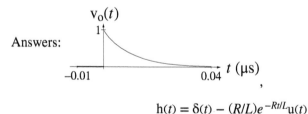

$h(t) = \delta(t) - (R/L)e^{-Rt/L}u(t)$

22. Graph the responses of the systems of Exercise 1 to a unit step.

Answers: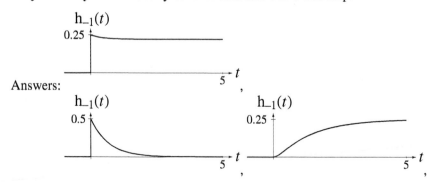

Stability

23. A continuous-time system has an impulse response rect(t) $* [\delta_8(t-1) - \delta_8(t-5)]$ u(t). Is it BIBO stable?

    Answer: No

24. Below are the impulse responses of some LTI systems. In each case determine whether or not the system is BIBO stable.

    (a) $h(t) = \sin(t)u(t)$
    (b) $h(t) = e^{1.2t}\sin(30\pi t)u(t)$
    (c) $h(t) = [\text{rect}(t) * \delta_2(t)]\text{rect}\left(\frac{t}{200}\right)$
    (d) $h(t) = \text{ramp}(t)$
    (e) $h(t) = \delta_1(t)e^{-t/10}u(t)$
    (f) $h(t) = [\delta_1(t) - \delta_1(t-1/2)]u(t)$

    Answers: 4 BIBO Unstable and 2 BIBO Stable

25. Find the impulse responses of the two systems in Figure E.25. Are these systems BIBO stable?

    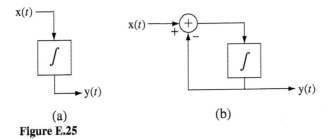

    Figure E.25

    Answers: One BIBO Stable and one BIBO Unstable

    $h(t) = u(t)$
    $h(t) = e^{-t}u(t)$

26. Find the impulse response of the system in Figure E.26. Is this system BIBO stable?

    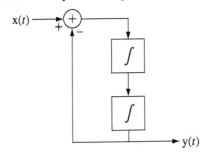

    Figure E.26
    A double-integrator system

    Answer: No

27. Find the impulse response of the system in Figure E.27 and evaluate its BIBO stability.

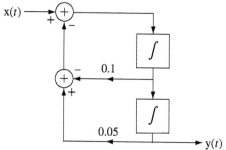

Figure E.27
A two-integrator system

Answers: $4.5893e^{0.05t}\sin(0.2179t)u(t)$. BIBO Unstable

Frequency Response

28. A continuous-time system is described by the differential equation

$$4y'''(t) - 2y''(t) + 3y'(t) - y(t) = 8x''(t) + x'(t) - 4x(t).$$

Its transfer function can be written in the standard form

$$H(s) = \frac{\sum_{k=0}^{M} b_k s^k}{\sum_{k=0}^{N} a_k s^k} = \frac{b_M s^M + b_{M-1} s^{M-1} + \cdots + b_2 s^2 + b_1 s + b_0}{a_N s^N + a_{N-1} s^{N-1} + \cdots + a_2 s^2 + a_1 s + a_0}$$

Find the values of M, N and all the a and b coefficients ($a_N \to a_0$ and $b_M \to b_0$).

Answers: $-1, 8, 1, -4, 3, -2, 3, 4, 2$

29. A continuous-time system is described by $2y'(t) + 4y(t) = -x(t)$, where x is the excitation and y is the response. If $x(t) = Xe^{j\omega t}$ and $y(t) = Ye^{j\omega t}$ and $H(j\omega) = \frac{Y}{X}$ find the numerical value of $H(j2)$.

Answer: $0.1768e^{j2.3562}$

Discrete Time

Impulse Response

30. Find the total numerical solution of this difference equation with initial conditions.

$$y[n] - 0.1y[n-1] - 0.2y[n-2] = 5, \ y[0] = 1, \ y[1] = 4$$

Answer: $y[n] = -6.2223(0.5)^n + 0.0794(-0.4)^n + 7.1429$

31. Find the first three numerical values, starting at time $n = 0$, of the impulse response of the discrete-time system described by the difference equation

$$9y[n] - 3y[n-1] + 2y[n-2] = x[n].$$

Answers: $1/9, 1/27, -1/81$

32. Find the impulse responses of the systems described by these equations.
    (a) $y[n] = x[n] - x[n-1]$
    (b) $25y[n] + 6y[n-1] + y[n-2] = x[n]$
    (c) $4y[n] - 5y[n-1] + y[n-2] = x[n]$
    (d) $2y[n] + 6y[n-2] = x[n] - x[n-2]$

    Answers: $h[n] = \dfrac{\cos(2.214n + 0.644)}{20(5)^n}$,
    $h[n] = \dfrac{(\sqrt{3})^n}{2}\cos(\pi n/2)(u[n] - (1/3)u[n-2])$, $h[n] = \delta[n] - \delta[n-1]$,
    $h[n] = [(1/3) - (1/12)(1/4)^n]u[n]$

33. A discrete-time system is described by the difference equation $y[n] - 0.95y[n-2] = x[n]$ where $x[n]$ is the excitation and $y[n]$ is the response.
    (a) Find these values, $h[0], h[1], h[2], h[3], h[4]$.
    (b) What is the numerical value of $h[64]$?

    Answers: 1, 0.95, 0, 0, 0.9025, 0.194

34. If a discrete-time system is described by $y[n] = \sum\limits_{m=-\infty}^{n-4} x[m]$, graph its impulse response $h[n]$.

    Answer:
    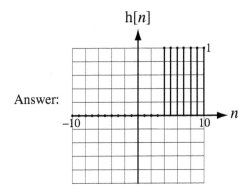

Convolution

35. Two discrete-time signals $x[n]$ and $h[n]$ are graphed in Figure E.35. If $y[n] = x[n] * h[n]$, graph $y[n]$.

    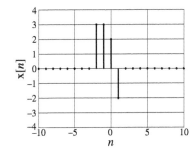
    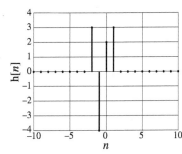

    Figure E.35

Answer: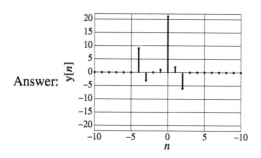

36. For each pair of signals $x_1[n]$ and $x_2[n]$, find the numerical value of $y[n] = x_1[n] * x_2[n]$ at the indicated value of n.

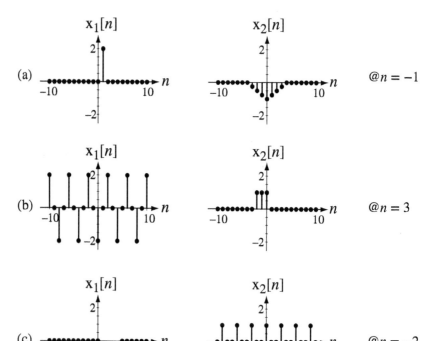

(d) $x_1[n] = -3u[n]$ and $x_2[n] = \text{ramp}[n-1]$ @$n = 1$

Answers: $-2, -4, 0, -1$

37. Find the numerical values of these functions.

(a) If $g[n] = 10\cos(2\pi n/12) * \delta[n+8]$ find $g[4]$
(b) If $g[n] = (u[n+2] - u[n-3]) * (\delta[n-1] - 2\delta[n-2])$ find $g[2]$
(c) If $g[n] = \text{ramp}[n] * u[n]$ find $g[3]$.
(d) If $g[n] = (u[n] - u[n-5]) * \delta_2[n]$ find $g[13]$.
(e) If $y[n] = x[n] * h[n]$ and $x[n] = \text{ramp}[n]$
(f) If $g[n] = 10\cos\left(\dfrac{2\pi n}{12}\right) * \delta[n+8]$ find $g[4]$

(g) If $g[n] = (u[2n + 2] - u[2n - 3]) * (\delta[n - 1] - 2\delta[n - 2])$ find $g[2]$
(h) If $y[n] = x[n] * h[n]$ and $x[n] = \text{ramp}[n]$

Answers: 2, 2, 10, 10, 2, −1, −1, 6

38. If $x[n] = (0.8)^n u[n] * u[n]$, what is the numerical value of $x[3]$?

    Answer: 2.952

39. Graph the convolution $y[n] = x[n] * h[n]$ where $x[n] = u[n] - u[n - 4]$ and $h[n] = \delta[n] - \delta[n - 2]$.

    Answer: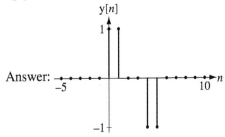

40. The impulse response $h[n]$ of an LTI system is illustrated in Figure E.40. Find the unit sequence response $h_{-1}[n]$ of that system over the same time range.

    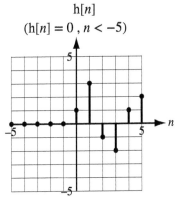

    Figure E.40

    Answer:

n	−5	−4	−3	−2	−1	0	1	2	3	4	5
$h_{-1}[n]$	0	0	0	0	0	1	4	3	1	2	4

41. Given the excitation $x[n] = \sin(2\pi n/32)$ and the impulse response $h[n] = (0.95)^n u[n]$, find a closed-form expression for and plot the system response $y[n]$.

    Answers: $y[n] = 5.0632 \sin(2\pi n/32 - 1.218)$,

    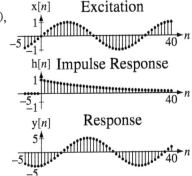

42. Given the excitations x[n] and the impulse responses h[n], use MATLAB to plot the system responses y[n].

   (a) $x[n] = u[n] - u[n-8]$, $h[n] = \sin(2\pi n/8)(u[n] - u[n-8])$

   (b) $x[n] = \sin(2\pi n/8)(u[n] - u[n-8])$, $h[n] = -\sin(2\pi n/8)(u[n] - u[n-8])$

Answers:

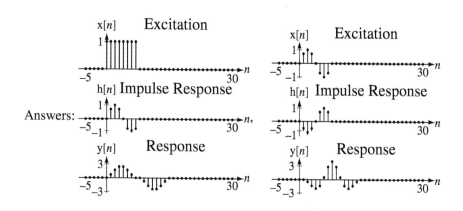

43. Two systems have impulse responses, $h_1[n] = (0.9)^n u[n]$ and $h_2[n] = \delta[n] - (0.9)^n u[n]$. When these two systems are connected in parallel what is the response y[n] of the overall system to the excitation $x[n] = u[n]$?

   Answer: $y[n] = u[n]$

44. A discrete-time system with impulse response $h[n] = 3(u[n] - u[n-4])$ is excited by the signal $x[n] = 2(u[n-2] - u[n-10])$ and the system response is y[n].

   (a) At what discrete time n does the first nonzero value of y[n] occur?
   (b) At what discrete time n does the last nonzero value of y[n] occur?
   (c) What is the maximum value of y[n] over all discrete time?
   (d) Find the signal energy of y[n].

   Answers: 2, 12, 24, 3888

45. Find and plot the unit-sequence responses of the systems in Figure E.43.

   (a)

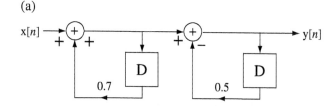

(b)

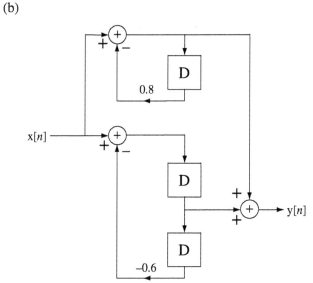

Figure E.43

Answers:

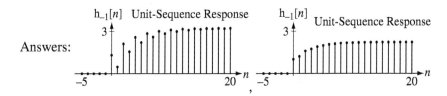

Stability

46. A discrete-time system is described by $y[n] + 1.8y[n-1] + 1.2y[n-2] = x[n]$. Is it BIBO stable?

    Answer: No

47. What numerical ranges of values of A and B make the system in Figure E.47 BIBO stable?

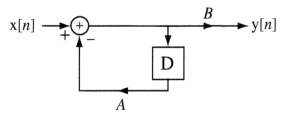

Figure E.47

Answers: $|A| < 1$, any B.

48. Below are the impulse responses of some LTI systems. In each case determine whether or not the system is BIBO stable.

(a) $h[n] = (1.1)^{-n}u[n]$
(b) $h[n] = u[n]$
(c) $h[n] = \text{tri}\left(\dfrac{n-4}{2}\right)$
(d) $h[n] = \delta_{10}[n]u[n]$
(e) $h[n] = \sin(2\pi n/6)u[n]$

Answers: 3 BIBO Unstable and 2 BIBO Stable

49. Which of the systems in Figure E.49 are BIBO stable?

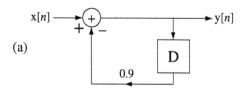

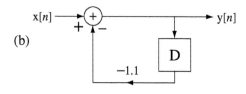

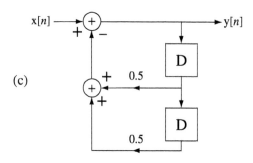

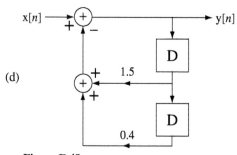

Figure E.49

Answers: 2 BIBO Stable and 2 BIBO Unstable

EXERCISES WITHOUT ANSWERS

Continuous Time

Impulse Response

50. Find the impulse responses of the systems described by these equations.
    (a) $4y''(t) = 2x(t) - x'(t)$
    (b) $y''(t) + 9y(t) = -6x'(t)$
    (c) $-y''(t) + 3y'(t) = 3x(t) + 5x''(t)$

51. Refer to the system of Figure E.51 with $a = 5$ and $b = 2$.

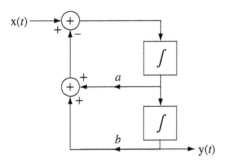

Figure E.51

   (a) Write the differential equation for the system.
   (b) The impulse response of this system can be written in the form, $h(t) = (K_1 e^{s_1 t} + K_2 e^{s_2 t})u(t)$. Find the values of K_1, K_2, s_2 and s_2.
   (c) Is this system BIBO stable?

52. Let $y(t) = x(t) * h(t)$ and let $x(t) = \text{rect}(t + 4) - \text{rect}(t - 1)$ and let $h(t) = \text{tri}(t - 2) + \text{tri}(t - 6)$.
    (a) What is the lowest value of t for which $y(t)$ is not zero?
    (b) What is the highest value of t for which $y(t)$ is not zero?

53. The impulse response of a continuous-time system described by a differential equation consists of a linear combination of one or more functions of the form Ke^{st} where s is an eigenvalue and, in some cases, an impulse of the form $K_\delta \delta(t)$. How many constants of each type are needed for each system?
    (a) $ay'(t) + by(t) = cx(t)$ a, b and c are constants.
    (b) $ay'(t) + by(t) = cx'(t)$ a, b and c are constants.
    (c) $ay'''(t) + by(t) = cx'(t)$ a, b and c are constants.
    (d) $ay''(t) + by(t) = cx''(t) + dx'(t)$ a, b, c and d are constants.

54. A rectangular voltage pulse which begins at $t = 0$, is 2 seconds wide and has a height of 0.5 V drives an RC lowpass filter in which $R = 10$ kΩ and $C = 100$ μF.
    (a) Graph the voltage across the capacitor versus time.
    (b) Change the pulse duration to 0.2 s and the pulse height to 5 V and repeat.

(c) Change the pulse duration to 2 ms and the pulse height to 500 V and repeat.
(d) Change the pulse duration to 2 ms and the pulse height to 500 kV and repeat.

Based on these results what do you think would happen if you let the input voltage be a unit impulse?

Convolution

55. (a) If $x(t) = \text{tri}(t/3) * \delta(t - 2)$ and $y(t) = x(2t)$, what is the numerical range of values of t for which $y(t)$ is not zero?

    (b) If $x(t) = \text{tri}(t/w) * \delta(t + t_0)$ and $y(t) = x(at)$, what is the range of values of t (in terms of w, t_0 and a) for which $y(t)$ is not zero?

56. What function convolved with $-2\cos(t)$ would produce $6\sin(t)$?

57. Graph these functions.

    (a) $g(t) = 3\cos(10\pi t) * 4\delta(t + 1/10)$

    (b) $g(t) = \text{tri}(2t) * \delta_1(t)$

    (c) $g(t) = 2[\text{tri}(2t) - \text{rect}(t - 1)] * \delta_2(t)$

    (d) $g(t) = 8[\text{tri}(t/4)\delta_1(t)] * \delta_8(t)$

    (e) $g(t) = e^{-2t}u(t) * [\delta_4(t) - \delta_4(t - 2)]$

58. For each graph in Figure E.58 select the corresponding signal or signals from the group, $x_1(t) \cdots x_8(t)$. (The corresponding signal may not be one of the choices A through E.)

$x_1(t) = \delta_2(t) * \text{rect}(t/2), \quad x_2(t) = 4\delta_2(t) * \text{rect}(t/2), \quad x_3(t) = (1/4)\delta_{1/2}(t) * \text{rect}(t/2)$

$x_4(t) = \delta_{1/2}(t) * \text{rect}(t/2), \quad x_5(t) = \delta_2(t) * \text{rect}(2t), \quad x_6(t) = 4\delta_2(t) * \text{rect}(2t)$

$x_7(t) = (1/4)\delta_{1/2}(t) * \text{rect}(2t), \quad x_8(t) = \delta_{1/2}(t) * \text{rect}(2t)$

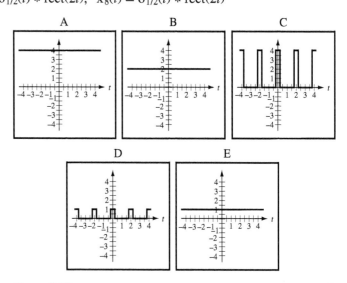

Figure E.58

59. Find the average signal power of these signals.

   (a) $x(t) = 4\,\text{rect}(t) * \delta_4(t)$ (b) $x(t) = 4\,\text{tri}(t) * \delta_4(t)$

60. If $x(t) = (u(t) - u(t-4)) * (\delta(t) + \delta(t-2))$, what is the signal energy of x?

61. A continuous-time system with impulse response $h(t) = 5\,\text{rect}(t)$ is excited by $x(t) = 4\,\text{rect}(2t)$.

   (a) Find the response $y(t)$ at time $t = 1/2$.
   (b) Change the excitation from part (a) to $x_a(t) = x(t-1)$ and keep the same impulse response. What is the new response $y_a(t)$ at time $t = 1/2$?
   (c) Change the excitation from part (a) to $x_b(t) = \dfrac{d}{dt}x(t)$ and keep the same impulse response. What is the new response $y_b(t)$ at time $t = 1/2$?

62. Write a functional description of the time-domain signal in Figure E.62 as the convolution of two functions of t.

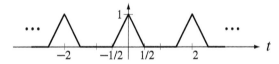

Figure E.62

63. In Figure E.63.1 are four continuouos-time functions, a, b, c and d. In Figure E.63.2 are 20 possible convolutions of pairs of these functions (including convolution of a functions with itself). For each convolution find the pair of signals convolved to produce it.

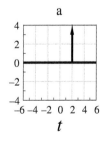 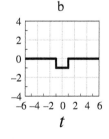 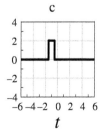 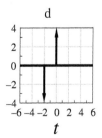

Figure E.63.1

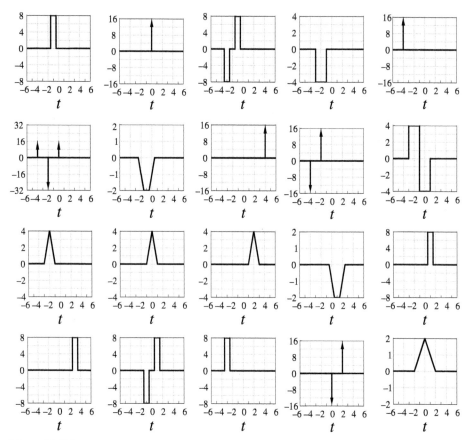

Figure E.63.2

Stability

64. Write the differential equations for the systems in Figure E.64, find their impulse responses and determine whether or not they are BIBO stable. For each system $a = 0.5$ and $b = -0.1$. Then comment on the effect on system BIBO stability of redefining the response.

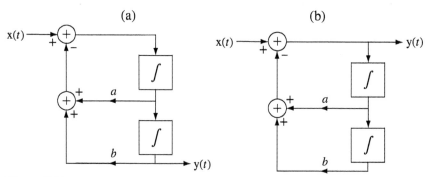

Figure E.64

65. Find the impulse response $h(t)$ of the continuous-time system described by $y'(t) + \beta y(t) = x(t)$. For what values of β is this system BIBO stable?

66. Find the impulse response of the system in Figure E.66 and evaluate its BIBO stability.

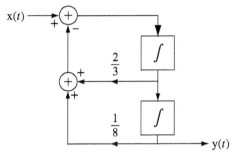

Figure E.66

67. For each impulse response indicate whether the LTI system it describes is BIBO stable or BIBO unstable.

   (a) $h[n] = \sin(2\pi n/6)u[n]$
   (b) $h(t) = \text{ramp}(t)$
   (c) $h(t) = \delta_1(t)e^{-t/10}u(t)$
   (d) $h(t) = [\delta_1(t) - \delta_1(t - 1/2)]u(t)$

Discrete Time

Impulse Response

68. A discrete-time system is described by the difference equation

$$7y[n] - 3y[n-1] + y[n-2] = 11.$$

   (a) The eigenvalues of this difference equation can be expressed in the polar form $Ae^{j\theta}$ where A is the magnitude and θ is the angle or phase. Find the values of A and θ.
   (b) The homogeneous solution approaches zero as $n \to \infty$. What value does $y[n]$ approach as $n \to \infty$?

69. Find the impulse responses of the systems described by these equations.

   (a) $3y[n] + 4y[n-1] + y[n-2] = x[n] + x[n-1]$
   (b) $(5/2)y[n] + 6y[n-1] + 10y[n-2] = x[n]$

Convolution

70. Graph $g[n]$. Verify with the MATLAB conv function.

   (a) $g[n] = (u[n+1] - u[n-2]) * \sin(2\pi n/9)$
   (b) $g[n] = (u[n+2] - u[n-3]) * \sin(2\pi n/9)$
   (c) $g[n] = (u[n+4] - u[n-5]) * \sin(2\pi n/9)$
   (d) $g[n] = (u[n+3] - u[n-4]) * (u[n+3] - u[n-4]) * \delta_{14}[n]$
   (e) $g[n] = (u[n+3] - u[n-4]) * (u[n+3] - u[n-4]) * \delta_7[n]$
   (f) $g[n] = 2\cos(2\pi n/7) * (7/8)^n u[n]$

71. If x[n] = (u[n + 4] − u[n − 3]) ∗ δ[n + 3] and y[n] = x[n − 4], what is the range of values for which y[n] is not zero?

72. Find the signal power of the discrete-time function, $4\,\text{sinc}(n) * \delta_3[n]$.

73. Given the function graphs 1 through 4 in Figure E.73.1 match each convolution expression a through j to one of the functions a through h in Figure E.73.2, if a match exists.

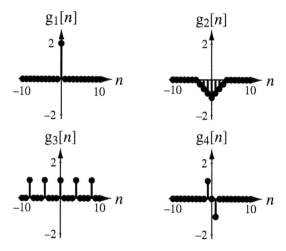

Figure E.73.1

(a) $g_1[n] * g_1[n]$ (b) $g_2[n] * g_2[n]$ (c) $g_3[n] * g_3[n]$
(d) $g_4[n] * g_4[n]$ (e) $g_1[n] * g_2[n]$ (f) $g_1[n] * g_3[n]$
(g) $g_1[n] * g_4[n]$ (h) $g_2[n] * g_3[n]$ (i) $g_2[n] * g_4[n]$
(j) $g_3[n] * g_4[n]$

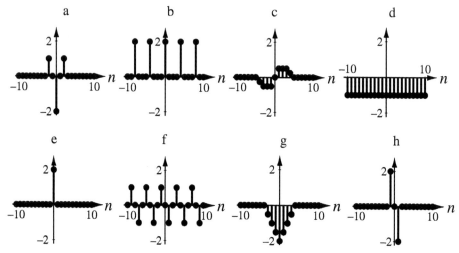

Figure E.73.2

74. In Figure E.74.1 are six discrete-time functions. All of them are zero outside the range graphed. In Figure E.74.2 are 15 candidate convolution results. All of them can be formed by convolving functions A–F in pairs. For each convolution result identify the two functions convolved to obtain it.

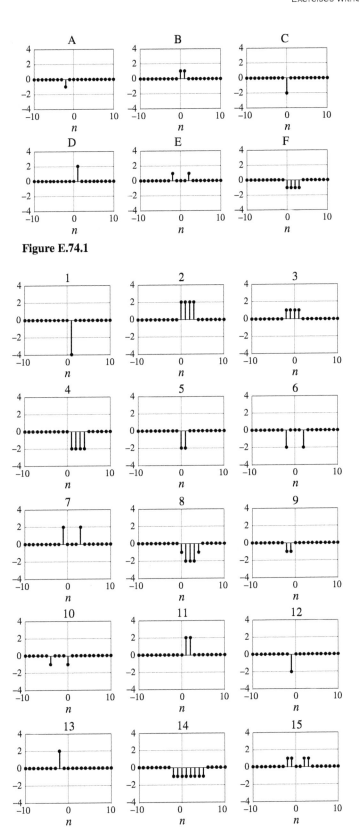

Figure E.74.1

Figure E.74.2

75. A system with an impulse response $h_1[n] = a^n u[n]$ is cascade connected with a second system with impulse response $h_2[n] = b^n u[n]$. If $a = 1/2$ and $b = 2/3$, and $h[n]$ is the impulse response of the overall cascade-connected system, find $h[3]$.

76. Find the impulse responses of the subsystems in Figure E.75 and then convolve them to find the impulse response of the cascade connection of the two subsystems.

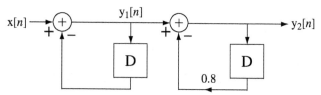

Figure E.75
Two cascaded subsystems

77. Given the excitations $x[n]$ and the impulse responses $h[n]$, find closed-form expressions for and graph the system responses $y[n]$.

(a) $x[n] = u[n]$, $h[n] = n(7/8)^n u[n]$
(b) $x[n] = u[n]$, $h[n] = (4/7)\delta[n] - (-3/4)^n u[n]$

Stability

78. A discrete-time LTI system has an impulse response of the form
$$h[n] = \begin{cases} 0, & n < 0 \\ a, 0, -a, 0, a, 0, -a, 0, \cdots, & n \geq 0 \end{cases}.$$
The pattern repeats forever in positive time. For what range of values of a is the system BIBO stable?

79. A system is excited by a unit-ramp function and the response is unbounded. From these facts alone, it is impossible to determine whether the system is BIBO stable or not. Why?

80. The impulse response of a system is zero for all negative time and, for $n \geq 0$, it is the alternating sequence, $1, -1, 1, -1, 1, -1, \cdots$ which continues forever. Is it BIBO stable?

CHAPTER 6

Continuous-Time Fourier Methods

6.1 INTRODUCTION AND GOALS

In Chapter 5 we learned how to find the response of an LTI system by expressing the excitation as a linear combination of impulses and expressing the response as a linear combination of impulse responses. We called that technique *convolution*. This type of analysis takes advantage of linearity and superposition and breaks one complicated analysis problem into multiple simpler analysis problems.

In this chapter we will also express an excitation as a linear combination of simple signals but now the signals will be sinusoids. The response will be a linear combination of the responses to those sinusoids. As we showed in Chapter 5, the response of an LTI system to a sinusoid is another sinusoid of the same frequency but with a generally different amplitude and phase. Expressing signals in this way leads to the **frequency domain** concept, thinking of signals as functions of frequency instead of time.

Analyzing signals as linear combinations of sinusoids is not as strange as it may sound. The human ear does something similar. When we hear a sound, what is the actual response of the brain? As indicated in Chapter 1, the ear senses a time variation of air pressure. Suppose this variation is a single-frequency tone like the sound of a person whistling. When we hear a whistled tone we are not aware of the (very fast) oscillation of air pressure with time. Rather, we are aware of three important characteristics of the sound, its pitch (a synonym for frequency), its intensity or amplitude, and its duration. The ear-brain system effectively parameterizes the signal into three simple descriptive parameters, pitch, intensity, and duration, and does not attempt to follow the rapidly changing (and very repetitive) air pressure in detail. In doing so, the ear-brain system has distilled the information in the signal down to its essence. The mathematical analysis of signals as linear combinations of sinusoids does something similar but in a more mathematically precise way. Looking at signals this way also lends new insight into the nature of systems and, for certain types of systems, greatly simplifies designing and analyzing them.

CHAPTER GOALS

1. To define the Fourier series as a way of representing periodic signals as linear combinations of sinusoids

2. To derive, using the concept of orthogonality, the methods for transforming signals back and forth between time and frequency descriptions
3. To determine the types of signals that can be represented by the Fourier series
4. To develop, and learn to use, the properties of the Fourier series
5. To generalize the Fourier series to the Fourier transform, which can represent aperiodic signals
6. To generalize the Fourier transform so it can apply to some very common useful signals
7. To develop, and learn to use, the properties of the Fourier transform
8. To see, through examples, some of the uses of the Fourier series and the Fourier transform

6.2 THE CONTINUOUS-TIME FOURIER SERIES

CONCEPTUAL BASIS

A common situation in signal and system analysis is an LTI system excited by a periodic signal. A very important result from Chapter 5 is that if an LTI system is excited by a sinusoid, the response is also a sinusoid, with the same frequency but generally a different amplitude and phase. This occurs because the complex exponential is the eigenfunction of the differential equations describing LTI systems and a sinusoid is a linear combination of complex exponentials.

An important result from Chapter 4 is that if an LTI system is excited by a sum of signals, the overall response is the sum of the responses to each of the signals individually. If we could find a way to express arbitrary signals as linear combinations of sinusoids we could use superposition to find the response of any LTI system to any arbitrary signal by summing the responses to the individual sinusoids. The representation of a periodic signal by a linear combination of sinusoids is called a **Fourier**[1] **series**. The sinusoids can be real sinusoids of the form $A\cos(2\pi t/T_0 + \theta)$ or they can be *complex* sinusoids of the form $Ae^{j2\pi t/T_0}$.

When first introduced to the idea of expressing *real* signals as linear combinations of *complex* sinusoids, students are often puzzled as to why we would want to introduce the extra (and seemingly unnecessary) dimension of imaginary numbers and functions. Euler's identity $e^{jx} = \cos(x) + j\sin(x)$ illustrates the very close relationship between real and *complex* sinusoids. It will turn out that, because of the compact notation that results, and because of certain mathematical simplifications that occur when using complex sinusoids, they are actually more convenient and powerful in analysis than real sinusoids. So the reader is encouraged to suspend disbelief for a while until the power of this method is revealed.

---

[1] Jean Baptiste Joseph Fourier was a French mathematician of the late 18th and early 19th centuries. (The name Fourier is commonly pronounced *fore-yay,* because of its similarity to the English word *four,* but the proper French pronunciation is *foor-yay,* where foor rhymes with tour.) Fourier lived in a time of great turmoil in France: the French Revolution and the reign of Napoleon Bonaparte. Fourier served as secretary of the Paris Academy of Science. In studying the propagation of heat in solids, Fourier developed the Fourier series and the Fourier integral. When he first presented his work to the great French mathematicians of the time, Laplace, LaGrange and LaCroix, they were intrigued by his theories but they (especially LaGrange) thought his theories lacked mathematical rigor. The publication of his paper at that time was denied. Some years later Dirichlet put the theories on a firmer foundation, explaining exactly what functions could and could not be expressed by a Fourier series. Then Fourier published his theories in what is now a classic text, *Theorie analytique de la chaleur*.

6.2 The Continuous-Time Fourier Series

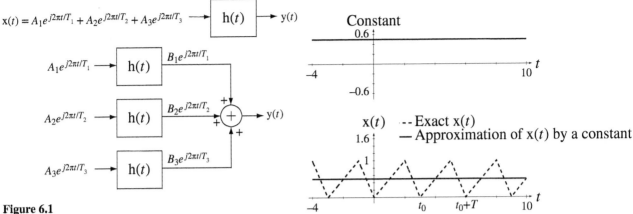

Figure 6.1
The equivalence of the response of an LTI system to an excitation signal and the sum of the system's responses to complex sinusoids whose sum is equivalent to the excitation

Figure 6.2
Signal approximated by a constant

If we are able to express an excitation signal as a linear combination of sinusoids, we can take advantage of linearity and superposition and apply each sinusoid to the system, one at a time, and then add the individual responses to obtain the overall response (Figure 6.1).

Consider an arbitrary original signal $x(t)$ that we would like to represent as a linear combination of sinusoids over a range of time from an initial time t_0 to a final time $t_0 + T$ as illustrated by the dashed line in Figure 6.2. In this illustration we will use real-valued sinusoids to make the visualization as simple as possible.

In Figure 6.2 the signal is approximated by a constant 0.5, which is the average value of the signal in the interval $t_0 \leq t < t_0 + T$. A constant is a special case of a sinusoid, a cosine of zero frequency. This is the best possible approximation of $x(t)$ by a constant. "Best" in this case means having the minimum mean-squared error between $x(t)$ and the approximation. Of course a constant, even the best one, is not a very good approximation to this signal. We can make the approximation better by adding to the constant a sinusoid whose fundamental period is the same as the fundamental period of $x(t)$ (Figure 6.3). This approximation is a big improvement on the previous one and is the best approximation that can be made using a constant and a single sinusoid of

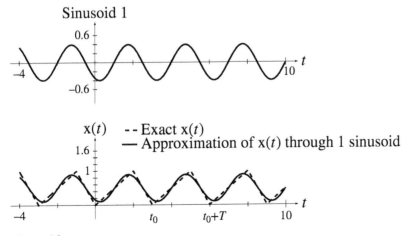

Figure 6.3
Signal approximated by a constant plus a single sinusoid

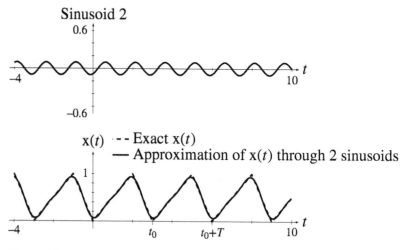

Figure 6.4
Signal approximated by a constant plus two sinusoids

the same fundamental frequency as x(*t*). We can improve the approximation further by adding a sinusoid at a frequency of twice the fundamental frequency of x(*t*) (Figure 6.4).

If we keep adding properly chosen sinusoids at higher integer multiples of the fundamental frequency of x(*t*), we can make the approximation better and better and, in the limit as the number of sinusoids approaches infinity, the approximation becomes exact (Figure 6.5 and Figure 6.6).

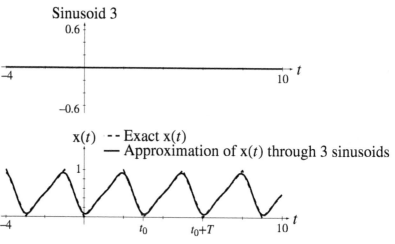

Figure 6.5
Signal approximated by a constant plus three sinusoids

The sinusoid added at three times the fundamental frequency of x(*t*) has an amplitude of zero, indicating that a sinusoid at that frequency does not improve the approximation. After the fourth sinusoid is added, the approximation is quite good, being hard to distinguish in Figure 6.6 from the exact x(*t*).

In this example the representation approaches the original signal in the representation time $t_0 \leq t < t_0 + T$, *and also for all time* because the fundamental period of the approximation is the same as the fundamental period of x(*t*). The most general

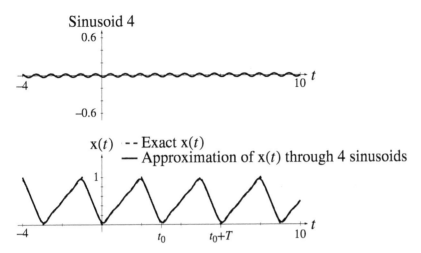

Figure 6.6
Signal approximated by a constant plus four sinusoids

application of Fourier series theory represents a signal in the interval $t_0 \leq t < t_0 + T$ but not necessarily outside that interval. But in signal and system analysis the representation is almost always of a periodic signal and the fundamental period of the representation is almost always chosen to also be a period of the signal so that the representation is valid for all time, not just in the interval $t_0 \leq t < t_0 + T$. In this example the signal and the representation have the same fundamental period but more generally the fundamental period of the representation can be chosen to be any period, fundamental or not, of the signal and the representation will still be valid everywhere.

Each of the sinusoids used in the approximations in the example above is of the form $A \cos(2\pi kt/T + \theta)$. Using the trigonometric identity

$$\cos(a + b) = \cos(a) \cos(b) - \sin(a) \sin(b)$$

we can express the sinusoid in the form

$$A \cos(2\pi kt/T + \theta) = A \cos(\theta) \cos(2\pi kt/T) - A \sin(\theta) \sin(2\pi kt/T).$$

This demonstrates that each phase-shifted cosine can be expressed also as the sum of an unshifted cosine and an unshifted sine of the same fundamental period, if the amplitudes are correctly chosen. The linear combination of all those sinusoids expressed as cosines and sines is called the **continuous-time Fourier series (CTFS)** and can be written in the form

$$x(t) = a_x[0] + \sum_{k=1}^{\infty} a_x[k] \cos(2\pi kt/T) + b_x[k] \sin(2\pi kt/T)$$

where $a_x[0]$ is the average value of the signal in the representation time, k is the **harmonic number** and $a_x[k]$ and $b_x[k]$ are functions of k called **harmonic functions**. Here we use the [·] notation to enclose the argument k because harmonic number is always an integer. The harmonic functions set the amplitudes of the sines and cosines and k determines the frequency. So the higher-frequency sines and cosines have frequencies that are integer multiples of the fundamental frequency and the multiple is k. The function $\cos(2\pi kt/T)$ is the kth-harmonic cosine. Its fundamental period is T/k and its fundamental cyclic frequency is k/T. Representing a signal this way as a linear combination of real-valued cosines and sines is called the **trigonometric** form of the CTFS.

Chapter 6 Continuous-Time Fourier Methods

For our purposes, it is important as a prelude to later work to see the equivalence of the complex form of the CTFS. Every real-valued sine and cosine can be replaced by a linear combination of complex sinusoids of the forms

$$\cos(2\pi kt/T) = \frac{e^{j2\pi kt/T} + e^{-j2\pi kt/T}}{2} \quad \text{and} \quad \sin(2\pi kt/T) = \frac{e^{j2\pi kt/T} - e^{-j2\pi kt/T}}{j2}.$$

If we add the cosine and sine with amplitudes $a_x[k]$ and $b_x[k]$, respectively, at any particular harmonic number k we get

$$a_x[k]\cos(2\pi kt/T) + b_x[k]\sin(2\pi kt/T) = \left\{\begin{array}{l} a_x[k]\dfrac{e^{j2\pi kt/T} + e^{-j2\pi kt/T}}{2} \\ +b_x[k]\dfrac{e^{j2\pi kt/T} - e^{-j2\pi kt/T}}{j2} \end{array}\right\}.$$

We can combine like complex-sinusoid terms on the right-hand side to form

$$a_x[k]\cos(2\pi kt/T) + b_x[k]\sin(2\pi kt/T) = \frac{1}{2}\left\{\begin{array}{l}(a_x[k] - jb_x[k])e^{j2\pi kt/T} \\ +(a_x[k] + jb_x[k])e^{-j2\pi kt/T}\end{array}\right\}.$$

Now if we define

$$c_x[0] = a_x[0], \quad c_x[k] = \frac{a_x[k] - jb_x[k]}{2}, \quad k > 0 \quad \text{and} \quad c_x[-k] = c_x^*[k]$$

we can write

$$a_x[k]\cos(2\pi kt/T) + b_x[k]\sin(2\pi kt/T) = c_x[k]e^{j2\pi kt/T} + c_x[-k]e^{j2\pi(-k)t/T}, \; k > 0$$

and we have the amplitudes $c_x[k]$ of the complex sinusoids $e^{j2\pi kt/T}$ at positive, and also negative, integer multiples of the fundamental cyclic frequency $1/T$. The sum of all these complex sinusoids and the constant $c_x[0]$ is equivalent to the original function, just as the sum of the sines and cosines and the constant was in the previous representation.

To include the constant term $c_x[0]$ in the general formulation of complex sinusoids we can let it be the zeroth ($k = 0$) harmonic of the fundamental. Letting k be zero, the complex sinusoid $e^{j2\pi kt/T}$ is just the number 1 and if we multiply it by a correctly chosen weighting factor $c_x[0]$ we can complete the complex CTFS representation. It will turn out in the material to follow that the same general formula for finding $c_x[k]$ for any nonzero k can also be used, without modification, to find $c_x[0]$, and that $c_x[0]$ is simply the average value in the representation time $t_0 \le t < t_0 + T$ of the function to be represented. $c_x[k]$ is the *complex* harmonic function of $x(t)$. The complex CTFS is more efficient than the trigonometric CTFS because there is only one harmonic function instead of two. The CTFS representation of the function can be written compactly in the form

$$\boxed{x(t) = \sum_{k=-\infty}^{\infty} c_x[k]e^{j2\pi kt/T}} \qquad (6.1)$$

So far we have asserted that the harmonic function exists but have not indicated how it can be found. That is the subject of the next section.

ORTHOGONALITY AND THE HARMONIC FUNCTION

In the Fourier series, the values of $c_x[k]$ determine the magnitudes and phases of complex sinusoids that are mutually **orthogonal**. Orthogonal means that the **inner product** of the two functions of time on some time interval is zero. An inner product is the integral of the product of one function and the complex conjugate of the other function over an interval, in this case the time interval T. For two functions x_1 and x_2 that are orthogonal on the interval $t_0 \leq t < t_0 + T$

$$\underbrace{(x_1(t), x_2(t))}_{\text{inner product}} = \int_{t_0}^{t_0+T} x_1(t) x_2^*(t)\, dt = 0.$$

We can show that the inner product of one complex sinusoid $e^{j2\pi kt/T}$ and another complex sinusoid $e^{j2\pi qt/T}$ on the interval $t_0 \leq t < t_0 + T$ is zero if k and q are integers and $k \neq q$. The inner product is

$$(e^{j2\pi kt/T}, e^{j2\pi qt/T}) = \int_{t_0}^{t_0+T} e^{j2\pi kt/T} e^{-j2\pi qt/T}\, dt = \int_{t_0}^{t_0+T} e^{j2\pi(k-q)t/T}\, dt.$$

Using Euler's identity

$$(e^{j2\pi kt/T}, e^{j2\pi qt/T}) = \int_{t_0}^{t_0+T} \left[\cos\left(2\pi \frac{k-q}{T} t\right) + j\sin\left(2\pi \frac{k-q}{T} t\right)\right] dt. \quad (6.2)$$

Since k and q are both integers, if $k \neq q$, the cosine and the sine in this integral are both being integrated over a period (an integer number of fundamental periods). The definite integral of any sinusoid (of nonzero frequency) over any period is zero. If $k = q$, the integrand is $\cos(0) + \sin(0) = 1$ and the inner product is T. If $k \neq q$, the inner product (6.2) is zero. So any two complex sinusoids with an integer number of fundamental periods on the interval $t_0 \leq t < t_0 + T$ are orthogonal, unless they have the same number of fundamental periods. Then we can conclude that functions of the form $e^{j2\pi kt/T}$, $-\infty < k < \infty$ constitute a countably infinite set of functions, all of which are mutually orthogonal on the interval $t_0 \leq t < t_0 + T$ where t_0 is arbitrary.

We can now take advantage of orthogonality by multiplying the expression for the Fourier series $x(t) = \sum_{k=-\infty}^{\infty} c_x[k] e^{j2\pi kt/T}$ through by $e^{-j2\pi qt/T}$ (q an integer) yielding

$$x(t) e^{-j2\pi qt/T} = \sum_{k=-\infty}^{\infty} c_x[k] e^{j2\pi kt/T} e^{-j2\pi qt/T} = \sum_{k=-\infty}^{\infty} c_x[k] e^{j2\pi(k-q)t/T}.$$

If we now integrate both sides over the interval $t_0 \leq t < t_0 + T$ we get

$$\int_{t_0}^{t_0+T} x(t) e^{-j2\pi qt/T}\, dt = \int_{t_0}^{t_0+T} \left[\sum_{k=-\infty}^{\infty} c_x[k] e^{j2\pi(k-q)t/T}\right] dt.$$

Since k and t are independent variables, the integral of the sum on the right side is equivalent to a sum of integrals. The equation can be written as

$$\int_{t_0}^{t_0+T} x(t) e^{-j2\pi qt/T}\, dt = \sum_{k=-\infty}^{\infty} c_x[k] \int_{t_0}^{t_0+T} e^{j2\pi(k-q)t/T}\, dt$$

and, using the fact that the integral is zero unless $k = q$, the summation

$$\sum_{k=-\infty}^{\infty} c_x[k] \int_{t_0}^{t_0+T} e^{j2\pi(k-q)t/T}\, dt$$

reduces to $c_x[q]T$ and

$$\int_{t_0}^{t_0+T} x(t)e^{-j2\pi qt/T}dt = c_x[q]T.$$

Solving for $c_x[q]$,

$$c_x[q] = \frac{1}{T}\int_{t_0}^{t_0+T} x(t)e^{-j2\pi qt/T}dt.$$

If this is a correct expression for $c_x[q]$, then $c_x[k]$ in the original Fourier series expression (6.1) must be

$$c_x[k] = \frac{1}{T}\int_{t_0}^{t_0+T} x(t)e^{-j2\pi kt/T}dt. \tag{6.3}$$

From this derivation we conclude that, if the integral in (6.3) converges, a periodic signal $x(t)$ can be expressed as

$$\boxed{x(t) = \sum_{k=-\infty}^{\infty} c_x[k]e^{j2\pi kt/T}} \tag{6.4}$$

where

$$\boxed{c_x[k] = \frac{1}{T}\int_T x(t)e^{-j2\pi kt/T}dt} \tag{6.5}$$

and the notation $\int_T$ means the same thing as $\int_{t_0}^{t_0+T}$ with t_0 arbitrarily chosen. Then $x(t)$ and $c_x[k]$ form a **CTFS pair**, which can be indicated by the notation

$$x(t) \xleftarrow[T]{\mathcal{FS}} c_x[k]$$

where the $\mathcal{FS}$ means "Fourier series" and the T means that $c_x[k]$ is computed with T as the fundamental period of the CTFS representation of $x(t)$.

This derivation was done on the basis of using a *period T* of the signal as the interval of orthogonality and also as the *fundamental period* of the CTFS representation. T could be any period of the signal, including its fundamental period T_0. In practice the most commonly used fundamental period of the representation is the fundamental period of the signal T_0. In that special case the CTFS relations become

$$x(t) = \sum_{k=-\infty}^{\infty} c_x[k]e^{j2\pi kt/T_0}$$

and

$$c_x[k] = \frac{1}{T_0}\int_{T_0} x(t)e^{-j2\pi kt/T_0}dt = f_0\int_{T_0} x(t)e^{-j2\pi kf_0 t}dt$$

where $f_0 = 1/T_0$ is the fundamental cyclic frequency of $x(t)$.

If the integral of a signal $x(t)$ over the time interval, $t_0 < t < t_0 + T$, diverges, a CTFS cannot be found for the signal. There are two other conditions on the applicability of the CTFS, which, together with the condition on the convergence of the integral, are called the *Dirichlet conditions*. The Dirichlet conditions are the following:

1. The signal must be absolutely integrable over the time, $t_0 < t < t_0 + T$. That is,

$$\int_{t_0}^{t_0+T} |x(t)| dt < \infty$$

2. The signal must have a finite number of maxima and minima in the time, $t_0 < t < t_0 + T$.
3. The signal must have a finite number of discontinuities, all of finite size, in the time, $t_0 < t < t_0 + T$.

There are hypothetical signals for which the Dirichlet conditions are not met, but they have no known engineering use.

THE COMPACT TRIGONOMETRIC FOURIER SERIES

Consider the trigonometric Fourier series.

$$x(t) = a_x[0] + \sum_{k=1}^{\infty} a_x[k] \cos(2\pi kt/T) + b_x[k] \sin(2\pi kt/T)$$

Now, using

$$A \cos(x) + B \sin(x) = \sqrt{A^2 + B^2} \cos(x - \tan^{-1}(B/A))$$

we have

$$x(t) = a_x[0] + \sum_{k=1}^{\infty} \sqrt{a_x^2[k] + b_x^2[k]} \cos\left(2\pi kt/T + \tan^{-1}\left(-\frac{b_x[k]}{a_x[k]}\right)\right)$$

or

$$x(t) = d_x[0] + \sum_{k=1}^{\infty} d_x[k] \cos(2\pi kt/T + \theta_x[k])$$

where

$$d_x[0] = a_x[0], \ d_x[k] = \sqrt{a_x^2[k] + b_x^2[k]}, \ k > 0$$

and

$$\theta_x[k] = \tan^{-1}\left(-\frac{b_x[k]}{a_x[k]}\right), \ k > 0$$

This is the so-called **compact trigonometric Fourier series**. It is also expressed in purely real-valued functions and coefficients and is a little more compact than the trigonometric form but it is still not as compact or efficient as the complex form $x(t) = \sum_{k=-\infty}^{\infty} c_x[k] e^{j2\pi kt/T}$. The trigonometric form is the one actually used by Jean Baptiste Joseph Fourier.

EXAMPLE 6.1

CTFS harmonic function of a rectangular wave

Find the complex CTFS harmonic function of $x(t) = A\text{rect}(t/w) * \delta_{T_0}(t)$, $w < T_0$ using its fundamental period as the representation time.

The fundamental period is T_0 so the CTFS harmonic function is

$$c_x[k] = (1/T_0)\int_{T_0} A\text{rect}(t/w) * \delta_{T_0}(t) e^{-j2\pi kt/T_0} dt$$

The integration interval can be anywhere in time as long as its length is T_0. For convenience, choose to integrate over the interval $-T_0/2 \leq t < T_0/2$. Then

$$c_x[k] = (A/T_0)\int_{-T_0/2}^{T_0/2} \text{rect}(t/w) * \delta_{T_0}(t) e^{-j2\pi kt/T_0} dt$$

Using $w < T_0$ and the fact that the interval contains only one rectangle function

$$c_x[k] = (A/T_0)\int_{-T_0/2}^{T_0/2} \text{rect}(t/w) e^{-j2\pi kt/T_0} dt = (A/T_0)\int_{-w/2}^{w/2} e^{-j2\pi kt/T_0} dt$$

$$c_x[k] = (A/T_0)\left[\frac{e^{-j2\pi kt/T_0}}{-j2\pi k/T_0}\right]_{-w/2}^{w/2} = A\left[\frac{e^{-j\pi kw/T_0} - e^{j\pi kw/T_0}}{-j2\pi k}\right] = A\frac{\sin(\pi kw/T_0)}{\pi k}$$

and finally

$$x(t) = A\text{rect}(t/w) * \delta_{T_0}(t) \xleftrightarrow[T_0]{\mathcal{FS}} c_x[k] = A\frac{\sin(\pi kw/T_0)}{\pi k}.$$

(Even though in this example we restricted w to be less than T_0 to simplify the analysis, the result is also correct for w greater than T_0.)

■

In Example 6.1 the harmonic function turned out to be $c_x[k] = A\frac{\sin(\pi kw/T_0)}{\pi k}$. This mathematical form of the sine of a quantity divided by the quantity itself occurs often enough in Fourier analysis to deserve its own name. We now define the unit-**sinc** function (Figure 6.7) as

$$\boxed{\text{sinc}(t) = \frac{\sin(\pi t)}{\pi t}} \quad (6.6)$$

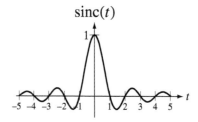

Figure 6.7
The unit sinc function

We can now express the harmonic function from Example 6.1 as

$$c_x[k] = (Aw/T_0)\text{sinc}(kw/T_0)$$

and the CTFS pair as

$$x(t) = A\,\text{rect}(t/w) * \delta_{T_0}(t) \xleftrightarrow[T_0]{\mathcal{FS}} c_x[k] = (Aw/T_0)\text{sinc}(wk/T_0).$$

The unit-sinc function is called a *unit function* because its height and area are both one.[2]

One common question when first encountering the sinc function is how to determine the value of sinc(0). When the independent variable t in $\sin(\pi t)/\pi t$ is zero, both the numerator $\sin(\pi t)$ and the denominator πt evaluate to zero, leaving us with an indeterminate form. The solution to this problem is to use L'Hôpital's rule. Then

$$\lim_{t\to 0}\text{sinc}(t) = \lim_{t\to 0}\frac{\sin(\pi t)}{\pi t} = \lim_{t\to 0}\frac{\pi\cos(\pi t)}{\pi} = 1.$$

So $\text{sinc}(t)$ is continuous at $t = 0$ and $\text{sinc}(0) = 1$.

CONVERGENCE

Continuous Signals

In this section we will examine how the CTFS summation approaches the signal it represents as the number of terms used in the sum approaches infinity. We do this by examining the partial sum

$$x_N(t) = \sum_{k=-N}^{N} c_x[k] e^{j2\pi kt/T}$$

for successively higher values of N. As a first example consider the CTFS representation of the continuous periodic signal in Figure 6.8. The CTFS pair is (using the signal's fundamental period as the fundamental period of the CTFS representation)

$$A\,\text{tri}(2t/T_0) * \delta_{T_0}(t) \xleftrightarrow[T_0]{\mathcal{FS}} (A/2)\text{sinc}^2(k/2)$$

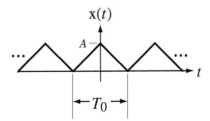

Figure 6.8
A continuous signal to be represented by a CTFS

and the partial-sum approximations $x_N(t)$ for $N = 1, 3, 5,$ and 59 are illustrated in Figure 6.9.

---
[2] The definition of the sinc function is generally, *but not universally*, accepted as $\text{sinc}(t) = \sin(\pi t)/\pi t$. In some books the sinc function is defined as $\text{sinc}(t) = \sin(t)/t$. In other books this second form is called the **Sa** function $\text{Sa}(t) = \sin(t)/t$. How the sinc function is defined is not really critical. As long as one definition is accepted and the sinc function is used in a manner consistent with that definition, signal and system analysis can be done with useful results.

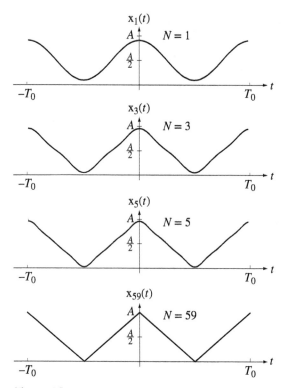

Figure 6.9
Successively closer approximations to a triangle wave

At $N = 59$ (and probably at lower values of N) it is impossible to distinguish the CTFS partial-sum approximation from the original signal by observing a graph on this scale.

Discontinuous Signals
Now consider a periodic signal with discontinuities

$$x(t) = A \operatorname{rect}\left(2\frac{t - T_0/4}{T_0}\right) * \delta_{T_0}(t)$$

(Figure 6.10). The CTFS pair is

$$A \operatorname{rect}\left(2\frac{t - T_0/4}{T_0}\right) * \delta_{T_0}(t) \xleftarrow[T_0]{\mathcal{FS}} (A/2)(-j)^k \operatorname{sinc}(k/2)$$

and the approximations $x_N(t)$ for $N = 1, 3, 5,$ and 59 are illustrated in Figure 6.11.

Although the mathematical derivation indicates that the original signal and its CTFS representation are equal everywhere, it is natural to wonder whether that is true after looking at Figure 6.11. There is an obvious overshoot and ripple near the discontinuities that does not appear to become smaller as N increases. In fact, the maximum vertical overshoot near a discontinuity does not decrease with N, even as N approaches infinity. This overshoot is called the *Gibbs phenomenon* in honor of Josiah Gibbs[3]

---

[3] Josiah Willard Gibbs, an American physicist, chemist, and mathematician, developed much of the theory for chemical thermodynamics and physical chemistry. He invented vector analysis (independently of Oliver Heaviside). He earned the first American Ph.D. in engineering from Yale in 1863 and he spent his entire career at Yale. In 1901, Gibbs was awarded the Copley Medal of the Royal Society of London for being "the first to apply the second law of thermodynamics to the exhaustive discussion of the relation between chemical, electrical, and thermal energy and capacity for external work."

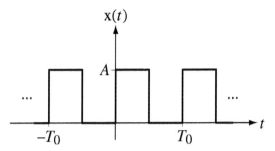

Figure 6.10
A discontinuous signal to be represented by a CTFS

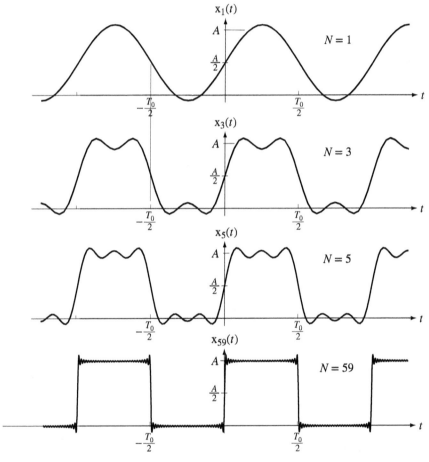

Figure 6.11
Successively closer approximations to a square wave

who first mathematically described it. But notice also that the ripple is also confined ever more closely in the vicinity of the discontinuity as N increases. In the limit as N approaches infinity the height of the overshoot is constant but its width approaches zero. The error in the partial-sum approximation is the difference between it and the original signal. In the limit as N approaches infinity the signal power of the error approaches zero because the zero-width difference at a point of discontinuity contains no signal energy. Also, at any particular value of t (except exactly at a discontinuity) the value of the CTFS representation approaches the value of the original signal as N approaches infinity.

At a discontinuity the functional value of the CTFS representation is always the average of the two limits of the original function approached from above and from below, for any N. Figure 6.12 is a magnified view of the CTFS representation at a discontinuity for three different values of N. Since the signal energy of the difference between the two signals is zero in any finite time interval, their effect on any real physical system is the same and they can be considered equal for any practical purpose.

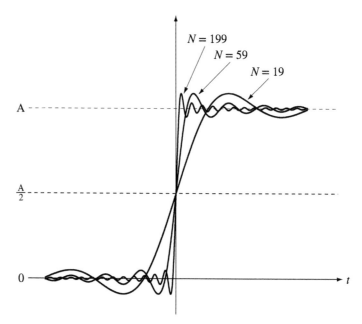

Figure 6.12
Illustration of the Gibbs phenomenon for increasing values of N

MINIMUM ERROR OF FOURIER-SERIES PARTIAL SUMS

The CTFS is an infinite summation of sinusoids. In general, for exact equality between an arbitrary original signal and its CTFS representation, infinitely many terms must be used. (Signals for which the equality is achieved with a finite number of terms are called **bandlimited** signals.) If a partial-sum approximation

$$x_N(t) = \sum_{k=-N}^{N} c_x[k] e^{j2\pi kt/T} \tag{6.7}$$

is made to a signal $x(t)$ by using only the first N harmonics of the CTFS, the difference between $x_N(t)$ and $x(t)$ is the approximation error $e_N(t) = x_N(t) - x(t)$. We know that in (6.7) when N goes to infinity the equality is valid at every point of continuity of $x(t)$. But when N is finite does the harmonic function $c_x[k]$ for $-N \leq k \leq N$ yield the best possible approximation to $x(t)$? In other words, could we have chosen a different harmonic function $c_{x,N}[k]$ that, if used in place of $c_x[k]$ in (6.7), would have been a better approximation to $x(t)$?

The first task in answering this question is to define what is meant by "best possible approximation." It is usually taken to mean that the signal energy of the error $e_N(t)$ over one period T is a minimum. Let's find the harmonic function $c_{x,N}[k]$ that minimizes the signal energy of the error.

$$e_N(t) = \underbrace{\sum_{k=-N}^{N} c_{x,N}[k]e^{j2\pi kt/T}}_{x_N(t)} - \underbrace{\sum_{k=-\infty}^{\infty} c_x[k]e^{j2\pi kt/T}}_{x(t)}$$

Let

$$c_y[k] = \begin{cases} c_{x,N}[k] - c_x[k], & |k| \leq N \\ -c_x[k], & |k| > N \end{cases}.$$

Then

$$e_N(t) = \sum_{k=-\infty}^{\infty} c_y[k]e^{j2\pi kt/T}.$$

The signal energy of the error over one period is

$$E_e = \frac{1}{T}\int_T |e_N(t)|^2 dt = \frac{1}{T}\int_T \left| \sum_{k=-\infty}^{\infty} c_y[k]e^{j2\pi kt/T} \right|^2 dt.$$

$$E_e = \frac{1}{T}\int_T \left(\sum_{k=-\infty}^{\infty} c_y[k]e^{j2\pi kt/T} \right) \left(\sum_{q=-\infty}^{\infty} c_y^*[q]e^{-j2\pi qt/T} \right) dt$$

$$E_e = \frac{1}{T}\int_T \left(\sum_{k=-\infty}^{\infty} c_y[k]c_y^*[k] + \sum_{k=-\infty}^{\infty} \sum_{\substack{q=-\infty \\ q \neq k}}^{\infty} c_y[k]c_y^*[q]e^{j2\pi(k-q)t/T} \right) dt$$

The integral of the double summation is zero for every combination of k and q for which $k \neq q$ because the integral over any period of $e^{j2\pi(k-q)t/T}$ is zero. Therefore

$$E_e = \frac{1}{T}\int_T \sum_{k=-\infty}^{\infty} c_y[k]c_y^*[k] dt = \frac{1}{T}\int_T \sum_{k=-\infty}^{\infty} |c_y[k]|^2 dt.$$

Substituting the definition of $c_y[k]$ we get

$$E_e = \frac{1}{T}\int_T \left(\sum_{k=-N}^{N} |c_{x,N}[k] - c_x[k]|^2 + \sum_{|k|>N} |-c_x[k]|^2 \right) dt$$

$$E_e = \sum_{k=-N}^{N} |c_{x,N}[k] - c_x[k]|^2 + \sum_{|k|>N} |c_x[k]|^2$$

All the quantities being summed are non-negative and, since the second summation is fixed, we want the first summation to be as small as possible. It is zero if $c_{x,N}[k] = c_x[k]$, proving that the harmonic function $c_x[k]$ gives the smallest possible mean-squared error in a partial sum approximation.

THE FOURIER SERIES OF EVEN AND ODD PERIODIC FUNCTIONS

Consider the case of representing a periodic even signal $x(t)$ with fundamental period T_0 with a complex CTFS. The CTFS harmonic function is

$$c_x[k] = \frac{1}{T}\int_T x(t)e^{-j2\pi kt/T} dt.$$

For periodic signals this integral over a period is independent of the starting point. Therefore we can rewrite the integral as

$$c_x[k] = \frac{1}{T}\int_{-T/2}^{T/2} x(t)e^{-j2\pi kt/T}\,dt = \frac{1}{T}\left[\int_{-T/2}^{T/2}\underbrace{\underbrace{x(t)}_{\text{even}}\underbrace{\cos(2\pi kt/T)}_{\text{even}}}_{\text{even}}\,dt - j\int_{-T/2}^{T/2}\underbrace{\underbrace{x(t)}_{\text{even}}\underbrace{\sin(2\pi kt/T)}_{\text{odd}}}_{\text{odd}}\,dt\right]$$

Using the fact that an odd function integrated over symmetrical limits about zero is zero, $c_x[k]$ must be real. By a similar argument, for a periodic *odd* function, $c_x[k]$ must be imaginary.

> For $x(t)$ even and real-valued, $c_x[k]$ is even and real-valued.
> For $x(t)$ odd and real-valued, $c_x[k]$ is odd and purely imaginary.

FOURIER-SERIES TABLES AND PROPERTIES

The properties of the CTFS are listed in Table 6.1. They can all be proven using the definition of the CTFS and the harmonic function

$$x(t) = \sum_{k=-\infty}^{\infty} c_x[k]e^{j2\pi kt/T} \xleftarrow{\mathscr{FS}}_{T} c_x[k] = (1/T)\int_T x(t)e^{-j2\pi kt/T}\,dt.$$

In the Multiplication–Convolution Duality property the integral

$$x(t) \circledast y(t) = \int_T x(\tau)y(t-\tau)\,d\tau$$

appears. It looks a lot like the convolution integral we have seen earlier except that the integration range is over the fundamental period T of the CTFS representation instead of from $-\infty$ to $+\infty$. This operation is called **periodic convolution**. Periodic convolution is always done with two periodic signals over a period T that is common to both of them. The convolution that was introduced in Chapter 5 is *aperiodic* convolution. Periodic convolution is equivalent to aperiodic convolution in the following way. Any periodic signal $x_p(t)$ with period T can be expressed as a sum of equally spaced aperiodic signals $x_{ap}(t)$ as

$$x_p(t) = \sum_{k=-\infty}^{\infty} x_{ap}(t - kT)$$

It can be shown that the periodic convolution of $x_p(t)$ with $y_p(t)$ is then

$$x_p(t) \circledast y_p(t) = x_{ap}(t) * y_p(t).$$

The function $x_{ap}(t)$ is not unique. It can be any function that satisfies $x_p(t) = \sum_{k=-\infty}^{\infty} x_{ap}(t - kT)$.

Table 6.2 shows some common CTFS pairs. All but one are based on the fundamental period T of the CTFS representation being mT_0, with m being a positive integer and T_0 being the fundamental period of the signal.

$$x(t) = \sum_{k=-\infty}^{\infty} c_x[k]e^{j2\pi kt/mT_0} \xleftarrow{\mathscr{FS}}_{mT_0} c_x[k] = \frac{1}{mT_0}\int_{mT_0} x(t)e^{-j2\pi kt/mT_0}\,dt$$

Table 6.1 CTFS properties

Linearity $\alpha x(t) + \beta y(t) \xleftrightarrow[T]{\mathscr{FS}} \alpha c_x[k] + \beta c_y[k]$

Time Shifting $x(t - t_0) \xleftrightarrow[T]{\mathscr{FS}} e^{-j2\pi k t_0/T} c_x[k]$

Frequency Shifting $e^{j2\pi k_0 t/T} x(t) \xleftrightarrow[T]{\mathscr{FS}} c_x[k - k_0]$

Conjugation $x^*(t) \xleftrightarrow[T]{\mathscr{FS}} c_x^*[-k]$

Time Differentiation $\frac{d}{dt}(x(t)) \xleftrightarrow[T]{\mathscr{FS}} (j2\pi k/T) c_x[k]$

Time Reversal $x(-t) \xleftrightarrow[T]{\mathscr{FS}} c_x[-k]$

Time Integration $\int_{-\infty}^{t} x(\tau) d\tau \xleftrightarrow[T]{\mathscr{FS}} \frac{c_x[k]}{j2\pi k/T}$, $k \neq 0$ if $c_x[0] = 0$

Parseval's Theorem $\frac{1}{T} \int_T |x(t)|^2 dt = \sum_{k=-\infty}^{\infty} |c_x[k]|^2$

Multiplication–Convolution Duality

$$x(t)y(t) \xleftrightarrow[T]{\mathscr{FS}} \sum_{m=-\infty}^{\infty} c_y[m] c_x[k - m] = c_x[k] * c_y[k]$$

$$x(t) \circledast y(t) = \int_T x(\tau) y(t - \tau) d\tau \xleftrightarrow[T]{\mathscr{FS}} T c_x[k] c_y[k]$$

Change of Period If $x(t) \xleftrightarrow[T]{\mathscr{FS}} c_x[k]$ and $x(t) \xleftrightarrow[mT]{\mathscr{FS}} c_{xm}[k]$, $c_{xm}[k] = \begin{cases} c_x[k/m], & k/m \text{ an integer} \\ 0, & \text{otherwise} \end{cases}$

Time Scaling If $x(t) \xleftrightarrow[T]{\mathscr{FS}} c_x[k]$ and $z(t) = x(mt) \xleftrightarrow[T]{\mathscr{FS}} c_z[k]$, $c_z[k] = \begin{cases} c_x[k/m], & k/m \text{ an integer} \\ 0, & \text{otherwise} \end{cases}$

Table 6.2 Some CTFS pairs

$$e^{j2\pi t/T_0} \xleftrightarrow[mT_0]{\mathscr{FS}} \delta[k - m]$$

$$\cos(2\pi k/T_0) \xleftrightarrow[mT_0]{\mathscr{FS}} (1/2)(\delta[k - m] + \delta[k + m])$$

$$\sin(2\pi k/T_0) \xleftrightarrow[mT_0]{\mathscr{FS}} (j/2)(\delta[k + m] - \delta[k - m])$$

$$1 \xleftrightarrow[T]{\mathscr{FS}} \delta[k], T \text{ is arbitrary}$$

$$\delta_{T_0}(t) \xleftrightarrow[mT_0]{\mathscr{FS}} (1/T_0)\delta_m[k]$$

$$\text{rect}(t/w) * \delta_{T_0}(t) \xleftrightarrow[mT_0]{\mathscr{FS}} (w/T_0)\text{sinc}(wk/mT_0)\delta_m[k]$$

$$\text{tri}(t/w) * \delta_{T_0}(t) \xleftrightarrow[mT_0]{\mathscr{FS}} (w/T_0)\text{sinc}^2(wk/mT_0)\delta_m[k]$$

$$\text{sinc}(t/w) * \delta_{T_0}(t) \xleftrightarrow[mT_0]{\mathscr{FS}} (w/T_0)\text{rect}(wk/mT_0)\delta_m[k]$$

$$t[u(t) - u(t - w)] * \delta_{T_0}(t) \xleftrightarrow[mT_0]{\mathscr{FS}} \frac{1}{T_0} \frac{[j(2\pi kw/mT_0) + 1]e^{-j(2\pi kw/mT_0)} - 1}{(2\pi k/mT_0)^2} \delta_m[k]$$

EXAMPLE 6.2

Periodic excitation and response of a continuous-time system

A continuous-time system is described by the differential equation

$$y''(t) + 0.04y'(t) + 1.58y(t) = x(t).$$

If the excitation is $x(t) = \text{tri}(t) * \delta_5(t)$, find the response $y(t)$.

The excitation can be expressed by a CTFS as

$$x(t) = \sum_{k=-\infty}^{\infty} c_x[k] e^{j2\pi kt/T_0}$$

where, from Table 6.2,

$$c_x[k] = (w/T_0)\text{sinc}^2(wk/mT_0)\delta_m[k]$$

with $w = 1$, $T_0 = 5$, and $m = 1$. Then

$$x(t) = \sum_{k=-\infty}^{\infty} (1/5)\text{sinc}^2(k/5)\delta_1[k] e^{j2\pi kt/5} = (1/5) \sum_{k=-\infty}^{\infty} \text{sinc}^2(k/5) e^{j2\pi kt/5}$$

We know that the CTFS expression for the excitation is a sum of complex sinusoids and the response to each of those sinusoids will be another sinusoid of the same frequency. Therefore, the response can be expressed in the form

$$y(t) = \sum_{k=-\infty}^{\infty} c_y[k] e^{j2\pi kt/5}$$

and each complex sinusoid in $y(t)$ with fundamental cyclic frequency $k/5$ is caused by the complex sinusoid in $x(t)$ of the same frequency. Substituting this form into the differential equation

$$\sum_{k=-\infty}^{\infty} (j2\pi k/5)^2 c_y[k] e^{j2\pi kt/5} + 0.04 \sum_{k=-\infty}^{\infty} (j2\pi k/5) c_y[k] e^{j2\pi kt/5} + 1.58 \sum_{k=-\infty}^{\infty} c_y[k] e^{j2\pi kt/5}$$

$$= \sum_{k=-\infty}^{\infty} c_x[k] e^{j2\pi kt/5}$$

Gathering terms and simplifying

$$\sum_{k=-\infty}^{\infty} [(j2\pi k/5)^2 + 0.04(j2\pi k/5) + 1.58] c_y[k] e^{j2\pi kt/5} = \sum_{k=-\infty}^{\infty} c_x[k] e^{j2\pi kt/5}.$$

Therefore, for any particular value of k, the excitation and response are related by

$$[(j2\pi k/5)^2 + 0.04(j2\pi k/5) + 1.58] c_y[k] = c_x[k]$$

and

$$\frac{c_y[k]}{c_x[k]} = \frac{1}{(j2\pi k/5)^2 + 0.04(j2\pi k/5) + 1.58}$$

The quantity $H[k] = \dfrac{c_y[k]}{c_x[k]}$ is analogous to frequency response and can logically be called **harmonic response**. The system response is

$$y(t) = (1/5) \sum_{k=-\infty}^{\infty} \frac{\text{sinc}^2(k/5)}{(j2\pi k/5)^2 + 0.04(j2\pi k/5) + 1.58} e^{j2\pi kt/5}.$$

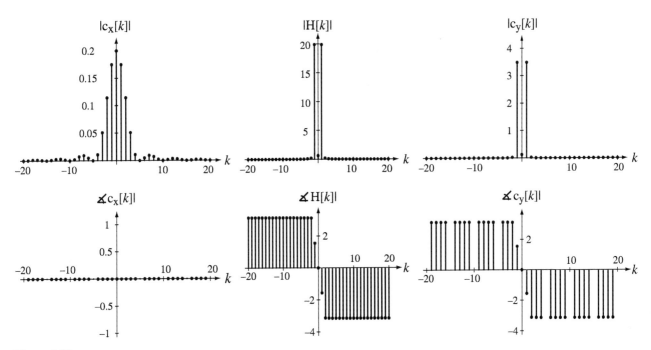

Figure 6.13
Excitation harmonic function, system harmonic response, and response harmonic function

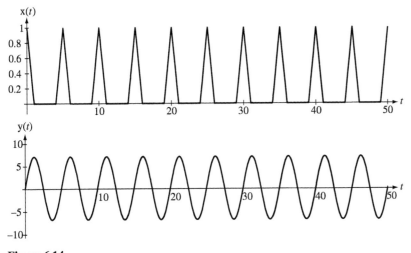

Figure 6.14
Excitation and response

This rather intimidating-looking expression can be easily programmed on a computer. The signals, their harmonic functions, and the harmonic response are illustrated in Figure 6.13 and Figure 6.14.

We can see from the harmonic response that the system responds strongly at harmonic number one, the fundamental. The fundamental period of x(t) is $T_0 = 5$ s. So y(t) should have a significant response at a frequency of 0.2 Hz. Looking at the response graph, we see a signal that looks like a sinusoid and its fundamental period is 5 s, so its fundamental frequency is 0.2 Hz. The magnitudes of all the other harmonics, including $k = 0$, are almost zero. That is why the average value of the response is practically zero and it looks like a sinusoid, a single-frequency signal. Also, notice the phase of the harmonic response at the fundamental. It is 1.5536 radians at $k = 1$, or almost $\pi/2$. That phase shift would convert a cosine into a sine. The excitation is an even function with only cosine components and the response is practically an odd function because of this phase shift.

We could compute x(t) and y(t) with a MATLAB program of the following form.

```
% Set up a vector of k's over a wide range. The ideal summation has an infinite
% range for k. We obviously cannot do that in MATLAB but we can make the range of
% k so large that making it any larger would not significantly change the
% computational results.

kmax = 1000 ;
T0 = 5 ;                % x(t) and y(t) have a fundamental period of five
dt = T0/100 ;           % Set the time increment for computing samples of x(t)
                        % and y(t)
t = -T0:dt:T0 ;         % Set the time vector for computing samples of x(t)
                        % and y(t)
x = 0*t ; y = x ;       % Initialize x and y each to a vector of zeros the same
                        % length as t
%   Compute samples of x(t) and y(t) in a for loop
for k = -kmax:kmax,
    % Compute samples of x for one k
    xk = sinc(k/5)^2*exp(j*2*pi*k*t/5)/5 ;
    % Add xk to previous x
    x = x + xk ;
    % Compute samples of y for one k
    yk = xk/((j*2*pi*k/5)^2 + 0.04*j*2*pi*k/5 + 1.58) ;
    % Add yk to previous y
    y = y + yk ;
end
```

NUMERICAL COMPUTATION OF THE FOURIER SERIES

Let's consider an example of a different kind of signal for which we might want to find the CTFS (Figure 6.15). This signal presents some problems. It is not at all obvious how to describe it, other than graphically. It is not sinusoidal, or any other obvious mathematical functional form. Up to this time in our study of the CTFS, in order to find a CTFS harmonic function of a signal, we needed a mathematical description of it. But just because we cannot describe a signal mathematically does not mean it does not have a CTFS description. Most real signals that we might want to analyze in practice do not have a known exact mathematical description. If we have a set of samples of the signal taken from one period, we can estimate the CTFS harmonic function *numerically*. The more samples we have, the better the estimate (Figure 6.16).

The harmonic function is

$$c_x[k] = \frac{1}{T}\int_T x(t) e^{-j2\pi k f t/T} dt.$$

6.2 The Continuous-Time Fourier Series

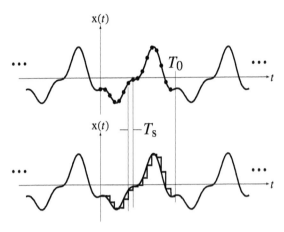

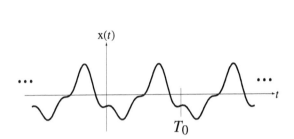

Figure 6.15
An arbitrary periodic signal

Figure 6.16
Sampling the arbitrary periodic signal to estimate its CTFS harmonic function

Since the starting point of the integral is arbitrary, for convenience set it to $t = 0$

$$c_x[k] = \frac{1}{T}\int_0^T x(t)e^{-j2\pi kt/T} dt.$$

We don't know the function $x(t)$ but if we have a set of N samples over one period starting at $t = 0$, the time between samples is $T_s = T/N$, and we can approximate the integral by the sum of several integrals, each covering a time of length T_s

$$c_x[k] \cong \frac{1}{T}\sum_{n=0}^{N-1}\left[\int_{nT_s}^{(n+1)T_s} x(nT_s)e^{-j2\pi knT_s/T} dt\right] \quad (6.8)$$

(In Figure 6.16, the samples extend over one fundamental period but they could extend over any period and the analysis would still be correct.) If the samples are close enough together $x(t)$ does not change much between samples and the integral (6.8) becomes a good approximation. The details of the integration process are in Web Appendix F where it is shown that, for harmonic numbers $|k| << N$, we can approximate the harmonic function as

$$c_x[k] \cong \frac{1}{N}\sum_{n=0}^{N-1} x(nT_s)e^{-j2\pi nk/N}. \quad (6.9)$$

The summation on the right side of (6.9)

$$\sum_{n=0}^{N-1} x(nT_s)e^{-j2\pi nk/N}$$

is a very important operation in signal processing called the **discrete Fourier transform (DFT)**. So (6.9) can be written as

$$\boxed{c_x[k] \cong (1/N)\mathscr{DFT}(x(nT_s)), \quad |k| << N}. \quad (6.10)$$

where

$$\mathscr{DFT}(x(nT_s)) = \sum_{n=0}^{N-1} x(nT_s)e^{-j2\pi nk/N}.$$

The DFT takes a set of samples representing a periodic function over one period and returns another set of numbers representing an approximation to its CTFS harmonic function, multiplied by the number of samples N. It is a built-in function in modern high-level programming languages like MATLAB. In MATLAB the name of the function is `fft`, which stands for **fast Fourier transform**. The fast Fourier transform is an efficient algorithm for computing the DFT. (The DFT and FFT are covered in more detail in Chapter 7.)

The simplest syntax of `fft` is `X = fft(x)` where `x` is a vector of N samples of a function indexed by n in the range $0 \leq n < N$ and `X` is a vector of N returned numbers indexed by k in the range $0 \leq k < N$.

The DFT

$$\sum_{n=0}^{N-1} x(nT_s)e^{-j2\pi nk/N}$$

is periodic in k with period N. This can be shown by finding $X[k+N]$.

$$X[k+N] = \frac{1}{N}\sum_{n=0}^{N-1} x(nT_s)e^{-j2\pi n(k+N)/N} = \frac{1}{N}\sum_{n=0}^{N-1} x(nT_s)e^{-j2\pi nk/N} \underbrace{e^{-j2\pi n}}_{=1} = X[k].$$

The approximation (6.9) is for $|k| << N$. This includes some negative values of k. But the `fft` function returns values of the DFT in the range $0 \leq k < N$. The values of the DFT for negative k are the same as the values of k in a positive range that are separated by one period. So, for example, to find $X[-1]$, find its periodic repetition $X[N-1]$, which is included in the range $0 \leq k < N$.

This numerical technique to find the CTFS harmonic function can also be useful in cases in which the functional form of $x(t)$ is known but the integral

$$c_x[k] = \frac{1}{T}\int_T x(t)e^{-j2\pi kt/T}dt$$

cannot be done analytically.

EXAMPLE 6.3

Using the DFT to approximate the CTFS

Find the approximate CTFS harmonic function of a periodic signal $x(t)$, one period of which is described by

$$x(t) = \sqrt{1-t^2}, \quad -1 \leq t < 1.$$

The fundamental period of this signal is 2. So we can choose any integer multiple of 2 as the time over which samples are taken (the representation time T). Choose 128 samples over one fundamental period. The following MATLAB program finds and graphs the CTFS harmonic function using the DFT.

```
% Program to approximate, using the DFT, the CTFS of a
% periodic signal described over one period by
% x(t) = sqrt(1-t^2), -1 < t < 1
N = 128 ;                       % Number of samples
T0 = 2 ;                        % Fundamental period
T = T0 ;                        % Representation time
Ts = T/N ;                      % Time between samples
```

6.2 The Continuous-Time Fourier Series

```
fs = 1/Ts ;                    % Sampling rate
n = [0:N-1]' ;                 % Time index for sampling
t = n*Ts ;                     % Sampling times

% Compute values of x(t) at the sampling times
x = sqrt(1-t.^2).*rect(t/2) +...
    sqrt(1-(t-2).^2).*rect((t-2)/2) +...
    sqrt(1-(t-4).^2).*rect((t-4)/2) ;
cx = fft(x)/N ;                % DFT of samples
k = [0:N/2-1]' ;               % Vector of harmonic numbers

% Graph the results
subplot(3,1,1) ;
p = plot(t,x,'k'); set(p,'LineWidth',2); grid on ; axis('equal');
axis([0,4,0,1.5]) ;
xlabel('Time, t (s)') ; ylabel('x(t)') ;
subplot(3,1,2) ;
p = stem(k,abs(cx(1:N/2)),'k') ; set(p,'LineWidth',2,'MarkerSize',4) ; grid on ;
xlabel('Harmonic Number, k') ; ylabel('|c_x[k]|') ;
subplot(3,1,3) ;
p = stem(k,angle(cx(1:N/2)),'k') ; set(p,'LineWidth',2,'MarkerSize',4) ;
grid on ;
xlabel('Harmonic Number, k') ; ylabel('Phase of c_x[k]') ;
```

Figure 6.17 is the graphical output of the program.

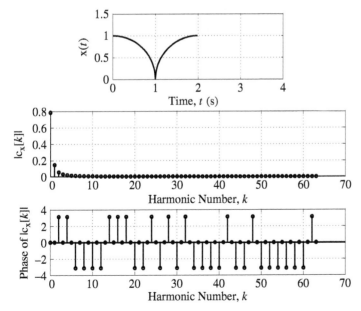

Figure 6.17
$x(t)$ and $c_x[k]$

Only three distinct values of phase occur in the phase graph, 0, π, and $-\pi$. The phases π and $-\pi$ are equivalent so they could all have been graphed as either π or $-\pi$. MATLAB computed the phase and, because of round-off errors in its computations, it sometimes chose a number very close to π and other times chose a number very close to $-\pi$.

The graphs of the magnitude and phase of $c_x[k]$ in Figure 6.17 are graphed only for k in the range $0 \leq k < N/2$. Since $c_x[k] = c_x^*[-k]$ this is sufficient to define $c_x[k]$ in the range $-N/2 \leq k < N/2$. It is often desirable to graph the harmonic function over the range $-N/2 \leq k < N/2$. That can be done by realizing that the numbers returned by the DFT are exactly one period of a periodic function. That being the case, the second half of these numbers covering the range $N/2 \leq k < N$ is exactly the same as the set of numbers that occur in the range $-N/2 \leq k < 0$. There is a function fftshift in MATLAB that swaps the second half of the set of numbers with the first half. Then the full set of N numbers covers the range $-N/2 \leq k < N/2$ instead of the range $0 \leq k < N$.

We can change the MATLAB program to analyze the signal over two fundamental periods instead of one by changing the line

```
T = T0 ;              %       Representation time
```
to
```
T = 2*T0 ;            %       Representation time
```

The results are illustrated in Figure 6.18.

Notice that now the CTFS harmonic function is zero for all odd values of k. That occurred because we used two fundamental periods of x(t) as the representation time T. The fundamental frequency of the CTFS representation is half the fundamental frequency of x(t). The signal power is at the fundamental frequency of x(t) and its harmonics, which are the even-numbered harmonics of this CTFS harmonic function. So only the even-numbered harmonics are nonzero. The kth harmonic in the previous analysis using one fundamental period as the representation time is the same as the ($2k$)th harmonic in this analysis.

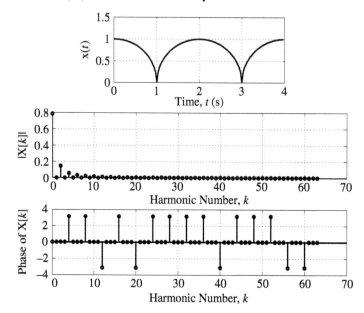

Figure 6.18
x(t) and X[k] using two fundamental periods as the representation time instead of one

EXAMPLE 6.4

Total harmonic distortion computation

A figure of merit for some systems is **total harmonic distortion (THD)**. If the excitation signal of the system is a sinusoid, the THD of the response signal is the total signal power in the response signal of all the harmonics other than the fundamental ($k = \pm 1$) divided by the total signal power in the response signal at the fundamental ($k = \pm 1$).

An audio amplifier with a nominal gain of 100 at 4 kHz is driven by a 4 kHz sine wave with a peak amplitude of 100 mV. The ideal response of the amplifier would be $x_i(t) = 10 \sin(8000\pi t)$ volts but the actual amplifier output signal $x(t)$ is limited to a range of ± 7 volts. So the actual response signal is correct for all voltages of magnitude less than 7 volts but for all ideal voltages greater than 7 in magnitude the response is "clipped" at ± 7 volts. Compute the THD of the response signal.

The CTFS harmonic function of $x(t)$ can be found analytically but it is a rather long, tedious, error-prone process. If we are only interested in a numerical THD we can find it numerically, using the DFT and a computer. That is done in the following MATLAB program and the results are illustrated in Figure 6.19.

```
f0 = 4000 ;            % Fundamental frequency of signal
T0 = 1/f0 ;            % Fundamental period of signal
N = 256 ;              % Number of samples to use in one period
Ts = T0/N ;            % Time between samples
fs = 1/Ts ;            % Sampling rate in samples/second
t = Ts*[0:N-1]' ;      % Time vector for graphing signals
```

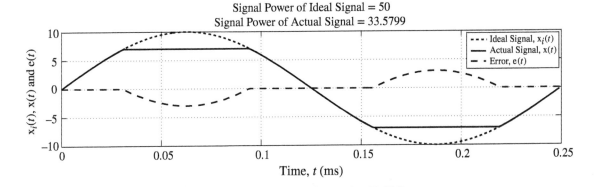

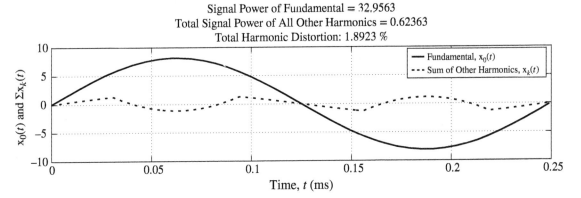

Figure 6.19
Results of THD computation

```
A = 10 ;                        % Ideal signal amplitude
xi = A*sin(2*pi*f0*t) ;         % Ideal signal
Pxi = A^2/2 ;                   % Signal power of ideal signal
x = min(xi,0.7*A) ;             % Clip ideal signal at 7 volts
x = max(x,-0.7*A) ;             % Clip ideal signal at -7 volts
Px = mean(x.^2) ;               % Signal power of actual signal
cx = fftshift(fft(x)/N);        % Compute harmonic function values up to k
                                % = +/- 128
k = [-N/2:N/2-1]' ;             % Vector of harmonic numbers
I0 = find(abs(k) == 1);         % Find harmonic function values at
                                % fundamental
P0 = sum(abs(cx(I0)).^2);       % Compute signal power of fundamental
Ik = find(abs(k) ~= 1) ;        % Find harmonic function values not at
                                % fundamental
Pk = sum(abs(cx(Ik)).^2);       % Compute signal power in harmonics
THD = Pk*100/P0 ;               % Compute total harmonic distortion

% Compute values of fundamental component of actual signal
x0 = 0*t ; for kk = 1:length(I0), x0 = x0 + cx(I0(kk))*exp(j*2*pi*
k(I0(kk))*f0*t) ; end

% Compute values of sum of signal components not at fundamental in
% actual signal
xk = 0*t ; for kk = 1:length(Ik), xk = xk + cx(Ik(kk))*exp(j*2*pi*
k(Ik(kk))*f0*t) ; end
x0 = real(x0);    % Remove any residual imaginary parts due to round-off
xk = real(xk);    % Remove any residual imaginary parts due to round-off

% Graph the results and report signal powers and THD

ttl = ['Signal Power of Ideal Signal = ',num2str(Pxi)] ;
ttl = str2mat(ttl,['Signal Power of Actual Signal = ', num2str(Px)]);
subplot(2,1,1) ;
ptr = plot(1000*t,xi,'k:',1000*t,x,'k',1000*t,x-xi,'k--') ; grid on ;
set(ptr,'LineWidth',2) ;
xlabel('Time, {\itt} (ms)','FontName','Times','FontSize',24) ;
ylabel('x_i({\itt}), x({\itt}) and e({\itt})','FontName','Times','
FontSize',24) ;
title(ttl,'FontName','Times','FontSize',24) ;
ptr = legend('Ideal Signal, x_i({\itt})','Actual Signal, x({\itt})',
'Error, e({\itt})') ;
set(ptr,'FontName','Times','FontSize',18) ;
set(gca,'FontSize',18) ;
subplot(2,1,2) ;
ttl = ['Signal Power of Fundamental = ',num2str(P0)] ;
ttl = str2mat(ttl,['Total Signal Power of All Other Harmonics = ',
num2str(Pk)]) ;
```

```
ttl = str2mat(ttl,['Total Harmonic Distortion: ',num2str(THD),'%']) ;
ptr = plot(1000*t,x0,'k',1000*t,xk,'k:') ; grid on ; set(ptr,'LineWidth',2) ;
xlabel('Time, {\itt} (ms)','FontName','Times','FontSize',24) ;
ylabel('x_0({\itt}) and \Sigma x_{\itk}({\itt})','FontName','Times',
'FontSize',24) ;
title(ttl,'FontName','Times','FontSize',24) ;
ptr = legend('Fundamental, x_0({\itt})','Sum of Other Harmonics, x_{\itk}
({\itt})') ;
set(ptr,'FontName','Times','FontSize',18) ;
set(gca,'FontSize',18) ;
```

The THD is 1.8923% even with this severe 30% clipping at each positive and negative peak. Therefore for good signal fidelity THD should generally be much smaller than 1%.

6.3 THE CONTINUOUS-TIME FOURIER TRANSFORM

The CTFS can represent any periodic signal with engineering usefulness over all time. Of course, some important signals are not periodic. So it would be useful to somehow extend the CTFS to also be able to represent aperiodic signals over all time. This can be done and the result is called the **Fourier transform**.

EXTENDING THE FOURIER SERIES TO APERIODIC SIGNALS

The salient difference between a periodic signal and an aperiodic signal is that a periodic signal repeats in a finite time T called the *period*. It has been repeating with that period forever and will continue to repeat with that period forever. An aperiodic signal does not have a finite period. An aperiodic signal may repeat a pattern many times within some finite time, but not over all time. The transition between the Fourier series and the Fourier transform is accomplished by finding the form of the Fourier series for a periodic signal and then letting the period approach infinity. Mathematically, saying that a function is aperiodic and saying that a function has an infinite period are saying the same thing.

Consider a time-domain signal x(t) consisting of rectangular pulses of height A and width w with fundamental period T_0 (Figure 6.20). This signal will illustrate the phenomena that occur in letting the fundamental period approach infinity for a general signal. Representing this pulse train with a complex CTFS, the harmonic function is found to be $c_x[k] = (Aw/T_0)\text{sinc}(kw/T_0)$ (with $T = T_0$).

Suppose $w = T_0/2$ (meaning the waveform is at A half the time and at zero the other half, a 50% duty cycle). Then $c_x[k] = (A/2)\text{sinc}(k/2)$ (Figure 6.21).

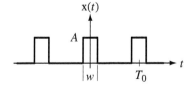

Figure 6.20
Rectangular-wave signal

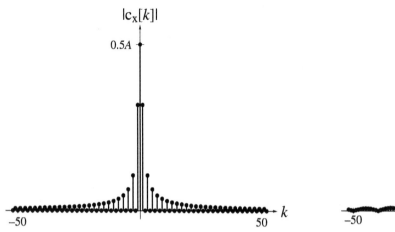

Figure 6.21
The magnitude of the CTFS harmonic function of a 50% duty-cycle rectangular-wave signal

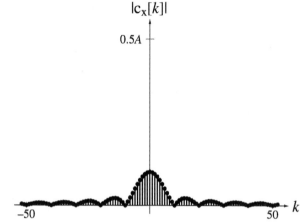

Figure 6.22
The magnitude of the CTFS harmonic function for a rectangular-wave signal with reduced duty cycle

Now let the fundamental period T_0 increase from 1 to 5 while w is unchanged. Then $c_x[0]$ becomes $1/10$ and the CTFS harmonic function is $c_x[k] = (1/10)\text{sinc}(k/10)$ (Figure 6.22).

The maximum harmonic amplitude magnitude is 5 times smaller than before because the average value of the function is 5 times smaller than before. As the fundamental period T_0 gets larger, the harmonic amplitudes lie on a wider sinc function whose amplitude goes down as T_0 increases. In the limit as T_0 approaches infinity, the original time-domain waveform $x(t)$ approaches a single rectangular pulse at the origin and the harmonic function approaches samples from an infinitely wide sinc function with zero amplitude. If we were to multiply $c_x[k]$ by T_0 before graphing it, the amplitude would not go to zero as T_0 approached infinity but would stay where it is and simply trace points on a widening sinc function. Also, graphing against $k/T_0 = kf_0$ instead of k would make the horizontal scale be frequency instead of harmonic number and the sinc function would remain the same width on that scale as T_0 increases (and f_0 decreases). Making those changes, the last two graphs would look like Figure 6.23.

Call this a "modified" harmonic function. For this modified harmonic function, $T_0 c_x[k] = Aw\text{sinc}(wkf_0)$. As T_0 increases without bound (making the pulse train a

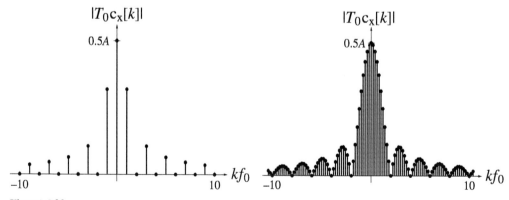

Figure 6.23
Magnitudes of the modified CTFS harmonic functions for rectangular-wave signals of 50% and 10% duty cycles

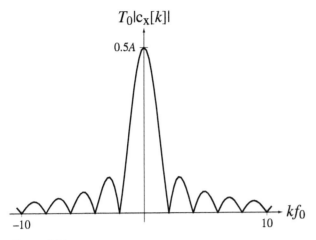

Figure 6.24
Limiting form of modified CTFS harmonic function for rectangular-wave signal

single pulse), f_0 approaches zero and the discrete variable kf_0 approaches a continuous variable (which we will call f). The modified CTFS harmonic function approaches the function illustrated in Figure 6.24. This modified harmonic function (with some notation changes) becomes the **continuous-time Fourier transform (CTFT)** of that single pulse.

The frequency difference between adjacent CTFS harmonic amplitudes is the same as the fundamental frequency of the CTFS representation, $f_0 = 1/T_0$. To emphasize its relationship to a frequency differential (which it will become in the limit as the fundamental period goes to infinity) let this spacing be called Δf. That is, let $\Delta f = f_0 = 1/T_0$. Then the complex CTFS representation of $x(t)$ can be written as

$$x(t) = \sum_{k=-\infty}^{\infty} c_x[k] e^{j2\pi k \Delta f t}.$$

Substituting the integral expression for $c_x[k]$,

$$x(t) = \sum_{k=-\infty}^{\infty} \left[\frac{1}{T_0} \int_{t_0}^{t_0+T_0} x(\tau) e^{-j2\pi k \Delta f \tau} d\tau \right] e^{j2\pi k \Delta f t}.$$

(The variable of integration is τ to distinguish it from the t in the function $e^{j2\pi k \Delta f t}$, which is outside the integral.) Since the starting point t_0 for the integral is arbitrary, let it be $t_0 = -T_0/2$. Then

$$x(t) = \sum_{k=-\infty}^{\infty} \left[\int_{-T_0/2}^{T_0/2} x(\tau) e^{-j2\pi k \Delta f \tau} d\tau \right] e^{j2\pi k \Delta f t} \Delta f$$

where Δf has replaced $1/T_0$. In the limit as T_0 approaches infinity, Δf approaches the differential df, $k\Delta f$ becomes a continuous variable f, the integration limits approach plus and minus infinity, and the summation becomes an integral

$$x(t) = \lim_{T_0 \to \infty} \left\{ \sum_{k=-\infty}^{\infty} \left[\int_{-T_0/2}^{T_0/2} x(\tau) e^{-j2\pi k \Delta f \tau} d\tau \right] e^{j2\pi k \Delta f t} \Delta f \right\}$$

$$= \int_{-\infty}^{\infty} \left[\int_{-\infty}^{\infty} x(\tau) e^{-j2\pi f \tau} d\tau \right] e^{j2\pi f t} df. \qquad (6.11)$$

The bracketed quantity on the right side of (6.11) is the CTFT of x(t)

$$X(f) = \int_{-\infty}^{\infty} x(t) e^{-j2\pi ft} dt \qquad (6.12)$$

and it follows that

$$x(t) = \int_{-\infty}^{\infty} X(f) e^{j2\pi ft} df. \qquad (6.13)$$

Here we have adopted the convention that the Fourier transform of a signal is indicated by the same letter of the alphabet but uppercase instead of lowercase. Notice that the Fourier transform is a function of cyclic frequency f and that the time dependence of the signal has been "integrated out" so the Fourier transform is not a function of time. The time-domain function (x) and its CTFT (X) form a "Fourier transform pair" that is often indicated by the notation $x(t) \xleftrightarrow{\mathscr{F}} X(f)$. Also conventional notation is $X(f) = \mathscr{F}(x(t))$ and $x(t) = \mathscr{F}^{-1}(X(f))$ where $\mathscr{F}(\cdot)$ is read "Fourier transform of" and $\mathscr{F}^{-1}(\cdot)$ is read "inverse Fourier transform of."

Another common form of the Fourier transform is defined by making the change of variable $f = \omega/2\pi$ where ω is radian frequency.

$$X(\omega/2\pi) = \int_{-\infty}^{\infty} x(t) e^{-j\omega t} dt \quad \text{and} \quad x(t) = \frac{1}{2\pi} \int_{-\infty}^{\infty} X(\omega/2\pi) e^{j\omega t} d\omega. \qquad (6.14)$$

This is the result we obtain by simply substituting $\omega/2\pi$ for f and $d\omega/2\pi$ for df. It is much more common in engineering literature to see this form written as

$$X(\omega) = \int_{-\infty}^{\infty} x(t) e^{-j\omega t} dt \quad \text{and} \quad x(t) = \frac{1}{2\pi} \int_{-\infty}^{\infty} X(\omega) e^{j\omega t} d\omega. \qquad (6.15)$$

In this second form, the strict mathematical meaning of the function "X" has changed and that could be a source of confusion if conversion between the two forms is necessary. To compound the confusion, it is also quite common to see the ω form written as

$$X(j\omega) = \int_{-\infty}^{\infty} x(t) e^{-j\omega t} dt \quad \text{and} \quad x(t) = \frac{1}{2\pi} \int_{-\infty}^{\infty} X(j\omega) e^{j\omega t} d\omega \qquad (6.16)$$

again changing the meaning of the function "X." The reason for including a j in the functional argument is to make the Fourier transform more directly compatible with the Laplace transform (Chapter 8).

Suppose we have used

$$X(f) = \int_{-\infty}^{\infty} x(t) e^{-j2\pi ft} dt$$

to form the Fourier pair

$$x(t) = e^{-\alpha t} u(t) \xleftrightarrow{\mathscr{F}} X(f) = \frac{1}{j2\pi f + \alpha}.$$

Ordinarily, in mathematical functional notation, if we were to then refer to a function $X(j\omega)$ we would mean

$$X(f) \xrightarrow{f \to j\omega} X(j\omega) = \frac{1}{j2\pi(j\omega) + \alpha} = \frac{1}{-2\pi\omega + \alpha}.$$

But in Fourier transform literature it is very common to say that if the cyclic-frequency form of a Fourier transform is

$$X(f) = \frac{1}{j2\pi f + \alpha}$$

then the radian-frequency form is

$$X(j\omega) = \frac{1}{j2\pi(\omega/2\pi) + \alpha} = \frac{1}{j\omega + \alpha}.$$

In going from $X(f)$ to $X(j\omega)$ what we have really done is to go from $X(f)$ to $x(t)$ using $x(t) = \int_{-\infty}^{\infty} X(f)e^{j2\pi ft}df$ and then, using $X(j\omega) = \int_{-\infty}^{\infty} x(t)e^{-j\omega t}dt$, find $X(j\omega)$. In other words, $X(f) \xrightarrow{\mathcal{F}^{-1}} x(t) \xrightarrow{\mathcal{F}} X(j\omega)$. This amounts to making the transition $X(f) \xrightarrow{f \to \omega/2\pi} X(j\omega)$ instead of $X(f) \xrightarrow{f \to j\omega} X(j\omega)$. In this text we will follow this traditional interpretation.

In any analysis it is important to choose a definition and then use it consistently. In this text the forms

$$x(t) = \int_{-\infty}^{\infty} X(f)e^{j2\pi ft}df \xleftarrow{\mathcal{F}} X(f) = \int_{-\infty}^{\infty} x(t)e^{-j2\pi ft}dt$$

$$x(t) = \frac{1}{2\pi}\int_{-\infty}^{\infty} X(j\omega)e^{j\omega t}d\omega \xleftarrow{\mathcal{F}} X(j\omega) = \int_{-\infty}^{\infty} x(t)e^{-j\omega t}dt$$

will be used for the f and ω forms because those are the two most often encountered in engineering literature. The Fourier transform, as introduced here, applies to continuous-time signals. The CTFT is widely used in the analysis of communication systems, filters, and Fourier optics.

The ω form and the f form of the CTFT are both widely used in engineering. Which one is used in any particular book or article depends on multiple factors including the traditional notation conventions in a particular field and the personal preference of the author. Since both forms are in common usage, in this text we will use whichever form seems to be the most convenient in any individual analysis. If, at any time, we need to change to the other form, that is usually easily done by simply replacing f by $\omega/2\pi$ or ω by $2\pi f$. (In addition to the definitions presented here there are also several other alternate definitions of the Fourier transform that can be found in engineering, mathematics, and physics books.)

Table 6.3 lists some CTFT pairs in the ω form derived directly from the definitions presented above. The ω form was used here because, for these functions, it is a little more compact.

Table 6.3 Some CTFT pairs

$$\delta(t) \xleftrightarrow{\mathcal{F}} 1$$

$$e^{-\alpha t}u(t) \xleftrightarrow{\mathcal{F}} 1/(j\omega + \alpha), \ \alpha > 0 \qquad -e^{-\alpha t}u(-t) \xleftrightarrow{\mathcal{F}} 1/(j\omega + \alpha), \ \alpha < 0$$

$$te^{-\alpha t}u(t) \xleftrightarrow{\mathcal{F}} 1/(j\omega + \alpha)^2, \ \alpha > 0 \qquad -te^{-\alpha t}u(-t) \xleftrightarrow{\mathcal{F}} 1/(j\omega + \alpha)^2, \ \alpha < 0$$

$$t^n e^{-\alpha t}u(t) \xleftrightarrow{\mathcal{F}} \frac{n!}{(j\omega + \alpha)^{n+1}}, \ \alpha > 0 \qquad -t^n e^{-\alpha t}u(-t) \xleftrightarrow{\mathcal{F}} \frac{n!}{(j\omega + \alpha)^{n+1}}, \ \alpha < 0$$

$$e^{-\alpha t}\sin(\omega_0 t)u(t) \xleftrightarrow{\mathcal{F}} \frac{\omega_0}{(j\omega + \alpha)^2 + \omega_0^2}, \ \alpha > 0 \qquad -e^{-\alpha t}\sin(\omega_0 t)u(-t) \xleftrightarrow{\mathcal{F}} \frac{\omega_0}{(j\omega + \alpha)^2 + \omega_0^2}, \ \alpha < 0$$

$$e^{-\alpha t}\cos(\omega_0 t)u(t) \xleftrightarrow{\mathcal{F}} \frac{j\omega + \alpha}{(j\omega + \alpha)^2 + \omega_0^2}, \ \alpha > 0 \qquad -e^{-\alpha t}\cos(\omega_0 t)u(-t) \xleftrightarrow{\mathcal{F}} \frac{j\omega + \alpha}{(j\omega + \alpha)^2 + \omega_0^2}, \ \alpha < 0$$

$$e^{-\alpha|t|} \xleftrightarrow{\mathcal{F}} \frac{2\alpha}{\omega^2 + \alpha^2}, \ \alpha > 0$$

THE GENERALIZED FOURIER TRANSFORM

There are some important practical signals that do not have Fourier transforms in the strict sense. Because these signals are so important, the Fourier transform has been "generalized" to include them. As an example of the generalized Fourier transform, let's find the CTFT of a very simple function $x(t) = A$, a constant. Using the CTFT definition

$$x(t) = \int_{-\infty}^{\infty} X(f)e^{+j2\pi ft} df \xleftrightarrow{\mathcal{F}} X(f) = \int_{-\infty}^{\infty} x(t)e^{-j2\pi ft} dt.$$

we obtain

$$X(f) = \int_{-\infty}^{\infty} Ae^{-j2\pi ft} dt = A \int_{-\infty}^{\infty} e^{-j2\pi ft} dt.$$

The integral does not converge. Therefore, strictly speaking, the Fourier transform does not exist. But we can avoid this problem by generalizing the Fourier transform with the following procedure. First we will find the CTFT of $x_\sigma(t) = Ae^{-\sigma|t|}$, $\sigma > 0$, a function that approaches the constant A as σ approaches zero. Then we will let σ approach zero *after* finding the transform. The factor $e^{-\sigma|t|}$ is a **convergence factor** that allows us to evaluate the integral (Figure 6.25).

The transform is

$$X_\sigma(f) = \int_{-\infty}^{\infty} Ae^{-\sigma|t|}e^{-j2\pi ft} dt = \int_{-\infty}^{0} Ae^{\sigma t}e^{-j2\pi ft} dt + \int_{0}^{\infty} Ae^{-\sigma t}e^{-j2\pi ft} dt$$

$$X_\sigma(f) = A\left[\int_{-\infty}^{0} e^{(\sigma - j2\pi f)t} dt + \int_{0}^{\infty} e^{(-\sigma - j2\pi f)t} dt\right] = A\frac{2\sigma}{\sigma^2 + (2\pi f)^2}$$

Now take the limit, as σ approaches zero, of $X_\sigma(f)$. For $f \neq 0$,

$$\lim_{\sigma \to 0} A\frac{2\sigma}{\sigma^2 + (2\pi f)^2} = 0.$$

Next find the area under the function $X_\sigma(f)$ as σ approaches zero.

$$\text{Area} = A\int_{-\infty}^{\infty} \frac{2\sigma}{\sigma^2 + (2\pi f)^2} df$$

Using

$$\int \frac{dx}{a^2 + (bx)^2} = \frac{1}{ab}\tan^{-1}\left(\frac{bx}{a}\right)$$

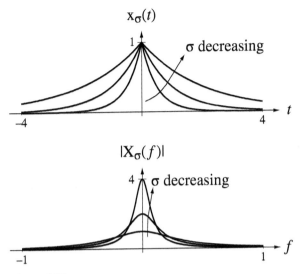

Figure 6.25
Effect of the convergence factor $e^{-\sigma|t|}$

we get

$$\text{Area} = A\left[\frac{2\sigma}{2\pi\sigma}\tan^{-1}\left(\frac{2\pi f}{\sigma}\right)\right]_{-\infty}^{\infty} = \frac{A}{\pi}\left(\frac{\pi}{2} + \frac{\pi}{2}\right) = A.$$

The area under the function is A and is *independent of* the value of σ. Therefore in the limit $\sigma \to 0$, the Fourier transform of the constant A is a function that is zero for $f \neq 0$ and has an area of A. This exactly describes an impulse of strength A occurring at $f = 0$. Therefore we can form the generalized Fourier-transform pair

$$A \xleftrightarrow{\mathcal{F}} A\delta(f).$$

The generalization of the CTFT extends it to other useful functions, including periodic functions. By similar reasoning the CTFT transform pairs

$$\cos(2\pi f_0 t) \xleftrightarrow{\mathcal{F}} (1/2)[\delta(f - f_0) + \delta(f + f_0)]$$

and

$$\sin(2\pi f_0 t) \xleftrightarrow{\mathcal{F}} (j/2)[\delta(f + f_0) - \delta(f - f_0)]$$

can be found. By making the substitution $f = \omega/2\pi$ and using the scaling property of the impulse, the equivalent radian-frequency forms of these transforms are found to be

$$A \xleftrightarrow{\mathcal{F}} 2\pi A\delta(\omega)$$
$$\cos(\omega_0 t) \xleftrightarrow{\mathcal{F}} \pi[\delta(\omega - \omega_0) + \delta(\omega + \omega_0)]$$
$$\sin(\omega_0 t) \xleftrightarrow{\mathcal{F}} j\pi[\delta(\omega + \omega_0) - \delta(\omega - \omega_0)].$$

The problem that caused the need for a generalized form of the Fourier transform is that these functions, constants, and sinusoids, are not absolutely integrable, even though they are bounded. The generalized Fourier transform can also be applied to other signals that are not absolutely integrable but are bounded, for example, the unit step and the signum.

Another way of finding the CTFT of a constant is to approach the problem from the other side by finding the inverse CTFT of an impulse $X(f) = A\delta(f)$ using the sampling property of the impulse.

$$x(t) = \int_{-\infty}^{\infty} X(f)e^{+j2\pi ft}df = A\int_{-\infty}^{\infty}\delta(f)e^{+j2\pi ft}df = Ae^0 = A$$

This is obviously a much quicker route to finding the forward transform of a constant than the preceding development. But the problem with this approach is that if we are trying to find the forward transform of a function we must first guess at the transform and then evaluate whether it is correct by finding the inverse transform.

EXAMPLE 6.5

CTFT of the signum and unit-step functions

Find the CTFT of $x(t) = \text{sgn}(t)$ and then use that result to find the CTFT of $x(t) = u(t)$.

Applying the integral formula directly we get

$$X(f) = \int_{-\infty}^{\infty} \text{sgn}(t)e^{-j2\pi ft}dt = -\int_{-\infty}^{0}e^{-j2\pi ft}dt + \int_{0}^{\infty}e^{-j2\pi ft}dt$$

and these integrals do not converge. We can use a convergence factor to find the generalized CTFT. Let $x_\sigma(t) = \text{sgn}(t)e^{-\sigma|t|}$ with $\sigma > 0$. Then

$$X_\sigma(f) = \int_{-\infty}^{\infty} \text{sgn}(t)e^{-\sigma|t|}e^{-j2\pi ft}dt = -\int_{-\infty}^{0}e^{(\sigma-j2\pi f)t}dt + \int_{0}^{\infty}e^{-(\sigma+j2\pi f)t}dt,$$

$$X_\sigma(f) = -\left.\frac{e^{(\sigma-j2\pi f)t}}{\sigma - j2\pi f}\right|_{-\infty}^{0} - \left.\frac{e^{-(\sigma+j2\pi f)t}}{\sigma + j2\pi f}\right|_{0}^{\infty} = -\frac{1}{\sigma - j2\pi f} + \frac{1}{\sigma + j2\pi f}$$

and

$$X(f) = \lim_{\sigma \to 0} X_\sigma(f) = 1/j\pi f$$

or in the radian-frequency form

$$X(j\omega) = 2/j\omega.$$

To find the CTFT of $x(t) = u(t)$, we observe that

$$u(t) = (1/2)[\text{sgn}(t) + 1]$$

So the CTFT is

$$U(f) = \int_{-\infty}^{\infty} (1/2)[\text{sgn}(t) + 1]e^{-j2\pi ft}dt = (1/2)\left[\underbrace{\int_{-\infty}^{\infty} \text{sgn}(t)e^{-j2\pi ft}dt}_{=\mathscr{F}(\text{sgn}(t)) = 1/j\pi f} + \underbrace{\int_{-\infty}^{\infty} e^{-j2\pi ft}dt}_{=\mathscr{F}(1) = \delta(f)}\right]$$

$$U(f) = (1/2)[1/j\pi f + \delta(f)] = 1/j2\pi f + (1/2)\delta(f)$$

or in the radian-frequency form

$$U(j\omega) = 1/j\omega + \pi\delta(\omega).$$

EXAMPLE 6.6

Verify that the inverse CTFT of $U(f) = 1/j2\pi f + (1/2)\delta(f)$ is indeed the unit-step function

If we apply the inverse Fourier transform integral to $U(f) = 1/j2\pi f + (1/2)\delta(f)$ we get

$$u(t) = \int_{-\infty}^{\infty} [1/j2\pi f + (1/2)\delta(f)] e^{j2\pi ft} df = \int_{-\infty}^{\infty} \frac{e^{j2\pi ft}}{j2\pi f} df + (1/2) \underbrace{\int_{-\infty}^{\infty} \delta(f) e^{j2\pi ft} df}_{=1 \text{ by the sampling property of the impulse}}$$

$$u(t) = 1/2 + \underbrace{\int_{-\infty}^{\infty} \frac{\cos(2\pi ft)}{j2\pi f} df}_{=0 \text{ (odd integrand)}} + \underbrace{\int_{-\infty}^{\infty} \frac{\sin(2\pi ft)}{2\pi f} df}_{\text{even integrand}} = 1/2 + 2\int_{0}^{\infty} \frac{\sin(2\pi ft)}{2\pi f} df$$

Case 1. $t = 0$.

$$u(t) = 1/2 + 2\int_{0}^{\infty} (0) d\omega = 1/2$$

Case 2. $t > 0$
Let $\lambda = 2\pi ft \Rightarrow d\lambda = 2\pi t\, df$.

$$u(t) = 1/2 + 2\int_{0}^{\infty} \frac{\sin(\lambda)}{\lambda/t} \frac{d\lambda}{2\pi t} = \frac{1}{2} + \frac{1}{\pi}\int_{0}^{\infty} \frac{\sin(\lambda)}{\lambda} d\lambda$$

Case 3. $t < 0$

$$u(t) = 1/2 + 2\int_{0}^{-\infty} \frac{\sin(\lambda)}{\lambda/t} \frac{d\lambda}{2\pi t} = \frac{1}{2} + \frac{1}{\pi}\int_{0}^{-\infty} \frac{\sin(\lambda)}{\lambda} d\lambda$$

The integrals in Case 2 and Case 3 are sine integrals defined by

$$Si(z) = \int_{0}^{z} \frac{\sin(\lambda)}{\lambda} d\lambda$$

and we can find in standard mathematical tables that

$$\lim_{z\to\infty} Si(z) = \pi/2, \quad Si(0) = 0, \quad \text{and} \quad Si(-z) = -Si(z)$$

(Abramowitz and Stegun, p. 231). Therefore

$$2\int_{0}^{\infty} \frac{\sin(2\pi ft)}{2\pi f} df = \begin{cases} 1/2, & t > 0 \\ 0, & t = 0 \\ -1/2, & t < 0 \end{cases}$$

and

$$u(t) = \begin{cases} 1, & t > 0 \\ 1/2, & t = 0. \\ 0, & t < 0 \end{cases}$$

This inverse CTFT shows that, for complete compatibility with Fourier transform theory, the value of u(0) should be defined as 1/2 as it was in Chapter 2. Defining the unit step this way is mathematically consistent and can occasionally have some engineering significance (see Chapter 14—digital filter design using the impulse invariant technique).

EXAMPLE 6.7

CTFT of the unit-rectangle function

Find the CTFT of the unit-rectangle function.
The CTFT of the unit-rectangle function is

$$\mathscr{F}(\text{rect}(t)) = \int_{-\infty}^{\infty} \text{rect}(t) e^{-j2\pi ft} dt = \int_{-1/2}^{1/2} [\cos(2\pi ft) + j\sin(2\pi ft)] dt$$

$$\mathscr{F}(\text{rect}(t)) = 2 \int_{0}^{1/2} \cos(2\pi ft) dt = \frac{\sin(\pi f)}{\pi f} = \text{sinc}(f)$$

We now have the CTFT pair $\text{rect}(t) \xleftrightarrow{\mathscr{F}} \text{sinc}(f)$. (In the ω form the pair is $\text{rect}(t) \xleftrightarrow{\mathscr{F}} \text{sinc}(\omega/2\pi)$. In this case the f form is simpler and more symmetrical than the ω form.) Recall the result of Example 6.1

$$A \, \text{rect}(t/w) * \delta_{T_0}(t) \xleftrightarrow[T_0]{\mathscr{FS}} (Aw/T_0) \text{sinc}(wk/T_0).$$

The CTFT of a rectangle function is a sinc function and the CTFS harmonic function of a periodically repeated rectangle function is a "sampled" sinc function. It is sampled because k only takes on integer values. This relationship between periodic repetition in time and sampling in frequency (harmonic number) will be important in the exploration of sampling in Chapter 10.

We can now extend the Fourier transform table to include several other functions that often occur in Fourier analysis. In Table 6.4 we used the cyclic frequency form of the CTFT because, for these functions, it is somewhat simpler and more symmetrical.

Table 6.4 More Fourier transform pairs

$\delta(t) \xleftrightarrow{\mathscr{F}} 1$	$1 \xleftrightarrow{\mathscr{F}} \delta(f)$
$\text{sgn}(t) \xleftrightarrow{\mathscr{F}} 1/j\pi f$	$u(t) \xleftrightarrow{\mathscr{F}} (1/2)\delta(f) + 1/j2\pi f$
$\text{rect}(t) \xleftrightarrow{\mathscr{F}} \text{sinc}(f)$	$\text{sinc}(t) \xleftrightarrow{\mathscr{F}} \text{rect}(f)$
$\text{tri}(t) \xleftrightarrow{\mathscr{F}} \text{sinc}^2(f)$	$\text{sinc}^2(t) \xleftrightarrow{\mathscr{F}} \text{tri}(f)$
$\delta_{T_0}(t) \xleftrightarrow{\mathscr{F}} f_0 \delta_{f_0}(f), \, f_0 = 1/T_0$	$T_0 \delta_{T_0}(t) \xleftrightarrow{\mathscr{F}} \delta_{f_0}(f), \, T_0 = 1/f_0$
$\cos(2\pi f_0 t) \xleftrightarrow{\mathscr{F}} (1/2)[\delta(f - f_0) + \delta(f + f_0)]$	$\sin(2\pi f_0 t) \xleftrightarrow{\mathscr{F}} (j/2)[\delta(f + f_0) - \delta(f - f_0)]$

FOURIER TRANSFORM PROPERTIES

Table 6.5 and **Table 6.6** illustrate some properties of the CTFT derived directly from the two definitions.

Table 6.5 Fourier transform properties, f form

Linearity	$\alpha g(t) + \beta h(t) \xleftrightarrow{\mathcal{F}} \alpha G(f) + \beta H(f)$
Time-Shifting	$g(t - t_0) \xleftrightarrow{\mathcal{F}} G(f)e^{-j2\pi f t_0}$
Frequency Shifting	$e^{j2\pi f_0 t} g(t) \xleftrightarrow{\mathcal{F}} G(f - f_0)$
Time Scaling	$g(at) \xleftrightarrow{\mathcal{F}} (1/|a|)G(f/a)$
Frequency Scaling	$(1/|a|)g(t/a) \xleftrightarrow{\mathcal{F}} G(af)$
Time Differentiation	$\frac{d}{dt}g(t) \xleftrightarrow{\mathcal{F}} j2\pi f G(f)$
Time Integration	$\int_{-\infty}^{t} g(\lambda)d\lambda \xleftrightarrow{\mathcal{F}} \frac{G(f)}{j2\pi f} + (1/2)G(0)\delta(f)$
Frequency Differentiation	$tg(t) \xleftrightarrow{\mathcal{F}} -\frac{j}{2\pi}\frac{d}{df}G(f)$
Multiplication–	$g(t) * h(t) \xleftrightarrow{\mathcal{F}} G(f)H(f)$
Convolution Duality	$g(t)h(t) \xleftrightarrow{\mathcal{F}} G(f) * H(f)$
Parseval's Theorem	$\int_{-\infty}^{\infty} |g(t)|^2 dt = \int_{-\infty}^{\infty} |G(f)|^2 df$
Total Area	$X(0) = \int_{-\infty}^{\infty} x(t)dt$ or $x(0) = \int_{-\infty}^{\infty} X(f)df$

Table 6.6 Fourier transform properties, ω form

Linearity	$\alpha g(t) + \beta h(t) \xleftrightarrow{\mathcal{F}} \alpha G(j\omega) + \beta H(j\omega)$
Time-Shifting	$g(t - t_0) \xleftrightarrow{\mathcal{F}} G(j\omega)e^{-j\omega t_0}$
Frequency Shifting	$e^{j\omega_0 t} g(t) \xleftrightarrow{\mathcal{F}} G(j(\omega - \omega_0))$
Time Scaling	$g(at) \xleftrightarrow{\mathcal{F}} (1/|a|)G(j\omega/a)$
Frequency Scaling	$(1/|a|)g(t/a) \xleftrightarrow{\mathcal{F}} G(ja\omega)$
Time Differentiation	$\frac{d}{dt}g(t) \xleftrightarrow{\mathcal{F}} j\omega G(j\omega)$
Time Integration	$\int_{-\infty}^{t} g(\lambda)d\lambda \xleftrightarrow{\mathcal{F}} \frac{G(j\omega)}{j\omega} + \pi G(0)\delta(\omega)$
Frequency Differentiation	$tg(t) \xleftrightarrow{\mathcal{F}} j\frac{d}{d\omega}G(j\omega)$
Multiplication–	$g(t) * h(t) \xleftrightarrow{\mathcal{F}} G(j\omega)H(j\omega)$
Convolution Duality	$g(t)h(t) \xleftrightarrow{\mathcal{F}} \frac{1}{2\pi}G(j\omega) * H(j\omega)$
Parseval's Theorem	$\int_{-\infty}^{\infty} |g(t)|^2 dt = \frac{1}{2\pi}\int_{-\infty}^{\infty} |G(j\omega)|^2 d\omega$
Total Area	$X(0) = \int_{-\infty}^{\infty} x(t)dt$ or $x(0) = \frac{1}{2\pi}\int_{-\infty}^{\infty} X(j\omega)dw$

Any periodic signal can be expressed as a Fourier series of the form

$$x(t) = \sum_{k=-\infty}^{\infty} c_x[k] e^{j2\pi kt/T}.$$

Using the frequency shifting property, we can find the CTFT as

$$X(f) = \sum_{k=-\infty}^{\infty} c_x[k] \delta(f - k/T).$$

So the CTFT of any periodic signal consists entirely of impulses. The strengths of those impulses at cyclic frequencies k/T are the same as the values of the CTFS harmonic function at harmonic number k.

EXAMPLE 6.8

CTFS harmonic function of a periodic signal using the CTFT

Use

$$X(f) = \sum_{k=-\infty}^{\infty} c_x[k] \delta(f - k/T)$$

to find the CTFS harmonic function of $x(t) = \text{rect}(2t) * \delta_1(t)$.

This is a convolution of two functions. Therefore, from the multiplication–convolution duality property, the CTFT of $x(t)$ is the product of the CTFTs of the individual functions,

$$X(f) = (1/2)\text{sinc}(f/2)\delta_1(f) = (1/2) \sum_{k=-\infty}^{\infty} \text{sinc}(k/2)\delta(f - k)$$

and the CTFS harmonic function must therefore be

$$c_x[k] = (1/2)\text{sinc}(k/2)$$

based on $T = T_0 = 1$.

EXAMPLE 6.9

CTFT of a modulated sinusoid

Find the CTFT of $x(t) = 24\cos(100\pi t)\sin(10{,}000\pi t)$.

This is the product of two functions. Therefore, using the multiplication–convolution duality property, the CTFT will be the convolution of their individual CTFTs. Using

$$\cos(2\pi f_0 t) \xleftrightarrow{\mathcal{F}} (1/2)[\delta(f - f_0) + \delta(f + f_0)]$$

and

$$\sin(2\pi f_0 t) \xleftrightarrow{\mathcal{F}} (j/2)[\delta(f + f_0) - \delta(f - f_0)]$$

we get

$$24\cos(100\pi t) \xleftrightarrow{\mathcal{F}} 12[\delta(f - 50) + \delta(f + 50)]$$

and

$$\sin(10,000\pi t) \overset{\mathcal{F}}{\longleftrightarrow} (j/2)[\delta(f+5000) - \delta(f-5000)]$$

Then the overall CTFT is

$$24\cos(100\pi t)\sin(10,000\pi t) \overset{\mathcal{F}}{\longleftrightarrow} j6\begin{bmatrix}\delta(f+4950) - \delta(f-5050) \\ + \delta(f+5050) - \delta(f-4920)\end{bmatrix}$$

■

The time-shifting property says that a shift in time corresponds to a phase shift in frequency. As an example of why the time-shifting property makes sense, let the time signal be the complex sinusoid $x(t) = e^{j2\pi t}$. Then $x(t - t_0) = e^{j2\pi(t-t_0)} = e^{j2\pi t}e^{-j2\pi t_0}$ (Figure 6.26).

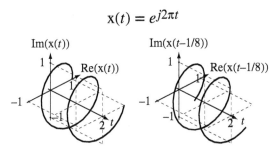

Figure 6.26
A complex exponential $x(t) = e^{j2\pi f_0 t}$ and a delayed version $x(t - 1/8) = e^{j2\pi f_0(t-1/8)}$

Shifting this signal in time corresponds to multiplying it by the complex number $e^{-j2\pi t_0}$. The CTFT expression

$$x(t) = \int_{-\infty}^{\infty} X(f)e^{+j2\pi ft}df$$

says that any signal that is Fourier transformable can be expressed as a linear combination of complex sinusoids over a continuum of frequencies f and, if $x(t)$ is shifted by t_0, each of those complex sinusoids is multiplied by the complex number $e^{-j2\pi ft_0}$. What happens to any complex number when it is multiplied by a complex exponential of the form e^{jx} where x is real? The magnitude of e^{jx} is one for any real x. Therefore multiplication by e^{jx} changes the phase, but not the magnitude, of the complex number. Changing the phase means changing its angle in the complex plane, which is a simple rotation of the vector representing the number. So multiplying a complex exponential function of time $e^{j2\pi t}$ by a complex constant $e^{-j2\pi t_0}$ rotates the complex exponential $e^{j2\pi t}$ with the time axis as the axis of rotation. Looking at Figure 6.26 it is apparent that, because of its unique helical shape, a rotation of a complex exponential function of time and a shift along the time axis have the same net effect.

The frequency-shifting property can be proven by starting with a frequency-shifted version of $X(f)$, $X(f - f_0)$ and using the inverse CTFT integral. The result is

$$x(t)e^{+j2\pi f_0 t} \overset{\mathcal{F}}{\longleftrightarrow} X(f - f_0).$$

Notice the similarity between the time-shifting and frequency-shifting properties. They both result in multiplying by a complex sinusoid in the other domain. However the sign of the exponent in the complex sinusoid is different. That occurs because of the signs in the forward and inverse CTFTs

$$X(f) = \int_{-\infty}^{\infty} x(t)e^{-j2\pi ft}dt, \quad x(t) = \int_{-\infty}^{\infty} X(f)e^{+j2\pi ft}df.$$

The frequency-shifting property is fundamental to understanding the effects of modulation in communication systems.

One consequence of the time-scaling and frequency-scaling properties is that a compression in one domain corresponds to an expansion in the other domain. One interesting way of illustrating that is through the function $x(t) = e^{-\pi t^2}$ whose CTFT is of the same functional form $e^{-\pi t^2} \xleftrightarrow{\mathcal{F}} e^{-\pi f^2}$. We can assign a characteristic width parameter w to these functions, the distance between inflection points (the time, or frequency between the points of maximum slope magnitude). Those points occur on $e^{-\pi t^2}$ at $t = \pm 1/\sqrt{2\pi}$, so $w = \sqrt{2/\pi}$. If we now time-scale through the transformation $t \to t/2$, for example, the transform pair becomes $e^{-\pi(t/2)^2} \xleftrightarrow{\mathcal{F}} 2e^{-\pi(2f)^2}$ (Figure 6.27) and the width parameter of the time function becomes $2\sqrt{2/\pi}$ while the width parameter of the frequency function becomes $\sqrt{2\pi}/2$.

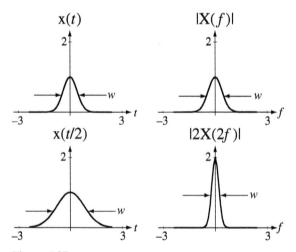

Figure 6.27
Time expansion and the corresponding frequency compression

The change of variable $t \to t/2$ causes a time expansion and the corresponding effect in the frequency domain is a frequency compression (accompanied by an amplitude scale factor). As the time-domain signal is expanded, it falls from its maximum of one at $t = 0$ more and more slowly as time departs from zero in either direction and, in the limit as the time expansion factor approaches infinity, it does not change at all and approaches the constant 1 ($w \to \infty$). As the time-domain signal is expanded by some factor, its CTFT is frequency-compressed and its height is multiplied by the same factor. In the limit as the time-domain expansion factor approaches infinity, the CTFT approaches an impulse

$$\lim_{a \to \infty} e^{-\pi(t/a)^2} = 1 \xleftrightarrow{\mathcal{F}} \lim_{a \to \infty}(1/|a|)e^{-\pi(af)^2} = \delta(f) \qquad (6.17)$$

(Figure 6.28) and $w \to 0$.

Figure 6.28
Constant and impulse as limits of time and frequency scaling $x(t) = e^{-\pi t^2}$ and its CTFT

The relation between compression in one domain and expansion in the other is the basis for an idea called the **uncertainty principle** of Fourier analysis. As $a \to \infty$ in (6.17), the signal energy of the time-domain function becomes less localized and the signal energy of the corresponding frequency-domain function becomes more localized. In that limit, the signal energy of the signal in the frequency domain is "infinitely localized" to a single frequency $f = 0$, while the time function's width becomes infinite, and therefore its signal energy is "infinitely unlocalized" in time. If we compress the time function instead, it becomes an impulse at time $t = 0$ and its signal energy occurs at one point while its CTFT becomes spread uniformly over the range $-\infty < f < \infty$ and its signal energy has no "locality" at all. As we know the location of the signal energy of one signal better, and better, we lose knowledge of the location of the signal energy of its transform counterpart. The name *uncertainty principle* comes from the principle in quantum mechanics of the same name.

If $x(t)$ is real valued, then $x(t) = x^*(t)$. The CTFT of $x(t)$ is $X(f)$ and the CTFT of $x^*(t)$ is

$$\mathcal{F}(x^*(t)) = \int_{-\infty}^{\infty} x^*(t) e^{-j2\pi ft} dt = \left[\int_{-\infty}^{\infty} x(t) e^{+j2\pi ft} dt \right]^* = X^*(-f)$$

Therefore, if $x(t) = x^*(t)$, $X(f) = X^*(-f)$. In words, if the time-domain signal is real valued, its CTFT has the property that the behavior for negative frequencies is the complex conjugate of the behavior for positive frequencies. This property is called Hermitian symmetry.

Let $x(t)$ be a real-valued signal. The square of the magnitude of $X(f)$ is $|X(f)|^2 = X(f)X^*(f)$. Then using $X(f) = X^*(-f)$ we can show that the square of the magnitude of $X(-f)$ is

$$|X(-f)|^2 = \underbrace{X(-f)}_{X^*(f)} \underbrace{X^*(-f)}_{X(f)} = X(f)X^*(f) = |X(f)|^2$$

proving that the magnitude of the CTFT of a real-valued signal is an even function of frequency. Using $X(f) = X^*(-f)$, we can also show that the phase of the CTFT of a real-valued signal can always be expressed as an odd function of frequency. (Since the phase of any complex function is multiple-valued, there are many equally correct ways of expressing phase. So we cannot say that the phase *is* an odd function, only that *it can always be expressed* as an odd function.) Often, in practical signal and system analysis the CTFT of a real-valued signal is only displayed for positive frequencies because, since $X(f) = X^*(-f)$, if we know the functional behavior for positive frequencies we also know it for negative frequencies.

Suppose a signal x(t) excites an LTI system with impulse response h(t) producing a response y(t). Then y(t) = x(t) ∗ h(t). Using the multiplication–convolution duality property Y(jω) = X(jω)H(jω). In words, the CTFT of x(t), X(jω), is a function of frequency and, when multiplied by H(jω), the CTFT of h(t), the result is Y(jω) = X(jω)H(jω), the CTFT of y(t). X(jω) describes the variation of the signal x(t) with radian frequency and Y(jω) performs the same function for y(t). So multiplication by H(jω) changes the frequency description of the excitation to the frequency description of the response. H(jω) is called the **frequency response** of the system. (This is the same frequency response first developed in Chapter 5.) When two LTI systems are cascaded the impulse response of the combined system is the convolution of the impulse responses of the two individual systems. Therefore, again using the multiplication-convolution duality property, the frequency response of the cascade of two LTI systems is the product of their individual frequency responses (Figure 6.29).

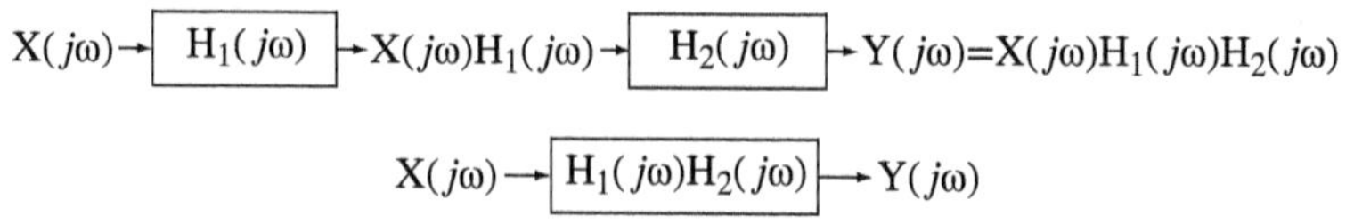

Figure 6.29
Frequency response of a cascade of two LTI systems

EXAMPLE 6.10

CTFT using the differentiation property

Find the CTFT of $x(t) = \text{rect}((t+1)/2) - \text{rect}((t-1)/2)$ using the differentiation property of the CTFT and the table entry for the CTFT of the triangle function (Figure 6.30).

The function x(t) is the derivative of a triangle function centered at zero with a base half-width of 2 and an amplitude of 2

$$x(t) = \frac{d}{dt}(2\text{tri}(t/2)).$$

In the table of CTFT pairs we find $\text{tri}(t) \xleftrightarrow{\mathcal{F}} \text{sinc}^2(f)$. Using the scaling and linearity properties, $2\text{tri}(t/2) \xleftrightarrow{\mathcal{F}} 4\text{sinc}^2(2f)$. Then, using the differentiation property, $x(t) \xleftrightarrow{\mathcal{F}} j8\pi f \text{sinc}^2(2f)$. If we find the CTFT of x(t) by using the table entry for the CTFT of a rectangle $\text{rect}(t) \xleftrightarrow{\mathcal{F}} \text{sinc}(f)$ and the time-scaling and time-shifting properties we get $x(t) \xleftrightarrow{\mathcal{F}} j4\text{sinc}(2f)\sin(2\pi f)$, which, using the definition of the sinc function, can be shown to be equivalent.

$$x(t) \xleftrightarrow{\mathcal{F}} j8\pi f \text{sinc}^2(2f) = j8\pi f \text{sinc}^2(2f)\frac{\sin(2\pi f)}{2\pi f} = j4\text{sinc}(2f)\sin(2\pi f)$$

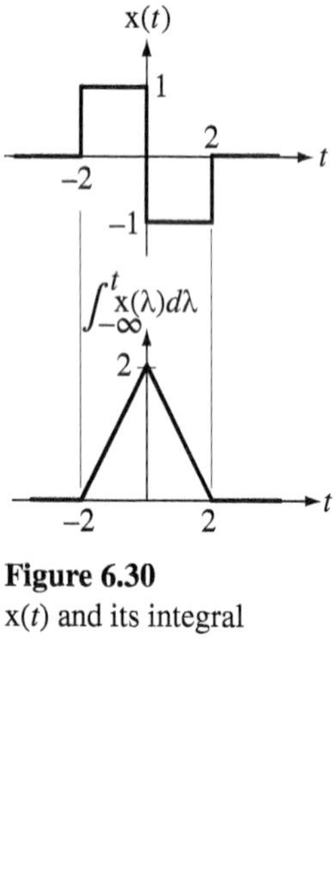

Figure 6.30
x(t) and its integral

Parseval's theorem says that we can find the energy of a signal either in the time or frequency domain.

$$\int_{-\infty}^{\infty} |x(t)|^2 dt = \int_{-\infty}^{\infty} |X(f)|^2 df. \quad (6.18)$$

(Marc-Antoine Parseval des Chênes, a French mathematician contemporary of Fourier of the late 18th and early 19th centuries, was born April 27, 1755, and died August 16, 1836.) The integrand $|X(f)|^2$ on the right-hand side of (6.18) is called **energy spectral density**. The name comes from the fact that its integral over all frequencies (the whole *spectrum* of frequencies) is the total signal energy of the signal. Therefore, to be consistent with the normal meaning of integration, $|X(f)|^2$ must be signal energy per unit cyclic frequency, a signal energy density. For example, suppose $x(t)$ represents a current in amperes (A). Then, from the definition of signal energy, the units of signal energy for this signal are $A^2 \cdot s$. The CTFT of $x(t)$ is $X(f)$ and its units are $A \cdot s$ or A/Hz. When we square this quantity we get the units

$$A^2/Hz^2 = \frac{A^2 \cdot s}{Hz} \quad \begin{array}{l} \leftarrow \text{signal energy} \\ \leftarrow \text{cyclic frequency} \end{array}$$

which confirm that the quantity $|X(f)|^2$ is signal energy per unit cyclic frequency.

EXAMPLE 6.11

Total area under a function using the CTFT

Find the total area under the function $x(t) = 10\,\text{sinc}((t+4)/7)$.

Ordinarily we would try to directly integrate the function over all t.

$$Area = \int_{-\infty}^{\infty} x(t)\,dt = \int_{-\infty}^{\infty} 10\,\text{sinc}\left(\frac{t+4}{7}\right)dt = \int_{-\infty}^{\infty} 10\,\frac{\sin(\pi(t+4)/7)}{\pi(t+4)/7}\,dt$$

This integral is a sine integral (first mentioned in Example 6.6) defined by

$$\text{Si}(z) = \int_0^z \frac{\sin(t)}{t}\,dt.$$

The sine integral can be found tabulated in mathematical tables books. However, evaluation of the sine integral is not necessary to solve this problem. We can use

$$X(0) = \int_{-\infty}^{\infty} x(t)\,dt.$$

First we find the CTFT of $x(t)$, which is $X(f) = 70\,\text{rect}(7f)e^{j8\pi f}$. Then $Area = X(0) = 70$.

EXAMPLE 6.12

CTFT of some time-scaled and time-shifted sines

If $x(t) = 10\sin(t)$, then find (a) the CTFT of $x(t)$, (b) the CTFT of $x(2(t-1))$, and (c) the CTFT of $x(2t-1)$.

(a) In this example the cyclic frequency of the sinusoid is $1/2\pi$ and the radian frequency is 1. Therefore the numbers will be simpler if we use the radian-frequency form of the CTFT. Using the linearity property and looking up the transform of the general sine form,

$$\sin(\omega_0 t) \xleftrightarrow{\mathcal{F}} j\pi[\delta(\omega + \omega_0) - \delta(\omega - \omega_0)]$$

$$\sin(t) \xleftrightarrow{\mathcal{F}} j\pi[\delta(\omega + 1) - \delta(\omega - 1)]$$

$$10\sin(t) \xleftrightarrow{\mathcal{F}} j10\pi[\delta(\omega + 1) - \delta(\omega - 1)]$$

(b) From part (a), $10\sin(t) \xleftrightarrow{\mathcal{F}} j10\pi[\delta(\omega+1) - \delta(\omega-1)]$. Using the time scaling property,

$$10\sin(2t) \xleftrightarrow{\mathcal{F}} j5\pi[\delta(\omega/2+1) - \delta(\omega/2-1)].$$

Then, using the time-shifting property,

$$10\sin(2(t-1)) \xleftrightarrow{\mathcal{F}} j5\pi[\delta(\omega/2+1) - \delta(\omega/2-1)]e^{-j\omega}.$$

Then, using the scaling property of the impulse,

$$10\sin(2(t-1)) \xleftrightarrow{\mathcal{F}} j10\pi[\delta(\omega+2) - \delta(\omega-2)]e^{-j\omega}$$

or

$$10\sin(2(t-1)) \xleftrightarrow{\mathcal{F}} j10\pi[\delta(\omega+2)e^{j2} - \delta(\omega-2)e^{-j2}]$$

(c) From part (a), $10\sin(t) \xleftrightarrow{\mathcal{F}} j10\pi[\delta(\omega+1) - \delta(\omega-1)]$.
Applying the time-shifting property first

$$10\sin(t-1) \xleftrightarrow{\mathcal{F}} j10\pi[\delta(\omega+1) - \delta(\omega-1)]e^{-j\omega}.$$

Then, applying the time-scaling property,

$$10\sin(2t-1) \xleftrightarrow{\mathcal{F}} j5\pi[\delta(\omega/2+1) - \delta(\omega/2-1)]e^{-j\omega/2}$$

Then, using the scaling property of the impulse,

$$10\sin(2t-1) \xleftrightarrow{\mathcal{F}} j10\pi[\delta(\omega+2) - \delta(\omega-2)]e^{-j\omega/2}$$

or

$$10\sin(2t-1) \xleftrightarrow{\mathcal{F}} j10\pi[\delta(\omega+2)e^{j} - \delta(\omega-2)e^{-j}]$$

EXAMPLE 6.13

CTFT of a scaled and shifted rectangle

Find the CTFT of $x(t) = 25\,\text{rect}((t-4)/10)$.

We can find the CTFT of the unit rectangle function in the table of Fourier transforms $\text{rect}(t) \xleftrightarrow{\mathcal{F}} \text{sinc}(f)$. First apply the linearity property $25\,\text{rect}(t) \xleftrightarrow{\mathcal{F}} 25\,\text{sinc}(f)$. Then apply the time-scaling property $25\,\text{rect}(t/10) \xleftrightarrow{\mathcal{F}} 250\,\text{sinc}(10f)$. Then apply the time-shifting property

$$25\,\text{rect}((t-4)/10) \xleftrightarrow{\mathcal{F}} 250\,\text{sinc}(10f)e^{-j8\pi f}.$$

EXAMPLE 6.14

CTFT of the convolution of some signals

Find the CTFT of the convolution of $10\sin(t)$ with $2\delta(t+4)$.
Method 1: Do the convolution first and find the CTFT of the result.

$$10\sin(t) * 2\delta(t+4) = 20\sin(t+4)$$

Apply the time-shifting property.

$$20\sin(t+4) \xleftrightarrow{\mathcal{F}} j20\pi[\delta(\omega+1) - \delta(\omega-1)]e^{j4\omega}$$

or

$$20\sin(t+4) \xleftrightarrow{\mathcal{F}} j10[\delta(f+1/2\pi) - \delta(f-1/2\pi)]e^{j8\pi f}$$

Method 2: Do the CTFT first to avoid the convolution.

$$10\sin(t) * 2\delta(t+4) \xleftrightarrow{\mathcal{F}} \mathcal{F}(10\sin(t))\mathcal{F}(2\delta(t+4)) = 2\mathcal{F}(10\sin(t))\mathcal{F}(\delta(t))e^{j4\omega}$$

$$10\sin(t) * 2\delta(t+4) \xleftrightarrow{\mathcal{F}} j20\pi[\delta(\omega+1) - \delta(\omega-1)]e^{j4\omega}$$

or

$$10\sin(t) * 2\delta(t+4) \xleftrightarrow{\mathcal{F}} \mathcal{F}(10\sin(t))\mathcal{F}(2\delta(t+4)) = 2\mathcal{F}(10\sin(t))\mathcal{F}(\delta(t))e^{j8\pi f}$$

$$10\sin(t) * 2\delta(t+4) \xleftrightarrow{\mathcal{F}} j10[\delta(f+1/2\pi) - \delta(f-1/2\pi)]e^{j8\pi f}.$$

NUMERICAL COMPUTATION OF THE FOURIER TRANSFORM

In cases in which the signal to be transformed is not readily describable by a mathematical function or the Fourier-transform integral cannot be done analytically, we can sometimes find an approximation to the CTFT *numerically* using the discrete Fourier transform (DFT), which was used to approximate the CTFS harmonic function. If the signal to be transformed is a causal energy signal, it can be shown (Web Appendix G) that we can approximate its CTFT (f form) at discrete frequencies by

$$X(kf_s/N) \cong T_s \sum_{n=0}^{N-1} x(nT_s)e^{-j2\pi kn/N} \cong T_s \times \mathcal{DFT}(x(nT_s)), \quad |k| << N \quad (6.19)$$

where $T_s = 1/f_s$ is chosen such that the signal x does not change much in that amount of time and N is chosen such that the time range 0 to NT_s covers all or practically all of the signal energy of the signal x (Figure 6.31).

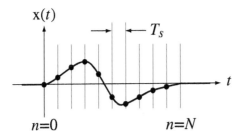

Figure 6.31
A causal energy signal sampled with T_s seconds between samples over a time NT_s

So if the signal to be transformed is a causal energy signal and we sample it over a time containing practically all of its energy and if the samples are close enough

Example 6.15

Using the DFT to approximate the CTFT

Using the DFT, find the approximate CTFT of

$$x(t) = \begin{cases} t(1-t), & 0 < t < 1 \\ 0, & \text{otherwise} \end{cases} = t(1-t)\text{rect}(t-1/2)$$

numerically by sampling it 32 times over the time interval $0 \leq t < 2$.

The following MATLAB program can be used to make this approximation.

```
% Program to demonstrate approximating the CTFT of t(1-t)*rect(t-1/2)
% by sampling it 32 times in the time interval 0 <= t < 2 seconds
% and using the DFT.
N = 32 ;                            % Sample 32 times
Ts = 2/N ;                          % Sample for two seconds
                                    % and set sampling interval
fs = 1/Ts ;                         % Set sampling rate
df = fs/N ;                         % Set frequency-domain resolution
n = [0:N-1]' ;                      % Vector of 32 time indices
t = Ts*n ;                          % Vector of times
x = t.*(1-t).*rect((t-1/2));        % Vector of 32 x(t) function values
X = Ts*fft(x) ;                     % Vector of 32 approx X(f) CTFT
                                    % values
k = [0:N/2-1]' ;                    % Vector of 16 frequency indices
% Graph the results
subplot(3,1,1) ;
p = plot(t,x,'k') ; set(p,'LineWidth',2) ; grid on ;
xlabel('Time, t (s)') ; ylabel('x(t)') ;
subplot(3,1,2) ;
p = plot(k*df,abs(X(1:N/2)),'k') ; set(p,'LineWidth',2) ; grid on;
xlabel('Frequency, f (Hz)') ; ylabel('|X(f)|') ;
subplot(3,1,3) ;
p = plot(k*df,angle(X(1:N/2)),'k') ; set(p,'LineWidth',2) ; grid on ;
xlabel('Frequency, f (Hz)') ; ylabel('Phase of X(f)') ;
```

This MATLAB program produces the graphs in Figure 6.32.

Notice that 32 samples are taken from the time-domain signal and the DFT returns a vector of 32 numbers. We only used the first 16 in these graphs. The DFT is periodic and the 32 points returned represent one period. Therefore the second 16 points are the same as the second 16 points occurring in the previous period and can be used to graph the DFT for negative frequencies. The

6.3 The Continuous-Time Fourier Transform

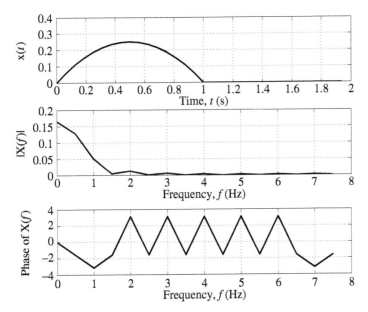

Figure 6.32
A signal and its approximate CTFT, found by using the DFT

MATLAB command `fftshift` is provided for just that purpose. Below is an example of using `fftshift` and graphing the approximate CTFT over equal negative and positive frequencies.

```
% Program to demonstrate approximating the CTFT of t(1-t)*rect(t-1/2)
% by sampling it 32 times in the time interval 0 < t < 2 seconds and
% using the DFT. The frequency domain graph covers equal negative
% and positive frequencies.

N = 32 ;                              % Sample 32 times
Ts = 2/N ;                            % Sample for two second
                                      % and set sampling interval
fs = 1/Ts ;                           % Set sampling rate
df = fs/N ;                           % Set frequency-domain resolution
n = [0:N-1]' ;                        % Vector of 32 time indices
t = Ts*n ;                            % Vector of times
x = t.*(1-t).*rect((t-1/2)) ;         % Vector of 32 x(t) function values
X = fftshift(Ts*fft(x)) ;             % Vector of 32 X(f) approx CTFT values
k = [-N/2:N/2-1]' ;                   % Vector of 32 frequency indices

% Graph the results

subplot(3,1,1) ;
p = plot(t,x,'k') ; set(p,'LineWidth',2) ; grid on ;
xlabel('Time, t (s)') ; ylabel('x(t)') ;
subplot(3,1,2) ;
p = plot(k*df,abs(X),'k') ; set(p,'LineWidth',2) ; grid on ;
xlabel('Frequency, f (Hz)') ; ylabel('|X(f)|') ;
subplot(3,1,3) ;
```

```
p = plot(k*dF,angle(X),'k') ; set(p,'LineWidth',2) ; grid on ;
xlabel('Frequency, f (Hz)') ; ylabel('Phase of X(f)') ;
```

Figure 6.33 and Figure 6.34 show the results of this MATLAB program with 32 points and 512 points.

This result is a rough approximation to the CTFT because only 32 points were used. If we use 512 points over a time period of 16 seconds we get an approximation with higher frequency-domain resolution and over a wider frequency range.

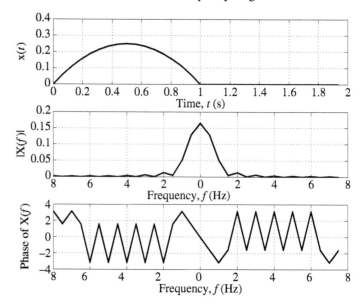

Figure 6.33
Approximate CTFT found by using the DFT graphed over equal negative and positive frequencies

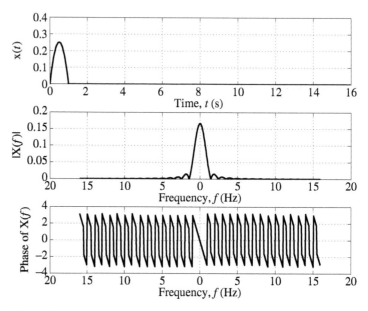

Figure 6.34
Approximate CTFT found by using the DFT with higher resolution

EXAMPLE 6.16

System analysis using the CTFT

A system described by the differential equation $y'(t) + 1000y(t) = 1000\,x(t)$ is excited by $x(t) = 4\mathrm{rect}(200t)$. Find and graph the response $y(t)$.

If we Fourier transform the differential equation we get

$$j2\pi f Y(f) + 1000 Y(f) = 1000 X(f)$$

which can be rearranged into

$$Y(f) = \frac{1000 X(f)}{j2\pi f + 1000}.$$

The CTFT of the excitation is $X(f) = 0.02\,\mathrm{sinc}(f/200)$. Therefore the CTFT of the response is

$$Y(f) = \frac{20\,\mathrm{sinc}(f/200)}{j2\pi f + 1000}$$

or, using the definition of the sinc function and the exponential definition of the sine function,

$$Y(f) = 20\frac{\sin(\pi f/200)}{(\pi f/200)(j2\pi f + 1000)} = 4000\frac{e^{j2\pi f/400} - e^{-j2\pi f/400}}{j2\pi f(j2\pi f + 1000)}.$$

To find the inverse CTFT, start with the CTFT pair $e^{-\alpha t}u(t) \xleftrightarrow{\mathcal{F}} 1/(j2\pi f + \alpha)$, $\alpha > 0$,

$$e^{-1000t}u(t) \xleftrightarrow{\mathcal{F}} \frac{1}{j2\pi f + 1000}.$$

Next use the integration property,

$$\int_{-\infty}^{t} g(\lambda)\,d\lambda \xleftrightarrow{\mathcal{F}} \frac{G(f)}{j2\pi f} + (1/2)G(0)\delta(f).$$

$$\int_{-\infty}^{t} e^{-1000\lambda}u(\lambda)\,d\lambda \xleftrightarrow{\mathcal{F}} \frac{1}{j2\pi f}\frac{1}{j2\pi f + 1000} + \frac{1}{2000}\delta(f).$$

Then apply the time-shifting property,

$$g(t - t_0) \xleftrightarrow{\mathcal{F}} G(f)e^{-j2\pi f t_0}.$$

$$\int_{0}^{t+1/400} e^{-1000\lambda}\,d\lambda \xleftrightarrow{\mathcal{F}} \frac{1}{j2\pi f}\frac{e^{j2\pi f/400}}{j2\pi f + 1000} + \underbrace{\frac{e^{j2\pi f/400}}{2000}\delta(f)}_{=\delta(f)/2000}$$

$$\int_{0}^{t-1/400} e^{-1000\lambda}\,d\lambda \xleftrightarrow{\mathcal{F}} \frac{1}{j2\pi f}\frac{e^{-j2\pi f/400}}{j2\pi f + 1000} + \underbrace{\frac{e^{-j2\pi f/400}}{2000}\delta(f)}_{=\delta(f)/2000}$$

Subtracting the second result from the first and multiplying through by 4000

$$4000\int_{-\infty}^{t+1/400} e^{-1000\lambda}u(\lambda)\,d\lambda - 4000\int_{-\infty}^{t-1/400} e^{-1000\lambda}u(\lambda)\,d\lambda$$

$$\xleftrightarrow{\mathcal{F}} \frac{4000}{j2\pi f}\frac{e^{j2\pi f/400} - e^{-j2\pi f/400}}{j2\pi f + 1000}$$

$$4000\left[\int_{-\infty}^{t+1/400} e^{-1000\lambda}u(\lambda)d\lambda - \int_{-\infty}^{t-1/400} e^{-1000\lambda}u(\lambda)d\lambda\right]$$

$$\xleftrightarrow{\mathcal{F}} \frac{4000}{j2\pi f} \frac{e^{j2\pi f/400} - e^{-j2\pi f/400}}{j2\pi f + 1000}$$

The two integral expressions can be simplified as follows.

$$\int_{-\infty}^{t+1/400} e^{-1000\lambda}u(\lambda)d\lambda = \begin{cases} (1/1000)(1 - e^{-1000(t+1/400)}), & t \geq -1/400 \\ 0, & t < -1/400 \end{cases}$$

$$= \frac{1}{1000}(1 - e^{-1000(t+1/400)})u(t + 1/400)$$

$$\int_{-\infty}^{t-1/400} e^{-1000\lambda}u(\lambda)d\lambda = \begin{cases} (1/1000)(1 - e^{-1000(t-1/400)}), & t \geq 1/400 \\ 0, & t < 1/400 \end{cases}$$

$$= \frac{1}{1000}(1 - e^{-1000(t-1/400)})u(t - 1/400)$$

Then

$$4\left[(1 - e^{-1000(t+1/400)})u(t + 1/400) - (1 - e^{-1000(t-1/400)})u(t - 1/400)\right]$$

$$\xleftrightarrow{\mathcal{F}} \frac{4000}{j2\pi f} \frac{e^{j2\pi f/400} - e^{-j2\pi f/400}}{j2\pi f + 1000}$$

Therefore the response is

$$y(t) = 4[(1 - e^{-1000(t+1/400)})u(t + 1/400) - (1 - e^{-1000(t-1/400)})u(t - 1/400)]$$

(Figure 6.35 and Figure 6.36).

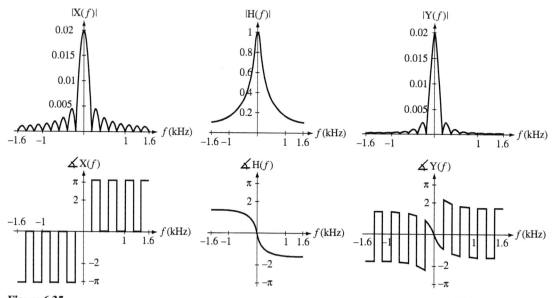

Figure 6.35
Magnitudes and phases of CTFTs of excitation and response, and of system frequency response

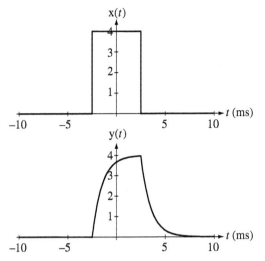

Figure 6.36
Rectangular pulse excitation and system response

EXAMPLE 6.17

System analysis using the CTFT

A system described by the differential equation $y'(t) + 1000y(t) = 1000x(t)$ is excited by $x(t) = 4\text{rect}(200t) * \delta_{0.01}(t)$. Find and graph the response $y(t)$.

From Example 6.16,

$$Y(f) \xrightarrow{f=\omega/2\pi} Y(j\omega) = \frac{1000X(j\omega)}{j\omega + 1000}.$$

The CTFT (f form) of the excitation is $X(f) = 2\text{sinc}(f/200)\delta_{100}(f)$ implying that $X(j\omega) = 2\text{sinc}(\omega/400\pi)\delta_{100}(\omega/2\pi)$. Using the scaling property of the periodic impulse,

$$X(j\omega) = 2\text{sinc}(\omega/400\pi) \times 2\pi\delta_{200\pi}(\omega) = 4\pi\text{sinc}(\omega/400\pi)\delta_{200\pi}(\omega)$$

Therefore the CTFT of the response is

$$Y(j\omega) = \frac{4000\pi\text{sinc}(\omega/400\pi)\delta_{200\pi}(\omega)}{j\omega + 1000}$$

or, using the definition of the periodic impulse,

$$Y(j\omega) = 4000\pi \sum_{k=-\infty}^{\infty} \frac{\text{sinc}(\omega/400\pi)\delta(\omega - 200\pi k)}{j\omega + 1000}.$$

Now, using the equivalence property of the impulse,

$$Y(j\omega) = 4000\pi \sum_{k=-\infty}^{\infty} \frac{\text{sinc}(k/2)\delta(\omega - 200\pi k)}{j200\pi k + 1000}$$

and the inverse CTFT yields the response

$$y(t) = 2000 \sum_{k=-\infty}^{\infty} \frac{\text{sinc}(k/2)}{j200\pi k + 1000} e^{j200\pi kt}.$$

If we separate the $k = 0$ term and pair each k and $-k$ this result can be written as

$$y(t) = 2 + \sum_{k=1}^{\infty} \left[\frac{\text{sinc}(k/2)}{j0.1\pi k + 0.5} e^{j200\pi kt} + \frac{\text{sinc}(-k/2)}{-j0.1\pi k + 0.5} e^{-j200\pi kt} \right].$$

Using the fact that the sinc function is even and combining the terms over one common denominator,

$$y(t) = 2 + \sum_{k=1}^{\infty} \text{sinc}(k/2) \frac{(-j0.1\pi k + 0.5)e^{j200\pi kt} + (j0.1\pi k + 0.5)e^{-j200\pi kt}}{(0.1\pi k)^2 + (0.5)^2}$$

$$y(t) = 2 + \sum_{k=1}^{\infty} \text{sinc}(k/2) \frac{\cos(200\pi kt) + 0.2\pi k \sin(200\pi kt)}{(0.1\pi k)^2 + (0.5)^2}$$

The response is a constant plus a linear combination of real cosines and sines at integer multiples of 100 Hz (Figure 6.37 and Figure 6.38).

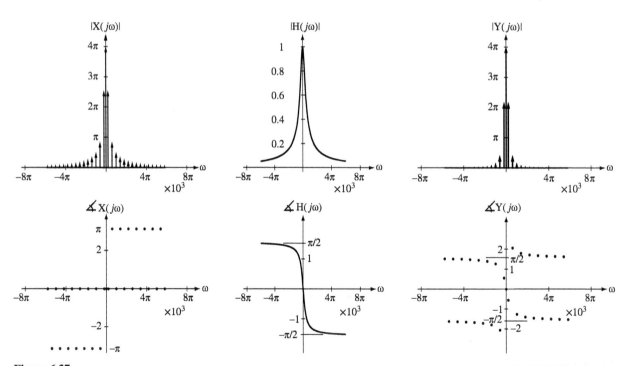

Figure 6.37
Magnitudes and phases of CTFTs of excitation and response, and of system frequency response

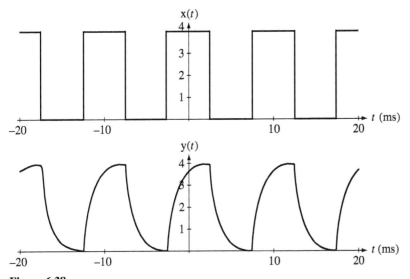

Figure 6.38
Excitation and response

6.4 SUMMARY OF IMPORTANT POINTS

1. The Fourier series is a method of representing any arbitrary signal with engineering usefulness as a linear combination of sinusoids, real-valued or complex.
2. The complex sinusoids used by the complex form of the Fourier series form a set of mutually orthogonal functions that can be combined in linear combinations to form any arbitrary periodic function with engineering usefulness.
3. The formula for finding the harmonic function of the Fourier series can be derived using the principle of orthogonality.
4. The Fourier series can be used to find the response of an LTI system to a periodic excitation.
5. The Fourier series can be extended to allow the representation of aperiodic signals and the extension is called the Fourier transform.
6. With a table of Fourier transform pairs and their properties the forward and inverse transforms of almost any signal of engineering significance, periodic or aperiodic, can be found.
7. The frequency response of a stable system is the Fourier transform of its impulse response.
8. The Fourier transform can be used to find the response of an LTI system to energy signals as well as to periodic signals.

EXERCISES WITH ANSWERS

(Answers to each exercise are in random order.)

Fourier Series

1. Using MATLAB graph each sum of complex sinusoids over the time period indicated.

(a) $x(t) = \dfrac{1}{10} \displaystyle\sum_{k=-30}^{30} \text{sinc}\left(\dfrac{k}{10}\right) e^{j200\pi kt}, \quad -15 \text{ ms} < t < 15 \text{ ms}$

(b) $x(t) = \dfrac{j}{4} \displaystyle\sum_{k=-9}^{9} \left[\text{sinc}\left(\dfrac{k+2}{2}\right) - \text{sinc}\left(\dfrac{k-2}{2}\right)\right] e^{j10\pi kt}, \quad -200 \text{ ms} < t < 200 \text{ ms}$

Answers:

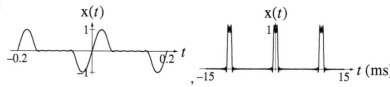

Orthogonality

2. Show by direct analytical integration that the integral of the function,
$$g(t) = A\sin(2\pi t)B\sin(4\pi t)$$
is zero over the interval, $-1/2 < t < 1/2$.

3. A periodic signal $x(t)$ with a period of 4 seconds is described over one period by
$$x(t) = 3 - t, \quad 0 < t < 4.$$
Graph the signal and find its CTFS description. Then graph on the same scale approximations to the signal $x_N(t)$ given by
$$x_N(t) = \sum_{k=-N}^{N} c_x[k]\, e^{j2\pi kt/T_0}$$
for $N = 1, 2,$ and 3. (In each case the time scale of the graph should cover at least two periods of the original signal.)

Answers: $c_x[k] = \dfrac{1}{4}\dfrac{2e^{-j2\pi k}(-2 - j\pi k) - j6\pi k + 4}{(\pi k)^2},$

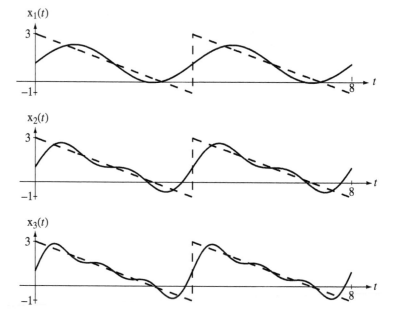

4. Using the CTFS table of transforms and the CTFS properties, find the CTFS harmonic function of each of these periodic signals using the representation time T indicated.

   (a) $x(t) = 10 \sin(20\pi t)$, $T = 1/10$

   (b) $x(t) = 2 \cos(100\pi(t - 0.005))$, $T = 1/50$

   (c) $x(t) = -4 \cos(500\pi t)$, $T = 1/50$

   (d) $x(t) = \frac{d}{dt}(e^{-j10\pi t})$, $T = 1/5$

   (e) $x(t) = \text{rect}(t) * 4\delta_4(t)$, $T = 4$

   (f) $x(t) = \text{rect}(t) * \delta_1(t)$, $T = 1$

   (g) $x(t) = \text{tri}(t) * \delta_1(t)$, $T = 1$

   Answers: $j5(\delta[k + 1] - \delta[k - 1])$, $-2(\delta[k - 5] + \delta[k + 5])$, $\text{sinc}(k/4)$,
   $\delta[k], \delta[k], j(\delta[k + 1] - \delta[k - 1]), -j10\pi\delta[k + 1]$

5. Given $x(t) \xleftrightarrow{\mathcal{FS}}_{12} \text{tri}((k - 1)/4) + \text{tri}((k + 1)/4)$ what is the average value of $x(t)$?

   Answer: 3/2

6. Given $x(t) \xleftrightarrow{\mathcal{FS}}_{8} 4(u[k + 3] - u[k - 4])$ what is the average signal power of $x(t)$?

   Answer: 112

7. The CTFS harmonic function of $x(t)$ based on one fundamental period is found to be

$$c_x[k] = \frac{1 - \cos(\pi k)}{(\pi k)^2}$$

   (a) Is the signal even, odd, or neither?
   (b) What is the average value of the signal?

   Answers: Even, 1/2

8. A signal $x(t) = [7\text{rect}(2t) - 5\text{rect}(4(t - 1))] * \delta_8(t)$ has a CTFS harmonic function $c_x[k]$. What is the numerical value of $c_x[0]$? (It is not necessary to find a general expression for $c_x[k]$ to answer this question.)

   Answer: 0.28125

9. For each function is its CTFS harmonic function purely real, purely imaginary, or neither?

   (a) $x(t) = 8 \cos(50\pi t) - 4 \sin(22\pi t)$

   (b) $x(t) = 32 \cos(50\pi t) \sin(22\pi t)$

   (c) $x(t) = \left[\text{tri}\left(\frac{t-1}{4}\right) - \text{tri}\left(\frac{t+1}{4}\right)\right] \sin(100\pi t)$

   (d) $x(t) = 100 \cos\left(4000\pi t + \frac{\pi}{2}\right)$

(e) $x(t) = -5\,\text{rect}(2t) * 2\delta_2(t)$

(f) $x(t) = -5t$, $-3 < t < 3$ and $x(t) = x(t + 6n)$, n any integer.

(g) $x(t) = 100\sin\left(4000\pi t + \frac{\pi}{4}\right)$

Answers: 2 Neither, 3 Purely Imaginary, 2 Purely Real

10. Find the numerical values of the literal constants in these CTFS pairs.

   (a) $10\sin(32\pi t) \xleftrightarrow[1/16]{\mathcal{FS}} A\delta[k-a] + B\delta[k-b]$

   (b) $3\cos(44\pi t) \xleftrightarrow[1/11]{\mathcal{FS}} A\delta[k-a] + B\delta[k-b]$

   (c) $A\,\text{rect}(at) * (1/b)\delta_{1/b}(t) \xleftrightarrow[2]{\mathcal{FS}} 30\,\text{sinc}(2k)$

   (d) $\dfrac{d}{dt}(2\,\text{rect}(4t) * \delta_1(t)) \xleftrightarrow[1]{\mathcal{FS}} Ak\,\text{sinc}(ak)$, $T_F = 1$

Answers: $\{\pi, 1/4\}$, $\{3/2, 2, -2\}$, $\{j5, -1, -j5, 1\}$, $\{15/2, 1/4, 1/2\}$

11. The harmonic function $c_x[k]$ for a periodic continuous-time signal $x(t)$ whose fundamental period is 4 is zero everywhere except at exactly two values of k, $k = \pm 8$. Its value is the same at those two points, $c_x[8] = c_x[-8] = 3$. What is the representation time used in calculating this harmonic function?

Answer: 32

12. For the following signals $x(t)$ the harmonic function is based on $T = T_0$, the fundamental period of $x(t)$. Find the numerical values of the literal constants.

   (a) $x(t) = \text{rect}(2t) * 5\delta_5(t)$ $\quad c_x[k] = A\,\text{sinc}(k/b)$

   (b) $x(t) = -2\sin(4\pi t) + 3\cos(12\pi t)$ $\quad c_x[k] = A(\delta[k+a] - \delta[k-a]) + B$
   $(\delta[k-b] + \delta[k+b])$

Answers: $\{-j, 1, 3/2, 3\}$, $\{1/2, 10\}$

13. The CTFS harmonic function $c_x[k]$ for the signal

$$x(t) = \text{rect}(2(t-1)) * 3\delta_3(t) = \text{rect}(2t) * 3\delta_3(t-1)$$

is of the form $c_x[k] = A\,\text{sinc}(ak)\,e^{-jb\pi k}$. Find A, a, and b using the fundamental period as the representation time.

Answers: 1/2, 1/6, 2/3

14. If $x(t) = 10\,\text{tri}(3t) * \delta_2(t)$ and $\dfrac{d}{dt}(x(t)) \xleftrightarrow[T_0]{\mathcal{FS}} Ak\,\text{sinc}^2(bk)$ find the numerical values of A and b.

Answers: $j5.236$, $1/6$

15. If $x(t) = A\,\text{tri}(bt) * \delta_c(t) \xleftrightarrow[8]{\mathcal{FS}} c_x[k] = 10\,\text{sinc}^2(k/3)$, find the numerical values of A, b, and c.

Answers: 30, 3/8, 8

16. In Figure E.16 is a graph of one fundamental period of a periodic function x(t). A CTFS harmonic function $c_x[k]$ is found based on the representation time being the same as the fundamental period T_0. If $A_1 = 4$, $A_2 = -3$, and $T_0 = 5$, what is the numerical value of $c_x[0]$?

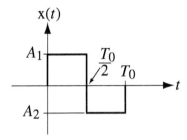

Figure E.16

If the representation period is changed to $3T_0$ what is the new numerical value of $c_x[0]$?

Answers: 0.5, 0.5

17. Find and graph the magnitude and phase of the CTFS harmonic function of $x(t) = \text{rect}(20t) * \delta_{1/5}(t)$ using a representation time $T = 1/5$ two ways and compare the graphs.

(a) Using the CTFS tables.

(b) Numerically with the time between points being $T_s = 1/2000$.

Answers:

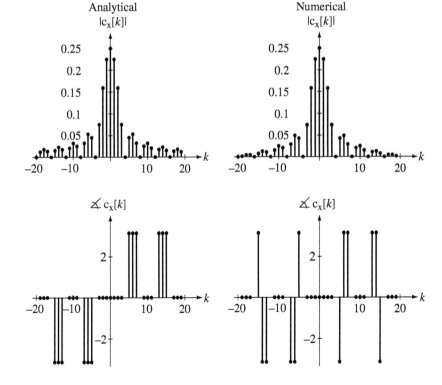

18. A quantizer accepts a continuous-time input signal and responds with a continuous-time output signal that only has a finite number of equally spaced values. If $x_{in}(t)$ is the input signal and $x_{out}(t)$ is the output signal and q is the difference between adjacent output levels, the value of $x_{out}(t)$ at any point in time can be computed by forming the ratio $x_{in}(t)/q$, rounding it off to the nearest integer and then multiplying the result by q. Let the range of the signal levels accepted by the quantizer be from -10 to $+10$ and let the number of quantization levels be 16. Find the numerical total harmonic distortion (see Example 6.4 in Chapter 6) of the quantizer output signal if the input signal is $x_{in}(t) = 10 \sin(2000\pi t)$.

    Answer: 0.2342%

Forward and Inverse Fourier Transforms

19. Let a signal be defined by

    $$x(t) = 2\cos(4\pi t) + 5\cos(15\pi t).$$

    Find the CTFTs of $x(t - 1/40)$ and $x(t + 1/20)$ and identify the resultant phase shift of each sinusoid in each case. Plot the phase of the CTFT and draw a straight line through the 4 phase points that result in each case. What is the general relationship between the slope of that line and the time delay?

    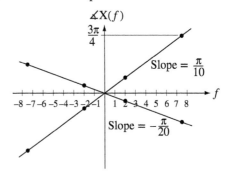

    Answers: The slope of the line is $-2\pi f$ times the delay.

20. If $x(t) = e^{-5t}u(t)$ and $y(t) = e^{-12t}u(t)$ and $x(t) \xleftrightarrow{\mathcal{F}} X(f)$ and $y(t) \xleftrightarrow{\mathcal{F}} Y(f)$ and $z(t) = x(t) * y(t)$ and $z(t) \xleftrightarrow{\mathcal{F}} Z(f)$ what is the numerical value of $Z(3)$?

    Answer: $0.0023\, e^{-j2.3154}$

21. Find the CTFT of $x(t) = \text{sinc}(t)$. Then make the change of scale $t \to 2t$ in $x(t)$ and find the CTFT of the time-scaled signal.

    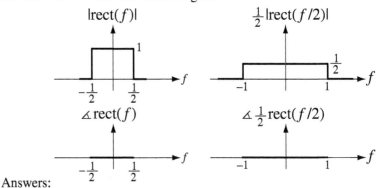

    Answers:

22. Using the multiplication–convolution duality of the CTFT, find an expression for y(t) which does not use the convolution operator ∗ and graph y(t).

(a) $y(t) = \text{rect}(t) * \cos(\pi t)$
(b) $y(t) = \text{rect}(t) * \cos(2\pi t)$
(c) $y(t) = \text{sinc}(t) * \text{sinc}(t/2)$
(d) $y(t) = \text{sinc}(t) * \text{sinc}^2(t/2)$
(e) $y(t) = e^{-t}u(t) * \sin(2\pi t)$

Answers:

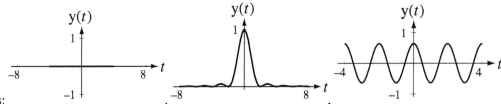
,
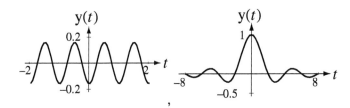
,

23. Find the following numerical values.

(a) $x(t) = 20\,\text{rect}(4t)$ $X(f)\big|_{f=2}$
(b) $x(t) = 2\,\text{sinc}(t/8) * \text{sinc}(t/4)$ $x(4)$
(c) $x(t) = 2\,\text{tri}(t/4) * \delta(t-2)$ $x(1)$ and $x(-1)$
(d) $x(t) = -5\,\text{rect}(t/2) * (\delta(t+1) + \delta(t))$ $x(1/2)$, $x(-1/2)$, and $x(-5/2)$
(e) $x(t) = 3\,\text{rect}(t-1)$ $X(f)\big|_{f=1/4}$
(f) $x(t) = 4\,\text{sinc}^2(3t)$ $X(j\omega)\big|_{\omega=4\pi}$
(g) $x(t) = \text{rect}(t) * \text{rect}(2t)$ $X(f)\big|_{f=1/2}$
(h) $X(f) = 10[\delta(f - 1/2) + \delta(f + 1/2)]$ $x(1)$
(i) $X(j\omega) = -2\,\text{sinc}(\omega/2\pi) * 3\,\text{sinc}(\omega/\pi)$ $x(0)$

Answers: -20, $4/9$, $1/2$, -3, 3.1831, $\{-5, -10, 0\}$, $-j2.70$, 0.287, 5.093

24. Find the following forward or inverse Fourier transforms. No final result should contain the convolution operator ∗.

(a) $\mathscr{F}(15\,\text{rect}((t+2)/7))$
(b) $\mathscr{F}^{-1}\left(2\,\text{tri}(f/2)e^{-j6\pi f}\right)$
(c) $\mathscr{F}(\sin(20\pi t)\cos(200\pi t))$

Answers: $105\,\text{sinc}(7f)\,e^{j4\pi f}$, $2\text{tri}(f/2)e^{-j6\pi f}$

$$(j/4)\begin{bmatrix}\delta(f-90)+\delta(f+110)\\-\delta(f-110)-\delta(f+90)\end{bmatrix}$$

25. Find the signal energy of

    (a) $x(t) = 28\,\text{sinc}(t/15)$ (b) $x(t) = -3\,\text{sinc}^2(2t)$

    Answers: 3, 11760

26. Let $y(t) = x(t) * h(t)$ and let $x(t) = e^{-t}u(t)$ and let $h(t) = x(-t)$.

    What is the numerical value of y(2)?

    Answer: 0.06765

27. Using Parseval's theorem, find the signal energy of these signals.

    (a) $x(t) = 4\,\text{sinc}(t/5)$ (b) $x(t) = 2\,\text{sinc}^2(3t)$

    Answers: 8/9, 80

28. What is the total area under the function, $g(t) = 100\,\text{sinc}((t-8)/30)$?

    Answer: 3000

29. Let a continuous-time signal x(t) have a CTFT $X(f) = \begin{cases} |f|, & |f| < 2 \\ 0, & |f| \geq 2 \end{cases}$.

    Let $y(t) = x(4(t-2))$. Find the numerical values of the magnitude and phase of Y(3) where $y(t) \xleftrightarrow{\mathcal{F}} Y(f)$.

    Answer: {3/16, 0}

30. Using the integration property, find the CTFT of these functions and compare with the CTFT found using other properties.

    (a) $g(t) = \begin{cases} 1, & |t| < 1 \\ 2 - |t|, & 1 < |t| < 2 \\ 0, & \text{elsewhere} \end{cases}$ (b) $g(t) = 8\,\text{rect}(t/3)$

    Answers: $3\,\text{sinc}(3f)\,\text{sinc}(f)$, $24\,\text{sinc}(3f)$

31. Graph the magnitudes and phases of the CTFTs of these signals in the f form.

    (a) $x(t) = \delta(t-2)$ (b) $x(t) = u(t) - u(t-1)$

    (c) $x(t) = 5\,\text{rect}\left(\dfrac{t+2}{4}\right)$ (d) $x(t) = 25\,\text{sinc}(10(t-2))$

    (e) $x(t) = 6\sin(200\pi t)$ (f) $x(t) = 2e^{-t}u(t)$

    (g) $x(t) = 4e^{-3t^2}$

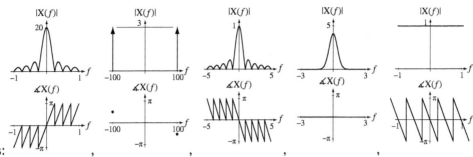

Answers:

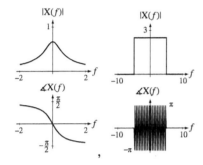

,

32. Graph the magnitudes and phases of the CTFTs of these signals in the ω form.

(a) $x(t) = \delta_2(t)$
(b) $x(t) = \text{sgn}(2t)$
(c) $x(t) = 10\,\text{tri}((t-4)/20)$
(d) $x(t) = (1/10)\,\text{sinc}^2((t+1)/3)$
(e) $x(t) = \dfrac{\cos(200\pi t - \pi/4)}{4}$
(f) $x(t) = 2e^{-3t}u(t) = 2e^{-3t}u(3t)$
(g) $x(t) = 7e^{-5|t|}$

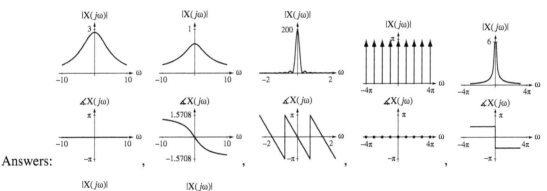

Answers:

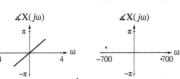

,

33. Graph the inverse CTFTs of these functions.

(a) $X(f) = -15\text{rect}(f/4)$

(b) $X(f) = \dfrac{\text{sinc}(-10f)}{30}$

(c) $X(f) = \dfrac{18}{9 + f^2}$

(d) $X(f) = \dfrac{1}{10 + jf}$

(e) $X(f) = (1/6)[\delta(f-3) + \delta(f+3)]$

(f) $X(f) = 8\delta(5f) = (8/5)\delta(f)$

(g) $X(f) = -3/j\pi f$

Answers:

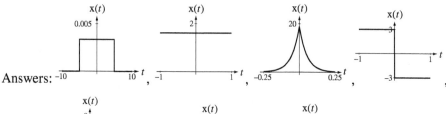

34. Graph the inverse CTFTs of these functions.

(a) $X(j\omega) = e^{-4\omega^2}$

(b) $X(j\omega) = 7\,\text{sinc}^2(\omega/\pi)$

(c) $X(j\omega) = j\pi[\delta(\omega + 10\pi) - \delta(\omega - 10\pi)]$

(d) $X(j\omega) = (\pi/20)\,\delta_{\pi/4}(\omega)$

(e) $X(j\omega) = 5\pi/j\omega + 10\pi\delta(\omega)$

(f) $X(j\omega) = \dfrac{6}{3 + j\omega}$

(g) $X(j\omega) = 20\,\text{tri}(8\omega)$

Answers:

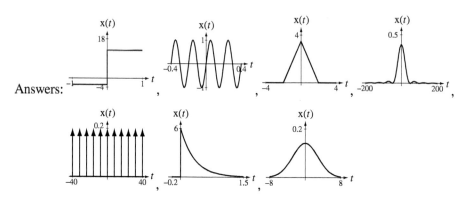

35. Graph the magnitudes and phases of these functions. Graph the inverse CTFT of the functions also.

(a) $X(j\omega) = \dfrac{10}{3 + j\omega} - \dfrac{4}{5 + j\omega}$

(b) $X(f) = 4\left[\text{sinc}\left(\dfrac{f-1}{2}\right) + \text{sinc}\left(\dfrac{f+1}{2}\right)\right]$

(c) $X(f) = \dfrac{j}{10}\left[\text{tri}\left(\dfrac{f+2}{8}\right) - \text{tri}\left(\dfrac{f-2}{8}\right)\right]$

(d) $X(f) = \begin{Bmatrix} \delta(f+1050) + \delta(f+950) \\ +\delta(f-950) + \delta(f-1050) \end{Bmatrix}$

(e) $X(f) = \begin{bmatrix} \delta(f+1050) + 2\delta(f+1000) \\ +\delta(f+950) + \delta(f-950) \\ +2\delta(f-1000) + \delta(f-1050) \end{bmatrix}$

Answers:

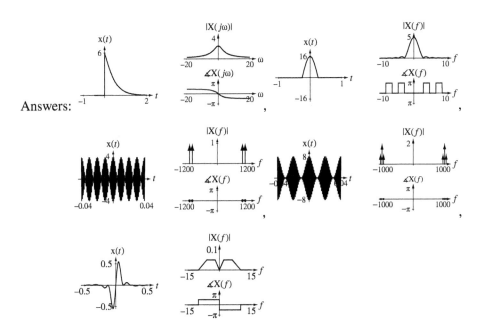

36. Graph these signals versus time. Graph the magnitudes and phases of the CTFTs of these signals in either the f or ω form, whichever is more convenient. In some cases the time graph may be conveniently done first. In other cases it may be more convenient to do the time graph after the CTFT has been found, by finding the inverse CTFT.

    (a) $x(t) = e^{-\pi t^2} \sin(20\pi t)$

    (b) $x(t) = \cos(400\pi t)(1/100)\delta_{1/100}(t) = (1/100) \sum_{n=-\infty}^{\infty} \cos(4\pi n)\delta(t - n/100)$

    (c) $x(t) = [1 + \cos(400\pi t)]\cos(4000\pi t)$

    (d) $x(t) = [1 + \text{rect}(100t) * \delta_{1/50}(t)]\cos(500\pi t)$

    (e) $x(t) = \text{rect}(t/7)\delta_1(t)$

Answers:

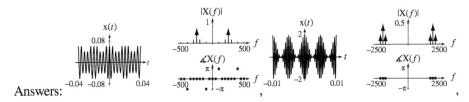

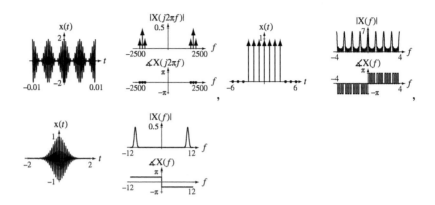

37. Graph the magnitudes and phases of these functions. Graph the inverse CTFTs of the functions also.

   (a) $X(f) = \text{sinc}(f/4)\,\delta_1(f)$

   (b) $X(f) = \left[\text{sinc}\left(\dfrac{f-1}{4}\right) + \text{sinc}\left(\dfrac{f+1}{4}\right)\right]\delta_1(f)$

   (c) $X(f) = \text{sinc}(f)\,\text{sinc}(2f)$

Answers:

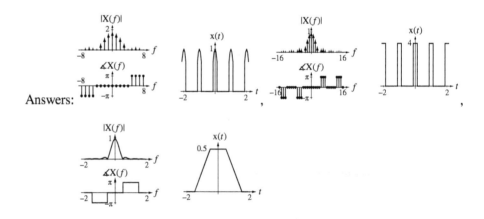

38. Graph these signals versus time and the magnitudes and phases of their CTFTs.

   (a) $x(t) = \displaystyle\int_{-\infty}^{t}\sin(2\pi\lambda)\,d\lambda$

   (b) $x(t) = \displaystyle\int_{-\infty}^{t}\text{rect}(\lambda)\,d\lambda = \begin{cases} 0, & t < -1/2 \\ t + 1/2, & |t| < 1/2 \\ 1, & t > 1/2 \end{cases}$

   (c) $x(t) = \displaystyle\int_{-\infty}^{t} 3\,\text{sinc}(2\lambda)\,d\lambda$

Answers: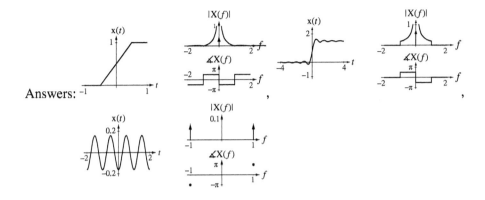

Relation of CTFS to CTFT

39. The transition from the CTFS to the CTFT is illustrated by the signal,
$$x(t) = \text{rect}(t/w) * \delta_{T_0}(t)$$
or
$$x(t) = \sum_{n=-\infty}^{\infty} \text{rect}\left(\frac{t - nT_0}{w}\right).$$

The complex CTFS for this signal is given by
$$c_x[k] = (Aw/T_0)\,\text{sinc}(kw/T_0).$$

Graph the modified CTFS,
$$T_0 c_x[k] = Aw\,\text{sinc}(w(kf_0)),$$
for $w = 1$ and $f_0 = 0.5, 0.1$, and 0.02 versus kf_0 for the range $-8 < kf_0 < 8$.

Answers: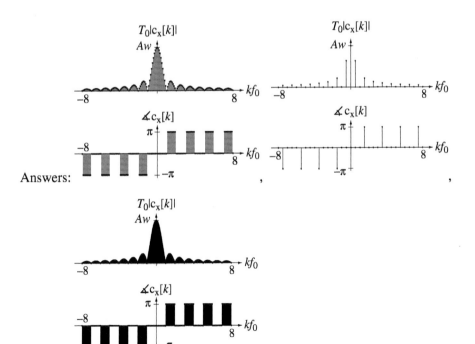

40. Find the CTFS and CTFT of these periodic functions and compare answers.

   (a) $x(t) = \text{rect}(t) * \delta_2(t)$ (b) $x(t) = \text{tri}(10t) * \delta_{1/4}(t)$

   Answers: $X(f) = \sum_{k=-\infty}^{\infty} (1/2)\text{sinc}(k/2)\delta(f - k/2) = \sum_{k=-\infty}^{\infty} c_x[k]\delta(f - kf_0)$,

   $X(f) = (2/5) \sum_{k=-\infty}^{\infty} \text{sinc}^2(2k/5)\delta(f - 4k) = \sum_{k=-\infty}^{\infty} c_x[k]\delta(f - 4k)$

Numerical CTFT

41. Find and graph the approximate magnitude and phase of the CTFT of

    $$x(t) = [4 - (t - 2)^2]\text{rect}((t - 2)/4)$$

    using the DFT to approximate the CTFT. Let the time between samples of x(t) be 1/16 and sample over the time range $0 \le t < 16$.

    Answer:

    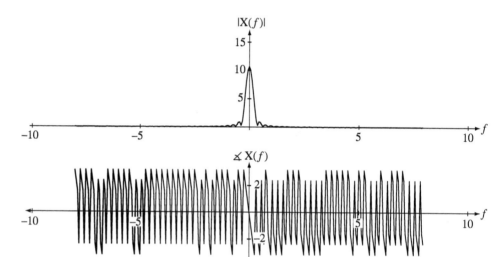

System Response

42. A system has a frequency response $H(j\omega) = \dfrac{100}{j\omega + 200}$.

    (a) If we apply the constant signal $x(t) = 12$ to this system the response is also a constant. What is the numerical value of the response constant?

    (b) If we apply the signal $x(t) = 3\sin(14\pi t)$ to the system the response is $y(t) = A\sin(14\pi t + \theta)$. What are the numerical values of A and θ (θ in radians)?

    Answers: $\{1.4649, -0.2165\}, 6$

EXERCISES WITHOUT ANSWERS

Fourier Series

43. Why can the function $c_x[k] = \begin{cases} 0, & k < -2 \\ k, & -2 \le k < 3 \\ 0, & k \ge 3 \end{cases}$ not be the harmonic function of a real signal?

44. A continuous-time signal x(t) consists of the periodic repetition of an even function. It is represented by a CTFS using its fundamental period as the representation time.

    (a) What is definitely known about its complex (exponential form) harmonic function?

    (b) If the representation time is doubled, what additional fact is known about the new complex harmonic function?

45. A periodic signal x(t) with a fundamental period of 2 seconds is described over one period by

$$x(t) = \begin{cases} \sin(2\pi t), & |t| < 1/2 \\ 0, & 1/2 < |t| < 1 \end{cases}.$$

Plot the signal and find its CTFS description. Then graph on the same scale approximations to the signal $x_N(t)$ given by

$$x_N(t) = \sum_{k=-N}^{N} X[k] e^{j2\pi kt/T_0}$$

for $N = 1, 2,$ and 3. (In each case the time scale of the graph should cover at least two periods of the original signal.)

46. Using MATLAB, graph the following signals over the time range, $-3 < t < 3$.

    (a) $x_0(t) = 1$
    (b) $x_1(t) = x_0(t) + 2\cos(2\pi t)$
    (c) $x_2(t) = x_1(t) + 2\cos(4\pi t)$
    (d) $x_{20}(t) = x_{19}(t) + 2\cos(40\pi t)$

    For each part, (a) through (d), numerically evaluate the area of the signal over the time range, $-1/2 < t < 1/2$.

47. Identify which of these functions has a complex CTFS $c_g[k]$ for which (1) $\text{Re}(c_g[k]) = 0$ for all k, or (2) $\text{Im}(c_g[k]) = 0$ for all k, or (3) neither of these conditions applies.

    (a) $g(t) = 18\cos(200\pi t) + 22\cos(240\pi t)$

    (b) $g(t) = -4\sin(10\pi t)\sin(2000\pi t)$

    (c) $g(t) = \text{tri}\left(\dfrac{t-1}{4}\right) * \delta_{10}(t)$

48. A continuous-time signal x(t) with fundamental period T_0 has a CTFS harmonic function, $c_x[k] = \text{tri}(k/10)$ using the representation time $T = T_0$. If $z(t) = x(2t)$ and $c_z[k]$ is its CTFS harmonic function, using the same representation time $T = T_0$, find the numerical values of $c_z[1]$ and $c_z[2]$.

49. Each signal in Figure E.49 is graphed over a range of exactly one fundamental period. Which of the signals have harmonic functions $c_x[k]$ that have a purely real value for every value of k? Which have a purely imaginary value for every value of k?

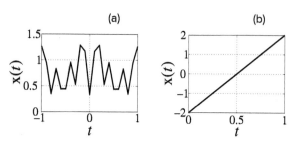

Figure E.49

50. Find the CTFS harmonic function $c_x[k]$, for the continuous-time function $x(t) = \operatorname{sinc}(8t) * (1/2)\delta_{1/2}(t)$ using its fundamental period as the representation time.

51. If $x(t) = 4\operatorname{sinc}(t/2) * \delta_9(t)$ and the representation time is one fundamental period, use Parseval's theorem to find the numerical value of the signal power of $x(t)$.

52. In Figure E.52 is graphed exactly one period of a periodic function $x(t)$. Its harmonic function $c_x[k]$ (with $T = T_0$) can be written as

$$c_x[k] = A g(bk) e^{jck}.$$

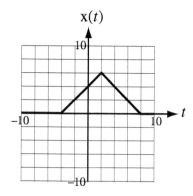

Figure E.52

What is the name of the function $g(\cdot)$?

What are the numerical values of A, b, and c?

53. Find the numerical values of the literal constants.

(a) $8\cos(30\pi t) \xleftarrow{\mathcal{FS}}_{1/15} A(\delta[k-m] + \delta[k+m])$ Find A and m.

(b) $5\cos(10\pi t) \xleftarrow{\mathcal{FS}}_{1} A(\delta[k-m] + \delta[k+m])$ Find A and m.

(c) $3\operatorname{rect}(t/6) * \delta_{18}(t) \xleftarrow{\mathcal{FS}}_{18} A\operatorname{sinc}(bk)$ Find A and b.

(d) $9\operatorname{rect}(3t) * \delta_2(t) \xleftarrow{\mathcal{FS}}_{4} A\operatorname{sinc}(bk)\delta_m[k]$ Find A, b, and m.

54. If $x(t) \xleftrightarrow[200 \text{ ms}]{\mathcal{FS}} 4(\delta[k-1] + 3\delta[k] + \delta[k+1])$

   (a) What is average value of $x(t)$?

   (b) Is $x(t)$ an even or an odd function or neither?

   (c) If $y(t) = \frac{d}{dt}(x(t))$, what is the value of $c_y[1]$?

55. If $c_x[k] = 3(\delta[k-1] + \delta[k+1])$ and $c_y[k] = j2(\delta[k+2] - \delta[k-2])$ and both are based on the same representation time and $z(t) = x(t)y(t)$, $c_z[k]$ can be written in the form

$$c_z[k] = A(\delta[k-a] - \delta[k-b] + \delta[k-c] - \delta[k-d])$$

   Find the numerical values of A, a, b, c, and d.

56. Let $x(t) \xleftrightarrow[10]{\mathcal{FS}} 4\operatorname{sinc}(k/5)$ and let $y(t) = \frac{d}{dt}(x(t))$ and let $y(t) \xleftrightarrow[10]{\mathcal{FS}} c_y[k]$.

   (a) What is the numerical value of $c_y[3]$?

   (b) Is $x(t)$ an even function, an odd function, neither, or impossible to know?

   (c) Is $y(t)$ an even function, an odd function, neither, or impossible to know?

57. If $x(t) \xleftrightarrow[T_0]{\mathcal{FS}} c_x[k] = u[k+3] - u[k-4]$ find the average signal power P_x of $x(t)$.

58. In Figure E.58 is a graph of one fundamental period of the product of an unshifted cosine, $x_1(t) = B\cos(2\pi f_x t)$, and an unshifted sine, $x_2(t) = B\sin(2\pi f_x t)$, of equal amplitude and frequency. If $A = 6$ and $T_0 = 10$, find B and f_x.

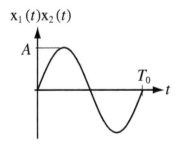

Figure E.58

59. Four continuous-time signals have the following descriptions.

$$c_{x1}[k] = 3\{\operatorname{sinc}(k/2) - \delta[k]\}, \quad T = T_0 = 1$$

$x_2(t)$ is periodic and one period of $x_2(t) = t\delta_1(t)\operatorname{rect}(t/5)$, $-2.5 < t < 2.5$ and $T = T_0 = 5$

$x_3(t)$ is periodic and one period of $x_3(t) = t + 1$, $-2 < t < 2$ and $T = T_0 = 4$

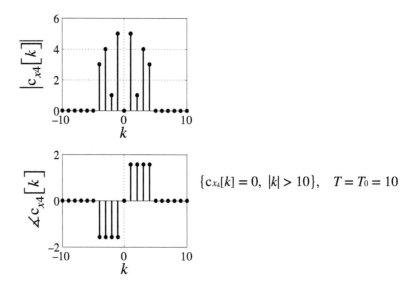

$\{c_{x4}[k] = 0, \; |k| > 10\}, \quad T = T_0 = 10$

Answer the following questions.

1. Which continuous-time signals are even functions?

2. Which continuous-time signals are not even but can be made even by adding or subtracting a constant?

3. Which continuous-time signals are odd functions?

4. Which continuous-time signals are not odd but can be made odd by adding or subtracting a constant?

5. Which continuous-time signals have an average value of zero?

6. Which of the continuous-time signals are square waves?

7. What is the average signal power of $x_1(t)$?

8. What is the average signal power of $x_4(t)$?

60. In some types of communication systems binary data are transmitted using a technique called binary phase-shift keying (BPSK) in which a "1" is represented by a "burst" of a continuous-time sine wave and a "0" is represented by a burst which is the exact negative of the burst that represents a "1." Let the sine frequency be 1 MHz and let the burst width be 10 periods of the sine wave. Find and graph the CTFS harmonic function for a periodic binary signal consisting of alternating "1's" and "0's" using its fundamental period as the representation time.

61. Match the CTFS magnitude graphs to the time functions. (In all cases $T = 4$.)

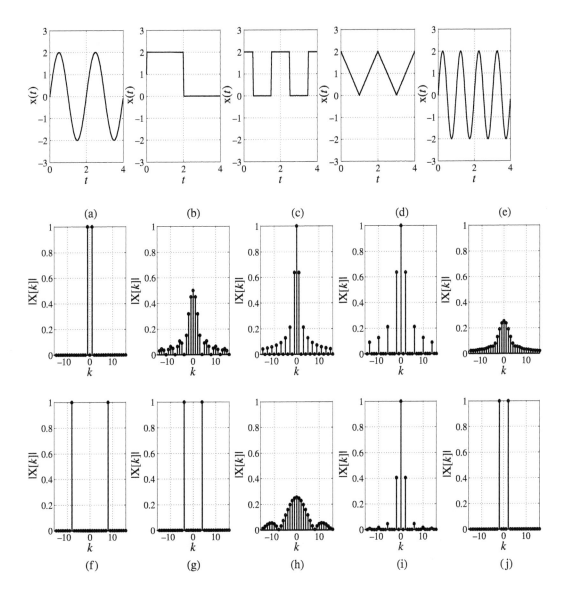

62. A continuous-time system is described by the differential equation
$$a_2 y''(t) + a_1 y'(t) + a_0 y(t) = b_2 x''(t) + b_1 x'(t) + b_0$$
and the system is excited by $x(t) = \text{rect}(t/w) * \delta_{T_0}(t)$.

(a) Let $a_2 = 1$, $a_1 = 20$, $a_0 = 250{,}100$, $b_2 = 1$, $b_1 = 0$, and $b_0 = 250{,}000$. Also let $T_0 = 3 \times \dfrac{2\pi}{\sqrt{b_0}}$ and let $w = T_0/2$. Graph the response $y(t)$ over the time range $0 \leq t < 2T_0$. At what harmonic number is the magnitude of the harmonic response a minimum? What cyclic frequency does that correspond to? Can you see in $y(t)$ the effect of this minimum magnitude response?

(b) Change T_0 to $\dfrac{2\pi}{\sqrt{b_0}}$ and repeat part (a).

Forward and Inverse Fourier Transforms

63. A system is excited by a signal x(t) = 4rect(t/2) and its response is

$$y(t) = 10\left[(1 - e^{-(t+1)})u(t+1) - (1 - e^{-(t-1)})u(t-1)\right].$$

   What is its impulse response?

64. Graph the magnitudes and phases of the CTFTs of the following functions.

   (a) $g(t) = 5\delta(4t)$
   (b) $g(t) = 4[\delta_4(t+1) - \delta_4(t-3)]$
   (c) $g(t) = u(2t) + u(t-1)$
   (d) $g(t) = \text{sgn}(t) - \text{sgn}(-t)$
   (e) $g(t) = \text{rect}\left(\frac{t+1}{2}\right) + \text{rect}\left(\frac{t-1}{2}\right)$
   (f) $g(t) = \text{rect}(t/4)$
   (g) $g(t) = 5\,\text{tri}(t/5) - 2\,\text{tri}(t/2)$
   (h) $g(t) = (3/2)\text{rect}(t/8) * \text{rect}(t/2)$

65. Graph the magnitudes and phases of the CTFTs of the following functions.

   (a) $\text{rect}(4t)$
   (b) $\text{rect}(4t) * 4\delta(t)$
   (c) $\text{rect}(4t) * 4\delta(t-2)$
   (d) $\text{rect}(4t) * 4\delta(2t)$
   (e) $\text{rect}(4t) * \delta_1(t)$
   (f) $\text{rect}(4t) * \delta_1(t-1)$
   (g) $(1/2)\text{rect}(4t) * \delta_{1/2}(t)$
   (h) $(1/2)\text{rect}(t) * \delta_{1/2}(t)$

66. Given that $y(t) = x(t-2) * x(t+2)$ and that $Y(f) = 3\,\text{sinc}^2(2f)$, find x(t).

67. An LTI system has a frequency response $H(j\omega) = \dfrac{1}{j\omega - j6} + \dfrac{1}{j\omega + j6}$.

   (a) Find an expression for its impulse response h(t) which does not contain the square root of minus one (j).
   (b) Is this system stable? Explain how you know.

68. A periodic signal has a fundamental period of 4 seconds.

   (a) What is the lowest positive frequency at which its CTFT could be nonzero?
   (b) What is the next-lowest positive frequency at which its CTFT could be nonzero?

69. A signal x(t) has a CTFT $X(f) = \dfrac{j2\pi f}{3 + jf/10}$.

   (a) What is the total net area under the signal x(t)?
   (b) Let y(t) be the integral of x(t), $y(t) = \int_{-\infty}^{t} x(\lambda)d\lambda$. What is the total net area under y(t)?
   (c) What is the numerical value of $|X(f)|$ in the limit as $f \to +\infty$?

70. Answer the following questions.

   (a) A signal $x_1(t)$ has a CTFT $X_1(f)$. If $x_2(t) = x_1(t+4)$, what is the relationship between $|X_1(f)|$ and $|X_2(f)|$?

(b) A signal $x_1(t)$ has a CTFT, $X_1(f)$. If $x_2(t) = x_1(t/5)$, what is the relationship between the maximum value of $|X_1(f)|$ and the maximum value of $|X_2(f)|$?

(c) A CTFT has the value, $e^{-j\pi/4}$ at a frequency, $f = 20$. What is the value of that same CTFT at a frequency of $f = -20$?

71. If $x(t) = \delta_{T_0}(t) * \text{rect}(t/4)$ and $x(t) \overset{\mathcal{F}}{\longleftrightarrow} X(f)$ find three different numerical values of T_0 for which $X(f) = A\delta(f)$ and the corresponding numerical impulse strengths A.

72. If $y(t) \overset{\mathcal{F}}{\longleftrightarrow} Y(f)$ and $\frac{d}{dt}(y(t)) \overset{\mathcal{F}}{\longleftrightarrow} 1 - e^{-j\pi f/2}$, find and graph $y(t)$.

73. Let a signal $x(t)$ have a CTFT $X(f) = \begin{cases} |f|, & |f| < 2 \\ 0, & |f| \geq 2 \end{cases}$. Let $y(t) = x(4(t-2))$.

Find the values of the magnitude and phase of $Y(3)$ where $y(t) \overset{\mathcal{F}}{\longleftrightarrow} Y(f)$.

74. Graph the magnitude and phase of the CTFT of each of the signals in Figure E.74 (ω form):

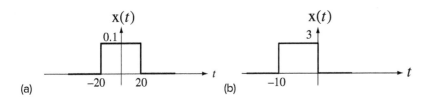
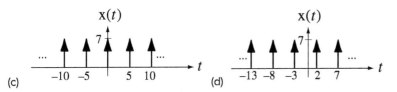

Figure E.74

75. Graph the inverse CTFTs of the functions in Figure E.75.

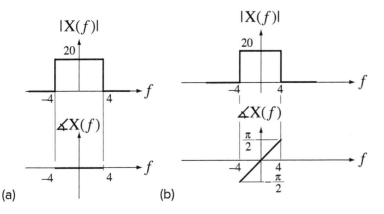

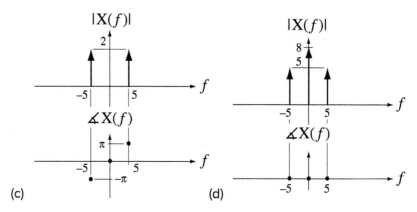

Figure E.75

76. If a CTFT in the f form is $24\delta_3(f)$ and in the ω form it is $A\delta_b(c\omega)$, what are the numerical values of A, b, and c?

77. Find the numerical values of the literal constants in the following CTFT pairs.

(a) $Ae^{at}u(t) \xleftrightarrow{\mathscr{F}} \dfrac{8}{j\omega + 2}$ (b) $Ae^{at}u(t) \xleftrightarrow{\mathscr{F}} \dfrac{1}{j8\omega + 2}$

(c) $3\delta_4(t) \xleftrightarrow{\mathscr{F}} A\delta_a(f)$ (d) $4\delta_3(5t) \xleftrightarrow{\mathscr{F}} A\delta_a(bf)$

(e) $4\delta_3(5t) \xleftrightarrow{\mathscr{F}} A\delta_a(f)$

(Hint: Use the definition of the continuous-time periodic impulse and continuous-time impulse properties.)

(f) $A\sin(200\pi t + \pi/3) \xleftrightarrow{\mathscr{F}} j8[\delta(f+a) - \delta(f-a)]e^{bf}$

(g) $A\mathrm{rect}(t/a) * \mathrm{rect}(t/b) \xleftrightarrow{\mathscr{F}} 40\,\mathrm{sinc}(2f)\mathrm{sinc}(4f)$

(h) $\dfrac{d}{dt}(3\mathrm{tri}(5t)) \xleftrightarrow{\mathscr{F}} Af\,\mathrm{sinc}^2(af)$

(i) $10\sin(20t) \xleftrightarrow{\mathscr{F}} A[\delta(\omega+b) - \delta(\omega-b)]$

(j) $-4\cos(6t - \pi/2) \xleftrightarrow{\mathscr{F}} A[\delta(f-b) + \delta(f+b)]e^{cf}$

(k) $\delta(t+2) - \delta(t-2) \xleftrightarrow{\mathscr{F}} A\sin(2\pi bf)$

(l) $u(t-5) - u(t+5) \xleftrightarrow{\mathscr{F}} A\,\mathrm{sinc}(b\omega)$

(m) $u(t+3) - u(t-6) \xleftrightarrow{\mathscr{F}} A\,\mathrm{sinc}(bf)e^{cf}$

(n) $A\,\mathrm{sinc}(bt)\cos(ct) \xleftrightarrow{\mathscr{F}} \mathrm{rect}\left(\dfrac{f-50}{4}\right) + \mathrm{rect}\left(\dfrac{f+50}{4}\right)$

(o) $A\cos(b(t-c)) \xleftrightarrow{\mathscr{F}} \delta(f-5)e^{j\pi/4} + \delta(f+5)e^{-j\pi/4}$

(p) $A\,\mathrm{rect}(bt) * \delta_c(t) \xleftrightarrow{\mathscr{F}} 9\,\mathrm{sinc}(f/10)\delta_1(f)$

78. Below are two lists, one of time-domain functions and one of frequency-domain functions. By writing the number of a time-domain function, match the frequency-domain functions to their inverse CTFTs in the list of time-domain functions.

(a)
Time Domain

1   $-(1/2)\delta_{1/8}(t)$

2   $5\,\mathrm{sinc}(2(t+2))$

3   $3\delta(3t-9)$

4   $-7\,\mathrm{sinc}^2(t/12)$

5   $5\,\mathrm{sinc}(2(t-2))$

6   $5\cos(200\pi t)$

7   $2\,\mathrm{tri}((t+5)/10)$

8   $3\delta(t-3)$

9   $-24[u(t+1)-u(t-3)]$

10 $-2\delta_{1/4}(-t)$

11 $9\,\mathrm{rect}((t-4)/20)$

12 $2\,\mathrm{tri}((t+10)/5)$

13 $-24[u(t+3)-u(t-1)]$

14 $10\cos(400\pi t)$

Frequency Domain

_____ $5[\delta(f-200)+\delta(f+200)]$

_____ $(5/2)\,\mathrm{rect}(f/2)\,e^{-j4\pi f}$

_____ $180\,\mathrm{sinc}(20f)\,e^{-j8\pi f}$

_____ $-84\,\mathrm{tri}(12f)$

_____ $-96\,\mathrm{sinc}(4f)\,e^{j2\pi f}$

_____ $-4\delta_8(-f)$

_____ $e^{-j6\pi f}$

_____ $10\,\mathrm{sinc}^2(5f)\,e^{j10\pi f}$

(b)
Time Domain

1   $3\delta(t-3)$

2   $3\,\mathrm{sinc}(8t+7)$

3   $-\mathrm{rect}((t+3)/6)$

4   $12[u(t-3)-u(t+5)]$

5   $4\,\mathrm{sinc}^2((t+1)/3)$

6   $10\sin(5\pi t)$

7   $-(1/2)\delta_{1/8}(t)$

8   $3\,\mathrm{sinc}(8(t+7))$

9   $3\delta(3t-9)$

10 $12[u(t+3)-u(t-5)]$

Frequency Domain

_____ $-4\delta_8(-f)$

_____ $0.375\,\mathrm{rect}(\omega/16\pi)\,e^{j7\omega}$

_____ $e^{j3\omega}$

_____ $12\,\mathrm{tri}(3f)\,e^{-j2\pi f}$

_____ $0.375\,\mathrm{rect}(f/8)\,e^{j7\pi f/4}$

_____ $j10\pi[\delta(\omega+10\pi)-\delta(\omega-10\pi)]$

_____ $-1.25\,\mathrm{sinc}^2(f/4)\,e^{-j4\pi f}$

_____ $3e^{-j3\omega}$

_____ $96\,\mathrm{sinc}(4\omega/\pi)\,e^{-j\omega}$

_____ $6\,\mathrm{sinc}(6f)\,e^{j6\pi f}$

11. $18\,\text{tri}(6(t+5))$ _____ $3\,\text{sinc}^2(3\omega/\pi)\,e^{j5\omega}$

12. $-5\,\text{tri}(4(t-2))$

13. $-2\delta_4(-t)$

14. $5\sin(10\pi t)$

79. Find the inverse CTFT of the real, frequency-domain function in Figure E.79 and graph it. (Let $A = 1$, $f_1 = 95$ kHz, and $f_2 = 105$ kHz.)

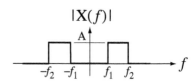

Figure E.79

80. In Figure E.80.1 are graphs of some continuous-time energy signals. In Figure E.80.2 are graphs of some magnitude CTFTs in random order. Find the CTFT magnitude that matches each energy signal.

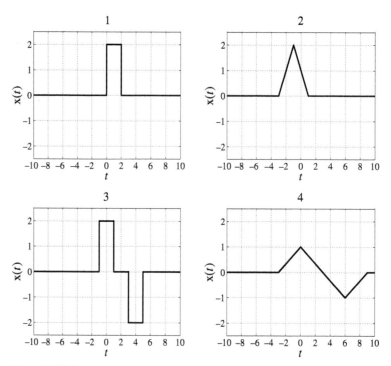

Figure E.80.1

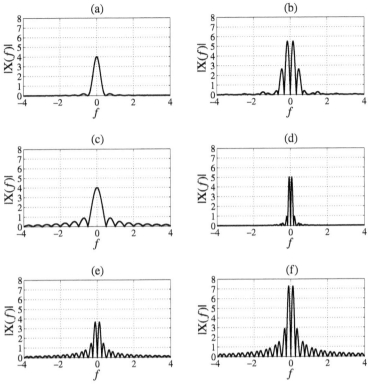

Figure E.80.2

81. In many communication systems a device called a "mixer" is used. In its simplest form a mixer is simply an analog multiplier. That is, its response signal y(t) is the product of its two excitation signals. If the two excitation signals are

$$x_1(t) = 10\text{sinc}(20t) \quad \text{and} \quad x_2(t) = 5\cos(2000\pi t)$$

graph the magnitude of the CTFT of y(t) Y(f) and compare it to the magnitude of the CTFT of $x_1(t)$. In simple terms what does a mixer do?

82. One major problem in real instrumentation systems is electromagnetic interference caused by the 60 Hz power lines. A system with an impulse response of the form, $h(t) = A(u(t) - u(t - t_0))$ can reject 60 Hz and all its harmonics. Find the numerical value of t_0 that makes this happen.

83. In electronics, one of the first circuits studied is the rectifier. There are two forms, the half-wave rectifier and the full-wave rectifier. The half-wave rectifier cuts off half of an excitation sinusoid and leaves the other half intact. The full-wave rectifier reverses the polarity of half of the excitation sinusoid and leaves the other half intact. Let the excitation sinusoid be a typical household voltage, 120 Vrms at 60 Hz, and let both types of rectifiers alter the negative half of the sinusoid while leaving the positive half unchanged. Find and graph the magnitudes of the CTFTs of the responses of both types of rectifiers (either form).

System Response

84. An LTI continuous-time system is described by the differential equation $2y'(t) + 5y(t) = x'(t)$.

Which is the correct description of this systems frequency response?

(a) The system attenuates low frequencies more than high frequencies.

(b) The system attenuates high frequencies more than low frequencies.

(c) The system has the same effect on all frequencies.

Relation of CTFS to CTFT

85. An aperiodic signal x(t) has a CTFT

$$X(f) = \begin{cases} 0, & |f| > 4 \\ 4 - |f|, & 2 < |f| < 4 \\ 2, & |f| < 2 \end{cases} = 4\text{tri}\left(\frac{f}{4}\right) - 2\text{tri}\left(\frac{f}{2}\right)$$

Let $x_p(t) = x(t) * \delta_2(t) = \sum_{k=-\infty}^{\infty} x(t - 2k)$ and let $X_p[k]$ be its CTFS harmonic function.

(a) Draw the magnitude of $X(f)$ vs. f.

(b) Find an expression for $c_{x_p}[k]$

(c) Find the numerical values of $c_{x_p}[3]$, $c_{x_p}[5]$, $c_{x_p}[10]$.

CHAPTER 7

Discrete-Time Fourier Methods

7.1 INTRODUCTION AND GOALS

In Chapter 6 we developed the continuous-time Fourier series as a method of representing periodic continuous-time signals and finding the response of a continuous-time LTI system to a periodic excitation. Then we extended the Fourier series to the Fourier transform by letting the period of the periodic signal approach infinity. In this chapter we take a similar path applied to discrete-time systems. Most of the basic concepts are the same but there are a few important differences.

CHAPTER GOALS

1. To develop methods of expressing discrete-time signals as linear combinations of sinusoids, real or complex

2. To explore the general properties of these ways of expressing discrete-time signals

3. To generalize the discrete-time Fourier series to include aperiodic signals by defining the discrete-time Fourier transform

4. To establish the types of signals that can or cannot be described by a discrete-time Fourier transform

5. To derive and demonstrate the properties of the discrete-time Fourier transform

6. To illustrate the interrelationships among the Fourier methods

7.2 THE DISCRETE-TIME FOURIER SERIES AND THE DISCRETE FOURIER TRANSFORM

LINEARITY AND COMPLEX-EXPONENTIAL EXCITATION

As was true in continuous-time, if a discrete-time LTI system is excited by a sinusoid, the response is also a sinusoid, with the same frequency but generally a different magnitude and phase. If an LTI system is excited by a sum of signals, the overall response is the sum of the responses to each of the signals individually. The **discrete-time Fourier series** (**DTFS**) expresses arbitrary periodic signals as linear combinations of sinusoids, real-valued or complex, so we can use superposition to find the response of any LTI system to any arbitrary signal by summing the responses to the individual sinusoids (Figure 7.1).

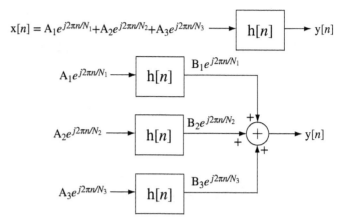

Figure 7.1
The equivalence of the response of an LTI system to a signal and the sum of its responses to complex sinusoids whose sum is equivalent to the signal

The sinusoids can be real-valued or complex. Real-valued sinusoids and complex sinusoids are related by

$$\cos(x) = \frac{e^{jx} + e^{-jx}}{2} \quad \text{and} \quad \sin(x) = \frac{e^{jx} - e^{-jx}}{j2}$$

and this relationship is illustrated in Figure 7.2.

Consider an arbitrary periodic signal $x[n]$ that we would like to represent as a linear combination of sinusoids as illustrated by the center graph in Figure 7.3. (Here we use real-valued sinusoids to simplify the visualization.)

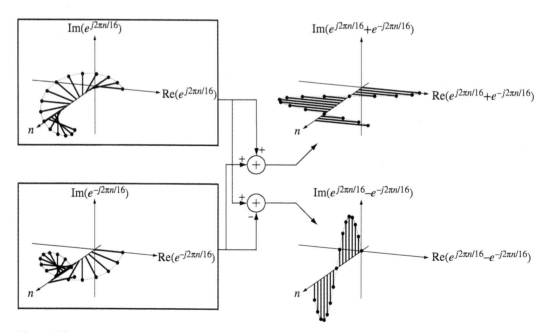

Figure 7.2
Addition and subtraction of $e^{j2\pi n/16}$ and $e^{-j2\pi n/16}$ to form $2\cos(2\pi n/16)$ and $j2\sin(2\pi n/16)$

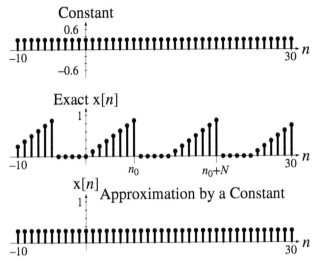

Figure 7.3
Signal approximated by a constant

In Figure 7.3, the signal is approximated by a constant 0.2197, which is the average value of the signal. A constant is a special case of a sinusoid, in this case $0.2197\cos(2\pi kn/N)$ with $k = 0$. This is the best possible approximation of $x[n]$ by a constant because the mean squared error between $x[n]$ and the approximation is a minimum. We can make this poor approximation better by adding to the constant a sinusoid whose fundamental period N is the fundamental period of $x[n]$ (Figure 7.4).

This is the best approximation that can be made using a constant and a single sinusoid of the same fundamental period as $x[n]$. We can improve the approximation further by adding a sinusoid at a frequency of twice the fundamental frequency of $x[n]$ (Figure 7.5).

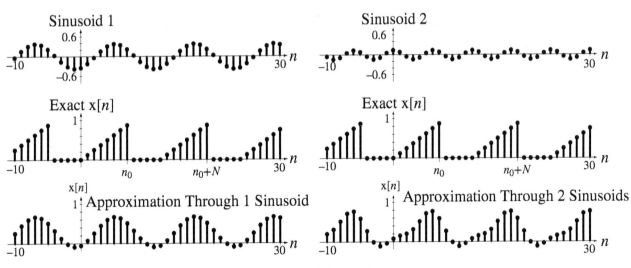

Figure 7.4
Signal approximated by a constant plus a single sinusoid

Figure 7.5
Signal approximated by a constant plus two sinusoids

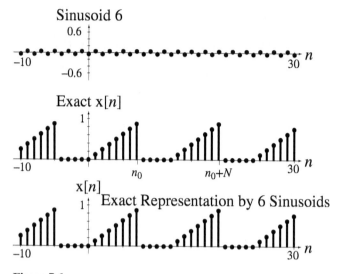

Figure 7.6
Signal represented by a constant plus six sinusoids

If we keep adding properly chosen sinusoids at higher integer multiples of the fundamental frequency of x[n], we can make the approximation better and better. Unlike the general case in continuous time, with a finite number of sinusoids the representation becomes exact (Figure 7.6).

This illustrates one significant difference between continuous-time and discrete-time Fourier series representations. In discrete time exact representation of a periodic signal is always achieved with a finite number of sinusoids.

Just as in the CTFS, k is called the *harmonic number* and all the sinusoids have frequencies that are k times the fundamental cyclic frequency, which, for the DTFS, is $1/N$. The DTFS represents a discrete-time periodic signal of fundamental period N_0 as a linear combination of complex sinusoids of the form

$$x[n] = \sum_{k=\langle N \rangle} c_x[k] e^{j2\pi kn/N}$$

where $N = mN_0$ (m an integer) and $c_x[k]$ is the DTFS harmonic function. The notation $\sum_{k=\langle N \rangle}$ is equivalent to $\sum_{k=n_0}^{n_0+N-1}$ where n_0 is arbitrary; in other words, a summation over any set of N consecutive values of k. Although the most commonly used value for N is the fundamental period of the signal N_0 ($m = 1$), N does not have to be N_0. N can be any period of the signal.

In discrete-time signal and system analysis, there is a very similar form of representation of discrete-time periodic signals using the **discrete Fourier transform (DFT)** first mentioned in Chapter 6. It also represents periodic discrete-time signals as a linear combination of complex sinusoids. The inverse DFT is usually written in the form

$$x[n] = \frac{1}{N} \sum_{k=0}^{N-1} X[k] e^{j2\pi kn/N}$$

where $X[k]$ is the DFT harmonic function of $x[n]$ and $X[k] = N\, c_x[k]$. The name ends with "transform" instead of "series" but, since it is a linear combination of sinusoids at a discrete set of frequencies, for terminological consistency it probably should have been called a series. The name "transform" probably came out of its use in digital signal processing in which it is often used to find a numerical approximation to the CTFT. The DFT is so widely used and so similar to the DTFS that in this text we will concentrate on the DFT knowing that conversion to the DTFS is very simple.

The formula $x[n] = \frac{1}{N}\sum_{k=0}^{N-1} X[k] e^{j2\pi kn/N}$ is the inverse DFT. It forms the time-domain function as a linear combination of complex sinusoids. The forward DFT is

$$X[k] = \sum_{n=0}^{N-1} x[n] e^{-j2\pi kn/N}$$

where N is any period of $x[n]$. It forms the harmonic function from the time-domain function.

As shown in Chapter 6, one important property of the DFT is that $X[k]$ is periodic

$$X[k] = X[k+N], \quad k \text{ any integer.}$$

So now it should be clear why the summation in the inverse DFT is over a finite range of k values. The harmonic function $X[k]$ is periodic with period N, and therefore has only N unique values. The summation needs only N terms to utilize all the unique values of $X[k]$. The formula for the inverse DFT is most commonly written as

$$x[n] = \frac{1}{N}\sum_{k=0}^{N-1} X[k] e^{j2\pi kn/N}$$

but, since $X[k]$ is periodic with period N, it can be written more generally as

$$x[n] = \frac{1}{N}\sum_{k=\langle N \rangle} X[k] e^{j2\pi kn/N}.$$

ORTHOGONALITY AND THE HARMONIC FUNCTION

We can find the forward DFT $X[k]$ of $x[n]$ by a process analogous to the one used for the CTFS. To streamline the notation in the equations to follow let

$$W_N = e^{j2\pi/N}. \tag{7.1}$$

Since the beginning point of the summation $\sum_{k=\langle N \rangle} X[k] e^{j2\pi kn/N}$ is arbitrary let it be $k = 0$. Then, if we write $e^{j2\pi kn/N}$ for each n in $n_0 \le n < n_0 + N$, using (7.1) we can write the matrix equation

$$\underbrace{\begin{bmatrix} x[n_0] \\ x[n_0+1] \\ \vdots \\ x[n_0+N-1] \end{bmatrix}}_{\mathbf{x}} = \frac{1}{N} \underbrace{\begin{bmatrix} W_N^0 & W_N^{n_0} & \cdots & W_N^{n_0(N-1)} \\ W_N^0 & W_N^{n_0+1} & \cdots & W_N^{(n_0+1)(N-1)} \\ \vdots & \vdots & \ddots & \vdots \\ W_N^0 & W_N^{n_0+N-1} & \cdots & W_N^{(n_0+N-1)(N-1)} \end{bmatrix}}_{\mathbf{W}} \underbrace{\begin{bmatrix} X[0] \\ X[1] \\ \vdots \\ X[N-1] \end{bmatrix}}_{\mathbf{X}} \tag{7.2}$$

or in the compact form $N\mathbf{x} = \mathbf{W}\mathbf{X}$. If $\mathbf{W}$ is nonsingular, we can directly find $\mathbf{X}$ as $\mathbf{X} = \mathbf{W}^{-1} N\mathbf{x}$. Equation (7.2) can also be written in the form

$$N\begin{bmatrix} x[n_0] \\ x[n_0+1] \\ \vdots \\ x[n_0+N-1] \end{bmatrix} = \underbrace{\begin{bmatrix} 1 \\ 1 \\ \vdots \\ 1 \end{bmatrix}}_{k=0} X[0] + \underbrace{\begin{bmatrix} W_N^{n_0} \\ W_N^{n_0+1} \\ \vdots \\ W_N^{n_0+N-1} \end{bmatrix}}_{k=1} X[1] + \cdots + \underbrace{\begin{bmatrix} W_N^{n_0(N-1)} \\ W_N^{(n_0+1)(N-1)} \\ \vdots \\ W_N^{(n_0+N-1)(N-1)} \end{bmatrix}}_{k=N-1} X[N-1]$$

(7.3)

or

$$N\mathbf{x} = \mathbf{w}_0 X[0] + \mathbf{w}_1 X[1] + \cdots + \mathbf{w}_{N-1} X[N-1] \tag{7.4}$$

where $\mathbf{W} = [\mathbf{w}_0 \mathbf{w}_1 \cdots \mathbf{w}_{N-1}]$. The elements of the first column vector $\mathbf{w}_0$ are all the constant one and can be thought of as the function values in a unit-amplitude complex sinusoid of zero frequency. The second column vector $\mathbf{w}_1$ consists of the function values from one cycle of a unit-amplitude complex sinusoid over the time period $n_0 \leq n < n_0 + N$. Each succeeding column vector consists of the function values from k cycles of a unit-amplitude complex sinusoid at the next higher harmonic number over the time period $n_0 \leq n < n_0 + N$.

Figure 7.7 illustrates these complex sinusoids for the case of $N = 8$ and $n_0 = 0$.

Notice that the sequence of complex sinusoid values versus n for $k = 7$ looks just like the sequence for $k = 1$, except rotating in the opposite direction. In fact, the sequence for $k = 7$ is the same as the sequence for $k = -1$. This had to happen because of the periodicity of the DFT.

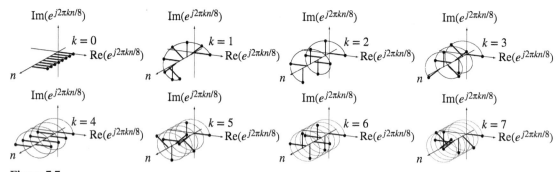

Figure 7.7
Illustration of a complete set of orthogonal basis vectors for $N = 8$ and $n_0 = 0$

These vectors form a family of **orthogonal basis vectors**. Recall from basic linear algebra or vector analysis that the projection $\mathbf{p}$ of a real vector $\mathbf{x}$ in the direction of another real vector $\mathbf{y}$ is

$$\mathbf{p} = \frac{\mathbf{x}^T \mathbf{y}}{\mathbf{y}^T \mathbf{y}} \mathbf{y} \tag{7.5}$$

and when that projection is zero, $\mathbf{x}$ and $\mathbf{y}$ are said to be orthogonal. That happens when the dot product (or scalar product or inner product) of $\mathbf{x}$ and $\mathbf{y}$, $\mathbf{x}^T \mathbf{y}$, is zero. If the vectors are complex-valued, the theory is practically the same except the dot product is $\mathbf{x}^H \mathbf{y}$ and the projection is

$$\mathbf{p} = \frac{\mathbf{x}^H \mathbf{y}}{\mathbf{y}^H \mathbf{y}} \mathbf{y} \tag{7.6}$$

where the notation $\mathbf{x}^H$ means the complex-conjugate of the transpose of $\mathbf{x}$. (This is such a common operation on complex-valued matrices that the transpose of a complex-valued matrix is often defined as including the complex-conjugation operation. This is true for transposing matrices in MATLAB.) A set of correctly chosen orthogonal vectors can form a **basis**. An orthogonal vector basis is a set of vectors that can be combined in linear combinations to form any arbitrary vector of the same dimension.

The dot product of the first two basis vectors in (7.4) is

$$\mathbf{w}_0^H \mathbf{w}_1 = \begin{bmatrix} 1 & 1 & \cdots & 1 \end{bmatrix} \begin{bmatrix} W_N^{n_0} \\ W_N^{n_0+1} \\ \vdots \\ W_N^{n_0+N-1} \end{bmatrix} = W_N^{n_0}(1 + W_N + \cdots + W_N^{N-1}). \quad (7.7)$$

The sum of a finite-length geometric series is

$$\sum_{n=0}^{N-1} r^n = \begin{cases} N, & r = 1 \\ \dfrac{1 - r^N}{1 - r}, & r \neq 1 \end{cases}.$$

Summing the geometric series in (7.7),

$$\mathbf{w}_0^H \mathbf{w}_1 = W^{n_0} \frac{1 - W_N^N}{1 - W_N} = W_N^{n_0} \frac{1 - e^{j2\pi}}{1 - e^{j2\pi/N}} = 0$$

proving that they are indeed orthogonal (if $N \neq 1$). In general, the dot product of the k_1-harmonic vector and the k_2-harmonic vector is

$$\mathbf{w}_{k_1}^H \mathbf{w}_{k_2} = \begin{bmatrix} W_N^{-n_0 k_1} & W_N^{-(n_0+1)k_1} & \cdots & W_N^{-(n_0+N-1)k_1} \end{bmatrix} \begin{bmatrix} W_N^{n_0 k_2} \\ W_N^{(n_0+1)k_2} \\ \vdots \\ W_N^{(n_0+N-1)k_2} \end{bmatrix}$$

$$\mathbf{w}_{k_1}^H \mathbf{w}_{k_2} = W_N^{n_0(k_2-k_1)} \left[1 + W_N^{(k_2-k_1)} + \cdots + W_N^{(N-1)(k_2-k_1)} \right]$$

$$\mathbf{w}_{k_1}^H \mathbf{w}_{k_2} = W_N^{n_0(k_2-k_1)} \frac{1 - [W_N^{(k_2-k_1)}]^N}{1 - W_N^{(k_2-k_1)}} = W_N^{n_0(k_2-k_1)} \frac{1 - e^{j2\pi(k_2-k_1)}}{1 - e^{j2\pi(k_2-k_1)/N}}$$

$$\mathbf{w}_{k_1}^H \mathbf{w}_{k_2} = \begin{cases} 0, & k_1 \neq k_2 \\ N, & k_1 = k_2 \end{cases} = N\delta[k_1 - k_2].$$

This result is zero for $k_1 \neq k_2$ because the numerator is zero and the denominator is not. The numerator is zero because both k_1 and k_2 are integers and therefore $e^{j2\pi(k_2-k_1)}$ is one. The denominator is not zero because both k_1 and k_2 are in the range $0 \leq k_1, k_2 < N$ and the ratio $(k_2 - k_1)/N$ cannot be an integer (if $k_1 \neq k_2$ and $N \neq 1$). So all the vectors in (7.4) are mutually orthogonal.

The fact that the columns of **W** are orthogonal leads to an interesting interpretation of how **X** can be calculated. If we premultiply all the terms in (7.4) by $\mathbf{w}_0^H$ we get

$$\mathbf{w}_0^H N\mathbf{x} = \underbrace{\mathbf{w}_0^H \mathbf{w}_0}_{=N} X[0] + \underbrace{\mathbf{w}_0^H \mathbf{w}_1}_{=0} X[1] + \cdots + \underbrace{\mathbf{w}_0^H \mathbf{w}_{N-1}}_{=0} X[N-1] = NX[0].$$

Then we can solve for X[0] as

$$X[0] = \frac{\mathbf{w}_0^H N\mathbf{x}}{\underbrace{\mathbf{w}_0^H \mathbf{w}_0}_{=N}} = \mathbf{w}_0^H \mathbf{x}.$$

The vector $X[0]\mathbf{w}_0$ is the projection of the vector $N\mathbf{x}$ in the direction of the basis vector $\mathbf{w}_0$. Similarly, each $X[k]\mathbf{w}_k$ is the projection of the vector $N\mathbf{x}$ in the direction of the basis vector $\mathbf{w}_k$. The value of the harmonic function $X[k]$ can be found at each harmonic number as

$$X[k] = \mathbf{w}_k^H \mathbf{x}$$

and we can summarize the entire process of finding the harmonic function as

$$\mathbf{X} = \begin{bmatrix} \mathbf{w}_0^H \\ \mathbf{w}_1^H \\ \vdots \\ \mathbf{w}_{N-1}^H \end{bmatrix} \mathbf{x} = \mathbf{W}^H \mathbf{x}. \tag{7.8}$$

Because of the orthogonality of the vectors $\mathbf{w}_{k_1}$ and $\mathbf{w}_{k_2}$ ($k_1 \neq k_2$), the product of **W** and its complex-conjugate transpose $\mathbf{W}^H$ is

$$\mathbf{W}\mathbf{W}^H = [\mathbf{w}_0 \mathbf{w}_1 \cdots \mathbf{w}_{N-1}] \begin{bmatrix} \mathbf{w}_0^H \\ \mathbf{w}_1^H \\ \vdots \\ \mathbf{w}_{N-1}^H \end{bmatrix} = \begin{bmatrix} N & 0 & \cdots & 0 \\ 0 & N & \cdots & 0 \\ \vdots & \vdots & \ddots & \vdots \\ 0 & 0 & \cdots & N \end{bmatrix} = N\mathbf{I}.$$

Dividing both sides by N,

$$\frac{\mathbf{W}\mathbf{W}^H}{N} = \begin{bmatrix} 1 & 0 & \cdots & 0 \\ 0 & 1 & \cdots & 0 \\ \vdots & \vdots & \ddots & \vdots \\ 0 & 0 & \cdots & 1 \end{bmatrix} = \mathbf{I}.$$

Therefore, the inverse of **W** is

$$\mathbf{W}^{-1} = \frac{\mathbf{W}^H}{N}$$

and, from $\mathbf{X} = \mathbf{W}^{-1}N\mathbf{x}$ we can solve for **X** as

$$\mathbf{X} = \mathbf{W}^H \mathbf{x} \tag{7.9}$$

which is the same as (7.8). Equations (7.8) and (7.9) can be written in a summation form

$$X[k] = \sum_{n=n_0}^{n_0+N-1} x[n]e^{-j2\pi kn/N}.$$

Now we have the forward and inverse DFT formulas as

$$X[k] = \sum_{n=n_0}^{n_0+N-1} x[n]e^{-j2\pi kn/N}, \quad x[n] = \frac{1}{N}\sum_{k=\langle N_0 \rangle} X[k]e^{j2\pi kn/N}. \quad (7.10)$$

If the time-domain function x[n] is bounded on the time $n_0 \leq n < n_0 + N$, the harmonic function can always be found and is itself bounded because it is a finite summation of bounded terms.

In most of the literature concerning the DFT the transform pair is written in this form.

$$\boxed{x[k] = \sum_{n=0}^{N-1} x[n]e^{-j2\pi kn/N}, \quad x[n] = \frac{1}{N}\sum_{k=0}^{N-1} X[k]e^{j2\pi kn/N}} \quad (7.11)$$

Here the beginning point for x[n] is taken as $n_0 = 0$ and the beginning point for X[k] is taken as $k = 0$. This is the form of the DFT that is implemented in practically all computer languages. So in using the DFT on a computer the user should be aware that the result returned by the computer is based on the assumption that the first element in the vector of N values of x sent to the DFT for processing is x[0]. If the first element is $x[n_0]$, $n_0 \neq 0$, then the DFT result will have an extra linear phase shift of $e^{j2\pi kn_0/N}$. This can be compensated for by multiplying the DFT result by $e^{-j2\pi kn_0/N}$. Similarly, if the first value of X[k] is not at $k = 0$, the inverse DFT result will be multiplied by a complex sinusoid.

DISCRETE FOURIER TRANSFORM PROPERTIES

In all the properties listed in Table 7.1, $x[n] \xleftrightarrow[N]{\mathcal{DFT}} X[k]$ and $y[n] \xleftrightarrow[N]{\mathcal{DFT}} Y[k]$.

If a signal x[n] is even and periodic with period N, then its harmonic function is

$$X[k] = \sum_{n=0}^{N-1} x[n]e^{-j2\pi kn/N}.$$

If N is an even number,

$$X[k] = x[0] + \sum_{n=1}^{N/2-1} x[n]e^{-j2\pi kn/N} + x[N/2]e^{-j\pi k} + \sum_{n=N/2+1}^{N-1} x[n]e^{-j2\pi kn/N}$$

$$X[k] = x[0] + \sum_{n=1}^{N/2-1} x[n]e^{-j2\pi kn/N} + \sum_{n=N-1}^{N/2+1} x[n]e^{-j2\pi kn/N} + (-1)^k x[N/2].$$

Knowing that x is periodic with period N, we can subtract N from n in the second summation yielding

$$X[k] = x[0] + \sum_{n=1}^{N/2-1} x[n]e^{-j2\pi kn/N} + \sum_{n=-1}^{-(N/2-1)} x[n]e^{-j2\pi k(n-N)/N} + (-1)^k x[N/2]$$

$$X[k] = x[0] + \sum_{n=1}^{N/2-1} x[n]e^{-j2\pi kn/N} + \underbrace{e^{j2\pi k}}_{=1} \sum_{n=-1}^{-(N/2-1)} x[n]e^{-j2\pi kn/N} + (-1)^k x[N/2]$$

Table 7.1 DFT properties

Linearity	$\alpha x[n] + \beta y[n] \xleftrightarrow{\mathcal{DFT}}_{N} \alpha X[k] + \beta Y[k]$
Time Shifting	$x[n - n_0] \xleftrightarrow{\mathcal{DFT}}_{N} X[k]e^{-j2\pi k n_0/N}$
Frequency Shifting	$x[n]e^{j2\pi k_0 n/N} \xleftrightarrow{\mathcal{DFT}}_{N} X[k - k_0]$
Time Reversal	$x[-n] = x[N - n] \xleftrightarrow{\mathcal{DFT}}_{N} X[-k] = X[N - k]$
Conjugation	$x^*[n] \xleftrightarrow{\mathcal{DFT}}_{N} X^*[-k] = X^*[N - k]$
$\vdots$	$x^*[-n] = x^*[N - n] \xleftrightarrow{\mathcal{DFT}}_{N} X^*[k]$
Time Scaling	$z[n] = \begin{cases} x[n/m], & n/m \text{ an integer} \\ 0, & \text{otherwise} \end{cases}$
$\vdots$	$N \to mN, \quad Z[k] = X[k]$
Change of Period	$N \to qN, \ q$ a positive integer
$\vdots$	$X_q[k] = \begin{cases} qX[k/q], & k/q \text{ an integer} \\ 0, & \text{otherwise} \end{cases}$
Multiplication-Convolution Duality	$x[n]y[n] \xleftrightarrow{\mathcal{DFT}}_{N} (1/N)Y[k]\circledast X[k]$
$\vdots$	$x[n]\circledast y[n] \xleftrightarrow{\mathcal{DFT}}_{N} Y[k]X[k]$
$\vdots$	where $x[n]\circledast y[n] = \sum_{m=\langle N \rangle} x[m]y[n - m]$
Parseval's Theorem	$\frac{1}{N}\sum_{n=\langle N \rangle}|x[n]|^2 = \frac{1}{N^2}\sum_{k=\langle N \rangle}|X[k]|^2$

$$X[k] = x[0] + \sum_{n=1}^{N/2-1} (x[n]e^{-j2\pi kn/N} + x[-n]e^{j2\pi kn/N}) + (-1)^k x[N/2]$$

Now, since $x[n] = x[-n]$,

$$X[k] = x[0] + 2\sum_{n=1}^{N/2-1} x[n] \cos(2\pi k/N) + (-1)^k x[N/2].$$

All these terms are real-valued, therefore $X[k]$ is also. A similar analysis shows that if N is an odd number, the result is the same; $X[k]$ has all real values. Also, if $x[n]$ is an odd periodic function with period N, all the values of $X[k]$ are purely imaginary.

EXAMPLE 7.1

DFT of a periodically repeated rectangular pulse 1

Find the DFT of $x[n] = (u[n] - u[n - n_x]) * \delta_{N_0}[n]$, $0 \leq n_x \leq N_0$, using N_0 as the representation time.

$$(u[n] - u[n - n_x]) * \delta_{N_0}[n] \xleftrightarrow{\mathcal{DFT}}_{N_0} \sum_{n=0}^{n_x-1} e^{-j2\pi kn/N_0}$$

Summing the finite-length geometric series,

$$(u[n] - u[n - n_x]) * \delta_{N_0}[n] \xleftrightarrow[N_0]{\mathscr{DFT}} \frac{1 - e^{-j2\pi k n_x/N_0}}{1 - e^{j2\pi k n/N_0}} = \frac{e^{-j\pi k n_x/N_0}}{e^{-j\pi k/N_0}} \frac{e^{j\pi k n_x/N_0} - e^{-j\pi k n_x/N_0}}{e^{-j\pi k/N_0} - e^{-j\pi k/N_0}}$$

$$(u[n] - u[n - n_x]) * \delta_{N_0}[n] \xleftrightarrow[N_0]{\mathscr{DFT}} e^{-j\pi k(n_x-1)/N_0} \frac{\sin(\pi k n_x/N_0)}{\sin(\pi k/N_0)}, \ 0 \le n_x \le N_0.$$

EXAMPLE 7.2

DFT of a periodically repeated rectangular pulse 2

Find the DFT of $x[n] = (u[n - n_0] - u[n - n_1]) * \delta_{N_0}[n]$, $0 \le n_1 - n_0 \le N_0$.

From Example 7.1 we already know the DFT pair

$$(u[n] - u[n - n_x]) * \delta_{N_0}[n] \xleftrightarrow[N_0]{\mathscr{DFT}} e^{-j\pi k(n_x-1)/N_0} \frac{\sin(\pi k n_x/N_0)}{\sin(\pi k/N_0)}, \ 0 \le n_x \le N_0.$$

If we apply the time-shifting property

$$x[n - n_y] \xleftrightarrow[N]{\mathscr{DFT}} X[k] e^{-j2\pi k n_y/N}$$

to this result we have

$$(u[n - n_y] - u[n - n_y - n_x]) * \delta_{N_0}[n] \xleftrightarrow[N_0]{\mathscr{DFT}} e^{-j\pi k(n_x-1)/N_0} e^{-j2\pi k n_y/N_0} \frac{\sin(\pi k n_x/N_0)}{\sin(\pi k/N_0)},$$

$$0 \le n_x \le N_0$$

$$(u[n - n_y] - u[n - (n_y + n_x)]) * \delta_{N_0}[n] \xleftrightarrow[N_0]{\mathscr{DFT}} e^{-j\pi k(n_x+2n_y-1)/N_0} \frac{\sin(\pi k n_x/N_0)}{\sin(\pi k/N_0)},$$

$$0 \le n_x \le N_0.$$

Now, let $n_0 = n_y$ and let $n_1 = n_y + n_x$.

$$(u[n - n_0] - u[n - n_1]) * \delta_{N_0}[n] \xleftrightarrow[N]{\mathscr{DFT}} e^{-j\pi k(n_0+n_1-1)/N} \frac{\sin(\pi k(n_1 - n_0)/N_0)}{\sin(\pi k/N_0)},$$

$$0 \le n_1 - n_0 \le N_0$$

Consider the special case in which $n_0 + n_1 = 1$. Then

$$u[n - n_0] - u[n - n_1] * \delta_{N_0}[n] \xleftrightarrow[N_0]{\mathscr{DFT}} \frac{\sin(\pi k(n_1 - n_0)/N_0)}{\sin(\pi k/N_0)}, \ n_0 + n_1 = 1.$$

This is the case of a rectangular pulse of width $n_1 - n_0 = 2n_1 - 1$, centered at $n = 0$. This is analogous to a continuous-time, periodically repeated pulse of the form

$$T_0 \text{rect}(t/w) * \delta_{T_0}(t).$$

Compare their harmonic functions.

$$T_0 \text{rect}(t/w) * \delta_{T_0}(t) \xleftrightarrow[T_0]{\mathscr{FS}} w \, \text{sinc}(wk/T_0) = \frac{\sin(\pi w k/T_0)}{\pi k/T_0}$$

$$u[n - n_0] - u[n - n_1] * \delta_{N_0}[n] \xleftrightarrow[N_0]{\mathscr{DFT}} \frac{\sin(\pi k(n_1 - n_0)/N_0)}{\sin(\pi k/N_0)}, \ n_0 + n_1 = 1$$

318 **Chapter 7** Discrete-Time Fourier Methods

The harmonic function of $T_0 \operatorname{rect}(t/w) * \delta_{T_0}(t)$ is a sinc function. Although it may not yet be obvious, the harmonic function of $(u[n - n_0] - u[n - n_1]) * \delta_{N_0}[n]$ is a *periodically repeated* sinc function.

The DFT harmonic function of a periodically repeated rectangular pulse can be found using this result. It can also be found numerically using the `fft` function in MATLAB. This MATLAB program computes the harmonic function both ways and graphs the results for comparison. The phase graphs are not identical, but they only differ when the phase is $\pm\pi$ radians and these two phases are equivalent See Figure 7.8. So the two methods of computing the harmonic function yield the same result.

```
N = 16 ;          %  Set fundamental period to 16
n0 = 2 ;          %  Turn on rectangular pulse at n=2
n1 = 7 ;          %  Turn off rectangular pulse at n=7
n = 0:N-1 ;       %  Discrete-time vector for computing x[n] over one
                          fundamental period
%  Compute values of x[n] over one fundamental period
x = usD(n-n0) - usD(n-n1) ;    %   usD is a user-written unit sequence function
X = fft(x) ;      %  Compute the DFT harmonic function X[k] of x[n] using "fft"
k = 0:N-1 ;       %  Harmonic number vector for graphing X[k]
%   Compute harmonic function X[k] analytically
Xa = exp(-j*pi*k*(n1+n0)/N)*(n1-n0).*drcl(k/N,n1-n0)./exp(-j*pi*k/N) ;

close all ; figure('Position',[20,20,1200,800]) ;

subplot(2,2,1) ;
ptr = stem(n,abs(X),'k','filled') ; grid on ;
set(ptr,'LineWidth',2,'MarkerSize',4) ;
xlabel('\itk','FontName','Times','FontSize',36) ;
ylabel('|X[{\itk}]|','FontName','Times','FontSize',36) ;
title('fft Result','FontName','Times','FontSize',36) ;
set(gca,'FontName','Times','FontSize',24) ;
subplot(2,2,3) ;
ptr = stem(n,angle(X),'k','filled') ; grid on ;
set(ptr,'LineWidth',2,'MarkerSize',4) ;
xlabel('\itk','FontName','Times','FontSize',36) ;
ylabel('Phase of X[{\itk}]','FontName','Times','FontSize',36) ;
set(gca,'FontName','Times','FontSize',24) ;

subplot(2,2,2) ;
ptr = stem(n,abs(Xa),'k','filled') ; grid on ;
set(ptr,'LineWidth',2,'MarkerSize',4) ;
xlabel('\itk','FontName','Times','FontSize',36) ;
ylabel('|X[{\itk}]|','FontName','Times','FontSize',36) ;
title('Analytical Result','FontName','Times','FontSize',36) ;
set(gca,'FontName','Times','FontSize',24) ;
subplot(2,2,4) ;
ptr = stem(n,angle(Xa),'k','filled') ; grid on ;
set(ptr,'LineWidth',2,'MarkerSize',4) ;
```

```
xlabel('\itk','FontName','Times','FontSize',36) ;
ylabel('Phase of X[{\itk}]','FontName','Times','FontSize',36) ;
set(gca,'FontName','Times','FontSize',24) ;
```

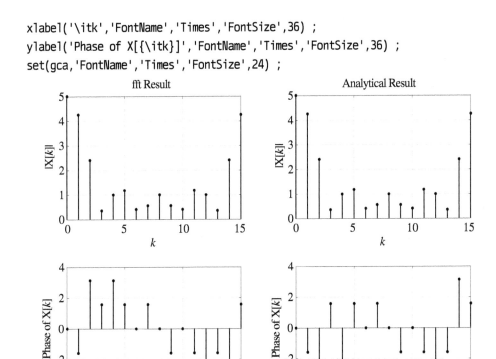

Figure 7.8
A comparison of the numerical and analytical DFT of a periodically-repeated discrete-time rectangular pulse

The functional form $\dfrac{\sin(\pi N x)}{N \sin(\pi x)}$ (see Example 7.2) appears commonly enough in the analysis of signals and systems to be given the name **Dirichlet** function (Figure 7.9).

$$\boxed{\mathrm{drcl}(t,N) = \frac{\sin(\pi N t)}{N \sin(\pi t)}}. \tag{7.12}$$

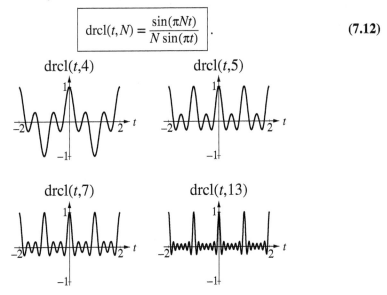

Figure 7.9
The Dirichlet function for $N = 4$, 5, 7, and 13

For N odd, the similarity to a sinc function is obvious; the Dirichlet function is an infinite sum of uniformly spaced sinc functions. The numerator $\sin(N\pi t)$ is zero when t is any integer multiple of $1/N$. Therefore, the Dirichlet function is zero at those points, *unless the denominator is also zero*. The denominator $N \sin(\pi t)$ is zero for every integer value of t. Therefore, we must use L'Hôpital's rule to evaluate the Dirichlet function at integer values of t.

$$\lim_{t \to m} \text{drcl}(t,N) = \lim_{t \to m} \frac{\sin(N\pi t)}{N \sin(\pi t)} = \lim_{t \to m} \frac{N\pi \cos(N\pi t)}{N\pi \cos(\pi t)} = \pm 1, \quad m \text{ an integer}$$

If N is even, the extrema of the Dirichlet function alternate between $+1$ and -1. If N is odd, the extrema are all $+1$. A version of the Dirichlet function is a part of the MATLAB signal toolbox with the function name `diric`. It is defined as

$$\text{diric}(x,N) = \frac{\sin(Nx/2)}{N \sin(x/2)}.$$

Therefore,

$$\text{drcl}(t,N) = \text{diric}(2\pi t, N).$$

```
%   Function to compute values of the Dirichlet function.
%   Works for vectors or scalars equally well.
%
%   x = sin(N*pi*t)/(N*sin(pi*t))
%
function x = drcl(t,N)
    x = diric(2*pi*t,N) ;
```

```
%   Function to implement the Dirichlet function without
%   using the
%   MATLAB diric function. Works for vectors or scalars
%   equally well.
%
%   x = sin(N*pi*t)/(N*sin(pi*t))
%
function x = drcl(t,N),
    num = sin(N*pi*t) ; den = N*sin(pi*t) ;
    I = find(abs(den) < 10*eps) ;
    num(I) = cos(N*pi*t(I)) ; den(I) = cos(pi*t(I)) ;
    x = num./den ;
```

Using the definition of the dirichlet function, the DFT pair from Example 7.2 can be written as

$$(\text{u}[n - n_0] - \text{u}[n - n_1]) * \delta_N[n] \xleftrightarrow[N]{\mathcal{DFT}} \frac{e^{-j\pi k(n_1+n_0)/N}}{e^{-j\pi k/N}} (n_1 - n_0) \text{drcl}(k/N, n_1 - n_0).$$

Table 7.2 shows several common DFT pairs.

Table 7.2 DFT pairs

(For each pair, m is a positive integer.)

$$e^{j2\pi n/N} \xleftrightarrow[mN]{\mathcal{DFT}} mN\delta_{mN}[k-m]$$

$$\cos(2\pi qn/N) \xleftrightarrow[mN]{\mathcal{DFT}} (mN/2)(\delta_{mN}[k-mq]+\delta_{mN}[k+mq])$$

$$\sin(2\pi qn/N) \xleftrightarrow[mN]{\mathcal{DFT}} (jmN/2)(\delta_{mN}[k+mq]-\delta_{mN}[k-mq])$$

$$\delta_N[n] \xleftrightarrow[mN]{\mathcal{DFT}} m\delta_{mN}[k]$$

$$1 \xleftrightarrow[N]{\mathcal{DFT}} N\delta_N[k]$$

$$(\mathrm{u}[n-n_0]-\mathrm{u}[n-n_1])*\delta_N[n] \xleftrightarrow[N]{\mathcal{DFT}} \frac{e^{-j\pi k(n_1+n_0)/N}}{e^{-j\pi k/N}}(n_1-n_0)\mathrm{drcl}(k/N, n_1-n_0)$$

$$\mathrm{tri}(n/N_w)*\delta_N[n] \xleftrightarrow[N]{\mathcal{DFT}} N_w\mathrm{drcl}^2(k/N, N_w),\ N_w\ \text{an integer}$$

$$\mathrm{sinc}(n/w)*\delta_N[n] \xleftrightarrow[N]{\mathcal{DFT}} w\mathrm{rect}(wk/N)*\delta_N[k]$$

THE FAST FOURIER TRANSFORM

The forward DFT is defined by

$$X[k] = \sum_{n=0}^{N-1} x[n]e^{-j2\pi nk/N}.$$

A straightforward way of computing the DFT would be by the following algorithm (written in MATLAB), which directly implements the operations indicated above:

```
.
.
.
%   (Acquire the input data in an array x with N elements.)
.
.
.
%
%   Initialize the DFT array to a column vector of zeros.
%
X = zeros(N,1) ;
%
%   Compute the X[k]'s in a nested, double for loop.
%
for k = 0:N-1
        for n = 0:N-1
                X(k+1) = X(k+1) + x(n+1)*exp(-j*2*pi*n*k/N) ;
        end
end
.
.
.
```

Table 7.3 Numbers of additions and multiplications and ratios for several N's

γ	$N = 2^\gamma$	A_{DFT}	M_{DFT}	A_{FFT}	M_{FFT}	A_{DFT}/A_{FFT}	M_{DFT}/M_{FFT}
1	2	2	4	2	1	1	4
2	4	12	16	8	4	1.5	4
3	8	56	64	24	12	2.33	5.33
4	16	240	256	64	32	3.75	8
5	32	992	1024	160	80	6.2	12.8
6	64	4032	4096	384	192	10.5	21.3
7	128	16,256	16,384	896	448	18.1	36.6
8	256	65,280	65,536	2048	1024	31.9	64
9	512	261,632	262,144	4608	2304	56.8	113.8
10	1024	1,047,552	1,048,576	10,240	5120	102.3	204.8

(One should never actually write this program in MATLAB because the DFT is already built in to MATLAB as an intrinsic function called fft.)

The computation of a DFT using this algorithm requires N^2 complex multiply-add operations. Therefore, the number of computations increases as the square of the number of elements in the input vector that is being transformed. In 1965 James Cooley[1] and John Tukey[2] popularized an algorithm that is much more efficient in computing time for large input arrays whose length is an integer power of 2. This algorithm for computing the DFT is the so-called **fast Fourier transform** or just FFT.

The reduction in calculation time for the fast Fourier transform algorithm versus the double-for-loop approach presented above is illustrated in Table 7.3 in which A is the number of complex-number additions required and M is the number of complex-number multiplications required, the subscript DFT indicates using the straightforward double-for-loop approach and FFT indicates the FFT algorithm.

As the number of points N in the transformation process is increased, the speed advantage of the FFT grows very quickly. *But these speed improvement factors do not apply if N is not an integer power of two.* For this reason, practically all actual DFT analysis is done with the FFT using a data vector length that is an integer power of 2. (In MATLAB if the input vector is an integer power of 2 in length, the algorithm used in the MATLAB function, fft, is the FFT algorithm just discussed. If it is not an integer power of 2 in length, the DFT is still computed but the speed suffers because a less efficient algorithm must be used.)

---

[1] James Cooley received his Ph.D. in applied mathematics from Columbia University in 1961. Cooley was a pioneer in the digital signal processing field, having developed, with John Tukey, the fast Fourier transform. He developed the FFT through mathematical theory and applications, and has helped make it more widely available by devising algorithms for scientific and engineering applications.

[2] John Tukey received his Ph.D. from Princeton in mathematics in 1939. He worked at Bell Labs from 1945 to 1970. He developed new techniques in data analysis, and graphing and plotting methods that now appear in standard statistics texts. He wrote many publications on time series analysis and other aspects of digital signal processing that are now very important in engineering and science. He developed, along with James Cooley, the fast Fourier transform algorithm. He is credited with having coined, as a contraction of "binary digit," the word "bit," the smallest unit of information used by a computer.

7.3 THE DISCRETE-TIME FOURIER TRANSFORM

EXTENDING THE DISCRETE FOURIER TRANSFORM TO APERIODIC SIGNALS

Consider a discrete-time rectangular-wave signal (Figure 7.10).

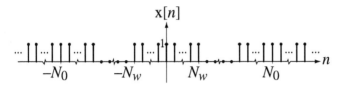

Figure 7.10
A general discrete-time rectangular-wave signal

The DFT harmonic function based on one fundamental period ($N = N_0$) is

$$X[k] = (2N_w + 1)\,\text{drcl}(k/N_0, 2N_w + 1),$$

a sampled Dirichlet function with maxima of $2N_w + 1$ and a period of N_0.

To illustrate the effects of different fundamental periods N_0 let $N_w = 5$ and graph the magnitude of $X[k]$ versus k for $N_0 = 22, 44,$ and 88 (Figure 7.11).

The effect on the DFT harmonic function of increasing the fundamental period of $x[n]$ is to spread it out as a function of harmonic number k. So in the limit as N_0 approaches infinity the period of the DFT harmonic function also approaches infinity. If the period of a function is infinite, it is no longer periodic. We can normalize by graphing the DFT harmonic function versus discrete-time cyclic frequency k/N_0 instead of harmonic number k. Then the fundamental period of the DFT harmonic function (as graphed) is always one, rather than N_0 (Figure 7.12).

As N_0 approaches infinity, the separation between points of $X[k]$ approaches zero and the discrete frequency graph becomes a continuous frequency graph (Figure 7.13).

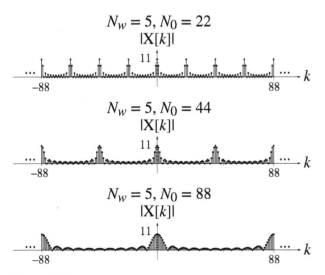

Figure 7.11
Effect of the fundamental period N_0 on the magnitude of the DFT harmonic function of a rectangular-wave signal

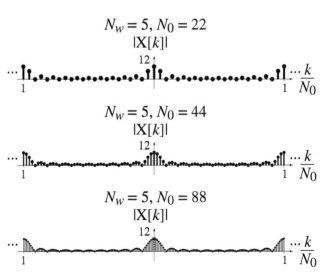

Figure 7.12
Magnitude of the DFT harmonic function of a rectangular-wave signal graphed versus k/N_0 instead of k

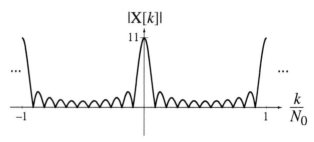

Figure 7.13
Limiting DFT harmonic function of a rectangular-wave signal

DERIVATION AND DEFINITION

To analytically extend the DFT to aperiodic signals, let $\Delta F = 1/N_0$, a finite increment in discrete-time cyclic frequency F. Then x$[n]$ can be written as the inverse DFT of X$[k]$,

$$\text{x}[n] = \frac{1}{N_0} \sum_{k=\langle N_0 \rangle} \text{X}[k] e^{j2\pi k n/N_0} = \Delta F \sum_{k=\langle N_0 \rangle} \text{X}[k] e^{j2\pi k \Delta F n}.$$

Substituting the summation expression for X$[k]$ in the DFT definition

$$\text{x}[n] = \Delta F \sum_{k=\langle N_0 \rangle} \left(\sum_{m=0}^{N_0-1} \text{x}[m] e^{-j2\pi k \Delta F m} \right) e^{j2\pi k \Delta F n}$$

(The index of summation n in the expression for X$[k]$ has been changed to m to avoid confusion with the n in the expression for x$[n]$ since they are independent variables.) Since x$[n]$ is periodic with fundamental period N_0, the inner summation can be over any period and the previous equation can be written as

$$\text{x}[n] = \sum_{k=\langle N_0 \rangle} \left(\sum_{m=\langle N_0 \rangle} \text{x}[m] e^{-j2\pi k \Delta F m} \right) e^{j2\pi k \Delta F n} \Delta F.$$

Let the range of the inner summation be $-N_0/2 \leq m < N_0/2$ for N_0 even or $-(N_0 - 1)/2 \leq m < (N_0 + 1)/2$ for N_0 odd. The outer summation is over any arbitrary range of k of width N_0 so let its range be $k_0 \leq k < k_0 + N_0$. Then

$$\text{x}[n] = \sum_{k=k_0}^{k_0+N_0-1} \left(\sum_{m=-N_0/2}^{N_0/2-1} \text{x}[m] e^{-j2\pi k \Delta F m} \right) e^{j2\pi k \Delta F n} \Delta F, \quad N_0 \text{ even} \qquad (7.13)$$

or

$$\text{x}[n] = \sum_{k=k_0}^{k_0+N_0-1} \left(\sum_{m=-(N_0-1)/2}^{(N_0-1)/2} \text{x}[m] e^{-j2\pi k \Delta F m} \right) e^{j2\pi k \Delta F n} \Delta F, \quad N_0 \text{ odd}. \qquad (7.14)$$

Now let the fundamental period N_0 of the DFT approach infinity. In that limit the following things happen:

1. ΔF approaches the differential discrete-time frequency dF.
2. $k\Delta F$ becomes discrete-time frequency F, a continuous independent variable, because ΔF is approaching dF.

3. The outer summation approaches an integral in $F = k\Delta F$. The summation covers a range of $k_0 \leq k < k_0 + N_0$. The equivalent range of (limits on) the integral it approaches can be found using the relationships $F = kdF = k/N_0$. Dividing the harmonic-number range $k_0 \leq k < k_0 + N_0$ by N_0 translates it to the discrete-time frequency range $F_0 < F < F_0 + 1$ where F_0 is arbitrary because k_0 is arbitrary. The inner summation covers an infinite range because N_0 is approaching infinity.

Then, in the limit, (7.13) and (7.14) both become

$$x[n] = \int_1 \underbrace{\left(\sum_{m=-\infty}^{\infty} x[m]e^{-j2\pi Fm}\right)}_{=\mathcal{F}(x[m])} e^{j2\pi Fn} dF.$$

The equivalent radian-frequency form is

$$x[n] = \frac{1}{2\pi} \int_{2\pi} \left(\sum_{m=-\infty}^{\infty} x[m]e^{-j\Omega m}\right) e^{j\Omega n} d\Omega$$

in which $\Omega = 2\pi F$ and $dF = d\Omega/2\pi$. These results define the **discrete-time Fourier transform (DTFT)** as

$$x[n] = \int_1 X(F)e^{j2\pi Fn} dF \xleftrightarrow{\mathcal{F}} X(F) = \sum_{n=-\infty}^{\infty} x[n]e^{-j2\pi Fn}$$

or

$$x[n] = (1/2\pi)\int_{2\pi} X(e^{j\Omega})e^{j\Omega n} d\Omega \xleftrightarrow{\mathcal{F}} X(e^{j\Omega}) = \sum_{n=-\infty}^{\infty} x[n]e^{-j\Omega n}.$$

Table 7.4 has some DTFT pairs for some typical simple signals.

Table 7.4 Some DTFT pairs derived directly from the definition

$$\delta[n] \xleftrightarrow{\mathcal{F}} 1$$

$$\alpha^n u[n] \xleftrightarrow{\mathcal{F}} \frac{e^{j\Omega}}{e^{j\Omega} - \alpha} = \frac{1}{1 - \alpha e^{-j\Omega}}, \quad |\alpha| < 1, \qquad -\alpha^n u[-n-1] \xleftrightarrow{\mathcal{F}} \frac{e^{j\Omega}}{e^{j\Omega} - \alpha} = \frac{1}{1 - \alpha e^{-j\Omega}}, \quad |\alpha| > 1$$

$$n\alpha^n u[n] \xleftrightarrow{\mathcal{F}} \frac{\alpha e^{j\Omega}}{(e^{j\Omega} - \alpha)^2} = \frac{\alpha e^{-j\Omega}}{(1 - \alpha e^{-j\Omega})^2}, \quad |\alpha| < 1, \qquad -n\alpha^n u[-n-1] \xleftrightarrow{\mathcal{F}} \frac{\alpha e^{j\Omega}}{(e^{j\Omega} - \alpha)^2} = \frac{\alpha e^{-j\Omega}}{(1 - \alpha e^{-j\Omega})^2}, \quad |\alpha| > 1$$

$$\alpha^n \sin(\Omega_0 n) u[n] \xleftrightarrow{\mathcal{F}} \frac{e^{j\Omega}\alpha \sin(\Omega_0)}{e^{j2\Omega} - 2\alpha e^{j\Omega}\cos(\Omega_0) + \alpha^2}, \quad |\alpha| < 1, \qquad -\alpha^n \sin(\Omega_0 n) u[-n-1] \xleftrightarrow{\mathcal{F}} \frac{e^{j\Omega}\alpha \sin(\Omega_0)}{e^{j2\Omega} - 2\alpha e^{j\Omega}\cos(\Omega_0) + \alpha^2}, \quad |\alpha| > 1$$

$$\alpha^n \cos(\Omega_0 n) u[n] \xleftrightarrow{\mathcal{F}} \frac{e^{j\Omega}[e^{j\Omega} - \alpha \cos(\Omega_0)]}{e^{j2\Omega} - 2\alpha e^{j\Omega}\cos(\Omega_0) + \alpha^2}, \quad |\alpha| < 1, \qquad -\alpha^n \cos(\Omega_0 n) u[-n-1] \xleftrightarrow{\mathcal{F}} \frac{e^{j\Omega}[e^{j\Omega} - \alpha \cos(\Omega_0)]}{e^{j2\Omega} - 2\alpha e^{j\Omega}\cos(\Omega_0) + \alpha^2}, \quad |\alpha| > 1$$

$$\alpha^{|n|} \xleftrightarrow{\mathcal{F}} \frac{e^{j\Omega}}{e^{j\Omega} - \alpha} - \frac{e^{j\Omega}}{e^{j\Omega} - 1/\alpha}, \quad |\alpha| < 1$$

Here we are faced with the same notational decision we encountered in deriving the CTFT in Chapter 6. $X(F)$ is defined by $X(F) = \sum_{n=-\infty}^{\infty} x[n]e^{-j2\pi Fn}$ and $X(e^{j\Omega})$ is defined by $X(e^{j\Omega}) = \sum_{n=-\infty}^{\infty} x[n]e^{-j\Omega n}$, but the two X's are actually mathematically different functions because $X(e^{j\Omega}) \neq X(F)_{F \to e^{j\Omega}}$. The decision here will be similar to the one reached in Chapter 6. We will use the forms $X(F)$ and $X(e^{j\Omega})$ for the same reasons. The use of $X(e^{j\Omega})$ instead of the simpler form $X(\Omega)$ is motivated by the desire to maintain consistency of functional definition between the DTFT and the z transform to be presented in Chapter 9.

THE GENERALIZED DTFT

Just as we saw in continuous time, in discrete time there are some important practical signals that do not have a DTFT in the strict sense. Because these signals are so important, the DTFT has been generalized to include them. Consider the DTFT of $x[n] = A$, a constant.

$$X(F) = \sum_{n=-\infty}^{\infty} Ae^{-j2\pi Fn} = A \sum_{n=-\infty}^{\infty} e^{-j2\pi Fn}$$

The series does not converge. Therefore, strictly speaking, the DTFT does not exist. We faced a similar situation with the CTFT and found that the generalized CTFT of a constant is an impulse at $f = 0$ or $\omega = 0$. Because of the close relationship between the CTFT and DTFT we might expect a similar result for the DTFT of a constant. But all DTFTs must be periodic. So a periodic impulse is the logical choice. Let a signal $x[n]$ have a DTFT of the form $A\delta_1(F)$. Then $x[n]$ can be found by finding the inverse DTFT of $A\delta_1(F)$.

$$x[n] = \int_1 A\delta_1(F)e^{j2\pi Fn}dF = A \int_{-1/2}^{1/2} \delta(F)e^{j2\pi Fn}dF = A.$$

This establishes the DTFT pairs

$$A \xleftrightarrow{\mathcal{F}} A\delta_1(F) \quad \text{or} \quad A \xleftrightarrow{\mathcal{F}} 2\pi A\delta_{2\pi}(\Omega).$$

If we now generalize to the form $A\delta_1(F - F_0)$, $-1/2 < F_0 < 1/2$ we get

$$x[n] = \int_1 A\delta_1(F - F_0)e^{j2\pi Fn}dF = A \int_{-1/2}^{1/2} \delta(F - F_0)e^{j2\pi Fn}dF = Ae^{j2\pi F_0 n}.$$

Then, if $x[n] = A\cos(2\pi F_0 n) = (A/2)(e^{j2\pi F_0 n} + e^{-j2\pi F_0 n})$ we get the DTFT pairs

$$A\cos(2\pi F_0 n) \xleftrightarrow{\mathcal{F}} (A/2)[\delta_1(F - F_0) + \delta_1(F + F_0)]$$

or

$$A\cos(\Omega_0 n) \xleftrightarrow{\mathcal{F}} \pi A[\delta_1(\Omega - \Omega_0) + \delta_1(\Omega + \Omega_0)].$$

By a similar process we can also derive the DTFT pairs

$$A\sin(2\pi F_0 n) \xleftrightarrow{\mathcal{F}} (jA/2)[\delta_1(F + F_0) - \delta_1(F - F_0)]$$

or

$$A \sin(\Omega_0 n) \xleftrightarrow{\mathcal{F}} j\pi A[\delta_1(\Omega + \Omega_0) - \delta_1(\Omega - \Omega_0)].$$

Now we can extend the table of DTFT pairs to include more useful functions (Table 7.5).

Table 7.5 More DTFT pairs

$$\delta[n] \xleftrightarrow{\mathcal{F}} 1$$

$$u[n] \xleftrightarrow{\mathcal{F}} \frac{1}{1 - e^{-j2\pi F}} + (1/2)\delta_1(F), \qquad u[n] \xleftrightarrow{\mathcal{F}} \frac{1}{1 - e^{-j\Omega}} + \pi\delta_1(\Omega)$$

$$\text{sinc}(n/w) \xleftrightarrow{\mathcal{F}} w\,\text{rect}(wF) * \delta_1(F), \qquad \text{sinc}(n/w) \xleftrightarrow{\mathcal{F}} w\,\text{rect}(w\Omega/2\pi) * \delta_{2\pi}(\Omega)$$

$$\text{tri}(n/w) \xleftrightarrow{\mathcal{F}} w\,\text{drcl}^2(F, w), \qquad \text{tri}(n/w) \xleftrightarrow{\mathcal{F}} w\,\text{drcl}^2(\Omega/2\pi, w)$$

$$1 \xleftrightarrow{\mathcal{F}} \delta_1(F), \qquad 1 \xleftrightarrow{\mathcal{F}} 2\pi\delta_{2\pi}(\Omega)$$

$$\delta_{N_0}[n] \xleftrightarrow{\mathcal{F}} (1/N_0)\delta_{1/N_0}(F), \qquad \delta_{N_0}[n] \xleftrightarrow{\mathcal{F}} (2\pi/N_0)\delta_{2\pi/N_0}(\Omega)$$

$$\cos(2\pi F_0 n) \xleftrightarrow{\mathcal{F}} (1/2)[\delta_1(F - F_0) + \delta_1(F + F_0)], \quad \cos(\Omega_0 n) \xleftrightarrow{\mathcal{F}} \pi[\delta_{2\pi}(\Omega - \Omega_0) + \delta_{2\pi}(\Omega + \Omega_0)]$$

$$\sin(2\pi F_0 n) \xleftrightarrow{\mathcal{F}} (j/2)[\delta_1(F + F_0) - \delta_1(F - F_0)], \quad \sin(\Omega_0 n) \xleftrightarrow{\mathcal{F}} j\pi[\delta_{2\pi}(\Omega + \Omega_0) - \delta_{2\pi}(\Omega - \Omega_0)]$$

$$u[n - n_0] - u[n - n_1] \xleftrightarrow{\mathcal{F}} \frac{e^{j2\pi F}}{e^{j2\pi F} - 1}(e^{-j2\pi n_0 F} - e^{-j2\pi n_1 F}) = \frac{e^{-j\pi F(n_0 + n_1)}}{e^{-j\pi F}}(n_1 - n_0)\text{drcl}(F, n_1 - n_0)$$

$$u[n - n_0] - u[n - n_1] \xleftrightarrow{\mathcal{F}} \frac{e^{j\Omega}}{e^{j\Omega} - 1}(e^{-jn_0\Omega} - e^{-jn_1\Omega}) = \frac{e^{-j\Omega(n_0 + n_1)/2}}{e^{-j\Omega/2}}(n_1 - n_0)\text{drcl}(\Omega/2\pi, n_1 - n_0)$$

CONVERGENCE OF THE DISCRETE-TIME FOURIER TRANSFORM

The condition for convergence of the DTFT is simply that the summation in

$$X(F) = \sum_{n=-\infty}^{\infty} x[n]e^{-j2\pi Fn} \quad \text{or} \quad X(e^{j\Omega}) = \sum_{n=-\infty}^{\infty} x[n]e^{-j\Omega n} \qquad (7.15)$$

actually converges. It will converge if

$$\sum_{n=-\infty}^{\infty} |x[n]| < \infty. \qquad (7.16)$$

If the DTFT function is bounded, the inverse transform,

$$x[n] = \int_1 X(F)e^{j2\pi Fn}dF \quad \text{or} \quad x[n] = \frac{1}{2\pi}\int_{2\pi} X(e^{j\Omega})e^{j\Omega n}d\Omega, \qquad (7.17)$$

will always converge because the integration interval is finite.

DTFT PROPERTIES

Let $x[n]$ and $y[n]$ be two signals whose DTFTs are $X(F)$ and $Y(F)$ or $X(e^{j\Omega})$ and $Y(e^{j\Omega})$. Then the properties in Table 7.6 apply.

Table 7.6 DTFT properties

$$\alpha x[n] + \beta y[n] \xleftrightarrow{\mathscr{F}} \alpha X(F) + \beta Y(F), \qquad \alpha x[n] + \beta y[n] \xleftrightarrow{\mathscr{F}} \alpha X(e^{j\Omega}) + \beta Y(e^{j\Omega})$$

$$x[n - n_0] \xleftrightarrow{\mathscr{F}} e^{-j2\pi F n_0} X(F), \qquad x[n - n_0] \xleftrightarrow{\mathscr{F}} e^{-j\Omega n_0} X(e^{j\Omega})$$

$$e^{j2\pi F_0 n} x[n] \xleftrightarrow{\mathscr{F}} X(F - F_0), \qquad e^{j\Omega_0 n} x[n] \xleftrightarrow{\mathscr{F}} X(e^{j(\Omega - \Omega_0)})$$

$$\text{If } z[n] = \begin{cases} x[n/m], & n/m \text{ an integer} \\ 0, & \text{otherwise} \end{cases} \text{ then } z[n] \xleftrightarrow{\mathscr{F}} X(mF) \text{ or } z[n] \xleftrightarrow{\mathscr{F}} X(e^{jm\Omega})$$

$$x^*[n] \xleftrightarrow{\mathscr{F}} X^*(-F), \qquad x^*[n] \xleftrightarrow{\mathscr{F}} X^*(e^{-j\Omega})$$

$$x[n] - x[n-1] \xleftrightarrow{\mathscr{F}} (1 - e^{-j2\pi F}) X(F), \qquad x[n] - x[n-1] \xleftrightarrow{\mathscr{F}} (1 - e^{-j\Omega}) X(e^{j\Omega})$$

$$\sum_{m=-\infty}^{n} x[m] \xleftrightarrow{\mathscr{F}} \frac{X(F)}{1 - e^{-j2\pi F}} + \frac{1}{2} X(0) \delta_1(F), \qquad \sum_{m=-\infty}^{n} x[m] \xleftrightarrow{\mathscr{F}} \frac{X(e^{j\Omega})}{1 - e^{-j\Omega}} + \pi X\left(e^{j0}\right) \delta_{2\pi}(\Omega)$$

$$x[-n] \xleftrightarrow{\mathscr{F}} X(-F), \qquad x[-n] \xleftrightarrow{\mathscr{F}} X(e^{-j\Omega})$$

$$x[n] * y[n] \xleftrightarrow{\mathscr{F}} X(F) Y(F), \qquad x[n] * y[n] \xleftrightarrow{\mathscr{F}} X(e^{j\Omega}) Y(e^{j\Omega})$$

$$x[n] y[n] \xleftrightarrow{\mathscr{F}} X(F) \circledast Y(F), \qquad x[n] y[n] \xleftrightarrow{\mathscr{F}} (1/2\pi) X(e^{j\Omega}) \circledast Y(e^{j\Omega})$$

$$\sum_{n=-\infty}^{\infty} e^{j2\pi F n} = \delta_1(F), \qquad \sum_{n=-\infty}^{\infty} e^{j\Omega n} = 2\pi \delta_{2\pi}(\Omega)$$

$$\sum_{n=-\infty}^{\infty} |x[n]|^2 = \int_1 |X(F)|^2 dF, \qquad \sum_{n=-\infty}^{\infty} |x[n]|^2 = (1/2\pi) \int_{2\pi} |X(e^{j\Omega})|^2 d\Omega$$

In the property

$$x[n] y[n] \xleftrightarrow{\mathscr{F}} (1/2\pi) X(e^{j\Omega}) \circledast Y(e^{j\Omega})$$

the $\circledast$ operator indicates periodic convolution, which was first introduced in Chapter 6. In this case

$$X(e^{j\Omega}) \circledast Y(e^{j\Omega}) = \int_{2\pi} X(e^{j\Phi}) Y(e^{j(\Omega - \Phi)}) d\Phi.$$

EXAMPLE 7.3

Inverse DTFT of two periodic shifted rectangles

Find and graph the inverse DTFT of
$$X(F) = [\text{rect}(50(F - 1/4)) + \text{rect}(50(F + 1/4))] * \delta_1(F)$$

(Figure 7.14).

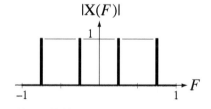

Figure 7.14
Magnitude of X(F)

We can start with the table entry $\text{sinc}(n/w) \xleftrightarrow{\mathcal{F}} w\,\text{rect}(wF) * \delta_1(F)$ or, in this case, $(1/50)\,\text{sinc}(n/50) \xleftrightarrow{\mathcal{F}} \text{rect}(50F) * \delta_1(F)$. Now apply the frequency-shifting property $e^{j2\pi F_0 n}\,x[n] \xleftrightarrow{\mathcal{F}} X(F - F_0)$,

$$e^{j\pi n/2}(1/50)\,\text{sinc}(n/50) \xleftrightarrow{\mathcal{F}} \text{rect}(50(F - 1/4)) * \delta_1(F) \qquad (7.18)$$

and

$$e^{-j\pi n/2}(1/50)\,\text{sinc}(n/50) \xleftrightarrow{\mathcal{F}} \text{rect}(50(F + 1/4)) * \delta_1(F) \qquad (7.19)$$

(Remember, when two functions are convolved, a shift of either one of them, but not both, shifts the convolution by the same amount.) Finally, combining (7.18) and (7.19) and simplifying,

$$(1/25)\,\text{sinc}(n/50)\cos(\pi n/2) \xleftrightarrow{\mathcal{F}} [\text{rect}(50(F - 1/4)) + \text{rect}(50(F + 1/4))] * \delta_1(F).$$

■

Time scaling in the DTFT is quite different from time scaling in the CTFT because of the differences between discrete time and continuous time. Let $z[n] = x[an]$. If a is not an integer, some values of $z[n]$ are undefined and a DTFT cannot be found for it. If a is an integer greater than one, some values of $x[n]$ will not appear in $z[n]$ because of decimation, and there cannot be a unique relationship between the DTFTs of $x[n]$ and $z[n]$ (Figure 7.15).

In Figure 7.15 the two signals $x_1[n]$ and $x_2[n]$ are different signals but have the same values at even values of n. Each of them, when decimated by a factor of 2, yields the same decimated signal $z[n]$. Therefore, the DTFT of a signal and the DTFT of a decimated version of that signal are not uniquely related and no time-scaling property can be found for that kind of time scaling. However, if $z[n]$ is a time-expanded version of $x[n]$, formed by inserting zeros between values of $x[n]$, there is a unique relationship between the DTFTs of $x[n]$ and $z[n]$. Let

$$z[n] = \begin{cases} x[n/m], & n/m \text{ an integer} \\ 0, & \text{otherwise} \end{cases}$$

where m is an integer. Then $Z(e^{j\Omega}) = X(e^{jm\Omega})$ and the time-scaling property of the DTFT is

$$\text{If } z[n] = \begin{cases} x[n/m], & n/m \text{ an integer} \\ 0, & \text{otherwise} \end{cases} \text{ then } \begin{cases} z[n] \xleftrightarrow{\mathcal{F}} X(mF) \\ z[n] \xleftrightarrow{\mathcal{F}} X(e^{jm\Omega}) \end{cases} \qquad (7.20)$$

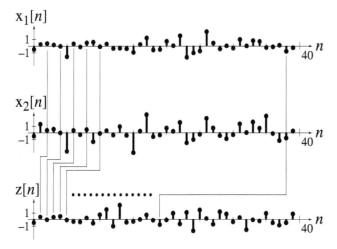

Figure 7.15
Two different signals which, when decimated by a factor of 2, yield the same signal

These results can also be interpreted as a frequency-scaling property. Given a DTFT $X(e^{j\Omega})$, if we scale Ω to $m\Omega$ where $m \geq 1$, the effect in the time domain is to insert $m-1$ zeros between the points in $x[n]$. The only scaling that can be done in the frequency domain is compression and only by a factor that is an integer. This is necessary because all DTFTs must have a period (not necessarily a fundamental period) of 2π in Ω.

EXAMPLE 7.4

General expression for the DTFT of a periodic impulse

Given the DTFT pair $1 \xleftrightarrow{\mathscr{F}} 2\pi\delta_{2\pi}(\Omega)$, use the time-scaling property to find a general expression for the DTFT of $\delta_{N_0}[n]$.

The constant 1 can be expressed as $\delta_1[n]$. The periodic impulse $\delta_{N_0}[n]$ is a time-scaled version of $\delta_1[n]$ scaled by the integer N_0. That is

$$\delta_{N_0}[n] = \begin{cases} \delta_1[n/N_0], & n/N_0 \text{ an integer} \\ 0, & \text{otherwise} \end{cases}$$

Therefore, from (7.20)

$$\delta_{N_0}[n] \xleftrightarrow{\mathscr{F}} 2\pi\delta_{2\pi}(N_0\Omega) = (2\pi/N_0)\delta_{2\pi/N_0}(\Omega).$$

■

The implications of multiplication-convolution duality for signal and system analysis are the same for discrete-time signals and systems as for continuous-time signals and systems. The response of a system is the convolution of the excitation with the impulse response. The equivalent statement in the frequency domain is that the DTFT of the response of a system is the product of the DTFT of the excitation and the frequency response, which is the DTFT of the impulse response (Figure 7.16).

The implications for cascade connections of systems are also the same (Figure 7.17).

If the excitation is a sinusoid of the form $x[n] = A\cos(2\pi n/N_0 + \theta)$, then

$$X(e^{j\Omega}) = \pi A[\delta_{2\pi}(\Omega - \Omega_0) + \delta_{2\pi}(\Omega + \Omega_0)]e^{j\theta\Omega/\Omega_0}$$

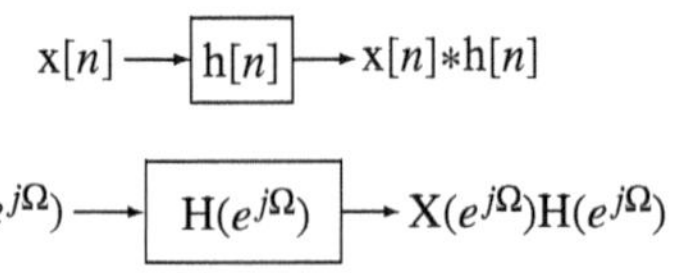

$$x[n] \rightarrow \boxed{h[n]} \rightarrow x[n]*h[n]$$

$$X(e^{j\Omega}) \rightarrow \boxed{H(e^{j\Omega})} \rightarrow X(e^{j\Omega})H(e^{j\Omega})$$

Figure 7.16
Equivalence of convolution in the time domain and multiplication in the frequency domain

$$X(e^{j\Omega}) \rightarrow \boxed{H_1(e^{j\Omega})} \rightarrow X(e^{j\Omega})H_1(e^{j\Omega}) \rightarrow \boxed{H_2(e^{j\Omega})} \rightarrow Y(e^{j\Omega}) = X(e^{j\Omega})H_1(e^{j\Omega})H_2(e^{j\Omega})$$

$$X(e^{j\Omega}) \rightarrow \boxed{H_1(e^{j\Omega})H_2(e^{j\Omega})} \rightarrow Y(e^{j\Omega})$$

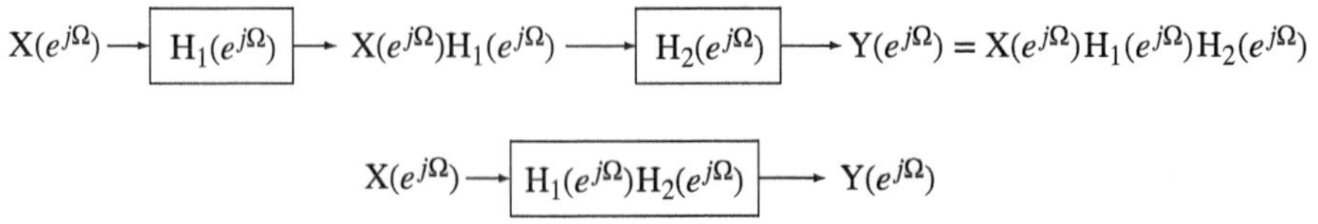

Figure 7.17
Cascade connection of systems

where $\Omega_0 = 2\pi/N_0$. Then

$$Y(e^{j\Omega}) = X(e^{j\Omega})H(e^{j\Omega}) = H(e^{j\Omega}) \times \pi A[\delta_{2\pi}(\Omega - \Omega_0) + \delta_{2\pi}(\Omega + \Omega_0)]e^{j\theta\Omega/\Omega_0}.$$

Using the equivalence property of the impulse, the periodicity of the DTFT, and the conjugation property of the CTFT,

$$Y(e^{j\Omega}) = \pi A \left[H(e^{j\Omega_0})\delta_{2\pi}(\Omega - \Omega_0) + \underbrace{H(e^{-j\Omega_0})}_{=H^*(e^{j\Omega_0})}\delta_{2\pi}(\Omega + \Omega_0) \right] e^{j\theta\Omega/\Omega_0}$$

$$Y(e^{j\Omega}) = \pi A \left\{ \begin{array}{l} \text{Re}(H(e^{j\Omega_0}))[\delta_{2\pi}(\Omega - \Omega_0) + \delta_{2\pi}(\Omega + \Omega_0)] \\ + j\,\text{Im}(H(e^{j\Omega_0}))[\delta_{2\pi}(\Omega - \Omega_0) - \delta_{2\pi}(\Omega + \Omega_0)] \end{array} \right\} e^{j\theta\Omega/\Omega_0}$$

$$y[n] = A[\text{Re}(H(e^{j\Omega_0}))\cos(2\pi n/N_0 + \theta) - \text{Im}(H(e^{j\Omega_0}))\sin(2\pi n/N_0 + \theta)]$$

$$y[n] = A|H(e^{j2\pi/N_0})|\cos(2\pi n/N_0 + \theta + \angle H(e^{j2\pi/N_0})).$$

EXAMPLE 7.5

Frequency response of a system

Graph the magnitude and phase of the frequency response of the system in Figure 7.18. If the system is excited by a signal $x[n] = \sin(\Omega_0 n)$, find and graph the response $y[n]$ for $\Omega_0 = \pi/4, \pi/2, 3\pi/4$.

The difference equation describing the system is $y[n] + 0.7y[n-1] = x[n]$ and the impulse response is $h[n] = (-0.7)^n u[n]$. The frequency response is the Fourier transform of the impulse response. We can use the DTFT pair

$$\alpha^n u[n] \xleftrightarrow{\mathcal{F}} \frac{1}{1 - \alpha e^{-j\Omega}}$$

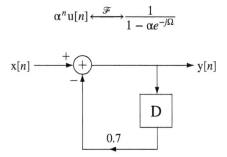

Figure 7.18
A discrete-time system

to get

$$h[n] = (-0.7)^n u[n] \xleftrightarrow{\mathcal{F}} H(e^{j\Omega}) = \frac{1}{1 + 0.7e^{-j\Omega}}.$$

Since the frequency response is periodic in Ω with period 2π, a range $-\pi \leq \Omega < \pi$ will show all the frequency-response behavior. At $\Omega = 0$ the frequency response is $H(e^{j0}) = 0.5882$. At $\Omega = \pm\pi$ the frequency response is $H(e^{\pm j\pi}) = 3.333$. The response at $\Omega = \Omega_0$ is

$$y[n] = |H(e^{j\Omega_0})|\sin(\Omega_0 n + \angle H(e^{j\Omega_0})),$$

(Figure 7.19).

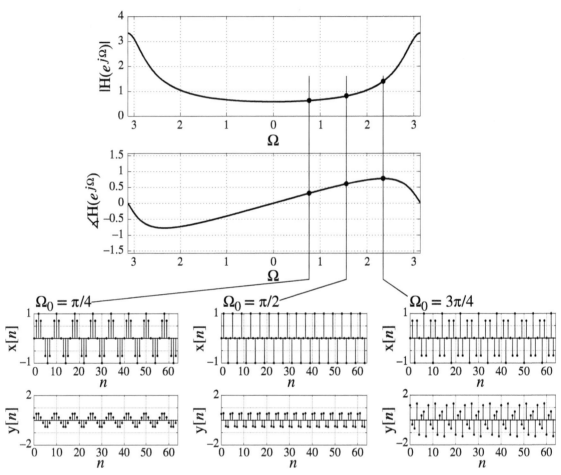

Figure 7.19
Frequency response and three sinusoidal signals and responses to them

EXAMPLE 7.6

Signal energy of a sinc signal

Find the signal energy of $x[n] = (1/5)\text{sinc}(n/100)$.
The signal energy of a signal is defined as

$$E_x = \sum_{n=-\infty}^{\infty} |x[n]|^2.$$

But we can avoid doing a complicated infinite summation by using Parseval's theorem. The DTFT of $x[n]$ can be found by starting with the Fourier pair

$$\text{sinc}(n/w) \xleftrightarrow{\mathcal{F}} w\text{rect}(wF) * \delta_1(F)$$

and applying the linearity property to form

$$(1/5)\text{sinc}(n/100) \xleftrightarrow{\mathcal{F}} 20\text{rect}(100F) * \delta_1(F).$$

Parseval's theorem is

$$\sum_{n=-\infty}^{\infty} |x[n]|^2 = \int_1 |X(F)|^2 dF.$$

So the signal energy is

$$E_x = \int_1 |20\text{rect}(100F) * \delta_1(F)|^2 dF = \int_{-\infty}^{\infty} |20\text{rect}(100F)|^2 dF$$

or

$$E_x = 400 \int_{-1/200}^{1/200} dF = 4.$$

∎

EXAMPLE 7.7

Inverse DTFT of a periodically repeated rectangle

Find the inverse DTFT of $X(F) = \text{rect}(wF) * \delta_1(F)$, $w > 1$ using the definition of the DTFT.

$$x[n] = \int_1 X(F)e^{j2\pi Fn} dF = \int_1 \text{rect}(wF) * \delta_1(F) e^{j2\pi Fn} dF$$

Since we can choose to integrate over any interval in F of width one, let's choose the simplest one

$$x[n] = \int_{-1/2}^{1/2} \text{rect}(wF) * \delta_1(F) e^{j2\pi Fn} dF$$

In this integration interval, there is exactly one rectangle function of width $1/w$ (because $w > 1$) and

$$x[n] = \int_{-1/2w}^{1/2w} e^{j2\pi Fn} dF = 2\int_0^{1/2w} \cos(2\pi Fn)dF = \frac{\sin(\pi n/w)}{\pi n} = \frac{1}{w}\text{sinc}\left(\frac{n}{w}\right). \quad (7.21)$$

From this result we can also establish the handy DTFT pair (which appears in the table of DTFT pairs),

$$\text{sinc}(n/w) \xleftrightarrow{\mathcal{F}} w\text{rect}(wF) * \delta_1(F), \quad w > 1$$

or

$$\text{sinc}(n/w) \xleftrightarrow{\mathcal{F}} w \sum_{k=-\infty}^{\infty} \text{rect}(w(F-k)), \quad w > 1$$

or, in radian-frequency form, using the convolution property,

$$y(t) = x(t) * h(t) \Rightarrow y(at) = |a|x(at) * h(at),$$

we get

$$\text{sinc}(n/w) \xleftrightarrow{\mathcal{F}} w\text{rect}(w\Omega/2\pi) * \delta_{2\pi}(\Omega), \quad w > 1$$

or

$$\text{sinc}(n/w) \xleftrightarrow{\mathcal{F}} w \sum_{k=-\infty}^{\infty} \text{rect}(w(\Omega - 2\pi k)/2\pi), \quad w > 1.$$

(Although these Fourier pairs we derived under the condition $w > 1$ to make the inversion integral (7.21) simpler, they are actually also correct for $w \leq 1$.)

NUMERICAL COMPUTATION OF THE DISCRETE-TIME FOURIER TRANSFORM

The DTFT is defined by $X(F) = \sum_{n=-\infty}^{\infty} x[n]e^{-j2\pi Fn}$ and the DFT is defined by $X[k] = \sum_{n=0}^{N-1} x[n]e^{-j2\pi kn/N}$. If the signal $x[n]$ is causal and time limited, the summation in the DTFT is over a finite range of n values beginning with $n = 0$. We can set the value of N by letting $N - 1$ be the last value of n needed to cover that finite range. Then

$$X(F) = \sum_{n=0}^{N-1} x[n]e^{-j2\pi Fn}.$$

If we now make the change of variable $F \to k/N$ we get

$$X(F)_{F \to k/N} = X(k/N) = \sum_{n=0}^{N-1} x[n]e^{-j2\pi kn/N} = X[k]$$

or in the radian-frequency form

$$X(e^{j\Omega})_{\Omega \to 2\pi k/N} = X(e^{j2\pi k/N}) = \sum_{n=0}^{N-1} x[n]e^{-j2\pi kn/N} = X[k].$$

So the DTFT of $x[n]$ can be found from the DFT of $x[n]$ at a discrete set of frequencies $F = k/N$ or equivalently $\Omega = 2\pi k/N$, k being any integer. If it is desired to increase the resolution of this set of discrete frequencies, we can just make N larger. The extra values of $x[n]$ corresponding to the larger value of N will all be zero. This technique for increasing the frequency-domain resolution is called **zero padding**.

The inverse DTFT is defined by

$$x[n] = \int_1 X(F)e^{j2\pi Fn} dF$$

and the inverse DFT is defined by

$$x[n] = \frac{1}{N} \sum_{k=0}^{N-1} X[k]e^{j2\pi kn/N}.$$

We can approximate the inverse DTFT by the sum of N integrals that together approximate the inverse DTFT integral.

$$x[n] \cong \sum_{k=0}^{N-1} \int_{k/N}^{(k+1)/N} X(k/N)e^{j2\pi Fn} dF = \sum_{k=0}^{N-1} X(k/N) \int_{k/N}^{(k+1)/N} e^{j2\pi Fn} dF$$

$$x[n] \cong \sum_{k=0}^{N-1} X(k/N) \frac{e^{j2\pi(k+1)n/N} - e^{j2\pi k/N}}{j2\pi n} = \frac{e^{j2\pi n/N} - 1}{j2\pi n} \sum_{k=0}^{N-1} X(k/N)e^{j2\pi kn/N}$$

$$x[n] \cong e^{j\pi n/N} \frac{j2 \sin(\pi n/N)}{j2\pi n} \sum_{k=0}^{N-1} X(k/N)e^{j2\pi kn/N} = e^{j\pi n/N} \operatorname{sinc}(n/N) \frac{1}{N} \sum_{k=0}^{N-1} X(k/N)e^{j2\pi kn/N}$$

7.3 The Discrete-Time Fourier Transform

For $n \ll N$,

$$x[n] \cong \frac{1}{N}\sum_{k=0}^{N-1} X(k/N)e^{j2\pi kn/N}$$

or in the radian-frequency form

$$x[n] \cong \frac{1}{N}\sum_{k=0}^{N-1} X(e^{j2\pi k/N})e^{j2\pi kn/N}$$

This is the inverse DFT with

$$X[k] = X(F)_{F \to k/N} = X(k/N) \quad \text{or} \quad X[k] = X(e^{j\Omega})_{\Omega \to 2\pi k/N} = X(e^{j2\pi k/N}).$$

EXAMPLE 7.8

Inverse DTFT using the DFT

Find the approximate inverse DTFT of

$$X(F) = [\text{rect}(50(F - 1/4)) + \text{rect}(50(F + 1/4))] * \delta_1(F)$$

using the DFT.

```
N = 512 ;      %      Number of pts to approximate X(F)
k = [0:N-1]' ; %      Harmonic numbers

%   Compute samples from X(F) between 0 and 1 assuming
%   periodic repetition with period 1

X = rect(50*(k/N - 1/4)) + rect(50*(k/N - 3/4)) ;

%   Compute the approximate inverse DTFT and
%   center the function on n = 0

xa = real(fftshift(ifft(X))) ;

n = [-N/2:N/2-1]' ;  %     Vector of discrete times for plotting
%   Compute exact x[n] from exact inverse DTFT

xe = sinc(n/50).*cos(pi*n/2)/25 ;

%   Graph the exact inverse DTFT

subplot(2,1,1) ; p = stem(n,xe,'k','filled') ; set(p,'LineWidth',1,
'MarkerSize',2) ;
axis([-N/2,N/2,-0.05,0.05]) ; grid on ;
xlabel('\itn','FontName','Times','FontSize',18) ;
ylabel('x[{\itn}]','FontName','Times','FontSize',18) ;
title('Exact','FontName','Times','FontSize',24) ;
```

```
%   Graph the approximate inverse DTFT

subplot(2,1,2) ; p = stem(n,xa,'k','filled') ; set(p,'LineWidth',1,
'MarkerSize',2) ;
axis([-N/2,N/2,-0.05,0.05]) ; grid on ;
xlabel('\itn','FontName','Times','FontSize',18) ;
ylabel('x[{\itn}]','FontName','Times','FontSize',18) ;
title('Approximation Using the DFT','FontName','Times','FontSize',24) ;
```

The exact and approximate inverse DTFT results are illustrated in Figure 7.20. Notice that the exact and approximate x[n] are practically the same near $n = 0$ but are noticeably different near $n = \pm 256$. This occurs because the approximate result is periodic and the overlap of the periodically repeated sinc functions causes these errors near plus or minus half a period.

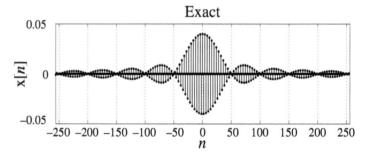
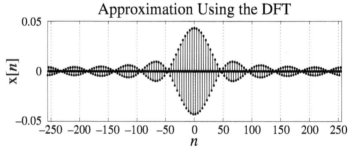

Figure 7.20
Exact and approximate inverse DTFT of X(F)

Example 7.9 illustrates a common analysis problem and a different kind of solution.

EXAMPLE 7.9

System response using the DTFT and the DFT

A system with frequency response $H(e^{j\Omega}) = \dfrac{e^{j\Omega}}{e^{j\Omega} - 0.7}$ is excited by $x[n] = \text{tri}((n-8)/8)$. Find the system response.

The DTFT of the excitation is $X(e^{j\Omega}) = 8\,\text{drcl}^2(\Omega/2\pi, 8)e^{-j8\Omega}$. So the DTFT of the response is

$$Y(e^{j\Omega}) = \frac{e^{j\Omega}}{e^{j\Omega} - 0.7} \times 8\,\text{drcl}^2(\Omega/2\pi, 8)e^{-j8\Omega}.$$

Here we have a problem. How do we find the inverse DTFT of $Y(e^{j\Omega})$? For an analytical solution it would probably be easier in this case to do the convolution in the time domain than to use transforms. But there is another way. We could use the inverse DFT to approximate the inverse DTFT and find $Y(e^{j\Omega})$ numerically.

When we compute the inverse DFT, the number of values of y[n] will be the same as the number of values of $Y(e^{j2\pi k/N})$ we use, N. To make this a good approximation we need a value of N large enough to cover the time range over which we expect y[n] to have values significantly different from zero. The triangle signal has a full base width of 16 and the impulse response of the system is $(0.7)^n$ u[n]. This is a decaying exponential, which approaches, but never reaches, zero. If we use the width at which its value goes below 1% of its initial value, we get a width of about 13. Since the convolution will be the sum of those two widths minus one, we need an N of at least 28. Also, remember that the approximation relies on the inequality $n \ll N$ for a good approximation. So let's use $N = 128$ in doing the computations and then use only the first 30 values. Below is a MATLAB program to find this inverse DTFT. Following that are the three graphs produced by the program (Figure 7.21).

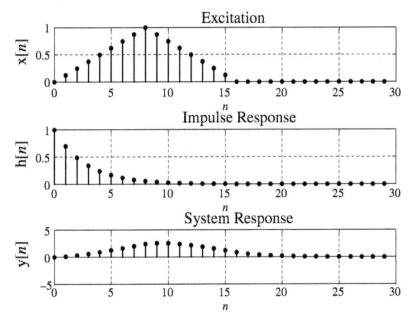

Figure 7.21
Excitation, impulse response, and system response

```
% Program to find an inverse DTFT using the inverse DFT

N = 128 ;                   % Number of points to use
k = [0:N-1]' ;              % Vector of harmonic numbers
n = k ;                     % Vector of discrete times
x = tri((n-8)/8) ;          % Vector of excitation signal values

% Compute the DTFT of the excitation
X = 8*drcl(k/N,8).^2.*exp(-j*16*pi*k/N) ;

% Compute the frequency response of the system
H = exp(j*2*pi*k/N)./(exp(j*2*pi*k/N) - 0.7) ;
```

```
h = 0.7.^n.*uD(n) ;     % Vector of impulse response values
Y = H.*X ;              % Compute the DTFT of the response

y = real(ifft(Y)) ; n = k ;    % Vector of system response values

% Graph the excitation, impulse response and response
n = n(1:30) ; x = x(1:30) ; h = h(1:30) ; y = y(1:30) ;
subplot(3,1,1) ;
ptr = stem(n,x,'k','filled') ; grid on ;
set(ptr,'LineWidth',2,'MarkerSize',4) ;
% xlabel('\itn','FontSize',24,'FontName','Times') ;
ylabel('x[{\itn}]','FontSize',24,'FontName','Times') ;
title('Excitation','FontSize',24,'FontName','Times') ;
set(gca,'FontSize',18,'FontName','Times') ;
subplot(3,1,2) ;
ptr = stem(n,h,'k','filled') ; grid on ;
set(ptr,'LineWidth',2,'MarkerSize',4) ;
% xlabel('\itn','FontSize',24,'FontName','Times') ;
ylabel('h[{\itn}]','FontSize',24,'FontName','Times') ;
title('Impulse Response','FontSize',24,'FontName','Times') ;
set(gca,'FontSize',18,'FontName','Times') ;
subplot(3,1,3) ;
ptr = stem(n,y,'k','filled') ; grid on ;
set(ptr,'LineWidth',2,'MarkerSize',4) ;
xlabel('\itn','FontSize',24,'FontName','Times') ;
ylabel('y[{\itn}]','FontSize',24,'FontName','Times') ;
title('System Response','FontSize',24,'FontName','Times') ;
set(gca,'FontSize',18,'FontName','Times') ;
```

EXAMPLE 7.10

Using the DFT to find a system response

A set of samples

n	0	1	2	3	4	5	6	7	8	9	10
$x[n]$	-9	-8	6	4	-4	9	-9	-1	-2	5	6

is taken from an experiment and processed by a smoothing filter whose impulse response is $h[n] = n(0.7)^n u[n]$. Find the filter response $y[n]$.

We can find a DTFT of $h[n]$ in the table. But $x[n]$ is not an identifiable functional form. We could find the transform of $x[n]$ by using the direct formula

$$X(e^{j\Omega}) = \sum_{z=0}^{10} x[n]e^{-j\Omega n}.$$

But this is pretty tedious and time consuming. If the nonzero portion of x[n] were much longer, this would become quite impractical. Instead, we can find the solution numerically using the relation derived above for approximating a DTFT with the DFT

$$X(e^{j2\pi k/N}) = \sum_{n=0}^{N-1} x[n]e^{-j2\pi kn/N}.$$

This problem could also be solved in the time domain using numerical convolution. But there are two reasons why using the DFT might be preferable. First, if the number of points used is an integer power of two, the fft algorithm that is used to implement the DFT on computers is very efficient and may have a significant advantage in a shorter computing time than time-domain convolution. Second, using the DFT method, the time scale for the excitation, impulse response, and system response are all the same. That is not true when using numerical time-domain convolution.

The following MATLAB program solves this problem numerically using the DFT. Figure 7.22 shows the graphs of the excitation, impulse response, and system response.

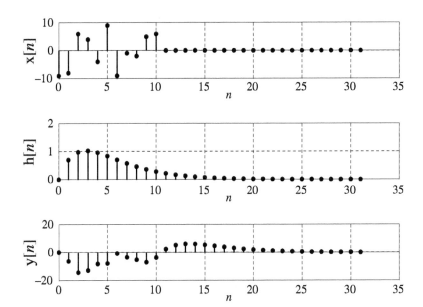

Figure 7.22
Excitation, impulse response, and system response

```
% Program to find a discrete-time system response using the DFT
N = 32 ;                % Use 32 points
n = [0:N-1]' ;          % Time vector
% Set excitation values
x = [[-9,-8,6,4,-4,9,-9,-1,-2,5,6],zeros(1,21)]' ;
h = n.*(0.7).^n.*uD(n) ; % Compute impulse response
X = fft(x) ;            % DFT of excitation
H = fft(h) ;            % DFT of impulse response
Y = X.*H ;              % DFT of system response
y = real(ifft(Y)) ;     % System response
% Graph the excitation, impulse response and system response
subplot(3,1,1) ;
```

```
ptr = stem(n,x,'k','filled') ; set(ptr,'LineWidth',2,'MarkerSize',4) ; grid on ;
xlabel('\itn','FontName','Times','FontSize',24) ;
ylabel('x[{\itn}]','FontName','Times','FontSize',24) ;
set(gca,'FontName','Times','FontSize',18) ;
subplot(3,1,2) ;
ptr = stem(n,h,'k','filled') ; set(ptr,'LineWidth',2,'MarkerSize',4) ; grid on ;
xlabel('\itn','FontName','Times','FontSize',24) ;
ylabel('h[{\itn}]','FontName','Times','FontSize',24) ;
set(gca,'FontName','Times','FontSize',18) ;
subplot(3,1,3) ;
ptr = stem(n,y,'k','filled') ; set(ptr,'LineWidth',2,'MarkerSize',4) ; grid on ;
xlabel('\itn','FontName','Times','FontSize',24) ;
ylabel('y[{\itn}]','FontName','Times','FontSize',24) ;
set(gca,'FontName','Times','FontSize',18) ;
```

7.4 FOURIER METHOD COMPARISONS

The DTFT completes the four Fourier analysis methods. These four methods form a "matrix" of methods for the four combinations of continuous and discrete time and continuous and discrete frequency (expressed as harmonic number) (Figure 7.23).

	Continuous Frequency	Discrete Frequency
Continuous Time	CTFT	CTFS
Discrete Time	DTFT	DFT

Figure 7.23
Fourier methods matrix

In Figure 7.24 are four rectangles or periodically repeated rectangles in both continuous and discrete time along with their Fourier transforms or harmonic functions. The CTFT of a single continuous-time rectangle is a single continuous-frequency sinc function. If that continuous-time rectangle is sampled to produce a single discrete-time rectangle, its DTFT is similar to the CTFT except that it is now periodically repeated. If the continuous-time rectangle is periodically repeated, its CTFS harmonic function is similar to the CTFT except that it has been sampled in frequency (harmonic number). If the original continuous-time rectangle is both periodically repeated and sampled, its DFT is also both periodically repeated and sampled. So, in general, periodic repetition in one domain, time or frequency, corresponds to sampling in the other domain, frequency or time, and

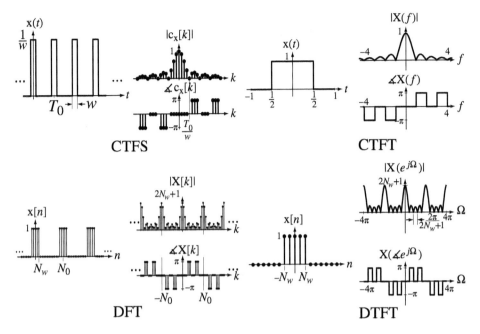

Figure 7.24
Fourier transform comparison for four related signals

sampling in one domain, time or frequency, corresponds to periodic repetition in the other domain, frequency or time. These relationships will be important in Chapter 10 on sampling.

7.5 SUMMARY OF IMPORTANT POINTS

1. Any discrete-time signal of engineering significance can be represented by a discrete-time Fourier series or inverse discrete Fourier transform (DFT), and the number of harmonics needed in the representation is the same as the fundamental period of the representation.
2. The complex sinusoids used in the DFT constitute a set of orthogonal basis functions.
3. The fast Fourier transform (FFT) is an efficient computer algorithm for computing the DFT if the representation time is an integer power of two.
4. The DFT can be extended to a discrete-time Fourier transform (DTFT) for aperiodic signals by letting the representation time approach infinity.
5. By allowing impulses in the transforms, the DTFT can be generalized to apply to some important signals.
6. The DFT and inverse DFT can be used to numerically approximate the DTFT and inverse DTFT under certain conditions.
7. With a table of discrete-time Fourier transform pairs and their properties, the forward and inverse transforms of almost any signal of engineering significance can be found.
8. The CTFS, CTFT, DFT, and DTFT are closely related analysis methods for periodic or aperiodic, continuous-time, or discrete-time signals.

EXERCISES WITH ANSWERS

(Answers to each exercise are in random order.)

Orthogonality

1. Without using a calculator or computer find the dot products of (a) $\mathbf{w}_1$ and $\mathbf{w}_{-1}$, (b) $\mathbf{w}_1$ and $\mathbf{w}_{-2}$, (c) $\mathbf{w}_{11}$ and $\mathbf{w}_{37}$, where

$$\mathbf{w}_k = \begin{bmatrix} W_4^0 \\ W_4^k \\ W_4^{2k} \\ W_4^{3k} \end{bmatrix} \text{ and } W_N = e^{j2\pi/N}$$

to show that they are orthogonal.

Answers: 0, 0, 0

2. Find the projection $\mathbf{p}$ of the vector $\mathbf{x} = \begin{bmatrix} 11 \\ 4 \end{bmatrix}$ in the direction of the vector $\mathbf{y} = \begin{bmatrix} -2 \\ 1 \end{bmatrix}$.

Answer: $18 \begin{bmatrix} 2/5 \\ -1/5 \end{bmatrix}$

3. Find the projection $\mathbf{p}$ of the vector $\mathbf{x} = \begin{bmatrix} 2 \\ -3 \\ 1 \\ 5 \end{bmatrix}$ in the direction of the vector $\mathbf{y} = \begin{bmatrix} 1 \\ j \\ -1 \\ -j \end{bmatrix}$. Then find the DFT of $\mathbf{x}$ and compare this result with $X[3]\mathbf{y}/4$.

Answer: $\begin{bmatrix} 1/4 - j2 \\ 2 + j/4 \\ -1/4 + j2 \\ -2 - j/4 \end{bmatrix}$

Discrete Fourier Transform

4. A periodic discrete-time signal with fundamental period $N = 3$ has the values $x[1] = 7$, $x[2] = -3$, $x[3] = 1$. If $x[n] \xleftrightarrow[3]{\mathcal{DFT}} X[k]$, find the magnitude and angle (in radians) of $X[1]$.

Answer: $8.7178 \angle -1.6858$

5. Using the direct summation formula find the DFT harmonic function of $\delta_{10}[n]$ with $N = 10$ and compare it with the DFT given in the table.

Answer: $\delta_1[k]$

6. Without using a computer, find the forward DFT of the following sequence of data and then find the inverse DFT of that sequence and verify that you get back the original sequence.

$$\{x[0], x[1], x[2], x[3]\} = \{3, 4, 1, -2\}$$

Answer: Forward DFT is $\{6, 2 - j6, 2, 2 + j6\}$

7. A signal x is sampled eight times. The samples are

$$\{x[0],\cdots,x[7]\} = \{a,b,c,d,e,f,g,h\}.$$

These samples are sent to a DFT algorithm and the output from that algorithm is X, a set of eight numbers $\{X[0],\cdots,X[7]\}$.

   (a) In terms of a,b,c,d,e,f,g, and h what is $X[0]$?
   (b) In terms of a,b,c,d,e,f,g, and h what is $X[4]$?
   (c) If $X[3] = 2 - j5$, what is the numerical value of $X[-3]$?
   (d) If $X[5] = 3e^{-j\pi/3}$, what is the numerical value of $X[-3]$?
   (e) If $X[5] = 9e^{j3\pi/4}$, what is the numerical value of $X[3]$?

   Answers: $9e^{-j3\pi/4}$, $a - b + c - d + e - f + g - h$, $3e^{-j\pi/3}$,
   $a + b + c + d + e + f + g + h$, $2 + j5$

8. A discrete-time periodic signal with fundamental period $N_0 = 6$ has the values
   $x[4] = 3$, $x[9] = -2$, $x[-1] = 1$, $x[14] = 5$, $x[24] = -3$, $x[7] = 9$.

   Also, $x[n] \xleftrightarrow[6]{\mathcal{DFT}} X[k]$.

   (a) Find $x[-5]$.
   (b) Find $x[322]$.
   (c) Find $X[2]$.

   Answers: 9, 3, $14.9332e^{-2.7862}$

9. Find the numerical values of the literal constants in

   (a) $8(u[n + 3] - u[n - 2]) * \delta_{12}[n] \xleftrightarrow[12]{\mathcal{DFT}} Ae^{bk}\,\mathrm{drcl}(ck, D)$
   (b) $5\delta_8[n - 2] \xleftrightarrow[8]{\mathcal{DFT}} Ae^{jak\pi}$
   (c) $\delta_4[n + 1] - \delta_4[n - 1] \xleftrightarrow[4]{\mathcal{DFT}} jA(\delta_4[k + a] - \delta_4[k - a])$

   Answers: $\{A = 5, a = -1/2\}$, $\{A = 2, a = 1\}$,
   $\{A = 40, b = j\pi/6, c = 1/12, D = 5\}$

10. The signal $x[n] = 1$ has a fundamental period $N_0 = 1$.

    (a) Find its DFT harmonic function using that fundamental period as the representation time.
    (b) Now let $z[n] = \begin{cases} x[n/4], & n/4 \text{ an integer} \\ 0, & \text{otherwise} \end{cases}$. Find the DFT harmonic function for $z[n]$ using its fundamental period as the representation time.
    (c) Verify that $z[0] = 1$ and that $z[1] = 0$ by using the DFT representation of $z[n]$.

    Answers: $N\delta_N[k]$, 0, $N\delta_N[k]$

11. If $x[n] = 5\cos(2\pi n/5) \xleftrightarrow[15]{\mathcal{DFT}} X[k]$, find the numerical values of $X[-11]$, $X[-33]$, $X[9]$, $X[12]$, $X[24]$, $X[48]$, and $X[75]$.

    Answers: 75/2, 75/2, 75/2, 0, 0, 0, 0

12. Find the DFT harmonic function of $x[n] = (u[n] - u[n-20]) * \delta_{20}[n]$ using its fundamental period as the representation time. There are at least two ways of computing X[k] and one of them is much easier than the other. Find the easy way.

    Answer: $20\delta_{20}[k]$

13. For each of these signals find the DFT over one fundamental period and show that $X[N_0/2]$ is real.

    (a) $x[n] = (u[n+2] - u[n-3]) * \delta_{12}[n]$
    (b) $x[n] = (u[n+3] - u[n-2]) * \delta_{12}[n]$
    (c) $x[n] = \cos(14\pi n/16)\cos(2\pi n/16)$
    (d) $x[n] = \cos(12\pi n/14)\cos\left(\dfrac{2\pi(n-3)}{14}\right)$

    Answers: 1, −1, 4, −7

Discrete-Time Fourier Transform Definition

14. From the summation definition, find the DTFT of
    $$x[n] = 10(u[n+4] - u[n-5])$$
    and compare with the DTFT table.

    Answer: $90\,\mathrm{drcl}(\Omega/2\pi, 9)$

15. From the definition, derive a general expression for the Ω form of the DTFT of functions of the form
    $$x[n] = \alpha^n \sin(\Omega_0 n)u[n],\ |\alpha| < 1.$$
    Compare with the DTFT table.

    Answer: $\dfrac{\alpha e^{j\Omega}\sin(\Omega_0)}{e^{j2\Omega} - 2\alpha e^{j\Omega}\cos(\Omega_0) + \alpha^2},\ |\alpha| < 1$

16. Given the DTFT pairs below convert them from the radian-frequency form to the cyclic frequency form using $\Omega = 2\pi F$ without doing any inverse DTFTs.

    (a) $\alpha^n \cos(\Omega_0 n) u[n] \xleftrightarrow{\mathcal{Z}} \dfrac{z[z - \alpha\cos(\Omega_0)]}{z^2 - 2\alpha z \cos(\Omega_0) + \alpha^2},\ |z| > |\alpha|$

    (b) $\cos(\Omega_0 n) \xleftrightarrow{\mathcal{F}} \pi[\delta_{2\pi}(\Omega - \Omega_0) + \delta_{2\pi}(\Omega + \Omega_0)]$

    Answers: $(1/2)[\delta_1(F - F_0) + \delta_1(F + F_0)]$,

    $\dfrac{1 - \alpha\cos(2\pi F_0)e^{-j2\pi F}}{1 - 2\alpha\cos(2\pi F_0)e^{-j2\pi F} + \alpha^2 e^{-j4\pi F}},\ |\alpha| < 1$

17. If $x[n] = n^2(u[n] - u[n-3])$ and $x[n] \xleftrightarrow{\mathcal{F}} X(e^{j\Omega})$, what is the value of $X(e^{j\Omega})\big|_{\Omega=0}$?

    Answer: 5

Exercises with Answers

Forward and Inverse Discrete-Time Fourier Transforms

18. A discrete-time signal is defined by

$$x[n] = \sin(\pi n/6).$$

Graph the magnitude and phase of the DTFT of $x[n-3]$ and $x[n+12]$.

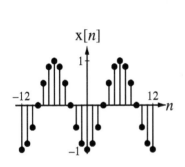

Answers:

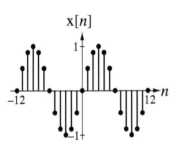

19. If $X(F) = 3[\delta_1(F - 1/4) + \delta_1(F + 1/4)] - j4[\delta_1(F + 1/9) - \delta(F - 1/9)]$ and $x[n] \xleftrightarrow{\mathcal{F}} X(F)$, what is the fundamental period of $x[n]$?

Answer: 36

20. If $X(F) = \delta_1(F - 1/10) + \delta_1(F + 1/10) + \delta_{1/16}(F)$ and $x[n] \xleftrightarrow{\mathcal{F}} X(F)$, what is the fundamental period of $x[n]$?

Answer: 80

21. Graph the magnitude and phase of the DTFT of

$$x[n] = (u[n+4] - u[n-5]) * \cos(2\pi n/6).$$

Then graph $x[n]$.

Answers:

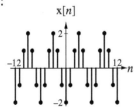 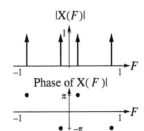

22. Graph the inverse DTFT of $X(F) = (1/2)[\text{rect}(4F) * \delta_1(F)] \circledast \delta_{1/2}(F)$.

Answer:

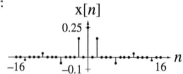

23. Let $X(e^{j\Omega}) = 4\pi - j6\pi \sin(2\Omega)$. Its inverse DTFT is x[n]. Find the numerical values of x[n] for $-3 \leq n < 3$.

Answers: 0, 0, −3/2, 3/2, 0, 2

24. A signal x[n] has a DTFT, $X(F) = 5 \text{ drcl}(F,5)$. What is its signal energy?

Answer: 20

25. Find the numerical values of the literal constants.

(a) $A(u[n + W] - u[n - W - 1])e^{jB\pi n} \xleftrightarrow{\mathcal{F}} 10\dfrac{\sin(5\pi(F + 1))}{\sin(\pi(F + 1))}$

(b) $2\delta_{15}[n - 3](u[n + 3] - u[n - 4]) \xleftrightarrow{\mathcal{F}} Ae^{jB\Omega}$

(c) $(2/3)^n u[n + 2] \xleftrightarrow{\mathcal{F}} \dfrac{Ae^{jB\Omega}}{1 - \alpha e^{-j\Omega}}$

(d) $4 \text{ sinc}(n/10) \xleftrightarrow{\mathcal{F}} A \text{ rect}(BF) * \delta_1(F)$

(e) $10 \cos\left(\dfrac{5\pi n}{14}\right) \xleftrightarrow{\mathcal{F}} A[\delta_1(F - a) + \delta_1(F + a)]$

(f) $4(\delta[n - 3] - \delta[n + 3]) \xleftrightarrow{\mathcal{F}} A \sin(aF)$

(g) $8(u[n + 3] - u[n - 2]) \xleftrightarrow{\mathcal{F}} Ae^{b\Omega} \text{drcl}(c\Omega, D)$

(h) $7\begin{pmatrix} u[n + 3] \\ -u[n - 4] \end{pmatrix} * 4 \sin(2\pi n/12) \xleftrightarrow{\mathcal{F}} A \text{ drcl}(a\Omega, b) \begin{bmatrix} \delta_{2\pi}(\Omega + c) \\ -\delta_{2\pi}(\Omega - c) \end{bmatrix}$

(i) $j42 \text{ drcl}(F, 5) \begin{bmatrix} \delta_1(F + 1/16) \\ -\delta_1(F - 1/16) \end{bmatrix} e^{j4\pi F} = A \begin{cases} \delta_1(F + 1/16) - \delta_1(F - 1/16) \\ -j[\delta_1(F + 1/16) + \delta_1(F - 1/16)] \end{cases}$

(j) $A \cos\left(\dfrac{2\pi(n - n_0)}{N_0}\right) \xleftrightarrow{\mathcal{F}} j\dfrac{36}{\sqrt{2}}\{(1 - j)\delta_1(F + 1/16) - (1 + j)\delta_1(F - 1/16)\}$

Answers: {A = 10, W = 2, B = −2}, {A = 40, B = 10},
{A = 5, a = 5/28 = 0.1786}, {A = j25.3148}, {A = 2, B = −3},
{A = 9/4, B = 2, α = 2/3}, {A = −j8, a = 6π = 18.85},
{A = j196π, a = 1/2π, b = 7, c = π/6}, {A = 72, N_0 = 16, n_0 = 2},
{A = 40, b = j, c = 1/2π, D = 5}

26. Given the DTFT pair $x[n] \xleftrightarrow{\mathcal{F}} \dfrac{10}{1 - 0.6e^{-j\Omega}}$ and

$$y[n] = \begin{cases} x[n/2], & n/2 \text{ an integer} \\ 0, & \text{otherwise} \end{cases},$$

find the magnitude and phase of $Y(e^{j\Omega})_{\Omega=\pi/4}$.

Answer: $8.575 \angle -0.5404$ radians

27. Let $x[n] \xleftrightarrow{\mathcal{F}} X(F) = 8\,\text{tri}(2F)e^{-j2\pi F} * \delta_1(F)$, a phase-shifted triangle in the range $-1/2 < F < 1/2$ that repeats that pattern periodically, with fundamental period one. Also let

$$y[n] = \begin{cases} x[n/3], & n/3 \text{ an integer} \\ 0, & \text{if } n/3 \text{ is not an integer}. \end{cases}$$

and let $y[n] \xleftrightarrow{\mathcal{F}} Y(F)$.

(a) Find the magnitude and angle (in radians) of $X(0.3)$.
(b) Find the magnitude and angle (in radians) of $X(2.2)$.
(c) What is the fundamental period of $Y(F)$?
(d) Find the magnitude and angle (in radians) of $Y(0.55)$.

Answers: $1/3$, $3.2 \angle -1.885$, $2.3984 \angle 2.1997$, $4.8 \angle -1.2566$

28. A signal $x[n]$ has a DTFT $X(F)$. Some of the values of $x[n]$ are

n	-2	-1	0	1	2	3	4	5	6
$x[n]$	-8	2	1	-5	7	9	8	2	3

Let $Y(F) = X(2F)$ with $y[n] \xleftrightarrow{\mathcal{F}} Y(F)$. Find the numerical values of $y[n]$ for $-2 \le n < 4$.

Answers: $1, -5, 0, 2, 0, 0$

29. Using the differencing property of the DTFT and the transform pair,

$$\text{tri}(n/2) \xleftrightarrow{\mathcal{F}} 1 + \cos(\Omega),$$

find the DTFT of $(1/2)(\delta[n+1] + \delta[n] - \delta[n-1] - \delta(n-2))$. Compare it with Fourier transform found using the table.

Answer: $(1/2)(e^{j\Omega} + 1 - e^{-j\Omega} - e^{-j2\Omega})$

30. A signal is described by

$$x[n] = \begin{cases} \ln(n+1), & 0 \le n < 10 \\ -\ln(-n+1), & -10 < n < 0 \\ 0, & \text{otherwise} \end{cases}$$

Graph the magnitude and phase of its DTFT over the range $-\pi \le \Omega < \pi$.

Answer:

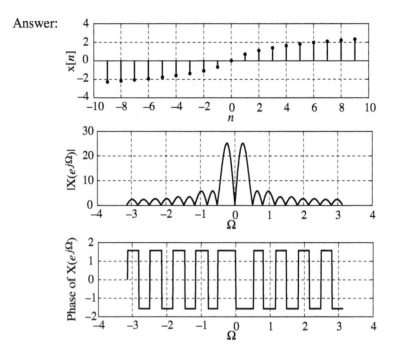

EXERCISES WITHOUT ANSWERS

Discrete Fourier Transform

31. If $x = \begin{bmatrix} 3 \\ -j \\ 3+j \\ 7 \end{bmatrix}$ and $y = \begin{bmatrix} 9 \\ -4+j \\ -5 \\ 2+j7 \end{bmatrix}$ find $x^H y$.

32. Fill in the blanks with correct numbers for this DFT harmonic function of a real-valued signal with $N = 8$.

k	0	1	2	3	4	5	6	7
X[k]	5	____	2−j7	4+j2	−3	____	____	9+j4

k	11	−9	26	−47
X[k]	____	____	____	____

33. A discrete-time signal x[n] is periodic with period 8. One period of its DFT harmonic function is

    $\{X[0], \cdots, X[7]\} = \{3, 4+j5, -4-j3, 1+j5, -4, 1-j5, -4+j3, 4-j5\}$.

    (a) What is the average value of x[n]?
    (b) What is the signal power of x[n]?
    (c) Is x[n] even, odd, or neither?

34. A set of samples x[n] from a signal is converted to a set of numbers X[k] by using the DFT.

(a) If $\{x[0], x[1], x[2], x[3], x[4]\} = \{2, 8, -3, 1, 9\}$, find the numerical value of X[0].
(b) If x[n] consists of 24 samples taken from exactly three periods of a sinusoid at a sampling rate which is exactly eight times the frequency of the sinusoid, two values of X[k] in the range $0 \leq k < 24$ are not zero. Which ones are they?
(c) If x[n] consists of seven samples, all of which are the same, -5, in the range $0 \leq k < 7$, which X[k] values are zero?
(d) If $\{x[0], x[1], x[2], x[3]\} = \{a, b, 0, b\}$ and

$$\{X[0], X[1], X[2], X[3]\} = \{A, B, 0, B^*\}$$

how are a and b related to each other? Express B in terms of a and b.

35. If $x_1[n] = 10\cos(2\pi n/8) \xleftrightarrow[8]{\mathscr{DFT}} X_1[k]$ and $x_2[n] \xleftrightarrow[32]{\mathscr{DFT}} X_1[k]$, find the numerical values of $x_2[2], x_2[4], x_2[8]$, and $x_2[204]$.

36. A periodic discrete-time signal x[n] is exactly described for all discrete time by its DFT

$$X[k] = 8(\delta_8[k-1] + \delta_8[k+1] + j2\delta_8[k+2] - j16\delta_8[k-2])e^{-j\pi k/4}.$$

(a) Write a correct analytical expression for x[n] in which $\sqrt{-1}$, "j," does not appear.
(b) What is the numerical value of x[n] at $n = -10$?

37. A discrete-time signal x[n] with fundamental period $N_0 = 4$ has a DFT X[k].

k	0	1	2	3
X[k]	4	2 − j3	1	2 + j3

(a) Find X[−5].
(b) Find X[22].
(c) Find x[0].
(d) Find x[3].

38. A signal x(t) is sampled four times and the samples are $\{x[0], x[1], x[2], x[3]\}$. Its DFT harmonic function is $\{X[0], X[1], X[2], X[3]\}$. X[3] can be written as $X[3] = ax[0] + bx[1] + cx[2] + dx[3]$. What are the numerical values of a, b, c, and d?

39. The DFT harmonic function X[k] of $x[n] = 5\cos(\pi n) \circledast 3\sin(\pi n/2)$ using $N = 4$ can be written in the form

$$X[k] = A(\delta_4[k-a] + \delta_4[k+a])(\delta_4[k+b] - \delta_4[k-b]).$$

Find the numerical values of A, a, and b.

40. Four data points $\{x[0], x[1], x[2], x[3]\}$ are converted by the DFT into four corresponding data points $\{X[0], X[1], X[2], X[3]\}$. If x[0] = 2, x[1] = −3, x[3] = 7, X[1] = −3 + j10, X[2] = 3, find the numerical values of the missing data, x[2], X[0], and X[3]. (Be careful to observe which symbols are lower and upper case.)

41. Demonstrate with a counterexample that the fundamental period of a discrete-time function and the fundamental period of its DFT harmonic function are not necessarily the same.

42. In Figure E.42 is graphed exactly one period of a periodic function x[n]. Its harmonic function X[k] (with $N = N_0$) can be written in the form

$$X[k] = A(Be^{jbk} - e^{jck} + De^{jdk})$$

Find the numerical values of the constants.

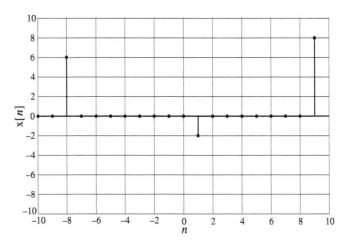

Figure E.42

43. Associate each discrete-time signal in Figure E.43.1 with its corresponding DFT magnitude in Figure E.43.2.

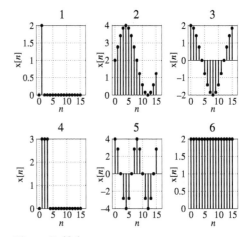

Figure E.43.1

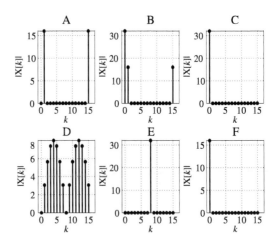

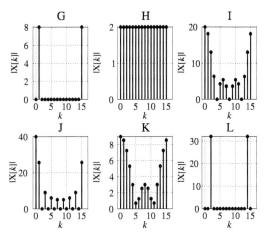

Figure E.43.2

44. Associate each discrete-time signal in Figure E.44.1 with its corresponding DFT magnitude in Figure E.44.2.

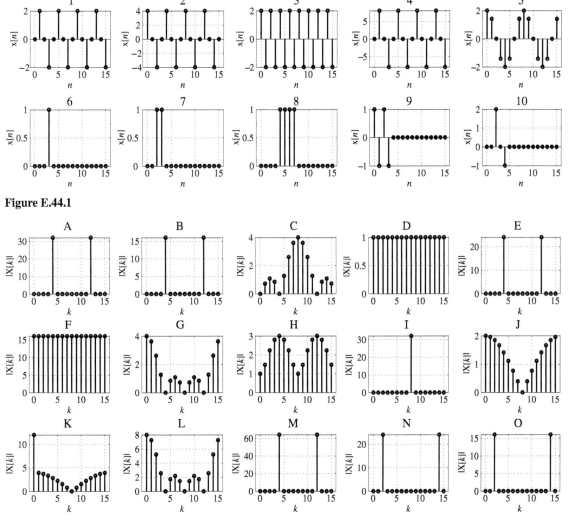

Figure E.44.1

Figure E.44.2

Forward and Inverse Discrete-Time Fourier Transforms

45. Given the DTFT pairs below convert them from the cyclic-frequency form to the radian-frequency form using $\Omega = 2\pi F$ without doing any inverse DTFTs.

    (a) $\delta_N[n] \xleftrightarrow{\mathcal{F}} (1/N)\delta_{1/N}(F)$

    (b) $\sin(2\pi F_0 n) \xleftrightarrow{\mathcal{F}} (j/2)[\delta_1(F + F_0) - \delta_1(F - F_0)]$

    (c) $\text{sinc}(n/w) \xleftrightarrow{\mathcal{F}} w\,\text{rect}(wF) * \delta_1(F)$

46. Find the DTFT of each of these signals:

    (a) $x[n] = (1/3)^n u[n-1]$

    (b) $x[n] = \sin(\pi n/4)(1/4)^n u[n-2]$

    (c) $x[n] = \text{sinc}(2\pi n/8) * \text{sinc}(2\pi(n-4)/8)$

    (d) $x[n] = \text{sinc}^2(2\pi n/8)$

47. Graph the inverse DTFTs of these functions:

    (a) $X(F) = \delta_1(F) - \delta_1(F - 1/2)$

    (b) $X(e^{j\Omega}) = j2\pi[\delta_{2\pi}(\Omega + \pi/4) - \delta_{2\pi}(\Omega - \pi/4)]$

    (c) $X(e^{j\Omega}) = 2\pi[\delta(\Omega - \pi/2) + \delta(\Omega - 3\pi/8) + \delta(\Omega - 5\pi/8)] * \delta_{2\pi}(2\Omega)$

48. A signal x[n] has a DTFT, $X(e^{j\Omega}) = 10\,\text{drcl}(\Omega/2\pi, 5)$. What is its signal energy?

49. Given this DTFT pair
$$x_1[n] \xleftrightarrow{\mathcal{F}} X_1(e^{j\Omega}) = e^{-j\Omega} + 2e^{-j2\Omega} + 3e^{-j3\Omega}$$
and the related pair $x_2[n] = 3x_1[n-2] \xleftrightarrow{\mathcal{F}} X_2(e^{j\Omega})$, find the value of $X_2(e^{j\Omega})\big|_{\Omega=\pi/4}$.

50. A signal x[n] has a DTFT,
$$X(e^{j\Omega}) = 2\pi[\delta_{2\pi}(\Omega - \pi/2) + \delta_{2\pi}(\Omega + \pi/2) + j\delta_{2\pi}(\Omega + 2\pi/3) - j\delta_{2\pi}(\Omega - 2\pi/3)].$$

    What is the fundamental period, N_0, of x[n]?

51. Let $x[n] = \text{tri}(n/4) * \delta_6[n]$ and let $x[n] \xleftrightarrow{\mathcal{F}} X(e^{j\Omega})$.

    (a) What is x[21]?

    (b) What is the lowest positive frequency Ω at which $X(e^{j\Omega}) \neq 0$?

52. Find an expression for the inverse DTFT x[n] of
$$X(F) = [\text{rect}(10F) * \delta_1(F)] \circledast (1/2)[\delta_1(F - 1/4) + \delta_1(F + 1/4)]$$
and evaluate it at the discrete time $n = 2$.

53. If $X(F) = \text{tri}(4F) * \delta_1(F) \circledast [\delta_1(F - 1/4) + \delta_1(F + 1/4)]$ then what is the numerical value of $X(11/8)$?

54. Find the numerical values of the literal constants.

    (a) $x[n] = 2\delta[n+3] - 3\delta[n-3] \xleftrightarrow{\mathcal{F}} X(e^{j\Omega}) = A\sin(b\Omega) + Ce^{d\Omega}$

    (b) $2\cos\left(\dfrac{2\pi n}{24}\right)\cos\left(\dfrac{2\pi n}{4}\right) \xleftrightarrow{\mathcal{F}} A\begin{bmatrix}\delta_1(F-a) + \delta_1(F+a) \\ +\delta_1(F-b) + \delta_1(F+b)\end{bmatrix}$

(c) $4(u[n-4] - u[n+5]) * \delta_9[n] \xleftrightarrow{\mathcal{F}} A\delta_b(\Omega)$

(d) $5\cos(2\pi n/14) \xleftrightarrow{\mathcal{F}} A[\delta_{2\pi}(\Omega - b) + \delta_{2\pi}(\Omega + b)]$

(e) $-2\sin(2\pi(n-3)/9) \xleftrightarrow{\mathcal{F}} A[\delta_1(F+b) - \delta_1(F-b)]e^{cF}$

(f) $7\text{sinc}(n/20) \xleftrightarrow{\mathcal{F}} A\text{rect}(bF) * \delta_{1/c}(cF)$

(g) $A + B\cos(n/c) \xleftrightarrow{\mathcal{F}} 4\pi\delta_\pi(\Omega)$

(h) $A\text{sinc}(n/b)\sin(n/c) \xleftrightarrow{\mathcal{F}} 12\begin{bmatrix}\text{rect}(5(F+1/5))\\-\text{rect}(5(F-1/5))\end{bmatrix} * \delta_1(F)$

(i) $A(\delta_b[n] - \delta_b[n-c]) \xleftrightarrow{\mathcal{F}} \pi\dfrac{1 - e^{-j2\Omega}}{3}\delta_{\pi/3}(\Omega)$

55. Let x[n] be a signal and let $y[n] = \sum_{m=-\infty}^{n} x[m]$. If $Y(e^{j\Omega}) = \cos(2\Omega)$, x[n] consists of exactly four discrete-time impulses. What are their strengths and locations?

CHAPTER 8

The Laplace Transform

8.1 INTRODUCTION AND GOALS

The continuous-time Fourier transform (CTFT) is a powerful tool for signal and system analysis but it has its limitations. There are some useful signals that do not have a CTFT, even in the generalized sense, which allows for impulses in the CTFT of a signal. The CTFT expresses signals as linear combinations of complex sinusoids. In this chapter we extend the CTFT to the Laplace transform, which expresses signals as linear combinations of complex exponentials, the eigenfunctions of the differential equations that describe continuous-time LTI systems. Complex sinusoids are a special case of complex exponentials. Some signals that do not have a CTFT do have a Laplace transform.

The impulse responses of LTI systems completely characterize them. Because the Laplace transform describes the impulse responses of LTI systems as linear combinations of the eigenfunctions of LTI systems, it directly encapsulates the characteristics of a system in a very useful way. Many system analysis and design techniques are based on the Laplace transform.

CHAPTER GOALS

1. To develop the Laplace transform, which is applicable to some signals that do not have a CTFT
2. To define the range of signals to which the Laplace transform applies
3. To develop a technique for realizing a system directly from its transfer function
4. To learn how to find forward and inverse Laplace transforms
5. To derive and illustrate the properties of the Laplace transform, especially those that do not have a direct counterpart in the Fourier transform
6. To define the unilateral Laplace transform and explore its unique features
7. To learn how to solve differential equations with initial conditions using the unilateral Laplace transform
8. To relate the pole and zero locations of a transfer function of a system directly to the frequency response of the system
9. To learn how MATLAB represents the transfer functions of systems

8.2 DEVELOPMENT OF THE LAPLACE TRANSFORM

When we extended the Fourier series to the Fourier transform we let the fundamental period of a periodic signal increase to infinity, making the discrete frequencies kf_0 in the CTFS merge into the continuum of frequencies f in the CTFT. This led to the two alternate definitions of the Fourier transform,

$$X(j\omega) = \int_{-\infty}^{\infty} x(t)e^{-j\omega t}\,dt, \quad x(t) = (1/2\pi)\int_{-\infty}^{\infty} X(j\omega)e^{+j\omega t}\,d\omega$$

and

$$X(f) = \int_{-\infty}^{\infty} x(t)e^{-j2\pi ft}\,dt, \quad x(t) = \int_{-\infty}^{\infty} X(f)e^{+j2\pi ft}\,df$$

There are two common approaches to introducing the Laplace transform. One approach is to conceive the Laplace transform as a generalization of the Fourier transform by expressing functions as linear combinations of complex exponentials instead of as linear combinations of the more restricted class of functions, complex sinusoids, used in the Fourier transform. The other approach is to exploit the unique nature of the complex exponential as the eigenfunction of the differential equations that describe linear systems and to realize that an LTI system excited by a complex exponential responds with another complex exponential. The relation between the excitation and response complex exponentials of an LTI system is the Laplace transform. We will consider both approaches.

GENERALIZING THE FOURIER TRANSFORM

If we simply generalize the forward Fourier transform by replacing complex sinusoids of the form $e^{j\omega t}$, ω a real variable, with complex exponentials e^{st}, s a complex variable, we get

$$\mathcal{L}(x(t)) = X(s) = \int_{-\infty}^{\infty} x(t)e^{-st}\,dt$$

which defines a forward Laplace[1] transform, where the notation $\mathcal{L}(\cdot)$ means "Laplace transform of."

Being a complex variable, s can have values anywhere in the complex plane. It has a real part called σ and an imaginary part called ω, so $s = \sigma + j\omega$. Then, for the special case in which σ is zero and the Fourier transform of the function $x(t)$ exists in the strict sense, the forward Laplace transform is equivalent to a forward Fourier transform.

$$X(j\omega) = X(s)_{s \to j\omega}.$$

This relationship between the Fourier and Laplace transforms is the reason for choosing in Chapter 6 the functional notation for the CTFT as $X(j\omega)$ instead of $X(\omega)$. This choice preserves the strict mathematical meaning of the function "X."

---

[1] Pierre Simon Laplace attended a Benedictine priory school until the age of 16 when he entered Caen University intending to study theology. But he soon realized that his real talent and love were in mathematics. He quit the university and went to Paris, where he was befriended by d'Alembert, who secured for him a teaching position in a military school. He produced in the next few years a sequence of many papers on various topics, all of high quality. He was elected to the Paris Academy in 1773 at the age of 23. He spent most of his career working in the areas of probability and celestial mechanics.

Using $s = \sigma + j\omega$ in the forward Laplace transform we get

$$X(s) = \int_{-\infty}^{\infty} x(t)e^{-(\sigma+j\omega)t}\,dt = \int_{-\infty}^{\infty} [x(t)e^{-\sigma t}]e^{-j\omega t}\,dt = \mathscr{F}[x(t)e^{-\sigma t}].$$

So one way of conceptualizing the Laplace transform is that it is equivalent to a Fourier transform of the product of the function x(t) and a real exponential **convergence factor** of the form $e^{-\sigma t}$ as illustrated in Figure 8.1.

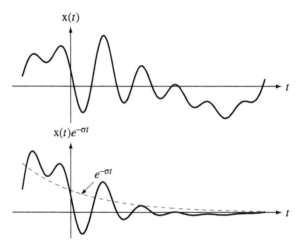

Figure 8.1
The effect of the decaying-exponential convergence factor on the original function

The convergence factor allows us, in some cases, to find transforms for which the Fourier transform cannot be found. As mentioned in an earlier chapter, the Fourier transforms of some functions do not (strictly speaking) exist. For example, the function $g(t) = A\,u(t)$ would have the Fourier transform

$$G(j\omega) = \int_{-\infty}^{\infty} A\,u(t)e^{-j\omega t}\,dt = A\int_{0}^{\infty} e^{-j\omega t}\,dt.$$

This integral does not converge. The technique used in Chapter 6 to make the Fourier transform converge was to multiply the signal by a convergence factor $e^{-\sigma|t|}$ where σ is positive real. Then the Fourier transform of the modified signal can be found and the limit taken as σ approaches zero. The Fourier transform found by this technique was called a generalized Fourier transform in which the impulse was allowed as a part of the transform. Notice that, for time $t > 0$ this convergence factor is the same in the Laplace transform and the generalized Fourier transform, but in the Laplace transform the limit as σ approaches zero is not taken. As we will soon see there are other useful functions that do not have even a generalized Fourier transform.

Now, to formally derive the forward and inverse Laplace transforms from the Fourier transform, we take the Fourier transform of $g_\sigma(t) = g(t)e^{-\sigma t}$ instead of the original function $g(t)$. That integral would then be

$$\mathscr{F}(g_\sigma(t)) = G_\sigma(j\omega) = \int_{-\infty}^{\infty} g_\sigma(t)e^{-j\omega t}\,dt = \int_{-\infty}^{\infty} g(t)e^{-(\sigma+j\omega)t}\,dt.$$

8.2 Development of the Laplace Transform

This integral may or may not converge, depending on the nature of the function g(t) and the choice of the value of σ. We will soon explore the conditions under which the integral converges. Using the notation $s = \sigma + j\omega$

$$\mathcal{F}(g_\sigma(t)) = \mathcal{L}(g(t)) = G_\mathcal{L}(s) = \int_{-\infty}^{\infty} g(t) e^{-st} dt.$$

This is the Laplace transform of g(t), *if the integral converges.*

The inverse Fourier transform would be

$$\mathcal{F}^{-1}(G_\sigma(j\omega)) = g_\sigma(t) = \frac{1}{2\pi} \int_{-\infty}^{\infty} G_\sigma(j\omega) e^{+j\omega t} d\omega = \frac{1}{2\pi} \int_{-\infty}^{\infty} G_\mathcal{L}(s) e^{+j\omega t} d\omega.$$

Using $s = \sigma + j\omega$ and $ds = jd\omega$ we get

$$g_\sigma(t) = \frac{1}{j2\pi} \int_{\sigma-j\infty}^{\sigma+j\infty} G_\mathcal{L}(s) e^{+(s-\sigma)t} ds = \frac{e^{-\sigma t}}{j2\pi} \int_{\sigma-j\infty}^{\sigma+j\infty} G_\mathcal{L}(s) e^{+st} ds$$

or, dividing both sides by $e^{-\sigma t}$,

$$g(t) = \frac{1}{j2\pi} \int_{\sigma-j\infty}^{\sigma+j\infty} G_\mathcal{L}(s) e^{+st} ds.$$

This defines an inverse Laplace transform. When we are dealing only with Laplace transforms, the $\mathcal{L}$ subscript will not be needed to avoid confusion with Fourier transforms, and the forward and inverse transforms can be written as

$$\boxed{X(s) = \int_{-\infty}^{\infty} x(t) e^{-st} dt \quad \text{and} \quad x(t) = \frac{1}{j2\pi} \int_{\sigma-j\infty}^{\sigma+j\infty} X(s) e^{+st} ds}. \quad (8.1)$$

This result shows that a function can be expressed as a linear combination of complex exponentials, a generalization of the Fourier transform in which a function is expressed as a linear combination of complex sinusoids. A common notational convention is

$$x(t) \xleftrightarrow{\mathcal{L}} X(s)$$

indicating that x(t) and X(s) form a **Laplace-transform pair.**

COMPLEX EXPONENTIAL EXCITATION AND RESPONSE

Another approach to the Laplace transform is to consider the response of an LTI system to a complex exponential excitation of the form $x(t) = Ke^{st}$ where $s = \sigma + j\omega$ and σ, ω, and K are all real-valued. Using convolution, the response y(t) of an LTI system with impulse response h(t) to x(t) is

$$y(t) = h(t) * Ke^{st} = K \int_{-\infty}^{\infty} h(\tau) e^{s(t-\tau)} d\tau = \underbrace{Ke^{st}}_{x(t)} \int_{-\infty}^{\infty} h(\tau) e^{-s\tau} d\tau.$$

The response of an LTI system to a complex-exponential excitation is that same excitation multiplied by the quantity $\int_{-\infty}^{\infty} h(\tau) e^{-s\tau} d\tau$ *if this integral converges.* This is the integral of the product of the impulse response h(τ) and a complex exponential $e^{-s\tau}$ over all τ and the result of this operation is a function of s only. This result is usually written as

$$\boxed{H(s) = \int_{-\infty}^{\infty} h(t)e^{-st} dt} \qquad (8.2)$$

and H(s) is called the Laplace transform of h(t). (The name of the variable of integration was changed from τ to t but that does not change the result H(s).)

For an LTI system, knowledge of h(t) is enough to completely characterize the system. H(s) also contains enough information to completely characterize the system, but the information is in a different form. The fact that this form is different can lend insight into the system's operation that is more difficult to see by examining h(t) alone. In the chapters to follow, we will see many examples of the advantage of viewing system properties and performance through H(s) in addition to h(t).

8.3 THE TRANSFER FUNCTION

Now let's find the Laplace transform Y(s) of the response y(t) of an LTI system with impulse response h(t) to an excitation x(t).

$$Y(s) = \int_{-\infty}^{\infty} y(t)e^{-st} dt = \int_{-\infty}^{\infty} [h(t) * x(t)]e^{-st} dt = \int_{-\infty}^{\infty} \left(\int h(\tau)x(t-\tau)d\tau \right) e^{-st} dt$$

Separating the two integrals,

$$Y(s) = \int_{-\infty}^{\infty} h(\tau)d\tau \int_{-\infty}^{\infty} x(t-\tau)e^{-st} dt$$

Let $\lambda = t - \tau \Rightarrow d\lambda = dt$. Then

$$Y(s) = \int_{-\infty}^{\infty} h(\tau)d\tau \int_{-\infty}^{\infty} x(\lambda)e^{-s(\lambda+\tau)}d\lambda = \underbrace{\int_{-\infty}^{\infty} h(\tau)e^{-s\tau}d\tau}_{= H(s)} \underbrace{\int_{-\infty}^{\infty} x(\lambda)e^{-s\lambda}d\lambda}_{= X(s)}.$$

The Laplace transform Y(s) of the response y(t) is

$$\boxed{Y(s) = H(s)X(s)}, \qquad (8.3)$$

the product of the Laplace transforms of the excitation and impulse response (if all the transforms exist). H(s) is called the **transfer function** of the system because it describes in the s domain how the system "transfers" the excitation to the response. This is a fundamental result in system analysis. In this new "s domain," time-convolution becomes s-domain multiplication just as it did using the Fourier transform.

$$y(t) = x(t) * h(t) \xleftarrow{\mathcal{L}} Y(s) = X(s)H(s)$$

8.4 CASCADE-CONNECTED SYSTEMS

If the response of one system is the excitation of another system, they are said to be cascade connected (Figure 8.2). The Laplace transform of the overall system response is then

$$Y(s) = H_2(s)[H_1(s)X(s)] = [H_1(s)H_2(s)]X(s)$$

and the cascade-connected systems are equivalent to a single system whose transfer function is $H(s) = H_1(s)H_2(s)$.

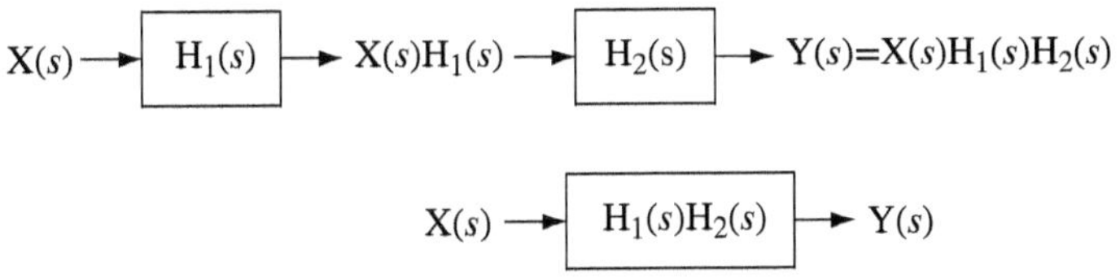

Figure 8.2
Cascade connection of systems

8.5 DIRECT FORM II REALIZATION

System realization is the process of putting together system components to form an overall system with a desired transfer function. In Chapter 5 we found that if a system is described by a linear differential equation of the form

$$\sum_{k=0}^{N} a_k y^{(k)}(t) = \sum_{k=0}^{N} b_k x^{(k)}(t)$$

its transfer function is a ratio of polynomials in s and the coefficients of the powers of s are the same as the coefficients of derivatives of x and y in the differential equation.

$$H(s) = \frac{Y(s)}{X(s)} = \frac{\sum_{k=0}^{N} b_k s^k}{\sum_{k=0}^{N} a_k s^k} = \frac{b_N s^N + b_{N-1} s^{N-1} + \cdots + b_1 s + b_0}{a_N s^N + a_{N-1} s^{N-1} + \cdots + a_1 s + a_0} \quad (8.4)$$

(Here the nominal orders of the numerator and denominator are both assumed to be N. If the numerator order is actually less than N, some of the higher-order b coefficients will be zero.) The denominator order must be N and a_N cannot be zero if this is an Nth-order system.

One standard form of system realization is called **Direct Form II**. The transfer function can be thought of as the product of two transfer functions

$$H_1(s) = \frac{Y_1(s)}{X(s)} = \frac{1}{a_N s^N + a_{N-1} s^{N-1} + \cdots + a_1 s + a_0} \quad (8.5)$$

and

$$H_2(s) = \frac{Y(s)}{Y_1(s)} = b_N s^N + b_{N-1} s^{N-1} + \cdots + b_1 s + b_0$$

(Figure 8.3) where the output signal of the first system $Y_1(s)$ is the input signal of the second system.

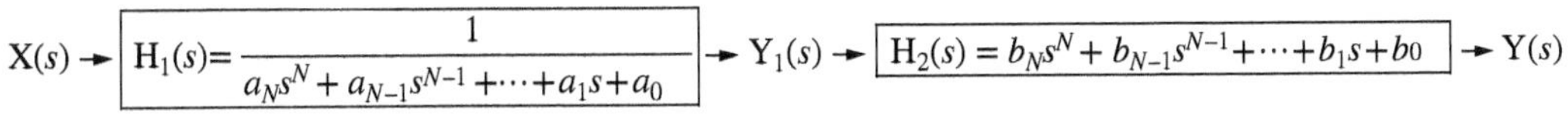

Figure 8.3
A system conceived as two cascaded systems

We can draw a block diagram of $H_1(s)$ by rewriting (8.5) as

$$X(s) = [a_N s^N + a_{N-1} s^{N-1} + \cdots + a_1 s + a_0] Y_1(s)$$

or

$$X(s) = a_N s^N Y_1(s) + a_{N-1} s^{N-1} Y_1(s) + \cdots + a_1 s Y_1(s) + a_0 Y_1(s)$$

or

$$s^N Y_1(s) = \frac{1}{a_N}\{X(s) - [a_{N-1} s^{N-1} Y_1(s) + \cdots + a_1 s Y_1(s) + a_0 Y_1(s)]\}$$

(Figure 8.4).

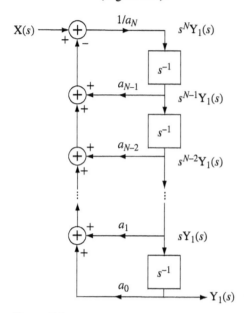

Figure 8.4
Realization of $H_1(s)$

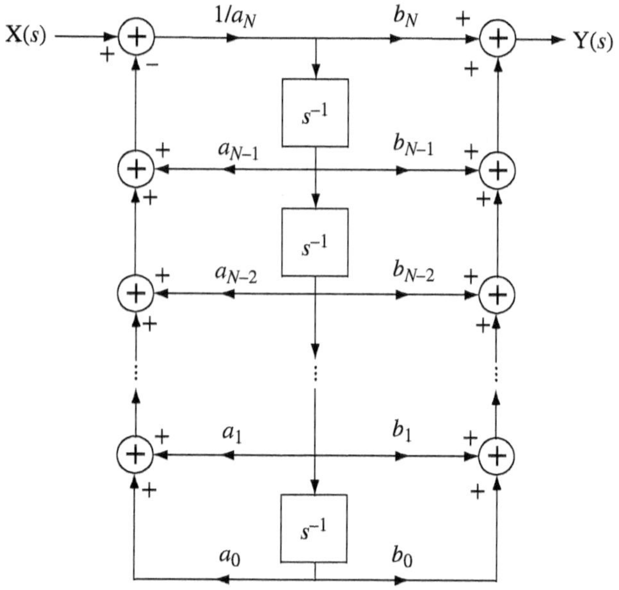

Figure 8.5
Overall Direct Form II system realization

Now we can immediately synthesize the overall response $Y(s)$ as a linear combination of the various powers of s multiplying $Y_1(s)$ (Figure 8.5).

8.6 THE INVERSE LAPLACE TRANSFORM

In the practical application of the Laplace transform we need a way to convert $Y(s)$ to $y(t)$, an **inverse Laplace transform**. It was shown in (8.1) that

$$y(t) = \frac{1}{j2\pi} \int_{\sigma-j\infty}^{\sigma+j\infty} Y(s) e^{st}\, ds$$

where σ is the real part of s. This is a contour integral in the complex s plane and is beyond the scope of this text. The inversion integral is rarely used in practical problem solving because the Laplace transforms of most useful signals have already been found and tabulated.

8.7 EXISTENCE OF THE LAPLACE TRANSFORM

We should now explore under what conditions the Laplace transform $X(s) = \int_{-\infty}^{\infty} x(t) e^{-st}\, dt$ actually exists. It exists if the integral converges, and whether or not the integral converges depend on $x(t)$ and s.

TIME-LIMITED SIGNALS

If $x(t) = 0$ for $t < t_0$ and $t > t_1$ (with t_0 and t_1 finite) it is called a **time-limited** signal. If $x(t)$ is also finite for all t, the Laplace-transform integral converges for any value of s and the Laplace transform of $x(t)$ exists (Figure 8.6).

RIGHT- AND LEFT-SIDED SIGNALS

If $x(t) = 0$ for $t < t_0$ it is called a **right-sided** signal and the Laplace transform becomes

$$X(s) = \int_{t_0}^{\infty} x(t) e^{-st} dt$$

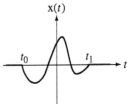

Figure 8.6
A finite, time-limited signal

(Figure 8.7(a)).
Consider the Laplace transform $X(s)$ of the right-sided signal $x(t) = e^{\alpha t} u(t - t_0)$, $\alpha \in \mathbb{R}$

$$X(s) = \int_{t_0}^{\infty} e^{\alpha t} e^{-st} dt = \int_{t_0}^{\infty} e^{(\alpha - \sigma)t} e^{-j\omega t} dt$$

(Figure 8.8(a)).

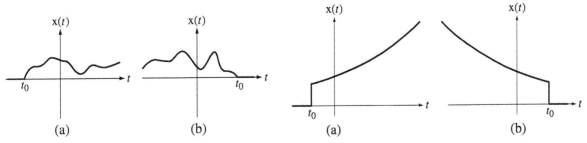

Figure 8.7
(a) A right-sided signal, (b) a left-sided signal

Figure 8.8
(a) $x(t) = e^{\alpha t} u(t - t_0)$, $\alpha \in \mathbb{R}$, (b) $x(t) = e^{\beta t} u(t_0 - t)$, $\beta \in \mathbb{R}$

If $\sigma > \alpha$ the integral converges. The inequality $\sigma > \alpha$ defines a region in the s plane called the **region of convergence (ROC)** (Figure 8.9(a)).

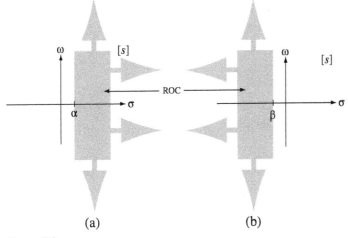

Figure 8.9
Regions of convergence for (a) the right-sided signal $x(t) = e^{\alpha t} u(t - t_0)$, $\alpha \in \mathbb{R}$ and (b) the left-sided signal $x(t) = e^{\beta t} u(t_0 - t)$, $\beta \in \mathbb{R}$

If $x(t) = 0$ for $t > t_0$ it is called a **left-sided** signal (Figure 8.7(b)). The Laplace transform becomes $X(s) = \int_{-\infty}^{t_0} x(t)e^{-st}dt$. If $x(t) = e^{\beta t}u(t_0 - t)$, $\beta \in \mathbb{R}$,

$$X(s) = \int_{-\infty}^{t_0} e^{\beta t}e^{-st}dt = \int_{-\infty}^{t_0} e^{(\beta-\sigma)t}e^{-j\omega t}dt$$

and the integral converges for any $\sigma < \beta$ (Figure 8.8(b) and Figure 8.9(b)).

Any signal can be expressed as the sum of a right-sided signal and a left-sided signal (Figure 8.10).

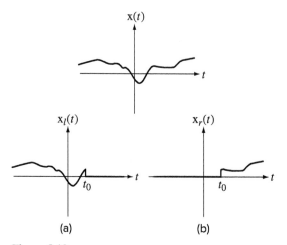

Figure 8.10
A signal divided into a left-sided part (a) and a right-sided part (b)

If $x(t) = x_r(t) + x_l(t)$ where $x_r(t)$ is the right-sided part and $x_l(t)$ is the left-sided part, and if $|x_r(t)| < |K_r e^{\alpha t}|$ and $|x_l(t)| < |K_l e^{\beta t}|$, (where K_r and K_l are constants), then the Laplace-transform integral converges and the Laplace transform exists for $\alpha < \sigma < \beta$. This implies that if $\alpha < \beta$ a Laplace transform can be found and the ROC in the s plane is the region $\alpha < \sigma < \beta$. If $\alpha > \beta$ the Laplace transform does not exist. For right-sided signals the ROC is always the region of the s plane to the right of α. For left-sided signals the ROC is always the region of the s plane to the left of β.

8.8 LAPLACE-TRANSFORM PAIRS

We can build a table of Laplace-transform pairs, starting with signals described by $\delta(t)$ and $e^{-\alpha t}\cos(\omega_0 t)u(t)$. Using the definition,

$$\delta(t) \overset{\mathcal{L}}{\longleftrightarrow} \int_{-\infty}^{\infty} \delta(t)e^{-st}dt = 1, \quad \text{All } s$$

$$e^{-\alpha t}\cos(\omega_0 t)u(t) \overset{\mathcal{L}}{\longleftrightarrow} \int_{-\infty}^{\infty} e^{-\alpha t}\cos(\omega_0 t)u(t)e^{-st}dt = \int_{0}^{\infty} \frac{e^{j\omega_0 t} + e^{-j\omega_0 t}}{2}e^{-(s+\alpha)t}dt, \quad \sigma > -\alpha$$

$$e^{-\alpha t}\cos(\omega_0 t)u(t) \overset{\mathcal{L}}{\longleftrightarrow} (1/2)\int_{0}^{\infty} (e^{-(s-j\omega_0+\alpha)t} + e^{-(s+j\omega_0+\alpha)t})dt, \quad \sigma > -\alpha$$

$$e^{-\alpha t}\cos(\omega_0 t)u(t) \xleftrightarrow{\mathcal{L}} (1/2)\left[\frac{1}{(s - j\omega_0 + \alpha)} + \frac{1}{(s + j\omega_0 + \alpha)}\right], \quad \sigma > -\alpha$$

$$e^{-\alpha t}\cos(\omega_0 t)u(t) \xleftrightarrow{\mathcal{L}} \frac{s + \alpha}{(s + \alpha)^2 + \omega_0^2}, \quad \sigma > -\alpha.$$

If $\alpha = 0$,

$$\cos(\omega_0 t)u(t) \xleftrightarrow{\mathcal{L}} \frac{s}{s^2 + \omega_0^2}, \quad \sigma > 0.$$

If $\omega_0 = 0$,

$$e^{-\alpha t}u(t) \xleftrightarrow{\mathcal{L}} \frac{1}{s + \alpha}, \quad \sigma > -\alpha.$$

If $\alpha = \omega_0 = 0$,

$$u(t) \xleftrightarrow{\mathcal{L}} 1/s, \quad \sigma > 0.$$

Using similar methods we can build a table of the most often used Laplace-transform pairs (Table 8.1).

To illustrate the importance of specifying not only the algebraic form of the Laplace transform, but also its ROC, consider the Laplace transforms of $e^{-\alpha t}u(t)$ and $-e^{-\alpha t}u(-t)$

$$e^{-\alpha t}u(t) \xleftrightarrow{\mathcal{L}} \frac{1}{s + \alpha}, \quad \sigma > -\alpha \quad \text{and} \quad -e^{-\alpha t}u(-t) \xleftrightarrow{\mathcal{L}} \frac{1}{s + \alpha}, \quad \sigma < -\alpha.$$

Table 8.1 Some common Laplace-transform pairs

$$\delta(t) \xleftrightarrow{\mathcal{L}} 1, \quad \text{All } \sigma$$

$u(t) \xleftrightarrow{\mathcal{L}} 1/s, \; \sigma > 0$	$-u(-t) \xleftrightarrow{\mathcal{L}} 1/s, \; \sigma < 0$
$\text{ramp}(t) = tu(t) \xleftrightarrow{\mathcal{L}} 1/s^2, \; \sigma > 0$	$\text{ramp}(-t) = -tu(-t) \xleftrightarrow{\mathcal{L}} 1/s^2, \; \sigma < 0$
$e^{-\alpha t}u(t) \xleftrightarrow{\mathcal{L}} 1/(s+\alpha), \; \sigma > -\alpha$	$-e^{-\alpha t}u(-t) \xleftrightarrow{\mathcal{L}} 1/(s+\alpha), \; \sigma < -\alpha$
$t^n u(t) \xleftrightarrow{\mathcal{L}} n!/s^{n+1}, \; \sigma > 0$	$-t^n u(-t) \xleftrightarrow{\mathcal{L}} n!/s^{n+1}, \; \sigma < 0$
$te^{-\alpha t}u(t) \xleftrightarrow{\mathcal{L}} 1/(s+\alpha)^2, \; \sigma > -\alpha$	$-te^{-\alpha t}u(-t) \xleftrightarrow{\mathcal{L}} 1/(s+\alpha)^2, \; \sigma < -\alpha$
$t^n e^{-\alpha t}u(t) \xleftrightarrow{\mathcal{L}} \dfrac{n!}{(s+\alpha)^{n+1}}, \; \sigma > -\alpha$	$-t^n e^{-\alpha t}u(-t) \xleftrightarrow{\mathcal{L}} \dfrac{n!}{(s+\alpha)^{n+1}}, \; \sigma < -\alpha$
$\sin(\omega_0 t)u(t) \xleftrightarrow{\mathcal{L}} \dfrac{\omega_0}{s^2 + \omega_0^2}, \; \sigma > 0$	$-\sin(\omega_0 t)u(-t) \xleftrightarrow{\mathcal{L}} \dfrac{\omega_0}{s^2 + \omega_0^2}, \; \sigma < 0$
$\cos(\omega_0 t)u(t) \xleftrightarrow{\mathcal{L}} \dfrac{s}{s^2 + \omega_0^2}, \; \sigma > 0$	$-\cos(\omega_0 t)u(-t) \xleftrightarrow{\mathcal{L}} \dfrac{s}{s^2 + \omega_0^2}, \; \sigma < 0$
$e^{-\alpha t}\sin(\omega_0 t)u(t) \xleftrightarrow{\mathcal{L}} \dfrac{\omega_0}{(s+\alpha)^2 + \omega_0^2}, \; \sigma > -\alpha$	$-e^{-\alpha t}\sin(\omega_0 t)u(-t) \xleftrightarrow{\mathcal{L}} \dfrac{\omega_0}{(s+\alpha)^2 + \omega_0^2}, \; \sigma < -\alpha$
$e^{-\alpha t}\cos(\omega_0 t)u(t) \xleftrightarrow{\mathcal{L}} \dfrac{s+\alpha}{(s+\alpha)^2 + \omega_0^2}, \; \sigma > -\alpha$	$-e^{-\alpha t}\cos(\omega_0 t)u(-t) \xleftrightarrow{\mathcal{L}} \dfrac{s+\alpha}{(s+\alpha)^2 + \omega_0^2}, \; \sigma < -\alpha$

$$e^{-\alpha|t|} \xleftrightarrow{\mathcal{L}} \frac{1}{s+\alpha} - \frac{1}{s-\alpha} = -\frac{2\alpha}{s^2 - \alpha^2}, \quad -\alpha < \sigma < \alpha$$

The algebraic expression for the Laplace transform is the same in each case but the ROCs are totally different, in fact mutually exclusive. That means that the Laplace transform of a linear combination of these two functions cannot be found because we cannot find a region in the s plane that is common to the ROCs of both $e^{-\alpha t}u(t)$ and $-e^{-\alpha t}u(-t)$.

An observant reader may have noticed that some very common signal functions do not appear in Table 8.1, for example, a constant. The function $x(t) = u(t)$ appears but $x(t) = 1$ does not. The Laplace transform of $x(t) = 1$ would be

$$X(s) = \int_{-\infty}^{\infty} e^{-st} dt = \underbrace{\int_{-\infty}^{0} e^{-\sigma t} e^{-j\omega t} dt}_{\text{ROC: } \sigma<0} + \underbrace{\int_{0}^{\infty} e^{-\sigma t} e^{-j\omega t} dt}_{\text{ROC: } \sigma>0}.$$

There is no ROC common to both of these integrals, therefore the Laplace transform does not exist. For the same reason $\cos(\omega_0 t)$, $\sin(\omega_0 t)$, $\text{sgn}(t)$, and $\delta_{T_0}(t)$ do not appear in the table although $\cos(\omega_0 t)u(t)$ and $\sin(\omega_0 t)u(t)$ do appear.

The Laplace transform $1/(s + \alpha)$ is finite at every point in the s plane *except* the point $s = -\alpha$. This unique point is called a **pole** of $1/(s + \alpha)$. In general, a pole of a Laplace transform is a value of s at which the transform tends to infinity. The opposite concept is a **zero** of a Laplace transform, a value of s at which the transform is zero. For $1/(s + \alpha)$ there is a single zero at infinity. The Laplace transform

$$\cos(\omega_0 t)u(t) \xleftrightarrow{\mathcal{L}} \frac{s}{s^2 + \omega_0^2}$$

has poles at $s = \pm j\omega_0$, a zero at $s = 0$, and a zero at infinity.

A useful tool in signal and system analysis is the pole-zero diagram in which an "x" marks a pole and an "o" marks a zero in the s plane (Figure 8.11).

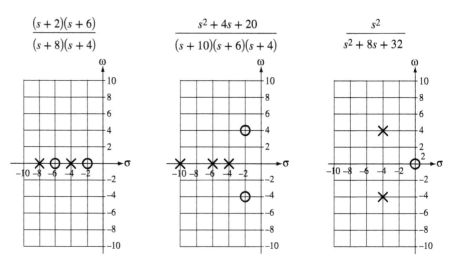

Figure 8.11
Example pole-zero diagrams

(The small "2" next to the zero in the rightmost pole-zero diagram in Figure 8.11 indicates that there is a double zero at $s = 0$.) As we will see in later material, the poles and zeros of the Laplace transform of a function contain much valuable information about the nature of the function.

EXAMPLE 8.1

Laplace transform of a noncausal exponential signal

Find the Laplace transform of $x(t) = e^{-t}u(t) + e^{2t}u(-t)$.

The Laplace transform of this sum is the sum of the Laplace transforms of the individual terms $e^{-t}u(t)$ and $e^{2t}u(-t)$. The ROC of the sum is the region in the s plane that is common to the two ROCs. From Table 8.1

$$e^{-t}u(t) \xleftrightarrow{\mathcal{L}} \frac{1}{s+1}, \quad \sigma > -1$$

and

$$e^{2t}u(-t) \xleftrightarrow{\mathcal{L}} -\frac{1}{s-2}, \quad \sigma < 2.$$

In this case, the region in the s plane that is common to both ROCs is $-1 < \sigma < 2$ and

$$e^{-t}u(t) + e^{2t}u(-t) \xleftrightarrow{\mathcal{L}} \frac{1}{s+1} - \frac{1}{s-2}, \quad -1 < \sigma < 2$$

(Figure 8.12). This Laplace transform has poles at $s = -1$ and $s = +2$ and two zeros at infinity.

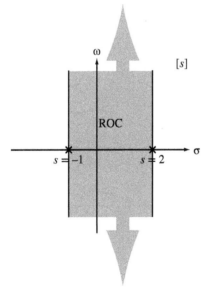

Figure 8.12
ROC for the Laplace transform of
$x(t) = e^{-t}u(t) + e^{2t}u(-t)$

EXAMPLE 8.2

Inverse Laplace transforms

Find the inverse Laplace transforms of

(a) $X(s) = \dfrac{4}{s+3} - \dfrac{10}{s-6}, \quad -3 < \sigma < 6$

(b) $X(s) = \dfrac{4}{s+3} - \dfrac{10}{s-6}$, $\sigma > 6$

(c) $X(s) = \dfrac{4}{s+3} - \dfrac{10}{s-6}$, $\sigma < -3$.

(a) $X(s)$ is the sum of two s-domain functions and the inverse Laplace transform must be the sum of two time-domain functions. $X(s)$ has two poles, one at $s = -3$ and one at $s = 6$. We know that for right-sided signals the ROC is always to the right of the pole and for left-sided signals the ROC is always to the left of the pole. Therefore, $\dfrac{4}{s+3}$ must inverse transform into a right-sided signal and $\dfrac{10}{s-6}$ must inverse transform into a left-sided signal. Then using

$$e^{-\alpha t}u(t) \xleftrightarrow{\mathcal{L}} \dfrac{1}{s+\alpha},\ \sigma > -\alpha \quad \text{and} \quad -e^{-\alpha t}u(-t) \xleftrightarrow{\mathcal{L}} \dfrac{1}{s+\alpha},\ \sigma < -\alpha$$

we get

$$x(t) = 4e^{-3t}u(t) + 10e^{6t}u(-t)$$

(Figure 8.13(a)).

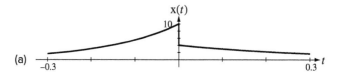
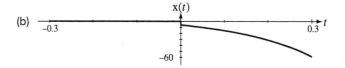
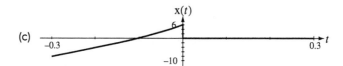

Figure 8.13
Three inverse Laplace transforms

(b) In this case the ROC is to the right of both poles and both time-domain signals must be right sided and, using $e^{-\alpha t}u(t) \xleftrightarrow{\mathcal{L}} \dfrac{1}{s+\alpha},\ \sigma > -\alpha$

$$x(t) = 4e^{-3t}u(t) - 10e^{6t}u(t)$$

(Figure 8.13(b)).

(c) In this case the ROC is to the left of both poles and both time-domain signals must be left sided and, using $-e^{-\alpha t}u(-t) \xleftrightarrow{\mathcal{L}} \dfrac{1}{s+\alpha},\ \sigma < -\alpha$

$$x(t) = -4e^{-3t}u(-t) + 10e^{6t}u(-t)$$

(Figure 8.13(c)).

8.9 PARTIAL-FRACTION EXPANSION

In Example 8.2 each s-domain expression was in the form of two terms, each of which can be found directly in Table 8.1. But what do we do when the Laplace-transform expression is in a more complicated form? For example, how do we find the inverse Laplace transform of

$$X(s) = \frac{s}{s^2 + 4s + 3} = \frac{s}{(s+3)(s+1)}, \quad \sigma > -1?$$

This form does not appear in Table 8.1. In a case like this, a technique called **partial-fraction expansion** becomes very useful. Using that technique it is possible to write $X(s)$ as

$$X(s) = \frac{3/2}{s+3} - \frac{1/2}{s+1} = \frac{1}{2}\left(\frac{3}{s+3} - \frac{1}{s+1}\right), \quad \sigma > -1.$$

Then the inverse transform can be found as

$$x(t) = (1/2)(3e^{-3t} - e^{-t})u(t).$$

The most common type of problem in signal and system analysis using Laplace methods is to find the inverse transform of a rational function in s of the form

$$G(s) = \frac{b_M s^M + b_{M-1} s^{M-1} + \cdots + b_1 s + b_0}{s^N + a_{N-1} s^{N-1} + \cdots a_1 s + a_0}$$

where the numerator and denominator coefficients a and b are constants. Since the orders of the numerator and denominator are arbitrary, this function does not appear in standard tables of Laplace transforms. But, using partial-fraction expansion, it can be expressed as a sum of functions that do appear in standard tables of Laplace transforms.

It is always possible (numerically, if not analytically) to factor the denominator polynomial and to express the function in the form

$$G(s) = \frac{b_M s^M + b_{M-1} s^{M-1} + \cdots + b_1 s + b_0}{(s - p_1)(s - p_2) \cdots (s - p_N)}$$

where the p's are the finite poles of $G(s)$. Let's consider, for now, the simplest case, that there are no repeated finite poles and that $N > M$, making the fraction proper in s. Once the poles have been identified we should be able to write the function in the partial-fraction form

$$G(s) = \frac{K_1}{s - p_1} + \frac{K_2}{s - p_2} + \cdots + \frac{K_N}{s - p_N},$$

if we can find the correct values of the K's. For this form of the function to be correct, the identity

$$\frac{b_M s^M + b_{M-1} s^{M-1} + \cdots b_1 s + b_0}{(s - p_1)(s - p_2) \cdots (s - p_N)} \equiv \frac{K_1}{s - p_1} + \frac{K_2}{s - p_2} + \cdots + \frac{K_N}{s - p_N} \quad (8.6)$$

must be satisfied for any arbitrary value of s. The K's can be found by putting the right side into the form of a single fraction with a common denominator that is the same as the left-side denominator, and then setting the coefficients of each power of s in the numerators equal and solving those equations for the K's. But there is another way that is often easier. Multiply both sides of (8.6) by $s - p_1$.

$$(s - p_1)\frac{b_M s^M + b_{M-1}s^{M-1} + \cdots + b_1 s + b_0}{(s - p_1)(s - p_2)\cdots(s - p_N)} = \left[\begin{array}{l}(s - p_1)\frac{K_1}{s - p_1} + (s - p_1)\frac{K_2}{s - p_2} + \cdots \\ + (s - p_1)\frac{K_N}{s - p_N}\end{array}\right]$$

or

$$\frac{b_M s^M + b_{M-1}s^{M-1} + \cdots + b_1 s + b_0}{(s - p_2)\cdots(s - p_N)} = K_1 + (s - p_1)\frac{K_2}{s - p_2} + \cdots + (s - p_1)\frac{K_N}{s - p_N} \quad (8.7)$$

Since (8.6) must be satisfied for any arbitrary value of s, let $s = p_1$. All the factors $(s - p_1)$ on the right side become zero, (8.7) becomes

$$K_1 = \frac{b_M p_1^M + b_{M-1}p_1^{M-1} + \cdots + b_1 p_1 + b_0}{(p_1 - p_2)\cdots(p_1 - p_N)}$$

and we immediately have the value of K_1. We can use the same technique to find all the other K's. Then, using the Laplace-transform pairs

$$e^{-\alpha t}u(t) \xleftrightarrow{\mathcal{L}} \frac{1}{s + \alpha}, \quad \sigma > -\alpha \quad \text{and} \quad -e^{-\alpha t}u(-t) \xleftrightarrow{\mathcal{L}} \frac{1}{s + \alpha}, \quad \sigma < -\alpha,$$

we can find the inverse Laplace transform.

EXAMPLE 8.3

Inverse Laplace transform using partial-fraction expansion

Find the inverse Laplace transform of $G(s) = \dfrac{10s}{(s + 3)(s + 1)}$, $\sigma > -1$.

We can expand this expression in partial fractions yielding

$$G(s) = \frac{\left[\frac{10s}{s + 1}\right]_{s=-3}}{s + 3} + \frac{\left[\frac{10s}{s + 3}\right]_{s=-1}}{s + 1}, \quad \sigma > -1$$

$$G(s) = \frac{15}{s + 3} - \frac{5}{s + 1}, \quad \sigma > -1.$$

Then, using

$$e^{-\alpha t}u(t) \xleftrightarrow{\mathcal{L}} \frac{1}{s + a}, \quad \sigma > -\alpha,$$

we get

$$g(t) = 5(3e^{-3t} - e^{-t})u(t).$$

The most common situation in practice is that there are no repeated poles, but let's see what happens if we have two poles that are identical,

$$G(s) = \frac{b_M s^M + b_{M-1}s^{M-1} + \cdots + b_1 s + b_0}{(s - p_1)^2(s - p_3)\cdots(s - p_N)}.$$

8.9 Partial-Fraction Expansion

If we try the same technique to find the partial-fraction form we get

$$G(s) = \frac{K_{11}}{s - p_1} + \frac{K_{12}}{s - p_1} + \frac{K_3}{s - p_3} + \cdots + \frac{K_N}{s - p_N}.$$

But this can be written as

$$G(s) = \frac{K_{11} + K_{12}}{s - p_1} + \frac{K_3}{s - p_3} + \cdots + \frac{K_N}{s - p_N} = \frac{K_1}{s - p_1} + \frac{K_3}{s - p_3} + \cdots + \frac{K_N}{s - p_N}$$

and we see that the sum of two arbitrary constants $K_{11} + K_{12}$ is really only a single arbitrary constant. There are really only $N - 1$ K's instead of N K's and when we form the common denominator of the partial-fraction sum, it is not the same as the denominator of the original function. We could change the form of the partial-fraction expansion to

$$G(s) = \frac{K_1}{(s - p_1)^2} + \frac{K_3}{s - p_3} + \cdots + \frac{K_N}{s - p_N}.$$

Then, if we tried to solve the equation by finding a common denominator and equating equal powers of s, we would find that we have N equations in $N - 1$ unknowns and there is no unique solution. The solution to this problem is to find a partial-fraction expansion in the form

$$G(s) = \frac{K_{12}}{(s - p_1)^2} + \frac{K_{11}}{s - p_1} + \frac{K_3}{s - p_3} + \cdots + \frac{K_N}{s - p_N}.$$

We can find K_{12} by multiplying both sides of

$$\frac{b_M s^M + b_{M-1} s^{M-1} + \cdots + b_1 s + b_0}{(s - p_1)^2 (s - p_3) \cdots (s - p_N)} = \frac{K_{12}}{(s - p_1)^2} + \frac{K_{11}}{s - p_1} + \frac{K_3}{s - p_3} + \cdots + \frac{K_N}{s - p_N} \quad (8.8)$$

by $(s - p_1)^2$, yielding

$$\frac{b_M s^M + b_{M-1} s^{M-1} + \cdots + b_1 s + b_0}{(s - p_3) \cdots (s - p_N)} = \left[\begin{array}{l} K_{12} + (s - p_1) K_{11} + (s - p_1)^2 \frac{K_3}{s - p_3} + \cdots \\ + (s - p_1)^2 \frac{K_N}{s - p_N} \end{array} \right]$$

and then letting $s = p_1$, yielding

$$K_{12} = \frac{b_M p_1^M + b_{M-1} p_1^{M-1} + \cdots + b_1 p_1 + b_0}{(p_1 - p_3) \cdots (p_1 - p_N)}.$$

But when we try to find K_{11} by the usual technique we encounter another problem.

$$(s - p_1) \frac{b_M s^M + b_{M-1} s^{M-1} + \cdots b_1 s + b_0}{(s - p_1)^2 (s - p_3) \cdots (s - p_N)} = \left[\begin{array}{l} (s - p_1) \frac{K_{12}}{(s - p_1)^2} + (s - p_1) \frac{K_{11}}{s - p_1} \\ + (s - p_1) \frac{K_3}{s - p_3} + \cdots + (s - p_1) \frac{K_N}{s - p_N} \end{array} \right]$$

or

$$\frac{b_M s^M + b_{M-1} s^{M-1} + \cdots + b_1 s + b_0}{(s - p_1)(s - p_3) \cdots (s - p_N)} = \frac{K_{12}}{s - p_1} + K_{11}.$$

Now if we set $s = p_1$ we get division by zero on both sides of the equation and we cannot directly solve it for K_{11}. But we can avoid this problem by multiplying (8.8) through by $(s - p_1)^2$, yielding

$$\frac{b_M s^M + b_{M-1} s^{M-1} + \cdots + b_1 s + b_0}{(s - p_3) \cdots (s - p_N)} = \left[\begin{array}{l} K_{12} + (s - p_1) K_{11} + \\ (s - p_1)^2 \dfrac{K_3}{s - p_3} + \cdots + (s - p_1)^2 \dfrac{K_N}{s - p_N} \end{array} \right],$$

differentiating with respect to s, yielding

$$\frac{d}{ds}\left[\frac{b_M s^M + b_{M-1} s^{M-1} + \cdots + b_1 s + b_0}{(s - p_3) \cdots (s - p_N)}\right] = \left[\begin{array}{l} K_{11} + \dfrac{(s - p_3)2(s - p_1) - (s - p_1)^2}{(s - p_3)^2} K_3 + \cdots \\ + \dfrac{(s - p_q)2(s - p_1) - (s - p_1)^2}{(s - p_N)^2} K_N \end{array} \right]$$

and then setting $s = p_1$ and solving for K_{11},

$$K_{11} = \frac{d}{ds}\left[\frac{b_M s^M + b_{M-1} s^{M-1} + \cdots + b_1 s + b_0}{(s - p_3) \cdots (s - p_N)}\right]_{s \to p_1} = \frac{d}{ds}\left[(s - p_1)^2 G(s)\right]_{s \to p_1}.$$

If there were a higher-order repeated pole such as a triple, quadruple, and so on (very unusual in practice), we could find the coefficients by extending this differentiation idea to multiple derivatives. In general, if H(s) is of the form

$$\text{H}(s) = \frac{b_M s^M + b_{M-1} s^{M-1} + \cdots + b_1 s + b_0}{(s - p_1)(s - p_2) \cdots (s - p_{N-1})(s - p_N)^m}$$

with $N - 1$ distinct finite poles and a repeated Nth pole of order m, it can be written as

$$\text{H}(s) = \frac{K_1}{s - p_1} + \frac{K_2}{s - p_2} + \cdots + \frac{K_{N-1}}{s - p_{N-1}} + \frac{K_{N,m}}{(s - p_N)^m} + \frac{K_{N,m-1}}{(s - p_N)^{m-1}} + \cdots + \frac{K_{N,1}}{s - p_N}$$

where the K's for the distinct poles are found as before and where the K for a repeated pole p_q of order m for the denominator of the form $(s - p_q)^{m-k}$ is

$$\boxed{K_{q,k} = \frac{1}{(m - k)!} \frac{d^{m-k}}{ds^{m-k}}\left[(s - p_q)^m \text{H}(s)\right]_{s \to p_q}, \quad k = 1, 2, \ldots, m} \qquad (8.9)$$

and it is understood that $0! = 1$.

EXAMPLE 8.4

Inverse Laplace transform using partial-fraction expansion

Find the inverse Laplace transform of

$$G(s) = \frac{s + 5}{s^2(s + 2)}, \quad \sigma > 0.$$

This function has a repeated pole at $s = 0$. Therefore, the form of the partial-fraction expansion must be

$$G(s) = \frac{K_{12}}{s^2} + \frac{K_{11}}{s} + \frac{K_3}{s + 2}, \quad \sigma > 0.$$

We find K_{12} by multiplying $G(s)$ by s^2, and setting s to zero in the remaining expression, yielding

$$K_{12} = [s^2 G(s)]_{s \to 0} = 5/2.$$

We find K_{11} by multiplying $G(s)$ by s^2, differentiating with respect to s and setting s to zero in the remaining expression, yielding

$$K_{11} = \frac{d}{ds}[s^2 G(s)]_{s \to 0} = \frac{d}{ds}\left[\frac{s+5}{s+2}\right]_{s \to 0} = \left[\frac{(s+2)-(s+5)}{(s+2)^2}\right]_{s \to 0} = -\frac{3}{4}.$$

We find K_3 by the usual method to be 3/4. So

$$G(s) = \frac{5}{2s^2} - \frac{3}{4s} + \frac{3}{4(s+2)}, \quad \sigma > 0$$

and the inverse transform is

$$g(t) = \left(\frac{5}{2}t - \frac{3}{4} + \frac{3}{4}e^{-2t}\right)u(t) = \frac{10t - 3(1-e^{-2t})}{4}u(t).$$

■

Let's now examine the effect of a violation of one of the assumptions in the original explanation of the partial-fraction expansion method, the assumption that

$$G(s) = \frac{b_M s^M + b_{M-1} s^{M-1} + \cdots + b_1 s + b_0}{(s-p_1)(s-p_2)\cdots(s-p_N)}$$

is a proper fraction in s. If $M \geq N$ we cannot expand in partial fractions because the partial-fraction expression is in the form

$$G(s) = \frac{K_1}{s-p_1} + \frac{K_2}{s-p_2} + \cdots + \frac{K_N}{s-p_N}.$$

Combining these terms over a common denominator,

$$G(s) = \frac{K_1 \prod_{\substack{k=1 \\ k \neq 1}}^{k=N}(s-p_k) + K_2 \prod_{\substack{k=1 \\ k \neq 2}}^{k=N}(s-p_k) + \cdots + K_2 \prod_{\substack{k=1 \\ k \neq N}}^{k=N}(s-p_k)}{(s-p_1)(s-p_2)\cdots(s-p_N)}.$$

The highest power of s in the numerator is $N-1$. Therefore, any ratio of polynomials in s that is to be expanded in partial fractions must have a numerator degree in s no greater than $N-1$ making it proper in s. This is not really much of a restriction because, if the fraction is improper in s, we can always synthetically divide the numerator by the denominator until we have a remainder that is of lower order than the denominator. Then we will have an expression consisting of the sum of terms with non-negative integer powers of s plus a proper fraction in s. The terms with non-negative powers of s have inverse Laplace transforms that are impulses and higher-order singularities.

EXAMPLE 8.5

Inverse Laplace transform using partial-fraction expansion

Find the inverse Laplace transform of $G(s) = \dfrac{10s^2}{(s+1)(s+3)}$, $\sigma > 0$.

This rational function is an improper fraction in s. Synthetically dividing the numerator by the denominator we get

$$s^2 + 4s + 3 \overline{\smash{\big)}\, 10s^2} \atop {\underline{10s^2 + 40s + 30} \atop -40s - 30}} 10 \qquad \Rightarrow \qquad \frac{10s^2}{(s+1)(s+3)} = 10 - \frac{40s + 30}{s^2 + 4s + 3}.$$

Therefore,

$$G(s) = 10 - \frac{40s + 30}{(s+1)(s+3)}, \quad \sigma > 0.$$

Expanding the (proper) fraction in s in partial fractions,

$$G(s) = 10 - 5\left(\frac{9}{s+3} - \frac{1}{s+1}\right), \quad \sigma > 0.$$

Then, using

$$e^{-at}\mathrm{u}(t) \xleftrightarrow{\mathcal{L}} \frac{1}{s+a} \quad \text{and} \quad \delta(t) \xleftrightarrow{\mathcal{L}} 1$$

we get

$$g(t) = 10\delta(t) - 5(9e^{-3t} - e^{-t})\mathrm{u}(t)$$

(Figure 8.14).

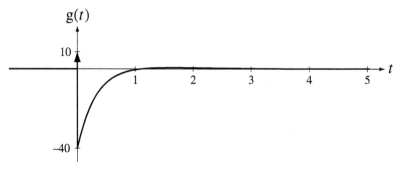

Figure 8.14
Inverse Laplace transform of $G(s) = \dfrac{10s^2}{(s+1)(s+3)}$

EXAMPLE 8.6

Inverse Laplace transform using partial-fraction expansion

Find the inverse Laplace transform of $G(s) = \dfrac{s}{(s-3)(s^2 - 4s + 5)}$, $\sigma < 2$.

If we take the usual route of finding a partial-fraction expansion we must first factor the denominator,

$$G(s) = \frac{s}{(s-3)(s-2+j)(s-2-j)}, \quad \sigma < 2$$

and find that we have a pair of complex-conjugate poles. The partial-fraction method still works with complex poles. Expanding in partial fractions,

$$G(s) = \frac{3/2}{s-3} - \frac{(3+j)/4}{s-2+j} - \frac{(3-j)/4}{s-2-j}, \quad \sigma < 2.$$

With complex poles like this we have a choice. We can either

1. Continue as though they were real poles, find a time-domain expression and then simplify it, or
2. Combine the last two fractions into one fraction with all real coefficients and find its inverse Laplace transform by looking up that form in a table.

Method 1:

$$g(t) = \left(-\frac{3}{2}e^{3t} + \frac{3+j}{4}e^{(2-j)t} + \frac{3-j}{4}e^{(2+j)t}\right)u(-t).$$

This is a correct expression for $g(t)$ but it is not in the most convenient form. We can manipulate it into an expression containing only real-valued functions. Finding a common denominator and recognizing trigonometric functions,

$$g(t) = \left(-\frac{3}{2}e^{3t} + \frac{3e^{(2-j)t} + 3e^{(2+j)t} + je^{(2-j)t} - je^{(2+j)t}}{4}\right)u(-t)$$

$$g(t) = \left(-\frac{3}{2}e^{3t} + e^{2t}\frac{3(e^{-jt} + e^{jt}) + j(e^{-jt} - e^{jt})}{4}\right)u(-t)$$

$$g(t) = (3/2)\{e^{2t}[\cos(t) + (1/3)\sin(t)] - e^{3t}\}u(-t).$$

Method 2:

$$G(s) = \frac{3/2}{s-3} - \frac{1}{4}\frac{(3+j)(s-2-j) + (3-j)(s-2+j)}{s^2 - 4s + 5}, \quad \sigma < 2.$$

When we simplify the numerator we have a first-degree polynomial in s divided by a second-degree polynomial in s.

$$G(s) = \frac{3/2}{s-3} - \frac{1}{4}\frac{6s - 10}{s^2 - 4s + 5} = \frac{3/2}{s-3} - \frac{6}{4}\frac{s - 5/3}{(s-2)^2 + 1}, \quad \sigma < 2$$

In the table of transforms we find

$$-e^{-\alpha t}\cos(\omega_0 t)u(-t) \xleftrightarrow{\mathcal{L}} \frac{s + \alpha}{(s+\alpha)^2 + \omega_0^2}, \quad \sigma < -\alpha$$

and

$$-e^{-\alpha t}\sin(\omega_0 t)u(-t) \xleftrightarrow{\mathcal{L}} \frac{\omega_0}{(s+\alpha)^2 + \omega_0^2}, \quad \sigma < -\alpha.$$

Our denominator form matches these denominators but the numerator form does not. But we can add and subtract numerator forms to form two rational functions whose numerator forms do appear in the table.

$$G(s) = \frac{3/2}{s-3} - \frac{3}{2}\left[\frac{s - 2}{(s-2)^2 + 1} + (1/3)\frac{1}{(s-2)^2 + 1}\right], \quad \sigma < 2 \quad (8.10)$$

Now we can directly find the inverse transform

$$g(t) = (3/2)\{e^{2t}[\cos(t) + (1/3)\sin(t)] - e^{3t}\}u(-t).$$

Realizing that there are two complex conjugate roots, we could have combined the two terms with the complex roots into one with a common denominator of the form

$$G(s) = \frac{A}{s-3} + \frac{K_2}{s-p_2} + \frac{K_3}{s-p_3} = \frac{A}{s-3} + \frac{s(K_2+K_3) - K_3 p_2 - K_2 p_3}{s^2 - 4s + 5}$$

or, since K_2 and K_3 are arbitrary constants,

$$G(s) = \frac{A}{s-3} + \frac{Bs+C}{s^2 - 4s + 5}$$

(Both B and C will be real numbers because K_2 and K_3 are complex conjugates and so are p_2 and p_3.) Then we can find the partial-fraction expansion in this form. A is found exactly as before to be 3/2. Since $G(s)$ and its partial-fraction expansion must be equal for any arbitrary value of s and

$$G(s) = \frac{s}{(s-3)(s^2 - 4s + 5)}$$

we can write

$$\left[\frac{s}{(s-3)(s^2 - 4s + 5)}\right]_{s=0} = \left[\frac{3/2}{s-3} + \frac{Bs+C}{s^2 - 4s + 5}\right]_{s=0}$$

or

$$0 = -1/2 + C/5 \Rightarrow C = 5/2.$$

Then

$$\frac{s}{(s-3)(s^2 - 4s + 5)} = \frac{3/2}{s-3} + \frac{Bs + 5/2}{s^2 - 4s + 5}$$

and we can find B by letting s be any convenient number, for example, one. Then

$$-\frac{1}{4} = -\frac{3}{4} + \frac{B + 5/2}{2} \Rightarrow B = -\frac{3}{2}$$

and

$$G(s) = \frac{3/2}{s-3} - \frac{3}{2}\frac{s - 5/3}{s^2 - 4s + 5}.$$

This result is identical to (8.10) and the rest of the solution is therefore the same. ∎

MATLAB has a function `residue` that can be used in finding partial-fraction expansions. The syntax is

$$[r,p,k] = \text{residue}(b,a)$$

where **b** is a vector of coefficients of descending powers of s in the numerator of the expression and **a** is a vector of coefficients of descending powers of s in the denominator of the expression, **r** is a vector of **residues**, **p** is a vector of finite pole locations, and **k** is a vector of so-called **direct terms**, which result when the degree of the numerator is equal to or greater than the degree of the denominator. The vectors **a** and **b** must always include all powers of s down through zero. The term *residue* comes from theories of closed-contour integration in the complex plane, a topic that is beyond the scope of this text. For our purposes, residues are simply the numerators in the partial-fraction expansion.

EXAMPLE 8.7

Partial-fraction expansion using MATLAB's **residue** function

Expand the expression

$$H(s) = \frac{s^2 + 3s + 1}{s^4 + 5s^3 + 2s^2 + 7s + 3}$$

in partial fractions.

In MATLAB,

```
»b = [1 3 1] ; a = [1 5 2 7 3] ;
»[r,p,k] = residue(b,a) ;
»r
r =
    -0.0856
     0.0496 - 0.2369i
     0.0496 + 0.2369i
    -0.0135
»p
p =
    -4.8587
     0.1441 + 1.1902i
     0.1441 - 1.1902i
    -0.4295
»k
k =
    []
»
```

There are four poles at -4.8587, $0.1441 + j1.1902$, $0.1441 - j1.1902$, and -0.4295 and the residues at those poles are -0.0856, $0.0496 - j0.2369$, $0.0496 + j0.2369$, and -0.0135, respectively. There are no direct terms because H(s) is a proper fraction in s. Now we can write H(s) as

$$H(s) = \frac{0.0496 - j0.2369}{s - 0.1441 - j1.1902} + \frac{0.0496 + j0.2369}{s - 0.1441 + j1.1902} - \frac{0.0856}{s + 4.8587} - \frac{0.0135}{s + 0.4295}$$

or, combining the two terms with complex poles and residues into one term with all real coefficients,

$$H(s) = \frac{0.0991s + 0.5495}{s^2 - 0.2883s + 1.437} - \frac{0.0856}{s + 0.48587} - \frac{0.0135}{s + 0.4295}.$$

∎

EXAMPLE 8.8

Response of an LTI system

Find the response y(t) of an LTI system
(a) With impulse response $h(t) = 5e^{-4t}u(t)$ if excited by $x(t) = u(t)$
(b) With impulse response $h(t) = 5e^{-4t}u(t)$ if excited by $x(t) = u(-t)$
(c) With impulse response $h(t) = 5e^{4t}u(-t)$ if excited by $x(t) = u(t)$

(d) With impulse response $h(t) = 5e^{4t}u(-t)$ if excited by $x(t) = u(-t)$

(a) $h(t) = 5e^{-4t}u(t) \xleftrightarrow{\mathcal{L}} H(s) = \dfrac{5}{s+4}, \quad \sigma > -4$

$$x(t) = u(t) \xleftrightarrow{\mathcal{L}} X(s) = 1/s, \quad \sigma > 0$$

Therefore

$$Y(s) = H(s)X(s) = \dfrac{5}{s(s+4)}, \quad \sigma > 0$$

$Y(s)$ can be expressed in the partial-fraction form

$$Y(s) = \dfrac{5/4}{s} - \dfrac{5/4}{s+4}, \quad \sigma > 0$$

$$y(t) = (5/4)(1 - e^{-4t})u(t) \xleftrightarrow{\mathcal{L}} Y(s) = \dfrac{5/4}{s} - \dfrac{5/4}{s+4}, \quad \sigma > 0$$

(Figure 8.15).

$h(t) = 5e^{-4t}u(t),\ x(t) = u(t)$ $h(t) = 5e^{-4t}u(t),\ x(t) = u(-t)$

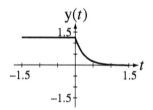

$h(t) = 5e^{4t}u(-t),\ x(t) = u(t)$ $h(t) = 5e^{4t}u(-t),\ x(t) = u(-t)$

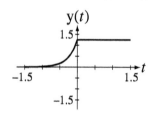 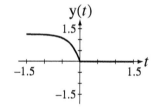

Figure 8.15
The four system responses

(b) $x(t) = u(-t) \xleftrightarrow{\mathcal{L}} X(s) = -1/s, \quad \sigma < 0$

$$Y(s) = H(s)X(s) = -\dfrac{5}{s(s+4)}, \quad -4 < \sigma < 0$$

$$Y(s) = -\dfrac{5/4}{s} + \dfrac{5/4}{s+4}, \quad -4 < \sigma < 0$$

$$y(t) = (5/4)[e^{-4t}u(t) + u(-t)] \xleftrightarrow{\mathcal{L}} Y(s) = -\dfrac{5/4}{s} + \dfrac{5/4}{s+4}, \quad -4 < \sigma < 0$$

(Figure 8.15).

(c) $h(t) = 5e^{4t}u(-t) \xleftrightarrow{\mathcal{L}} H(s) = -\dfrac{5}{s-4}, \quad \sigma < 4$

$$Y(s) = H(s)X(s) = -\dfrac{5}{s(s-4)}, \quad 0 < \sigma < 4$$

$$Y(s) = \dfrac{5/4}{s} - \dfrac{5/4}{s-4}, \quad 0 < \sigma < 4$$

$$y(t) = (5/4)[u(t) + e^{4t}u(-t)] \xleftrightarrow{\mathcal{L}} Y(s) = \dfrac{5/4}{s} - \dfrac{5/4}{s+4}, \quad 0 < \sigma < 4$$

(Figure 8.15).

(d) $Y(s) = H(s)X(s) = \dfrac{5}{s(s-4)}, \quad \sigma < 0$

$$Y(s) = -\dfrac{5/4}{s} + \dfrac{5/4}{s-4}, \quad \sigma < 0$$

$$y(t) = (5/4)[u(-t) - e^{4t}u(-t)] \xleftrightarrow{\mathcal{L}} Y(s) = -\dfrac{5/4}{s} + \dfrac{5/4}{s-4}, \quad \sigma < 4$$

(Figure 8.15).

8.10 LAPLACE-TRANSFORM PROPERTIES

Let $g(t)$ and $h(t)$ have Laplace transforms $G(s)$ and $H(s)$ with regions of convergence ROC_G and ROC_H, respectively. Then it can be shown that the following properties apply (Table 8.2):

Table 8.2 Laplace-transform properties

Linearity	$\alpha g(t) + \beta h(t - t_0) \xleftrightarrow{\mathcal{L}} \alpha G(s) + \beta H(s),$	$\text{ROC} \supseteq \text{ROC}_G \cap \text{ROC}_H$
Time-Shifting	$g(t - t_0) \xleftrightarrow{\mathcal{L}} G(s)e^{-st_0},$	$\text{ROC} = \text{ROC}_G$
s-Domain Shift	$e^{s_0 t}g(t) \xleftrightarrow{\mathcal{L}} G(s - s_0),$	$\text{ROC} = \text{ROC}_G$ shifted by s_0
		(s is in ROC if $s - s_0$ is in ROC_G)
Time Scaling	$g(at) \xleftrightarrow{\mathcal{L}} (1/|a|)G(s/a),$	$\text{ROC} = \text{ROC}_G$ scaled by a
		(s is in ROC if s/a is in ROC_G)
Time Differentiation	$\dfrac{d}{dt}g(t) \xleftrightarrow{\mathcal{L}} sG(s),$	$\text{ROC} \supseteq \text{ROC}_G$
s-Domain Differentiation	$-tg(\tau) \xleftrightarrow{\mathcal{L}} \dfrac{d}{ds}G(s),$	$\text{ROC} = \text{ROC}_G$
Time Integration	$\displaystyle\int_{-\infty}^{t} g(\tau)d\tau \xleftrightarrow{\mathcal{L}} G(s)/s,$	$\text{ROC} \supseteq \text{ROC}_G \cap (\sigma > 0)$
Convolution in Time	$g(t) * h(t) \xleftrightarrow{\mathcal{L}} G(s)H(s),$	$\text{ROC} \supseteq \text{ROC}_G \cap \text{ROC}_H$

If $g(t) = 0, \ t < 0$ and there are no impulses or higher-order singularities at $t = 0$ then

Initial Value Theorem: $\quad g(0^+) = \lim_{s \to \infty} sG(s)$

Final Value Theorem: $\quad \lim_{t \to \infty} g(t) = \lim_{s \to 0} sG(s)$ if $\lim_{t \to \infty} g(t)$ exists

The final value theorem applies *if* the limit $\lim_{t \to \infty} g(t)$ exists. The limit $\lim_{s \to 0} sG(s)$ may exist even if the limit $\lim_{t \to \infty} g(t)$ does not exist. For example, if $X(s) = \dfrac{s}{s^2 + 4}$ then $\lim_{s \to 0} sG(s) = \lim_{s \to 0} \dfrac{s^2}{s^2 + 4} = 0$. But $x(t) = \cos(4t)u(t)$ and $\lim_{t \to \infty} g(t) = \lim_{t \to \infty} \cos(4t)u(t)$ does not exist. Therefore the conclusion that the final value is zero is wrong. It can be shown that For the final value theorem to apply to a function $G(s)$ all the finite poles of the function $sG(s)$ must lie in the open left half of the s plane.

EXAMPLE 8.9

Use of the s-domain shifting property

If $X_1(s) = \dfrac{1}{s+5}$, $\sigma > -5$. and $X_2(s) = X_1(s - j4) + X_1(s + j4)$, $\sigma > -5$ find $x_2(t)$.

$$e^{-5t}u(t) \xleftrightarrow{\mathcal{L}} \dfrac{1}{s+5}, \quad \sigma > -5$$

Using the s-domain shifting property

$$e^{-(5-j4)t}u(t) \xleftrightarrow{\mathcal{L}} \dfrac{1}{s-j4+5}, \quad \sigma > -5 \quad \text{and} \quad e^{-(5+j4)t}u(t) \xleftrightarrow{\mathcal{L}} \dfrac{1}{s+j4+5}, \quad \sigma > -5.$$

Therefore,

$$x_2(t) = e^{-(5-j4)t}u(t) + e^{-(5+j4)t}u(t) = e^{-5t}(e^{j4t} + e^{-j4t})u(t) = 2e^{-5t}\cos(4t)u(t).$$

The effect of shifting equal amounts in opposite directions parallel to the ω axis in the s domain and adding corresponds to multiplication by a causal cosine in the time domain. The overall effect is double-sideband suppressed carrier modulation, which will be discussed in Web Chapter 15.

EXAMPLE 8.10

Laplace transforms of two time-scaled rectangular pulses

Find the Laplace transforms of $x(t) = u(t) - u(t - a)$ and $x(2t) = u(2t) - u(2t - a)$.

We have already found the Laplace transform of $u(t)$, which is $1/s$, $\sigma > 0$. Using the linearity and time-shifting properties,

$$u(t) - u(t - a) \xleftrightarrow{\mathcal{L}} \dfrac{1 - e^{-as}}{s}, \quad \text{all } \sigma.$$

Now, using the time-scaling property,

$$u(2t) - u(2t - a) \xleftrightarrow{\mathcal{L}} \dfrac{1}{2}\dfrac{1 - e^{-as/2}}{s/2} = \dfrac{1 - e^{-as/2}}{s}, \quad \text{all } \sigma.$$

This result is sensible when we consider that $u(2t) = u(t)$ and $u(2t - a) = u(2(t - a/2)) = u(t - a/2)$.

EXAMPLE 8.11

Using s-domain differentiation to derive a transform pair

Using s-domain differentiation and the basic Laplace transform $u(t) \xleftrightarrow{\mathcal{L}} 1/s$, $\sigma > 0$, find the inverse Laplace transform of $1/s^2$, $\sigma > 0$.

$$u(t) \xleftrightarrow{\mathcal{L}} 1/s, \quad \sigma > 0$$

Using $-t g(t) \xleftrightarrow{\mathcal{L}} \dfrac{d}{ds}(G(s))$

$$-t u(t) \xleftrightarrow{\mathcal{L}} -1/s^2, \quad \sigma > 0.$$

Therefore,
$$\text{ramp}(t) = tu(t) \xleftrightarrow{\mathcal{L}} 1/s^2, \quad \sigma > 0.$$

By induction we can extend this to the general case.

$$\frac{d}{ds}\left(\frac{1}{s}\right) = -\frac{1}{s^2}, \frac{d^2}{ds^2}\left(\frac{1}{s}\right) = \frac{2}{s^3}, \frac{d^3}{ds^3}\left(\frac{1}{s}\right) = -\frac{6}{s^4}, \frac{d^4}{ds^4}\left(\frac{1}{s}\right) = \frac{24}{s^5}, \cdots, \frac{d^n}{ds^n}\left(\frac{1}{s}\right) = (-1)^n \frac{n!}{s^{n+1}}$$

The corresponding transform pairs are

$$tu(t) \xleftrightarrow{\mathcal{L}} \frac{1}{s^2}, \quad \sigma > 0, \quad \frac{t^2}{2}u(t) \xleftrightarrow{\mathcal{L}} \frac{1}{s^3}, \quad \sigma > 0$$

$$\frac{t^3}{6}u(t) \xleftrightarrow{\mathcal{L}} \frac{1}{s^4}, \quad \sigma > 0, \cdots, \frac{t^n}{n!}u(t) \xleftrightarrow{\mathcal{L}} \frac{1}{s^{n+1}}, \quad \sigma > 0$$

■

EXAMPLE 8.12

Using the time integration property to derive a transform pair

In Example 8.11 we used complex-frequency differentiation to derive the Laplace-transform pair

$$tu(t) \xleftrightarrow{\mathcal{L}} 1/s^2, \quad \sigma > 0.$$

Derive the same pair from $u(t) \xleftrightarrow{\mathcal{L}} 1/s$, $\sigma > 0$ using the time integration property instead.

$$\int_{-\infty}^{t} u(\tau)d\tau = \begin{cases} \int_{0^-}^{t} d\tau = t, & t \geq 0 \\ 0, & t < 0 \end{cases} = tu(t).$$

Therefore,

$$tu(t) \xleftrightarrow{\mathcal{L}} \frac{1}{s} \times \frac{1}{s} = \frac{1}{s^2}, \quad \sigma > 0.$$

Successive integrations of $u(t)$ yield

$$tu(t), \frac{t^2}{2}u(t), \frac{t^3}{6}u(t)$$

and these can be used to derive the general form

$$\frac{t^n}{n!}u(t) \xleftrightarrow{\mathcal{L}} \frac{1}{s^{n+1}}, \quad \sigma > 0.$$

■

8.11 THE UNILATERAL LAPLACE TRANSFORM

DEFINITION

In the introduction to the Laplace transform it was apparent that if we consider the full range of possible signals to transform, sometimes a region of convergence can be found and sometimes it cannot be found. If we leave out some pathological functions like t^t or e^{t^2}, which grow faster than an exponential (and have no known engineering usefulness) and restrict ourselves to functions that are zero before or after time $t = 0$, the Laplace transform and its ROC become considerably simpler. The quality that made the functions $g_1(t) = Ae^{\alpha t}u(t)$, $\alpha > 0$ and $g_2(t) = Ae^{-\alpha t}u(-t)$, $\alpha > 0$ Laplace transformable was that each of them was restricted by the unit-step function to be zero over a semi-infinite range of time.

Even a function as benign as $g(t) = A$, which is bounded for all t, causes problems because a single convergence factor that makes the Laplace transform converge for all time cannot be found. But the function $g(t) = Au(t)$ *is* Laplace transformable. The presence of the unit step allows the choice of a convergence factor for positive time that makes the Laplace-transform integral converge. For this reason (and other reasons), a modification of the Laplace transform that avoids many convergence issues is usually used in practical analysis.

Let us now redefine the Laplace transform as $G(s) = \int_{0^-}^{\infty} g(t)e^{-st}\,dt$. Only the lower limit of integration has changed. The Laplace transform defined by $G(s) = \int_{-\infty}^{\infty} g(t)e^{-st}\,dt$ is conventionally called the **two-sided** or **bilateral** Laplace transform. The Laplace transform defined by $G(s) = \int_{0^-}^{\infty} g(t)e^{-st}\,dt$ is conventionally called the **one-sided** or **unilateral** Laplace transform. The unilateral Laplace transform is restrictive in the sense that it excludes the negative-time behavior of functions. But since, in the analysis of any real system, a time origin can be chosen to make all signals zero before that time, this is not really a practical problem and actually has some advantages. Since the lower limit of integration is $t = 0^-$, any functional behavior of $g(t)$ before time $t = 0$ is irrelevant to the transform. This means that any other function that has the same behavior at or after time $t = 0$ will have the same transform. Therefore, for the transform to be unique to one time-domain function, it should only be applied to functions that are zero before time $t = 0$.[2]

The inverse unilateral Laplace transform is exactly the same as derived above for the bilateral Laplace transform

$$g(t) = \frac{1}{j2\pi}\int_{\sigma-j\infty}^{\sigma+j\infty} G(s)e^{+st}\,ds.$$

It is common to see the Laplace-transform pair defined by

$$\boxed{\mathcal{L}(g(t)) = G(s) = \int_{0^-}^{\infty} g(t)e^{-st}\,dt, \quad \mathcal{L}^{-1}(G(s)) = g(t) = \frac{1}{j2\pi}\int_{\sigma-j\infty}^{\sigma+j\infty} G(s)e^{+st}\,ds}. \quad (8.11)$$

The unilateral Laplace transform has a simple ROC. It is always the region of the s plane to the right of all the finite poles of the transform (Figure 8.16).

---

[2] Even for times $t > 0$ the transform is not actually unique to a single time-domain function. As mentioned in Chapter 2 in the discussion of the definition of the unit-step function, all the definitions have exactly the signal energy over any finite time range and yet their values are different at the discontinuity time $t > 0$. This is a mathematical point without any real engineering significance. Their effects on any real system will be identical because there is no signal energy in a signal at a point (unless there is an impulse at the point) and real systems respond to the energy of input signals. Also, if two functions differ in value at a finite number of points, the Laplace-transform integral will yield the same transform for the two functions because the area under a point is zero.

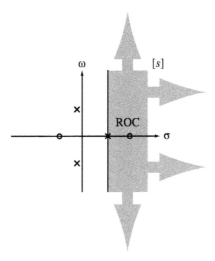

Figure 8.16
ROC for a unilateral Laplace transform

PROPERTIES UNIQUE TO THE UNILATERAL LAPLACE TRANSFORM

Most of the properties of the unilateral Laplace transform are the same as the properties of the bilateral Laplace transform, but there are a few differences. If $g(t) = 0$ for $t < 0$ and $h(t) = 0$ for $t < 0$ and

$$\mathcal{L}(g(t)) = G(s) \quad \text{and} \quad \mathcal{L}(h(t)) = H(s)$$

then the properties in Table 8.3 that are different for the unilateral Laplace transform can be shown to apply.

Table 8.3 Unilateral Laplace-transform properties that differ from bilateral Laplace-transform properties

Time Shifting	$g(t - t_0) \xleftrightarrow{\mathcal{L}} G(s)e^{-st_0}, \ t_0 > 0$
Time Scaling	$g(at) \xleftrightarrow{\mathcal{L}} (1/|a|)G(s/a), \ a > 0$
First Time Derivative	$\dfrac{d}{dt}g(t) \xleftrightarrow{\mathcal{L}} sG(s) - g(0^-)$
Nth Time Derivative	$\dfrac{d^N}{dt^N}(g(t)) \xleftrightarrow{\mathcal{L}} s^N G(s) - \sum_{n=1}^{N} s^{N-n} \left[\dfrac{d^{n-1}}{dt^{n-1}}(g(t)) \right]_{t=0^-}$
Time Integration	$\int_{0^-}^{t} g(\tau)d\tau \xleftrightarrow{\mathcal{L}} G(s)/s$

The time-shifting property is now only valid for time shifts to the right (time delays) because only for delayed signals is the entire nonzero part of the signal still guaranteed to be included in the integral from 0^- to infinity. If a signal were shifted to the left (advanced in time), some of it might occur before time $t = 0$ and not be included within the limits of the Laplace-transform integral. That would destroy the unique relation between the transform of the signal and the transform of its shifted version, making it impossible to relate them in any general way (Figure 8.17).

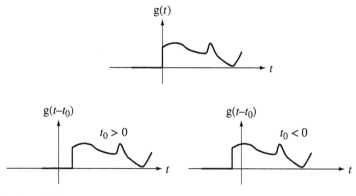

Figure 8.17
Shifts of a causal function

Similarly, in the time-scaling and frequency-scaling properties, the constant a cannot be negative because that would turn a causal signal into a noncausal signal, and the unilateral Laplace transform is only valid for causal signals.

The time derivative properties are important properties of the unilateral Laplace transform. These are the properties that make the solution of differential equations with initial conditions systematic. When using the differentiation properties in solving differential equations, the initial conditions are automatically called for in the proper form as an inherent part of the transform process. Table 8.4 has several commonly used, unilateral Laplace transforms.

Table 8.4 Common unilateral Laplace-transform pairs

$$\delta(t) \xleftrightarrow{\mathcal{L}} 1, \quad \text{All } s$$

$$u(t) \xleftrightarrow{\mathcal{L}} 1/s, \quad \sigma > 0$$

$$u_{-n}(t) = \underbrace{u(t) * \cdots u(t)}_{(n-1)\,\text{convolutions}} \xleftrightarrow{\mathcal{L}} 1/s^n, \quad \sigma > 0$$

$$\text{ramp}(t) = tu(t) \xleftrightarrow{\mathcal{L}} 1/s^2, \quad \sigma > 0$$

$$e^{-\alpha t}u(t) \xleftrightarrow{\mathcal{L}} \frac{1}{s+\alpha}, \quad \sigma > -\alpha$$

$$t^n u(t) \xleftrightarrow{\mathcal{L}} n!/s^{n+1}, \quad \sigma > 0$$

$$te^{-\alpha t}u(t) \xleftrightarrow{\mathcal{L}} \frac{1}{(s+\alpha)^2}, \quad \sigma > -\alpha$$

$$t^n e^{-\alpha t}u(t) \xleftrightarrow{\mathcal{L}} \frac{n!}{(s+\alpha)^{n+1}}, \quad \sigma > -\alpha$$

$$\sin(\omega_0 t)u(t) \xleftrightarrow{\mathcal{L}} \frac{\omega_0}{s^2+\omega_0^2}, \quad \sigma > 0$$

$$\cos(\omega_0 t)u(t) \xleftrightarrow{\mathcal{L}} \frac{s}{s^2+\omega_0^2}, \quad \sigma > 0$$

$$e^{-\alpha t}\sin(\omega_0 t)u(t) \xleftrightarrow{\mathcal{L}} \frac{\omega_0}{(s+\alpha)^2+\omega_0^2}, \quad \sigma > -\alpha$$

$$e^{-\alpha t}\cos(\omega_0 t)u(t) \xleftrightarrow{\mathcal{L}} \frac{s+\alpha}{(s+\alpha)^2+\omega_0^2}, \quad \sigma > -\alpha$$

SOLUTION OF DIFFERENTIAL EQUATIONS WITH INITIAL CONDITIONS

The power of the Laplace transform lies in its use in the analysis of linear system dynamics. This comes about because linear continuous-time systems are described by linear differential equations and, after Laplace transformation, differentiation is represented by multiplication by s. Therefore, the solution of the differential equation is transformed into the solution of an algebraic equation. The unilateral Laplace transform is especially convenient for transient analysis of systems whose excitation begins at an initial time, which can be identified as $t = 0$ and of unstable systems or systems driven by forcing functions that are unbounded as time increases.

EXAMPLE 8.13

Solution of a differential equation with initial conditions using the unilateral Laplace transform

Solve the differential equation

$$x''(t) + 7x'(t) + 12x(t) = 0$$

for times $t > 0$ subject to the initial conditions

$$x(0^-) = 2 \quad \text{and} \quad \frac{d}{dt}(x(t))_{t=0^-} = -4.$$

First, Laplace transform both sides of the equation.

$$s^2 X(s) - sx(0^-) - \frac{d}{dt}(x(t))_{t=0^-} + 7[sX(s) - x(0^-)] + 12X(s) = 0$$

Then solve for $X(s)$.

$$X(s) = \frac{sx(0^-) + 7x(0^-) + \frac{d}{dt}(x(t))_{t=0^-}}{s^2 + 7s + 12}$$

or

$$X(s) = \frac{2s + 10}{s^2 + 7s + 12}.$$

Expanding $X(s)$ in partial fractions,

$$X(s) = \frac{4}{s+3} - \frac{2}{s+4}.$$

From the Laplace-transform table,

$$e^{-\alpha t} u(t) \xleftrightarrow{\mathcal{L}} \frac{1}{s + \alpha}.$$

Inverse Laplace transforming, $x(t) = (4e^{-3t} - 2e^{-4t})u(t)$. Substituting this result into the original differential equation, for times $t \geq 0$

$$\frac{d^2}{dt^2}[4e^{-3t} - 2e^{-4t}] + 7\frac{d}{dt}[4e^{-3t} - 2e^{-4t}] + 12[4e^{-3t} - 2e^{-4t}] = 0$$

$$36e^{-3t} - 32e^{-4t} - 84e^{-3t} + 56e^{-4t} + 48e^{-3t} - 24e^{-4t} = 0$$

$$0 = 0$$

proving that the x(t) found actually solves the differential equation. Also

$$x(0^-) = 4 - 2 = 2 \quad \text{and} \quad \frac{d}{dt}(x(t))_{t=0^-} = -12 + 8 = -4$$

which verifies that the solution also satisfies the stated initial conditions.

Example 8.14

Response of a bridged-T network

In Figure 8.18 the excitation voltage is $v_i(t) = 10u(t)$ volts. Find the zero-state response $v_{R_L}(t)$.

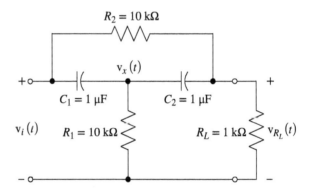

Figure 8.18
Bridged-T network

We can write nodal equations.

$$C_1\frac{d}{dt}[v_x(t) - v_i(t)] + C_2\frac{d}{dt}[v_x(t) - v_{R_L}(t)] + G_1v_x(t) = 0$$

$$C_2\frac{d}{dt}[v_{R_L}(t) - v_x(t)] + G_Lv_{R_L}(t) + G_2[v_{R_L}(t) - v_i(t)] = 0$$

where $G_1 = 1/R_1 = 10^{-4}$ S, $G_2 = 1/R_2 = 10^{-4}$ S, and $G_L = 10^{-3}$ S. Laplace transforming the equations

$$C_1\{sV_x(s) - v_x(0^-) - [sV_i(s) - v_i(0^-)]\} + C_2\{sV_x(s) - v_x(0^-) - [sV_{R_L}(s) - v_{R_L}(0^-)]\}$$
$$+ G_1V_x(s) = 0$$

$$C_2\{sV_{R_L}(s) - v_{R_L}(0^-) - [sV_x(s) - v_x(0^-)]\} + G_LV_{R_L}(s) + G_2[V_{R_L}(s) - V_i(s)] = 0.$$

Since we seek the zero-state response, all the initial conditions are zero and the equations simplify to

$$sC_1[V_x(s) - V_i(s)] + sC_2[V_x(s) - V_{R_L}(s)] + G_1V_x(s) = 0$$

$$sC_2[V_{R_L}(s) - V_x(s)] + G_LV_{R_L}(s) + G_2[V_{R_L}(s) - V_i(s)] = 0$$

The Laplace transform of the excitation is $V_i(s) = 10/s$. Then

$$\begin{bmatrix} s(C_1 + C_2) + G_1 & -sC_2 \\ -sC_2 & sC_2 + (G_L + G_2) \end{bmatrix} \begin{bmatrix} V_x(s) \\ V_{R_L}(s) \end{bmatrix} = \begin{bmatrix} 10C_1 \\ 10G_2/s \end{bmatrix}.$$

The determinant of the 2 by 2 matrix is

$$\Delta = [s(C_1 + C_2) + G_1][sC_2 + (G_L + G_2)] - s^2 C_2^2$$
$$= s^2 C_1 C_2 + s[G_1 C_2 + (G_L + G_2)(C_1 + C_2)] + G_1(G_L + G_2)$$

and, by Cramer's rule, the solution for the Laplace transform of the response is

$$V_{R_L}(s) = \frac{\begin{vmatrix} s(C_1 + C_2) + G_1 & 10C_1 \\ -sC_2 & 10G_2/s \end{vmatrix}}{s^2 C_1 C_2 + s[G_1 C_2 + (G_L + G_2)(C_1 + C_2)] + G_1(G_L + G_2)}$$

$$V_{R_L}(s) = 10 \frac{s^2 C_1 C_2 + sG_2(C_1 + C_2) + G_1 G_2}{s\{s^2 C_1 C_2 + s[G_1 C_2 + (G_L + G_2)(C_1 + C_2)] + G_1(G_L + G_2)\}}$$

or

$$V_{R_L}(s) = 10 \frac{s^2 + sG_2(C_1 + C_2)/C_1 C_2 + G_1 G_2/C_1 C_2}{s\{s^2 + s[G_1/C_1 + (G_L + G_2)(C_1 + C_2)/C_1 C_2] + G_1(G_L + G_2)/C_1 C_2\}}.$$

Using the component numerical values,

$$V_{R_L}(s) = 10 \frac{s^2 + 200s + 10{,}000}{s(s^2 + 2300s + 110{,}000)}.$$

Expanding in partial fractions,

$$V_{R_L}(s) = \frac{0.9091}{s} - \frac{0.243}{s + 48.86} + \frac{9.334}{s + 2251}.$$

Inverse Laplace transforming,

$$v_{R_L}(t) = [0.9091 - 0.243 e^{-48.86t} + 9.334 e^{-2251t}] u(t).$$

As a partial check on the correctness of this solution the response approaches 0.9091 as $t \to \infty$. This is exactly the voltage found using voltage division between the two resistors, considering the capacitors to be open circuits. So the final value looks right. The initial response at time $t = 0^+$ is 10 V. The capacitors are initially uncharged so, at time $t = 0^+$, their voltages are both zero and the excitation and response voltages must be the same. So the initial value also looks right. These two checks on the solution do not guarantee that it is correct for all time, but they are very good checks on the reasonableness of the solution and will often detect an error.

■

8.12 POLE-ZERO DIAGRAMS AND FREQUENCY RESPONSE

In practice, the most common kind of transfer function is one that can be expressed as a ratio of polynomials in s

$$H(s) = \frac{N(s)}{D(s)}.$$

This type of transfer function can be factored into the form

$$H(s) = A \frac{(s - z_1)(s - z_2) \cdots (s - z_M)}{(s - p_1)(s - p_2) \cdots (s - p_N)}.$$

Then the frequency response of the system is

$$H(j\omega) = A\frac{(j\omega - z_1)(j\omega - z_2)\cdots(j\omega - z_M)}{(j\omega - p_1)(j\omega - p_2)\cdots(j\omega - p_N)}.$$

To illustrate a graphical interpretation of this result with an example, let the transfer function be

$$H(s) = \frac{3s}{s+3}.$$

This transfer function has a zero at $s = 0$ and a pole at $s = -3$ (Figure 8.19).
Converting the transfer function to a frequency response,

$$H(j\omega) = 3\frac{j\omega}{j\omega + 3}.$$

The frequency response is three times the ratio of $j\omega$ to $j\omega + 3$. The numerator and denominator can be conceived as vectors in the s plane as illustrated in Figure 8.20 for an arbitrary choice of ω.

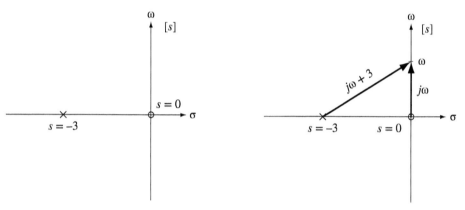

Figure 8.19
Pole-zero plot for H(s) = 3s/(s + 3)

Figure 8.20
Diagram showing the vectors, $j\omega$ and $j\omega + 3$

As the frequency ω is changed, the vectors change also. The magnitude of the frequency response at any particular frequency is three times the magnitude of the numerator vector divided by the magnitude of the denominator vector.

$$|H(j\omega)| = 3\frac{|j\omega|}{|j\omega + 3|}$$

The phase of the frequency response at any particular frequency is the phase of the constant +3 (which is zero), plus the phase of the numerator $j\omega$ (a constant $\pi/2$ radians for positive frequencies and a constant $-\pi/2$ radians for negative frequencies), minus the phase of the denominator $j\omega + 3$.

$$\sphericalangle H(j\omega) = \underbrace{\sphericalangle 3}_{=0} + \sphericalangle j\omega - \sphericalangle(j\omega + 3).$$

At frequencies approaching zero from the positive side, the numerator vector length approaches zero and the denominator vector length approaches a minimum value of 3, making the overall frequency response magnitude approach zero. In that same limit, the phase of $j\omega$ is $\pi/2$ radians and the phase of $j\omega + 3$ approaches zero so that the overall frequency response phase approaches $\pi/2$ radians,

$$\lim_{\omega \to 0^+} |H(j\omega)| = \lim_{\omega \to 0^+} 3\frac{|j\omega|}{|j\omega + 3|} = 0$$

and

$$\lim_{\omega \to 0^+} \angle H(j\omega) = \lim_{\omega \to 0^+} \angle j\omega - \lim_{\omega \to 0^+} \angle(j\omega + 3) = \pi/2 - 0 = \pi/2.$$

At frequencies approaching zero from the negative side, the numerator vector length approaches zero and the denominator vector length approaches a minimum value of 3, making the overall frequency response magnitude approach zero, as before. In that same limit, the phase of $j\omega$ is $-\pi/2$ radians and the phase of $j\omega + 3$ approaches zero so that the overall frequency response phase approaches $-\pi/2$ radians,

$$\lim_{\omega \to 0^-} |H(j\omega)| = \lim_{\omega \to 0^-} 3\frac{|j\omega|}{|j\omega + 3|} = 0$$

and

$$\lim_{\omega \to 0^-} \angle H(j\omega) = \lim_{\omega \to 0^-} \angle j\omega - \lim_{\omega \to 0^-} \angle(j\omega + 3) = -\pi/2 - 0 = -\pi/2.$$

At frequencies approaching positive infinity, the two vector lengths approach the same value and the overall frequency response magnitude approaches 3. In that same limit, the phase of $j\omega$ is $\pi/2$ radians and the phase of $j\omega + 3$ approach $\pi/2$ radians so that the overall frequency-response phase approaches zero,

$$\lim_{\omega \to +\infty} |H(j\omega)| = \lim_{\omega \to +\infty} 3\frac{|j\omega|}{|j\omega + 3|} = 3$$

and

$$\lim_{\omega \to +\infty} \angle H(j\omega) = \lim_{\omega \to +\infty} \angle j\omega - \lim_{\omega \to +\infty} \angle(j\omega + 3) = \pi/2 - \pi/2 = 0.$$

At frequencies approaching negative infinity, the two vector lengths approach the same value and the overall frequency response magnitude approaches 3, as before. In that same limit, the phase of $j\omega$ is $-\pi/2$ radians and the phase of $j\omega + 3$ approach $-\pi/2$ radians so that the overall frequency-response phase approaches zero,

$$\lim_{\omega \to -\infty} |H(j\omega)| = \lim_{\omega \to -\infty} 3\frac{|j\omega|}{|j\omega + 3|} = 3$$

and

$$\lim_{\omega \to -\infty} \angle H(j\omega) = \lim_{\omega \to -\infty} \angle j\omega - \lim_{\omega \to -\infty} \angle(j\omega + 3) = -\pi/2 - (-\pi/2) = 0.$$

These attributes of the frequency response inferred from the pole-zero plot are borne out by a graph of the magnitude and phase frequency response (Figure 8.21). This system attenuates low frequencies relative to higher frequencies. A system with this type of frequency response is often called a **highpass** filter because it generally lets high frequencies pass through and generally blocks low frequencies.

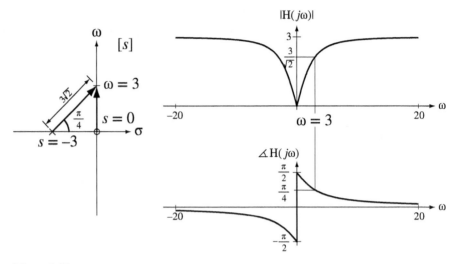

Figure 8.21
Magnitude and phase frequency response of a system whose transfer function is $H(s) = 3s/(s+3)$

EXAMPLE 8.15

Frequency response of a system from its pole-zero diagram

Find the magnitude and phase frequency response of a system whose transfer function is

$$H(s) = \frac{s^2 + 2s + 17}{s^2 + 4s + 104}.$$

This can be factored into

$$H(s) = \frac{(s+1-j4)(s+1+j4)}{(s+2-j10)(s+2+j10)}.$$

So the poles and zeros of this transfer function are $z_1 = -1+j4$, $z_2 = -1-j4$, $p_1 = -2+j10$, and $p_2 = -2-j10$ as illustrated in Figure 8.22.

Converting the transfer function to a frequency response,

$$H(j\omega) = \frac{(j\omega + 1 - j4)(j\omega + 1 + j4)}{(j\omega + 2 - j10)(j\omega + 2 + j10)}.$$

The magnitude of the frequency response at any particular frequency is the product of the numerator-vector magnitudes divided by the product of the denominator-vector magnitudes

$$|H(j\omega)| = \frac{|j\omega + 1 - j4||j\omega + 1 + j4|}{|j\omega + 2 - j10||j\omega + 2 + j10|}.$$

The phase of the frequency response at any particular frequency is the sum of the numerator-vector angles minus the sum of the denominator-vector angles

$$\angle H(j\omega) = \angle(j\omega + 1 - j4) + \angle(j\omega + 1 + j4) - [\angle(j\omega + 2 - j10) + \angle(j\omega + 2 + j10)].$$

This transfer function has no poles or zeros on the ω axis. Therefore its frequency response is neither zero nor infinite at any real frequency. But the finite poles and finite zeros are *near* the real axis and, because of that proximity, will strongly influence the frequency response for real frequencies near those poles and zeros. For a real frequency ω near the pole p_1 the denominator factor $j\omega + 2 - j10$ becomes very small and that makes the overall frequency response magnitude become very large. Conversely, for a real frequency ω near the zero z_1, the numerator

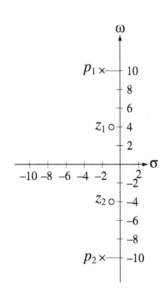

Figure 8.22
Pole-zero plot
of $H(s) = \dfrac{s^2 + 2s + 17}{s^2 + 4s + 104}$

factor $j\omega + 1 - j4$ becomes very small and that makes the overall frequency response magnitude become very small. So, not only does the frequency response magnitude go to zero at zeros and to infinity at poles, it also becomes small near zeros and it becomes large near poles.

The frequency response magnitude and phase are illustrated in Figure 8.23.

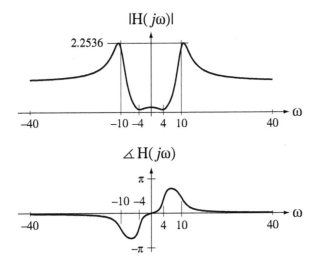

Figure 8.23
Magnitude and phase frequency response of a system whose transfer function is $H(s) = \dfrac{s^2 + 2s + 17}{s^2 + 4s + 104}$

Frequency response can be graphed using the MATLAB control toolbox command bode, and pole-zero diagrams can be plotted using the MATLAB control toolbox command pzmap.

By using graphical concepts to interpret pole-zero plots one can, with practice, perceive approximately how the frequency response looks. There is one aspect of the transfer function that is not evident in the pole-zero plot. The frequency-independent gain A has no effect on the pole-zero plot and therefore cannot be determined by observing it. But all the dynamic behavior of the system is determinable from the pole-zero plot, to within a gain constant.

Below is a sequence of illustrations of how frequency response and step response change as the number and locations of the finite poles and zeros of a system are changed. In Figure 8.24 is a pole-zero diagram of a system with one finite pole and no finite zeros. Its frequency response emphasizes low frequencies relative to high frequencies,

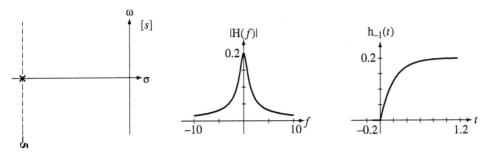

Figure 8.24
One-finite-pole lowpass filter

making it a lowpass filter, and its step response reflects that fact by not jumping discontinuously at time $t = 0$ and approaching a nonzero final value. The continuity of the step response at time $t = 0$ is a consequence of the fact that the high-frequency content of the unit step has been attenuated so that the response cannot change discontinuously.

In Figure 8.25 a zero at zero has been added to the system in Figure 8.24. This changes the frequency response to that of a highpass filter. This is reflected in the step response in the fact that it jumps discontinuously at time $t = 0$ and approaches a final value of zero. The final value of the step response must be zero because the filter completely blocks the zero-frequency content of the input signal. The jump at $t = 0$ is discontinuous because the high-frequency content of the unit step has been retained.

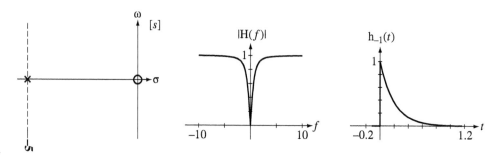

Figure 8.25
One-finite-pole, one-finite-zero highpass filter

In Figure 8.26 is a lowpass filter with two real finite poles and no finite zeros. The step response does not jump discontinuously at time $t = 0$ and approaches a nonzero final value. The response is similar to that in Figure 8.24 but the attenuation of high-frequency content is stronger, as can be seen in the fact that the frequency response falls faster with increasing frequency than the response in Figure 8.24. The step response is also slightly different, starting at time $t = 0$ with a zero slope instead of the nonzero slope in Figure 8.24.

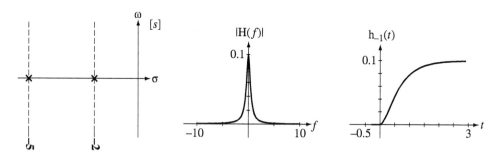

Figure 8.26
Two-finite pole system

In Figure 8.27 a zero at zero has been added to the system of Figure 8.26. The step response does not jump discontinuously at time $t = 0$ and approaches a final value of zero because the system attenuates both the high-frequency content and the low-frequency content relative to the mid-range frequencies. A system with this general form of frequency response is called a **bandpass** filter. Attenuating the high-frequency content makes the step response continuous and attenuating the low-frequency content makes the final value of the step response zero.

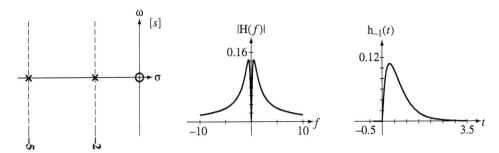

Figure 8.27
Two-finite-pole, one-finite-zero bandpass filter

In Figure 8.28 another zero at zero has been added to the filter of Figure 8.27 making it a highpass filter. The step response jumps discontinuously at time $t = 0$ and the response approaches a final value of zero. The low-frequency attenuation is stronger than the system of Figure 8.25 and that also affects the step response, making it undershoot zero before settling to zero.

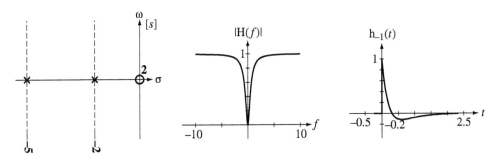

Figure 8.28
Two-finite-pole, two-finite-zero highpass filter

In Figure 8.29 is another two-finite-pole lowpass filter but with a frequency response that is noticeably different from the system in Figure 8.26 because the poles are now complex conjugates instead of real. The frequency response increases and reaches a peak at frequencies near the two poles before it falls at high frequencies. A system with this general form of frequency response is said to be **underdamped**. In an underdamped system, the step response overshoots its final value and "rings" before settling.

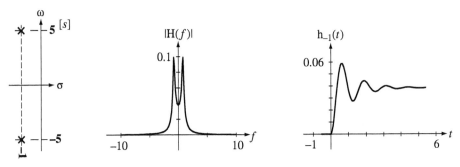

Figure 8.29
Two-finite-pole underdamped lowpass filter

The step response is still continuous everywhere and still approaches a nonzero final value but in a different way than in Figure 8.26.

In Figure 8.30 a zero at zero has been added to the system of Figure 8.29. This changes it from lowpass to bandpass but now, because of the complex-conjugate pole locations, the response is underdamped as is seen in the peaking in the frequency response and the ringing in the step response as compared with the system in Figure 8.27.

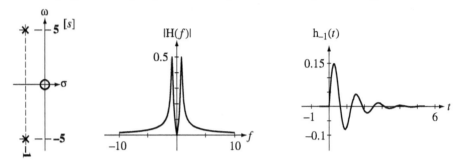

Figure 8.30
Two-finite-pole, one-finite-zero underdamped bandpass filter

In Figure 8.31 another zero at zero has been added to the system of Figure 8.30 making it a highpass filter. It is still underdamped as is evident in the peaking of the frequency response and the ringing in the step response.

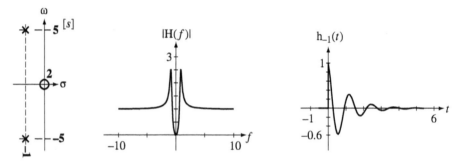

Figure 8.31
Two-finite-pole, two-finite-zero underdamped highpass filter

We see in these examples that moving the poles nearer to the ω axis decreases the damping, makes the step response "ring" for a longer time, and makes the frequency response peak to a higher value. What would happen if we put the poles *on* the ω axis? Having two finite poles on the ω axis (and no finite zeros) means that there are poles at $s = \pm j\omega_0$, the transfer function is of the form $H(s) = \dfrac{K\omega_0}{s^2 + \omega_0^2}$ and the impulse response is of the form $h(t) = K\sin(\omega_0 t)u(t)$. The response to an impulse is equal to a sinusoid after $t = 0$ and oscillates with stable amplitude forever thereafter. The frequency response is $H(j\omega) = \dfrac{K\omega_0}{(j\omega)^2 - \omega_0^2}$. So if the system is excited by a sinusoid $x(t) = A\sin(\omega_0 t)$, the response is infinite, an unbounded response to a bounded excitation. If the system were excited by a sinusoid applied at $t = 0$, $x(t) = A\sin(\omega_0 t)u(t)$, the response would be

$$y(t) = \frac{KA}{2}\left[\frac{\sin(\omega_0 t)}{\omega_0} - t\cos(\omega_0 t)\right]u(t).$$

This contains a sinusoid starting at time $t = 0$ and growing in amplitude linearly forever in positive time. Again this is an unbounded response to a bounded excitation indicating an unstable system. Undamped resonance is never achieved in a real passive system, but it can be achieved in a system with active components that can compensate for energy losses and drive the damping ratio to zero.

8.13 MATLAB SYSTEM OBJECTS

The MATLAB `control` toolbox contains many helpful commands for the analysis of systems. They are based on the idea of a **system object**, a special type of variable in MATLAB for the description of systems. One way of creating a system description in MATLAB is through the use of the `tf` (transfer function) command. The syntax for creating a system object with `tf` is

$$\text{sys = tf(num,den)}.$$

This command creates a system object `sys` from two vectors `num` and `den`. The two vectors are all the coefficients of s (including any zeros), in descending order, in the numerator and denominator of a transfer function. For example, let the transfer function be

$$H_1(s) = \frac{s^2 + 4}{s^5 + 4s^4 + 7s^3 + 15s^2 + 31s + 75}.$$

In MATLAB we can form $H_1(s)$ with

```
»num = [1 0 4] ;
»den = [1 4 7 15 31 75] ;
»H1 = tf(num,den) ;
»H1

Transfer function:
         s^2 + 4
-----------------------------------------
s^5 + 4 s^4 + 7 s^3 + 15 s^2 + 31 s + 75
```

Alternately we can form a system description by specifying the finite zeros, finite poles, and a gain constant of a system using the `zpk` command. The syntax is

$$\text{sys = zpk(z,p,k)},$$

where `z` is a vector of finite zeros of the system, `p` is a vector of finite poles of the system, and `k` is the gain constant. For example, suppose we know that a system has a transfer function

$$H_2(s) = 20 \frac{s + 4}{(s + 3)(s + 10)}.$$

We can form the system description with

```
»z = [-4] ;
»p = [-3 -10] ;
»k = 20 ;
»H2 = zpk(z,p,k) ;
»H2
Zero/pole/gain:
   20 (s+4)
------------
(s+3) (s+10)
```

Another way of forming a system object in MATLAB is to first define s as the independent variable of the Laplace transform with the command

```
»s = tf('s') ;
```

Then we can simply write a transfer function like $H_3(s) = \dfrac{s(s+3)}{s^2 + 2s + 8}$ in the same way we would on paper.

```
»H3 = s*(s+3)/(s^2+2*s+8)

Transfer function:
  s^2 + 3 s
---------------
s^2 + 2 s + 8
```

We can convert one type of system description to the other type.

```
»tf(H2)
Transfer function:
  20 s + 80
---------------
s^2 + 13 s + 30

»zpk(H1)
Zero/pole/gain:
                          (s^2 + 4)
---------------------------------------------------------
(s+3.081) (s^2 + 2.901s + 5.45) (s^2 - 1.982s + 4.467)
```

We can get information about systems from their descriptions using the two commands, tfdata and zpkdata. For example,

```
»[num,den] = tfdata(H2,'v') ;
»num
num =
    0 20 80
»den
den =
    1 13 30
»[z,p,k] = zpkdata(H1,'v') ;
»z
z =
        0 + 2.0000i
        0 - 2.0000i
»p
p =
    -3.0807
    -1.4505 + 1.8291i
    -1.4505 - 1.8291i
     0.9909 + 1.8669i
     0.9909 - 1.8669i
»k
k =
        1
```

(The 'v' argument in these commands indicates that the answers should be returned in vector form.) This last result indicates that the transfer function $H_1(s)$ has zeros at $\pm j2$ and poles at -3.0807, $-1.4505 \pm j1.829$, and $0.9909 \pm j1.8669$.

MATLAB has some handy functions for doing frequency-response analysis in the `control` toolbox. The command

$$H = \text{freqs}(\text{num},\text{den},\text{w});$$

accepts the two vectors `num` and `den` and interprets them as the coefficients of the powers of s in the numerator and denominator of the transfer function $H(s)$, starting with the highest power and going all the way to the zero power, not skipping any. It returns in H the complex frequency response at the radian frequencies in the vector w.

8.14 SUMMARY OF IMPORTANT POINTS

1. The Laplace transform can be used to determine the transfer function of an LTI system and the transfer function can be used to find the response of an LTI system to an arbitrary excitation.
2. The Laplace transform exists for signals whose magnitudes do not grow any faster than an exponential in either positive or negative time.
3. The region of convergence of the Laplace transform of a signal depends on whether the signal is right or left sided.
4. Systems described by ordinary, linear, constant-coefficient differential equations have transfer functions in the form of a ratio of polynomials in s.
5. Pole-zero diagrams of a system's transfer function encapsulate most of its properties and can be used to determine its frequency response to within a gain constant.
6. MATLAB has an object defined to represent a system transfer function and many functions to operate on objects of this type.
7. With a table of Laplace transform pairs and properties, the forward and inverse transforms of almost any signal of engineering significance can be found.
8. The unilateral Laplace transform is commonly used in practical problem solving because it does not require any involved consideration of the region of convergence and is, therefore, simpler than the bilateral form.

EXERCISES WITH ANSWERS

(Answers to each exercise are in random order.)

Laplace-Transform Definition

1. Starting with the definition of the Laplace transform,

$$\mathcal{L}(g(t)) = G(s) = \int_{0^-}^{\infty} g(t)e^{-st}\,dt,$$

find the Laplace transforms of these signals:

(a) $x(t) = e^t u(t)$ (b) $x(t) = e^{2t}\cos(200\pi t)\,u(-t)$

(c) $x(t) = \text{ramp}(t)$ (d) $x(t) = te^t u(t)$

Answers: $\dfrac{1}{s-1},\ \sigma > 1,\ \dfrac{1}{(s-1)^2},\ \sigma > 1,\ -\dfrac{s-2}{(s-2)^2 + (200\pi)^2},\ \sigma < 2,$

$\dfrac{1}{s^2},\ \sigma > 0$

Direct Form II System Realization

2. Draw Direct Form II system diagrams of the systems with these transfer functions:

   (a) $H(s) = \dfrac{1}{s+1}$

   (b) $H(s) = 4\dfrac{s+3}{s+10}$

   Answers:

   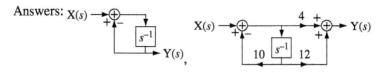

Forward and Inverse Laplace Transforms

3. Using the complex-frequency-shifting property, find and graph the inverse Laplace transform of
$$X(s) = \frac{1}{(s+j4)+3} + \frac{1}{(s-j4)+3}, \quad \sigma > -3.$$

   Answer:

   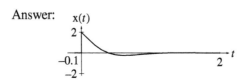

4. Using the time-scaling property, find the Laplace transforms of these signals:

   (a) $x(t) = \delta(4t)$

   (b) $x(t) = u(4t)$

   Answers: $\dfrac{1}{s}$, $\sigma > 0$, $1/4$, All s

5. Using the convolution in time property, find the Laplace transforms of these signals and graph the signals versus time:

   (a) $x(t) = e^{-t}u(t) * u(t)$
   (b) $x(t) = e^{-t}\sin(20\pi t)u(t) * u(-t)$
   (c) $x(t) = 8\cos(\pi t/2)u(t) * [u(t) - u(t-1)]$
   (d) $x(t) = 8\cos(2\pi t)u(t) * [u(t) - u(t-1)]$

   Answers:

   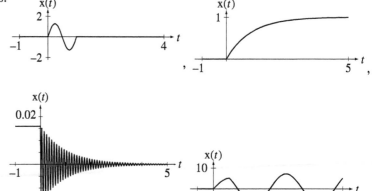

6. A system impulse response h(t) has a unilateral Laplace transform,
$$H(s) = \frac{s(s-4)}{(s+3)(s-2)}.$$
   (a) What is the region of convergence for H(s)?
   (b) Could the CTFT of h(t) be found directly from H(s)?
       If not, why not?

Answers: $\sigma > 2$, No

7. Find the inverse Laplace transforms of these functions:

   (a) $X(s) = \dfrac{24}{s(s+8)}$, $\sigma > 0$

   (b) $X(s) = \dfrac{20}{s^2 + 4s + 3}$, $\sigma < -3$

   (c) $X(s) = \dfrac{5}{s^2 + 6s + 73}$, $\sigma > -3$

   (d) $X(s) = \dfrac{10}{s(s^2 + 6s + 73)}$, $\sigma > 0$

   (e) $X(s) = \dfrac{4}{s^2(s^2 + 6s + 73)}$, $\sigma > 0$

   (f) $X(s) = \dfrac{2s}{s^2 + 2s + 13}$, $\sigma < -1$

   (g) $X(s) = \dfrac{s}{s+3}$, $\sigma > -3$

   (h) $X(s) = \dfrac{s}{s^2 + 4s + 4}$, $\sigma > -2$

   (i) $X(s) = \dfrac{s^2}{s^2 - 4s + 4}$, $\sigma < 2$

   (j) $X(s) = \dfrac{10s}{s^4 + 4s^2 + 4}$, $\sigma > -2$

Answers: $(5/\sqrt{2})t\sin(\sqrt{2}t)u(t)$, $\delta(t) - 4e^{2t}(t+1)u(-t)$, $0.137[1 - e^{-3t}(\cos(8t) + 0.375\sin(8t))]u(t)$, $\dfrac{1}{(73)^2}\left[292t - 24 + 24e^{-3t}\left(\cos(8t) - \dfrac{55}{48}\sin(8t)\right)\right]u(t)$, $(5/8)e^{-3t}\sin(8t)u(t)$, $e^{-2t}(1-2t)u(t)$, $10(e^{-3t} - e^{-t})u(-t)$, $3(1 - e^{-8t})u(t)$, $\delta(t) - 3e^{-3t}u(t)$, $2e^{-t}[(1/\sqrt{12})\sin(\sqrt{12}t) - \cos\sqrt{12}t]u(-t)$

8. Using the initial and final value theorems, find the initial and final values (if possible) of the signals whose Laplace transforms are these functions:

   (a) $X(s) = \dfrac{10}{s+8}$, $\sigma > -8$

   (b) $X(s) = \dfrac{s+3}{(s+3)^2 + 4}$, $\sigma > -3$

   (c) $X(s) = \dfrac{s}{s^2 + 4}$, $\sigma > 0$

   (d) $X(s) = \dfrac{10s}{s^2 + 10s + 300}$, $\sigma < -5$

   (e) $X(s) = \dfrac{8}{s(s+20)}$, $\sigma > 0$

   (f) $X(s) = \dfrac{8}{s^2(s+20)}$, $\sigma > 0$

   (g) $X(s) = \dfrac{s-3}{s(s+5)}$

   (h) $X(s) = \dfrac{s^2 + 7}{s^2 + 4}$

Answers: {0, Does not apply}, {0, 2/5}, {10, 0},

{Does not apply, Does not apply}, {1, Does not apply}, {1, 0},

{1, −0.6}, {Does not apply, Does not apply}

9. A system has a transfer function $H(s) = \dfrac{s^2 + 2s + 3}{s^2 + s + 1}$. Its impulse response is $h(t)$ and its unit step response is $h_{-1}(t)$. Find the final values of each of them.

   Answers: 0, 3

10. Find the numerical values of the literal constants in the form $\dfrac{b_2 s^2 + b_1 s + b_0}{a_2 s^2 + a_1 s + a_0} e^{-s t_0}$ of the bilateral Laplace transforms of these functions:

    (a) $3e^{-8(t-1)} u(t-1)$
    (b) $4\cos(32\pi t) u(t)$
    (c) $4e^{t+2}\sin(32\pi(t+2)) u(t+2)$

    Answers: $\{0, 0, 3, 1, 0, 8, 1\}$, $\{0, 4, 0, 1, 0, (32\pi)^2, 0\}$,
    $\{0, 0, 128\pi, 1, -2, (32\pi)^2, -2\}$

11. Let the function $x(t)$ be defined by $x(t) \xleftrightarrow{\mathcal{L}} \dfrac{s(s+5)}{s^2 + 16}$, $\sigma > 0$. $x(t)$ can be written as the sum of three functions, two of which are causal sinusoids.

    (a) What is the third function?

    (b) What is the cyclic frequency of the causal sinusoids?

    Answers: Impulse, 0.637 Hz

12. A system has a transfer function, $H(s) = \dfrac{s(s-1)}{(s+2)(s+a)}$, which can be expanded in partial fractions in the form $H(s) = A + \dfrac{B}{s+2} + \dfrac{C}{s+a}$. If $a \neq 2$ and $B = \dfrac{3}{2}$, find the values of a, A, and C.

    Answers: 6, 1, −10.5

13. Find the numerical values of the literal constants.

    (a) $4e^{-5t}\cos(25\pi t) u(t) \xleftrightarrow{\mathcal{L}} A \dfrac{s+a}{s^2 + bs + c}$

    (b) $\dfrac{6}{(s+4)(s+a)} = \dfrac{2}{s+4} + \dfrac{b}{s+a}$

    (c) $[A\sin(at) + B\cos(at)] u(t) \xleftrightarrow{\mathcal{L}} 3 \dfrac{3s+4}{s^2+9}$

    (d) $Ae^{-at}[\sin(bt) + B\cos(bt)] u(t) \xleftrightarrow{\mathcal{L}} \dfrac{35s + 325}{s^2 + 18s + 85}$

    (e) $A\delta(t) + (Bt - C) u(t) \xleftrightarrow{\mathcal{L}} 3 \dfrac{(s-1)(s-2)}{s^2}$

    (f) $[At + B(1 - e^{-bt})] u(t) \xleftrightarrow{\mathcal{L}} \dfrac{s-1}{s^2(s+3)}$

    (g) $3e^{-2t}\cos(8t - 24) u(t-3) \xleftrightarrow{\mathcal{L}} A \dfrac{s+a}{s^2 + bs + c} e^{ds}$

    (h) $4e^{-at} u(t) * Ae^{-t/2} u(t) \xleftrightarrow{\mathcal{L}} \dfrac{36}{s^2 + bs + 3}$, $\sigma > -1/2 \cap \sigma > -a$

    Answers: $\{9, 6, 6.5\}$, $\{3, 6, -9\}$, $\{4, 5, 10, 6193.5\}$, $\{4, 9, 3\}$, $\{5, 9, 2\}$,
    $\{0.0074, 2, 4, 68, -3\}$, $\{7, -2\}$, $\{-1/3, 4/9, 3\}$

Unilateral Laplace-Transform Integral

14. Starting with the definition of the unilateral Laplace transform,

$$\mathcal{L}(g(t)) = G(s) = \int_{0^-}^{\infty} g(t)e^{-st}\,dt,$$

find the Laplace transforms of these signals:

(a) $x(t) = e^{-t}u(t)$

(b) $x(t) = e^{2t}\cos(200\pi t)u(t)$

(c) $x(t) = u(t+4)$

(d) $x(t) = u(t-4)$

Answers: $\dfrac{1}{s+1},\ \sigma > 1,\ \dfrac{1}{s},\ \sigma > 0,\ \dfrac{e^{-4s}}{s},\ \sigma > 0,\ \dfrac{s-2}{(s-2)^2 + (200\pi)^2},\ \sigma > 2$

Solving Differential Equations

15. Using the differentiation properties of the unilateral Laplace transform, write the Laplace transform of the differential equation

$$x''(t) - 2x'(t) + 4x(t) = u(t)$$

Answer: $s^2 X(s) - sx(0^-) - \left(\dfrac{d}{dt}(x(t))\right)_{t=0^-} - 2[sX(s) - x(0^-)] + 4X(s) = \dfrac{1}{s}$

16. Using the unilateral Laplace transform, solve these differential equations for $t \geq 0$:

(a) $x'(t) + 10x(t) = u(t)$, $\qquad x(0^-) = 1$

(b) $x''(t) - 2x'(t) + 4x(t) = u(t)$, $\qquad x(0^-) = 0,\ \left[\dfrac{d}{dt}x(t)\right]_{t=0^-} = 4$

(c) $x'(t) + 2x(t) = \sin(2\pi t)u(t)$, $\qquad x(0^-) = -4$

Answers: $(1/4)(1 - e^t\cos(\sqrt{3}t) + (17/\sqrt{3})e^t\sin(\sqrt{3}t))u(t),$

$\left[\dfrac{2\pi e^{-2t} - 2\pi\cos(2\pi t) + 2\sin(2\pi t)}{4 + (2\pi)^2} - 4e^{-2t}\right]u(t),\ \dfrac{1 + 9e^{-10t}}{10}u(t)$

17. Write the differential equations describing the systems in Figure E.17 and find and graph the indicated responses.

(a) $x(t) = u(t)$, $y(t)$ is the response, $y(0^-) = 0$

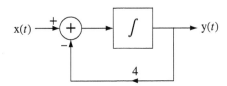

(b) $v(0^-) = 10$, $v(t)$ is the response

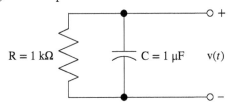

Figure E.17

Answers:

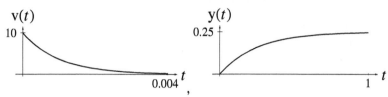

EXERCISES WITHOUT ANSWERS

Region of Convergence

18. Find the regions of convergence of the Laplace transforms of the following functions:

    (a) $3e^{2t}u(t)$
    (b) $-10e^{t/2}u(-t)$
    (c) $10e^{2t}u(t)$
    (d) $5\delta(3t-1)$
    (e) $[-(1/2)+(3/2)e^{-2t}]u(t)$
    (f) $(1/2)u(-t)+(3/2)e^{-2t}u(t)$
    (g) $-5e^{2t}u(t)+7e^{-t}u(t)$
    (h) $\delta(t)-2e^{-2t}u(t)$

Existence of the Laplace Transform

19. Graph the pole-zero plot and region of convergence (if it exists) for these signals:

    (a) $x(t) = e^{-t}u(-t) - e^{-4t}u(t)$
    (b) $x(t) = e^{-2t}u(-t) - e^{-t}u(t)$

Direct Form II System Realization

20. Draw Direct Form II system diagrams of the systems with these transfer functions:

    (a) $H(s) = 10\dfrac{s^2+8}{s^3+3s^2+7s+22}$
    (b) $H(s) = 10\dfrac{s+20}{(s+4)(s+8)(s+14)}$

21. Fill in the blanks in the block diagram in Figure E.21 with numbers for a system whose transfer function is $H(s) = \dfrac{7s^2}{s(s+4)}$.

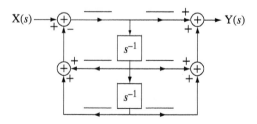

Figure E.21

Forward and Inverse Laplace Transforms

22. Using a table of Laplace transforms and the properties, find the Laplace transforms of the following functions:

    (a) $g(t) = 5 \sin(2\pi(t-1))u(t-1)$
    (b) $g(t) = 5 \sin(2\pi t)u(t-1)$
    (c) $g(t) = 2 \cos(10\pi t)\cos(100\pi t)u(t)$
    (d) $g(t) = \frac{d}{dt}(u(t-2))$
    (e) $g(t) = \int_{0^-}^{t} u(\tau)d\tau$
    (f) $g(t) = \frac{d}{dt}(5e^{-\frac{t-\tau}{2}}u(t-\tau)), \quad \tau > 0$
    (g) $g(t) = 2e^{-5t}\cos(10\pi t)u(t)$
    (h) $x(t) = 5\sin(\pi t - \pi/8)u(-t)$

23. Given
$$g(t) \xleftrightarrow{\mathscr{L}} \frac{s+1}{s(s+4)}, \quad \sigma > 0$$

    find the Laplace transforms of
    (a) $g(2t)$
    (b) $\frac{d}{dt}(g(t))$
    (c) $g(t-4)$
    (d) $g(t) * g(t)$

24. Find the time-domain functions that are the inverse Laplace transforms of these functions. Then, using the initial and final value theorems verify that they agree with the time-domain functions.

    (a) $G(s) = \frac{4s}{(s+3)(s+8)}, \quad \sigma > -3$
    (b) $G(s) = \frac{4}{(s+3)(s+8)}, \quad \sigma > -3$
    (c) $G(s) = \frac{s}{s^2 + 2s + 2}, \quad \sigma > -1$
    (d) $G(s) = \frac{e^{-2s}}{s^2 + 2s + 2}, \quad \sigma > -1$

25. Given
$$e^{4t}u(-t) \xleftrightarrow{\mathscr{L}} G(s)$$

    find the inverse Laplace transforms of
    (a) $G(s/3), \quad \sigma < 4$
    (b) $G(s-2) + G(s+2), \quad \sigma < 4$
    (c) $G(s)/s, \quad \sigma < 4$

26. Given the Laplace-transform pair $g(t) \xleftrightarrow{\mathscr{L}} \frac{3s(s+5)}{(s-2)(s^2+2s+8)}$ and the fact that $g(t)$ is continuous at $t = 0$, complete the following Laplace-transform pairs:

    (a) $\frac{d}{dt}g(t) \xleftrightarrow{\mathscr{L}}$

    (b) $g(t-3) \xleftrightarrow{\mathscr{L}}$

27. For each time-domain function in the column on the left, find its Laplace transform in the column on the right. (It may or may not be there.)

    1. $e^{-3(t-1)}u(t-1)$ A $\frac{e^{-1}}{s+3}$

    2. $e^{-(3t+1)}u(t-1)$ B $\frac{e^{2-s}}{s+3}$

3. $e^{-t/3-1}u(t-1)$ C $\dfrac{e^{-3-s}}{s+3}$

4. $e^{-3t}u(t-1)$ D $\dfrac{e^{4-s}}{s+1/3}$

5. $e^{-t/3}u(t-1)$ E $\dfrac{e^{-(s+1/3)}}{s+1/3}$

6. $e^{-(3t+1)}u(t)$ F $\dfrac{e^{-s}}{s+3}$

28. A causal signal x(t) has a Laplace transform $X(s) = \dfrac{s}{s^4-16}$. If $y(t) = 5x(3t)$, $Y(s)$ can be expressed in the form $Y(s) = \dfrac{As}{s^4-a^4}$. What are the numerical values of A and a?

29. Find the numerical values of the literal constants.

(a) $A\delta(t) - (Be^{at} - e^{bt})u(t) \xleftrightarrow{\mathcal{L}} \dfrac{3s^2}{s^2+5s+4}$

(b) $Ae^{at}(1-bt)u(t) \xleftrightarrow{\mathcal{L}} \dfrac{5s}{(s+1)^2}$

(c) $-2\sin(10\pi t)u(-t) \xleftrightarrow{\mathcal{L}} \dfrac{A}{s^2+as+b}, \quad \sigma < 0$

(d) $7\cos(4t-1)u(t-1/4) \xleftrightarrow{\mathcal{L}} \dfrac{As}{s^2+as+b}e^{cs}, \quad \sigma > 0$

(e) $7e^{3t}\sin(2\pi t)u(-t) \xleftrightarrow{\mathcal{L}} \dfrac{A}{s^2+as+b}, \quad \sigma < c$

(f) $[A\,\text{ramp}(at) + B + Ce^{ct}]u(t) \xleftrightarrow{\mathcal{L}} 12\dfrac{s-1}{s^2(s+3)}, \quad \sigma > 0$

(g) $A\delta(t) + [Be^{-bt} + Ce^{-ct}]u(t) \xleftrightarrow{\mathcal{L}} \dfrac{3s(s-3)}{s^2+4s+3}$

(h) $[At + B + Ce^{-ct}]u(t) \xleftrightarrow{\mathcal{L}} \dfrac{1}{s^2(s+8)}$

(i) $A\big(B\,\text{ramp}(t-t_b) + C + De^{d(t-t_d)}\big)u(t-t_e) \xleftrightarrow{\mathcal{L}} \dfrac{10e^{-3s}}{s^2(s+1)}$

(j) $(K_1 te^{at} + K_2 e^{bt} + K_3 e^{ct})u(t) \xleftrightarrow{\mathcal{L}} \dfrac{2(s-3)}{s^2(s+5)}$

(k) $Ae^{-at}\sin(3t)u(t) \xleftrightarrow{\mathcal{L}} \dfrac{12}{s^2+6s+b}$

(l) $3e^{-2t}\cos(8t-24)u(t-3) \xleftrightarrow{\mathcal{L}} A\dfrac{s+a}{s^2+bs+c}e^{ds}$

30. Find the numerical values of the constants $K_0, K_1, K_2, p_1,$ and p_2.

$$\dfrac{s^2+3}{3s^2+s+9} = K_0 + \dfrac{K_1}{s-p_1} + \dfrac{K_2}{s-p_2}$$

31. Given that $g(t) * u(t) \xleftrightarrow{\mathcal{L}} \dfrac{8(s+2)}{s(s+4)}$ find the function $g(t)$.

32. Given that $g(t) * u(t) \xleftrightarrow{\mathcal{L}} \dfrac{5}{s+12}$ find the function $g(t)$.

Solution of Differential Equations

33. Write the differential equations describing the systems in Figure E.33 and find and graph the indicated responses.

   (a) $x(t) = u(t)$, $y(t)$ is the response $y(0^-) = -5$, $\left[\dfrac{d}{dt}(y(t))\right]_{t=0^-} = 10$

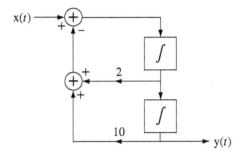

   (b) $i_s(t) = u(t)$, $v(t)$ is the response, No initial energy storage

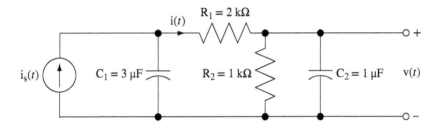

   (c) $i_s(t) = \cos(2000\pi t)$, $v(t)$ is the response, No initial energy storage

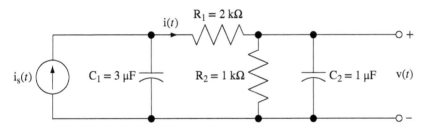

   Figure E.33

Pole-Zero Diagrams and Frequency Response

34. Draw pole-zero diagrams of these transfer functions

   (a) $H(s) = \dfrac{(s+3)(s-1)}{s(s+2)(s+6)}$

   (b) $H(s) = \dfrac{s}{s^2+s+1}$

   (c) $H(s) = \dfrac{s(s+10)}{s^2+11s+10}$

   (d) $H(s) = \dfrac{1}{(s+1)(s^2+1.618s+1)(s^2+0.618s+1)}$

35. In Figure E.35 are some pole-zero plots of transfer functions of systems of the general form $H(s) = A\dfrac{(s-z_1)\cdots(s-z_N)}{(s-p_1)\cdots(s-p_D)}$ in which $A = 1$, the z's are the zeros and the p's are the poles. Answer the following questions:

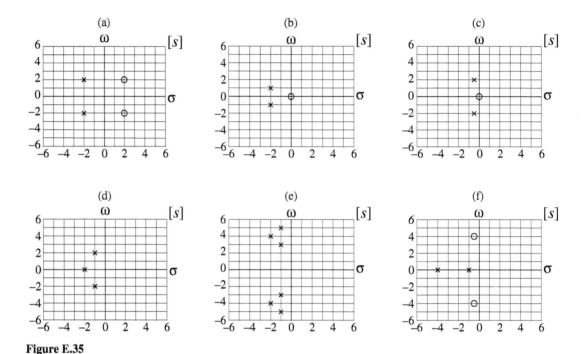

Figure E.35

(a) Which one(s) have a magnitude frequency response that is nonzero at $\omega = 0$?

(b) Which one(s) have a magnitude frequency response that is nonzero as $\omega \to \infty$?

(c) There are two which have a bandpass frequency response (zero at zero and zero at infinity). Which one is more underdamped (higher Q)?

(d) Which one has a magnitude frequency response whose shape is closest to being a bandstop filter?

(e) Which one(s) have a magnitude frequency response that approaches K/ω^6 at very high frequencies (K is a constant)?

(f) Which one has a magnitude frequency response that is constant?

(g) Which one has a magnitude frequency response whose shape is closest to an ideal lowpass filter?

(h) Which one(s) have a phase frequency response that is discontinuous at $\omega = 0$?

36. For each of the pole-zero plots in Figure E.36 determine whether the frequency response is that of a practical lowpass, bandpass, highpass, or bandstop filter.

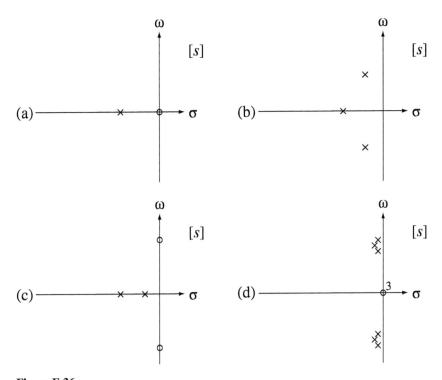

Figure E.36

9 CHAPTER

The z Transform

9.1 INTRODUCTION AND GOALS

Every analysis method used in continuous time has a corresponding method in discrete time. The counterpart to the Laplace transform is the z transform, which expresses signals as linear combinations of discrete-time complex exponentials. Although the transform methods in discrete time are very similar to those used in continuous time, there are a few important differences.

This material is important because in modern system designs digital signal processing is being used more and more. An understanding of discrete-time concepts is needed to grasp the analysis and design of systems that process both continuous-time and discrete-time signals and convert back and forth between them with sampling and interpolation.

CHAPTER GOALS

The chapter goals in this chapter parallel those of Chapter 8 but as applied to discrete-time signals and systems.

1. To develop the z transform as a more general analysis technique for systems than the discrete-time Fourier transform and as a natural result of the convolution process when a discrete-time system is excited by its eigenfunction

2. To define the z transform and its inverse and to determine for what signals it exists

3. To define the transfer function of discrete-time systems and learn a way of realizing a discrete-time system directly from a transfer function

4. To build tables of z transform pairs and properties and learn how to use them with partial-fraction expansion to find inverse z transforms

5. To define a unilateral z transform

6. To solve difference equations with initial conditions using the unilateral z transform

7. To relate the pole and zero locations of a transfer function of a system directly to the frequency response of the system

8. To learn how MATLAB represents the transfer functions of systems

9. To compare the usefulness and efficiency of different transform methods in some typical problems

9.2 GENERALIZING THE DISCRETE-TIME FOURIER TRANSFORM

The Laplace transform is a generalization of the continuous-time Fourier transform (CTFT), which allows consideration of signals and impulse responses that do not have a CTFT. In Chapter 8 we saw how this generalization allowed analysis of signals and systems that could not be analyzed with the Fourier transform and also how it gives insight into system performance through analysis of the locations of the poles and zeros of the transfer function in the s-plane. The z transform is a generalization of the discrete-time Fourier transform (DTFT) with similar advantages. The z transform is to discrete-time signal and system analysis what the Laplace transform is to continuous-time signal and system analysis.

There are two approaches to deriving the z transform that are analogous to the two approaches taken to the derivation of the Laplace transform, generalizing the DTFT and exploiting the unique properties of complex exponentials as the eigenfunctions of LTI systems.

The DTFT is defined by

$$x[n] = \frac{1}{2\pi}\int_{2\pi} X(e^{j\Omega}) e^{j\Omega n} d\Omega \xleftrightarrow{\mathcal{F}} X(e^{j\Omega}) = \sum_{n=-\infty}^{\infty} x[n] e^{-j\Omega n}$$

or

$$x[n] = \int_1 X(F) e^{j2\pi F n} dF \xleftrightarrow{\mathcal{F}} X(F) = \sum_{n=-\infty}^{\infty} x[n] e^{-j2\pi F n}.$$

The Laplace transform generalizes the CTFT by changing complex sinusoids of the form $e^{j\omega t}$ where ω is a real variable, to complex exponentials of the form e^{st} where s is a complex variable. The independent variable in the DTFT is discrete-time radian frequency Ω. The exponential function $e^{j\Omega n}$ appears in both the forward and inverse transforms (as $e^{-j\Omega n} = 1/e^{j\Omega n}$ in the forward transform). For real Ω, $e^{j\Omega n}$ is a discrete time complex sinusoid and has a magnitude of one for any value of discrete time n, which is real. By analogy with the Laplace transform, we could generalize the DTFT by replacing the real variable Ω with a complex variable S and thereby replace $e^{j\Omega n}$ with e^{Sn}, a complex exponential. For complex S, e^S can lie anywhere in the complex plane. We can simplify the notation by letting $z = e^S$ and expressing discrete-time signals as linear combinations of z^n instead of e^{Sn}. Replacing $e^{j\Omega n}$ with z^n in the DTFT leads directly to the conventional definition of a forward z transform

$$\boxed{X(z) = \sum_{n=-\infty}^{\infty} x[n] z^{-n}} \tag{9.1}$$

and $x[n]$ and $X(z)$ are said to form a z-transform pair

$$x[n] \xleftrightarrow{\mathcal{Z}} X(z).$$

The fact that z can range anywhere in the complex plane means that we can use discrete-time complex exponentials instead of just discrete-time complex sinusoids in representing a discrete-time signal. Some signals cannot be represented by linear combinations of complex sinusoids but can be represented by a linear combination of complex exponentials.

9.3 COMPLEX EXPONENTIAL EXCITATION AND RESPONSE

Let the excitation of a discrete-time LTI system be a complex exponential of the form Kz^n where z is, in general, complex and K is any constant. Using convolution, the response $y[n]$ of an LTI system with impulse response $h[n]$ to a complex exponential excitation $x[n] = Kz^n$ is

$$y[n] = h[n] * Kz^n = K \sum_{m=-\infty}^{\infty} h[m] z^{n-m} = \underbrace{Kz^n}_{=x[n]} \sum_{m=-\infty}^{\infty} h[m] z^{-m}.$$

So the response to a complex exponential is that same complex exponential, multiplied by $\sum_{m=-\infty}^{\infty} h[m] z^{-m}$ *if the series converges*. This is identical to (9.1).

9.4 THE TRANSFER FUNCTION

If an LTI system with impulse response $h[n]$ is excited by a signal $x[n]$, the z transform $Y(z)$ of the response $y[n]$ is

$$Y(z) = \sum_{n=-\infty}^{\infty} y[n] z^{-n} = \sum_{n=-\infty}^{\infty} (h[n] * x[n]) z^{-n} = \sum_{n=-\infty}^{\infty} \sum_{m=-\infty}^{\infty} h[m] x[n-m] z^{-n}.$$

Separating the two summations,

$$Y(z) = \sum_{m=-\infty}^{\infty} h[m] \sum_{n=-\infty}^{\infty} x[n-m] z^{-n}.$$

Let $q = n - m$. Then

$$Y(z) = \sum_{m=-\infty}^{\infty} h[m] \sum_{q=-\infty}^{\infty} x[q] z^{-(q+m)} = \underbrace{\sum_{m=-\infty}^{\infty} h[m] z^{-m}}_{=H(z)} \underbrace{\sum_{q=-\infty}^{\infty} x[q] z^{-q}}_{=X(z)}.$$

So, in a manner similar to the Laplace transform, $Y(z) = H(z) X(z)$ and $H(z)$ is called the **transfer function** of the discrete-time system, just as first introduced in Chapter 5.

9.5 CASCADE-CONNECTED SYSTEMS

The transfer functions of components in the cascade connection of discrete-time systems combine in the same way they do in continuous-time systems (Figure 9.1).

The overall transfer function of two systems in cascade connection is the product of their individual transfer functions.

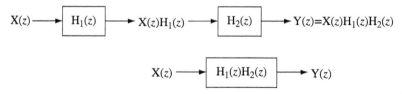

Figure 9.1
Cascade connection of systems

9.6 DIRECT FORM II SYSTEM REALIZATION

In engineering practice the most common form of description of a discrete-time system is a difference equation or a system of difference equations. We showed in Chapter 5 that for a discrete-time system described by a difference equation of the form

$$\sum_{k=0}^{N} a_k y[n-k] = \sum_{k=0}^{M} b_k x[n-k], \qquad (9.2)$$

the transfer function is

$$\boxed{H(z) = \frac{\sum_{k=0}^{M} b_k z^{-k}}{\sum_{k=0}^{N} a_k z^{-k}} = \frac{b_0 + b_1 z^{-1} + b_2 z^{-2} + \cdots + b_M z^{-M}}{a_0 + a_1 z^{-1} + a_2 z^{-2} + \cdots + a_N z^{-N}}} \qquad (9.3)$$

or, alternately,

$$\boxed{H(z) = \frac{\sum_{k=0}^{M} b_k z^{-k}}{\sum_{k=0}^{N} a_k z^{-k}} = z^{N-M} \frac{b_0 z^M + b_1 z^{M-1} + \cdots + b_{M-1} z + b_M}{a_0 z^N + a_1 z^{N-1} + \cdots + a_{N-1} z + a_N}}. \qquad (9.4)$$

The Direct Form II, realization of discrete-time systems, is directly analogous to Direct Form II in continuous time. The transfer function

$$H(z) = \frac{Y(z)}{X(z)} = \frac{b_0 + b_1 z^{-1} + \cdots + b_N z^{-N}}{a_0 + a_1 z^{-1} + \cdots + a_N z^{-N}} = \frac{b_0 z^N + b_1 z^{N-1} + \cdots + b_N}{a_0 z^N + a_1 z^{N-1} + \cdots + a_N}$$

can be separated into the cascade of two subsystem transfer functions

$$H_1(z) = \frac{Y_1(z)}{X(z)} = \frac{1}{a_0 z^N + a_1 z^{N-1} + \cdots + a_N} \qquad (9.5)$$

and

$$H_2(z) = \frac{Y(z)}{Y_1(z)} = b_0 z^N + b_1 z^{N-1} + \cdots + b_N.$$

(Here the order of the numerator and denominator are both indicated as N. If the order of the numerator is actually less than N, some of the b's will be zero. But a_0 must not be zero.) From (9.5),

$$z^N Y_1(z) = (1/a_0)\{X(z) - [a_1 z^{N-1} Y_1(z) + \cdots + a_N Y_1(z)]\}$$

(Figure 9.2).

All the terms of the form $z^k Y_1(z)$ that are needed to form $H_2(z)$ are available in the realization of $H_1(z)$. Combining them in a linear combination using the b coefficients, we get the Direct Form II realization of the overall system (Figure 9.3).

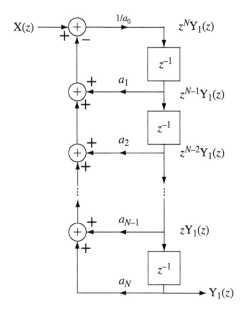

Figure 9.2
Direct Form II, canonical realization of $H_1(z)$

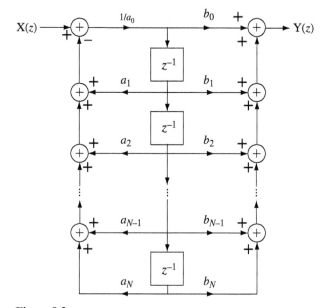

Figure 9.3
Overall Direct Form II canonical system realization

9.7 THE INVERSE z TRANSFORM

The conversion from H(z) to h[n] is the **inverse z transform** and can be done by the direct formula

$$h[n] = \frac{1}{j2\pi} \oint_C H(z) z^{n-1} \, dz.$$

This is a contour integral around a circle in the complex z plane and is beyond the scope of this text. Most practical inverse z transforms are done using a table of z-transform pairs and the properties of the z transform.

9.8 EXISTENCE OF THE z TRANSFORM

TIME-LIMITED SIGNALS

The conditions for existence of the z transform are analogous to the conditions for existence of the Laplace transform. If a discrete-time signal is time limited and bounded, the z-transform summation is finite and its z transform exists for any finite, nonzero value of z (Figure 9.4).

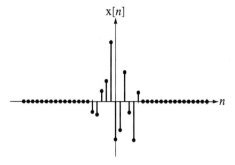

Figure 9.4
A time-limited discrete-time signal

An impulse $\delta[n]$ is a very simple, bounded, time-limited signal and its z transform is

$$\sum_{n=-\infty}^{\infty} \delta[n] z^{-n} = 1.$$

This z transform has no zeros or poles. For any nonzero value of z, the transform of this impulse exists. If we shift the impulse in time in either direction, we get a slightly different result.

$$\delta[n-1] \xleftrightarrow{\mathcal{Z}} z^{-1} \Rightarrow \text{pole at zero}$$
$$\delta[n+1] \xleftrightarrow{\mathcal{Z}} z \Rightarrow \text{pole at infinity}$$

So the z transform of $\delta[n-1]$ exists for every nonzero value of z and the z transform of $\delta[n+1]$ exists for every finite value of z.

RIGHT- AND LEFT-SIDED SIGNALS

A right-sided signal $x_r[n]$ is one for which $x_r[n] = 0$ for any $n < n_0$ and a left-sided signal $x_l[n]$ is one for which $x_l[n] = 0$ for any $n > n_0$ (Figure 9.5).

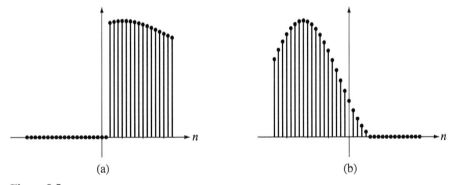

Figure 9.5
(a) Right-sided discrete-time signal, (b) left-sided discrete-time signal

Consider the right-sided signal $x[n] = \alpha^n u[n - n_0]$, $\alpha \in \mathbb{C}$ (Figure 9.6(a)). Its z transform is

$$X(z) = \sum_{n=-\infty}^{\infty} \alpha^n u[n - n_0] z^{-n} = \sum_{n=n_0}^{\infty} (\alpha z^{-1})^n$$

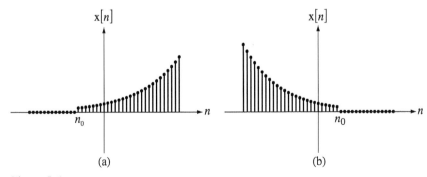

Figure 9.6
(a) $x[n] = \alpha^n u[n - n_0]$, $\alpha \in \mathbb{C}$, (b) $x[n] = \beta^n u[n_0 - n]$, $\beta \in \mathbb{C}$

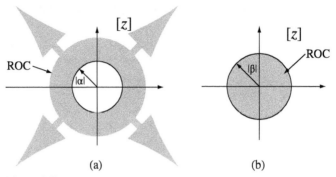

Figure 9.7
Region of convergence for (a) the right-sided signal $x[n] = \alpha^n u[n - n_0]$, $\alpha \in \mathbb{C}$, and (b) the left-sided signal $x[n] = \beta^n u[n_0 - n]$, $\beta \in \mathbb{C}$

if the series converges, and the series converges if $|\alpha/z| < 1$ or $|z| > |\alpha|$. This region of the z plane is called the **region of convergence (ROC)** (Figure 9.7(a)).

If $x[n] = 0$ for $n > n_0$ it is called a **left-sided** signal (Figure 9.6(b)).
If $x[n] = \beta^n u[n_0 - n]$, $\beta \in \mathbb{C}$,

$$X(z) = \sum_{n=-\infty}^{n_0} \beta^n z^{-n} = \sum_{n=-\infty}^{n_0} (\beta z^{-1})^n = \sum_{n=-n_0}^{\infty} (\beta^{-1} z)^n$$

and the summation converges for $|\beta^{-1} z| < 1$ or $|z| < |\beta|$ (Figure 9.7(b)).

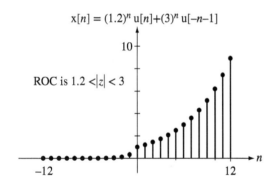
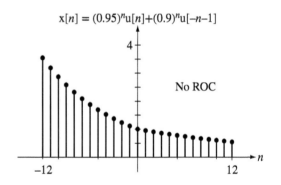
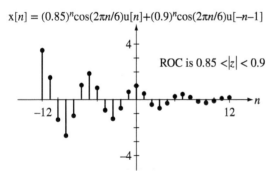
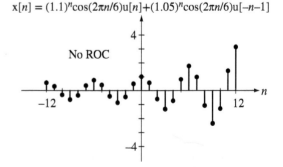

Figure 9.8
Some noncausal signals and their ROCs (if they exist)

Just as in continuous time, any discrete-time signal can be expressed as a sum of a right-sided signal and a left-sided signal. If $x[n] = x_r[n] + x_l[n]$ and if $|x_r[n]| < |K_r \alpha^n|$ and $|x_l[n]| < |K_l \beta^n|$ (where K_r and K_l are constants), then the summation converges and the z transform exists for $|\alpha| < |z| < |\beta|$. This implies that if $|\alpha| < |\beta|$, a z transform can be found and the ROC in the z plane is the region $|\alpha| < |z| < |\beta|$. If $|\alpha| > |\beta|$, the z transform does not exist (Figure 9.8).

EXAMPLE 9.1

z transform of a noncausal signal

Find the z transform of $x[n] = K\alpha^{|n|}$, $\alpha \in \mathbb{R}$
Its variation with n depends on the value of α (Figure 9.9). It can be written as

$$x[n] = K(\alpha^n u[n] + \alpha^{-n} u[-n] - 1).$$

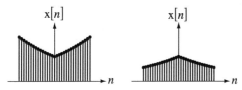

Figure 9.9
(a) $x[n] = K\alpha^{|n|}$, $\alpha > 1$, (b) $x[n] = K\alpha^{|n|}$, $\alpha < 1$

If $|\alpha| \geq 1$ then $|\alpha| \geq |\alpha^{-1}|$, no ROC can be found and it does not have a z transform. If $|\alpha| < 1$ then $|\alpha| < |\alpha^{-1}|$, the ROC is $|\alpha| < z < |\alpha^{-1}|$ and the z transform is

$$K\alpha^{|n|} \xleftrightarrow{\mathcal{Z}} K \sum_{n=-\infty}^{\infty} \alpha^{|n|} z^{-n} = K \left[\sum_{n=0}^{\infty} (\alpha z^{-1})^n + \sum_{n=-\infty}^{0} (\alpha^{-1} z^{-1})^n - 1 \right], |\alpha| < z < |\alpha^{-1}|$$

$$K\alpha^{|n|} \xleftrightarrow{\mathcal{Z}} K \left[\sum_{n=0}^{\infty} (\alpha z^{-1})^n + \sum_{n=0}^{\infty} (\alpha z)^n - 1 \right], |\alpha| < z < |\alpha^{-1}|.$$

This consists of two summations plus a constant. Each summation is a geometric series of the form $\sum_{n=0}^{\infty} r^n$ and the series converges to $1/(1-r)$ if $|r| < 1$.

$$K\alpha^{|n|} \xleftrightarrow{\mathcal{Z}} K \left(\frac{1}{1 - \alpha z^{-1}} + \frac{1}{1 - \alpha z} - 1 \right) = K \left(\frac{z}{z - \alpha} - \frac{z}{z - \alpha^{-1}} \right), |\alpha| < z < |\alpha^{-1}|$$

9.9 z-TRANSFORM PAIRS

We can start a useful table of z transforms with the impulse $\delta[n]$ and the damped cosine $\alpha^n \cos(\Omega_0 n) u[n]$. As we have already seen, $\delta[n] \xleftrightarrow{\mathcal{Z}} 1$. The z transform of the damped cosine is

$$\alpha^n \cos(\Omega_0 n) u[n] \xleftrightarrow{\mathcal{Z}} \sum_{n=-\infty}^{\infty} \alpha^n \cos(\Omega_0 n) u[n] z^{-n}$$

$$\alpha^n \cos(\Omega_0 n)\, u[n] \xleftrightarrow{\mathcal{Z}} \sum_{n=0}^{\infty} \alpha^n \frac{e^{j\Omega_0 n} + e^{-j\Omega_0 n}}{2} z^{-n}$$

$$\alpha^n \cos(\Omega_0 n)\, u[n] \xleftrightarrow{\mathcal{Z}} (1/2) \sum_{n=0}^{\infty} \left[(\alpha e^{j\Omega_0} z^{-1})^n + (\alpha e^{-j\Omega_0} z^{-1})^n \right].$$

For convergence of this z transform $|z| > |\alpha|$ and

$$\alpha^n \cos(\Omega_0 n)\, u[n] \xleftrightarrow{\mathcal{Z}} (1/2) \left[\frac{1}{1 - \alpha e^{j\Omega_0} z^{-1}} + \frac{1}{1 - \alpha e^{-j\Omega_0} z^{-1}} \right], \quad |z| > |\alpha|.$$

This can be simplified to either of the two alternate forms

$$\alpha^n \cos(\Omega_0 n)\, u[n] \xleftrightarrow{\mathcal{Z}} \frac{1 - \alpha \cos(\Omega_0) z^{-1}}{1 - 2\alpha \cos(\Omega_0) z^{-1} + \alpha^2 z^{-2}}, \quad |z| > |\alpha|$$

or

$$\alpha^n \cos(\Omega_0 n)\, u[n] \xleftrightarrow{\mathcal{Z}} \frac{z[z - \alpha \cos(\Omega_0)]}{z^2 - 2\alpha \cos(\Omega_0) z + \alpha^2}, \quad |z| > |\alpha|.$$

If $\alpha = 1$, then

$$\cos(\Omega_0 n)\, u[n] \xleftrightarrow{\mathcal{Z}} \frac{z[z - \cos(\Omega_0)]}{z^2 - 2\cos(\Omega_0) z + 1} = \frac{1 - \cos(\Omega_0) z^{-1}}{1 - 2\cos(\Omega_0) z^{-1} + z^{-2}}, \quad |z| > 1.$$

If $\Omega_0 = 0$, then

$$\alpha^n u[n] \xleftrightarrow{\mathcal{Z}} \frac{z}{z - \alpha} = \frac{1}{1 - \alpha z^{-1}}, \quad |z| > |\alpha|.$$

If $\alpha = 1$ and $\Omega_0 = 0$, then

$$u[n] \xleftrightarrow{\mathcal{Z}} \frac{z}{z - 1} = \frac{1}{1 - z^{-1}}, \quad |z| > 1.$$

9.9 z-Transform Pairs   **415**

Table 9.1 shows the z-transform pairs for several commonly used functions.

Table 9.1 Some z-transform pairs

$$\delta[n] \xleftrightarrow{\mathcal{Z}} 1, \quad \text{All } z$$

$$u[n] \xleftrightarrow{\mathcal{Z}} \frac{z}{z-1} = \frac{1}{1-z^{-1}}, \quad |z|>1, \qquad -u[-n-1] \xleftrightarrow{\mathcal{Z}} \frac{z}{z-1}, \quad |z|<1$$

$$\alpha^n u[n] \xleftrightarrow{\mathcal{Z}} \frac{z}{z-\alpha} = \frac{1}{1-\alpha z^{-1}}, \quad |z|>|\alpha|, \qquad -\alpha^n u[-n-1] \xleftrightarrow{\mathcal{Z}} \frac{z}{z-\alpha} = \frac{1}{1-\alpha z^{-1}}, \quad |z|<|\alpha|$$

$$n u[n] \xleftrightarrow{\mathcal{Z}} \frac{z}{(z-1)^2} = \frac{z^{-1}}{(1-z^{-1})^2}, \quad |z|>1, \qquad -n u[-n-1] \xleftrightarrow{\mathcal{Z}} \frac{z}{(z-1)^2} = \frac{z^{-1}}{(1-z^{-1})^2}, \quad |z|<1$$

$$n^2 u[n] \xleftrightarrow{\mathcal{Z}} \frac{z(z+1)}{(z-1)^3} = \frac{1+z^{-1}}{z(1-z^{-1})^3}, \quad |z|>1, \qquad -n^2 u[-n-1] \xleftrightarrow{\mathcal{Z}} \frac{z(z+1)}{(z-1)^3} = \frac{1+z^{-1}}{z(1-z^{-1})^3}, \quad |z|<1$$

$$n\alpha^n u[n] \xleftrightarrow{\mathcal{Z}} \frac{\alpha z}{(z-\alpha)^2} = \frac{\alpha z^{-1}}{(1-\alpha z^{-1})^2}, \quad |z|>|\alpha|, \qquad -n\alpha^n u[-n-1] \xleftrightarrow{\mathcal{Z}} \frac{\alpha z}{(z-\alpha)^2} = \frac{\alpha z^{-1}}{(1-\alpha z^{-1})^2}, \quad |z|<|\alpha|$$

$$\sin(\Omega_0 n) u[n] \xleftrightarrow{\mathcal{Z}} \frac{z \sin(\Omega_0)}{z^2 - 2z\cos(\Omega_0) + 1}, \quad |z|>1, \qquad -\sin(\Omega_0 n) u[-n-1] \xleftrightarrow{\mathcal{Z}} \frac{z \sin(\Omega_0)}{z^2 - 2z\cos(\Omega_0) + 1}, \quad |z|<1$$

$$\cos(\Omega_0 n) u[n] \xleftrightarrow{\mathcal{Z}} \frac{z[z - \cos(\Omega_0)]}{z^2 - 2z\cos(\Omega_0) + 1}, \quad |z|>1, \qquad -\cos(\Omega_0 n) u[-n-1] \xleftrightarrow{\mathcal{Z}} \frac{z[z - \cos(\Omega_0)]}{z^2 - 2z\cos(\Omega_0) + 1}, \quad |z|<1$$

$$\alpha^n \sin(\Omega_0 n) u[n] \xleftrightarrow{\mathcal{Z}} \frac{z\alpha \sin(\Omega_0)}{z^2 - 2\alpha z\cos(\Omega_0) + \alpha^2}, \quad |z|>|\alpha|, \qquad -\alpha^n \sin(\Omega_0 n) u[-n-1] \xleftrightarrow{\mathcal{Z}} \frac{z\alpha \sin(\Omega_0)}{z^2 - 2\alpha z\cos(\Omega_0) + \alpha^2}, \quad |z|<|\alpha|$$

$$\alpha^n \cos(\Omega_0 n) u[n] \xleftrightarrow{\mathcal{Z}} \frac{z[z - \alpha\cos(\Omega_0)]}{z^2 - 2\alpha z\cos(\Omega_0) + \alpha^2}, \quad |z|>|\alpha|, \qquad -\alpha^n \cos(\Omega_0 n) u[-n-1] \xleftrightarrow{\mathcal{Z}} \frac{z[z - \alpha\cos(\Omega_0)]}{z^2 - 2\alpha z\cos(\Omega_0) + \alpha^2}, \quad |z|<|\alpha|$$

$$\alpha^{|n|} \xleftrightarrow{\mathcal{Z}} \frac{z}{z-\alpha} - \frac{z}{z-\alpha^{-1}}, \quad |\alpha|<|z|<|\alpha^{-1}|$$

$$u[n-n_0] - u[n-n_1] \xleftrightarrow{\mathcal{Z}} \frac{z}{z-1}(z^{-n_0} - z^{-n_1}) = \frac{z^{n_1-n_0-1} + z^{n_1-n_0-2} + \cdots + z + 1}{z^{n_1-1}}, \quad |z|>0$$

EXAMPLE 9.2

Inverse z transforms

Find the inverse z transforms of

(a) $X(z) = \dfrac{z}{z-0.5} - \dfrac{z}{z+2}, \quad 0.5 < |z| < 2$

(b) $X(z) = \dfrac{z}{z-0.5} - \dfrac{z}{z+2}, \quad |z| > 2$

(c) $X(z) = \dfrac{z}{z-0.5} - \dfrac{z}{z+2}, \quad |z| < 0.5$

(a) Right-sided signals have ROCs that are outside a circle and left-sided signals have ROCs that are inside a circle. Therefore, using

$$\alpha^n u[n] \xleftrightarrow{\mathcal{Z}} \frac{z}{z-\alpha} = \frac{1}{1-\alpha z^{-1}}, \quad |z|>|\alpha|$$

and

$$-\alpha^n u[-n-1] \xleftrightarrow{\mathcal{Z}} \frac{z}{z-\alpha} = \frac{1}{1-\alpha z^{-1}}, \quad |z| < |\alpha|$$

we get

$$(0.5)^n u[n] - (-(-2)^n u[-n-1]) \xleftrightarrow{\mathcal{Z}} X(z) = \frac{z}{z-0.5} - \frac{z}{z+2}, \quad 0.5 < |z| < 2$$

or

$$(0.5)^n u[n] + (-2)^n u[-n-1] \xleftrightarrow{\mathcal{Z}} X(z) = \frac{z}{z-0.5} - \frac{z}{z+2}, \quad 0.5 < |z| < 2.$$

(b) In this case both signals are right sided.

$$[(0.5)^n - (-2)^n] u[n] \xleftrightarrow{\mathcal{Z}} X(z) = \frac{z}{z-0.5} - \frac{z}{z+2}, \quad |z| > 2$$

(c) In this case both signals are left sided.

$$-[(0.5)^n - (-2)^n] u[-n-1] \xleftrightarrow{\mathcal{Z}} X(z) = \frac{z}{z-0.5} - \frac{z}{z+2}, \quad |z| < 0.5$$

■

9.10 z-TRANSFORM PROPERTIES

Given the z-transform pairs $g[n] \xleftrightarrow{\mathcal{Z}} G(z)$ and $h[n] \xleftrightarrow{\mathcal{Z}} H(z)$ with ROCs of ROC_G and ROC_H, respectively, the properties of the z transform are listed in Table 9.2.

Table 9.2 z-transform properties

Property	
Linearity	$\alpha g[n] + \beta h[n] \xleftrightarrow{\mathcal{Z}} \alpha G(z) + \beta H(z)$. ROC = $\text{ROC}_G \cap \text{ROC}_H$
Time Shifting	$g[n - n_0] \xleftrightarrow{\mathcal{Z}} z^{-n_0} G(z)$, ROC = ROC_G except perhaps $z = 0$ or $z \to \infty$
Change of Scale in z	$\alpha^n g[n] \xleftrightarrow{\mathcal{Z}} G(z/\alpha)$, ROC = $|\alpha| \text{ROC}_G$
Time Reversal	$g[-n] \xleftrightarrow{\mathcal{Z}} G(z^{-1})$, ROC = $1/\text{ROC}_G$
Time Expansion	$\begin{cases} g[n/k], & n/k \text{ an integer} \\ 0, & \text{otherwise} \end{cases} \xleftrightarrow{\mathcal{Z}} G(z^k)$, ROC = $(\text{ROC}_G)^{1/k}$
Conjugation	$g^*[n] \xleftrightarrow{\mathcal{Z}} G^*(z^*)$, ROC = ROC_G
z-Domain Differentiation	$-n g[n] \xleftrightarrow{\mathcal{Z}} z \frac{d}{dz} G(z)$, ROC = ROC_G
Convolution	$g[n] * h[n] \xleftrightarrow{\mathcal{Z}} H(z)G(z)$
First Backward Difference	$g[n] - g[n-1] \xleftrightarrow{\mathcal{Z}} (1 - z^{-1})G(z)$, ROC $\supseteq \text{ROC}_G \cap |z| > 0$
Accumulation	$\sum_{m=-\infty}^{n} g[m] \xleftrightarrow{\mathcal{Z}} \frac{z}{z-1} G(z)$, ROC $\supseteq \text{ROC}_G \cap |z| > 1$
Initial Value Theorem	If $g[n] = 0, n < 0$ then $g[0] = \lim_{z \to \infty} G(z)$
Final Value Theorem	If $g[n] = 0, n < 0$, $\lim_{n \to \infty} g[n] = \lim_{z \to 1}(z-1)G(z)$ if $\lim_{n \to \infty} g[n]$ exists.

9.11 INVERSE z-TRANSFORM METHODS

SYNTHETIC DIVISION

For rational functions of z of the form

$$H(z) = \frac{b_M z^M + b_{M-1} z^{M-1} + \cdots + b_1 z + b_0}{a_N z^N + a_{N-1} z^{N-1} + \cdots + a_1 z + a_0}$$

we can always synthetically divide the numerator by the denominator and get a sequence of powers of z. For example, if the function is

$$H(z) = \frac{(z-1.2)(z+0.7)(z+0.4)}{(z-0.2)(z-0.8)(z+0.5)}, \quad |z| > 0.8$$

or

$$H(z) = \frac{z^3 - 0.1z^2 - 1.04z - 0.336}{z^3 - 0.5z^2 - 0.34z + 0.08}, \quad |z| > 0.8$$

the synthetic division process produces

$$
\begin{array}{r}
1 + 0.4z^{-1} + 0.5z^{-2} \cdots \\
z^3 - 0.5z^2 - 0.34z + 0.08 \overline{\smash{)}\, z^3 - 0.1z^2 - 1.04z - 0.336} \\
\underline{z^3 - 0.5z^2 - 0.34z + 0.08} \\
0.4z^2 - 0.7z - 0.256 \\
\underline{0.4z^2 - 0.2z - 0.136 - 0.032z^{-1}} \\
0.5z - 0.12 + 0.032z^{-1} \\
\vdots \quad \vdots \quad \vdots
\end{array}
$$

Then the inverse z transform is

$$h[n] = \delta[n] + 0.4\delta[n-1] + 0.5\delta[n-2]\cdots$$

There is an alternate form of synthetic division.

$$
\begin{array}{r}
-4.2 - 30.85z - 158.613z^2 \cdots \\
0.08 - 0.34z - 0.5z^2 + z^3 \overline{\smash{)}\, -0.336 - 1.04z - 0.1z^2 + z^3} \\
\underline{-0.336 + 1.428z + 2.1z^2 - 4.2z^3} \\
-2.468z - 2.2z^2 + 5.2z^3 \\
\underline{-2.468z + 10.489z^2 + 15.425z^3 - 30.85z^4} \\
-12.689z^2 - 10.225z^3 + 30.85z^4 \\
\vdots \quad \vdots \quad \vdots
\end{array}
$$

From this result, we might conclude that the inverse z transform would be

$$-4.2\delta[n] - 30.85\delta[n+1] - 158.613\delta[n+2]\cdots$$

It is natural at this point to wonder why these two results are different and which one is correct. The key to knowing which one is correct is the ROC, $|z| > 0.8$. This implies a right-sided inverse transform and the first synthetic division result is of that form. That series converges for $|z| > 0.8$. The second series converges for $|z| < 0.2$ and would be the correct answer if the ROC were $|z| < 0.2$.

PARTIAL-FRACTION EXPANSION

The technique of partial-fraction expansion to find the inverse z transform is algebraically identical to the method used to find inverse Laplace transforms with the variable s replaced by the variable z. But there is a situation in inverse z transforms that deserves mention. It is very common to have z-domain functions in which the number of finite zeros equals the number of finite poles (making the expression improper in z), with at least one zero at $z = 0$.

$$H(z) = \frac{z^{N-M}(z - z_1)(z - z_2)\cdots(z - z_M)}{(z - p_1)(z - p_2)\cdots(z - p_N)}, \quad N > M.$$

We cannot immediately expand $H(z)$ in partial fractions because it is an improper rational function of z. In a case like this it is convenient to divide both sides of the equation by z.

$$\frac{H(z)}{z} = \frac{z^{N-M-1}(z - z_1)(z - z_2)\cdots(z - z_M)}{(z - p_1)(z - p_2)\cdots(z - p_N)}$$

$H(z)/z$ is a proper fraction in z and can be expanded in partial fractions.

$$\frac{H(z)}{z} = \frac{K_1}{z - p_1} + \frac{K_2}{z - p_2} + \cdots + \frac{K_N}{z - p_N}$$

Then both sides can be multiplied by z and the inverse transform can be found.

$$H(z) = \frac{zK_1}{z - p_1} + \frac{zK_2}{z - p_2} + \cdots + \frac{zK_N}{z - p_N}$$

$$h[n] = K_1 p_1^n u[n] + K_2 p_2^n u[n] + \cdots + K_N p_N^n u[n]$$

Just as we did in finding inverse Laplace transforms, we could have solved this problem using synthetic division to obtain a proper remainder. But this new technique is often simpler.

EXAMPLES OF FORWARD AND INVERSE z TRANSFORMS

The time-shifting property is very important in converting z-domain transfer-function expressions into actual systems and, other than the linearity property, is probably the most often-used property of the z transform.

EXAMPLE 9.3

System block diagram from a transfer function using the time-shifting property

A system has a transfer function

$$H(z) = \frac{Y(z)}{X(z)} = \frac{z - 1/2}{z^2 - z + 2/9}, \quad |z| > 2/3.$$

Draw a system block diagram using delays, amplifiers, and summing junctions.

We can rearrange the transfer-function equation into

$$Y(z)(z^2 - z + 2/9) = X(z)(z - 1/2)$$

or

$$z^2 Y(z) = zX(z) - (1/2)X(z) + zY(z) - (2/9)Y(z).$$

Multiplying this equation through by z^{-2} we get

$$Y(z) = z^{-1} X(z) - (1/2) z^{-2} X(z) + z^{-1} Y(z) + (2/9) z^{-2} Y(z).$$

Now, using the time-shifting property, if $x[n] \xleftrightarrow{\mathcal{Z}} X(z)$ and $y[n] \xleftrightarrow{\mathcal{Z}} Y(z)$, then the inverse z transform is

$$y[n] = x[n-1] - (1/2)x[n-2] + y[n-1] - (2/9)y[n-2].$$

This is called a **recursion** relationship between $x[n]$ and $y[n]$ expressing the value of $y[n]$ at discrete time n as a linear combination of the values of both $x[n]$ and $y[n]$ at discrete times n, $n-1$, $n-2$, $\cdots$. From it we can directly synthesize a block diagram of the system (Figure 9.10).

This system realization uses four delays, two amplifiers, and two summing junctions. This block diagram was drawn in a "natural" way by directly implementing the recursion relation in the diagram. Realized in Direct Form II, the realization uses two delays, three amplifiers, and three summing junctions (Figure 9.11). There are multiple other ways of realizing the system (see Chapter 13).

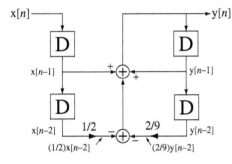

Figure 9.10
Time-domain system block diagram for the transfer function $H(z) = \dfrac{z - 1/2}{z^2 - z + 2/9}$

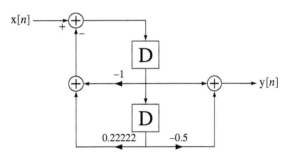

Figure 9.11
Direct Form II realization of $H(z) = \dfrac{z - 1/2}{z^2 - z + 2/9}$

A special case of the change-of-scale-in-z property

$$\alpha^n g[n] \xleftrightarrow{\mathcal{Z}} G(z/\alpha)$$

is of particular interest. Let the constant α be $e^{j\Omega_0}$ with Ω_0 real. Then

$$e^{j\Omega_0 n} g[n] \xleftrightarrow{\mathcal{Z}} G(z e^{-j\Omega_0}).$$

Every value of z is changed to $z e^{-j\Omega_0}$. This accomplishes a counterclockwise rotation of the transform $G(z)$ in the z plane by the angle Ω_0 because $e^{-j\Omega_0}$ has a magnitude of one and a phase of $-\Omega_0$. This effect is a little hard to see in the abstract. An example will illustrate it better. Let

$$G(z) = \frac{z - 1}{(z - 0.8 e^{-j\pi/4})(z - 0.8 e^{+j\pi/4})}$$

and let $\Omega_0 = \pi/8$. Then

$$G(ze^{-j\Omega_0}) = G(ze^{-j\pi/8}) = \frac{ze^{-j\pi/8} - 1}{(ze^{-j\pi/8} - 0.8e^{-j\pi/4})(ze^{-j\pi/8} - 0.8e^{+j\pi/4})}$$

or

$$G(ze^{-j\pi/8}) = \frac{e^{-j\pi/8}(z - e^{j\pi/8})}{e^{-j\pi/8}(z - 0.8e^{-j\pi/8})e^{-j\pi/8}(z - 0.8e^{+j3\pi/8})}$$

$$= e^{j\pi/8}\frac{z - e^{j\pi/8}}{(z - 0.8e^{-j\pi/8})(z - 0.8e^{+j3\pi/8})}.$$

The original function has finite poles at $z = 0.8e^{\pm j\pi/4}$ and a zero at $z = 1$. The transformed function $G(ze^{-j\pi/8})$ has finite poles at $z = 0.8e^{-j\pi/8}$ and $0.8e^{+j3\pi/8}$ and a zero at $z = e^{j\pi/8}$. So the finite pole and zero locations have been rotated counterclockwise by $\pi/8$ radians (Figure 9.12).

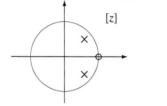

Figure 9.12
Illustration of the frequency-scaling property of the z transform for the special case of a scaling by $e^{j\Omega_0}$

A multiplication by a complex sinusoid of the form $e^{j\Omega_0 n}$ in the time domain corresponds to a rotation of its z transform.

EXAMPLE 9.4

z transforms of a causal exponential and a causal exponentially damped sinusoid

Find the z transforms of $x[n] = e^{-n/40}u[n]$ and $x_m[n] = e^{-n/40}\sin(2\pi n/8)u[n]$ and draw pole-zero diagrams for $X(z)$ and $X_m(z)$.

Using

$$\alpha^n u[n] \xleftrightarrow{\mathcal{Z}} \frac{z}{z - \alpha} = \frac{1}{1 - \alpha z^{-1}}, \quad |z| > |\alpha|$$

we get

$$e^{-n/40}u[n] \xleftrightarrow{\mathcal{Z}} \frac{z}{z - e^{-1/40}}, \quad |z| > |e^{-1/40}|.$$

Therefore,

$$X(z) = \frac{z}{z - e^{-1/40}}, \quad |z| > |e^{-1/40}|.$$

We can rewrite $x_m[n]$ as

$$x_m[n] = e^{-n/40}\frac{e^{j2\pi n/8} - e^{-j2\pi n/8}}{j2}u[n]$$

or

$$x_m[n] = -\frac{j}{2}[e^{-n/40}e^{j2\pi n/8} - e^{-n/40}e^{-j2\pi n/8}]u[n].$$

Then, starting with

$$e^{-n/40}u[n] \xleftrightarrow{\mathcal{Z}} \frac{z}{z - e^{-1/40}}, \quad |z| > |e^{-1/40}|$$

and, using the change-of-scale property $\alpha^n g[n] \xleftrightarrow{\mathcal{Z}} G(z/\alpha)$, we get

$$e^{j2\pi n/8}e^{-n/40}u[n] \xleftrightarrow{\mathcal{Z}} \frac{ze^{-j2\pi/8}}{ze^{-j2\pi/8} - e^{-1/40}}, \quad |z| > |e^{-1/40}|$$

and

$$e^{-j2\pi n/8}e^{-n/40}u[n] \xleftrightarrow{\mathcal{Z}} \frac{ze^{j2\pi/8}}{ze^{j2\pi/8} - e^{-1/40}}, \quad |z| > |e^{-1/40}|$$

and

$$-\frac{j}{2}[e^{-n/40}e^{j2\pi n/8} - e^{-n/40}e^{-j2\pi n/8}]u[n] \xleftrightarrow{\mathcal{Z}}$$

$$-\frac{j}{2}\left[\frac{ze^{-j2\pi/8}}{ze^{-j2\pi/8} - e^{-1/40}} - \frac{ze^{j2\pi/8}}{ze^{j2\pi/8} - e^{-1/40}}\right], \quad |z| > |e^{-1/40}|$$

or

$$X_m(z) = -\frac{j}{2}\left[\frac{ze^{-j2\pi/8}}{ze^{-j2\pi/8} - e^{-1/40}} - \frac{ze^{j2\pi/8}}{ze^{j2\pi/8} - e^{-1/40}}\right]$$

$$= \frac{ze^{-1/40}\sin(2\pi/8)}{z^2 - 2ze^{-1/40}\cos(2\pi/8) + e^{-1/20}}, \quad |z| > |e^{-1/40}|$$

or

$$X_m(z) = \frac{0.6896z}{z^2 - 1.3793z + 0.9512}$$

$$= \frac{0.6896z}{(z - 0.6896 - j0.6896)(z - 0.6896 + j0.6896)}, \quad |z| > |e^{-1/40}|$$

(Figure 9.13).

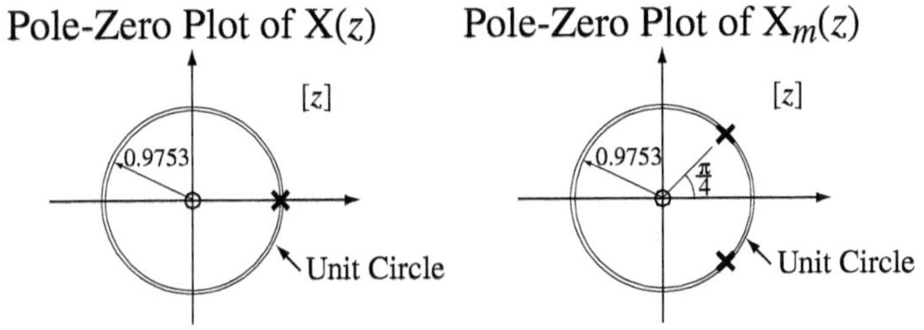

Pole-Zero Plot of X(z)   Pole-Zero Plot of $X_m(z)$

Figure 9.13
Pole-zero plots of X(z) and $X_m(z)$

EXAMPLE 9.5

z transform using the differentiation property

Using the z-domain differentiation property, show that the z transform of $n\,\mathrm{u}[n]$ is $\frac{z}{(z-1)^2}$, $|z| > 1$.
Start with

$$\mathrm{u}[n] \xleftrightarrow{\mathcal{Z}} \frac{z}{z-1}, \quad |z| > 1.$$

Then, using the z-domain differentiation property,

$$-n\,\mathrm{u}[n] \xleftrightarrow{\mathcal{Z}} z\frac{d}{dz}\left(\frac{z}{z-1}\right) = -\frac{z}{(z-1)^2}, \quad |z| > 1$$

or

$$n\,\mathrm{u}[n] \xleftrightarrow{\mathcal{Z}} \frac{z}{(z-1)^2}, \quad |z| > 1.$$

EXAMPLE 9.6

z transform using the accumulation property

Using the accumulation property, show that the z transform of $n\,\mathrm{u}[n]$ is $\frac{z}{(z-1)^2}$, $|z| > 1$.
First express $n\,\mathrm{u}[n]$ as an accumulation

$$n\,\mathrm{u}[n] = \sum_{m=0}^{n} \mathrm{u}[m-1].$$

Then, using the time-shifting property, find the z transform of $\mathrm{u}[n-1]$,

$$\mathrm{u}[n-1] \xleftrightarrow{\mathcal{Z}} z^{-1}\frac{z}{z-1} = \frac{1}{z-1}, \quad |z| > 1.$$

Then, applying the accumulation property,

$$n\,\mathrm{u}[n] = \sum_{m=0}^{n} \mathrm{u}[m-1] \xleftrightarrow{\mathcal{Z}} \left(\frac{z}{z-1}\right)\frac{1}{z-1} = \frac{z}{(z-1)^2}, \quad |z| > 1.$$

As was true for the Laplace transform, the final value theorem applies *if* the limit $\lim_{n\to\infty} g[n]$ exists. The limit $\lim_{z\to 1}(z-1)G(z)$ may exist even if the limit $\lim_{n\to\infty} g[n]$ does not. For example, if

$$X(z) = \frac{z}{z-2}, \quad |z| > 2$$

then

$$\lim_{z\to 1}(z-1)X(z) = \lim_{z\to 1}(z-1)\frac{z}{z-2} = 0.$$

But $\mathrm{x}[n] = 2^n\mathrm{u}[n]$ and the limit $\lim_{n\to\infty} \mathrm{x}[n]$ does not exist. Therefore, the conclusion that the final value is zero is wrong.

In a manner similar to the analogous proof for Laplace transforms, the following can be shown:

> For the final value theorem to apply to a function $G(z)$, all the finite poles of the function $(z-1)G(z)$ must lie in the open interior of the unit circle of the z plane.

EXAMPLE 9.7

z transform of an anticausal signal

Find the z transform of $x[n] = 4(-0.3)^{-n}u[-n]$.

Using $-\alpha^n u[-n-1] \xleftrightarrow{\mathcal{Z}} \dfrac{z}{z-\alpha} = \dfrac{1}{1-\alpha z^{-1}}$, $|z| < |\alpha|$

Identify α as -0.3^{-1}. Then

$$-(-0.3^{-1})^n u[-n-1] \xleftrightarrow{\mathcal{Z}} \dfrac{z}{z+0.3^{-1}}, \quad |z| < |-0.3^{-1}|$$

$$-(-10/3)^n u[-n-1] \xleftrightarrow{\mathcal{Z}} \dfrac{z}{z+10/3}, \quad |z| < |10/3|.$$

Use the time-shifting property.

$$-(-10/3)^{n-1} u[-(n-1)-1] \xleftrightarrow{\mathcal{Z}} z^{-1}\dfrac{z}{z+10/3} = \dfrac{1}{z+10/3}, \quad |z| < |10/3|$$

$$-(-3/10)(-10/3)^n u[-n] \xleftrightarrow{\mathcal{Z}} \dfrac{1}{z+10/3}, \quad |z| < |10/3|$$

$$(3/10)(-10/3)^n u[-n] \xleftrightarrow{\mathcal{Z}} \dfrac{1}{z+10/3}, \quad |z| < |10/3|$$

Using the linearity property, multiply both sides by $4/(3/10)$ or $40/3$.

$$4(-0.3)^{-n}u[-n] \xleftrightarrow{\mathcal{Z}} \dfrac{40/3}{z+10/3} = \dfrac{40}{3z+10}, \quad |z| < |10/3|$$

9.12 THE UNILATERAL z TRANSFORM

The unilateral Laplace transform proved convenient for continuous-time functions and the unilateral z transform is convenient for discrete-time functions for the same reasons. We can define a unilateral z transform, which is only valid for functions that are zero before discrete time $n = 0$ and avoid, in most practical problems, any complicated consideration of the region of convergence.

The unilateral z transform is defined by

$$\boxed{X(z) = \sum_{n=0}^{\infty} x[n] z^{-n}}. \tag{9.6}$$

The region of convergence of the unilateral z transform is always the open exterior of a circle, centered at the origin of the z plane whose radius is the largest finite pole magnitude.

PROPERTIES UNIQUE TO THE UNILATERAL z TRANSFORM

The properties of the unilateral z transform are very similar to the properties of the bilateral z transform. The time-shifting property is a little different. Let $g[n] = 0$, $n < 0$. Then, for the unilateral z transform,

$$g[n-n_0] \xleftrightarrow{\mathcal{Z}} \begin{cases} z^{-n_0} G(z), & n_0 \geq 0 \\ z^{-n_0}\left\{ G(z) - \sum_{m=0}^{-(n_0+1)} g[m] z^{-m} \right\}, & n_0 < 0. \end{cases}$$

This property must be different for shifts to the left because when a causal function is shifted to the left, some nonzero values may no longer lie in the summation range of the unilateral z transform, which begins at $n = 0$. The extra terms

$$-\sum_{m=0}^{-(n_0+1)} g[m]z^{-m}$$

account for any function values that are shifted into the $n < 0$ range.

The accumulation property for the unilateral z transform is

$$\sum_{m=0}^{n} g[m] \xleftrightarrow{\mathcal{Z}} \frac{z}{z-1} G(z).$$

Only the lower summation limit has changed. Actually the bilateral form

$$\sum_{m=-\infty}^{n} g[m] \xleftrightarrow{\mathcal{Z}} \frac{z}{z-1} G(z)$$

would still work because, for a causal signal $g[n]$,

$$\sum_{m=-\infty}^{n} g[m] = \sum_{m=0}^{n} g[m].$$

The unilateral z transform of any causal signal is exactly the same as the bilateral z transform of that signal. So the table of bilateral z transforms can be used as a table of unilateral z transforms.

SOLUTION OF DIFFERENCE EQUATIONS

One way of looking at the z transform is that it bears a relationship to difference equations analogous to the relationship of the Laplace transform to differential equations. A linear difference equation with initial conditions can be converted by the z transform into an algebraic equation. Then it is solved and the solution in the time domain is found by an inverse z transform.

EXAMPLE 9.8

Solution of a difference equation with initial conditions using the z transform

Solve the difference equation

$$y[n+2] - (3/2)y[n+1] + (1/2)y[n] = (1/4)^n, \quad \text{for } n \geq 0$$

with initial conditions $y[0] = 10$ and $y[1] = 4$.

Initial conditions for a second-order differential equation usually consist of a specification of the initial value of the function and its first derivative. Initial conditions for a second-order difference equation usually consist of the specification of the initial two values of the function (in this case $y[0]$ and $y[1]$).

Taking the z transform of both sides of the difference equation (using the time-shifting property of the z transform),

$$z^2(Y(z) - y[0] - z^{-1}y[1]) - (3/2)z(Y(z) - y[0]) + (1/2)Y(z) = \frac{z}{z - 1/4}$$

Solving for $Y(z)$,

$$Y(z) = \frac{\frac{z}{z - 1/4} + z^2 y[0] + z y[1] - (3/2)z y[0]}{z^2 - (3/2)z + 1/2}$$

$$Y(z) = z\frac{z^2 y[0] - z(7y[0]/4 - y[1]) - y[1]/4 + 3y[0]/8 + 1}{(z - 1/4)(z^2 - (3/2)z + 1/2)}$$

Substituting in the numerical values of the initial conditions,

$$Y(z) = z\frac{10z^2 - (27/2)z + 15/4}{(z - 1/4)(z - 1/2)(z - 1)}$$

Dividing both sides by z,

$$\frac{Y(z)}{z} = \frac{10z^2 - (27/2)z + 15/4}{(z - 1/4)(z - 1/2)(z - 1)}.$$

This is a proper fraction in z and can therefore be expanded in partial fractions as

$$\frac{Y(z)}{z} = \frac{16/3}{z - 1/4} + \frac{4}{z - 1/2} + \frac{2/3}{z - 1} \Rightarrow Y(z) = \frac{16z/3}{z - 1/4} + \frac{4z}{z - 1/2} + \frac{2z/3}{z - 1}.$$

Then using

$$\alpha^n u[n] \xleftrightarrow{\mathcal{Z}} \frac{z}{z - \alpha}$$

and taking the inverse z transform, $y[n] = [5.333(0.25)^n + 4(0.5)^n + 0.667]u[n]$. Evaluating this expression for $n = 0$ and $n = 1$ yields

$$y[0] = 5.333(0.25)^0 + 4(0.5)^0 + 0.667 = 10$$

$$y[1] = 5.333(0.25)^1 + 4(0.5)^1 + 0.667 = 1.333 + 2 + 0.667 = 4$$

which agree with the initial conditions. Substituting the solution into the difference equation,

$$\begin{cases} 5.333(0.25)^{n+2} + 4(0.5)^{n+2} + 0.667 \\ -1.5[5.333(0.25)^{n+1} + 4(0.5)^{n+1} + 0.667] \\ +0.5[5.333(0.25)^n + 4(0.5)^n + 0.667] \end{cases} = (0.25)^n, \text{ for } n \geq 0$$

or

$$0.333(0.25)^n + (0.5)^n + 0.667 - 2(0.25)^n - 3(0.5)^n - 1 + 2.667(0.25)^n$$

$$+ 2(0.5)^n + 0.333 = (0.25)^n, \text{ for } n \geq 0$$

or

$$(0.25)^n = (0.25)^n, \text{ for } n \geq 0$$

which proves that the solution does indeed solve the difference equation. ∎

9.13 POLE-ZERO DIAGRAMS AND FREQUENCY RESPONSE

To examine the frequency response of discrete-time systems we can specialize the z transform to the DTFT through the transformation $z \to e^{j\Omega}$ with Ω being a real variable representing discrete-time radian frequency. The fact that Ω is real means that in determining frequency response the only values of z that we are now considering are those on the unit circle in the z plane because $|e^{j\Omega}| = 1$ for any real Ω. This is directly analogous to determining the frequency response of a continuous-time system by examining the behavior of its s-domain transfer function as s moves along the ω axis in the s-plane, and a similar graphical technique can be used.

Suppose the transfer function of a system is

$$H(z) = \frac{z}{z^2 - z/2 + 5/16} = \frac{z}{(z - p_1)(z - p_2)}$$

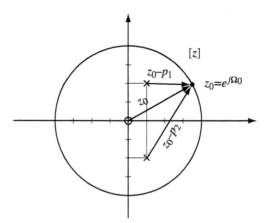

Figure 9.14
z-domain pole-zero diagram of a system transfer function

where

$$p_1 = \frac{1+j2}{4} \quad \text{and} \quad p_2 = \frac{1-j2}{4}.$$

The transfer function has a zero at zero and two complex-conjugate finite poles (Figure 9.14).

The frequency response of the system at any particular radian frequency Ω_0 is determined (to within a multiplicative constant) by the vectors from the finite poles and finite zeros of the transfer function to the z-plane point $z_0 = e^{j\Omega_0}$. The magnitude of the frequency response is the product of the magnitudes of the zero vectors divided by the product of the magnitudes of the pole vectors. In this case,

$$\left|\mathrm{H}(e^{j\Omega})\right| = \frac{\left|e^{j\Omega}\right|}{\left|e^{j\Omega} - p_1\right|\left|e^{j\Omega} - p_2\right|}. \tag{9.7}$$

It is apparent that as $e^{j\Omega}$ approaches a pole, p_1 for example, the magnitude of the difference $e^{j\Omega} - p_1$ becomes small, making the magnitude of the denominator small and tending to make the magnitude of the transfer function larger. The opposite effect occurs when $e^{j\Omega}$ approaches a zero.

The phase of the frequency response is the sum of the angles of the zero vectors minus the sum of the angles of the pole vectors. In this case, $\sphericalangle\mathrm{H}(e^{j\Omega}) = \sphericalangle e^{j\Omega} - \sphericalangle(e^{j\Omega} - p_1) - \sphericalangle(e^{j\Omega} - p_2)$ (Figure 9.15).

The maximum magnitude frequency response occurs at approximately $z = e^{\pm j1.11}$, which are the points on the unit circle at the same angle as the finite poles of the transfer function and, therefore, the points on the unit circle at which the denominator factors $e^{j\Omega_0} - p_1$ and $e^{j\Omega_0} - p_2$ in (9.7) reach their minimum magnitudes.

An important difference between the frequency response of continuous-time and discrete-time systems is that, for discrete-time systems, the frequency response is always periodic, with period 2π in Ω. That difference can be seen directly in this graphical technique because as Ω moves from zero in a positive direction, it traverses the entire unit circle in a counterclockwise direction and then, on its second traversal of the unit circle, retraces its previous positions, repeating the same frequency responses found on the first traversal.

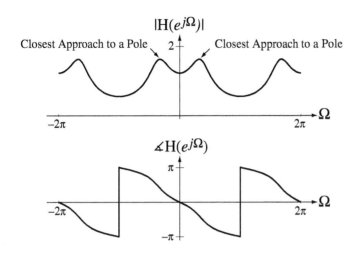

Figure 9.15
Magnitude and phase frequency response of the system whose transfer function is $H(z) = \dfrac{z}{z^2 - z/2 + 5/16}$

EXAMPLE 9.9

Pole-zero plot and frequency response from a transfer function 1

Draw the pole-zero plot and graph the frequency response for the system whose transfer function is

$$H(z) = \frac{z^2 - 0.96z + 0.9028}{z^2 - 1.56z + 0.8109}.$$

The transfer function can be factored into

$$H(z) = \frac{(z - 0.48 + j0.82)(z - 0.48 - j0.82)}{(z - 0.78 + j0.45)(z - 0.78 - j0.45)}.$$

The pole-zero diagram is in Figure 9.16.

The magnitude and phase frequency responses of the system are illustrated in Figure 9.17.

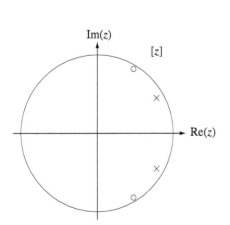

Figure 9.16
Pole-zero diagram of the transfer function,
$H(z) = \dfrac{z^2 - 0.96z + 0.9028}{z^2 - 1.56z + 0.8109}$

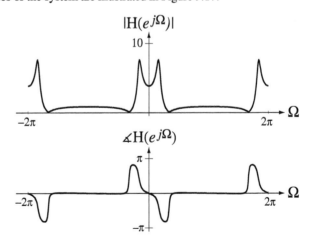

Figure 9.17
Magnitude and phase frequency response of the system whose transfer function is $H(z) = \dfrac{z^2 - 0.96z + 0.9028}{z^2 - 1.56z + 0.8109}$

EXAMPLE 9.10

Pole-zero plot and frequency response from a transfer function 2

Draw the pole-zero plot and graph the frequency response for the system whose transfer function is

$$H(z) = \frac{0.0686}{(z^2 - 1.087z + 0.3132)(z^2 - 1.315z + 0.6182)}.$$

This transfer function can be factored into

$$H(z) = \frac{0.0686}{(z - 0.5435 + j0.1333)(z - 0.5435 - j0.1333)(z - 0.6575 + j0.4312)(z - 0.6575 - j0.4312)}.$$

The pole-zero diagram is illustrated in Figure 9.18. The magnitude and phase frequency responses of the system are illustrated in Figure 9.19.

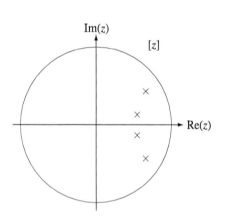

Figure 9.18
Pole-zero diagram for the transfer function,
$$H(z) = \frac{0.0686}{(z^2 - 1.087z + 0.3132)(z^2 - 1.315z + 0.6182)}$$

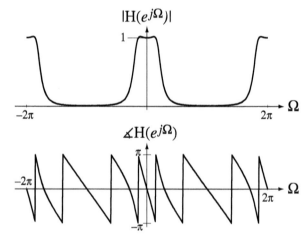

Figure 9.19
Magnitude and phase frequency response of the system whose transfer function is $H(z) = \dfrac{0.0686}{(z^2 - 1.087z + 0.3132)(z^2 - 1.315z + 0.6182)}$

9.14 MATLAB SYSTEM OBJECTS

Discrete-time system objects can be created and used in much the same way as continuous-time system objects. The syntax for creating a system object with `tf` is almost the same

```
sys = tf(num,den,Ts)
```

but with the extra argument `Ts`, the time between samples, on the assumption that the discrete-time signals were created by sampling continuous-time signals. For example, let the transfer function be

$$H_1(z) = \frac{z^2(z - 0.8)}{(z + 0.3)(z^2 - 1.4z + 0.2)} = \frac{z^3 - 0.8z^2}{z^3 - 1.1z^2 - 0.22z + 0.06}.$$

IN MATLAB

```
»num = [1 -0.8 0 0] ;
»den = [1 -1.1 -0.22 0.06] ;
»Ts = 0.008 ;
»H1 = tf(num,den,Ts) ;
»H1
Transfer function:
 z^3 - 0.8 z^2
-----------------------------
z^3 - 1.1 z^2 - 0.22 z + 0.06
Sampling time: 0.008
```

We can also use zpk.

```
»z = [0.4] ;
»p = [0.7 -0.6] ;
»k = 3 ;
»H2 = zpk(z,p,k,Ts) ;
»H2
Zero/pole/gain:
 3 (z-0.4)
---------------
(z-0.7) (z+0.6)
Sampling time: 0.008
```

We can also define z as the independent variable of the z transform with the command.

```
»z = tf('z',Ts) ;
»H3 = 7*z/(z^2+0.2*z+0.8) ;
»H3

Transfer function:
 7 z
-----------------
z^2 + 0.2 z + 0.8
Sampling time: 0.008
```

We are not required to specify the sampling time.

```
>> z = tf('z');
>> H3 = 7*z/(z^2+0.2*z+0.8);
>> H3
Transfer function:
       7 z
-----------------
z^2 + 0.2 z + 0.8
Sampling time: unspecified
```

The command

$$H = \text{freqz(num,den,W)} ;$$

accepts the two vectors num and den and interprets them as the coefficients of the powers of z in the numerator and denominator of the transfer function H(z). It returns in H the complex frequency response at the discrete-time radian frequencies in the vector W.

9.15 TRANSFORM METHOD COMPARISONS

Each type of transform has a niche in signal and system analysis where it is particularly convenient. If we want to find the total response of a discrete-time system to a causal or anticausal excitation, we would probably use the z transform. If we are interested in the frequency response of a system, the DTFT is convenient. If we want to find the forced response of a system to a periodic excitation, we might use the DTFT or the discrete Fourier transform, depending on the type of analysis needed and the form in which the excitation is known (analytically or numerically).

EXAMPLE 9.11

Total system response using the z transform and the DTFT

A system with transfer function $H(z) = \dfrac{z}{(z-0.3)(z+0.8)}$, $|z| > 0.8$ is excited by a unit sequence. Find the total response.

The z transform of the response is

$$Y(z) = H(z)X(z) = \frac{z}{(z-0.3)(z+0.8)} \times \frac{z}{z-1}, \quad |z| > 1.$$

Expanding in partial fractions,

$$Y(z) = \frac{z^2}{(z-0.3)(z+0.8)(z-1)} = -\frac{0.1169}{z-0.3} + \frac{0.3232}{z+0.8} + \frac{0.7937}{z-1}, \quad |z| > 1.$$

Therefore, the total response is

$$y[n] = [-0.1169(0.3)^{n-1} + 0.3232(-0.8)^{n-1} + 0.7937]u[n-1].$$

This problem can also be analyzed using the DTFT but the notation is significantly clumsier, mainly because the DTFT of a unit sequence is

$$\frac{1}{1-e^{-j\Omega}} + \pi\delta_{2\pi}(\Omega).$$

The system frequency response is

$$H(e^{j\Omega}) = \frac{e^{j\Omega}}{(e^{j\Omega}-0.3)(e^{j\Omega}+0.8)}.$$

The DTFT of the system response is

$$Y(e^{j\Omega}) = H(e^{j\Omega})X(e^{j\Omega}) = \frac{e^{j\Omega}}{(e^{j\Omega}-0.3)(e^{j\Omega}+0.8)} \times \left(\frac{1}{1-e^{-j\Omega}} + \pi\delta_{2\pi}(\Omega)\right)$$

or

$$Y(e^{j\Omega}) = \frac{e^{j2\Omega}}{(e^{j\Omega}-0.3)(e^{j\Omega}+0.8)(e^{j\Omega}-1)} + \pi\frac{e^{j\Omega}}{(e^{j\Omega}-0.3)(e^{j\Omega}+0.8)}\delta_{2\pi}(\Omega).$$

Expanding in partial fractions

$$Y(e^{j\Omega}) = \frac{-0.1169}{e^{j\Omega}-0.3} + \frac{0.3232}{e^{j\Omega}+0.8} + \frac{0.7937}{e^{j\Omega}-1} + \frac{\pi}{(1-0.3)(1+0.8)}\delta_{2\pi}(\Omega).$$

Using the equivalence property of the impulse and the periodicity of both $\delta_{2\pi}(\Omega)$ and $e^{j\Omega}$

$$Y(e^{j\Omega}) = \frac{-0.1169 e^{-j\Omega}}{1 - 0.3 e^{-j\Omega}} + \frac{0.3232 e^{-j\Omega}}{1 + 0.8 e^{-j\Omega}} + \frac{0.7937 e^{-j\Omega}}{1 - e^{-j\Omega}} + 2.4933 \delta_{2\pi}(\Omega).$$

Then, manipulating this expression into a form for which the inverse DTFT is direct

$$Y(e^{j\Omega}) = \frac{-0.1169 e^{-j\Omega}}{1 - 0.3 e^{-j\Omega}} + \frac{0.3232 e^{-j\Omega}}{1 + 0.8 e^{-j\Omega}} + 0.7937 \left(\frac{e^{-j\Omega}}{1 - e^{-j\Omega}} + \pi \delta_{2\pi}(\Omega) \right)$$

$$\underbrace{-0.7937\pi\delta_{2\pi}(\Omega) + 2.4933\delta_{2\pi}(\Omega)}_{=0}$$

$$Y(e^{j\Omega}) = \frac{-0.1169 e^{-j\Omega}}{1 - 0.3 e^{-j\Omega}} + \frac{0.3232 e^{-j\Omega}}{1 + 0.8 e^{-j\Omega}} + 0.7937 \left(\frac{e^{-j\Omega}}{1 - e^{-j\Omega}} + \pi \delta_{2\pi}(\Omega) \right).$$

And, finally, taking the inverse DTFT

$$y[n] = [-0.1169 (0.3)^{n-1} + 0.3232 (-0.8)^{n-1} + 0.7937] u[n-1].$$

The result is the same but the effort and the probability of error are considerably greater.

■

EXAMPLE 9.12

System response to a sinusoid

A system with transfer function $H(z) = \dfrac{z}{z - 0.9}$, $|z| > 0.9$ is excited by the sinusoid $x[n] = \cos(2\pi n/12)$. Find the response.

The excitation is the pure sinusoid $x[n] = \cos(2\pi n/12)$, not the *causal* sinusoid $x[n] = \cos(2\pi n/12) u[n]$. Pure sinusoids do not appear in the table of z transforms. Since the excitation is a pure sinusoid, we are finding the forced response of the system and we can use the DTFT pairs

$$\cos(\Omega_0 n) \xleftrightarrow{\mathcal{F}} \pi[\delta_{2\pi}(\Omega - \Omega_0) + \delta_{2\pi}(\Omega + \Omega_0)]$$

and

$$\delta_{N_0}[n] \xleftrightarrow{\mathcal{F}} (2\pi/N_0) \delta_{2\pi/N_0}(\Omega)$$

and the duality of multiplication and convolution

$$x[n] * y[n] \xleftrightarrow{\mathcal{F}} X(e^{j\Omega}) Y(e^{j\Omega}).$$

The DTFT of the response of the system is

$$Y(e^{j\Omega}) = \frac{e^{j\Omega}}{e^{j\Omega} - 0.9} \times \pi[\delta_{2\pi}(\Omega - \pi/6) + \delta_{2\pi}(\Omega + \pi/6)]$$

$$Y(e^{j\Omega}) = \pi \left[e^{j\Omega} \frac{\delta_{2\pi}(\Omega - \pi/6)}{e^{j\Omega} - 0.9} + e^{j\Omega} \frac{\delta_{2\pi}(\Omega + \pi/6)}{e^{j\Omega} - 0.9} \right].$$

Using the equivalence property of the impulse and the fact that both $e^{j\Omega}$ and $\delta_{2\pi}(\Omega)$ have a fundamental period of 2π

$$Y(e^{j\Omega}) = \pi \left[e^{j\pi/6} \frac{\delta_{2\pi}(\Omega - \pi/6)}{e^{j\pi/6} - 0.9} + e^{-j\pi/6} \frac{\delta_{2\pi}(\Omega + \pi/6)}{e^{-j\pi/6} - 0.9} \right].$$

Finding a common denominator and simplifying,

$$Y(e^{j\Omega}) = \pi \frac{\delta_{2\pi}(\Omega - \pi/6)(1 - 0.9e^{j\pi/6}) + \delta_{2\pi}(\Omega + \pi/6)(1 - 0.9e^{-j\pi/6})}{1.81 - 1.8\cos(\pi/6)}$$

$$Y(e^{j\Omega}) = \pi \frac{0.2206\,[\delta_{2\pi}(\Omega - \pi/6) + \delta_{2\pi}(\Omega + \pi/6)] + j0.45[\delta_{2\pi}(\Omega + \pi/6) - \delta_{2\pi}(\Omega - \pi/6)]}{0.2512}$$

$$Y(e^{j\Omega}) = 2.7589[\delta_{2\pi}(\Omega - \pi/6) + \delta_{2\pi}(\Omega + \pi/6)] + j5.6278\,[\delta_{2\pi}(\Omega + \pi/6) - \delta_{2\pi}(\Omega - \pi/6)].$$

Recognizing the DTFTs of a cosine and a sine,

$$y[n] = 0.8782\cos(2\pi n/12) + 1.7914\sin(2\pi n/12).$$

Using $A\cos(x) + B\sin(x) = \sqrt{A^2 + B^2}\cos(x - \tan^{-1}(B/A))$

$$y[n] = 1.995\cos(2\pi n/12 - 1.115).$$

We did not use the z transform because there is no entry in the table of z-transform pairs for a sinusoid. But there is an entry for a sinusoid multiplied by a unit sequence.

$$\cos(\Omega_0 n)\,\mathrm{u}[n] \xleftrightarrow{\mathcal{Z}} \frac{z[z - \cos(\Omega_0)]}{z^2 - 2z\cos(\Omega_0) + 1},\quad |z| > 1$$

It is instructive to find the response of the system to this different, but similar, excitation. The transfer function is

$$\mathrm{H}(z) = \frac{z}{z - 0.9},\quad |z| > 0.9.$$

The z transform of the response is

$$Y(z) = \frac{z}{z - 0.9} \times \frac{z[z - \cos(\pi/6)]}{z^2 - 2z\cos(\pi/6) + 1},\quad |z| > 1.$$

Expanding in partial fractions,

$$Y(z) = \frac{0.1217z}{z - 0.9} + \frac{0.8783z^2 + 0.1353z}{z^2 - 1.732z + 1},\quad |z| > 1.$$

To find the inverse z transform we need to manipulate the expressions into forms similar to the table entries. The first fraction form appears directly in the table. The second fraction has a denominator of the same form as the z transforms of $\cos(\Omega_0 n)\,\mathrm{u}[n]$ and $\sin(\Omega_0 n)\,\mathrm{u}[n]$ but the numerator is not in exactly the right form. But by adding and subtracting the right amounts in the numerator we can express $Y(z)$ in the form

$$Y(z) = \frac{0.1217}{z - 0.9} + 0.8783\left[\frac{z(z - 0.866)}{z^2 - 1.732z + 1} + 2.04\frac{0.5z}{z^2 - 1.732z + 1}\right],\quad |z| > 1$$

$$y[n] = 0.1217(0.9)^n\,\mathrm{u}[n] + 0.8783[\cos(2\pi n/12) + 2.04\sin(2\pi n/12)]\,\mathrm{u}[n]$$

$$y[n] = 0.1217(0.9)^n\,\mathrm{u}[n] + 1.995\cos(2\pi n/12 - 1.115)\,\mathrm{u}[n].$$

Notice that the response consists of two parts, a transient response $0.1217(0.9)^n\,\mathrm{u}[n]$ and a forced response $1.995\cos(2\pi n/12 - 1.115)\,\mathrm{u}[n]$ that, except for the unit sequence factor, is exactly the same as the forced response we found using the DTFT. So, even though we do not have a z transform of a sinusoid in the z transform table we can use the z transforms of $\cos(\Omega_0 n)\,\mathrm{u}[n]$ and $\sin(\Omega_0 n)\,\mathrm{u}[n]$ to find the forced response to a sinusoid.

The analysis in Example 9.12, a system excited by a sinusoid, is very common in some types of signal and system analysis. It is important enough to generalize the process. If the transfer function of the system is

$$H(z) = \frac{N(z)}{D(z)},$$

the response of the system to $\cos(\Omega_0 n)\,u[n]$ is

$$Y(z) = \frac{N(z)}{D(z)} \frac{z[z - \cos(\Omega_0)]}{z^2 - 2z\cos(\Omega_0) + 1}.$$

The poles of this response are the poles of the transfer function plus the roots of $z^2 - 2z\cos(\Omega_0) + 1 = 0$, which are the complex conjugate pair $p_1 = e^{j\Omega_0}$ and $p_2 = e^{-j\Omega_0}$. Therefore, $p_1 = p_2^*$, $p_1 + p_2 = 2\cos(\Omega_0)$, $p_1 - p_2 = j2\sin(\Omega_0)$, and $p_1 p_2 = 1$. Then if $\Omega_0 \neq m\pi$, m an integer and, if there is no pole-zero cancellation, these poles are distinct. The response can be written in partial-fraction form as

$$Y(z) = z\left[\frac{N_1(z)}{D(z)} + \frac{1}{p_1 - p_2}\frac{H(p_1)(p_1 - \cos(\Omega_0))}{z - p_1} + \frac{1}{p_2 - p_1}\frac{H(p_2)(p_2 - \cos(\Omega_0))}{z - p_2}\right]$$

or, after simplification,

$$Y(z) = z\left[\left\{\frac{N_1(z)}{D(z)} + \left[\frac{H_r(p_1)(z - p_{1r}) - H_i(p_1)p_{1i}}{z^2 - z(2p_{1r}) + 1}\right]\right\}\right]$$

where $p_1 = p_{1r} + jp_{1i}$ and $H(p_1) = H_r(p_1) + jH_i(p_1)$. This can be written in terms of the original parameters as

$$Y(z) = \left\{z\frac{N_1(z)}{D(z)} + \left[\begin{array}{l}\mathrm{Re}(H(\cos(\Omega_0) + j\sin(\Omega_0)))\dfrac{z^2 - z\cos(\Omega_0)}{z^2 - z(2\cos(\Omega_0)) + 1} \\ -\mathrm{Im}(H(\cos(\Omega_0) + j\sin(\Omega_0)))\dfrac{z\sin(\Omega_0)}{z^2 - z(2\cos(\Omega_0)) + 1}\end{array}\right]\right\}.$$

The inverse z transform is

$$y[n] = \mathcal{Z}^{-1}\left(z\frac{N_1(z)}{D(z)}\right) + \left[\begin{array}{l}\mathrm{Re}(H(\cos(\Omega_0) + j\sin(\Omega_0)))\cos(\Omega_0 n) \\ -\mathrm{Im}(H(\cos(\Omega_0) + j\sin(\Omega_0)))\sin(\Omega_0 n)\end{array}\right]u[n]$$

or, using

$$\mathrm{Re}(A)\cos(\Omega_0 n) - \mathrm{Im}(A)\sin(\Omega_0 n) = |A|\cos(\Omega_0 n + \angle A),$$

$$y[n] = \mathcal{Z}^{-1}\left(z\frac{N_1(z)}{D(z)}\right) + |H(\cos(\Omega_0) + j\sin(\Omega_0))|\cos(\Omega_0 n +$$

$$\angle H(\cos(\Omega_0) + j\sin(\Omega_0)))\,u[n]$$

or finally

$$\boxed{y[n] = \mathcal{Z}^{-1}\left(z\frac{N_1(z)}{D(z)}\right) + |H(p_1)|\cos(\Omega_0 n + \angle H(p_1))u[n]}.\qquad(9.8)$$

If the system is stable, the term

$$Z^{-1}\left(z\frac{N_1(z)}{D(z)}\right)$$

(the natural or transient response) decays to zero with discrete time and the term $|H(p_1)|\cos(\Omega_0 n + \angle H(p_1))u[n]$ is equal to a sinusoid after discrete time $n = 0$ and persists forever.

Using this result we could now solve the problem in Example 9.12 much more quickly. The response to $x[n] = \cos(2\pi n/12)u[n]$ is

$$y[n] = Z^{-1}\left(z\frac{N_1(z)}{D(z)}\right) + |H(p_1)|\cos(\Omega_0 n + \angle H(p_1))u[n]$$

and the response to $x[n] = \cos(2\pi n/12)$ is

$$y_f[n] = |H(p_1)|\cos(\Omega_0 n + \angle H(p_1))$$

where $H(z) = \dfrac{z}{z - 0.9}$ and $p_1 = e^{j\pi/6}$. Therefore,

$$H(e^{j\pi/6}) = \frac{e^{j\pi/6}}{e^{j\pi/6} - 0.9} = 0.8783 - j1.7917 = 1.995\angle -1.115$$

and

$$y_f[n] = 1.995\cos(\Omega_0 n - 0.115).$$

9.16 SUMMARY OF IMPORTANT POINTS

1. The z transform can be used to determine the transfer function of a discrete-time LTI system and the transfer function can be used to find the response of a discrete-time LTI system to an arbitrary excitation.
2. The z transform exists for discrete-time signals whose magnitudes do not grow any faster than an exponential in either positive or negative time.
3. The region of convergence of the z transform of a signal depends on whether the signal is right or left sided.
4. Systems described by ordinary, linear, constant-coefficient difference equations have transfer functions in the form of a ratio of polynomials in z and the systems can be realized directly from the transfer function.
5. With a table of z transform pairs and z-transform properties the forward and inverse transforms of almost any signal of engineering significance can be found.
6. The unilateral z transform is commonly used in practical problem solving because it does not require any involved consideration of the region of convergence and is, therefore, simpler than the bilateral form.
7. Pole-zero diagrams of a system's transfer function encapsulate most of its properties and can be used to determine its frequency response.
8. MATLAB has an object defined to represent a discrete-time system transfer function and many functions to operate on objects of this type.

EXERCISES WITH ANSWERS

Direct-Form II System Realization

1. Draw a Direct Form II block diagram for each of these system transfer functions:

   (a) $H(z) = \dfrac{z(z-1)}{z^2 + 1.5z + 0.8}$

   (b) $H(z) = \dfrac{z^2 - 2z + 4}{(z - 1/2)(2z^2 + z + 1)}$

   Answers:

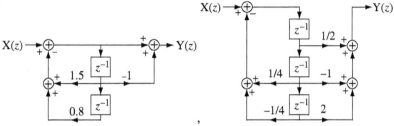

Existence of the z Transform

2. Find the region of convergence (if it exists) in the z plane, of the z transform of these signals:

   (a) $x[n] = u[n] + u[-n]$

   (b) $x[n] = u[n] - u[n - 10]$

   (c) $x[n] = 4n\,u[n + 1]$

   (Hint: Express the time-domain function as the sum of a causal function and an anticausal function, combine the z-transform results over a common denominator, and simplify.)

   (d) $x[n] = 4n\,u[n - 1]$

   (e) $x[n] = 12(0.85)^n \cos(2\pi n/10)\,u[-n - 1] + 3(0.4)^{n+2}\,u[n + 2]$

   Answers: $|z| > 1$, $|z| > 0$, $|z| > 1$, Does Not Exist, $0.4 < |z| < 0.85$

Forward and Inverse z Transforms

3. Using the time-shifting property, find the z transforms of these signals:

   (a) $x[n] = u[n - 5]$ (b) $x[n] = u[n + 2]$

   (c) $x[n] = (2/3)^n u[n + 2]$

   Answers: $\dfrac{z^{3}}{z-1}$, $|z| > 1$, $\dfrac{9}{4}\dfrac{z^{3}}{z - 2/3}$, $|z| > 2/3$, $\dfrac{z^{-4}}{z - 1}$, $|z| > 1$

4. Using the change-of-scale property, find the z transform of $x[n] = \sin(2\pi n/32)\cos(2\pi n/8)\,u[n]$.

   Answer:

   $$\sin(2\pi n/32)\cos(2\pi n/8)u[n] \overset{\mathcal{Z}}{\longleftrightarrow} z\dfrac{0.1379 z^2 - 0.3827z + 0.1379}{z^4 - 2.7741 z^3 + 3.8478 z^2 - 2.7741 z + 1},\ |z| > 1$$

5. Using the z-domain-differentiation property, find the z transform of
$$x[n] = n(5/8)^n u[n].$$

Answer: $n(5/8)^n u[n] \xleftrightarrow{\mathcal{Z}} \dfrac{5z/8}{(z-5/8)^2},\ |z| > 5/8$

6. Using the convolution property, find the z transforms of these signals:

   (a) $x[n] = (0.9)^n u[n] * u[n]$
   (b) $x[n] = (0.9)^n u[n] * (0.6)^n u[n]$

Answers: $\dfrac{z^2}{z^2 - 1.9z + 0.9},\ |z| > 1,\ \dfrac{z^2}{z^2 - 1.5z + 0.54},\ |z| > 0.9$

7. Using the differencing property and the z transform of the unit sequence, find the z transform of the unit impulse and verify your result by checking the z-transform table.

8. Find the z transform of
$$x[n] = u[n] - u[n-10]$$

and, using that result and the differencing property, find the z transform of
$$x[n] = \delta[n] - \delta[n-10].$$

Compare this result with the z transform found directly by applying the time-shifting property to an impulse.

Answer: $1 - z^{-10}$, All z

9. Using the accumulation property, find the z transforms of these signals.

   (a) $x[n] = \text{ramp}[n]$
   (b) $x[n] = \displaystyle\sum_{m=-\infty}^{n} (u[m+5] - u[m])$

Answers: $\dfrac{z^2(z^5 - 1)}{(z-1)^2},\ |z| > 1,\ \dfrac{z}{(z-1)^2},\ |z| > 1$

10. A discrete-time signal y[n] is related to another discrete-time signal x[n] by
$$y[n] = \sum_{m=0}^{n} x[m].$$
If $y[n] \xleftrightarrow{\mathcal{Z}} \dfrac{1}{(z-1)^2}$, what are the values of x[−1], x[0], x[1], and x[2]?

Answers: 1, 0, 0, 0

11. Using the final-value theorem, find the final value of functions that are the inverse z transforms of these functions (if the theorem applies).

   (a) $X(z) = \dfrac{z}{z-1},\ |z| > 1$
   (b) $X(z) = z\dfrac{2z - 7/4}{z^2 - 7/4z + 3/4},\ |z| > 1$

   (c) $X(z) = \dfrac{z^3 + 2z^2 - 3z + 7}{(z-1)(z^2 - 1.8z + 0.9)},\ |z| > 1$

Answers: 1, 1, 70

12. A discrete-time signal x[n] has a z transform, $X(z) = \dfrac{5z^2 - z + 3}{2z^2 - \frac{1}{2}z + \frac{1}{4}}$. What is the numerical value of x[0]?

    Answer: 2.5

13. Find the inverse z transforms of these functions in series form by synthetic division.

    (a) $X(z) = \dfrac{z}{z - 1/2}$, $|z| > 1/2$

    (b) $X(z) = \dfrac{z - 1}{z^2 - 2z + 1}$, $|z| > 1$

    (c) $X(z) = \dfrac{z}{z - 1/2}$, $|z| < 1/2$

    (d) $X(z) = \dfrac{z + 2}{4z^2 - 2z + 3}$, $|z| < \sqrt{3}/2$

    Answers: $\delta[n - 1] + \delta[n - 2] + \cdots + \delta[n - k] + \cdots$,

    $\delta[n] + (1/2)\delta[n - 1] + \cdots + (1/2^k)\delta[n - k] + \cdots$,

    $0.667\delta[n] + 0.778\delta[n + 1] - 0.3704\delta[n + 2] + \cdots$,

    $-2\delta[n + 1] - 4\delta[n + 2] - 8\delta[n + 3] - \cdots - 2^k\delta[n + k] - \cdots$

14. Find the inverse z transforms of these functions in closed form using partial-fraction expansions, a z-transform table, and the properties of the z transform.

    (a) $X(z) = \dfrac{1}{z(z - 1/2)}$, $|z| > 1/2$

    (b) $X(z) = \dfrac{z^2}{(z - 1/2)(z - 3/4)}$, $|z| < 1/2$

    (c) $X(z) = \dfrac{z^2}{z^2 + 1.8z + 0.82}$, $|z| > 0.9055$

    (d) $X(z) = \dfrac{z - 1}{3z^2 - 2z + 2}$, $|z| < 0.8165$

    (e) $X(z) = 2\dfrac{z^2 - 0.1488z}{z^2 - 1.75z + 1}$

    Answers: $[2(1/2)^n - 3(3/4)^n]u[-n - 1]$, $2\{(1/2)^{n-1} - \delta[n - 1]\}u[n - 1]$,

    $(0.9055)^n[\cos(3.031n) - 9.03 \sin(3.031n)]u[n]$,

    $0.4472(0.8165)^n \left\{ \begin{array}{l} 1.2247 \sin(1.1503(n - 1))u[-n] \\ -\sin(1.1503n)u[-n - 1] \end{array} \right\}$,

    $[2\cos(0.5054n) + 3\sin(0.5054n)]u[n]$

15. A discrete-time system has a transfer function (the z transform of its impulse response) $H(z) = \dfrac{z}{z^2 + z + 0.24}$. If a unit sequence u[n] is applied as an excitation to this system, what are the numerical values of the responses y[0], y[1], and y[2]?

    Answers: 0, 0, 1

16. The z transform of a discrete-time signal x[n] is $X(z) = \dfrac{z^4}{z^4 + z^2 + 1}$. What are the numerical values of x[0], x[1], and x[2]?

    Answer: −1, 0, 1

17. If $H(z) = \dfrac{z^2}{(z - 1/2)(z + 1/3)}$, $|z| > 1/2$, then, by finding the partial-fraction expansion of this improper fraction in z two different ways, its inverse z transform, h[n] can be written in two different forms,

    $$h[n] = [A(1/2)^n + B(-1/3)^n]u[n]$$

    and

    $h[n] = \delta[n] + [C(1/2)^{n-1} + D(-1/3)^{n-1}]u[n-1]$. Find the numerical values of A, B, C and D.

    Answers: 0.6, 0.4, 0.3, −0.1333

Unilateral z-Transform Properties

18. Using the time-shifting property, find the unilateral z transforms of these signals:

    (a) $x[n] = u[n - 5]$

    (b) $x[n] = u[n + 2]$

    (c) $x[n] = (2/3)^n u[n + 2]$

    Answers: $\dfrac{z^{-4}}{z-1}$, $|z| > 1$, $\dfrac{z}{z - 2/3}$, $|z| > 2/3$, $\dfrac{z}{z-1}$, $|z| > 1$

19. If the unilateral z transform of x[n] is $X(z) = \dfrac{z}{z-1}$, what are the z transforms of x[n − 1] and x[n + 1]?

    Answers: $\dfrac{1}{z-1}, \dfrac{z}{z-1}$

20. The unilateral z transform of $x[n] = 5(0.7)^{n+1} u[n + 1]$ can be written in the form, $X(z) = A\dfrac{z}{z+a}$. Find the numerical values of A and a.

    Answers: 3.5, −0.7

Solution of Difference Equations

21. Using the z transform, find the total solutions to these difference equations with initial conditions, for discrete time, $n \geq 0$.

    (a) $2y[n + 1] - y[n] = \sin(2\pi n/16)u[n]$, $y[0] = 1$

    (b) $5y[n + 2] - 3y[n + 1] + y[n] = (0.8)^n u[n]$, $y[0] = -1$, $y[1] = 10$

    Answers: $y[n] = 0.2934(1/2)^{n-1} u[n-1] + (1/2)^n u[n]$

    $- 0.2934[\cos((\pi/8)(n-1)) - 2.812 \sin((\pi/8)(n-1))]u[n-1],$

$$y[n] = 0.4444\,(0.8)^n\,u[n]$$
$$-\left\{\delta[n] - 9.5556\,(0.4472)^{n-1}\begin{bmatrix}\cos(0.8355(n-1))\\+0.9325\sin(0.8355(n-1))\end{bmatrix}u[n-1]\right\}$$

22. For each block diagram in Figure E.22, write the difference equation and find and graph the response $y[n]$ of the system for discrete times $n \geq 0$ assuming no initial energy storage in the system and impulse excitation $x[n] = \delta[n]$.

(a)
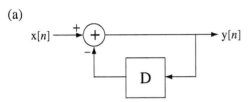

(b)
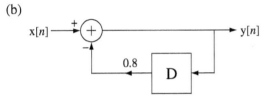

(c)
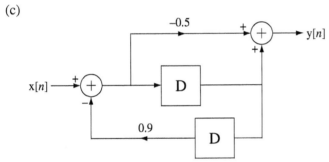

Figure E.22

Answers:
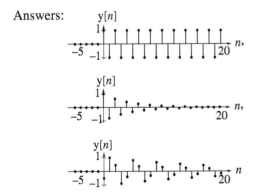

Pole-Zero Diagrams and Frequency Response

23. Sketch the magnitude frequency response of the systems in Figure E.23 from their pole-zero diagrams.

(a)
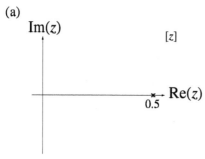

(b)
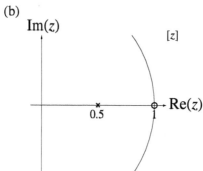

(c)
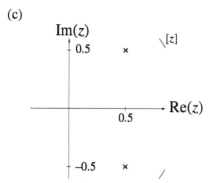

Figure E.23

Answers:

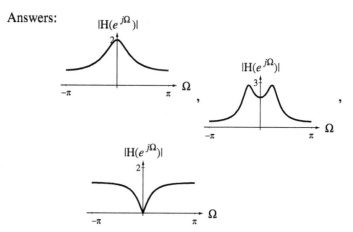

24. Where are the finite poles and zeros of $H(z) = \dfrac{z^2(z-2)}{6z^3 - 4z^2 + 3z}$?

    Answers: 2, $0.333 \pm j0.6236$, 0

EXERCISES WITHOUT ANSWERS

Direct Form II System Realization

25. Draw a Direct Form II block diagram for each of these system transfer functions:

    (a) $H(z) = \dfrac{z^2}{2z^4 + 1.2z^3 - 1.06z^2 + 0.08z - 0.02}$

    (b) $H(z) = \dfrac{z^2(z^2 + 0.8z + 0.2)}{(2z^2 + 2z + 1)(z^2 + 1.2z + 0.5)}$

Existence of the z Transform

26. Graph the region of convergence (if it exists) in the z plane, of the z transform of these discrete-time signals.

    (a) $x[n] = (1/2)^n u[n]$ (b) $x[n] = (5/4)^n u[n] + (10/7)^n u[-n]$

Forward and Inverse z-Transforms

27. Find the inverse z transform of $H(z) = \dfrac{z^2}{z + 1/2}$. Is a system with this transfer function causal? Why or why not?

28. Find the forward z transforms of these discrete-time functions:

    (a) $4\cos(2\pi n/4)u[n]$ (b) $4\cos(2\pi(n-4)/4)u[n-4]$

    (c) $4\cos(2\pi n/4)u[n-4]$ (d) $4\cos(2\pi(n-4)/4)u[n]$

29. Using the time-shifting property, find the z transforms of these signals:

    (a) $x[n] = (2/3)^{n-1} u[n-1]$ (b) $x[n] = (2/3)^n u[n-1]$

    (c) $x[n] = \sin\left(\dfrac{2\pi(n-1)}{4}\right) u[n-1]$

30. If the z transform of $x[n]$ is $X(z) = \dfrac{1}{z - 3/4}$, $|z| > 3/4$, and

    $$Y(z) = j[X(e^{j\pi/6} z) - X(e^{-j\pi/6} z)]$$

    what is $y[n]$?

31. Using the convolution property, find the z transforms of these signals.

    (a) $x[n] = \sin(2\pi n/8) u[n] * u[n]$
    (b) $x[n] = \sin(2\pi n/8) u[n] * (u[n] - u[n-8])$

32. A digital filter has an impulse response, $h[n] = \dfrac{\delta[n] + \delta[n-1] + \delta[n-2]}{10}$.

    (a) How many finite poles and finite zeros are there in its transfer function and what are their numerical locations?

    (b) If the excitation $x[n]$ of this system is a unit sequence, what is the final numerical value of the response $\lim\limits_{n \to \infty} y[n]$?

33. The forward z transform $h[n] = (4/5)^n u[n] * u[n]$ can be expressed in the general form, $H(z) = \dfrac{b_2 z^2 + b_1 z + b_0}{a_2 z^2 + a_1 z + a_0}$. Find the numerical values of b_2, b_1, b_0, a_2, a_1, and a_0.

34. Find the inverse z transforms of these functions in closed form using partial-fraction expansions, a z-transform table, and the properties of the z transform.

    (a) $X(z) = \dfrac{z-1}{z^2 + 1.8z + 0.82}, \ |z| > 0.9055$

    (b) $X(z) = \dfrac{z-1}{z(z^2 + 1.8z + 0.82)}, \ |z| > 0.9055$

    (c) $X(z) = \dfrac{z^2}{z^2 - z + 1/4}, \ |z| < 0.5$

    (d) $X(z) = \dfrac{z + 0.3}{z^2 + 0.8z + 0.16} = \dfrac{z + 0.3}{(z + 0.4)^2} = \dfrac{-0.1}{(z + 0.4)^2} + \dfrac{1}{z + 0.4}, \ |z| > 0.4$

    (e) $X(z) = \dfrac{z^2 - 0.8z + 0.3}{z^3} = z^{-1} - 0.8z^{-2} + 0.3z^{-3}, \ |z| > 0$

    (f) $X(z) = \dfrac{2z}{z^2 - z + 0.74}, \ |z| > 0.86$

    (g) $X(z) = \dfrac{z^3}{(z-1)^2 \left(z - \frac{1}{2}\right)}, \ |z| > 1$

35. The z transform of a signal $x[n]$ is $X(z) = \dfrac{z^{-4}}{z^4 + z^2 + 1}, \ |z| < 1$. What are the numerical values of $x[-2], x[-1], x[0], x[1], x[2], x[3]$, and $x[4]$?

36. Find the numerical values of the literal constants.

    (a) $10(-0.4)^n \sin(\pi n/8) u[n] \xleftrightarrow{\mathcal{Z}} \dfrac{b_2 z^2 + b_1 z + b_0}{z^2 + a_1 z + a_0}$

    (b) $A a^n \sin(bn) u[n] \xleftrightarrow{\mathcal{Z}} \dfrac{12z}{z^2 + 0.64}$

    (c) $(A a^n + B b^n) u[n] \xleftrightarrow{\mathcal{Z}} \dfrac{z(z - 0.4)}{z^2 + 1.5z + 0.3}$

    (d) $A \text{tri}((n - n_0)/b) \xleftrightarrow{\mathcal{Z}} \dfrac{z^2 + 2z + 1}{z^2}$

    (e) $(\delta[n] - 2\delta[n - 2]) * (0.7)^n u[n] \xleftrightarrow{\mathcal{Z}} A \dfrac{z + B z^{-1}}{z + C}$

    (f) $A a^n [\cos(bn) + B \sin(bn)] u[n] \xleftrightarrow{\mathcal{Z}} \dfrac{z^2}{z^2 + z + 0.8}$

    (g) $4u[n + 1] \xleftrightarrow{\mathcal{Z}} \dfrac{Az}{z - B}$

(h) $\left(\frac{4}{5}\right)^n u[n] * u[n] \xleftrightarrow{\mathcal{Z}} \frac{b_2 z^2 + b_1 z + b_0}{a_2 z^2 + a_1 z + a_0}$

(i) $\left(1 - e^{j\frac{2\pi n}{8}}\right) u[n] \xleftrightarrow{\mathcal{Z}} \frac{b_2 z^2 + b_1 z + b_0}{a_2 z^2 + a_1 z + a_0}$

(j) $A\alpha^n [\cos(\Omega_0 n) + B \sin(\Omega_0 n)] u[n] \xleftrightarrow{\mathcal{Z}} 6 \frac{z(z + 1/2)}{z^2 + 4/9}$

(k) $(2/3)^n u[n - 2] \xleftrightarrow{\mathcal{Z}} \frac{A z^b}{z - c}$

(l) $6(0.8)^{n-1} \cos(\pi(n-1)/12) u[n-1] \xleftrightarrow{\mathcal{Z}} \frac{b_2 z^2 + b_1 z + b_0}{a_2 z^2 + a_1 z + a_0}$

(m) $A b^n u[-n-1] + C d^n u[n] \xleftrightarrow{\mathcal{Z}} \frac{z^2}{(z+0.5)(z-0.2)}, \; 0.2 < |z| < 0.5$

(n) $(0.3)^n u[-n+8] \xleftrightarrow{\mathcal{Z}} A \frac{z^b}{z+a}, \; |z| < |c|$

(o) $A\alpha^{n-c} \begin{bmatrix} \cos(2\pi b(n-c)) \\ + B \sin(2\pi b(n-c)) \end{bmatrix} u[n-c] \xleftrightarrow{\mathcal{Z}} z^{-1} \frac{1.5625 z - 1}{z^2 + 0.64}, \; |z| > 0.8$

(p) $3(u[n-1] - u[n+1]) \xleftrightarrow{\mathcal{Z}} A z^a + B z^b$.

Pole-Zero Diagrams and Frequency Response

37. A digital filter has a transfer function $H(z) = 0.9525 \frac{z^2 - z + 1}{z^2 - 0.95 z + 0.9025}$.

    (a) What are the numerical locations of its poles and zeros?
    (b) Find the numerical frequency response magnitude at the radian frequencies $\Omega = 0$ and $\Omega = \pi/3$.

38. A filter has an impulse response $h[n] = \frac{\delta[n] + \delta[n-1]}{2}$. A sinusoid $x[n]$ is created by sampling, at $f_s = 10$ Hz, a continuous-time sinusoid with cyclic frequency f_0. What is the minimum positive numerical value of f_0 for which the forced filter response is zero?

39. Find the magnitude of the transfer function of the systems with the pole-zero plots in Figure E.39 at the specified frequencies. (In each case assume the transfer function is of the general form, $H(z) = K \frac{(z - z_1)(z - z_2) \cdots (z - z_N)}{(z - p_1)(z - p_2) \cdots (z - p_D)}$, where the z's are the zeros and the p's are the poles, and let $K = 1$.)

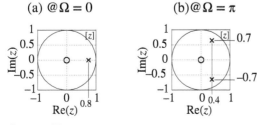

Figure E.39

40. For each of the systems with the pole-zero plots in Figure E.40, find the discrete-time radian frequencies, Ω_{max} and Ω_{min}, in the range, $-\pi \leq \Omega \leq \pi$ for which the transfer function magnitude is a maximum and a minimum. If there is more than one value of Ω_{max} or Ω_{min}, find all such values.

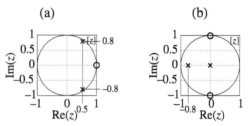

Figure E.40

41. Sketch the magnitude frequency response of the systems in Figure E.41 from their pole-zero diagrams.

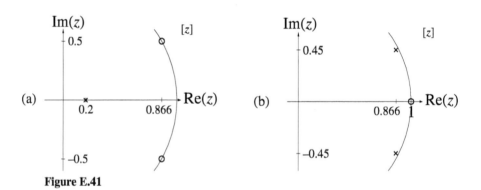

Figure E.41

42. Match the pole-zero plots in Figure E.42 to the corresponding magnitude frequency responses.

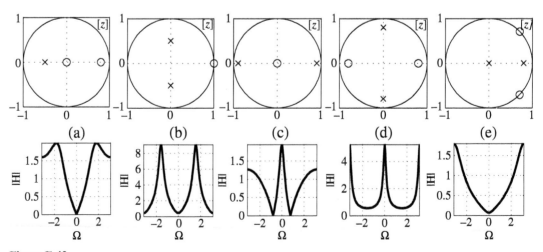

Figure E.42

43. Using the following definitions of lowpass, highpass, bandpass, and bandstop, classify the systems whose transfer functions have the pole-zero diagrams in Figure E.43. (Some may not be classifiable.) In each case the transfer function is H(z).

LP: $H(1) \neq 0$ and $H(-1) = 0$ HP: $H(1) = 0$ and $H(-1) \neq 0$
BP: $H(1) = 0$ and $H(-1) = 0$ and $H(z) \neq 0$ for some range of $|z| = 1$
BS: $H(1) \neq 0$ and $H(-1) \neq 0$ and $H(z) = 0$ for at least one $|z| = 1$

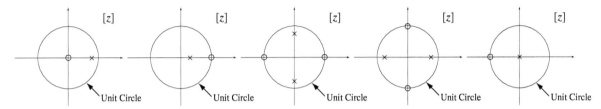

Figure E.43

44. For each magnitude frequency response and each unit sequence response in Figure E.44, find the corresponding pole-zero diagram.

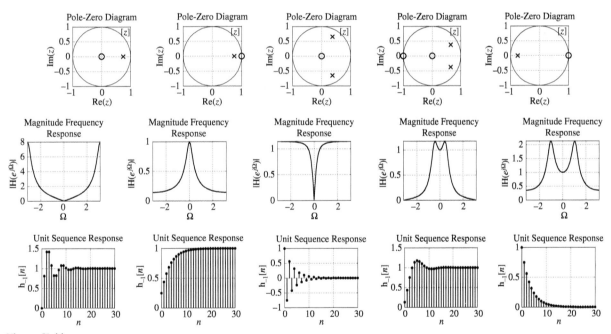

Figure E.44

10 CHAPTER

Sampling and Signal Processing

10.1 INTRODUCTION AND GOALS

In the application of signal processing to real signals in real systems, we often do not have a mathematical description of the signals. We must measure and analyze them to discover their characteristics. If the signal is unknown, the process of analysis begins with the **acquisition** of the signals, measuring and recording the signals over a period of time. This could be done with a tape recorder or other **analog** recording device but the most common technique of acquiring signals today is by **sampling**. (The term *analog* refers to continuous-time signals and systems.) Sampling converts a continuous-time signal into a discrete-time signal. In previous chapters we have explored ways of analyzing continuous-time signals and discrete-time signals. In this chapter we investigate the relationships between them.

Much signal processing and analysis today is done using **digital signal processing** (**DSP**). A DSP system can acquire, store, and perform mathematical calculations on numbers. A computer can be used as a DSP system. Since the memory and mass storage capacity of any DSP system are finite, it can only handle a finite number of numbers. Therefore, if a DSP system is to be used to analyze a signal, it can only be sampled for a finite time. The salient question addressed in this chapter is, "To what extent do the samples accurately describe the signal from which they are taken?" We will see that whether, and how much, information is lost by sampling depends on the way the samples are taken. We will find that under certain circumstances practically all of the signal information can be stored in a finite number of numerical samples.

Many filtering operations that were once done with analog filters now use digital filters, which operate on samples from a signal, instead of the original continuous-time signal. Modern cellular telephone systems use DSP to improve voice quality, separate channels, and switch users between cells. Long-distance telephone communication systems use DSP to efficiently use long trunk lines and microwave links. Television sets use DSP to improve picture quality. Robotic vision is based on signals from cameras that digitize (sample) an image and then analyze it with computation techniques to recognize features. Modern control systems in automobiles, manufacturing plants, and scientific instrumentation usually have embedded processors that analyze signals and make decisions using DSP.

CHAPTER GOALS

1. To determine how a continuous-time signal must be sampled to retain most or all of its information

2. To learn how to reconstruct a continuous-time signal from its samples
3. To apply sampling techniques to discrete-time signals and to see the similarities with continuous-time sampling

10.2 CONTINUOUS-TIME SAMPLING

SAMPLING METHODS

Sampling of electrical signals, occasionally currents but usually voltages, is most commonly done with two devices, the **sample-and-hold** (**S/H**) and the **analog-to-digital converter** (**ADC**). The excitation of the S/H is the analog voltage at its input. When the S/H is clocked, it responds with that voltage at its output and holds that voltage until it is clocked to acquire another voltage (Figure 10.1).

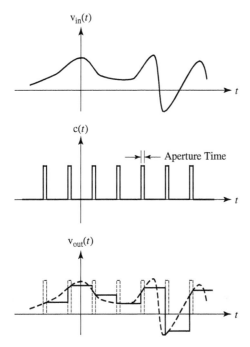

Figure 10.1
Operation of a sample-and-hold

In Figure 10.1 the signal $c(t)$ is the clock signal. The acquisition of the input voltage signal of the S/H occurs during the **aperture time**, which is the width of a clock pulse. During the clock pulse the output voltage signal very quickly moves from its previous value to track the excitation. At the end of the clock pulse the output voltage signal is held at a fixed value until the next clock pulse occurs.

An ADC accepts an analog voltage at its input and responds with a set of binary bits (often called a **code**). The ADC response can be serial or a parallel. If the ADC has a serial response, it produces on one output pin a single output voltage signal that is a timed sequence of high and low voltages representing the 1's and 0's of the set of binary bits. If the ADC has a parallel response, there is a response voltage for each bit and each bit appears simultaneously on a dedicated output pin of the ADC as a high

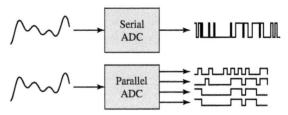

Figure 10.2
Serial and parallel ADC operation

or low voltage representing a 1 or a 0 in the set of binary bits (Figure 10.2). An ADC may be preceded by a S/H to keep its excitation constant during the conversion time.

The excitation of the ADC is a continuous-time signal and the response is a discrete-time signal. Not only is the response of the ADC discrete-time, but is also **quantized** and **encoded**. The number of binary bits produced by the ADC is finite. Therefore, the number of unique bit patterns it can produce is also finite. If the number of bits the ADC produces is n, the number of unique bit patterns it can produce is 2^n. **Quantization** is the effect of converting a continuum of (infinitely many) excitation values into a finite number of response values. Since the response has an error due to quantization, it is as though the signal has noise on it, and this noise is called **quantization noise**. If the number of bits used to represent the response is large enough, quantization noise is often negligible in comparison with other noise sources. After quantization the ADC encodes the signal also. Encoding is the conversion from an analog voltage to a binary bit pattern. The relation between the excitation and response of an ADC whose input voltage range is $-V_0 < v_{in}(t) < +V_0$ is illustrated in Figure 10.3 for a 3-bit ADC. (A 3-bit ADC is rarely, if ever, actually used, but it does illustrate the quantization effect nicely because the number of unique bit patterns is small and the quantization noise is large.)

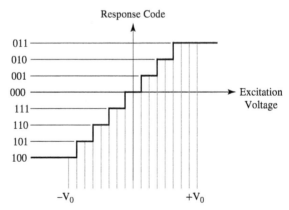

Figure 10.3
ADC excitation-response relationship

The effects of quantization are easy to see in a sinusoid quantized by a 3-bit ADC (Figure 10.4). When the signal is quantized to 8 bits, the quantization error is much smaller (Figure 10.5).

The opposite of analog-to-digital conversion is obviously digital-to-analog conversion done by a **digital-to-analog converter** (**DAC**). A DAC accepts binary bit

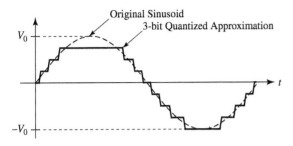
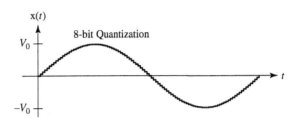

Figure 10.4
Sinusoid quantized to 3 bits

Figure 10.5
Sinusoid quantized to 8 bits

patterns as its excitation and produces an analog voltage as its response. Since the number of unique bit patterns it can accept is finite, the DAC response signal is an analog voltage that is quantized. The relation between excitation and response for a 3-bit DAC is shown in Figure 10.6.

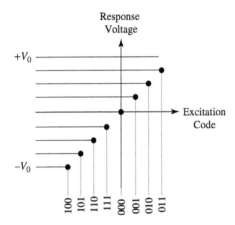

Figure 10.6
DAC excitation-response relationship

In the material to follow, the effects of quantization will not be considered. The model for analyzing the effects of sampling will be that the sampler is ideal in the sense that the response signal's quantization noise is zero.

THE SAMPLING THEOREM

Qualitative Concepts

If we are to use samples from a continuous-time signal, instead of the signal itself, the most important question to answer is how to sample the signal so as to retain the information it carries. If the signal can be exactly reconstructed from the samples, then the samples contain all the information in the signal. We must decide how fast to sample the signal and how long to sample it. Consider the signal $x(t)$ (Figure 10.7(a)). Suppose this signal is sampled at the sampling rate illustrated in Figure 10.7(b). Most people would probably intuitively say that there are enough samples here to describe the signal adequately by drawing a smooth curve through the points. How about the sampling rate in Figure 10.7(c)? Is this sampling rate adequate? How about the rate in Figure 10.7(d)? Most people would probably agree that the sampling rate in Figure 10.7(d) is inadequate.

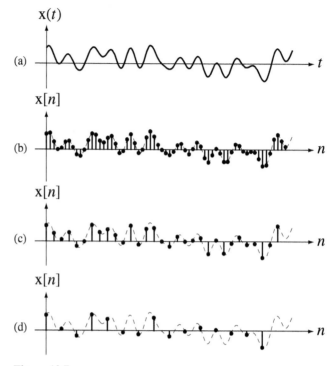

Figure 10.7
(a) A continuous-time signal, (b)–(d) discrete-time signals formed by sampling the continuous-time signal at different rates

A naturally drawn smooth curve through the last sample set would not look very much like the original curve. Although the last sampling rate was inadequate for this signal, it might be just fine for another signal (Figure 10.8). It seems adequate for the signal of Figure 10.8 because it is much smoother and more slowly varying.

Figure 10.8
A discrete-time signal formed by sampling a slowly varying signal

The minimum rate at which samples can be taken while retaining the information in the signal depends on how fast the signal varies with time, the frequency content of the signal. The question of how fast samples have to be taken to describe a signal was answered definitively by the sampling theorem. Claude Shannon[1] of Bell Labs was a major contributor to theories of sampling.

---

[1] Claude Shannon arrived as a graduate student at the Massachusetts Institute of Technology in 1936. In 1937 he wrote a thesis on the use of electrical circuits to make decisions based on Boolean logic. In 1948, while working at Bell Labs, he wrote "A Mathematical Theory of Communication," which outlined what we now call information theory. This work has been called the "Magna Carta" of the information age. He was appointed a professor of communication sciences and mathematics at MIT in 1957 but remained a consultant to Bell Labs. He was often seen in the corridors of MIT on a unicycle, sometimes juggling at the same time. He also devised one of the first chess-playing programs.

Sampling Theorem Derivation

Let the process of sampling a continuous-time signal $x(t)$ be to multiply it by a periodic pulse train $p(t)$. Let the amplitude of each pulse be one, let the width of each pulse be w, and let the fundamental period of the pulse train be T_s (Figure 10.9).

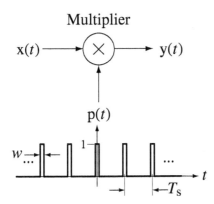

Figure 10.9
Pulse train

The pulse train can be mathematically described by $p(t) = \text{rect}(t/w) * \delta_{T_s}(t)$. The output signal is

$$y(t) = x(t)p(t) = x(t)[\text{rect}(t/w) * \delta_{T_s}(t)].$$

The average of the signal $y(t)$ over the width of the pulse centered at $t = nT_s$ can be considered an approximate sample of $x(t)$ at time $t = nT_s$. The continuous-time Fourier transform (CTFT) of $y(t)$ is $Y(f) = X(f) * w\,\text{sinc}(wf)f_s\,\delta_{f_s}(f)$ where $f_s = 1/T_s$ is the pulse repetition rate (pulse train fundamental frequency) and

$$Y(f) = X(f) * \left[wf_s \sum_{k=-\infty}^{\infty} \text{sinc}(wkf_s)\delta(f - kf_s) \right]$$

$$Y(f) = wf_s \sum_{k=-\infty}^{\infty} \text{sinc}(wkf_s)X(f - kf_s).$$

The CTFT $Y(f)$ of the response is a set of replicas of the CTFT of the input signal $x(t)$ repeated periodically at integer multiples of the pulse repetition rate f_s and also multiplied by the value of a sinc function whose width is determined by the pulse width w (Figure 10.10). Replicas of the spectrum of the input signal occur multiple times in the spectrum of the output signal, each centered at an integer multiple of the pulse repetition rate and multiplied by a different constant.

As we make each pulse shorter, its average value approaches the exact value of the signal at its center. The approximation of ideal sampling improves as w approaches zero. In the limit as w approaches zero,

$$y(t) = \lim_{w \to 0} \sum_{n=-\infty}^{\infty} x(t)\,\text{rect}((t - nT_s)/w).$$

In that limit, the signal power of $y(t)$ approaches zero. But if we now modify the sampling process to compensate for that effect by making the *area* of each sampling pulse

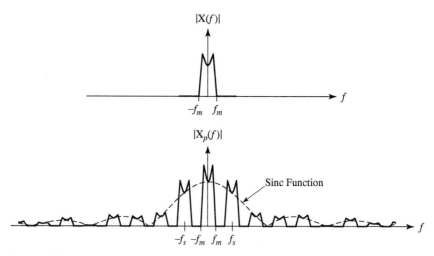

Figure 10.10
Magnitude CTFT of input and output signals

one instead of the *height*, we get the new pulse train

$$p(t) = (1/w)\text{rect}(t/w) * \delta_{T_s}(t)$$

and now y(t) is

$$y(t) = \sum_{n=-\infty}^{\infty} x(t)(1/w)\text{rect}((t - nT_s)/w).$$

Let the response in this limit as w approaches zero be designated $x_\delta(t)$. In that limit, the rectangular pulses $(1/w)\text{rect}((t - nT_s)/w)$ approach unit impulses and

$$x_\delta(t) = \lim_{w \to 0} y(t) = \sum_{n=-\infty}^{\infty} x(t)\delta(t - nT_s) = x(t)\delta_{T_s}(t).$$

This operation is called **impulse sampling** or sometimes **impulse modulation**. Of course, as a practical matter this kind of sampling is impossible because we cannot generate impulses. But the analysis of this hypothetical type of sampling is still useful because it leads to relationships between the values of a signal at discrete points and the values of the signal at all other times. Notice that in this model of sampling, the response of the sampler is still a continuous-time signal, but one whose value is zero except at the sampling instants.

It is revealing to examine the CTFT of the newly defined response $x_\delta(t)$. It is

$$X_\delta(f) = X(f) * (1/T_s)\delta_{1/T_s}(f) = f_s X(f) * \delta_{f_s}(f).$$

This is the sum of equal-size replicas of the CTFT $X(f)$ of the original signal $x(t)$, each shifted by a different integer multiple of the sampling frequency f_s, and multiplied by f_s (Figure 10.11). These replicas are called **aliases**. In Figure 10.11 the dashed lines represent the aliases of the original signal's CTFT magnitude and the solid line represents the magnitude of the sum of those aliases. Obviously the shape of the original signal's CTFT magnitude is lost in the overlapping process. But if $X(f)$ is zero for all $|f| > f_m$ and if $f_s > 2f_m$, the aliases do not overlap (Figure 10.12).

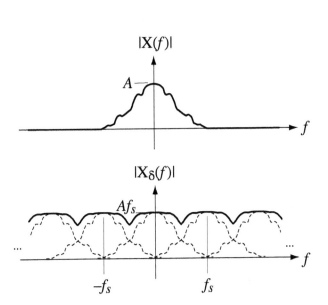

Figure 10.11
CTFT of an impulse-sampled signal

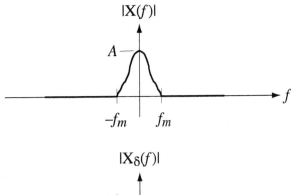

Figure 10.12
CTFT of a bandlimited signal impulse-sampled above twice its bandlimit

Signals for which $X(f)$ is zero for all $|f| > f_m$ are called **strictly bandlimited** or, more often, just **bandlimited** signals. If the aliases do not overlap, then, at least in principle, the original signal could be recovered from the impulse-sampled signal by filtering out the aliases centered at $f \pm f_s, \pm 2f_s, \pm 3f_s, \ldots$ with a lowpass filter whose frequency response is

$$H(f) = \begin{cases} T_s, & |f| < f_c \\ 0, & \text{otherwise} \end{cases} = T_s \, \text{rect}\left(\frac{f}{2f_c}\right)$$

where

$$f_m < f_c < f_s - f_m,$$

an "ideal" lowpass filter. This fact forms the basis for what is commonly known as the **sampling theorem**.

> If a continuous-time signal is sampled for all time at a rate f_s that is more than twice the bandlimit f_m of the signal, the original continuous-time signal can be recovered exactly from the samples.

If the highest frequency present in a signal is f_m, the sampling rate must be above $2f_m$ and the frequency $2f_m$ is called the **Nyquist**[2] **rate**. The words *rate* and *frequency* both describe something that happens periodically. In this text, the word *frequency* will refer to the frequencies present in a signal and the word *rate* will refer to the way a signal is sampled. A signal sampled at greater than its Nyquist rate is said to be **oversampled** and a

---

[2] Harry Nyquist received his Ph.D. from Yale in 1917. From 1917 to 1934 he was employed by Bell Labs where he worked on transmitting pictures using telegraphy and on voice transmission. He was the first to quantitatively explain thermal noise. He invented the vestigial sideband transmission technique still widely used in the transmission of television signals. He invented the Nyquist diagram for determining the stability of feedback systems.

signal sampled at less than its Nyquist rate is said to be **undersampled**. When a signal is sampled at a rate f_s the frequency $f_s/2$ is called the **Nyquist frequency**. Therefore, if a signal has any signal power at or above the Nyquist frequency the aliases will overlap.

Another sampling model that we have used in previous chapters is the creation of a discrete-time signal x[n] from a continuous-time signal x(t) through x[n] = x(nT_s) where T_s is the time between consecutive samples. This may look like a more realistic model of practical sampling, and in some ways it is, but instantaneous sampling at a point in time is also not possible practically. We will refer to this sampling model as simply "sampling" instead of "impulse sampling."

Recall that the DTFT of any discrete-time signal is always periodic. The CTFT of an impulse-sampled signal is also periodic. The CTFT of an impulse-sampled continuous-time signal $x_\delta(t)$ and the DTFT of a discrete-time signal $x_s[n]$ formed by sampling that same continuous-time signal are similar (Figure 10.13). (The s subscript on $x_s[n]$ is there to help avoid confusion between the different transforms that follow.) The waveshapes are the same. The only difference is that the DTFT is based on normalized frequency F or Ω and the CTFT on actual frequency f or ω. The sampling theorem can be derived using the DTFT instead of the CTFT and the result is the same.

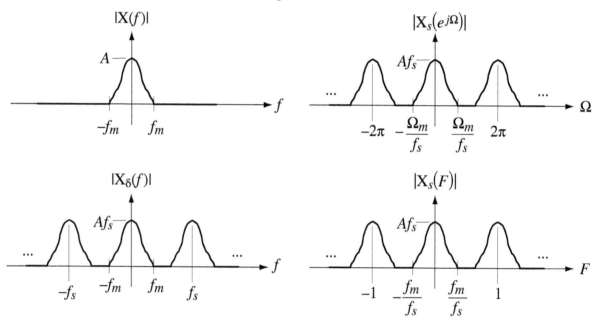

Figure 10.13
Comparison between the CTFT of an impulse-sampled signal and the DTFT of a sampled signal

ALIASING

The phenomenon of aliasing (overlapping of aliases) is not an exotic mathematical concept that is outside the experience of ordinary people. Almost everyone has observed aliasing, but probably without knowing what to call it. A very common experience that illustrates aliasing sometimes occurs while watching television. Suppose you are watching a Western movie on television and there is a picture of a horse-drawn wagon with spoked wheels. If the wheels on the wagon gradually rotate faster and faster, a point is reached at which the wheels appear to stop rotating forward and begin to appear to rotate backward even though the wagon is obviously moving forward. If the speed of rotation were increased further, the wheels would eventually appear to stop and then rotate forward again. This is an example of the phenomenon of aliasing.

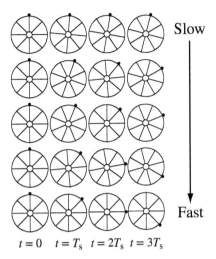

Figure 10.14
Wagon wheel angular positions at four sampling times

Although it is not apparent to the human eye, the image on a television screen is flashed upon the screen 30 times per second (under the NTSC video standard). That is, the image is effectively sampled at a rate of 30 samples/second. Figure 10.14 shows the positions of a spoked wheel at four sampling instants for several different rotational velocities, starting with a lower rotational velocity at the top and progressing toward a higher rotational velocity at the bottom. (A small index dot has been added to the wheel to help in seeing the actual rotation of the wheel, as opposed to the apparent rotation.)

This wheel has eight spokes, so upon rotation by one-eighth of a complete revolution the wheel looks exactly the same as in its initial position. Therefore the image of the wheel has an angular period of $\pi/4$ radians or $45°$, the angular spacing between spokes. If the rotational velocity of the wheel is f_0 revolutions/second (Hz) the image fundamental frequency is $8f_0$ Hz. The image repeats exactly eight times in one complete wheel rotation.

Let the image be sampled at 30 Hz ($T_s = 1/30$ s). On the top row the wheel is rotating clockwise at $-5°/T_s$ ($-150°/s$ or -0.416 rev/s) so that in the top row the spokes have rotated by $0°$, $5°$, $10°$, and $15°$ clockwise. The eye and brain of the observer interpret the succession of images to mean that the wheel is rotating clockwise because of the progression of angles at the sampling instants. In this case the wheel appears to be (and is) rotating at an image rotational frequency of $-150°/s$.

In the second row, the rotational speed is four times faster than in the top row and the angles of rotation at the sampling instants are $0°$, $20°$, $40°$, and $60°$ clockwise. The wheel still (correctly) appears to be rotating clockwise at its actual rotational frequency of $-600°/s$. In the third row, the rotational speed is $-675°/s$. Now the ambiguity caused by sampling begins. If the index dot were not there, it would be impossible to determine whether the wheel is rotating $-22.5°$ per sample or $+22.5°$ per sample because the image samples are identical for those two cases. It is impossible, by simply looking at the sample images, to determine whether the rotation is clockwise or counterclockwise. In the fourth row the wheel is rotating at $-1200°/s$. Now (ignoring the index dot) the wheel definitely appears to be rotating at $+5°$ per sample instead of the actual rotational frequency of $-40°$ per sample. The perception of the human brain would be that the wheel is rotating $5°$ counterclockwise per sample instead of $40°$ clockwise. In the bottom row the wheel rotation is −1350°/s or clockwise 45° per sample. Now the wheel appears to be standing still even though it is rotating clockwise. Its angular velocity seems to be zero because it is being sampled at a rate exactly equal to the image fundamental frequency.

EXAMPLE 10.1

Finding Nyquist rates of signals

Find the Nyquist rate for each of the following signals:

(a) $x(t) = 25\cos(500\pi t)$

$$X(f) = 12.5[\delta(f-250) + \delta(f+250)]$$

The highest frequency (and the only frequency) present in this signal is $f_m = 250$ Hz. The Nyquist rate is 500 samples/second.

(b) $x(t) = 15\text{rect}(t/2)$

$$X(f) = 30\text{sinc}(2f)$$

Since the sinc function never goes to zero and stays there, at a finite frequency, the highest frequency in the signal is infinite and the Nyquist rate is also infinite. The rectangle function is not bandlimited.

(c) $x(t) = 10\text{sinc}(5t)$

$$X(f) = 2\text{rect}(f/5)$$

The highest frequency present in x(t) is the value of f at which the rect function has its discontinuous transition from one to zero $fm = 2.5$ Hz. Therefore, the Nyquist rate is 5 samples/second.

(d) $x(t) = 2\text{sinc}(5000t)\sin(500,000\pi t)$

$$X(f) = \frac{1}{2500}\text{rect}\left(\frac{f}{5000}\right) * \frac{j}{2}[\delta(f+250,000) - \delta(f-250,000)]$$

$$X(f) = \frac{j}{5000}\left[\text{rect}\left(\frac{f+250,000}{5000}\right) - \text{rect}\left(\frac{f-250,000}{5000}\right)\right]$$

The highest frequency in x(t) is $f_m = 252.5$ kHz. Therefore, the Nyquist rate is 505,000 samples/second.

∎

EXAMPLE 10.2

Analysis of an *RC* filter as an anti-aliasing filter

Suppose a signal that is to be acquired by a data acquisition system is known to have an amplitude spectrum that is flat out to 100 kHz and drops suddenly there to zero. Suppose further that the fastest rate at which our data acquisition system can sample the signal is 60 kHz. Design an *RC*, lowpass, anti-aliasing filter that will reduce the signal's amplitude spectrum at 30 kHz to less than 1% of its value at very low frequencies so that aliasing will be minimized.

The frequency response of a unity-gain *RC* lowpass filter is

$$H(f) = \frac{1}{j2\pi fRC + 1}.$$

The squared magnitude of the frequency response is

$$|H(f)|^2 = \frac{1}{(2\pi f RC)^2 + 1}$$

and its value at very low frequencies approaches one. Set the RC time constant so that at 30 kHz, the squared magnitude of $H(f)$ is $(0.01)^2$.

$$|H(30{,}000)|^2 = \frac{1}{(2\pi \times 30{,}000 \times RC)^2 + 1} = (0.01)^2$$

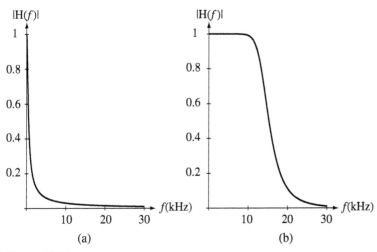

Figure 10.15
(a) Magnitude frequency response of the anti-aliasing RC lowpass filter,
(b) magnitude frequency response of a 6th-order Butterworth anti-aliasing lowpass filter

Solving for RC, $RC = 0.5305$ ms. The corner frequency (-3 dB frequency) of this RC lowpass filter is 300 Hz, which is 100 times lower than the Nyquist frequency of 30 kHz (Figure 10.15). It must be set this low to meet the specification using a single-pole filter because its frequency response rolls off so slowly. For this reason most practical anti-aliasing filters are designed as higher-order filters with much faster transitions from the pass band to the stop band. Figure 10.15(b) shows the frequency response of a 6th-order Butterworth lowpass filter. (Butterworth filters are covered in Chapter 14.) The higher-order filter preserves much more of the signal than the RC filter. ∎

TIME-LIMITED AND BANDLIMITED SIGNALS

Recall that the original mathematical statement of the way a signal is sampled is $x_s[n] = x(nT_s)$. This equation holds true for any integer value of n and that implies that the signal $x(t)$ is sampled *for all time*. Therefore, infinitely many samples are needed to describe $x(t)$ exactly from the information in $x_s[n]$. The sampling theorem is predicated on sampling this way. So, even though the Nyquist rate has been found, and may be finite, one must (in general) still take infinitely many samples to exactly reconstruct the original signal from its samples, *even if it is bandlimited and we oversample.*

It is tempting to think that if a signal is **time limited** (having nonzero values only over a finite time), one could then sample only over that time, knowing all the other samples are zero, and have all the information in the signal. The problem with that idea is that no time-limited signal can also be bandlimited, and therefore no finite sampling rate is adequate.

The fact that a signal cannot be simultaneously time limited and bandlimited is a fundamental law of Fourier analysis. The validity of this law can be demonstrated by the following argument. Let a signal x(t) have no nonzero values outside the time range $t_1 < t < t_2$. Let its CTFT be X(f). If x(t) is time limited to the time range $t_1 < t < t_2$, it can be multiplied by a rectangle function whose nonzero portion covers this same time range, without changing the signal. That is,

$$\text{x}(t) = \text{x}(t)\text{rect}\left(\frac{t - t_0}{\Delta t}\right) \tag{10.1}$$

where $t_0 = (t_1 + t_2)/2$ and $\Delta t = t_2 - t_1$ (Figure 10.16).

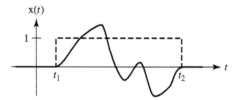

Figure 10.16
A time-limited function and a rectangle time-limited to the same time

Finding the CTFT of both sides of (10.1) we obtain $\text{X}(f) = \text{X}(f) * \Delta t\,\text{sinc}(\Delta t f)e^{-j2\pi f t_0}$. This last equation says that X(f) is unaffected by being convolved with a sinc function. Since sinc($\Delta t f$) has an infinite nonzero extent in f, if it is convolved with an X(f) that has a finite nonzero extent in f, the convolution of the two will have an infinite nonzero extent in f. Therefore, the last equation cannot be satisfied by any X(f) that has a finite nonzero extent in f, proving that if a signal is time limited it cannot be bandlimited. The converse, that a bandlimited signal cannot be time limited, can be proven by a similar argument.

> A signal can be unlimited in both time and frequency but it cannot be limited in both time and frequency.

INTERPOLATION

Ideal Interpolation

The description given above on how to recover the original signal indicated that we could filter the impulse-sampled signal to remove all the aliases except the one centered at zero frequency. If that filter were an ideal lowpass filter with a constant gain of $T_s = 1/f_s$ in its passband and bandwidth f_c where $f_m < f_c < f_s - f_m$ that operation in the frequency domain would be described by

$$\text{X}(f) = T_s\text{rect}(f/2f_c) \times \text{X}_\delta(f) = T_s\text{rect}(f/2f_c) \times f_s\text{X}(f) * \delta_{f_s}(f).$$

If we inverse transform this expression we get

$$\text{x}(t) = \underbrace{T_s f_s}_{=1} 2f_c\,\text{sinc}(2f_c t) * \underbrace{\text{x}(t)(1/f_s)\delta_{T_s}(t)}_{=(1/f_s)\sum_{n=-\infty}^{\infty}\text{x}(nT_s)\delta(t - nT_s)}$$

or

$$x(t) = 2(f_c/f_s)\operatorname{sinc}(2f_c t) * \sum_{n=-\infty}^{\infty} x(nT_s)\delta(t - nT_s)$$

$$x(t) = 2(f_c/f_s)\sum_{n=-\infty}^{\infty} x(nT_s)\operatorname{sinc}(2f_c(t - nT_s)). \quad (10.2)$$

By pursuing an admittedly impractical idea, impulse sampling, we have arrived at a result that allows us to fill in the values of a signal for all time, given its values at equally spaced points in time. There are no impulses in (10.2), only the sample values, which are the strengths of the impulses that would have been created by impulse sampling. The process of filling in the missing values between the samples is called **interpolation**.

Consider the special case $f_c = f_s/2$. In this case the interpolation process is described by the simpler expression

$$x(t) = \sum_{n=-\infty}^{\infty} x(nT_s)\operatorname{sinc}((t - nT_s)/T_s).$$

Now interpolation consists simply of multiplying each sinc function by its corresponding sample value and then adding all the scaled and shifted sinc functions as illustrated in Figure 10.17.

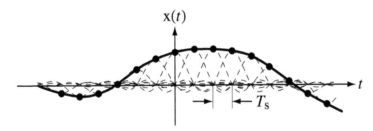

Figure 10.17
Interpolation process for an ideal lowpass filter corner frequency set to half the sampling rate

Referring to Figure 10.17, notice that each sinc function peaks at its sample time and is zero at every other sample time. So the interpolation is obviously correct at the sample times. The derivation above shows that it is also correct at all the points between sample times.

Practical Interpolation
The interpolation method in the previous section reconstructs the signal exactly but it is based on an assumption that is never justified in practice, the availability of infinitely many samples. The interpolated value at any point is the sum of contributions from infinitely many weighted sinc functions, each of infinite time extent. But since, as a practical matter, we cannot acquire infinitely many samples, much less process them, we must approximately reconstruct the signal using a finite number of samples. Many techniques can be used. The selection of the one to be used in any given situation depends on what accuracy of reconstruction is required and how oversampled the signal is.

Zero-Order Hold Probably the simplest approximate reconstruction idea is to simply let the reconstruction always be the value of the most recent sample (Figure 10.18). This is a simple technique because the samples, in the form of numerical codes, can be the input signal to a DAC that is clocked to produce a new output signal with every clock pulse. The signal produced by this technique has a "stair-step" shape that follows the original signal. This type of signal reconstruction can be modeled by impulse sampling the signal and letting the impulse-sampled signal excite a system called a **zero-order hold** whose impulse response is

$$h(t) = \begin{cases} 1, & 0 < t < T_s \\ 0, & \text{otherwise} \end{cases} = \text{rect}\left(\frac{t - T_s/2}{T_s}\right)$$

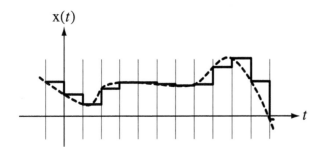

Figure 10.18
Zero-order hold signal reconstruction

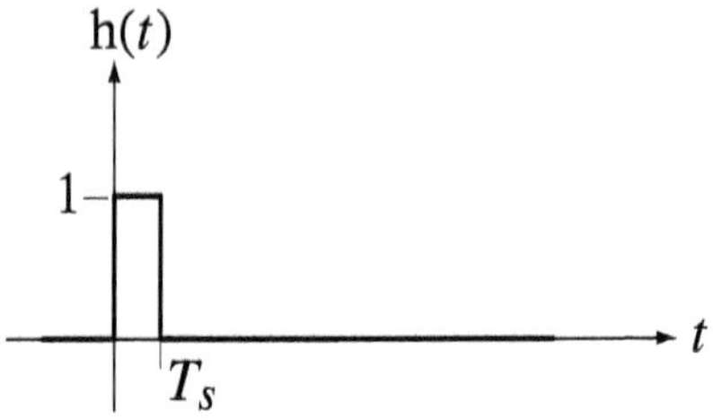

Figure 10.19
Impulse response of a zero-order hold

(Figure 10.19).

One popular way of further reducing the effects of the aliases is to follow the zero-order hold with a practical lowpass filter that smooths out the steps caused by the zero-order hold. The zero-order hold inevitably causes a delay relative to the original signal because it is causal and any practical lowpass smoothing filter will add still more delay.

First-Order Hold Another natural idea is to interpolate between samples with straight lines (Figure 10.20). This is obviously a better approximation to the original signal but it is a little harder to implement. As drawn in Figure 10.20, the value of the interpolated signal at any time depends on the value of the previous sample and the value of the next sample. This cannot be done in real time because the value of the next sample is not known in real time. But if we are willing to delay the reconstructed signal by one sample time T_s we can make the reconstruction process occur in real time. The reconstructed signal would appear as shown in Figure 10.21.

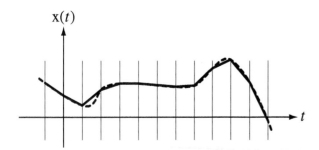

Figure 10.20
Signal reconstruction by straight-line interpolation

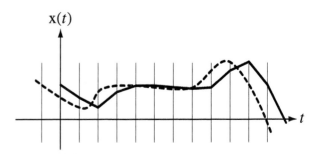

Figure 10.21
Straight-line signal reconstruction delayed by one sample time

This interpolation can be accomplished by following the zero-order hold by an identical zero-order hold. This means that the impulse response of such an interpolation system would be the convolution of the zero-order hold impulse response with itself

$$h(t) = \text{rect}\left(\frac{t - T_s/2}{T_s}\right) * \text{rect}\left(\frac{t - T_s/2}{T_s}\right) = \text{tri}\left(\frac{t - T_s}{T_s}\right)$$

(Figure 10.22). This type of interpolation system is called a **first-order hold**.

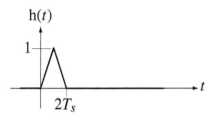

Figure 10.22
Impulse response of a first-order hold

One very familiar example of the use of sampling and signal reconstruction is the playback of an audio compact disk (CD). A CD stores samples of a musical signal that have been taken at a rate of 44,100 samples/second. Half of that sampling rate is 22.05 kHz. The frequency response of a young, healthy human ear is conventionally taken to span from about 20 Hz to about 20 kHz with some variability in that range. So the sampling rate is a little more than twice the highest frequency the human ear can detect.

SAMPLING BANDPASS SIGNALS

The sampling theorem, as stated above, was based on a simple idea. If we sample fast enough, the aliases do not overlap and the original signal can be recovered by an ideal lowpass filter. We found that if we sample faster than twice the highest frequency in the signal, we can recover the signal from the samples. That is true for all signals, but for some signals, the minimum sampling rate can be reduced.

In making the argument that we must sample at a rate greater than twice the highest frequency in the signal, we were implicitly assuming that if we sampled at any lower rate the aliases would overlap. In the spectra used above to illustrate the ideas, the aliases would overlap. But that is not true of all signals. For example, let a continuous-time signal have a narrow bandpass spectrum that is nonzero only for $15 \text{ kHz} < |f| < 20 \text{ kHz}$. Then the bandwidth of this signal is 5 kHz (Figure 10.23).

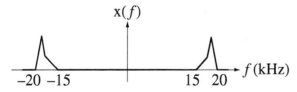

Figure 10.23
A narrow-bandpass signal spectrum

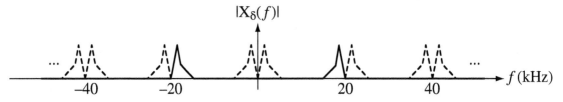

Figure 10.24
The spectrum of a bandpass signal impulse-sampled at 20 kHz

If we impulse sample this signal at 20 kHz we would get the aliases illustrated in Figure 10.24. These aliases do not overlap. Therefore, it must be possible, with knowledge of the original signal's spectrum and the right kind of filtering, to recover the signal from the samples. We could even sample at 10 kHz, half the highest frequency, get the aliases in Figure 10.25, and still recover the original signal (theoretically) with that same filter. But if we sampled at any lower rate the aliases would definitely overlap and we could not recover the original signal. Notice that this minimum sampling rate is not twice the *highest frequency* in the signal but rather twice the *bandwidth* of the signal.

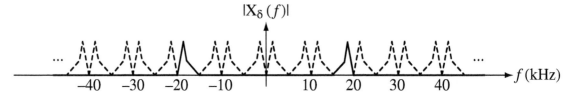

Figure 10.25
The spectrum of a bandpass signal impulse-sampled at 10 kHz

In this example the ratio of the highest frequency to the bandwidth of the signal was an integer. When that ratio is not an integer it becomes more difficult to find the minimum sampling rate that avoids aliasing (Figure 10.26).

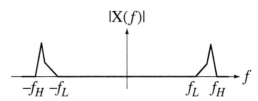

Figure 10.26
Magnitude spectrum of a general bandpass signal

The aliases occur at shifts of integer multiples of the sampling rate. Let the integer k index the aliases. Then the $(k-1)$th alias must lie wholly below f_L and the kth alias must lie wholly above f_H. That is

$$(k-1)f_s + (-f_L) < f_L \Rightarrow (k-1)f_s < 2f_L$$

and

$$kf_s + (-f_H) > f_H \Rightarrow kf_s > 2f_H.$$

Rearranging these two inequalities we get

$$(k-1)f_s < 2(f_H - B)$$

where B is the bandwidth $f_H - f_L$ and

$$\frac{1}{f_s} < \frac{k}{2f_H}.$$

Now set the product of the left sides of these inequalities less than the product of the right sides of these inequalities

$$k - 1 < (f_H - B)\frac{k}{f_H} \Rightarrow k < \frac{f_H}{B}$$

Since k must be an integer, that means that the real limit on k is

$$k_{\max} = \left\lfloor \frac{f_H}{B} \right\rfloor$$

the greatest integer in f_H/B. So the two conditions,

$$k_{\max} = \left\lfloor \frac{f_H}{B} \right\rfloor \quad \text{and} \quad k_{\max} > \frac{2f_H}{f_{s,\min}}$$

or the single condition

$$f_{s,\min} > \frac{2f_H}{\lfloor f_H/B \rfloor}$$

determine the minimum sampling rate for which aliasing does not occur.

EXAMPLE 10.3

Minimum sampling rate to avoid aliasing

Let a signal have no nonzero spectral components outside the range $34\,\text{kHz} < |f| < 47\,\text{kHz}$. What is the minimum sampling rate that avoids aliasing?

$$f_{s,\min} > \frac{2f_H}{\lfloor f_H/B \rfloor} = \frac{94\,\text{kHz}}{\lfloor 47\,\text{kHz}/13\,\text{kHz} \rfloor} = 31{,}333 \text{ samples/second}$$

EXAMPLE 10.4

Minimum sampling rate to avoid aliasing

Let a signal have no nonzero spectral components outside the range $0 < |f| < 580\,\text{kHz}$. What is the minimum sampling rate that avoids aliasing?

$$f_{s,\min} > \frac{2f_H}{\lfloor f_H/B \rfloor} = \frac{1160\,\text{kHz}}{\lfloor 580\,\text{kHz}/580\,\text{kHz} \rfloor} = 1{,}160{,}000 \text{ samples/second}$$

This is a lowpass signal and the minimum sampling rate is twice the highest frequency as originally determined in the sampling theorem.

In most real engineering design situations, choosing the sampling rate to be more than twice the highest frequency in the signal is the practical solution. As we will soon see, that rate is usually considerably above the Nyquist rate in order to simplify some of the other signal processing operations.

SAMPLING A SINUSOID

The whole point of Fourier analysis is that any signal can be decomposed into sinusoids (real or complex). Therefore, let's explore sampling by looking at some real sinusoids sampled above, below, and at the Nyquist rate. In each example a sample occurs at time $t = 0$. This sets a definite phase relationship between an exactly described mathematical signal and the way it is sampled. (This is arbitrary, but there must always be a sampling-time reference and, when we get to sampling for finite times, the first sample will always be at time $t = 0$ unless otherwise stated. Also, in the usual use of the DFT in DSP, the first sample is normally assumed to occur at time $t = 0$.)

Case 1. A cosine sampled at a rate that is four times its frequency or at twice its Nyquist rate (Figure 10.27).

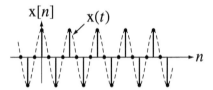

Figure 10.27
Cosine sampled at twice its Nyquist rate

It is clear here that the sample values and the knowledge that the signal is sampled fast enough are adequate to uniquely describe this sinusoid. No other sinusoid of this, or any other frequency, below the Nyquist frequency could pass exactly through all the samples in the full time range $-\infty < n < +\infty$. In fact no other signal *of any kind* that is bandlimited to below the Nyquist frequency could pass exactly through all the samples.

Case 2. A cosine sampled at twice its frequency or at its Nyquist rate (Figure 10.28)

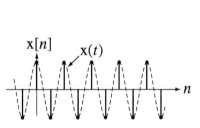

Figure 10.28
Cosine sampled at its Nyquist rate

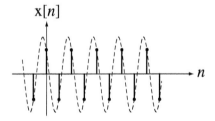

Figure 10.29
Sinusoid with same samples as a cosine sampled at its Nyquist rate

Is this sampling adequate to uniquely determine the signal? No. Consider the sinusoidal signal in Figure 10.29, which is of the same frequency and passes exactly through the same samples.

This is a special case that illustrates the subtlety mentioned earlier in the sampling theorem. To be sure of exactly reconstructing any general signal from its samples, the sampling rate must be *more than* the Nyquist rate instead of *at least* the Nyquist rate. In earlier examples, it did not matter because the signal power at exactly the Nyquist frequency was zero (no impulse in the amplitude spectrum there). If there is a sinusoid in a signal, exactly at its bandlimit, the sampling must exceed the Nyquist rate for exact reconstruction, in general. Notice that there is no ambiguity about the frequency of the signal. But there is ambiguity about the amplitude and phase as illustrated above. If the sinc-function-interpolation procedure derived earlier were applied to the samples in Figure 10.29, the cosine in Figure 10.28 that was sampled at its peaks would result.

Any sinusoid at some frequency can be expressed as the sum of an unshifted cosine of some amplitude at the same frequency and an unshifted sine of some amplitude at the same frequency. The amplitudes of the unshifted sine and cosine depend on the phase of the original sinusoid. Using a trigonometric identity,

$$A\cos(2\pi f_0 t + \theta) = A\cos(2\pi f_0 t)\cos(\theta) - A\sin(2\pi f_0 t)\sin(\theta).$$

$$A\cos(2\pi f_0 t + \theta) = \underbrace{A\cos(\theta)}_{A_c}\cos(2\pi f_0 t) + \underbrace{[-A\sin(\theta)]}_{A_s}\sin(2\pi f_0 t)$$

$$A\cos(2\pi f_0 t + \theta) = A_c\cos(2\pi f_0 t) + A_s\sin(2\pi f_0 t).$$

When a sinusoid is sampled at exactly the Nyquist rate the sinc-function interpolation always yields the cosine part and drops the sine part, an effect of aliasing. The cosine part of a general sinusoid is often called the **in-phase** part and the sine part is often called the **quadrature** part. The dropping of the quadrature part of a sinusoid can easily be seen in the time domain by sampling an unshifted sine function at exactly the Nyquist rate. All the samples are zero (Figure 10.30).

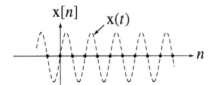

Figure 10.30
Sine sampled at its Nyquist rate

If we were to add a sine function of any amplitude at exactly this frequency to any signal and then sample the new signal, the samples would be the same as if the sine function were not there because its value is zero at each sample time (Figure 10.31). Therefore, the quadrature or sine part of a signal that is at exactly the Nyquist frequency is lost when the signal is sampled.

Case 3. A sinusoid sampled at slightly above the Nyquist rate (Figure 10.32).

Now, because the sampling rate is higher than the Nyquist rate, the samples do not all occur at zero crossings and there is enough information in the samples to reconstruct the signal. There is only one sinusoid whose frequency is less than the Nyquist frequency, of a unique amplitude, phase, and frequency that passes exactly through all these samples.

Case 4. Two sinusoids of different frequencies sampled at the same rate with the same sample values (Figure 10.33).

In this case, the lower-frequency sinusoid is oversampled and the higher-frequency sinusoid is undersampled. This illustrates the ambiguity caused by undersampling. If we only had access to the samples from the higher-frequency sinusoid and we believed that the signal had been properly sampled according to the sampling theorem, we would interpret them as having come from the lower-frequency sinusoid.

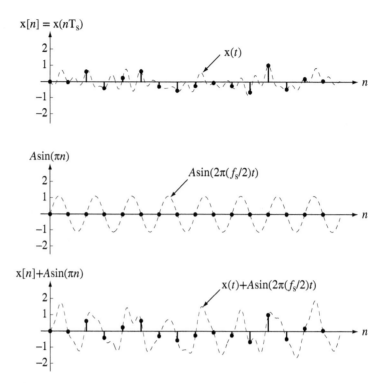

Figure 10.31
Effect on samples of adding a sine at the Nyquist frequency

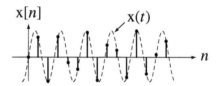

Figure 10.32
Sine sampled at slightly above its Nyquist rate

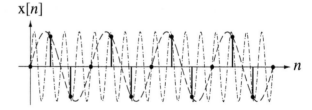

Figure 10.33
Two sinusoids of different frequencies that have the same sample values

If a sinusoid $x_1(t) = A\cos(2\pi f_0 t + \theta)$ is sampled at a rate f_s, the samples will be the same as the samples from another sinusoid $x_2(t) = A\cos(2\pi(f_0 + kf_s)t + \theta)$, where k is any integer (including negative integers). This can be shown by expanding the argument of $x_2(t) = A\cos(2\pi f_0 t + 2\pi(kf_s)t + \theta)$. The samples occur at times nT_s where n is an integer. Therefore, the nth sample values of the two sinusoids are

$$x_1(nT_s) = A\cos(2\pi f_0 nT_s + \theta) \quad \text{and} \quad x_2(nT_s) = A\cos(2\pi f_0 nT_s + 2\pi(kf_s)nT_s + \theta)$$

and, since $f_s T_s = 1$, the second equation simplifies to $x_2(nT_s) = A\cos(2\pi f_0 nT_s + 2k\pi n + \theta)$. Since kn is the product of integers and therefore also an integer, and since adding an integer multiple of 2π to the argument of a sinusoid does not change its value,

$$x_2(nT_s) = A\cos(2\pi f_0 nT_s + 2k\pi n + \theta) = A\cos(2\pi f_0 nT_s + \theta) = x_1(nT_s).$$

BANDLIMITED PERIODIC SIGNALS

In a previous section we saw what the requirements were for adequately sampling a signal. We also learned that, in general, for perfect reconstruction of the signal, infinitely many samples are required. Since any DSP system has a finite storage capability, it is important to explore methods of signal analysis using a finite number of samples.

There is one type of signal that can be completely described by a finite number of samples, a bandlimited, periodic signal. Knowledge of what happens in one period is sufficient to describe all periods and one period is finite in duration (Figure 10.34).

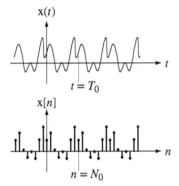

Figure 10.34
A bandlimited, periodic, continuous-time signal and a discrete-time signal formed by sampling it eight times per fundamental period

Therefore, a finite number of samples over one period of a bandlimited, periodic signal taken at a rate that is above the Nyquist rate *and is also an integer multiple of the fundamental frequency* is a complete description of the signal. Making the sampling rate an integer multiple of the fundamental frequency ensures that the samples from any fundamental period are exactly the same as the samples from any other fundamental period.

Let the signal formed by sampling a bandlimited, periodic signal $x(t)$ above its Nyquist rate be the periodic signal $x_s[n]$ and let an impulse-sampled version of $x(t)$, sampled at the same rate, be $x_\delta(t)$ (Figure 10.35).

Only one fundamental period of samples is shown in Figure 10.35 to emphasize that one fundamental period of samples is enough to completely describe the bandlimited periodic signal. We can find the appropriate Fourier transforms of these signals (Figure 10.36).

The CTFT of $x(t)$ consists only of impulses because it is periodic and it consists of a finite number of impulses because it is bandlimited. So a finite number of numbers completely characterizes the signal in both the time and frequency domains. If we multiply the impulse strengths in $X(f)$ by the sampling rate f_s, we get the impulse strengths in the same frequency range of $X_\delta(f)$.

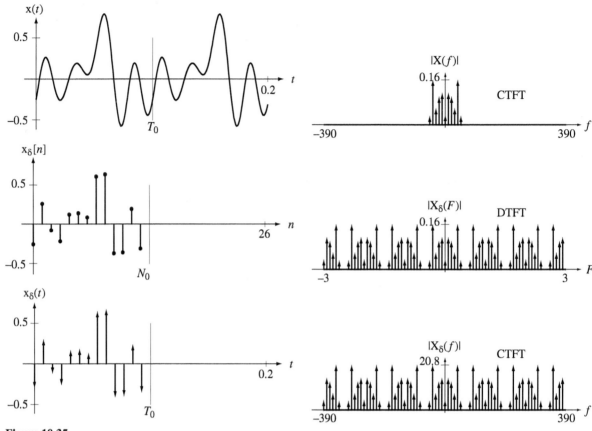

Figure 10.35
A bandlimited periodic continuous-time signal, a discrete-time signal, and a continuous-time impulse signal created by sampling it above its Nyquist rate

Figure 10.36
Magnitudes of the Fourier transforms of the three time-domain signals of Figure 10.35

EXAMPLE 10.5

Finding a CTFS harmonic function from a DFT harmonic function

Find the CTFS harmonic function for the signal $x(t) = 4 + 2\cos(20\pi t) - 3\sin(40\pi t)$ by sampling above the Nyquist rate at an integer multiple of the fundamental frequency over one fundamental period and finding the DFT harmonic function of the samples.

There are exactly three frequencies present in the signal, 0 Hz, 10 Hz, and 20 Hz. Therefore, the highest frequency present in the signal is $f_m = 20$ Hz and the Nyquist rate is 40 samples/second. The fundamental frequency is the greatest common divisor of 10 Hz and 20 Hz, which is 10 Hz. So we must sample for 1/10 second. If we were to sample at the Nyquist rate for exactly one fundamental period, we would get four samples. If we are to sample above the Nyquist rate at an integer multiple of the fundamental frequency, we must take five or more samples in one fundamental period. To keep the calculations simple we will sample eight times in one fundamental period, a sampling rate of 80 samples/second. Then, beginning the sampling at time $t = 0$, the samples are

$$\{x[0], x[1], \ldots x[7]\} = \{6, 1 + \sqrt{2}, 4, 7 - \sqrt{2}, 2, 1 - \sqrt{2}, 4, 7 + \sqrt{2}\}.$$

Using the formula for finding the DFT harmonic function of a discrete-time function,

$$X[k] = \sum_{n=\langle N_0 \rangle} x[n] e^{-j2\pi kn/N_0}$$

we get

$$\{X[0], X[1], \ldots, X[7]\} = \{32, 8, j12, 0, 0, 0, -j12, 8\}.$$

The right-hand side of this equation is one fundamental period of the DFT harmonic function $X[k]$ of the function $x[n]$. Finding the CTFS harmonic function of $x(t) = 4 + 2\cos(20\pi t) - 3\sin(40\pi t)$ directly using

$$c_x[k] = (1/T_0)\int_{T_0} x(t) e^{-j2\pi k t/T_0} dt$$

we get

$$\{c_x[-4], c_x[-3], \ldots, c_x[4]\} = \{0, 0, -j3/2, 1, 4, 1, j3/2, 0, 0\}.$$

From the two results, $1/N$ times the values $\{X[0], X[1], X[2], X[3], X[4]\}$ in the DFT harmonic function and the CTFS harmonic values $\{c_x[0], c_x[1], c_x[2], c_x[3], c_x[4]\}$ are the same and, using the fact that $X[k]$ is periodic with fundamental period 8, $(1/8)\{X[-4], X[-3], X[-2], X[-1]\}$ and $\{c_x[-4], c_x[-3], c_x[-2], c_x[-1]\}$ are the same also.

Now let's violate the sampling theorem by sampling at the Nyquist rate. In this case, there are four samples in one fundamental period

$$\{x[0], x[1], x[2], x[3]\} = \{6, 4, 2, 4\}$$

and one period of the DFT harmonic function is

$$\{X[0], X[1], X[2], X[3]\} = \{16, 4, 0, 4\}.$$

The nonzero values of the CTFS harmonic function are the set

$$\{c_x[-2], c_x[-1], \ldots, c_x[2]\} = \{-j3/2, 1, 4, 1, j3/2\}.$$

The $j3/2$ for $c_x[2]$ is missing from the DFT harmonic function because $X[2] = 0$. This is the amplitude of the sine function at 40 Hz. This is a demonstration that when we sample a sine function at exactly the Nyquist rate, we don't see it in the samples because we sample it exactly at its zero crossings.

∎

A thoughtful reader may have noticed that the description of a signal based on samples in the time domain from one fundamental period consists of a finite set of numbers $x_s[n], n_0 \leq n < n_0 + N$, which contains N independent *real* numbers, and the corresponding DFT harmonic-function description of the signal in the frequency domain consists of the finite set of numbers $X_s[k], k_0 \leq k < k_0 + N$, which contains N *complex* numbers and therefore $2N$ real numbers (two real numbers for each complex number, the real and imaginary parts). So it might seem that the description in the time domain is more efficient than in the frequency domain since it is accomplished with fewer real numbers. But how can this be when the set $X_s[k], k_0 \leq k < k_0 + N$ is calculated directly from the set $x_s[n], n_0 \leq n < n_0 + N$ with no extra information? A closer examination of the relationship between the two sets of numbers will reveal that this apparent difference is an illusion.

As first discussed in Chapter 7, $X_s[0]$ is always real. It can be computed by the DFT formula as

$$X_s[0] = \sum_{n=\langle N \rangle} x_s[n].$$

Since all the $x_s[n]$'s are real, $X_s[0]$ must also be real because it is simply the sum of all the $x_s[n]$'s. So this number never has a nonzero imaginary part. There are two cases to consider next, N even and N odd.

Case 1. N even

For simplicity, and without loss of generality, in

$$X_s[k] = \sum_{n=\langle N \rangle} x_s[n] e^{-j\pi kn/N} = \sum_{n=k_0}^{k_0+N-1} x_s[n] e^{-j\pi kn/N}$$

let $k_0 = -N/2$. Then

$$X_s[k_0] = X_s[-N/2] = \sum_{n=\langle N \rangle} x_s[n] e^{j\pi n} = \sum_{n=\langle N \rangle} x_s[n](-1)^n$$

and $X_s[-N/2]$ is guaranteed to be real. All the DFT harmonic function values in one period, other than $X_s[0]$ and $X_s[-N/2]$, occur in pairs $X_s[k]$ and $X_s[-k]$. Next recall that for any real $x_s[n]$, $X_s[k] = X_s^*[-k]$. That is, once we know $X_s[k]$ we also know $X_s[-k]$. So, even though each $X_s[k]$ contains two real numbers, and each $X_s[-k]$ does also, $X_s[-k]$ does not add any information since we already know that $X_s[k] = X_s^*[-k]$. $X_s[-k]$ is not independent of $X_s[k]$. So now we have, as independent numbers, $X_s[0]$, $X_s[N/2]$, and $X_s[k]$ for $1 \leq k < N/2$. All the $X_s[k]$'s from $k = 1$ to $k = N/2 - 1$ yield a total of $2(N/2 - 1) = N - 2$ independent real numbers. Add the two guaranteed-real values $X_s[0]$ and $X_s[N/2]$ and we finally have a total of N independent real numbers in the frequency-domain description of this signal.

Case 2: N odd

For simplicity, and without loss of generality, let $k_0 = -(N-1)/2$. In this case, we simply have $X_s[0]$ plus $(N-1)/2$ complex conjugate pairs $X_s[k]$ and $X_s[-k]$. We have already seen that $X_s[k] = X_s^*[-k]$. So we have the real number $X_s[0]$ and two independent real numbers per complex conjugate pair or $N - 1$ independent real numbers for a total of N independent real numbers.

The information content in the form of independent real numbers is conserved in the process of converting from the time to the frequency domain.

SIGNAL PROCESSING USING THE DFT

CTFT-DFT Relationship

In the following development of the relationship between the CTFT and the DFT, all the processing steps from the CTFT of the original function to the DFT will be illustrated by an example signal. Then several uses of the DFT are developed for signal processing operations. We will use the F form of the DTFT because the transform relationships are a little more symmetrical than in the Ω form.

Let a signal $x(t)$ be sampled and let the total number of samples taken be N where $N = Tf_s$, T is the total sampling time, and f_s is the sampling frequency. Then the time between samples is $T_s = 1/f_s$. Below is an example of an original signal in both the time and frequency domains (Figure 10.37).

The first processing step in converting from the CTFT to the DFT is to sample the signal $x(t)$ to form a signal $x_s[n] = x(nT_s)$. The frequency-domain counterpart of the discrete-time function is its DTFT. In the next section we will look at the relation between these two transforms.

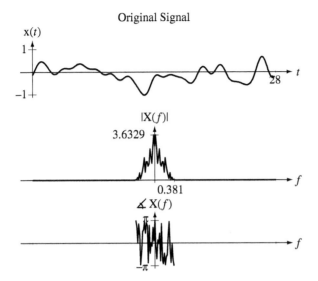

Figure 10.37
An original signal and its CTFT

CTFT-DTFT Relationship The CTFT is the Fourier transform of a continuous-time function and the DTFT is the Fourier transform of a discrete-time function. If we multiply a continuous-time function $x(t)$ by a periodic impulse of period T_s, we create the continuous-time impulse function

$$x_\delta(t) = x(t)\delta_{T_s}(t) = \sum_{n=-\infty}^{\infty} x(nT_s)\delta(t - nT_s). \tag{10.3}$$

If we now form a function $x_s[n]$ whose values are the values of the original continuous-time function $x(t)$ at integer multiples of T_s (and are therefore also the strengths of the impulses in the continuous-time impulse function $x_\delta(t)$), we get the relationship $x_s[n] = x(nT_s)$. The two functions $x_s[n]$ and $x_\delta(t)$ are described by the same set of numbers (the impulse strengths) and contain the same information. If we now find the CTFT of (10.3) we get

$$X_\delta(f) = X(f) * f_s \delta_{f_s}(f) = \sum_{n=-\infty}^{\infty} x(nT_s)e^{-j2\pi f nT_s}$$

where $f_s = 1/T_s$ and $x(t) \xleftrightarrow{\mathcal{F}} X(f)$ or

$$X_\delta(f) = f_s \sum_{k=-\infty}^{\infty} X(f - kf_s) = \sum_{n=-\infty}^{\infty} x_s[n]e^{-j2\pi fn/f_s}.$$

If we make the change of variable $f \to f_s F$ we get

$$X_\delta(f_s F) = f_s \sum_{k=-\infty}^{\infty} X(f_s(F - k)) = \underbrace{\sum_{n=-\infty}^{\infty} x_s[n]e^{-j2\pi nF}}_{=X_s(F)}.$$

The last expression is exactly the definition of the DTFT of $x_s[n]$, which is $X_s(F)$. Summarizing, if $x_s[n] = x(nT_s)$ and $x_\delta(t) = \sum_{n=-\infty}^{\infty} x_s[n]\delta(t - nT_s)$ then

$$\boxed{X_s(F) = X_\delta(f_s F)} \tag{10.4}$$

or

$$X_\delta(f) = X_s(f/f_s).\tag{10.5}$$

Also

$$X_s(F) = f_s \sum_{k=-\infty}^{\infty} X(f_s(F-k))\tag{10.6}$$

(Figure 10.38).

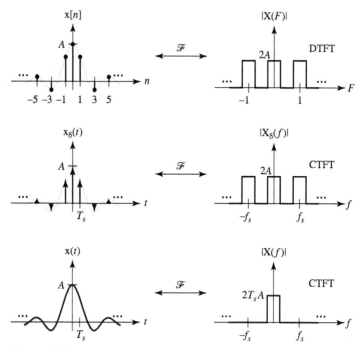

Figure 10.38
Fourier spectra of original signal, impulse-sampled signal, and sampled signal

Now we can write the DTFT of $x_s[n]$ which is $X_s(F)$ in terms of the CTFT of $x(t)$, which is $X(f)$. It is

$$X_s(F) = f_s X(f_s F) * \delta_1(F) = f_s \sum_{k=-\infty}^{\infty} X(f_s(F-k))$$

a frequency-scaled and periodically repeated version of $X(f)$ (Figure 10.39).

Next, we must limit the number of samples to those occurring in the total discrete-time sampling time N. Let the time of the first sample be $n = 0$. (This is the default assumption in the DFT. Other time references could be used but the effect of a different time reference is simply a phase shift that varies linearly with frequency.) This can be accomplished by multiplying $x_s[n]$ by a **window** function

$$w[n] = \begin{cases} 1, & 0 \leq n < N \\ 0, & \text{otherwise} \end{cases}$$

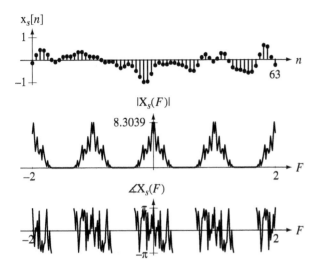

Figure 10.39
Original signal, time-sampled to form a discrete-time signal, and the DTFT of the discrete-time signal

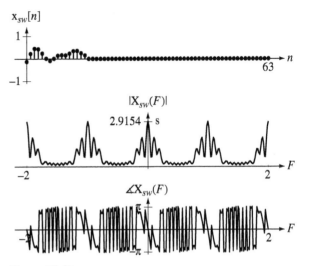

Figure 10.40
Original signal, time-sampled, and windowed to form a discrete-time signal, and the DTFT of that discrete-time signal

as illustrated in Figure 10.40. This window function has exactly N nonzero values, the first one being at discrete time $n = 0$. Call the sampled-and-windowed signal $x_{sw}[n]$. Then

$$x_{sw}[n] = w[n]x_s[n] = \begin{cases} x_s[n], & 0 \leq n < N \\ 0, & \text{otherwise} \end{cases}.$$

The process of limiting a signal to the finite range N in discrete time is called *windowing*, because we are considering only that part of the sampled signal that can be seen through a "window" of finite length. The window function need not be a rectangle. Other window shapes are often used in practice to minimize an effect called **leakage** (described below) in the frequency domain. The DTFT of $x_{sw}[n]$ is the periodic convolution of the DTFT of the signal $x_s[n]$ and the DTFT of the window function $w[n]$, which is $X_{sw}(F) = W(F) \circledast X_s(F)$. The DTFT of the window function is

$$W(F) = e^{-j\pi F(N-1)} N \, \text{drcl}(F, N).$$

Then

$$X_{sw}(F) = e^{-j\pi F(N-1)} N \, \text{drcl}(F, N) \circledast f_s \sum_{k=-\infty}^{\infty} X(f_s(F - k))$$

or, using the fact that periodic convolution with a periodic signal is equivalent to aperiodic convolution with any aperiodic signal that can be periodically repeated to form the periodic signal,

$$X_{sw}(F) = f_s[e^{-j\pi F(N-1)} N \, \text{drcl}(F, N)] * X(f_s F). \quad (10.7)$$

So the effect in the frequency domain of windowing in discrete-time is that the Fourier transform of the time-sampled signal has been periodically convolved with

$$W(F) = e^{-j\pi F(N-1)} N \, \text{drcl}(F, N)$$

(Figure 10.41).

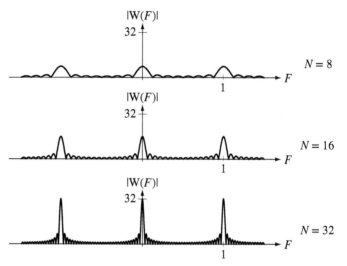

Figure 10.41
Magnitude of the DTFT of the rectangular window function
$$w[n] = \begin{cases} 1, & 0 \le n < N \\ 0, & \text{otherwise} \end{cases}$$ for three different window widths

The convolution process will tend to spread $X_s(F)$ in the frequency domain, which causes the power of $X_s(F)$ at any frequency to "leak" over into nearby frequencies in $X_{sw}(F)$. This is where the term *leakage* comes from. The use of a different window function whose DTFT is more confined in the frequency domain reduces (but can never completely eliminate) leakage. As can be seen in Figure 10.41, as the number of samples N increases, the width of the main lobe of each fundamental period of this function decreases, reducing leakage. So another way to reduce leakage is to use a larger set of samples.

At this point in the process we have a finite sequence of numbers from the sampled-and-windowed signal, but the DTFT of the windowed signal is a periodic function in continuous frequency F and therefore not appropriate for computer storage and manipulation. The fact that the time-domain function has become time limited by the windowing process and the fact that the frequency-domain function is periodic allow us to sample now *in the frequency domain* over one fundamental period to completely describe the frequency-domain function. It is natural at this point to wonder how a frequency-domain function must be sampled to be able to reconstruct it from its samples. The answer is almost identical to the answer for sampling time-domain signals except that time and frequency have exchanged roles. The relations between the time and frequency domains are almost identical because of the duality of the forward and inverse Fourier transforms.

Sampling and Periodic-Repetition Relationship The inverse DFT of a periodic function x[n] with fundamental period N is defined by

$$\text{x}[n] = \frac{1}{N} \sum_{k=\langle N \rangle} \text{X}[k] e^{j2\pi kn/N}. \tag{10.8}$$

Taking the DTFT of both sides, using the DTFT pair $e^{j2\pi F_0 n} \xleftrightarrow{\mathcal{F}} \delta_1(F - F_0)$, we can find the DTFT of x[n], yielding

$$\text{X}(F) = \frac{1}{N} \sum_{k=\langle N \rangle} \text{X}[k] \, \delta_1(F - k/N) \tag{10.9}$$

Then

$$X(F) = \frac{1}{N} \sum_{k=\langle N \rangle} X[k] \sum_{q=-\infty}^{\infty} \delta(F - k/N - q) = \frac{1}{N} \sum_{k=-\infty}^{\infty} X[k]\delta(F - k/N). \quad (10.10)$$

This shows that, for periodic functions, the DFT is simply a scaled special case of the DTFT. If a function $x[n]$ is periodic, its DTFT consists only of impulses occurring at k/N with strengths $X[k]/N$ (Figure 10.42).

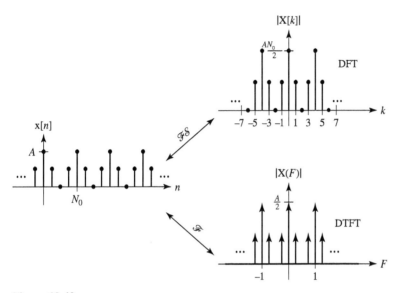

Figure 10.42
Harmonic function and DTFT of $x[n] = (A/2)[1 + \cos(2\pi n/4)]$

Summarizing, for a periodic function $x[n]$ with fundamental period N

$$\boxed{X(F) = \frac{1}{N} \sum_{k=-\infty}^{\infty} X[k]\delta(F - k/N)}. \quad (10.11)$$

Let $x[n]$ be an aperiodic function with DTFT $X(F)$. Let $x_p[n]$ be a periodic extension of $x[n]$ with fundamental period N_p such that

$$x_p[n] = \sum_{m=-\infty}^{\infty} x[n - mN_p] = x[n] * \delta_{N_p}[n] \quad (10.12)$$

(Figure 10.43).
Using the multiplication–convolution duality of the DTFT, and finding the DTFT of (10.12)

$$X_p(F) = X(F)(1/N_p)\delta_{1/N_p}(F) = (1/N_p) \sum_{k=-\infty}^{\infty} X(k/N_p)\delta(F - k/N_p). \quad (10.13)$$

Using (10.11) and (10.13),

$$\boxed{X_p[k] = X(k/N_p)}. \quad (10.14)$$

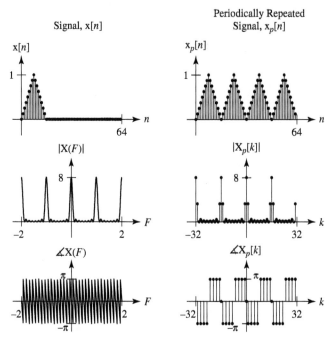

Figure 10.43
A signal and its DTFT and the periodic repetition of the signal and its DFT harmonic function

where $X_p[k]$ is the DFT of $x_p[n]$. If an aperiodic signal $x[n]$ is periodically repeated with fundamental period N_p to form a periodic signal $x_p[n]$ the values of its DFT harmonic function $X_p[k]$ can be found from $X(F)$, which is the DTFT of $x[n]$, evaluated at the discrete frequencies k/N_p. This forms a correspondence between sampling in the frequency domain and periodic repetition in the time domain.

If we now form a periodic repetition of $x_{sw}[n]$

$$x_{swp}[n] = \sum_{m=-\infty}^{\infty} x_{sw}[n - mN],$$

with fundamental period N, its DFT is

$$X_{swp}[k] = X_{sw}(k/N), \quad k \text{ an integer}$$

or, from (10.7),

$$X_{swp}[k] = f_s[e^{-j\pi F(N-1)} N \operatorname{drcl}(F, N) * X(f_s F)]_{F \to k/N}.$$

The effect of the last operation, sampling in the frequency domain, is sometimes called **picket fencing** (Figure 10.44).

Since the nonzero length of $x_{sw}[n]$ is exactly N, $x_{swp}[n]$ is a periodic repetition of $x_{sw}[n]$ with a fundamental period equal to its length so the multiple replicas of $x_{sw}[n]$ do not overlap but instead just touch. Therefore, $x_{sw}[n]$ can be recovered from $x_{swp}[n]$ by simply isolating one fundamental period of $x_{swp}[n]$ in the discrete-time range $0 \leq n < N$.

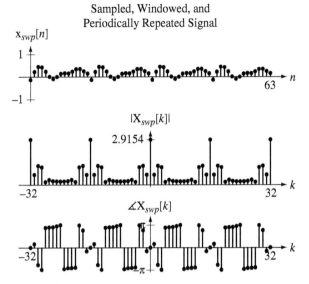

Figure 10.44
Original signal, time-sampled, windowed, and periodically repeated, to form a periodic discrete-time signal and the DFT of that signal

The result

$$X_{swp}[k] = f_s[e^{-j\pi F(N-1)} N \operatorname{drcl}(F,N) * X(f_s F)]_{F \to k/N}$$

is the DFT of a periodic extension of the discrete-time signal formed by sampling the original signal over a finite time.

In summary, in moving from the CTFT of a continuous-time signal to the DFT of samples of the continuous-time signal taken over a finite time, we do the following. In the time domain:

1. Sample the continuous time signal.
2. Window the samples by multiplying them by a window function.
3. Periodically repeat the nonzero samples from step 2.

In the frequency domain:

1. Find the DTFT of the sampled signal, which is a scaled and periodically repeated version of the CTFT of the original signal.
2. Periodically convolve the DTFT of the sampled signal with the DTFT of the window function.
3. Sample in frequency the result of step 2.

The DFT and inverse DFT, being strictly numerical operations, form an exact correspondence between a set of N real numbers and a set of N complex numbers. If the set of real numbers is a set of N signal values over exactly one period of a periodic discrete-time signal $x[n]$, then the set of N complex numbers is a set of complex amplitudes over one period of the DFT $X[k]$ of that discrete-time signal. These are the complex amplitudes of complex discrete-time sinusoids which, when added, will produce the periodic discrete-time signal $N\,x[n]$.

If the set of N real numbers is a set of samples from one period of a bandlimited periodic continuous-time signal sampled above its Nyquist rate and at a rate that is an integer multiple of its fundamental frequency, the numbers returned by the DFT can be scaled and interpreted as complex amplitudes of continuous-time complex sinusoids which, when added, will recreate the periodic continuous-time signal.

So when using the DFT in the analysis of periodic discrete-time signals or bandlimited periodic continuous-time signals, we can obtain results that can be used to exactly compute the DTFS or DTFT or CTFS or CTFT of the periodic signal. When we use the DFT in the analysis of aperiodic signals, we are inherently making an approximation because the DFT and inverse DFT are only exact for periodic signals.

If the set of N real numbers represents all, or practically all, the nonzero values of an aperiodic discrete-time energy signal, we can find an approximation to the DTFT of that signal at a set of discrete frequencies using the results returned by the DFT. If the set of N real numbers represents samples from all, or practically all, the nonzero range of an aperiodic continuous-time signal, we can find an approximation to the CTFT of that continuous-time signal at a set of discrete frequencies using the results returned by the DFT.

Computing the CTFS Harmonic Function with the DFT

It can be shown that if a signal $x(t)$ is periodic with fundamental frequency f_0, and if it is sampled at a rate f_s that is above the Nyquist rate, and if the ratio of the sampling rate to the fundamental frequency f_s/f_0 is an integer, that the DFT of the samples $X[k]$ is related to the CTFS harmonic function of the signal $c_x[k]$ by

$$X[k] = N c_x[k] * \delta_N[k].$$

In this special case the relationship is exact.

Approximating the CTFT with the DFT

Forward CTFT In cases in which the signal to be transformed is not readily describable by a mathematical function or the Fourier-transform integral cannot be done analytically, we can sometimes find an approximation to the CTFT numerically using the DFT. If the signal to be transformed is a causal energy signal, it can be shown that we can approximate its CTFT at discrete frequencies kf_s/N by

$$X(kf_s/N) \cong T_s \sum_{n=0}^{N-1} x(nT_s) e^{-j2\pi kn/N} \cong T_s \times \mathscr{DFT}(x(nT_s)), \quad |k| << N \quad (10.15)$$

where $T_s = 1/f_s$ and N is chosen such that the time range 0 to NT_s covers all or practically all of the signal energy of the signal x (Figure 10.45). So if the signal to be transformed

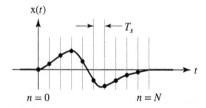

Figure 10.45
A causal energy signal sampled with T_s seconds between samples over a time NT_s

is a causal energy signal and we sample it over a time containing practically all of its energy, the approximation in (10.15) becomes accurate for $|k| \ll N$.

Inverse CTFT The inverse CTFT is defined by $x(t) = \int_{-\infty}^{\infty} X(f)e^{j2\pi ft}df$. If we know $X(kf_s/N)$ in the range $-N \ll -k_{max} \leq k \leq k_{max} \ll N$ and if the magnitude of $X(kf_s/N)$ is negligible outside that range, then it can be shown that for $n \ll N$,

$$x(nT_s) \cong f_s \times \mathcal{DFT}^{-1}(X_{ext}(kf_s/N))$$

where

$$X_{ext}(kf_s/N) = \begin{cases} X(kf_s/N), & -k_{max} \leq k \leq k_{max} \\ 0, & k_{max} < |k| \leq N/2 \end{cases} \text{ and}$$

$$X_{ext}(kf_s/N) = X_{ext}((k+mN)f_s/N).$$

Approximating the DTFT with the DFT

The numerical approximation of the DTFT using the DFT was derived in Chapter 7. The DTFT of $x[n]$ computed at frequencies $F = k/N$ or $\Omega = 2\pi k/N$ is

$$\boxed{X(k/N) \cong \mathcal{DFT}(x[n])}. \qquad (10.16)$$

Approximating Continuous-Time Convolution with the DFT

Aperiodic Convolution Another common use of the DFT is to approximate the convolution of two continuous-time signals using samples from them. Suppose we want to convolve two aperiodic energy signals $x(t)$ and $h(t)$ to form $y(t)$. It can be shown that for $|n| \ll N$,

$$\boxed{y(nT_s) \cong T_s \times \mathcal{DFT}^{-1}(\mathcal{DFT}(x(nT_s)) \times \mathcal{DFT}(h(nT_s)))}. \qquad (10.17)$$

Periodic Convolution Let $x(t)$ and $h(t)$ be two periodic continuous-time signals with a common period T and sample them over exactly that time at a rate f_s above the Nyquist rate, taking N samples of each signal. Let $y(t)$ be the periodic convolution of $x(t)$ with $h(t)$. Then it can be shown that

$$\boxed{y(nT_s) \cong T_s \times \mathcal{DFT}^{-1}(\mathcal{DFT}(x(nT_s)) \times \mathcal{DFT}(h(nT_s)))}. \qquad (10.18)$$

Discrete-Time Convolution with the DFT

Aperiodic Convolution If $x[n]$ and $h[n]$ are energy signals and most or all of their energy occurs in the time range $0 \leq n < N$, then it can be shown that for $|n| \ll N$,

$$\boxed{y[n] \cong \mathcal{DFT}^{-1}(\mathcal{DFT}(x[n]) \times \mathcal{DFT}(h[n]))}. \qquad (10.19)$$

Periodic Convolution Let $x[n]$ and $h[n]$ be two periodic signals with a common period N. Let $y[n]$ be the periodic convolution of $x[n]$ with $h[n]$. Then it can be shown that

$$\boxed{y[n] = \mathcal{DFT}^{-1}(\mathcal{DFT}(x[n]) \times \mathcal{DFT}(h[n]))}. \qquad (10.20)$$

Summary of signal processing using the DFT

CTFS	$c_x[k] \cong e^{-j\pi k/N} \dfrac{\text{sinc}(k/N)}{N} X[k], \; |k| \ll N$
CTFS	$X[k] = N\, c_x[k] * \delta_N[k]$ if $f_s > f_{Nyq}$ and f_s/f_0 is an integer
CTFT	$X(kf_s/N) \cong T_s \times \mathcal{DFT}(x(nT_s))$
DTFT	$X(k/N) \cong \mathcal{DFT}(x[n])$
Continuous-Time Aperiodic Convolution	$[x(t) * h(t)]_{t \to nT_s} \cong T_s \times \mathcal{DFT}^{-1}(\mathcal{DFT}(x(nT_s)) \times \mathcal{DFT}(h(nT_s)))$
Discrete-Time Aperiodic Convolution	$x[n] * h[n] \cong \mathcal{DFT}^{-1}(\mathcal{DFT}(x[n]) \times \mathcal{DFT}(h[n]))$
Continuous-Time Periodic Convolution	$[x(t) \circledast h(t)]_{t \to nT_s} \cong T_s \times \mathcal{DFT}^{-1}(\mathcal{DFT}(x(nT_s)) \times \mathcal{DFT}(h(nT_s)))$
Discrete-Time Periodic Convolution	$x[n] \circledast h[n] = \mathcal{DFT}^{-1}(\mathcal{DFT}(x[n]) \times \mathcal{DFT}(h[n]))$

A typical use of the DFT is to estimate the CTFT of a continuous-time signal using only a finite set of samples taken from it. Suppose we sample a continuous-time signal $x(t)$ 16 times at a 1000 samples/second rate and acquire the samples $x[n]$ illustrated in Figure 10.46.

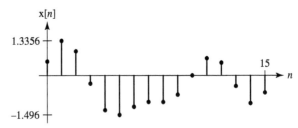

Figure 10.46
Sixteen samples taken from a continuous-time signal

What do we know so far? We know the value of $x(t)$ at 16 points in time, over a time span of 16 ms. We don't know what signal values preceded or followed $x(t)$. We also don't know what values occurred between the samples we acquired. So, to draw any reasonable conclusions about $x(t)$ and its CTFT we will need more information.

Suppose we know that $x(t)$ is bandlimited to less than 500 Hz. If it is bandlimited it cannot be time limited, so we know that outside the time over which we acquired the data the signal, values were not all zero. In fact, they cannot be any constant because, if they were, we could subtract that constant from the signal, creating a time-limited signal, which cannot be bandlimited. The signal values outside the 16 ms time range could vary in many different ways or could repeat in a periodic pattern. If they repeat in a periodic pattern, with this set of 16 values as the fundamental period, then $x(t)$ is a bandlimited, periodic signal and is unique. It is the only bandlimited signal with that fundamental period that could have produced the samples. The samples and the DFT of the samples form a DFT pair

$$x[n] \xleftrightarrow{\mathcal{DFT}}_{16} X[k].$$

The CTFS harmonic function $c_x[k]$ can be found from the DFT through

$$X[k] = N c_x[k] * \delta_N[k] \text{ if } f_s > f_{Nyq} \text{ and } f_s/f_0 \text{ is an integer}$$

and x(t) can therefore be recovered exactly. Also, the CTFT is a set of impulses spaced apart by the signal's fundamental frequency whose strengths are the same as the values of the CTFS harmonic function.

Now let's make a different assumption about what happened outside the 16-ms time of the sample set. Suppose we know that x(t) is zero outside the 16-ms range over which we sampled. Then it is time limited and cannot be bandlimited so we cannot exactly satisfy the sampling theorem. But if the signal is smooth enough and we sample fast enough it is possible that the amount of signal energy in the CTFT above the Nyquist frequency is negligible and we can compute good approximations of the CTFT of x(t) at a discrete set of frequencies using

$$X(kf_s/N) \cong T_s \times \mathcal{DFT}(x(nT_s)).$$

10.3 DISCRETE-TIME SAMPLING

PERIODIC-IMPULSE SAMPLING

In the previous sections all the signals that were sampled were continuous-time signals. Discrete-time signals can also be sampled. Just as in sampling continuous-time signals, the main concern in sampling discrete-time signals is whether the information in the signal is preserved by the sampling process. There are two complementary processes used in discrete-time signal processing to change the sampling rate of a signal, **decimation** and **interpolation**. Decimation is a process of reducing the number of samples and interpolation is a process of increasing the number of samples. We will consider decimation first.

We impulse-sampled a continuous-time signal by multiplying it by a continuous-time periodic impulse. Analogously, we can sample a discrete-time signal by multiplying it by a discrete-time periodic impulse. Let the discrete-time signal to be sampled be x[n]. Then the sampled signal would be

$$x_s[n] = x[n]\delta_{N_s}[n]$$

where N_s is the discrete time between samples (Figure 10.47).

The DTFT of the sampled signal is

$$X_s(F) = X(F) \circledast F_s \delta_{F_s}(F), \quad F_s = 1/N_s$$

(Figure 10.48).

The similarity of discrete-time sampling to continuous-time sampling is obvious. In both cases, if the aliases do not overlap, the original signal can be recovered from the samples and there is a minimum sampling rate for recovery of the signals. The sampling rate must satisfy the inequality $F_s > 2F_m$ where F_m is the discrete-time cyclic frequency above which the DTFT of the original discrete-time signal is zero (in the base fundamental period, $|F| < 1/2$). That is, for $F_m < |F| < 1 - F_m$ the DTFT of the original signal is zero. A discrete-time signal that satisfies this requirement is bandlimited in the discrete-time sense.

Just as with continuous-time sampling, if a signal is properly sampled we can reconstruct it from the samples using interpolation. The process of recovering the original signal is described in the discrete-time-frequency domain as a lowpass filtering operation,

$$X(F) = X_s(F)[(1/F_s)\text{rect}(F/2F_c) * \delta_1(F)]$$

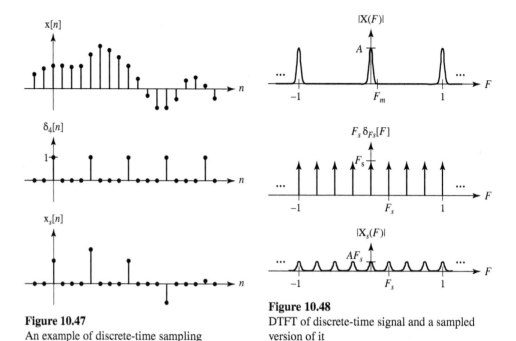

Figure 10.47
An example of discrete-time sampling

Figure 10.48
DTFT of discrete-time signal and a sampled version of it

where F_c is the cutoff discrete-time frequency of the ideal lowpass discrete-time filter. The equivalent operation in the discrete-time domain is a discrete-time convolution,

$$x[n] = x_s[n] * (2F_c/F_s)\operatorname{sinc}(2F_c n).$$

In the practical application of sampling discrete-time signals, it does not make much sense to retain all those zero values between the sampling points because we already know they are zero. Therefore, it is common to create a new signal $x_d[n]$, which has only the values of the discrete-time signal $x_s[n]$ at integer multiples of the sampling interval N_s. The process of forming this new signal is called **decimation**. Decimation was briefly discussed in Chapter 3. The relations between the signals are given by

$$x_d[n] = x_s[N_s n] = x[N_s n].$$

This operation is discrete-time time scaling which, for $N_s > 1$, causes discrete-time time compression and the corresponding effect in the discrete-time-frequency domain is discrete-time frequency expansion. The DTFT of $x_d[n]$ is

$$X_d(F) = \sum_{n=-\infty}^{\infty} x_d[n] e^{-j2\pi Fn} = \sum_{n=-\infty}^{\infty} x_s[N_s n] e^{-j2\pi Fn}.$$

We can make a change of variable $m = N_s n$ yielding

$$X_d(F) = \sum_{\substack{m=-\infty \\ m=\text{integer} \\ \text{multiple of } N_s}}^{\infty} x_s[m] e^{-j2\pi Fm/N_s}.$$

Now, taking advantage of the fact that all the values of $x_s[n]$ between the allowed values, $m = $ integer multiple of N_s, are zero, we can include the zeros in the summation, yielding

$$X_d(F) = \sum_{m=-\infty}^{\infty} x_s[m] e^{-j2\pi (F/N_s)m} = X_s(F/N_s).$$

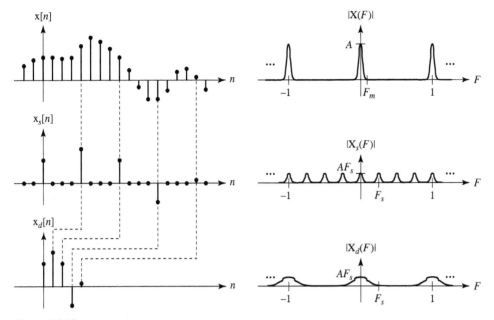

Figure 10.49
Comparison of the discrete-time-domain and discrete-time-frequency domain effects of sampling and decimation

So the DTFT of the decimated signal is a discrete-time-frequency-scaled version of the DTFT of the sampled signal (Figure 10.49).

Observe carefully that the DTFT of the decimated signal is not a discrete-time-frequency scaled version of the DTFT of the original signal, but rather a discrete-time-frequency scaled version of the DTFT of the discrete-time-sampled original signal,

$$X_d(F) = X_s(F/N_s) \neq X(F/N_s).$$

The term **downsampling** is sometimes used instead of decimation. This term comes from the idea that the discrete-time signal was produced by sampling a continuous-time signal. If the continuous-time signal was oversampled by some factor then the discrete-time signal can be decimated by the same factor without losing information about the original continuous-time signal, thus reducing the effective sampling rate or downsampling.

INTERPOLATION

The opposite of decimation is interpolation or **upsampling**. The process is simply the reverse of decimation. First extra zeros are placed between samples, then the signal so created is filtered by an ideal discrete-time lowpass filter. Let the original discrete-time signal be $x[n]$ and let the signal created by adding $N_s - 1$ zeros between samples be $x_s[n]$. Then

$$x_s[n] = \begin{cases} x[n/N_s], & n/N_s \text{ an integer} \\ 0, & \text{otherwise} \end{cases}.$$

This discrete-time expansion of $x[n]$ to form $x_s[n]$ is the exact opposite of the discrete-time compression of $x_s[n]$ to form $x_d[n]$ in decimation, so we should expect the effect in the discrete-time-frequency domain to be the opposite also. A discrete-time expansion by a factor of N_s creates a discrete-time-frequency compression by the same factor,

$$X_s(F) = X(N_s F)$$

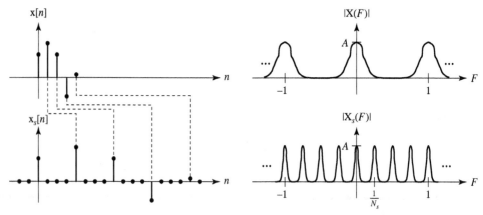

Figure 10.50
Effects, in both the discrete-time and discrete-time-frequency domains, of inserting $N_s - 1$ zeros between samples

(Figure 10.50).
 The signal $x_s[n]$ can be lowpass filtered to interpolate between the nonzero values. If we use an ideal unity-gain lowpass filter with a transfer function

$$H(F) = \text{rect}(N_s F) * \delta_1(F),$$

we get an interpolated signal

$$X_i(F) = X_s(F)[\text{rect}(N_s F) * \delta_1(F)]$$

and the equivalent in the discrete-time domain is

$$x_i[n] = x_s[n] * (1/N_s)\text{sinc}(n/N_s)$$

(Figure 10.51).

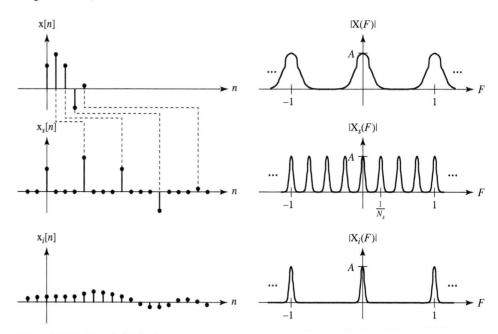

Figure 10.51
Comparison of the discrete-time-domain and discrete-time-frequency domain effects of expansion and interpolation

Notice that the interpolation using the unity-gain ideal lowpass filter introduced a gain factor of $1/N_s$, reducing the amplitude of the interpolated signal $x_i[n]$ relative to the original signal $x[n]$. This can be compensated for by using an ideal lowpass filter with a gain of N_s

$$H(F) = N_s \operatorname{rect}(N_s F) * \delta_1(F)$$

instead of unity gain.

EXAMPLE 10.6

Sample the signal

$$x(t) = 5 \sin(2000\pi t) \cos(20{,}000\pi t)$$

at 80 kHz over one fundamental period to form a discrete-time signal $x[n]$. Take every fourth sample of $x[n]$ to form $x_s[n]$, and decimate $x_s[n]$ to form $x_d[n]$. Then upsample $x_d[n]$ by a factor of eight to form $x_i[n]$ (Figure 10.52 and Figure 10.53).

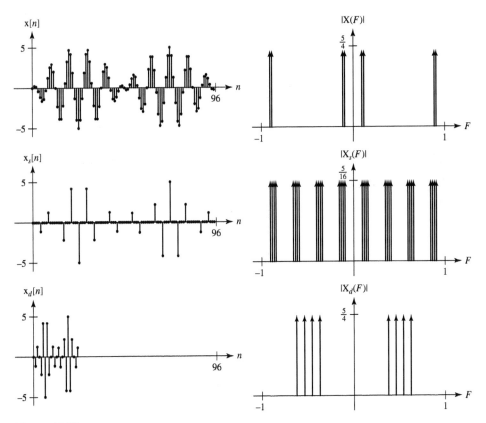

Figure 10.52
Original, sampled, and decimated discrete-time signals and their DTFTs

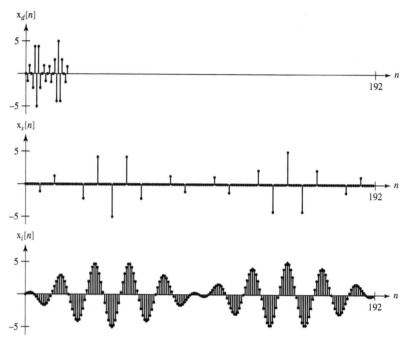

Figure 10.53
Original, upsampled, and discrete-time-lowpass-filtered discrete-time signals

10.4 SUMMARY OF IMPORTANT POINTS

1. A sampled or impulse-sampled signal has a Fourier spectrum that is a periodically repeated version of the spectrum of the signal sampled. Each repetition is called an alias.
2. If the aliases in the spectrum of the sampled signal do not overlap, the original signal can be recovered from the samples.
3. If the signal is sampled at a rate more than twice its highest frequency, the aliases will not overlap.
4. A signal cannot be simultaneously time limited and bandlimited.
5. The ideal interpolating function is the sinc function but since it is noncausal, other methods must be used in practice.
6. A bandlimited periodic signal can be completely described by a finite set of numbers.
7. The CTFT of a signal and the DFT of samples from it are related through the operations sampling in time, windowing, and sampling in frequency.
8. The DFT can be used to approximate the CTFT, the CTFS, and other common signal-processing operations and as the sampling rate and/or number of samples are increased, the approximation gets better.
9. The techniques used in sampling a continuous-time signal can be used in almost the same way in sampling discrete-time signals. There are analogous concepts of bandwidth, minimum sampling rate, aliasing, and so on.

EXERCISES WITH ANSWERS

(Answers to each exercise are in random order.)

Pulse Amplitude Modulation

1. Sample the signal

$$x(t) = 10\operatorname{sinc}(500t)$$

   by multiplying it by the pulse train

$$p(t) = \operatorname{rect}(10^4 t) * \delta_{1\text{ms}}(t)$$

   to form the signal $x_p(t)$. Graph the magnitude of the CTFT $X_p(f)$ of $x_p(t)$.

   Answer:

   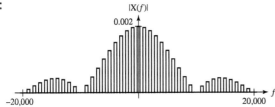

2. Let

$$x(t) = 10\operatorname{sinc}(500t)$$

   and form a signal,

$$x_p(t) = [x(t)\delta_{1\text{ms}}(t)] * \operatorname{rect}(10^4 t).$$

   Graph the magnitude of the CTFT $X_p(f)$ of $x_p(t)$.

   Answer:

   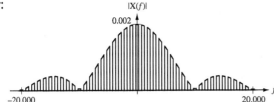

Sampling

3. A signal $x(t) = 25\sin(200\pi t)$ is sampled at 300 samples/second with the first sample being taken at time $t = 0$. What is the value of the fifth sample?

   Answer: 21.651

4. The signal $x(t) = 30\cos(2000\pi t)\sin(50\pi t)$ is sampled at a rate $f_s = 10^4$ samples/second with the first sample occurring at time $t = 0$. What is the value of the third sample?

   Answer: 0.2912

5. A continuous-time signal $x_1(t) = 20\sin(100\pi t)$ is undersampled at $f_s = 40$ samples/second to form a discrete-time signal $x[n]$. If the samples in $x[n]$ had

been instead taken from a continuous-time signal $x_2(t)$ at a rate more than twice the highest frequency in $x_2(t)$, what would be a correct mathematical description of $x_2(t)$?

Answer: $x_2(t) = 20\sin(20\pi t)$

6. A continuous-time signal $x(t) = 10\,\text{sinc}(25t)$ is sampled at $f_s = 20$ samples/second to form $x[n]$ and $x[n] \xleftrightarrow{\mathscr{F}} X(e^{j\Omega})$.

   (a) Find an expression for $X(e^{j\Omega})$

   (b) What is the maximum numerical magnitude of $X(e^{j\Omega})$?

   Answers: $8\,\text{rect}(\Omega/2.5\pi) * \delta_{2\pi}(\Omega)$, 16

7. Given a signal $x(t) = \text{tri}(100t)$ form a signal $x[n]$ by sampling $x(t)$ at a rate $f_s = 800$ and form an information-equivalent impulse signal $x_\delta(t)$ by multiplying $x(t)$ by a periodic sequence of unit impulses whose fundamental frequency is the same $f_0 = f_s = 800$. Graph the magnitude of the DTFT of $x[n]$ and the CTFT of $x_\delta(t)$. Change the sampling rate to $f_s = 5000$ and repeat.

   Answers:

8. Given a bandlimited signal $x(t) = \text{sinc}(t/4)\cos(2\pi t)$ form a signal $x[n]$ by sampling $x(t)$ at a rate $f_s = 4$ and form an information-equivalent impulse signal $x_\delta(t)$ by multiplying $x(t)$ by a periodic sequence of unit impulses whose fundamental frequency is the same $f_0 = f_s = 4$. Graph the magnitude of the DTFT of $x[n]$ and the CTFT of $x_\delta(t)$. Change the sampling rate to $f_s = 2$ and repeat.

   Answers:

Impulse Sampling

9. Let $x_\delta(t) = K\,\text{tri}(t/4)\,\delta_4(t - t_0)$ and let $x_\delta(t) \xleftrightarrow{\mathcal{F}} X_\delta(f)$. For $t_0 = 0$, $X_\delta(f) = X_{\delta 0}(f)$ and for $t_0 = 2$, $X_\delta(f) = G(f)X_{\delta 0}(f)$. What is the function $G(f)$?

   Answer: $\cos(4\pi f)$

10. Let $x(t) = 8\,\text{rect}(t/5)$, let an impulse-sampled version be $x_\delta(t) = x(t)\delta_{T_s}(t)$ and let $x_\delta(t) \xleftrightarrow{\mathcal{F}} X_\delta(f)$. The functional behavior of $X_\delta(f)$ generally depends on T_s but, for this signal, for all values of T_s above some minimum value $T_{s,\min}$ $X_\delta(f)$ is the same. What is the numerical value of $T_{s,\min}$?

    Answer: 2.5

11. For each signal $x(t)$, impulse sample it at the rate specified by multiplying it by a periodic impulse $\delta_{T_s}(t)$ ($T_s = 1/f_s$) and graph the impulse-sampled signal $x_\delta(t)$ over the time range specified and the magnitude and phase of its CTFT $X_\delta(f)$ over the frequency range specified.

    (a) $x(t) = \text{rect}(100t), f_s = 1100$
    $-20\text{ ms} < t < 20\text{ ms}, -3\text{ kHz} < f < 3\text{ kHz}$

    (b) $x(t) = \text{rect}(100t), f_s = 110 \Rightarrow T_s = 1/110 = 9.091\text{ ms}$
    $-20\text{ ms} < t < 20\text{ ms}, -3\text{ kHz} < f < 3\text{ kHz}$

    (c) $x(t) = \text{tri}(45t), f_s = 180$
    $-100\text{ ms} < t < 100\text{ ms}, -400 < f < 400$

    Answers:

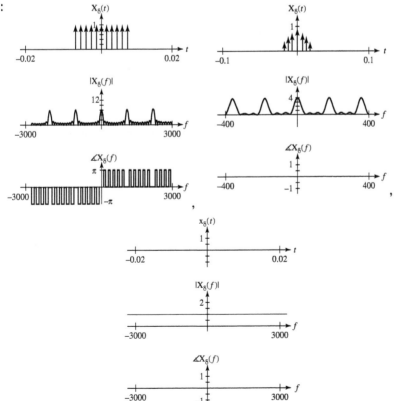

12. Given a signal $x(t) = \text{tri}(200t) * \delta_{0.05}(t)$, impulse sample it at the rate f_s specified by multiplying it by a periodic impulse of the form $\delta_{T_s}(t)$ ($T_s = 1/f_s$). Then filter the impulse-sampled signal $x_\delta(t)$ with an ideal lowpass filter whose gain is T_s in its passband and whose corner frequency is the Nyquist frequency. Graph the signal $x(t)$ and the response of the lowpass filter $x_f(t)$ over the time range -60 ms $< t < 60$ ms.

   (a) $f_s = 1000$ (b) $f_s = 200$ (c) $f_s = 100$

   Answers:

   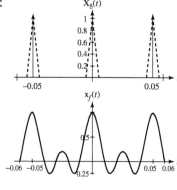 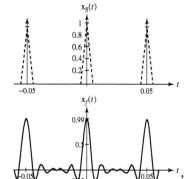

   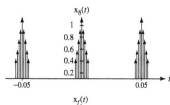

   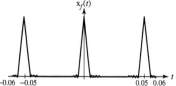

13. Given a signal $x(t) = 8\cos(24\pi t) - 6\cos(104\pi t)$, impulse sample it at the rate specified by multiplying it by a periodic impulse of the form $\delta_{T_s}(t)$ ($T_s = 1/f_s$). Then filter the impulse-sampled signal with an ideal lowpass filter whose gain is T_s in its passband and whose corner frequency is the Nyquist frequency. Graph the signal $x(t)$ and the response of the lowpass filter $x_i(t)$ over two fundamental periods of $x_i(t)$.

   (a) $f_s = 100$ (b) $f_s = 50$ (c) $f_s = 40$

   Answers:

   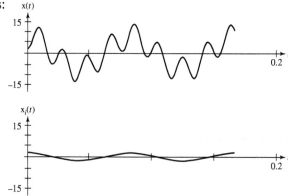

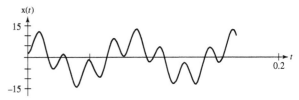
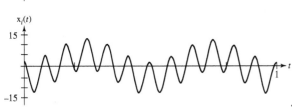

Nyquist Rates

14. Find the Nyquist rates for these signals:

    (a) $x(t) = \text{sinc}(20t)$
    (b) $x(t) = 4\,\text{sinc}^2(100t)$
    (c) $x(t) = 8\sin(50\pi t)$
    (d) $x(t) = 4\sin(30\pi t) + 3\cos(70\pi t)$
    (e) $x(t) = \text{rect}(300t)$
    (f) $x(t) = -10\sin(40\pi t)\cos(300\pi t)$
    (g) $x(t) = \text{sinc}(t/2) * \delta_{10}(t)$
    (h) $x(t) = \text{sinc}(t/2)\,\delta_{0.1}(t)$
    (i) $x(t) = 8\,\text{tri}((t-4)/12)$
    (j) $x(t) = 13\,e^{-20t}\cos(70\pi t)u(t)$
    (k) $x(t) = u(t) - u(t-5)$

    Answers: 70, Infinite, 200, 20, Infinite, 0.4, 340, Infinite, Infinite, Infinite, 50

15. Let $x(t) = 10\cos(4\pi t)$.

    (a) Is $x(t)$ bandlimited? Explain your answer. If it is bandlimited, what is its Nyquist rate?

    (b) If we multiply $x(t)$ by rect(t) to form $y(t)$, is $y(t)$ bandlimited? Explain your answer. If it is bandlimited, what is its Nyquist rate?

    (c) If we multiply $x(t)$ by sinc(t) to form $y(t)$, is $y(t)$ bandlimited? Explain your answer. If it is bandlimited, what is its Nyquist rate?

    (d) If we multiply $x(t)$ by $e^{-\pi t^2}$ to form $y(t)$, is $y(t)$ bandlimited? Explain your answer. If it is bandlimited, what is its Nyquist rate?

    Answers: No, No, {Yes, 4 Hz}, {Yes, 5 Hz}

16. Two sinusoids, one at 40 Hz and the other at 150 Hz, are combined to form a single signal $x(t)$.

    (a) If they are added, what is the Nyquist rate for $x(t)$?

    (b) If they are multiplied, what is the Nyquist rate for $x(t)$?

    Answers: 380 Hz, 300 Hz

Time-Limited and Bandlimited Signals

17. A continuous-time signal x(*t*) is described by x(*t*) = 4 cos(2π*t*)sin(20π*t*). If x(*t*) is filtered by a unity-gain ideal lowpass filter with a bandwidth of 10 Hz, the response is a sinusoid. What are the amplitude and frequency of that sinusoid?

    Answers: 2, 9 Hz

18. Graph these time-limited signals and find and graph the magnitude of their CTFTs and confirm that they are not bandlimited.

    (a) x(*t*) = 5 rect(*t*/100)
    (b) x(*t*) = 10 tri(5*t*)
    (c) x(*t*) = rect(*t*)[1 + cos(2π*t*)]
    (d) x(*t*) = rect(*t*)[1 + cos(2π*t*)] cos(16π*t*)

    Answers:

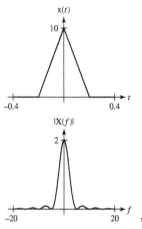 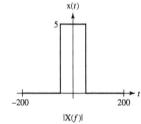 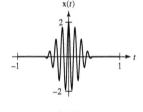 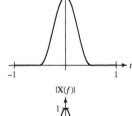
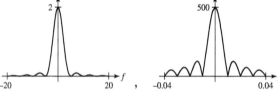 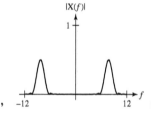 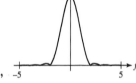

19. Graph the magnitudes of these bandlimited-signal CTFTs and find and graph their inverse CTFTs and confirm that they are not time limited.

    (a) X(*f*) = rect(*f*)$e^{-j4\pi f}$
    (b) X(*f*) = tri(100*f*) $e^{j\pi f}$
    (c) X(*f*) = δ(*f* − 4) + δ(*f* + 4)
    (d) X(*f*) = *j*[δ(*f* + 4) − δ(*f* − 4)] ∗ rect(8*f*)

    Answers:

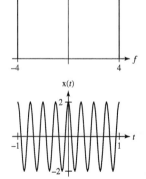 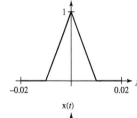 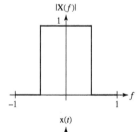 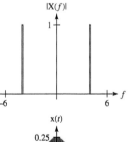
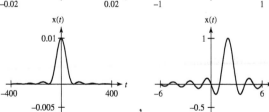 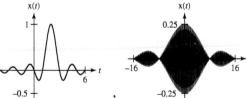

Interpolation

20. Sample the signal $x(t) = \sin(2\pi t)$ at a sampling rate f_s. Then, using MATLAB, graph the interpolation between samples in the time range $-1 < t < 1$ using the approximation,

$$x(t) \cong 2(f_c/f_s) \sum_{n=-N}^{N} x(nT_s) \operatorname{sinc}(2f_c(t - nT_s)),$$

with these combinations of f_s, f_c, and N.

(a) $f_s = 4$, $f_c = 2$, $N = 1$
(b) $f_s = 4$, $f_c = 2$, $N = 2$
(c) $f_s = 8$, $f_c = 4$, $N = 4$
(d) $f_s = 8$, $f_c = 2$, $N = 4$
(e) $f_s = 16$, $f_c = 8$, $N = 8$
(f) $f_s = 16$, $f_c = 8$, $N = 16$

Answers:

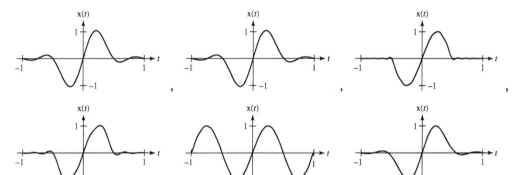

21. For each signal and specified sampling rate, graph the original signal and an interpolation between samples of the signal using a zero-order hold, over the time range $-1 < t < 1$. (The MATLAB function **stairs** could be useful here.)

(a) $x(t) = \sin(2\pi t)$, $f_s = 8$
(b) $x(t) = \sin(2\pi t)$, $f_s = 32$
(c) $x(t) = \operatorname{rect}(t)$, $f_s = 8$
(d) $x(t) = \operatorname{tri}(t)$, $f_s = 8$

Answers:

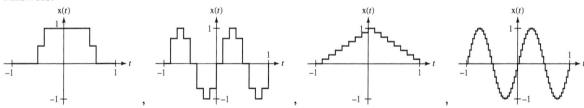

22. Repeat Exercise 21 except use a first-order hold instead of a zero-order hold.

Answers:

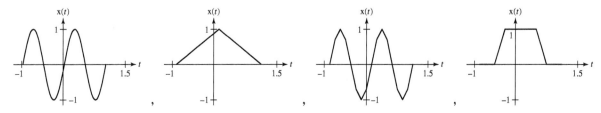

Chapter 10 Sampling and Signal Processing

23. Sample each signal x(t) N times at the rate f_s creating the signal x[n]. Graph x(t) versus t and x[n] versus nT_s over the time range $0 < t < NT_s$. Find the DFT X[k] of the N samples. Then graph the magnitude and phase of X(f) versus f and $T_s X[k]$ versus $k\Delta f$ over the frequency range $-f_s/2 < f < f_s/2$, where $\Delta f = f_s/N$. Graph $T_s X[k]$ as a continuous function of $k\Delta f$ using the MATLAB plot command.

(a) $x(t) = 5\,\text{rect}(2(t-2))$, $f_s = 16$, $N = 64$

(b) $x(t) = 3\,\text{sinc}\left(\dfrac{t-20}{5}\right)$, $f_s = 1$, $N = 40$

(c) $x(t) = 2\,\text{rect}(t-2)\sin(8\pi t)$, $f_s = 32$, $N = 128$

(d) $x(t) = 10\left[\text{tri}\left(\dfrac{t-2}{2}\right) - \text{tri}\left(\dfrac{t-6}{2}\right)\right]$, $f_s = 8$, $N = 64$

(e) $x(t) = 5\cos(2\pi t)\cos(16\pi t)$, $f_s = 64$, $N = 128$

Answers:

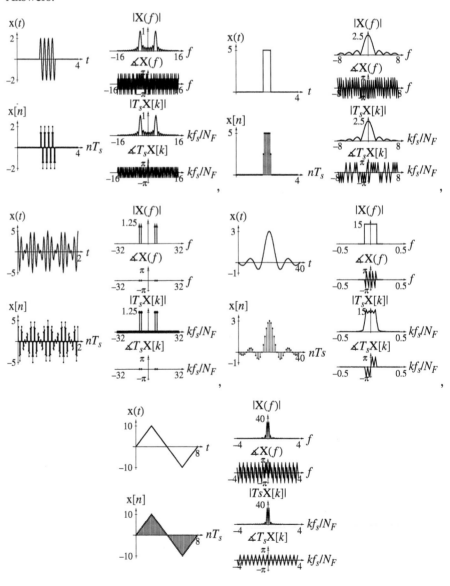

Aliasing

24. For each pair of signals below, sample at the specified rate and find the DTFT of the sampled signals. In each case, explain, by examining the DTFTs of the two signals, why the samples are the same.

    (a) $x_1(t) = 4\cos(16\pi t)$ and $x_2(t) = 4\cos(76\pi t)$, $f_s = 30$

    (b) $x_1(t) = 6\,\text{sinc}(8t)$ and $x_2(t) = 6\,\text{sinc}(8t)\cos(400\pi t)$, $f_s = 100$

    (c) $x_1(t) = 9\cos(14\pi t)$ and $x_2(t) = 9\cos(98\pi t)$, $f_s = 56$

25. For each sinusoid, find the two other sinusoids whose frequencies are nearest to the frequency of the given sinusoid and which, when sampled at the specified rate, have exactly the same samples.

    (a) $x(t) = 4\cos(8\pi t)$, $f_s = 20$ (b) $x(t) = 4\sin(8\pi t)$, $f_s = 20$

    (c) $x(t) = 2\sin(-20\pi t)$, $f_s = 50$ (d) $x(t) = 2\cos(-20\pi t)$, $f_s = 50$

    (e) $x(t) = 5\cos(30\pi t + \pi/4)$, $f_s = 50$

    Answers: $4\cos(48\pi t)$ and $4\cos(32\pi t)$, $-2\sin(-80\pi t)$ and $2\sin(-120\pi t)$, $5\cos(130\pi t + \pi/4)$ and $5\cos(-70\pi t + \pi/4)$, $2\cos(-80\pi t)$ and $2\cos(-120\pi t)$, $4\sin(48\pi t)$ and $-4\sin(32\pi t)$

Bandlimited Periodic Signals

26. Sample the following signals $x(t)$ to form signals $x[n]$. Sample at the Nyquist rate and then at the next higher rate for which f_s/f_0 is an integer (which implies that the total sampling time divided by the time between samples is also an integer). Graph the signals and the magnitudes of the CTFTs of the continuous-time signals and the DTFTs of the discrete-time signals.

    (a) $x(t) = 2\sin(30\pi t) + 5\cos(18\pi t)$ (b) $x(t) = 6\sin(6\pi t)\cos(24\pi t)$

    Answers:

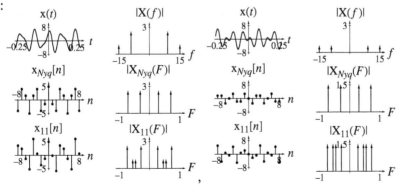

CTFT-CTFS-DFT Relationships

27. Start with a signal $x(t) = 8\cos(30\pi t)$ and sample, window, and periodically repeat it using a sampling rate of $f_s = 60$ and a window width of $N = 32$. For each signal in the process, graph the signal and its transform, either CTFT or DTFT.

Answers:

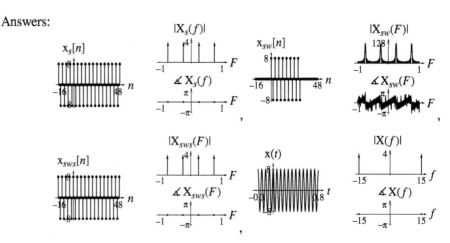

28. Sample each signal x(t) N times at the rate f_s creating the signal x[n]. Graph x(t) versus t and x[n] versus nT_s over the time range $0 < t < NT_s$. Find the DFT X[k] of the N samples. Then graph the magnitude and phase of X(f) versus f and X[k]/N versus $k\Delta f$ over the frequency range $-f_s/2 < f < f_s/2$, where $\Delta f = f_s/N$. Graph X[k]/N as an *impulse* function of $k\Delta f$ using the MATLAB stem command to represent the impulses.

(a) $x(t) = 4\cos(200\pi t)$, $f_s = 800$, $N = 32$

(b) $x(t) = 6\,\text{rect}(2t) * \delta_1(t)$, $f_s = 16$, $N = 128$

(c) $x(t) = 6\,\text{sinc}(4t) * \delta_1(t)$, $f_s = 16$, $N = 128$

(d) $x(t) = 5\cos(2\pi t)\cos(16\pi t)$, $f_s = 64$, $N = 128$

Answers:

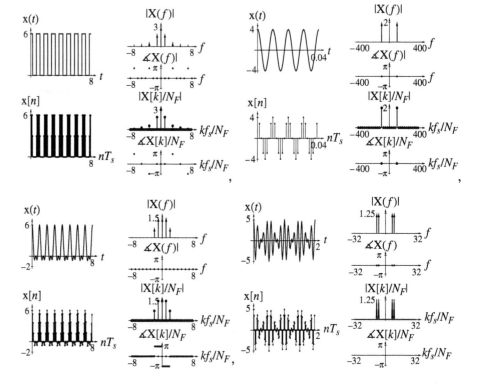

Windows

29. Sometimes window shapes other than a rectangle are used. Using MATLAB, find and graph the magnitudes of the DFTs of these window functions, with $N = 32$.

    (a) von Hann or Hanning

    $$w[n] = \frac{1}{2}\left[1 - \cos\left(\frac{2\pi n}{N-1}\right)\right], \quad 0 \le n < N$$

    (b) Bartlett

    $$w[n] = \begin{cases} \dfrac{2n}{N-1}, & 0 \le n \le \dfrac{N-1}{2} \\ 2 - \dfrac{2n}{N-1}, & \dfrac{N-1}{2} \le n < N \end{cases}$$

    (c) Hamming

    $$w[n] = 0.54 - 0.46\cos\left(\frac{2\pi n}{N-1}\right), \quad 0 \le n < N$$

    (d) Blackman

    $$w[n] = 0.42 - 0.5\cos\left(\frac{2\pi n}{N-1}\right) + 0.08\cos\left(\frac{4\pi n}{N-1}\right), \quad 0 \le n < N$$

Answers:

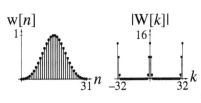 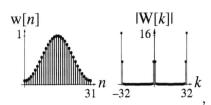

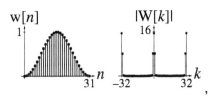 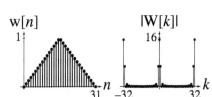

DFT

30. A signal x(t) is sampled four times to produce the signal x[n] and the sample values are

    $$\{x[0], x[1], x[2], x[3]\} = \{7, 3, -4, a\}.$$

    This set of four numbers is the set of input data to the DFT which returns the set $\{X[0], X[1], X[2], X[3]\}$.

    (a) What numerical value of a makes $X[-1]$ a purely real number?

    (b) Let $a = 9$. What is the numerical value of $X[29]$?

    (c) If $X[15] = 9 - j2$, what is the numerical value of $X[1]$?

    Answers: 3, $11 + j6$, $9 + j2$

31. Sample the following signals at the specified rates for the specified times and graph the magnitudes and phases of the DFTs versus harmonic number in the range $-N/2 < k < (N/2) - 1$.

(a) $x(t) = \text{tri}(t - 1)$, $f_s = 2$, $N = 16$

(b) $x(t) = \text{tri}(t - 1)$, $f_s = 8$, $N = 16$

(c) $x(t) = \text{tri}(t - 1)$, $f_s = 16$, $N = 256$

(d) $x(t) = \text{tri}(t) + \text{tri}(t - 4)$, $f_s = 2$, $N = 8$

(e) $x(t) = \text{tri}(t) + \text{tri}(t - 4)$, $f_s = 8$, $N = 32$

(f) $x(t) = \text{tri}(t) + \text{tri}(t - 4)$, $f_s = 64$, $N = 256$

Answers:

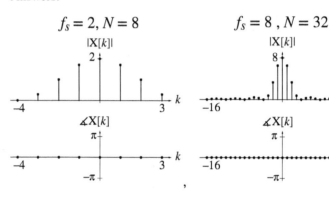

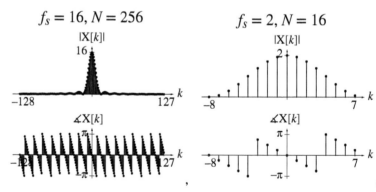

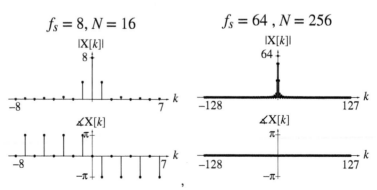

32. For each signal, graph the original signal and the decimated signal for the specified sampling interval. Also graph the magnitudes of the DTFT's of both signals.

(a) $x[n] = \text{tri}\left(\frac{n}{10}\right)$, $N_s = 2$ $x_d[n] = \text{tri}\left(\frac{n}{5}\right)$

(b) $x[n] = (0.95)^n \sin\left(\frac{2\pi n}{10}\right) u[n]$, $N_s = 2$

(c) $x[n] = \cos\left(\frac{2\pi n}{8}\right)$, $N_s = 7$

Answers:

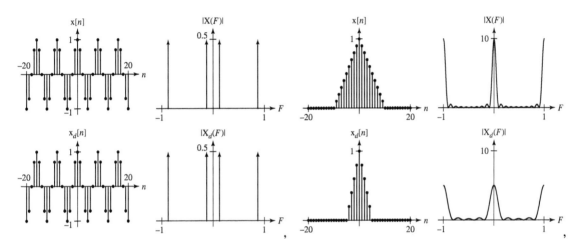

,

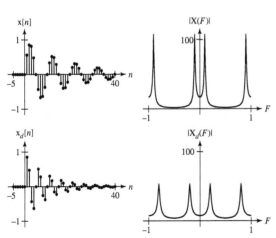

33. For each signal in Exercise 32, insert the specified number of zeros between samples, lowpass discrete-time filter the signals with the specified cutoff frequency, and graph the resulting signal and the magnitude of its DTFT.

(a) Insert 1 zero between points. Cutoff frequency is $F_c = 0.1$.

(b) Insert 4 zeros between points. Cutoff frequency is $F_c = 0.2$.

(c) Insert 4 zeros between points. Cutoff frequency is $F_c = 0.02$.

Answers: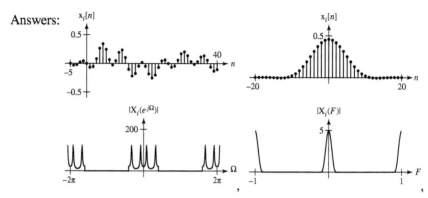

No graph needed

EXERCISES WITHOUT ANSWERS

Sampling

34. The theoretically perfect interpolation function is the sinc function. But we cannot actually use it in practice. Why not?

35. A continuous-time signal with a fundamental period of 2 seconds is sampled at a rate of 6 samples/second. Some selected values of the discrete-time signal that result are

$$x[0] = 3, \ x[13] = 1, \ x[7] = -7, \ x[33] = 0, \ x[17] = -3$$

Find the following numerical values, if it is possible to do so. If it is impossible, explain why.

(a) x[24]   (b) x[18]
(c) x[21]   (d) x[103]

36. The signal $x(t) = 5\text{tri}(t - 1) * \delta_2(t)$ is sampled at a rate of 8 samples/second with the first sample (sample number 1) occurring at time $t = 0$.

(a) What is the numerical value of sample number 6?

(b) What is the numerical value of sample number 63?

37. Using MATLAB (or an equivalent mathematical computer tool) graph the signal,

$$x(t) = 3\cos(20\pi t) - 2\sin(30\pi t)$$

over a time range of $0 < t < 400$ ms. Also graph the signal formed by sampling this function at the following sampling intervals: (a) $T_s = 1/120$ s, (b) $T_s = 1/60$ s, (c) $T_s = 1/30$ s, and (d) $T_s = 1/15$ s. Based on what you observe what can you say about how fast this signal should be sampled so that it could be reconstructed from the samples?

38. A signal, $x(t) = 20\cos(1000\pi t)$ is impulse sampled at a sampling rate of 2000 samples/second. Graph two periods of the impulse-sampled signal $x_\delta(t)$. (Let the one sample be at time, $t = 0$.) Then graph four periods, centered at 0 Hz, of the CTFT $X_\delta(f)$ of the impulse-sampled signal $x_\delta(t)$. Change the sampling rate to 500 samples/second and repeat.

39. A signal $x(t) = 10\,\text{rect}(t/4)$ is impulse sampled at a sampling rate of 2 samples/second. Graph the impulse-sampled signal $x_\delta(t)$ on the interval $-4 < t < 4$. Then

graph three periods, centered at $f = 0$, of the CTFT $X_\delta(f)$ of the impulse-sampled signal $x_\delta(t)$. Change the sampling rate to 1/2 samples/second and repeat.

40. A signal $x(t) = 4\operatorname{sinc}(10t)$ is impulse sampled at a sampling rate of 20 samples/second. Graph the impulse-sampled signal $x_\delta(t)$ on the interval $-0.5 < t < 0.5$. Then graph three periods, centered at $f = 0$, of the CTFT $X_\delta(f)$ of the impulse-sampled signal $x_\delta(t)$. Change the sampling rate to 4 samples/second and repeat.

41. A signal $x[n]$ is formed by sampling a signal $x(t) = 20\cos(8\pi t)$ at a sampling rate of 20 samples/second. Graph $x[n]$ over 10 periods versus discrete time. Then do the same for sampling frequencies of 8 samples/second and 6 samples/second.

42. A signal $x[n]$ is formed by sampling a signal $x(t) = -4\sin(200\pi t)$ at a sampling rate of 400 samples/second. Graph $x[n]$ over 10 periods versus discrete time. Then do the same for sampling frequencies of 200 samples/second and 60 samples/second.

43. A signal $x(t)$ is sampled above its Nyquist rate to form a signal $x[n]$ and is also impulse sampled at the same rate to form an impulse signal $x_\delta(t)$. The DTFT of $x[n]$ is $X(F) = 10\operatorname{rect}(5F) * \delta_1(F)$ or $X(e^{j\Omega}) = 10\operatorname{rect}(5\Omega/2\pi) * \delta_{2\pi}(\Omega)$.

    (a) If the sampling rate is 100 samples/second, what is the highest frequency at which the CTFT of $x(t)$ is nonzero?

    (b) What is the lowest positive frequency greater than the highest frequency in $x(t)$ at which the CTFT of $x_\delta(t)$ is nonzero?

    (c) If the original signal $x(t)$ is to be recovered from the impulse sampled signal $x_\delta(t)$, by using an ideal lowpass filter with impulse response $h(t) = A\operatorname{sinc}(wt)$, what is the maximum possible value of w?

44. Below is a graph of some samples taken from a sinusoid.

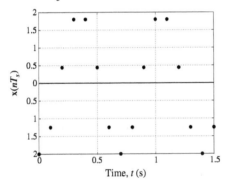

    (a) What is the sampling rate f_s?

    (b) If these samples have been taken properly (according to Shannon's sampling theorem), what is the fundamental frequency f_0 of the sinusoid?

    (c) The sinusoid from which the samples came can be expressed in the form $A\cos(2\pi f_0 t)$. What is the numerical value of A?

    (d) Specify the fundamental frequencies f_{01} and f_{02} of two other cosines of the same amplitude which, when sampled at the same rate, would yield the same set of samples.

45. A bandlimited periodic continuous-time signal is sampled at twice its Nyquist rate over exactly one fundamental period at $f_s = 220$ Hz with the first sample occurring at $t = 0$ and the samples are $\{1, -2, -4, 6, 3, 5, 9, 7\}$.

    (a) What is the maximum frequency at which the continuous-time signal could have any signal power?

(b) What is the fundamental period T_0 of the continuous-time signal?

(c) If sample 1 occurs at time $t = 0$ and the sampling continued indefinitely, what would be the numerical value of sample 317 and at what time would it occur?

46. A wagon wheel has eight spokes. It is rotating at a constant angular velocity. Four snapshots of the wheel are taken at the four times illustrated in Figure E.46. Let the sampling interval (time between snapshots) be 10 ms.

   (a) In Case 1, what are the three lowest positive angular velocities (in revolutions per second, rps) at which it could be rotating?

   (b) In Case 2, what are the three lowest positive angular velocities (in revolutions per second, rps) at which it could be rotating?

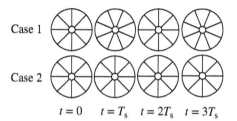

Figure E.46

Impulse Sampling

47. A signal $x(t)$ is impulse-sampled at 100 samples/second (which is above its Nyquist rate) to form the impulse signal $x_\delta(t)$. The CTFTs of the two signals are $X(f)$ and $X_\delta(f)$. What is the numerical ratio, $\dfrac{X_\delta(0)}{X(0)}$?

48. A sinusoidal signal of frequency f_0 is impulse sampled at a rate of f_s. The impulse-sampled signal is the excitation of an ideal lowpass filter with corner (cutoff) frequency of f_c. For each set of parameters below what frequencies are present in the response of the filter? (List only the non-negative frequencies, including zero if present.)

   (a) $f_0 = 20$, $f_s = 50$, $f_c = 210$

   (b) $f_0 = 20$, $f_s = 50$, $f_c = 40$

   (c) $f_0 = 20$, $f_s = 15$, $f_c = 35$

   (d) $f_0 = 20$, $f_s = 5$, $f_c = 22$

   (e) $f_0 = 20$, $f_s = 20$, $f_c = 50$

49. Each signal x below is sampled to form x_s by being multiplied by a periodic impulse of the form $\delta_{T_s}(t)$ for continuous-time signals or $\delta_{N_s}[n]$ for discrete-time signals and $f_s = 1/T_s$ and $N_s = 1/F_s$.

   (a) $x(t) = 4\cos(20\pi t)$, $f_s = 40$. What is the first positive frequency above 10 Hz at which $X_s(f)$ is not zero?

   (b) $x[n] = 3\sin\left(\dfrac{2\pi n}{10}\right)$, $N_s = 3$. If the sampled signal is filtered by an ideal lowpass discrete-time filter, what is the maximum corner frequency that would produce a pure sinusoidal response from the filter? What is the maximum corner frequency that would produce no response at all from the filter?

(c) $x(t) = 10\,\text{tri}(t)$, $f_s = 4$. If the sampled signal is interpolated by simply always holding the last sample value, what would be the value of the interpolated signal at time $t = 0.9$?

(d) $x[n] = 20$, $N_s = 4$. The DTFT of the sampled signal consists entirely of impulses, all of the same strength.

50. A continuous-time signal $x(t)$ is sampled above its Nyquist rate to form a discrete-time signal $x[n]$ and is also impulse sampled at the same rate to form a continuous-time impulse signal $x_\delta(t)$. The DTFT of $x[n]$ is $X(F) = 10\,\text{rect}(5F) * \delta_1(F)$ or $X(e^{j\Omega}) = 10\,\text{rect}\!\left(\dfrac{5\Omega}{2\pi}\right) * \delta_{2\pi}(\Omega)$.

    (a) If the sampling rate is 100 samples/second, what is the highest frequency in $x(t)$?

    (b) What is the lowest positive frequency greater than the highest frequency of $x(t)$ at which the CTFT of $x_\delta(t)$ is nonzero?

    (c) If the original signal $x(t)$ is to be recovered from the impulse sampled signal $x_\delta(t)$ by using an ideal lowpass filter with impulse response $h(t) = A\,\text{sinc}(wt)$, what is the maximum possible value of w?

51. For each signal $x(t)$, impulse sample it at the rate specified by multiplying it by a periodic impulse $\delta_{T_s}(t)$ ($T_s = 1/f_s$) and graph the impulse-sampled signal $x_\delta(t)$ over the time range specified and the magnitude and phase of its CTFT $X_\delta(f)$ over the frequency range specified.

    (a) $x(t) = 5(1 + \cos(200\pi t))\,\text{rect}(100t)$, $f_s = 1600$
    -20 ms $< t < 20$ ms, $-2000 < f < 2000$

    (b) $x(t) = e^{-t^2/2}$, $f_s = 5$
    $-5 < t < 5$, $-15 < f < 15$

    (c) $x(t) = 10e^{-t/20}u(t)$, $f_s = 1$
    $-10 < t < 100$, $-3 < f < 3$

52. Given a signal $x(t) = \text{rect}(20t) * \delta_{0.1}(t)$ and an ideal lowpass filter whose frequency response is $T_s\,\text{rect}(f/f_s)$, process $x(t)$ in two different ways.

    Process 1: Filter the signal and multiply it by f_s.

    Process 2: Impulse sample the signal at the rate specified, then filter the impulse-sampled signal.

    For each sampling rate, graph the original signal $x(t)$ and the processed signal $y(t)$ over the time range $-0.5 < t < 0.5$. In each case, by examining the CTFTs of the signals, explain why the two signals do, or do not, look the same.

    (a) $f_s = 1000$ (b) $f_s = 200$ (c) $f_s = 50$

    (d) $f_s = 20$ (e) $f_s = 10$ (f) $f_s = 4$

    (g) $f_s = 2$

53. Sample the signal

$$x(t) = \begin{cases} 4\sin(20\pi t), & -0.2 < t < 0.2 \\ 0, & \text{otherwise} \end{cases} = 4\sin(20\pi t)\operatorname{rect}(t/0.4)$$

over the time range $-0.5 < t < 0.5$ at the specified sampling rates and approximately reconstruct the signal by using the sinc-function technique

$$x(t) = 2(f_c/f_s)\sum_{n=-\infty}^{\infty} x(nT_s)\operatorname{sinc}(2f_c(t - nT_s))$$

except with a finite set of samples and with the specified filter cutoff frequency. That is, use

$$x(t) = 2(f_c/f_s)\sum_{n=-N}^{N} x(nT_s)\operatorname{sinc}(2f_c(t - nT_s))$$

where $N = 0.5/T_s$. Graph the reconstructed signal in each case.

(a) $f_s = 20, f_c = 10$ $f_s = 20 \Rightarrow T_s = 0.05 \Rightarrow N = 10$

(b) $f_s = 40, f_c = 10$ $f_s = 40 \Rightarrow T_s = 0.025 \Rightarrow N = 20$

(c) $f_s = 40, f_c = 20$ $f_s = 40 \Rightarrow T_s = 0.025 \Rightarrow N = 20$

(d) $f_s = 100, f_c = 10$ $f_s = 100 \Rightarrow T_s = 0.01 \Rightarrow N = 50$

(e) $f_s = 100, f_c = 20$ $f_s = 100 \Rightarrow T_s = 0.01 \Rightarrow N = 50$

(f) $f_s = 100, f_c = 50$ $f_s = 100 \Rightarrow T_s = 0.01 \Rightarrow N = 50$

Nyquist Rates

54. Find the Nyquist rates for these signals:

(a) $x(t) = 15\operatorname{rect}(300t)\cos(10^4\pi t)$

(b) $x(t) = 7\operatorname{sinc}(40t)\cos(150\pi t)$

(c) $x(t) = 15[\operatorname{rect}(500t) * \delta_{1/100}(t)]\cos(10^4\pi t)$

(d) $x(t) = 4[\operatorname{sinc}(500t) * \delta_{1/200}(t)]$

(e) $x(t) = -2[\operatorname{sinc}(500t) * \delta_{1/200}(t)]\cos(10^4\pi t)$

(f) $x(t) = \begin{cases} |t|, & |t| < 10 \\ 0, & |t| \geq 10 \end{cases}$

(g) $x(t) = -8\operatorname{sinc}(101t) + 4\cos(200\pi t)$

(h) $x(t) = -32\operatorname{sinc}(101t)\cos(200\pi t)$

(i) $x(t) = 7\operatorname{sinc}(99t) * \delta_1(t)$

(j) $x(t) = 6\operatorname{tri}(100t)\cos(20000\pi t)$

(k) $x(t) = \begin{cases} 2 + \cos(2\pi t), & |t| \leq 1/2 \\ 1, & |t| > 1/2 \end{cases}$

(l) $x(t) = [4\operatorname{sinc}(20t) * \delta_3(t)]\operatorname{tri}(t/10)$

55. A signal $x_1(t) = 5\sin(20\pi t)$ is sampled at four times its Nyquist rate. Another signal $x_2(t) = 5\sin(2\pi f_0 t)$ is sampled at the same rate. What is the smallest value of f_0 which is greater than 10 and for which the samples of $x_2(t)$ are exactly the same as the samples of $x_1(t)$?

56. A signal $x(t) = 4\cos(200\pi t) - 7\sin(200\pi t)$ is sampled at its Nyquist rate with one of the samples occurring at time $t = 0$. If an attempt is made to reconstruct this signal from these samples by ideal sinc-function interpolation, what signal will actually be created by this interpolation process?

Aliasing

57. Graph the signal $x[n]$ formed by sampling the signal $x(t) = 10\sin(8\pi t)$ at twice the Nyquist rate and $x(t)$ itself. Then on the same graph at least two other continuous-time sinusoids which would yield exactly the same samples if sampled at the same times.

58. A cosine $x(t)$ and a sine signal $y(t)$ of the same frequency are added to form a composite signal $z(t)$. The signal $z(t)$ is then sampled at exactly its Nyquist rate with the usual assumption that a sample occurs at time $t = 0$. Which of the two signals $x(t)$ or $y(t)$ would, if sampled by itself, produce exactly the same set of samples?

59. Each signal x below is impulse sampled to form x_s by being multiplied by a periodic impulse function of the form $\delta_{T_s}(t)$ and $f_s = 1/T_s$.

    (a) $x(t) = 4\cos(20\pi t), f_s = 40$, What is the first positive frequency above 10 Hz at which $X_s(f)$ is not zero?

    (b) $x(t) = 10\,\text{tri}(t), f_s = 4$, If the sampled signal is interpolated by simply always holding the last sample value, what would be the value of the interpolated signal at time $t = 0.9$?

Practical Sampling

60. Graph the magnitude of the CTFT of $x(t) = 25\,\text{sinc}^2(t/6)$. What is the minimum sampling rate required to exactly reconstruct $x(t)$ from its samples? Infinitely many samples would be required to exactly reconstruct $x(t)$ from its samples. If one were to make a practical compromise in which one sampled over the minimum possible time which could contain 99% of the energy of this waveform, how many samples would be required?

61. Graph the magnitude of the CTFT of $x(t) = 8\,\text{rect}(3t)$. This signal is not bandlimited so it cannot be sampled adequately to exactly reconstruct the signal from the samples. As a practical compromise, assume that a bandwidth which contains 99% of the energy of $x(t)$ is great enough to practically reconstruct $x(t)$ from its samples. What is the minimum required sampling rate in this case?

$$X(f) = (8/3)\,\text{sinc}(f/3)$$

Bandlimited Periodic Signals

62. A discrete-time signal $x[n]$ is formed by sampling a continuous-time sinusoid sinusoid $x(t) = A\cos\left(2\pi f_0 t - \frac{\pi}{3} + \theta\right)$ at exactly its Nyquist rate with one of the samples occurring exactly at time, $t = 0$.

    (a) What value of θ_{max} in the range $-\frac{\pi}{2} \leq \theta_{max} \leq \frac{\pi}{2}$ will maximize the signal power of $x[n]$, and, in terms of A, what is that maximum signal power?

(b) What value of θ_{min} in the range $-\frac{\pi}{2} \leq \theta_{min} \leq \frac{\pi}{2}$ will minimize the signal power of x[n], and, in terms of A, what is that minimum signal power?

63. How many sample values are required to yield enough information to exactly describe these bandlimited periodic signals?

   (a) $x(t) = 8 + 3\cos(8\pi t) + 9\sin(4\pi t)$, $f_m = 4, f_{Nyq} = 8$

   (b) $x(t) = 8 + 3\cos(7\pi t) + 9\sin(4\pi t)$, $f_m = 3.5, f_{Nyq} = 7$

64. Sample the signal $x(t) = 15\,[\text{sinc}(5t) * \delta_2(t)]\sin(32\pi t)$ to form the signal x[n]. Sample at the Nyquist rate and then at the next higher rate for which the number of samples per cycle is an integer. Graph the signals and the magnitude of the CTFT of the continuous-time signal and the DTFT of the discrete-time signal.

65. A signal x(t) is periodic and one period of the signal is described by

$$x(t) \begin{cases} 3t, & 0 < t < 5.5 \\ 0, & 5.5 < t < 8 \end{cases}.$$

Find the samples of this signal over one period sampled at a rate of 1 samples/second (beginning at time, t = 0). Then graph, on the same scale, two periods of the original signal and two periods of a periodic signal which is bandlimited to 0.5 Hz or less that would have these same samples.

DFT

66. A signal, x(t), is sampled four times and the samples are {x[0], x[1], x[2], x[3]}. Its DFT is {X[0], X[1], X[2], X[3]}. X[3] can be written as X[3] = ax[0] + bx[1] + cx[2] + dx[3]. What are the numerical values of a, b, c, and d?

67. Sample the bandlimited periodic signal, $x(t) = 8\cos(50\pi t) - 12\sin(80\pi t)$ at exactly its Nyquist rate over exactly one period of x(t). Find the DFT of those samples. From the DFT find the CTFS. Graph the CTFS representation of the signal that results and compare it with x(t). Explain any differences. Repeat for a sampling rate of twice the Nyquist rate.

68. An arbitrary real-valued signal is sampled 32 times to form a set of numbers, {x[0], x[1], $\cdots$, x[31]}. These are fed into an fft algorithm on a computer and it returns the set of numbers, {X[0], X[1], $\cdots$, X[31]}. Which of these returned numbers are guaranteed to be purely real?

69. A bandlimited periodic signal x(t) whose highest frequency is 25 Hz is sampled at 100 samples/second over exactly one fundamental period to form the signal x[n]. The samples are

$$\{x[0], x[1], x[2], x[3]\} = \{a, b, c, d\}.$$

   (a) Let one period of the DFT of those samples be {X[0], X[1], X[2], X[3]}. What is the value of X[1] in terms of a, b, c, and d?

   (b) What is the average value of x(t) in terms of a, b, c, and d?

(c) One of the numbers {X[0], X[1], X[2], X[3]} must be zero. Which one is it and why?

(d) Two of the numbers {X[0], X[1], X[2], X[3]} must be real numbers. Which ones are they and why?

(e) If $X[1] = 2 + j3$, what is the numerical value of $X[3]$ and why?

70. Using MATLAB,

   (a) Generate a pseudo-random sequence of 256 data points in a vector x, using the **randn** function which is built in to MATLAB.

   (b) Find the DFT of that sequence of data and put it in a vector X.

   (c) Set a vector **X1pf** equal to X.

   (d) Change all the values in **X1pf** to zero except the first 8 points and the last 8 points.

   (e) Take the real part of the inverse DFT of **X1pf** and put it in a vector **x1pf**.

   (f) Generate a set of 256 sample times t which begin with 0 and are uniformly separated by 1.

   (g) Graph x and **x1pf** versus t on the same scale and compare. What kind of effect does this operation have on a set of data? Why is the output array called **x1pf**?

71. In Figure E.71 match functions to their DFT magnitudes.

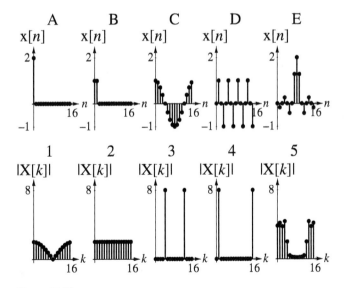

Figure E.71

72. For each x[n] in a–h in Figure E.72, find the DFT magnitude |X[k]| corresponding to it.

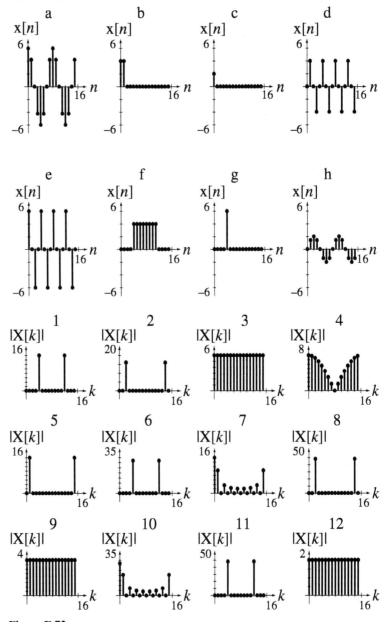

Figure E.72

Discrete-Time Sampling

73. A discrete-time signal, $x[n] = \text{sinc}\left(\frac{2n}{19}\right)$, is sampled by multiplying it by $\delta_{N_s}[n]$. What is the maximum value of N_s for which the original signal can be theoretically reconstructed exactly from the samples?

74. A discrete-time signal is passed through a discrete-time lowpass filter with frequency response $H(F) = \text{rect}(8.5F) * \delta_1(F)$. If every Nth point of the filter's output signal is sampled, what is the maximum numerical value of N for which all the information in the original discrete-time signal is preserved?

CHAPTER 11

Frequency Response Analysis

11.1 INTRODUCTION AND GOALS

Up to this point in this text the material has been highly mathematical and abstract. We have seen some occasional examples of the use of these signal and system analysis techniques but no really in-depth exploration of their use. We are now at the point at which we have enough analytical tools to attack some important types of signals and systems and demonstrate why frequency-domain methods are so popular and powerful in analysis of many systems. Once we have developed a real facility and familiarity with frequency-domain methods we will understand why many professional engineers spend practically their whole careers "in the frequency domain," creating, designing and analyzing systems with transform methods.

Every linear, time-invariant (LTI) system has an impulse response and, through the Fourier transform, a frequency response, and through the Laplace transform a transfer function. We will analyze systems called *filters* that are designed to have a certain frequency response. We will define the term *ideal filter* and we will see ways of approximating the ideal filter. Since frequency response is so important in the analysis of systems, we will develop efficient methods of finding the frequency responses of complicated systems.

CHAPTER GOALS

1. To demonstrate the use of transform methods in the analysis of some systems with practical engineering applications
2. To develop an appreciation of the power of signal and system analysis done directly in the frequency domain

11.2 FREQUENCY RESPONSE

Probably the most familiar example of frequency response in everyday life is the response of the human ear to sounds. Figure 11.1 illustrates the variation of the perception by the average healthy human ear of the loudness of a single sinusoidal frequency of a constant mid-level intensity as a function of frequency from 20 Hz to 20 kHz. This range of frequencies is commonly called the **audio range**.

This frequency response is a result of the structure of the human ear. A system designed with the ear's response in mind is a home-entertainment audio system. This is an

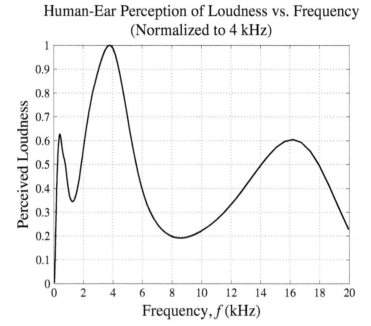

Figure 11.1
Average human ear's perception of the loudness of a constant-amplitude audio tone as a function of frequency

example of a system that is designed without knowing exactly what signals it will process or exactly how they should be processed. But it is known that the signals will lie in the audio frequency range. Since different people have different tastes in music and how it should sound, such a system should have some flexibility. An audio system typically has an amplifier that is capable of adjusting the relative loudness of one frequency versus another through tone controls like bass adjustment, treble adjustment, loudness compensation or a graphic equalizer. These controls allow any individual user of the system to adjust its frequency response for the most pleasing sound with any kind of music.

Audio-amplifier controls are good examples of systems designed in the frequency domain. Their purpose is to shape the frequency response of the amplifier. The term **filter** is commonly used for systems whose main purpose is to shape a frequency response. We have already seen a few examples of filters characterized as lowpass, highpass, bandpass or bandstop. What does the word *filter* mean in general? It is a device for separating something desirable from something undesirable. A coffee filter separates the desirable coffee from the undesirable coffee grounds. An oil filter removes undesirable particulates. In signal and system analysis, a filter separates the desirable part of a signal from the undesirable part. A filter is conventionally defined in signal and system analysis as a device that emphasizes the power of a signal in one frequency range while deemphasizing the power in another frequency range.

11.3 CONTINUOUS-TIME FILTERS

EXAMPLES OF FILTERS

Filters have **passbands** and **stopbands**. A passband is a frequency range in which the filter allows the signal power to pass through relatively unaffected. A stopband is a frequency range in which the filter significantly attenuates the signal power, allowing very little to pass through. The four basic types of filters are **lowpass**, **highpass**, **bandpass** and **bandstop** filters. In a lowpass filter the passband is a region of low frequency and the stopband is a region of high frequency. In a highpass filter those bands are reversed. Low frequencies are attenuated and high frequencies are not. A bandpass filter has a passband in a mid-range of frequencies and stopbands at both low and high frequencies. A bandstop filter reverses the pass and stop bands of the bandpass filter.

Simple adjustments of the bass and treble (low and high frequencies) volume in an audio amplifier could be accomplished by using lowpass and highpass filters with adjustable corner frequencies. We have seen a circuit realization of a lowpass filter. We can also make a lowpass filter using standard continuous-time system building blocks, integrators, amplifiers and summing junctions (Figure 11.2(a)).

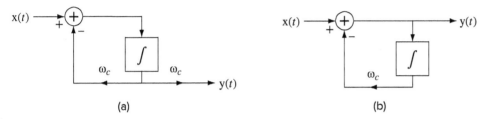

Figure 11.2
Simple filters, (a) lowpass, (b) highpass

The system in Figure 11.2(a) is a lowpass filter with a corner frequency of ω_c (in radians/second) and a frequency response magnitude that approaches one at low frequencies. This is a very simple Direct Form II system. The transfer function is

$$H(s) = \frac{\omega_c}{s + \omega_c}$$

therefore the frequency response is

$$H(j\omega) = H(s)_{s \to j\omega} = \frac{\omega_c}{j\omega + \omega_c} \quad \text{or} \quad H(f) = H(s)_{s \to j2\pi f} = \frac{2\pi f_c}{j2\pi f + 2\pi f_c} = \frac{f_c}{jf + f_c}$$

where $\omega_c = 2\pi f_c$. The system in Figure 11.2(b) is a highpass filter with a corner frequency of ω_c. Its transfer function and frequency response are

$$H(s) = \frac{s}{s + \omega_c}, \quad H(j\omega) = \frac{j\omega}{j\omega + \omega_c}, \quad H(f) = \frac{jf}{jf + f_c}.$$

In either filter, if ω_c can be varied, the relative power of the signals at low and high frequencies can be adjusted. These two systems can be cascade connected to form a bandpass filter (Figure 11.3). The transfer function and frequency response of the bandpass filter are

$$H(s) = \frac{s}{s + \omega_{ca}} \times \frac{\omega_{cb}}{s + \omega_{cb}} = \frac{\omega_{cb} s}{s^2 + (\omega_{ca} + \omega_{cb})s + \omega_{ca}\omega_{cb}}$$

$$H(j\omega) = \frac{j\omega\omega_{cb}}{(j\omega)^2 + j\omega(\omega_{ca} + \omega_{cb}) + \omega_{ca}\omega_{cb}}$$

$$H(f) = \frac{jff_{cb}}{(jf)^2 + jf(f_{ca} + f_{cb}) + f_{ca}f_{cb}}$$

$$f_{ca} = \omega_{ca}/2\pi, \quad f_{cb} = \omega_{cb}/2\pi.$$

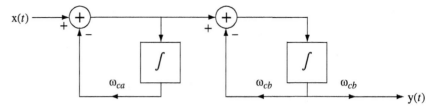

Figure 11.3
A bandpass filter formed by cascading a highpass filter and a lowpass filter

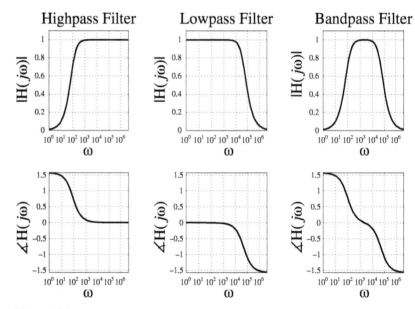

Figure 11.4
High, low and bandpass filter frequency responses

As an example, let $\omega_{ca} = 100$ and let $\omega_{cb} = 50,000$. Then the frequency responses of the lowpass, highpass and bandpass filters are as illustrated in Figure 11.4.

A bandstop filter can be made by parallel connecting a lowpass and highpass filter if the corner frequency of the lowpass filter is lower than the corner frequency of the highpass filter (Figure 11.5).

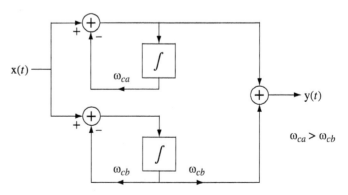

Figure 11.5
A bandstop filter formed by parallel connecting a lowpass filter and a highpass filter

The transfer function and frequency response of the bandstop filter are

$$H(s) = \frac{s^2 + 2\omega_{cb}s + \omega_{ca}\omega_{cb}}{s^2 + (\omega_{ca} + \omega_{cb})s + \omega_{ca}\omega_{cb}}$$

$$H(j\omega) = \frac{(j\omega)^2 + j2\omega\omega_{cb} + \omega_{ca}\omega_{cb}}{(j\omega)^2 + j\omega(\omega_{ca} + \omega_{cb}) + \omega_{ca}\omega_{cb}}$$

$$H(f) = \frac{(jf)^2 + j2ff_{cb} + f_{ca}f_{cb}}{(jf)^2 + jf(f_{ca} + f_{cb}) + f_{ca}f_{cb}}$$

$$f_{ca} = \omega_{ca}/2\pi, f_{cb} = \omega_{cb}/2\pi.$$

If, for example, $\omega_{ca} = 50{,}000$ and $\omega_{cb} = 100$, the frequency response would look like Figure 11.6.

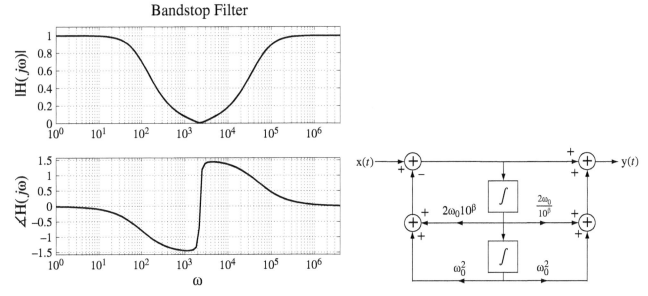

Figure 11.6
Bandstop filter frequency response

Figure 11.7
A biquadratic system

A graphic equalizer is a little more complicated than a simple lowpass, highpass or bandpass filter. It has several cascaded filters, each of which can increase or decrease the frequency response of the amplifier in a narrow range of frequencies. Consider the system in Figure 11.7. Its transfer function and frequency response are

$$H(s) = \frac{s^2 + 2\omega_0 s/10^\beta + \omega_0^2}{s^2 + 2\omega_0 s \times 10^\beta + \omega_0^2}.$$

$$H(j\omega) = \frac{(j\omega)^2 + j2\omega_0\omega/10^\beta + \omega_0^2}{(j\omega)^2 + j2\omega_0\omega \times 10^\beta + \omega_0^2}$$

This transfer function is **biquadratic** in s, a ratio of two quadratic polynomials. If we graph the frequency response magnitude with $\omega_0 = 1$ for several values of the parameter β, we can see how this system could be used as one filter in a graphic equalizer (Figure 11.8).

It is apparent that, with proper selection of the parameter β, this filter can either emphasize or deemphasize signals near its center frequency ω_0 and has a frequency

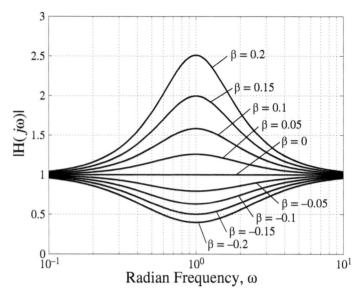

Figure 11.8
Frequency response magnitude for $H(j\omega) = \dfrac{(j\omega)^2 + j2\omega/10^\beta + 1}{(j\omega)^2 + j2\omega \times 10^\beta + 1}$

response approaching one for frequencies far from its center frequency. A set of cascaded filters of this type, each with a different center frequency, can be used to emphasize or deemphasize multiple bands of frequencies and thereby to tailor the frequency response to almost any shape a listener might desire (Figure 11.9).

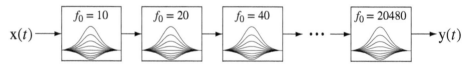

Figure 11.9
Conceptual block diagram of a graphic equalizer

With all the filters set to emphasize their frequency range the magnitude frequency responses of the subsystems could look like Figure 11.10. The center frequencies of these filters are 20 Hz, 40 Hz, 80 Hz, ..., 20,480 Hz. The filters are spaced at **octave** intervals in frequency. An octave is a factor-of-two change in frequency. That makes the individual-filter center frequencies be uniformly spaced on a logarithmic scale, and the bandwidths of the filters are also uniform on a logarithmic scale.

Another example of a system designed to handle unknown signals would be an instrumentation system measuring pressure, temperature, flow and so on in an industrial process. We do not know exactly how these process parameters vary. But they normally lie within some known range and can vary no faster than some maximum rate because of the physical limitations of the process. Again, this knowledge allows us to design a signal processing system appropriate for these types of signals.

Even though a signal's exact characteristics may be unknown, we usually know something about it. We often know its approximate **power spectrum**. That is, we have an approximate description of the signal power of the signal in the frequency domain. If we could not mathematically calculate the power spectrum, we could estimate it based on knowledge of the physics of the system that created it or we could measure it. One way to measure it would be through the use of filters.

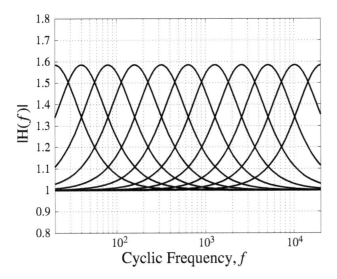

Figure 11.10
Frequency response magnitudes for 11 filters spanning the audio range

IDEAL FILTERS

Distortion

An **ideal** lowpass filter would pass all signal power at frequencies below some maximum, without distorting the signal at all in that frequency range, and completely stop or block all signal power at frequencies above that maximum. It is important here to define precisely what is meant by **distortion**. Distortion is commonly construed in signal and system analysis to mean changing the shape of a signal. This does not mean that if we change the signal we necessarily distort it. Multiplication of the signal by a constant, or a time shift of the signal, are changes that are not considered to be distortion.

Suppose a signal $x(t)$ has the shape illustrated at the top of Figure 11.11(a). Then the signal at the bottom of Figure 11.11(a) is an undistorted version of that signal. Figure 11.11(b) illustrates one type of distortion.

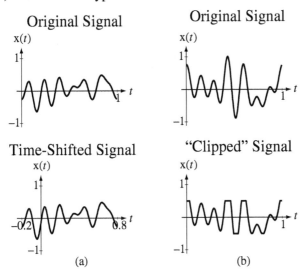

Figure 11.11
(a) An original signal and a changed, but undistorted, version of it, (b) an original signal and a distorted version of it

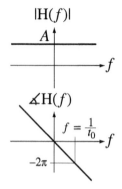

Figure 11.12
Magnitude and phase of a distortionless system

The response of any LTI system is the convolution of its excitation with its impulse response. Any signal convolved with a unit impulse at the origin is unchanged, $x(t) * \delta(t) = x(t)$. If the impulse has a strength other than one, the signal is multiplied by that strength but the shape is still unchanged, $x(t) * A\delta(t) = Ax(t)$. If the impulse is time shifted, the convolution is time shifted also, but without changing the shape, $x(t) * A\delta(t - t_0) = Ax(t - t_0)$. Therefore, the impulse response of a filter that does not distort would be an impulse, possibly with strength other than one and possibly shifted in time. The most general form of impulse response of a distortionless system would be $h(t) = A\delta(t - t_0)$. The corresponding frequency response would be the CTFT of the impulse response $H(f) = Ae^{-j2\pi f t_0}$. The frequency response can be characterized by its magnitude and phase $|H(f)| = A$ and $\angle H(f) = -2\pi f t_0$. Therefore a distortionless system has a frequency response magnitude that is constant with frequency and a phase that is linear with frequency (Figure 11.12).

It should be noted here that a distortionless impulse response or frequency response is a concept that cannot actually be realized in any real physical system. No real system can have a frequency response that is constant all the way to an infinite frequency. Therefore the frequency responses of all real physical systems must approach zero as frequency approaches infinity.

Filter Classifications

Since the purpose of a filter is to remove the undesirable part of a signal and leave the rest, no filter, not even an ideal one, is distortionless because its magnitude is not constant with frequency. But an ideal filter is distortionless *within its passband*. Its frequency response magnitude is constant within the passband and its frequency-response phase is linear within the passband.

We can now define the four basic types of ideal filter. In the following descriptions, f_m, f_L and f_H are all positive and finite.

An ideal lowpass filter passes signal power for frequencies $0 < |f| < f_m$ without distortion and stops signal power at other frequencies.

An ideal highpass filter stops signal power for frequencies $0 < |f| < f_m$ and passes signal power at other frequencies without distortion.

An ideal bandpass filter passes signal power for frequencies $f_L < |f| < f_H$ without distortion and stops signal power at other frequencies.

An ideal bandstop filter stops signal power for frequencies $f_L < |f| < f_H$ and passes signal power at other frequencies without distortion.

Ideal Filter Frequency Responses

Figure 11.13 and Figure 11.14 illustrate typical magnitude and phase frequency responses of the four basic types of ideal filters.

It is appropriate here to define a word that is very commonly used in signal and system analysis, **bandwidth**. The term *bandwidth* is applied to both signals and systems. It generally means "a range of frequencies." This could be the range of frequencies present in a signal or the range of frequencies a system passes or stops. For historical reasons, it is usually construed to mean a range of frequencies in positive frequency space. For example, an ideal lowpass filter with corner frequencies of $\pm f_m$ as illustrated in Figure 11.13 is said to have a bandwidth of f_m, even though the width of the filter's nonzero magnitude frequency response is obviously $2f_m$. The ideal bandpass filter has a bandwidth of $f_H - f_L$, which is the width of the passband in positive frequency space.

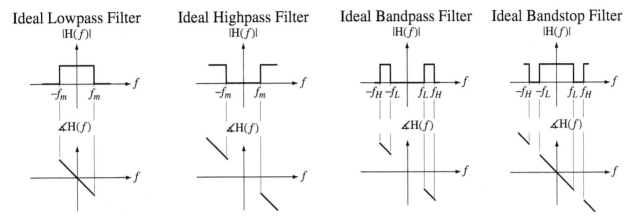

Figure 11.13
Magnitude and phase frequency responses of ideal lowpass and highpass filters

Figure 11.14
Magnitude and phase frequency responses of ideal bandpass and bandstop filters

There are many different kinds of definitions of bandwidth, absolute bandwidth, half-power bandwidth, null bandwidth and so on (Figure 11.15). Each of them is a range of frequencies but defined in different ways. For example, if a signal has no signal power at all below some minimum positive frequency and above some maximum positive frequency, its absolute bandwidth is the difference between those two frequencies. If a signal has a finite absolute bandwidth it is said to be **strictly bandlimited** or, more commonly, just **bandlimited**. Most real signals are not known to be bandlimited so other definitions of bandwidth are needed.

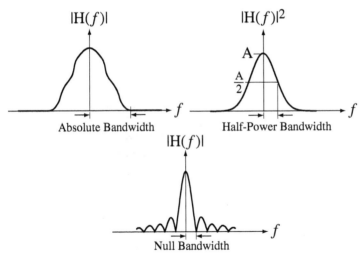

Figure 11.15
Examples of bandwidth definitions

Impulse Responses and Causality

The impulse responses of ideal filters are the inverse transforms of their frequency responses. The impulse and frequency responses of the four basic types of ideal filter are summarized in Figure 11.16.

These descriptions are general in the sense that they involve an arbitrary gain constant A and an arbitrary time delay t_0. Notice that the ideal highpass filter and the ideal bandstop filter have frequency responses extending all the way to infinity. This is

Ideal Filter Type	Frequency Response
Lowpass	$H(f) = A\text{rect}(f/2f_m)e^{-j2\pi f t_0}$
Highpass	$H(f) = A[1 - \text{rect}(f/2f_m)]e^{-j2\pi f t_0}$
Bandpass	$H(f) = A[\text{rect}((f-f_0)/\Delta f) + \text{rect}((f+f_0)/\Delta f)]e^{-j2\pi f t_0}$
Bandstop	$H(f) = A[1 - \text{rect}((f-f_0)/\Delta f) - \text{rect}((f+f_0)/\Delta f)]e^{-j2\pi f t_0}$

Ideal Filter Type	Impulse Response
Lowpass	$h(t) = 2Af_m \,\text{sinc}(2f_m(t-t_0))$
Highpass	$h(t) = A\delta(t-t_0) - 2Af_m \,\text{sinc}(2f_m(t-t_0))$
Bandpass	$h(t) = 2A\Delta f \,\text{sinc}(\Delta f(t-t_0))\cos(2\pi f_0(t-t_0))$
Bandstop	$h(t) = A\delta(t-t_0) - 2A\Delta f \,\text{sinc}(\Delta f(t-t_0))\cos(2\pi f_0(t-t_0))$

$$\Delta f = f_H - f_L, \quad f_0 = (f_H + f_L)/2$$

Figure 11.16
Frequency responses and impulse responses of the four basic types of ideal filter

impossible in any real physical system. Therefore, practical approximations to the ideal highpass and bandstop filter allow higher-frequency signals to pass but only up to some high, not infinite, frequency. "High" is a relative term and, as a practical matter, usually means beyond the frequencies of any signals actually expected to occur in the system.

In Figure 11.17 are some typical shapes of impulse responses for the four basic types of ideal filter.

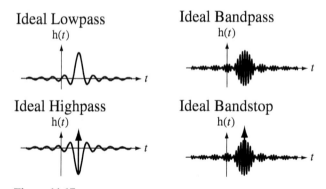

Figure 11.17
Typical impulse responses of ideal lowpass, highpass, bandpass and bandstop filters

As mentioned above, one reason ideal filters are called ideal is that they cannot physically exist. The reason is not simply that perfect system components with ideal characteristics do not exist (although that would be sufficient). It is more fundamental than that. Consider the impulse responses depicted in Figure 11.17. They are the responses of the filters to a unit impulse applied at time $t = 0$. Notice that all of the impulse responses of these ideal filters have a nonzero response *before the impulse is applied* at time $t = 0$. In fact, all of *these particular* impulse responses begin at *an infinite time* before time $t = 0$. It should be intuitively obvious that a real system cannot look into the future and anticipate the application of the excitation and start responding before it occurs. All ideal filters are noncausal.

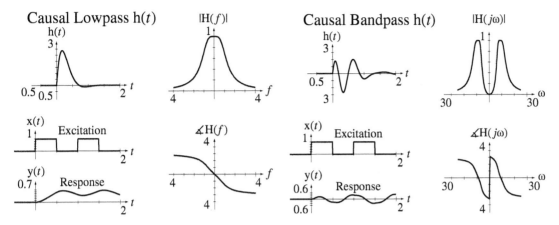

Figure 11.18
Impulse responses, frequency responses and responses to square waves of causal lowpass and bandpass filters

Although ideal filters cannot be built, useful approximations to them can be built. In Figure 11.18 and Figure 11.19 are some examples of the impulse responses, frequency responses and responses to square waves of some nonideal, causal filters that approximate the four common types of ideal filters.

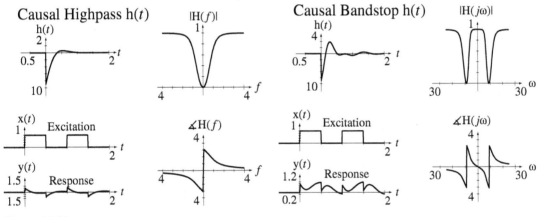

Figure 11.19
Impulse responses, frequency responses and responses to square waves of causal highpass and bandstop filters

The lowpass filter smooths the square wave by removing high-frequency signal power from it but leaves the low-frequency signal power (including zero frequency), making the average values of the input and output signals the same (because the frequency response at zero frequency is one). The bandpass filter removes high-frequency signal power, smoothing the signal, and removes low-frequency power (including zero frequency), making the average value of the response zero.

The highpass filter removes low-frequency signal power from the square wave, making the average value of the response zero. But the high-frequency signal power that defines the sharp discontinuities in the square wave is preserved. The bandstop filter removes signal power in a small range of frequencies and leaves the very low-frequency and very high-frequency signal power. So the discontinuities and the average value of the square wave are both preserved but some of the mid-frequency signal power is removed.

The Power Spectrum

One purpose of launching into filter analysis was to explain one way of determining the power spectrum of a signal by measuring it. That could be accomplished by the system illustrated in Figure 11.20. The signal is routed to multiple bandpass filters, each with the same bandwidth but a unique center frequency. Each filter's response is that part of the signal lying in the frequency range of the filter. Then the output signal from each filter is the input signal of a squarer and its output signal is the input signal of a time averager. A squarer simply takes the square of the signal. This is not a linear operation, so this is not a linear system. The output signal from any squarer is that part of the instantaneous signal power of the original $x(t)$ that lies in the passband of the bandpass filter. Then the time averager forms the time-average signal power. Each output response $P_x(f_n)$ is a measure of the signal power of the original $x(t)$ in a narrow band of frequencies centered at f_n. Taken together, the P's are an indication of the variation of the signal power with frequency, the power spectrum.

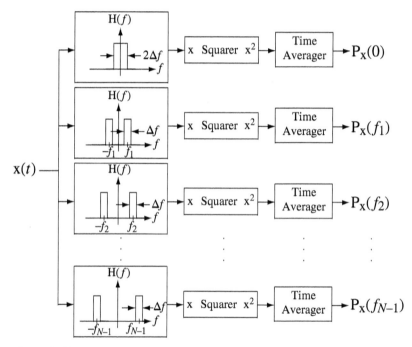

Figure 11.20
A system to measure the power spectrum of a signal

It is unlikely that an engineer today would actually build a system like this to measure the power spectrum of a signal. A better way to measure it is to use an instrument called a **spectrum analyzer**. But this illustration is useful because it reinforces the concept of what a filter does and what the term *power spectrum* means.

Noise Removal

Every useful signal always has another undesirable signal called **noise** added to it. One very important use of filters is in removing noise from a signal. The sources of noise are many and varied. By careful design, noise can often be greatly reduced but can never be completely eliminated. As an example of filtering, suppose the signal power is confined to a range of low frequencies and the noise power is spread over a much

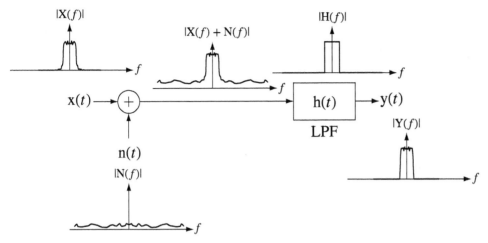

Figure 11.21
Partial removal of noise by a lowpass filter

wider range of frequencies (a very common situation). We can filter the signal plus noise with a lowpass filter and reduce the noise power without having much effect on the signal power (Figure 11.21).

The ratio of the signal power of the desired signal to the signal power of the noise is called the **signal-to-noise ratio**, often abbreviated **SNR**. Probably the most fundamental consideration in communication system design is to maximize the SNR, and filtering is a very important technique in maximizing SNR.

BODE DIAGRAMS

The Decibel

In graphing frequency response, the magnitude of the frequency response is often converted to a logarithmic scale using a unit called the **decibel (dB)**. If the frequency response magnitude is

$$|H(j\omega)| = \left|\frac{Y(j\omega)}{X(j\omega)}\right|,$$

then that magnitude, expressed in decibels, is

$$\boxed{|H(j\omega)|_{dB} = 20\log_{10}|H(j\omega)| = 20\log_{10}\left|\frac{Y(j\omega)}{X(j\omega)}\right| = |Y(j\omega)|_{dB} - |X(j\omega)|_{dB}}. \quad (11.1)$$

The name decibel comes from the original unit defined by Bell Telephone engineers, the **bel (B)**, named in honor of Alexander Graham Bell,[1] the inventor of the telephone. The bel is defined as the common logarithm (base 10) of a power ratio. For example, if the response signal power of a system is 100 and the input signal power

---

[1] Alexander Graham Bell was born in Scotland in a family specializing in elocution. In 1864 he became a resident master in Elgin's Weston House Academy in Scotland where he studied sound and first thought of transmitting speech with electricity. He moved to Canada in 1870 to recuperate from tuberculosis and later settled in Boston. There he continued working on transmitting sound over wires and on March 7, 1876, he was granted a patent for the telephone, arguably the most valuable patent ever issued. He became independently wealthy as a result of the income derived from this patent. In 1898 he became president of the National Geographic Society.

(expressed in the same units) is 20, the signal-power gain of the system, expressed in bels would be

$$\log_{10}(P_Y/P_X) = \log_{10}(100/20) \cong 0.699 \text{ B}.$$

Since the prefix **deci** is the international standard for one-tenth, a decibel is one-tenth of a bel and that same power ratio expressed in dB would be 6.99 dB. So the power gain, expressed in dB, would be $10\log_{10}(P_Y/P_X)$. Since signal power is proportional to the square of the signal itself, the ratio of powers, expressed directly in terms of the signals, would be

$$10\log_{10}(P_Y/P_X) = 10\log_{10}(Y^2/X^2) = 10\log_{10}[(Y/X)^2] = 20\log_{10}(Y/X).$$

In a system in which multiple subsystems are cascaded, the overall frequency response is the product of the individual frequency responses, but the overall frequency response expressed in dB is the sum of the individual frequency responses expressed in dB because of the logarithmic definition of the dB. Also, use of decibels may reveal frequency response behavior that is difficult to see on a linear graph.

Before considering the frequency responses of practical filters it is useful to become familiar with a very helpful and common way of displaying frequency response. Often linear graphs of frequency response, although accurate, do not reveal important system behavior. As an example, consider the graphs of the two quite different-looking frequency responses,

$$H_1(j\omega) = \frac{1}{j\omega + 1} \quad \text{and} \quad H_2(j\omega) = \frac{30}{30 - \omega^2 + j31\omega},$$

(Figure 11.22).

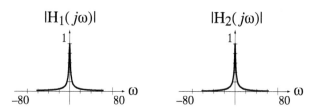

Figure 11.22
Comparison of the magnitudes of two apparently different frequency responses

Graphed this way, the two magnitude frequency response graphs look identical, yet we know the frequency responses are different. One way of seeing small differences between frequency responses is to graph them in dB. The decibel is defined logarithmically. A logarithmic graph deemphasizes large values and emphasizes small values. Then small differences between frequency responses can be more easily seen (Figure 11.23).

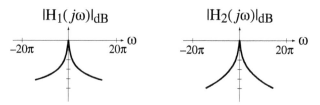

Figure 11.23
Log-magnitude graphs of the two frequency responses

In the linear graphs, the behavior of the magnitude frequency response looked identical because at very small values, the two graphs look the same. In a dB graph, the difference between the two magnitude frequency responses at very small values can be seen.

Although this type of graph is used sometimes, a more common way of displaying frequency response is the **Bode[2] diagram** or **Bode plot**. Like the log-magnitude graph, the Bode diagram reveals small differences between frequency responses but it is also a systematic way of quickly sketching or estimating the overall frequency response of a system that may contain multiple cascaded frequency responses. A log-magnitude graph is logarithmic in one dimension. A magnitude Bode diagram is logarithmic in both dimensions. A magnitude frequency response Bode diagram is a graph of the frequency response magnitude in dB against a logarithmic frequency scale. Since the frequency scale is now logarithmic, only positive frequencies can be used in a graph. That is not a loss of information since, for frequency responses of real systems, the value of the frequency response at any negative frequency is the complex conjugate of the value at the corresponding positive frequency.

Returning now to the two different system frequency responses

$$H_1(j\omega) = \frac{1}{j\omega + 1} \quad \text{and} \quad H_2(j\omega) = \frac{30}{30 - \omega^2 + j31\omega},$$

if we make a Bode diagram of each of them, their difference becomes more evident (Figure 11.24). The dB scale makes the behavior of the two magnitude frequency responses at the higher frequencies distinguishable.

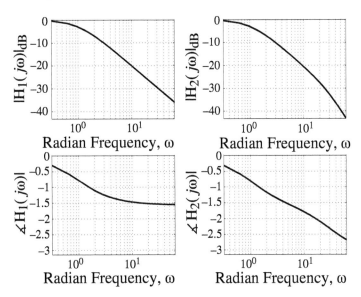

Figure 11.24
Bode diagrams of the two example of frequency responses

---

[2] Hendrik Bode received a B.A. in 1924 and an M.A. in 1926 from Ohio State University. In 1926 he started work at Bell Telephone Laboratories and worked with electronic filters and equalizers. While employed at Bell Labs, he went to Columbia University Graduate School and received his Ph.D. in 1935. In 1938 Bode used the magnitude and phase frequency response plots of a complex function. He investigated closed-loop stability using the notions of gain and phase margin. These Bode plots are used extensively with many electronic systems. He published *Network Analysis and Feedback Amplifier Design*, considered to be a very important book in this field. Bode retired in October 1967 and was elected Gordon Mckay Professor of Systems Engineering at Harvard University.

Although the fact that differences between low levels of magnitude frequency response can be better seen with a Bode diagram is a good reason to use it, it is by no means the only reason. It is not even the main reason. The fact that system gains in dB add instead of multiplying when systems are cascaded makes the quick graphical estimation of overall system gain characteristics easier using Bode diagrams than using linear graphs.

Most LTI systems are described by linear differential equations with constant coefficients. The most general form of such an equation is

$$\sum_{k=0}^{N} a_k \frac{d^k}{dt^k} y(t) = \sum_{k=0}^{M} b_k \frac{d^k}{dt^k} x(t) \qquad (11.2)$$

where $x(t)$ is the excitation and $y(t)$ is the response. From Chapter 5 we know that the transfer function is

$$H(s) = \frac{b_M s^M + b_{M-1} s^{M-1} + \cdots + b_1 s + b_0}{a_N s^N + a_{N-1} s^{N-1} + \cdots + a_1 s + b_0}.$$

The numerator and denominator polynomials can be factored, putting the transfer function into the form

$$H(s) = A \frac{(1 - s/z_1)(1 - s/z_2) \cdots (1 - s/z_M)}{(1 - s/p_1)(1 - s/p_2) \cdots (1 - s/p_N)}$$

where the z's and p's are the zeros and poles.

For real systems the coefficients a and b in (11.2) are all real and all the finite p's and z's in the factored forms must either be real or must occur in complex conjugate pairs, so that when the factored numerator and denominator are multiplied out to obtain the ratio-of-polynomials form, all the coefficients of the powers of s are real.

From the factored form, the system transfer function can be considered as being the cascade of a frequency-independent gain A and multiple subsystems, each having a transfer function with one finite pole or one finite zero. If we now convert the transfer function to a frequency response through $s \to j\omega$, we can think of the overall frequency response as resulting from the cascade of multiple components, each with a simple frequency response (Figure 11.25).

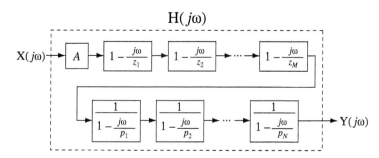

Figure 11.25
A system represented as a cascade of simpler systems

Each component system will have a Bode diagram and, because the magnitude Bode diagrams are graphed in dB, the overall magnitude Bode diagram is the sum of the individual magnitude Bode diagrams. Phase is graphed linearly as before (against a logarithmic frequency scale) and the overall phase Bode diagram is the sum of all the phases contributed by the components.

11.3 Continuous-Time Filters

The One-Real-Pole System Consider the frequency response of a subsystem with a single real pole at $s = p_k$ and no finite zeros,

$$H(s) = \frac{1}{1 - s/p_k} \Rightarrow H(j\omega) = \frac{1}{1 - j\omega/p_k}. \tag{11.3}$$

Before proceeding, first consider the inverse CTFT of $H(j\omega)$. We can use the CTFT pair

$$e^{-at}u(t) \xleftrightarrow{\mathcal{F}} \frac{1}{a + j\omega}, \quad \text{Re}(a) > 0,$$

and rewrite (11.3) as

$$H(j\omega) = -\frac{p_k}{j\omega - p_k}.$$

Then it follows that

$$-p_k e^{p_k t} u(t) \xleftrightarrow{\mathcal{F}} -\frac{p_k}{j\omega - p_k}, \quad p_k < 0. \tag{11.4}$$

This shows that the pole must have a negative real value for the frequency response to have meaning. If it is positive, we cannot do the inverse CTFT to find the corresponding time function. If p_k is negative, the exponential in (11.4) decays to zero in positive time. If it were positive that would indicate a growing exponential in positive time and the system would be unstable. The Fourier transform of a growing exponential does not exist. Also, frequency response has no practical meaning for an unstable system because it could never actually be tested.

The magnitudes and phases of $H(j\omega) = 1/(1 - j\omega/p_k)$ versus frequency are graphed in Figure 11.26. For frequencies $\omega \ll |p_k|$ the frequency response approaches $H(j\omega) = 1$, the magnitude response is approximately zero dB, and the phase response is approximately zero radians. For frequencies $\omega \gg |p_k|$ the frequency response is approximately $H(j\omega) = -p_k/j\omega$, the magnitude frequency response approaches a linear slope of -6 dB per octave or -20 dB per decade, and the phase response approaches a constant $-\pi/2$ radians. (An octave is a factor-of-2 change in frequency and a decade is a factor-of-10 change in frequency.) These limiting behaviors for extreme frequencies define magnitude and phase **asymptotes**. The intersection of the two magnitude asymptotes occurs at $\omega = |p_k|$, which is called the **corner frequency**. At the corner frequency $\omega = |p_k|$ the frequency response is

$$H(j\omega) = \frac{1}{1 - j|p_k|/p_k} = \frac{1}{1 + j}, \quad p_k < 0$$

and its magnitude is $1/\sqrt{2} \cong 0.707$. We can convert this to decibels.

$$(0.707)_{\text{dB}} = 20 \log_{10}(0.707) = -3 \text{ dB}$$

At that point the actual Bode diagram is 3 dB below the corner formed by the asymptotes. This is the point of largest deviation of this magnitude Bode diagram from its asymptotes. The phase Bode diagram goes through $-\pi/4$ radians at the corner frequency and approaches zero radians below and $-\pi/2$ radians above the corner frequency.

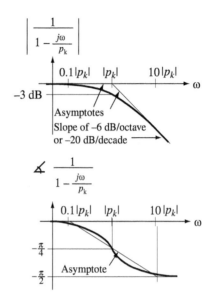

Figure 11.26
The magnitude and phase frequency response of a single-negative-real-pole subsystem

EXAMPLE 11.1

Bode diagram of frequency response of an *RC* lowpass filter

Draw magnitude and phase Bode diagrams for an *RC* lowpass filter frequency response with a time constant of 50 μs.

The form of the *RC* lowpass filter transfer function is

$$H(s) = \frac{1}{sRC + 1}.$$

The time constant is *RC*. Therefore

$$H(s) = \frac{1}{50 \times 10^{-6} s + 1} = \frac{1}{s/20{,}000 + 1}.$$

Setting the denominator equal to zero and solving for the pole location we get a pole at $s = -20{,}000$. Then we can write the frequency response in the standard one-negative-real-pole form,

$$H(j\omega) = H(s)_{s \to j\omega} = \frac{1}{1 - j\omega/(-20{,}000)}.$$

The corresponding corner frequency on the Bode diagram is at $\omega = 20{,}000$ (Figure 11.27).

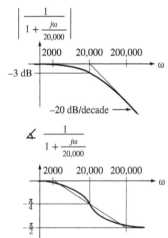

Figure 11.27
Magnitude and phase Bode diagram for the *RC* lowpass filter frequency response

The One-Real-Zero System An analysis similar to the one-real-pole system analysis yields the magnitude and phase Bode diagrams for a subsystem with a single *negative*-real zero and no finite poles.

$$H(s) = 1 - s/z_k \Rightarrow H(j\omega) = 1 - j\omega/z_k, \quad z_k < 0$$

(Figure 11.28).

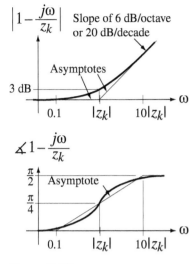

Figure 11.28
The magnitude and phase frequency response of a single-negative-real-zero subsystem

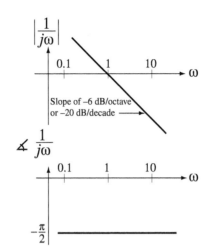

Figure 11.29
The magnitude and phase frequency response of a single pole at $s = 0$

The diagrams are very similar to those for the single-negative-real pole except that the magnitude asymptote above the corner frequency has a slope of +6 dB per octave or +20 dB per decade and the phase approaches $+\pi/2$ radians instead of $-\pi/2$ radians. They are basically the single-negative-real-pole Bode diagrams "turned upside down."

For a subsystem with a single-*positive*-real zero and no finite poles, of the form

$$\mathrm{H}(j\omega) = 1 - j\omega/z_k, \quad z_k > 0$$

the magnitude graph is the same as in Figure 11.28 but the phase approaches $-\pi/2$ instead of $+\pi/2$ at frequencies above the corner frequency.

Integrators and Differentiators We must also consider a pole or a zero at zero frequency (Figure 11.29 and Figure 11.30). A system component with a single pole at $s = 0$ is called an *integrator* because its transfer function is $\mathrm{H}(s) = 1/s$ and division by s corresponds to integration in the time domain.

A system component with a single zero at $s = 0$ is called a *differentiator* because its transfer function is $\mathrm{H}(s) = s$ and multiplication by s corresponds to differentiation in the time domain.

Frequency-Independent Gain The only remaining type of simple system component is a frequency-independent gain (Figure 11.31). In Figure 11.31, the gain constant A is assumed to be positive. That is why the phase is zero. If A is negative the phase is $\pm\pi$ radians.

The asymptotes are helpful in drawing the actual Bode diagram and they are especially helpful in sketching the overall Bode diagram for a more complicated system. The asymptotes can be quickly sketched from knowledge of a few simple rules and added together. Then the magnitude Bode diagram can be sketched approximately by drawing a smooth curve that approaches the asymptotes and deviates at the corners by ± 3 dB.

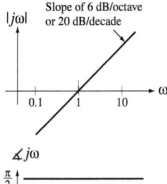
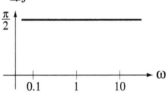
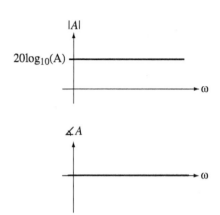

Figure 11.30
The magnitude and phase frequency response of a single zero at $s = 0$

Figure 11.31
The magnitude and phase frequency response of a frequency-independent gain A

EXAMPLE 11.2

Bode diagram of the frequency response of an *RC* circuit

Graph the Bode diagram for the voltage frequency response of the circuit in Figure 11.32, where $C_1 = 1$ F, $C_2 = 2$ F, $R_s = 4$ Ω, $R_1 = 2$ Ω, $R_2 = 3$ Ω.

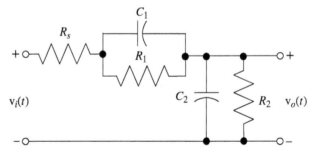

Figure 11.32
An *RC* Circuit

The transfer function is

$$H(s) = \frac{1}{R_s C_2} \frac{s + 1/R_1 C_1}{s^2 + \left(\frac{C_1 + C_2}{R_s C_1 C_2} + \frac{R_1 C_1 + R_2 C_2}{R_1 R_2 C_1 C_2}\right)s + \frac{R_1 + R_2 + R_s}{R_1 R_2 R_s C_1 C_2}}.$$

Substituting $s \to j\omega$ and using numerical values for the components, the frequency response is

$$H(j\omega) = 3 \frac{j2\omega + 1}{48(j\omega)^2 + j50\omega + 9} = 0.125 \frac{j\omega + 0.5}{(j\omega + 0.2316)(j\omega + 0.8104)}$$

$$H(j\omega) = 0.333 \frac{1 - \dfrac{j\omega}{(-0.5)}}{\left[1 - \dfrac{j\omega}{(-0.2316)}\right]\left[1 - \dfrac{j\omega}{(-0.8104)}\right]} = A \frac{1 - j\omega/z_1}{(1 - j\omega/p_1)(1 - j\omega/p_2)}$$

where $A = 0.333$, $z_1 = -0.5$, $p_1 = -0.2316$, $p_2 = -0.8104$.

So this frequency response has two finite poles, one finite zero and one frequency-independent gain. We can quickly construct an overall asymptotic Bode diagram by adding the asymptotic Bode diagrams for the four individual components of the overall frequency response (Figure 11.33).

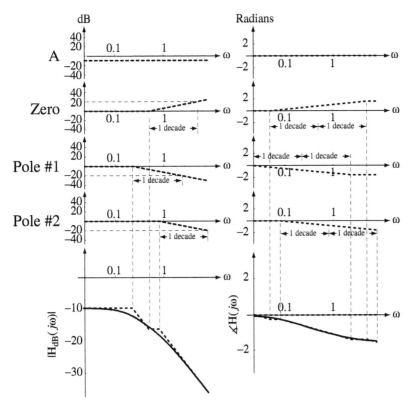

Figure 11.33
Individual asymptotic and overall asymptotic and exact Bode magnitude and phase diagrams for the circuit voltage frequency response

The following MATLAB program demonstrates some techniques for drawing Bode diagrams.

```
%   Set up a logarithmic vector of radian frequencies
%   for graphing the Bode diagram from 0.01 to 10 rad/sec
w = logspace(-2,1,200) ;

%   Set the gain, zero and pole values
A = 0.3333 ; z1 = -0.5 ; p1 = -0.2316 ; p2 = -0.8104

%   Compute the complex frequency response
H = A*(1-j*w/z1)./((1-j*w/p1).*(1-j*w/p2)) ;

%   Graph the magnitude Bode diagram
subplot(2,1,1) ; p = semilogx(w,20*log10(abs(H)),'k') ;
set(p,'LineWidth',2) ; grid on ;
xlabel('\omega','FontSize',18,'FontName','Times') ;
```

```
ylabel('|H({\itj}\omega)|_d_B','FontSize',18,'FontName','Times') ;
title('Magnitude','FontSize',24,'FontName','Times') ;
set(gca,'FontSize',14,'FontName','Times') ;

%   Graph the phase Bode diagram
subplot(2,1,2) ; p = semilogx(w,angle(H),'k') ;
set(p,'LineWidth',2) ; grid on ;
xlabel('\omega','FontSize',18,'FontName','Times') ;
ylabel('Phase of H({\itj}\omega)','FontSize',18,'FontName','Times') ;
title('Phase','FontSize',24,'FontName','Times') ;
set(gca,'FontSize',14,'FontName','Times') ;
```

The resulting magnitude and phase Bode diagrams are illustrated in Figure 11.34.

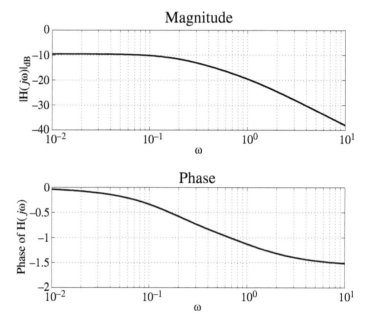

Figure 11.34
Magnitude and phase Bode diagrams of the frequency response of the filter

***

Complex Pole and Zero Pairs Now consider the more complicated case of complex poles and zeros. For real system functions, they always occur in complex conjugate pairs. So a complex conjugate pair of poles with no finite zeros would form a subsystem transfer function

$$H(s) = \frac{1}{(1 - s/p_1)(1 - s/p_2)} = \frac{1}{1 - (1/p_1 + 1/p_1^*)s + s^2/p_1 p_1^*}$$

and frequency response

$$H(j\omega) = \frac{1}{(1 - j\omega/p_1)(1 - j\omega/p_2)} = \frac{1}{1 - j\omega(1/p_1 + 1/p_1^*) + (j\omega)^2/p_1 p_1^*}$$

or

$$H(j\omega) = \cfrac{1}{1 - j\omega\cfrac{2\text{Re}(p_1)}{|p_1|^2} + \cfrac{(j\omega)^2}{|p_1|^2}}.$$

From the table of Fourier pairs, we find the pair

$$e^{-\omega_n\zeta t}\sin\left(\omega_n\sqrt{1-\zeta^2}\,t\right)u(t) \xleftrightarrow{\mathcal{F}} \frac{\omega_n\sqrt{1-\zeta^2}}{(j\omega)^2 + j\omega(2\zeta\omega_n) + \omega_n^2},$$

in the ω domain, which can be expressed in the form

$$\omega_n\frac{e^{-\omega_n\zeta t}\sin\left(\omega_n\sqrt{1-\zeta^2}\,t\right)}{\sqrt{1-\zeta^2}}u(t) \xleftrightarrow{\mathcal{F}} \cfrac{1}{1 + j\omega\cfrac{2\zeta\omega_n}{\omega_n^2} + \cfrac{(j\omega)^2}{\omega_n^2}}$$

whose right side is of the same functional form as

$$H(j\omega) = \cfrac{1}{1 - j\omega\cfrac{2\text{Re}(p_1)}{|p_1|^2} + \cfrac{(j\omega)^2}{|p_1|^2}}.$$

This is a standard form of a second-order underdamped system response where the natural radian frequency is ω_n and the damping ratio is ζ. Therefore, for this type of subsystem,

$$\omega_n^2 = |p_1|^2 = p_1 p_2 \quad \text{and} \quad \zeta = -\frac{\text{Re}(p_1)}{\omega_n} = -\frac{p_1 + p_2}{2\sqrt{p_1 p_2}}.$$

The Bode diagram for this subsystem is illustrated in Figure 11.35.

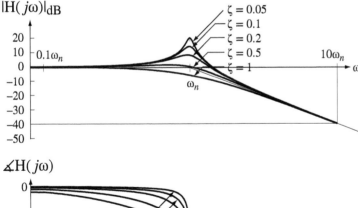
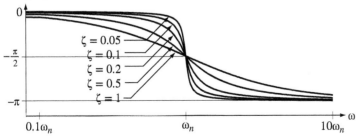

Figure 11.35
Magnitude and phase Bode diagram for a second-order complex pole pair

A complex pair of zeros would form a subsystem frequency response of the form,

$$H(j\omega) = \left(1 - \frac{j\omega}{z_1}\right)\left(1 - \frac{j\omega}{z_2}\right) = 1 - j\omega\left(\frac{1}{z_1} + \frac{1}{z_1^*}\right) + \frac{(j\omega)^2}{z_1 z_1^*} = 1 - j\omega\frac{2\,\text{Re}(z_1)}{|z_1|^2} + \frac{(j\omega)^2}{|z_1|^2}.$$

In this type of subsystem we can identify the natural radian frequency and the damping ratio as

$$\omega_n^2 = |z_1|^2 = z_1 z_2 \quad \text{and} \quad \zeta = -\frac{\text{Re}(z_1)}{\omega_n} = -\frac{z_1 + z_2}{2\sqrt{z_1 z_2}}.$$

The Bode diagram for this subsystem is illustrated in Figure 11.36.

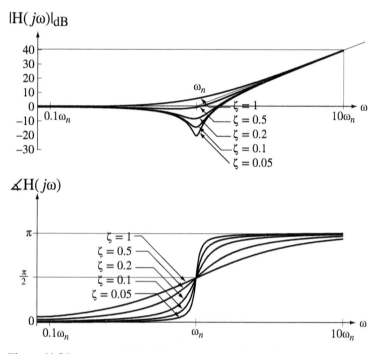

Figure 11.36
Magnitude and phase Bode diagram for a second-order complex zero pair

PRACTICAL FILTERS

Passive Filters

The Lowpass Filter Approximations to the ideal lowpass and bandpass filters can be made with certain types of circuits. The simplest approximation to the ideal lowpass filter is the one we have already analyzed more than once, the so-called *RC* lowpass filter (Figure 11.37). We have found its response to a step and to a sinusoid. Let us now analyze it directly in the frequency domain.

The differential equation describing this circuit is $RCv'_{out}(t) + v_{out}(t) = v_{in}(t)$. Laplace transforming both sides (assuming no initial charge on the capacitor), $sRCV_{out}(s) + V_{out}(s) = V_{in}(s)$. We can now solve directly for the transfer function,

$$H(s) = \frac{V_{out}(s)}{V_{in}(s)} = \frac{1}{sRC + 1}.$$

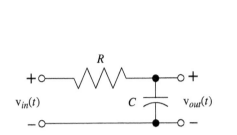

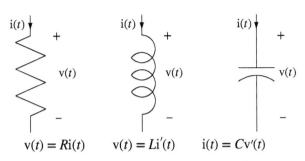

Figure 11.37
Practical *RC* lowpass filter

Figure 11.38
Defining equations for resistors, capacitors and inductors

The method commonly used in elementary circuit analysis to solve for the frequency response is based on the phasor and impedance concepts. Impedance is a generalization of the idea of resistance to apply to inductors and capacitors. Recall the voltage-current relationships for resistors, capacitors and inductors (Figure 11.38).

If we Laplace transform these relationships we get

$$V(s) = RI(s), \quad V(s) = sLI(s) \quad \text{and} \quad I(s) = sCV(s).$$

The impedance concept comes from the similarity of the inductor and capacitor equations to Ohm's law for resistors. If we form the ratios of voltage to current we get

$$\frac{V(s)}{I(s)} = R, \quad \frac{V(s)}{I(s)} = sL \quad \text{and} \quad \frac{V(s)}{I(s)} = \frac{1}{sC}$$

For resistors this ratio is called *resistance*. In the generalization this ratio is called **impedance**. Impedance is conventionally symbolized by Z. Using that symbol,

$$Z_R(s) = R, \quad Z_L(s) = sL \quad \text{and} \quad Z_C(s) = 1/sC.$$

This allows us to apply many of the techniques of resistive circuit analysis to circuits that contain inductors and capacitors and are analyzed in the frequency domain. In the case of the *RC* lowpass filter we can view it as a voltage divider (Figure 11.39).

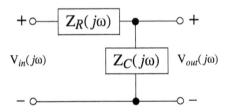

Figure 11.39
Impedance voltage divider representation of the *RC* lowpass filter

Then we can directly write the transfer function in the frequency domain

$$H(s) = \frac{V_{out}(s)}{V_{in}(s)} = \frac{Z_c(s)}{Z_c(s) + Z_f(s)} = \frac{1/sC}{1/sC + R} = \frac{1}{sRC + 1}$$

and the frequency response as

$$H(j\omega) = \frac{1}{j\omega RC + 1} \quad \text{or} \quad H(f) = \frac{1}{j2\pi fRC + 1},$$

arriving at the same result as before without a direct reference to the time domain. The magnitude and phase of the RC lowpass filter frequency response are illustrated in Figure 11.40.

The impulse response of the RC lowpass filter is the inverse CTFT of its frequency response

$$h(t) = \frac{e^{-t/RC}}{RC} u(t)$$

as illustrated in Figure 11.41. For this physically realizable filter the impulse response is zero before time $t = 0$. The filter is causal.

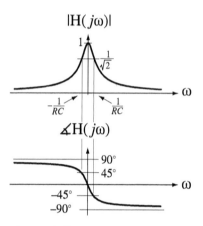

Figure 11.40
Magnitude and phase frequency responses of an RC lowpass filter

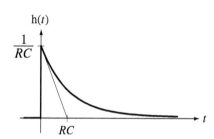

Figure 11.41
Impulse response of an RC lowpass filter

At very low frequencies (approaching zero) the capacitor's impedance is much greater in magnitude than the resistor's impedance, the voltage division ratio approaches one, and the output voltage signal and input voltage signal are about the same. At very high frequencies the capacitor's impedance becomes much smaller in magnitude than the resistor's impedance and the voltage division ratio approaches zero. Thus we can say approximately that low frequencies "pass through" and high frequencies "get stopped." This qualitative analysis of the circuit agrees with the mathematical form of the frequency response,

$$H(j\omega) = \frac{1}{j\omega RC + 1}.$$

At low frequencies,

$$\lim_{\omega \to 0} H(j\omega) = 1$$

and at high frequencies

$$\lim_{\omega \to \infty} H(j\omega) = 0.$$

The RC lowpass filter is lowpass only because the excitation is defined as the voltage at the input and the response is defined as the voltage at the output. If the response had been defined as the current, the nature of the filtering process would change completely. In that case the frequency response would become

$$H(j\omega) = \frac{I(j\omega)}{V_{in}(j\omega)} = \frac{1}{Z_R(j\omega) + Z_c(j\omega)} = \frac{1}{1/j\omega C + R} = \frac{j\omega C}{j\omega RC + 1}.$$

With this definition, at low frequencies the capacitor impedance is very large, blocking current flow so the response approaches zero. At high frequencies the capacitor impedance approaches zero so the circuit responds as though the capacitor were a perfect conductor and the current flow is determined by the resistance R. Mathematically the response approaches zero at low frequencies and approaches the constant $1/R$ at high frequencies. This defines a highpass filter.

$$\lim_{\omega \to 0} H(j\omega) = 0 \quad \text{and} \quad \lim_{\omega \to \infty} H(j\omega) = 1/R$$

Another (much less common) form of lowpass filter is illustrated in Figure 11.42.

$$H(s) = \frac{V_{out}(s)}{V_{in}(s)} = \frac{R}{sL + R} \Rightarrow H(j\omega) = \frac{R}{j\omega L + R}.$$

Using the impedance and voltage divider ideas, can you explain in words why this circuit is a lowpass filter?

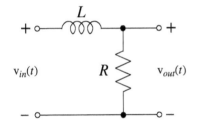

Figure 11.42
Alternate form of a practical lowpass filter

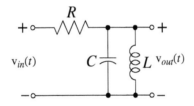

Figure 11.43
An *RLC* practical bandpass filter

The Bandpass Filter One of the simplest forms of a practical bandpass filter is illustrated in Figure 11.43.

$$H(s) = \frac{V_{out}(s)}{V_{in}(s)} = \frac{s/RC}{s^2 + s/RC + 1/LC} \Rightarrow H(j\omega) = \frac{j\omega/RC}{(j\omega)^2 + j\omega/RC + 1/LC}$$

At very low frequencies, the capacitor is an open circuit and the inductor is a perfect conductor. Therefore at very low frequencies, the output voltage signal is practically zero. At very high frequencies, the inductor is an open circuit and the capacitor is a perfect conductor, again making the output voltage signal zero. The impedance of the parallel inductor-capacitor combination is

$$Z_{LC}(s) = \frac{sL/sC}{sL + 1/sC} = \frac{sL}{s^2 LC + 1}.$$

For $s^2 LC + 1 = 0 \Rightarrow s = \pm j\sqrt{1/LC} \Rightarrow \omega = \pm 1/\sqrt{LC}$, the impedance is infinite. This frequency is called the **resonant** frequency. So at the resonant frequency of the parallel-*LC* circuit, the impedance of that parallel combination of inductor and capacitor goes to infinity and the output voltage signal is the same as the input voltage signal. The overall behavior of the circuit is to approximately pass frequencies near the resonant frequency and block other frequencies, hence it is a practical bandpass filter.

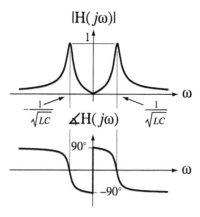

Figure 11.44
Magnitude and phase frequency responses of a practical *RLC* bandpass filter

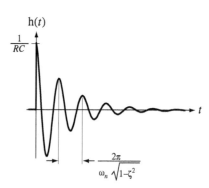

Figure 11.45
Impulse response of a practical *RLC* bandpass filter

A graph of the magnitude and phase of the frequency response (Figure 11.44) (for a particular choice of component values) reveals its bandpass nature.

The impulse response of the *RLC* bandpass filter is

$$h(t) = 2\zeta\omega_n e^{-\zeta\omega_n t}\left[\cos(\omega_c t) - \frac{\zeta}{\sqrt{1-\zeta^2}}\sin(\omega_c t)\right]u(t)$$

where

$$2\zeta\omega_n = 1/RC,\ \omega_n^2 = 1/LC \text{ and } \omega_c = \omega_n\sqrt{1-\zeta^2}$$

(Figure 11.45). Notice that the impulse response of this physically realizable filter is causal.

All physical systems are filters in the sense that each of them has a response that has a characteristic variation with frequency. This is what gives a musical instrument and each human voice its characteristic sound. To see how important this is, try playing just the mouthpiece of any wind instrument. The sound is very unpleasant until the instrument is attached, then it becomes very pleasant (when played by a good musician). The sun periodically heats the earth as it rotates and the earth acts like a lowpass filter, smoothing out the daily variations and responding with a lagging seasonal variation of temperature. In prehistoric times people tended to live in caves because the thermal mass of the rock around them smoothed out the seasonal variation of temperature and allowed them to be cooler in the summer and warmer in the winter, another example of lowpass filtering. Industrial foam-rubber ear plugs are designed to allow lower frequencies through so that people wearing them can converse but to block intense high-frequency sounds that may damage the ear. The list of examples of systems that we are familiar with in daily life that perform filtering operations is endless.

Active Filters

All the practical filters we have examined so far have been **passive** filters. The term *passive* means they contained no devices with the capability of producing an output signal with more actual power (not signal power) than the input signal. Many modern filters are **active** filters. They contain active devices like transistors and/or operational amplifiers and require an external source of power to operate properly. With the use of active devices the actual output power can be greater than the actual input power.

The subject of active filters is a large one and only the simplest forms of active filters will be introduced here.[3]

Operational Amplifiers There are two commonly used forms of operational amplifier circuits, the inverting amplifier form and the noninverting amplifier form (Figure 11.46). The analysis here will use the simplest possible model for the operational amplifier, the **ideal operational amplifier**. An ideal operational amplifier has infinite input impedance, zero output impedance, infinite gain and infinite bandwidth.

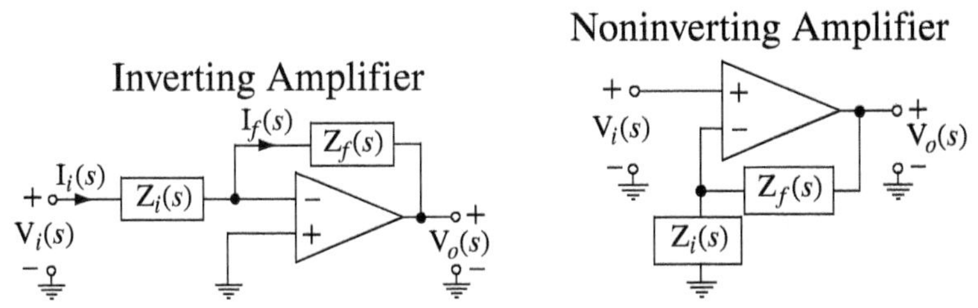

Figure 11.46
Two common forms of amplifiers utilizing operational amplifiers

For each type of amplifier there are two impedances $Z_i(s)$ and $Z_f(s)$ that control the transfer function. The transfer function of the inverting amplifier can be derived by observing that, since the operational amplifier input impedance is infinite, the current flowing into either input terminal is zero and therefore

$$I_f(s) = I_i(s). \tag{11.5}$$

Also, since the output voltage is finite and the operational amplifier gain is infinite, the voltage difference between the two input terminals must be zero. Therefore

$$I_i(s) = \frac{V_i(s)}{Z_i(s)} \tag{11.6}$$

and

$$I_f(s) = -\frac{V_f(s)}{Z_f(s)}. \tag{11.7}$$

Equating (11.6) and (11.7) according to (11.5), and solving for the transfer function,

$$\boxed{H(s) = \frac{V_o(s)}{V_i(s)} = -\frac{Z_f(s)}{Z_i(s)}}. \tag{11.8}$$

Similarly it can be shown that the noninverting amplifier transfer function is

$$\boxed{H(s) = \frac{V_o(s)}{V_i(s)} = \frac{Z_f(s) + Z_i(s)}{Z_i(s)} = 1 + \frac{Z_f(s)}{Z_i(s)}}. \tag{11.9}$$

---

[3] In some passive circuits, there is voltage gain at some frequencies. The output voltage signal can be larger than the input voltage signal. Therefore the output signal power, as defined previously, would be greater than the input signal power. But this is not actual power gain because that higher output voltage signal is across a higher impedance.

Chapter 11 Frequency Response Analysis

The Integrator Probably the most common and simplest form of active filter is the active **integrator** (Figure 11.47). Using the inverting amplifier gain formula (11.8) for the transfer function,

$$H(s) = -\frac{Z_f(s)}{Z_i(s)} = -\frac{1/sC}{R} = -\frac{1}{sRC} \Rightarrow H(f) = -\frac{1}{j2\pi fRC}.$$

The action of the integrator is easier to see if the frequency response is rearranged to the form

$$V_o(f) = -\frac{1}{RC}\frac{V_i(f)}{j2\pi f} \quad \text{or} \quad V_o(j\omega) = -\frac{1}{RC}\frac{V_i(j\omega)}{j\omega}.$$

The integrator integrates the signal but, at the same time, multiplies it by $-1/RC$. Notice that we did not introduce a practical passive integrator. The passive RC lowpass filter acts much like an integrator for frequencies well above its corner frequency but at a low enough frequency its response is not like an integrator. So the active device (the operational amplifier in this case) has given the filter designer another degree of freedom in design.

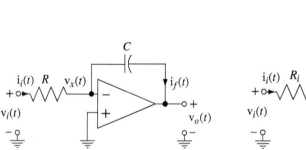

Figure 11.47
An active integrator

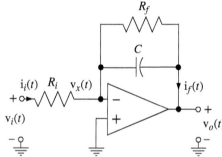

Figure 11.48
An active RC lowpass filter

The Lowpass Filter The integrator is easily changed to a lowpass filter by the addition of a single resistor (Figure 11.48). For this circuit,

$$H(s) = \frac{V_0(s)}{V_i(s)} = -\frac{R_f}{R_i}\frac{1}{sCR_f + 1} \Rightarrow H(j\omega) = \frac{V_0(j\omega)}{V_i(j\omega)} = -\frac{R_f}{R_i}\frac{1}{j\omega CR_f + 1}.$$

This frequency response has the same functional form as the passive RC lowpass filter except for the factor $-R_f/R_i$. So this is a filter with gain. It filters and amplifies the signal simultaneously. In this case the voltage gain is negative.

EXAMPLE 11.3

Bode diagram of the frequency response of a two-stage active filter

Graph the Bode magnitude and phase diagrams for the two-stage active filter in Figure 11.49.
The transfer function of the first stage is

$$H_1(s) = -\frac{Z_{f1}(s)}{Z_{i1}(s)} = -\frac{R_{f1}}{R_{i1}}\frac{1}{1 + sC_{f1}R_{f1}}.$$

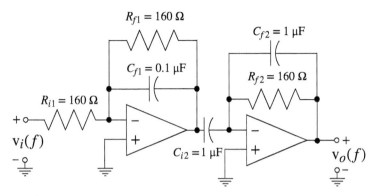

Figure 11.49
A two-stage active filter

The transfer function of the second stage is

$$H_2(s) = -\frac{Z_{f1}(s)}{Z_{i1}(s)} = -\frac{sR_{f2}C_{i2}}{1 + sR_{f2}C_{f2}}.$$

Since the output impedance of an ideal operational amplifier is zero, the second stage does not load the first stage and the overall transfer function is simply the product of the two transfer functions,

$$H(s) = \frac{R_{f1}}{R_{i1}} \frac{sR_{f2}C_{i2}}{(1 + sC_{f1}R_{f1})(1 + sR_{f2}C_{f2})}.$$

Substituting in parameter values and letting $s \to j2\pi f$, we get the frequency response

$$H(f) = \frac{j1000f}{(1000 + jf/10)(1000 + jf)}$$

(Figure 11.50). This is a practical bandpass filter.

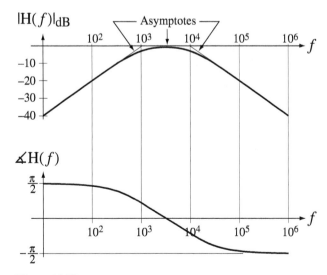

Figure 11.50
Bode diagram of the frequency response of the two-stage active filter

Example 11.4

Design of an active highpass filter

Design an active filter that attenuates signals at 60 Hz and below by more than 40 dB and amplifies signals at 10 kHz and above with a positive gain that deviates from 20 dB by no more than 2 dB.

This specifies a highpass filter. The gain must be positive. A positive gain and some highpass filtering can be accomplished by one noninverting amplifier. However, looking at the transfer function and frequency response for the noninverting amplifier

$$H(s) = \frac{V_o(s)}{V_i(s)} = \frac{Z_f(s) + Z_i(s)}{Z_i(s)} \Rightarrow H(j\omega) = \frac{Z_f(j\omega) + Z_i(j\omega)}{Z_i(j\omega)}$$

we see that if the two impedances consist of only resistors and capacitors its gain is never less than one and we need attenuation (a gain less than one) at low frequencies. (If we were to use both inductors and capacitors, we could make the magnitude of the sum $Z_f(j\omega) + Z_i(j\omega)$ be less than the magnitude of $Z_i(j\omega)$ at some frequencies and achieve a gain less than one. But we could not make that occur for all frequencies below 60 Hz, and the use of inductors is generally avoided in practical design unless absolutely necessary. There are other practical difficulties with this idea also using real, as opposed to ideal, operational amplifiers.)

If we use one inverting amplifier we have a negative gain. But we could follow it with another inverting amplifier making the overall gain positive. (Gain is the opposite of attenuation. If the attenuation is 60 dB the gain is −60 dB.) If the gain at 60 Hz is −40 dB and the response is that of a single-pole highpass filter, the Bode diagram asymptote on the magnitude frequency response would pass through −20 dB of gain at 600 Hz, 0 dB of gain at 6 kHz and 20 dB of gain at 60 kHz. But we need 20 dB of gain at 10 kHz so a single-pole filter is inadequate to meet the specifications. We need a two-pole highpass filter. We can achieve that with a cascade of two single-pole highpass filters, meeting the requirements for attenuation and positive gain simultaneously.

Now we must choose $Z_f(j\omega)$ and $Z_i(j\omega)$ to make the inverting amplifier a highpass filter. Figure 11.48 illustrates an active lowpass filter. That filter is lowpass because the gain is $-Z_f(j\omega)/Z_i(j\omega)$, $Z_i(j\omega)$ is constant and $Z_f(j\omega)$ has a larger magnitude at low frequencies than at high frequencies. There is more than one way to make a highpass filter using the same inverting amplifier configuration. We could make the magnitude of $Z_f(j\omega)$ be small at low frequencies and larger at high frequencies. That requires the use of an inductor but, again for practical reasons, inductors should be avoided unless really needed. We could make $Z_f(j\omega)$ constant and make the magnitude of $Z_i(j\omega)$ large at low frequencies and small at high frequencies. That general goal can be accomplished by either a parallel or series combination of a resistor and a capacitor (Figure 11.51).

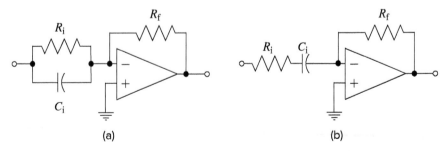

Figure 11.51
Two ideas for a highpass filter using only capacitors and resistors

If we just think about the limiting behavior of these two design ideas at very low and very high frequencies, we immediately see that only one of them meets the specifications of this design. Design (a) has a finite gain at very low frequencies and a gain that rises with frequency at higher frequencies, never approaching a constant. Design (b) has a gain that falls with frequency at low frequencies, approaching zero at zero frequency and approaching a constant gain at high frequencies. Design (b) can be used to meet our specification. So now the design is a cascade of two inverting amplifiers (Figure 11.52).

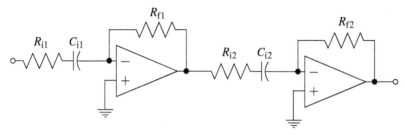

Figure 11.52
Cascade of two inverting highpass active filters

At this point we must select the resistor and capacitor values to meet the attenuation and gain requirements. There are many ways of doing that. The design is not unique. We can begin by selecting the resistors to meet the high-frequency gain requirement of 20 dB. That is an overall high-frequency gain of 10, which we can apportion any way we want between the two amplifiers. Let's let the two stage gains be approximately the same. Then the resistor ratios in each stage should be about 3.16. We should choose resistors large enough not to load the outputs of the operational amplifiers but small enough that stray capacitances don't cause problems. Resistors in the range of 500 Ω to 50 kΩ are usually good choices. But unless we are willing to pay a lot, we cannot arbitrarily choose a resistor value. Resistors come in standard values, typically in a sequence of

$$1, 1.2, 1.5, 1.8, 2.2, 2.7, 3.3, 3.9, 4.7, 5.6, 6.8, 8.2 \times 10^n$$

where n sets the decade of the resistance value. Some ratios that are very near 3.16 are

$$\frac{3.9}{1.2} = 3.25, \ \frac{4.7}{1.5} = 3.13, \ \frac{5.6}{1.8} = 3.11, \ \frac{6.8}{2.2} = 3.09, \ \frac{8.2}{2.7} = 3.03.$$

To set the overall gain very near 10 we can choose the first-stage ratio to be $3.9/1.2 = 3.25$ and the second-stage ratio to be $6.8/2.2 = 3.09$ and achieve an overall high-frequency gain of 10.043. So we set

$$R_{f1} = 3.9 \text{ k}\Omega, \ R_{i1} = 1.2 \text{ k}\Omega, \ R_{f2} = 6.8 \text{ k}\Omega, \ R_{i2} = 2.2 \text{ k}\Omega.$$

Now we must choose the capacitor values to achieve the attenuation at 60 Hz and below and the gain at 10 kHz and above. To simplify the design let's set the two corner frequencies of the two stages at the same (or very nearly the same) value. With a two-pole low-frequency rolloff of 40 dB per decade and a high-frequency gain of approximately 20 dB, we get a 60-dB difference between the frequency response magnitude at 60 Hz and 10 kHz. If we were to set the gain at 60 Hz to be exactly −40 dB, then at 600 Hz we would have approximately 0 dB gain and at 6 kHz we would have a gain of 40 dB, and it would be higher at 10 kHz. This does not meet the specification.

We can start at the high-frequency end and set the gain at 10 kHz to be approximately 10, meaning that the corner for the low-frequency rolloff should be well below 10 kHz. If we put it at 1 kHz, the approximate gain at 100 Hz based on asymptotic approximations will be −20 dB and at 10 Hz it will be −60 dB. We need −40 dB at 60 Hz. But we only get about −29 dB at 60 Hz. So we need to put the corner frequency a little higher, say 3 kHz. If we put the corner frequency at 3 kHz, the calculated capacitor values will be $C_{i1} = 46$ nF and $C_{i2} = 24$ nF. Again, we cannot arbitrarily choose a capacitor value. Standard capacitor values are typically arrayed at the same intervals as standard resistor values

$$1, 1.2, 1.5, 1.8, 2.2, 2.7, 3.3, 3.9, 4.7, 5.6, 6.8, 8.2 \times 10^n.$$

There is some leeway in the location of the corner frequency so we probably don't need a really precise value of capacitance. We can choose $C_{i1} = 0.47$ nF and $C_{i2} = 22$ nF, making one a little high and one a little low. This will separate the poles slightly but will still create the desired 40 dB per decade low-frequency rolloff. This looks like a good design but we need to verify its performance by drawing a Bode diagram (Figure 11.53).

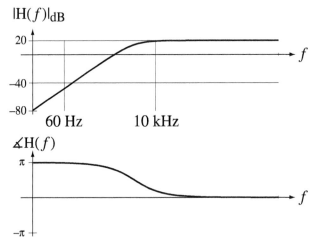

Figure 11.53
Bode diagram for two-stage active highpass filter design

It is apparent from the diagram that the attenuation at 60 Hz is adequate. Calculation of the gain at 10 kHz yields about 19.2 dB, which also meets specifications.

These results are based on exact values of resistors and capacitors. In reality all resistors and capacitors are typically chosen based on their nominal values but their actual values may differ from the nominal by a few percent. So any good design should have some tolerance in the specifications to allow for small deviations of component values from the design values.

EXAMPLE 11.5

Sallen-Key bandpass filter

A popular filter design that can be found in many books on electronics or filters is the two-pole, single-stage, **Sallen-Key** or **constant-K** bandpass filter (Figure 11.54).

The triangle symbol with the K inside represents an ideal noninverting amplifier with a finite voltage gain K, an infinite input impedance, a zero output impedance and infinite

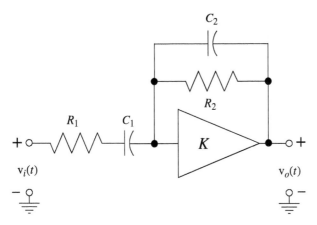

Figure 11.54
Sallen-Key or constant-K bandpass filter

bandwidth (not an operational amplifier). The overall bandpass-filter transfer function and frequency response are

$$H(s) = \frac{V_o(s)}{V_i(s)} = \frac{s\frac{K}{(1-K)}\frac{1}{R_1C_2}}{s^2 + \left[\frac{1}{R_1C_1} + \frac{1}{R_2C_2} + \frac{1}{R_1C_2(1-K)}\right]s + \frac{1}{R_1R_2C_1C_2}}$$

and

$$H(j\omega) = \frac{V_o(j\omega)}{V_i(j\omega)} = \frac{j\omega\frac{K}{(1-K)}\frac{1}{R_1C_2}}{(j\omega)^2 + j\omega\left[\frac{1}{R_1C_1} + \frac{1}{R_2C_2} + \frac{1}{R_1C_2(1-K)}\right] + \frac{1}{R_1R_2C_1C_2}}.$$

The frequency response is of the form

$$H(j\omega) = H_0 \frac{j2\zeta\omega_0^2}{(j\omega)^2 + 2\zeta\omega_0(j\omega) + \omega_0^2} = \frac{j\omega A}{(j\omega)^2 + 2\zeta\omega_0(j\omega) + \omega_0^2}$$

where

$$A = \frac{K}{(1-K)}\frac{1}{R_1C_2}, \quad \omega_0^2 = \frac{1}{R_1R_2C_1C_2}$$

$$\zeta = \frac{R_1C_1 + R_2C_2 + \frac{R_2C_1}{1-K}}{2\sqrt{R_1R_2C_1C_2}}, \quad Q = \frac{1}{2\zeta} = \frac{\sqrt{R_1R_2C_1C_2}}{R_1C_1 + R_2C_2 + \frac{R_2C_1}{1-K}}$$

and

$$H_0 = \frac{K}{1 + (1-K)\left(\frac{C_2}{C_1} + \frac{R_1}{R_2}\right)}.$$

The recommended design procedure is to choose the Q, and the resonant frequency $f_0 = \omega_0/2\pi$, choose $C_1 = C_2 = C$ as some convenient value and then calculate

$$R_1 = R_2 = \frac{1}{2\pi f_0 C} \quad \text{and} \quad K = \frac{3Q-1}{2Q-1} \quad \text{and} \quad |H_0| = 3Q - 1.$$

Also, it is recommended that Q should be less than 10 for this design. Design a filter of this type with a Q of 5 and a center frequency of 50 kHz.

We can pick convenient values of capacitance, so let $C_1 = C_2 = C = 10$ nF. Then $R_1 = R_2 = 318\ \Omega$ and $K = 1.556$ and $|H_0| = 14$. That makes the frequency response

$$H(j\omega) = -\frac{j\omega(8.792 \times 10^5)}{(j\omega)^2 + (6.4 \times 10^4)j\omega + 9.86 \times 10^{10}}$$

or, written as a function of cyclic frequency

$$H(f) = -\frac{j2\pi f(8.792 \times 10^5)}{(j2\pi f)^2 + (6.4 \times 10^4)j2\pi f + 9.86 \times 10^{10}}$$

(Figure 11.55).

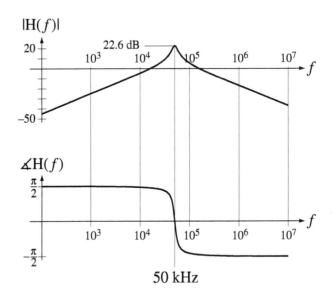

Figure 11.55
Bode diagram of the Sallen-Key bandpass filter frequency response

As in the previous example, we cannot choose the component values to be exactly those calculated, but we can come close. We would probably have to use nominal 330 Ω resistors and that would alter the frequency response slightly, depending on their actual values and the actual values of the capacitors.

EXAMPLE 11.6

Biquadratic RLC active filter

The biquadratic filter introduced in Section 11.2 can be realized as an active filter (Figure 11.56). Under the assumption of an ideal operational amplifier, the transfer function can be found using standard circuit analysis techniques. It is

$$H(s) = \frac{V_o(s)}{V_i(s)} = \frac{s^2 + \dfrac{R(R_1 + R_2) + R_1(R_f + R_2)}{L(R_1 + R_2)}s + \dfrac{1}{LC}}{s^2 + \dfrac{R(R_1 + R_2) + R_2(R_s + R_1)}{L(R_1 + R_2)}s + \dfrac{1}{LC}}$$

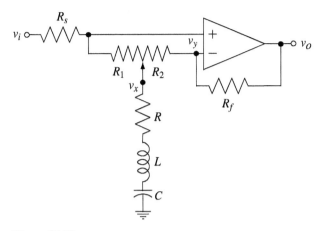

Figure 11.56
Active *RLC* realization of a biquadratic filter

Consider the two cases, $R_1 \neq 0, R_2 = 0$ and $R_1 = 0, R_2 \neq 0$. If $R_1 \neq 0, R_2 = 0$, then the frequency response is

$$H(j\omega) = \frac{(j\omega)^2 + j\omega(R + R_f)/L + 1/LC}{(j\omega)^2 + j\omega R/L + 1/LC}$$

The natural radian frequency is $\omega_n = 1/\sqrt{LC}$. There are poles at

$$j\omega = -(R/2L) \pm \sqrt{(R/2L)^2 - 1/LC}$$

and zeros at

$$j\omega = -\frac{R + R_f}{2L} \pm \sqrt{\left(\frac{R + R_f}{2L}\right)^2 - \frac{1}{LC}}$$

and, at low and high frequencies and at resonance,

$$\lim_{\omega \to 0} H(j\omega) = 1, \quad \lim_{\omega \to \infty} H(j\omega) = 1, \quad H(j\omega_n) = \frac{R + R_f}{R} > 1.$$

If $R < 2\sqrt{L/C}$ and $R + R_f \gg 2\sqrt{L/C}$, the poles are complex and the zeros are real and the dominant effect near ω_n is an increase in the frequency response magnitude. Notice that in this case the frequency response does not depend on R_1. This condition is just like having the *RLC* resonant circuit in the feedback with the potentiometer removed.

If $R_1 = 0, R_2 \neq 0$, then

$$H(j\omega) = \frac{(j\omega)^2 + j\omega\frac{R}{L} + \frac{1}{LC}}{(j\omega)^2 + j\omega\frac{R + R_s}{L} + \frac{1}{LC}}$$

The natural radian frequency is $\omega_n = 1/\sqrt{LC}$. There are zeros at

$$j\omega = -\frac{R}{2L} \pm \sqrt{\left(\frac{R}{2L}\right)^2 - \frac{1}{LC}}$$

and poles at

$$j\omega = -\frac{R + R_s}{2L} \pm \sqrt{\left(\frac{R + R_s}{2L}\right)^2 - \frac{1}{LC}}$$

and, at low and high frequencies and at resonance,

$$\lim_{\omega \to 0} H(j\omega) = 1, \quad \lim_{\omega \to \infty} H(j\omega) = 1, \quad H(j\omega_n) = \frac{R}{R+R_s} < 1.$$

If $R < 2\sqrt{L/C}$ and $R + R_s \gg 2\sqrt{L/C}$, the zeros are complex and the poles are real and the dominant effect near ω_n is a decrease in the frequency response magnitude. Notice that in this case the frequency response does not depend on R_2. This condition is just like having the RLC resonant circuit on the input of the amplifier with the potentiometer removed. If $R_1 = R_2$ and $R_f = R_s$, the frequency response is $H(j\omega) = 1$ and the output signal is the same as the input signal.

So one potentiometer can determine whether the frequency response magnitude is increased or decreased near a resonant frequency. The graphic equalizer of Section 11.2 could be realized with a cascade connection of 9 to 11 such biquadratic filters with their resonant frequencies spaced apart by octaves. But it can also be realized with only one operational amplifier as illustrated in Figure 11.57. Because of the interaction of the passive RLC networks, the operation of this circuit is not identical to that of multiple cascade-connected biquadratic filters, but it accomplishes the same goal with fewer parts.

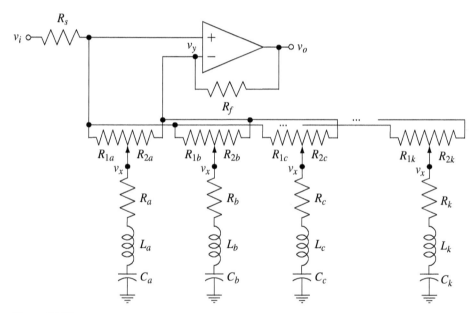

Figure 11.57
A circuit realization of a graphic equalizer with only one operational amplifier

11.4 DISCRETE-TIME FILTERS

NOTATION

The discrete-time Fourier transform (DTFT) was derived from the z transform by making the change of variable $z \to e^{j2\pi F}$ or $z \to e^{j\Omega}$ where F and Ω are both real variables representing frequency, cyclic and radian. In the literature on discrete-time (digital) systems the most commonly used variable for frequency is radian frequency Ω. So in the following sections on discrete-time filters we will also use Ω predominantly.[4]

IDEAL FILTERS

The analysis and design of discrete-time filters have many parallels with the analysis and design of continuous-time filters. In this and the next section we will explore the characteristics of discrete-time filters using many of the techniques and much of the terminology developed for continuous-time filters.

Distortion

The term *distortion* means the same thing for discrete-time filters as it does for continuous-time filters, changing the shape of a signal. Suppose a signal x[n] has the shape illustrated at the top of Figure 11.58(a). Then the signal at the bottom of Figure 11.58(a) is an undistorted version of that signal. Figure 11.58(b) illustrates one type of distortion.

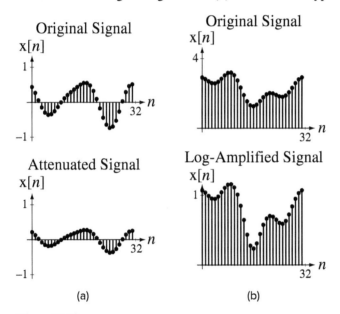

Figure 11.58
(a) An original signal and a changed, but undistorted, version of it, (b) an original signal and a distorted version of it

Just as was true for continuous-time filters, the impulse response of a filter that does not distort is an impulse, possibly with a strength other than one and possibly shifted in time. The most general form of an impulse response of a distortionless system is $h[n] = A\delta[n - n_0]$. The corresponding frequency response is the DTFT of the impulse response $H(e^{j\Omega}) = Ae^{-j\Omega n_0}$. The frequency response can be characterized by its magnitude and phase $|H(e^{j\Omega})| = A$ and $\sphericalangle H(e^{j\Omega}) = -\Omega n_0$. Therefore a distortionless

---

[4] The reader should be aware that notation varies widely among books and papers in this area. The DTFT of a discrete-time function x[n] might be written in any of the forms

$$X(e^{j2\pi f}), X(e^{j\Omega}), X(\Omega), X(e^{j\omega}), X(\omega).$$

Some authors use the same symbol ω for radian frequency in both continuous and discrete time. Some authors use ω and f in discrete time and Ω and F in continuous time. Some authors preserve the meaning of "X" as the z transform of "x" by replacing z by $e^{j\Omega}$ or $e^{j\omega}$. Other authors redefine the function "X" and the DTFT by using Ω or ω as the independent variable. All notation forms have advantages and disadvantages.

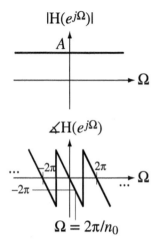

Figure 11.59
Magnitude and phase of a distortionless system

system has a frequency response magnitude that is constant with frequency and a phase that is linear with frequency (Figure 11.59).

The magnitude frequency response of a distortionless system is constant and the phase frequency response is linear over the range $-\pi < \Omega < \pi$ and repeats periodically outside that range. Since n_0 is an integer, the magnitude and phase of a distortionless filter are guaranteed to repeat every time Ω changes by 2π.

Filter Classifications

The terms *passband* and *stopband* have the same significance for discrete-time filters as they do for continuous-time filters. The descriptions of ideal discrete-time filters are similar in concept but have to be modified slightly because of the fact that all discrete-time systems have periodic frequency responses. They are periodic because, in the signal $A\cos(\Omega_0 n)$, if Ω_0 is changed by adding $2\pi m$, m an integer, the signal becomes $A\cos((\Omega_0 + 2\pi m)n)$ and the signal is unchanged because

$$A\cos(\Omega_0 n) = A\cos((\Omega_0 + 2\pi m)n) = A\cos(\Omega_0 n + 2\pi mn), \ m \text{ an integer.}$$

Therefore, a discrete-time filter is classified by its frequency response over the base period $-\pi < \Omega < \pi$.

> An ideal lowpass filter passes signal power for frequencies $0 < |\Omega| < \Omega_m < \pi$ without distortion and stops signal power at other frequencies in the range $-\pi < \Omega < \pi$.

> An ideal highpass filter stops signal power for frequencies $0 < |\Omega| < \Omega_m < \pi$ and passes signal power at other frequencies in the range $-\pi < \Omega < \pi$ without distortion.

> An ideal bandpass filter passes signal power for frequencies $0 < \Omega_L < |\Omega| < \Omega_H < \pi$ without distortion and stops signal power at other frequencies in the range $-\pi < \Omega < \pi$.

> An ideal bandstop filter stops signal power for frequencies $0 < \Omega_L < |\Omega| < \Omega_H < \pi$ and passes signal power at other frequencies in the range $-\pi < \Omega < \pi$ without distortion.

Frequency Responses

In Figure 11.60 and Figure 11.61 are the magnitude and phase frequency responses of the four basic types of ideal filters.

Impulse Responses and Causality

The impulse responses of ideal filters are the inverse transforms of their frequency responses. The impulse and frequency responses of the four basic types of ideal filter are summarized in Figure 11.62. These descriptions are general in the sense that they involve an arbitrary gain constant A and an arbitrary time delay n_0.

In Figure 11.63 are some typical shapes of impulse responses for the four basic types of ideal filter.

The consideration of causality is the same for discrete-time filters as for continuous-time filters. Like ideal continuous-time filters, ideal discrete-time filters have noncausal impulse responses and are therefore physically impossible to build.

11.4 Discrete-Time Filters

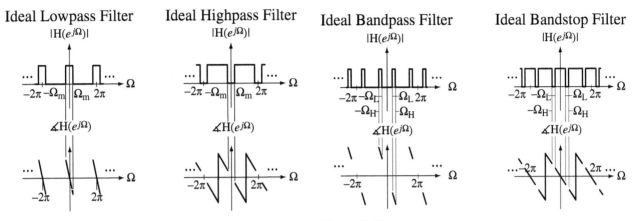

Figure 11.60
Magnitude and phase frequency responses of ideal lowpass and highpass filters

Figure 11.61
Magnitude and phase frequency responses of ideal bandpass and bandstop filters

Filter Type	Frequency Response
Lowpass	$H(e^{j\Omega}) = A\,\text{rect}(\Omega/2\Omega_m)e^{-j\Omega n_0} * \delta_{2\pi}(\Omega)$
Highpass	$H(e^{j\Omega}) = Ae^{-j\Omega n_0}[1 - \text{rect}(\Omega/2\Omega_m) * \delta_{2\pi}(\Omega)]$
Bandpass	$H(e^{j\Omega}) = A\left[\text{rect}\left(\dfrac{\Omega - \Omega_0}{\Delta\Omega}\right) + \text{rect}\left(\dfrac{\Omega + \Omega_0}{\Delta\Omega}\right)\right]e^{-j\Omega n_0} * \delta_{2\pi}(\Omega)$
Bandstop	$H(e^{j\Omega}) = Ae^{-j\Omega n_0}\left\{1 - \left[\text{rect}\left(\dfrac{\Omega - \Omega_0}{\Delta\Omega}\right) + \text{rect}\left(\dfrac{\Omega + \Omega_0}{\Delta\Omega}\right)\right] * \delta_{2\pi}(\Omega)\right\}$

Filter Type	Impulse Response
Lowpass	$h[n] = (A\Omega_m/\pi)\,\text{sinc}(\Omega_m(n - n_0)/\pi)$
Highpass	$h[n] = A\delta[n - n_0] - (A\Omega_m/\pi)\,\text{sinc}(\Omega_m(n - n_0)/\pi)$
Bandpass	$h[n] = 2A\Delta f\,\text{sinc}(\Delta f(t - t_0))\cos(2\pi f_0(t - t_0))$
Bandstop	$h[n] = A\delta[n - n_0] - (A\Delta\Omega/\pi)\,\text{sinc}(\Delta\Omega(n - n_0)/2\pi)\cos(\Omega_0(n - n_0))$

$$\Delta\Omega = \Omega_H - \Omega_L, \quad \Omega_0 = (\Omega_H + \Omega_L)/2$$

Figure 11.62
Frequency responses and impulse responses of the four basic types of ideal filter

In Figure 11.64 and Figure 11.65 are some examples of the impulse responses, frequency responses and responses to rectangular waves of some nonideal, causal filters that approximate the four common types of ideal filters. In each case the frequency response is graphed only over the base period $-\pi < \Omega < \pi$.

The effects of these practical filters on the rectangular waves are similar to those shown for the corresponding continuous-time filters.

Filtering Images

One interesting way to demonstrate what filters do is to filter an image. An image is a "two-dimensional signal." Images can be acquired in various ways. A film camera exposes light-sensitive film to a scene through a lens system, which puts an optical

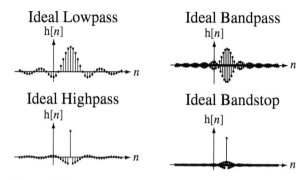

Figure 11.63
Typical impulse responses of ideal lowpass, highpass, bandpass and bandstop filters

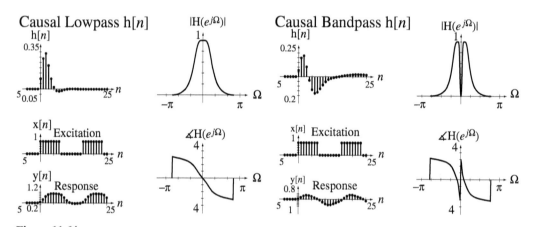

Figure 11.64
Impulse responses, frequency responses and responses to rectangular waves of causal lowpass and bandpass filters

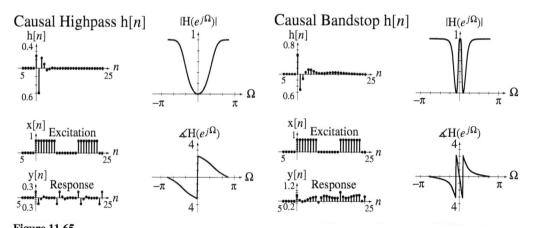

Figure 11.65
Impulse responses, frequency responses and responses to rectangular waves of causal highpass and bandstop filters

11.4 Discrete-Time Filters

image of the scene on the film. The photograph could be a color photograph or a black-and-white (monochrome) photograph. This discussion will be confined to monochrome images. A digital camera acquires an image by imaging the scene on a (usually) rectangular array of detectors, which convert light energy to electrical charge. Each detector, in effect, sees a very tiny part of the image called a **pixel** (short for picture element). The image acquired by the digital camera then consists of an array of numbers, one for each pixel, indicating the light intensity at that point (again assuming a monochrome image).

A photograph is a **continuous-space** function of two spatial coordinates conventionally called x and y. An acquired digital image is a **discrete-space** function of two discrete-space coordinates n_x and n_y. In principle a photograph could be directly filtered. In fact, there are optical techniques that do just that. But by far the most common type of image filtering is done digitally, meaning that an acquired digital image is filtered by a computer using numerical methods.

The techniques used to filter images are very similar to the techniques used to filter time signals, except that they are done in two dimensions. Consider the very simple example image in Figure 11.66.

One technique for filtering an image is to treat one row of pixels as a one-dimensional signal and filter it just like a discrete-time signal. Figure 11.67 is a graph of the brightness of the pixels in the top row of the image versus horizontal discrete-space n_x.

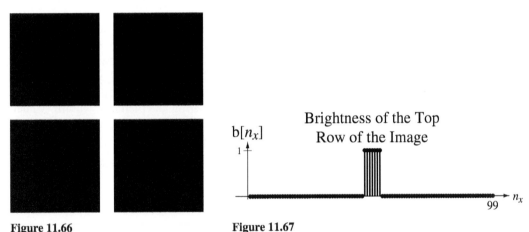

Figure 11.66
A white cross on a black background

Figure 11.67
Brightness of the top row of pixels in the white-cross image

If the signal were actually a function of discrete-time and we were filtering in real time (meaning we would not have future values available during the filtering process), the lowpass-filtered signal might look like Figure 11.68.

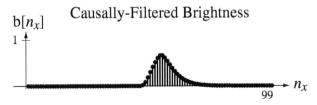

Figure 11.68
Brightness of the top row of pixels after being lowpass filtered by a causal lowpass filter

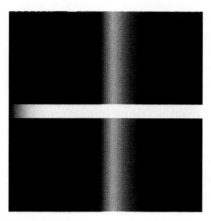

Figure 11.69
White-cross image after all rows have been lowpass filtered by a causal lowpass filter

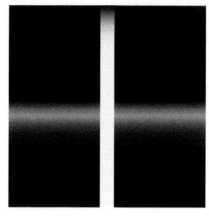

Figure 11.70
White-cross image after all columns have been lowpass filtered by a causal lowpass filter

After lowpass filtering all the rows in the image would look smeared or smoothed in the horizontal direction and unaltered in the vertical direction (Figure 11.69). If we had filtered the columns instead of the rows, the effect would have been as illustrated in Figure 11.70.

One nice thing about image filtering is that usually causality is not relevant to the filtering process. Usually the entire image is acquired and then processed. Following the analogy between time and space, during horizontal filtering "past" signal values would lie to the left and "future" values to the right. In real-time filtering of time signals we cannot use future values because we don't yet know what they are. In image filtering we have the entire image before we begin filtering and therefore "future" values are available. If we horizontally filtered the top row of the image with a "noncausal" lowpass filter, the effect might look as illustrated in Figure 11.71.

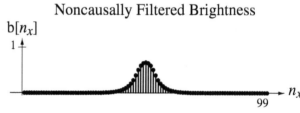

Figure 11.71
Brightness of top row of pixels after being lowpass filtered by a "noncausal" lowpass filter

If we horizontally lowpass filtered the entire image with a "noncausal" lowpass filter, the result would look like Figure 11.72. The overall effect of this type of filtering can be seen in Figure 11.73 where both the rows and columns of the image have been filtered by a lowpass filter.

Of course, the filter referred to above as "noncausal" is actually causal because all the image data are acquired before the filtering process begins. Knowledge of the future is never required. It is only called noncausal because if a space coordinate were instead time, and we were doing real-time filtering, the filtering would be noncausal. Figure 11.74 illustrates some other images and other filtering operations.

11.4 Discrete-Time Filters   553

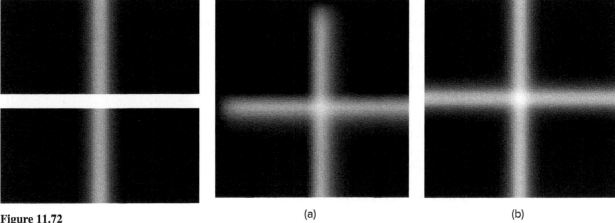

Figure 11.72
White-cross image after all rows have been lowpass filtered by a "noncausal" lowpass filter

Figure 11.73
White-cross image filtered by a lowpass filter, (a) causal, (b) "noncausal"

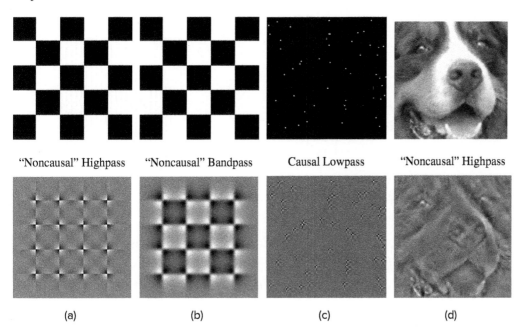

Figure 11.74
Examples of different types of image filtering
(d) © M. J. Roberts

In each image in Figure 11.74 the pixel values range from black to white with gray levels in between. To grasp the filtering effects think of a black pixel as having a value of 0 and a white pixel as having a value of +1. Then medium gray would have a pixel value of 0.5.

Image (a) is a checkerboard pattern filtered by a highpass filter in both dimensions. The effect of the highpass filter is to emphasize the edges and to deemphasize the constant values between the edges. The edges contain the "high-spatial-frequency" information in the image. So the highpass-filtered image has an average value of 0.5 (medium gray) and the black and white squares, which were very different in the original image, look about the same in the filtered image. The checkerboard in (b) is filtered by a bandpass filter. This type of filter smooths edges because it has little response at high frequencies.

It also attenuates the average values because it also has little response at very low frequencies including zero. Image (c) is a random dot pattern filtered by a causal lowpass filter. We can see that it is a causal filter because the smoothing of the dots always occurs to the right and below the dots, which would be "later" times if the signals were time signals. The response of a filter to a very small point of light in an image is called its **point spread function**. The point spread function is analogous to the impulse response in time-domain systems. A small dot of light approximates a two-dimensional impulse and the point spread function is the approximate two-dimensional impulse response. The last image (d) is of the face of a dog. It is highpass filtered. The effect is to form an image that looks like an "outline" of the original image because it emphasizes sudden changes (edges) and deemphasizes the slowly varying parts of the image.

PRACTICAL FILTERS

Comparison with Continuous-Time Filters

Figure 11.75 is an example of an LTI lowpass filter. Its unit-sequence response is $[5 - 4(0.8)^n]u[n]$ (Figure 11.76).

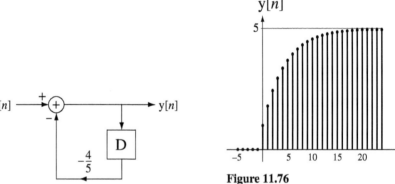

Figure 11.75
A lowpass filter

Figure 11.76
Unit-sequence response of the lowpass filter

The impulse response of any discrete-time system is the first backward difference of its unit-sequence response. In this case that is

$$h[n] = [5 - 4(4/5)^n]u[n] - [5 - 4(4/5)^{n-1}]u[n - 1]$$

which reduces to $h[n] = (0.8)^n u[n]$ (Figure 11.77). The transfer function and frequency response are

$$H(z) = \frac{z}{z - 0.8} \Rightarrow H(e^{j\Omega}) = \frac{e^{j\Omega}}{e^{j\Omega} - 0.8}$$

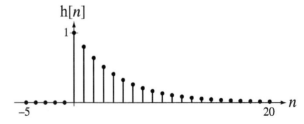

Figure 11.77
Impulse response of the lowpass filter

(Figure 11.78).

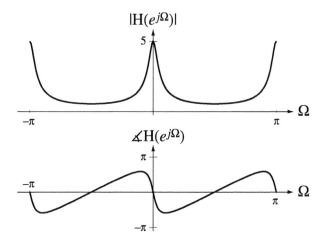

Figure 11.78
Frequency response of the lowpass filter

It is instructive to compare the impulse and frequency responses of this lowpass filter and the impulse and frequency responses of the *RC* lowpass filter. The impulse response of the discrete-time lowpass filter looks like a sampled version of the impulse response of the *RC* lowpass filter (Figure 11.79). Their frequency responses also have some similarities (Figure 11.80).

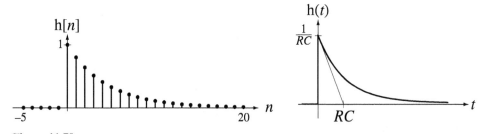

Figure 11.79
A comparison of the impulse responses of a discrete-time and an *RC* lowpass filter

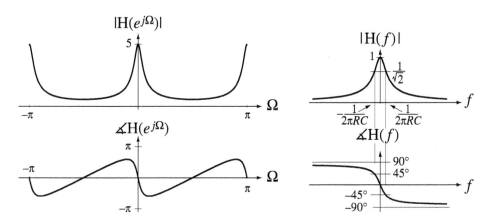

Figure 11.80
Frequency responses of discrete-time and continuous-time lowpass filters

If we compare the shapes of the magnitudes and phases of these frequency responses over the frequency range $-\pi < \Omega < \pi$, they look very much alike (magnitudes more than phases). But a discrete-time frequency response is always periodic and can never be lowpass in the same sense as the frequency response of the *RC* lowpass filter. The name *lowpass* applies accurately to the behavior of the frequency response in the range $-\pi < \Omega < \pi$ and that is the only sense in which the designation lowpass is correctly used for discrete-time systems.

Highpass, Bandpass and Bandstop Filters

Of course, we can have highpass and bandpass discrete-time filters also (Figure 11.81 through Figure 11.83). The transfer functions and frequency responses of these filters are

$$H(z) = \frac{z-1}{z+\alpha} \Rightarrow H(e^{j\Omega}) = \frac{e^{j\Omega}-1}{e^{j\Omega}+\alpha}$$

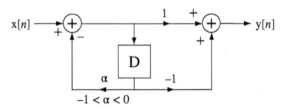

Figure 11.81
A highpass filter

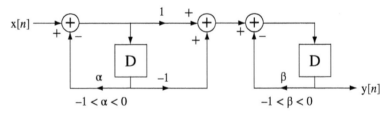

Figure 11.82
A bandpass filter

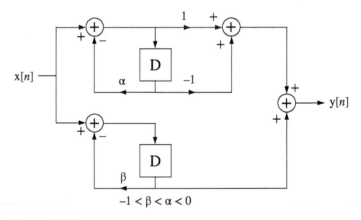

Figure 11.83
A bandstop filter

for the highpass filter,

$$H(e^{j\Omega}) = \frac{z(z-1)}{z^2 + (\alpha+\beta)z + \alpha\beta} \Rightarrow H(e^{j\Omega}) = \frac{e^{j\Omega}(e^{j\Omega}-1)}{e^{j2\Omega} + (\alpha+\beta)e^{j\Omega} + \alpha\beta}.$$

for the bandpass filter and

$$H(e^{j\Omega}) = \frac{2z^2 - (1-\beta-\alpha)z - \beta}{z^2 + (\alpha+\beta)z + \alpha\beta} \Rightarrow H(e^{j\Omega}) = \frac{2e^{j2\Omega} - (1-\beta-\alpha)e^{j\Omega} - \beta}{e^{j2\Omega} + (\alpha+\beta)e^{j\Omega} + \alpha\beta}, -1 < \beta < \alpha < 0$$

for the bandstop filter.

EXAMPLE 11.7

Response of a highpass filter to a sinusoid

A sinusoidal signal $x[n] = 5\sin(2\pi n/18)$ excites a highpass filter with transfer function

$$H(z) = \frac{z-1}{z-0.7}.$$

Graph the response $y[n]$.

The filter's frequency response is $H(e^{j\Omega}) = \frac{e^{j\Omega} - 1}{e^{j\Omega} - 0.7}$. The DTFT of the excitation is $X(e^{j\Omega}) = j5\pi[\delta_{2\pi}(\Omega + \pi/9) - \delta_{2\pi}(\Omega - \pi/9)]$. The DTFT of the response is the product of these two

$$Y(e^{j\Omega}) = \frac{e^{j\Omega} - 1}{e^{j\Omega} - 0.7} \times j5\pi[\delta_{2\pi}(\Omega + \pi/9) - \delta_{2\pi}(\Omega - \pi/9)].$$

Using the equivalence property of the impulse and the fact that both are periodic with period 2π,

$$Y(e^{j\Omega}) = j5\pi\left[\delta_{2\pi}(\Omega + \pi/9)\frac{e^{-j\pi/9} - 1}{e^{-j\pi/9} - 0.7} - \delta_{2\pi}(\Omega - \pi/9)\frac{e^{j\pi/9} - 1}{e^{j\pi/9} - 0.7}\right]$$

$$Y(e^{j\Omega}) = j5\pi\left[\frac{(e^{-j\pi/9}-1)(e^{j\pi/9} - 0.7)\delta_{2\pi}(\Omega + \pi/9) - (e^{j\pi/9}-1)(e^{-j\pi/9} - 0.7)\delta_{2\pi}(\Omega - \pi/9)}{(e^{-j\pi/9} - 0.7)(e^{j\pi/9} - 0.7)}\right]$$

$$Y(e^{j\Omega}) = j5\pi\left[\frac{(1.7 - e^{j\pi/9} - 0.7e^{-j\pi/9})\delta_{2\pi}(\Omega + \pi/9) - (1.7 - 0.7e^{j\pi/9} - e^{-j\pi/9})\delta_{2\pi}(\Omega - \pi/9)}{1.49 - 1.4\cos(\pi/9)}\right]$$

$$Y(e^{j\Omega}) = j28.67\pi\begin{cases} 1.7[\delta_{2\pi}(\Omega + \pi/9) - \delta_{2\pi}(\Omega - \pi/9)] \\ +0.7e^{j\pi/9}\delta_{2\pi}(\Omega - \pi/9) - e^{j\pi/9}\delta_{2\pi}(\Omega + \pi/9) \\ +e^{-j\pi/9}\delta_{2\pi}(\Omega - \pi/9) - 0.7e^{-j\pi/9}\delta_{2\pi}(\Omega + \pi/9) \end{cases}$$

$$Y(e^{j\Omega}) = j28.67\pi\begin{cases} 1.7[\delta_{2\pi}(\Omega + \pi/9) - \delta_{2\pi}(\Omega - \pi/9)] \\ +(0.7\cos(\pi/9) + j0.7\sin(\pi/9))\delta_{2\pi}(\Omega - \pi/9) \\ -(\cos(\pi/9) + j\sin(\pi/9))\delta_{2\pi}(\Omega + \pi/9) \\ +(\cos(\pi/9) - j\sin(\pi/9))\delta_{2\pi}(\Omega - \pi/9) \\ -(0.7\cos(\pi/9) - j0.7\sin(\pi/9))\delta_{2\pi}(\Omega + \pi/9) \end{cases}$$

$$Y(e^{j\Omega}) = j28.67\pi\begin{cases} 1.7(1 - \cos(\pi/9))[\delta_{2\pi}(\Omega + \pi/9) - \delta_{2\pi}(\Omega - \pi/9)] \\ -j0.3\sin(\pi/9)[\delta_{2\pi}(\Omega - \pi/9) + \delta_{2\pi}(\Omega + \pi/9)] \end{cases}.$$

Inverse transforming,

$$y[n] = 28.67 \times 1.7(1 - \cos(\pi/9))\sin(2\pi n/18) + 28.67 \times 0.3\sin(\pi/9)\cos(2\pi n/18)$$
$$y[n] = 2.939\sin(2\pi n/18) + 2.9412\cos(2\pi n/18) = 4.158\sin(2\pi n/18 + 0.786)$$

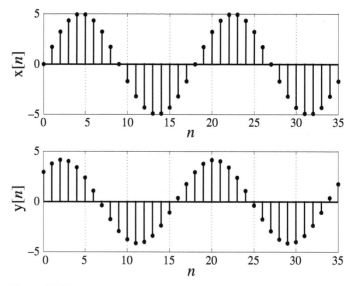

Figure 11.84
Excitation and response of a highpass filter

Figure 11.84 shows the excitation and response of the filter.

EXAMPLE 11.8

Effects of filters on example signals

Test the filter in Figure 11.85 with a unit impulse, a unit sequence and a random signal to show the filtering effects at all three outputs.

$$H_{LP}(e^{j\Omega}) = \frac{Y_{LP}(e^{j\Omega})}{X(e^{j\Omega})} = \frac{0.1}{1 - 0.9e^{-j\Omega}}$$

$$H_{HP}(e^{j\Omega}) = \frac{Y_{HP}(e^{j\Omega})}{X(e^{j\Omega})} = 0.95\frac{1 - e^{-j\Omega}}{1 - 0.9e^{-j\Omega}}$$

$$H_{BP}(e^{j\Omega}) = \frac{Y_{BP}(e^{j\Omega})}{X(e^{j\Omega})} = 0.2\frac{1 - e^{-j\Omega}}{1 - 1.8e^{-j\Omega} + 0.81e^{-j2\Omega}}$$

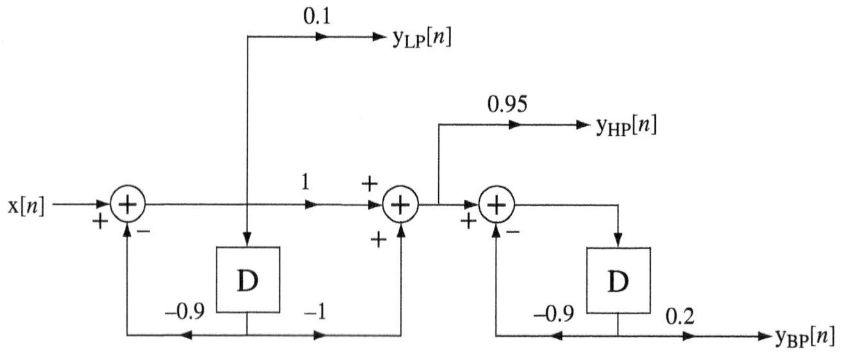

Figure 11.85
Filter with lowpass, highpass and bandpass outputs

Notice in Figure 11.86 that sums of the highpass and bandpass impulse responses are zero because the frequency response is zero at $\Omega = 0$.

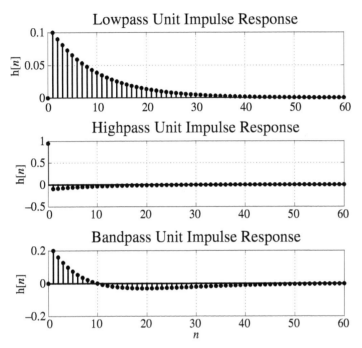

Figure 11.86
Impulse responses at the three outputs

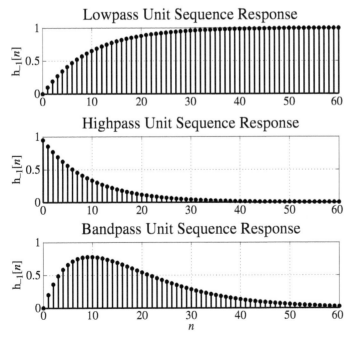

Figure 11.87
Unit-sequence responses at the three outputs

The lowpass filter's response to a unit sequence (Figure 11.87) approaches a nonzero final value because the filter passes the average value of the unit sequence. The unit-sequence responses of the highpass and bandpass filters both approach zero. Also, the unit-sequence response of the highpass filter jumps suddenly at the application of the unit sequence but the lowpass and bandpass filters both respond much more slowly, indicating that they do not allow high-frequency signals to pass through.

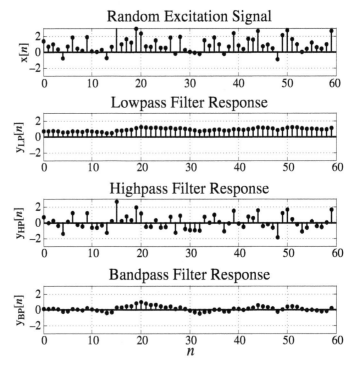

Figure 11.88
Responses at the three outputs to a random signal

The lowpass-filter output signal (Figure 11.88) is a smoothed version of the input signal. The rapidly changing (high-frequency) content has been removed by the filter. The highpass-filter response has an average value of zero and all the rapid changes in the input signal appear as rapid changes in the output signal. The bandpass filter removes the average value of the signal and also smooths it to some extent because it removes both the very low and very high frequencies.

The Moving Average Filter

A very common type of lowpass filter that will illustrate some principles of discrete-time filter design and analysis is the **moving-average** filter (Figure 11.89). The difference equation describing this filter is

$$y[n] = \frac{x[n] + x[n-1] + x[n-2] + \cdots + x[n-(N-1)]}{N}$$

and its impulse response is

$$h[n] = (u[n] - u[n-N])/N$$

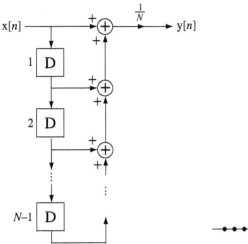

Figure 11.89
A moving-average filter

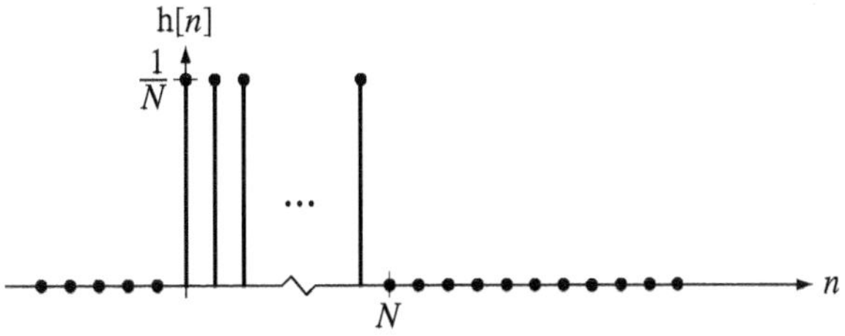

Figure 11.90
Impulse response of a moving-average filter

(Figure 11.90).

Its frequency response is

$$H(e^{j\Omega}) = \frac{e^{-j(N-1)\Omega/2}}{N}\frac{\sin(N\Omega/2)}{\sin(\Omega/2)} = e^{-j(N-1)\Omega/2}\,\text{drcl}(\Omega/2\pi, N)$$

(Figure 11.91).

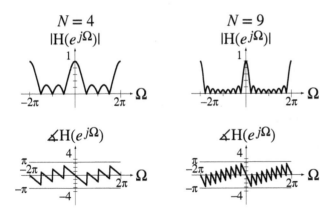

Figure 11.91
Frequency response of a moving-average filter for two different averaging times

This filter is usually described as a smoothing filter because it generally attenuates higher frequencies. That designation would be consistent with being a lowpass filter. However, observing the nulls in the frequency response magnitude, one might be tempted to call it a "multiple bandstop" filter. This illustrates that classification of a filter as lowpass, highpass, bandpass or bandstop is not always clear. However, because of the traditional use of this filter to smooth a set of data, it is usually classified as lowpass.

EXAMPLE 11.9

Filtering a pulse with a moving-average filter

Filter the signal x[n] = u[n] − u[n − 9]

(a) with a moving-average filter with $N = 6$
(b) with the bandpass filter in Figure 11.82 with $\alpha = 0.8$ and $\beta = 0.5$.

Using MATLAB, graph the zero-state response y[n] from each filter.

The zero-state response is the convolution of the impulse response with the excitation. The impulse response for the moving-average filter is

$$h[n] = (1/6)(u[n] - u[n-6]).$$

The frequency response of the bandpass filter is

$$H(e^{j\Omega}) = \frac{Y(e^{j\Omega})}{X(e^{j\Omega})} = \frac{1 - e^{-j\Omega}}{1 - 1.3e^{-j\Omega} + 0.4e^{-j2\Omega}} = \frac{1}{1 - 0.8e^{-j\Omega}} \times \frac{1 - e^{-j\Omega}}{1 - 0.5e^{-j\Omega}}.$$

Therefore its impulse response is

$$h[n] = (0.8)^n u[n] * \{(0.5)^n u[n] - (0.5)^{n-1} u[n-1]\}.$$

The MATLAB program has a main script file. It calls a function convD to do the discrete-time convolutions.

```
% Program to graph the response of a moving average filter
% and a discrete-time bandpass filter to a rectangular pulse

close all ;                              % Close all open figure windows
figure('Position',[20,20,800,600]) ;     % Open a new figure window

n = [-5:30]' ;                           % Set up a time vector for the
                                         % responses

x = uD(n) - uD(n-9) ;                    % Excitation vector

% Moving average filter response

h = uD(n) - uD(n-6) ;                    % Moving average filter impulse
                                         % response

[y,n] = convDT(x,n,h,n,n) ;              % Response of moving average
                                         % filter

% Graph the response

subplot(2,1,1) ; p = stem(n,y,'k','filled') ;
set(p,'LineWidth',2,'MarkerSize',4) ; grid on ;
xlabel('\itn','FontName','Times','FontSize',18) ;
ylabel('y[{\itn}]','FontName','Times','FontSize',18) ;
title('Moving-Average Filter','FontName','Times','FontSize',24) ;

% Bandpass filter response

% Find bandpass filter impulse response
```

```
h1 = 0.8.^n.*uD(n) ; h2 = 0.5.^n.*uD(n) - 0.5.^(n-1).*uD(n-1) ;
[h,n] = convD(h1,n,h2,n,n) ;

[y,n] = convD(x,n,h,n,n) ;              % Response of bandpass filter

% Graph the response

subplot(2,1,2) ; p = stem(n,y,'k','filled') ; set(p,'LineWidth',2,'
MarkerSize',4) ; grid on ;
xlabel('\itn','FontName','Times','FontSize',18) ;
ylabel('y[{\itn}]','FontName','Times','FontSize',18) ;
title('Bandpass Filter','FontName','Times','FontSize',24) ;

%   Function to perform a discrete-time convolution on two signals
%   and return their convolution at specified discrete times. The two
%   signals are in column vectors, x1 and x2, and their times
%   are in column vectors, n1 and n2. The discrete times at which
%   the convolution is desired are in the column, n12. The
%   returned convolution is in column vector, x12, and its
%   time is in column vector, n12. If n12 is not included
%   in the function call it is generated in the function as the
%   total time determined by the individual time vectors
%
%   [x12,n12] = convD(x1,n1,x2,n2,n12)

function [x12,n12] = convD(x1,n1,x2,n2,n12)

% Convolve the two vectors using the MATLAB conv command
    xtmp = conv(x1,x2) ;

% Set a temporary vector of times for the convolution
% based on the input time vectors

    ntmp = n1(1) + n2(1) + [0:length(n1)+length(n2)-2]' ;

% Set the first and last times in temporary vector

    nmin = ntmp(1) ; nmax = ntmp(length(ntmp)) ;

    if nargin < 5, % If no input time vector is specified use ntmp

            x12 = xtmp ; n12 = ntmp ;

    else
%           If an input time vector is specified, compute the
%           convolution at those times

            x12 = 0*n12 ; % Initialize output convolution to zero

%           Find the indices of the desired times which are between
%           the minimum and maximum of the temporary time vector

            I12intmp = find(n12 >= nmin & n12 <= nmax) ;
```

```
%           Translate them to the indices in the temporary time vector
            Itmp = (n12(I12intmp) - nmin) + 1 ;
%           Replace the convolution values for those times
%           in the desired time vector
            x12(I12intmp) = xtmp(Itmp) ;
    end
```

The graphs created are in Figure 11.92.

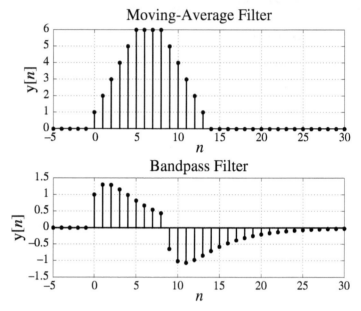

Figure 11.92
Two filter responses

The Almost Ideal Lowpass Filter

If we want to approach the frequency-domain performance of the ideal lowpass filter, we must design a discrete-time filter with an impulse response that closely approaches the inverse DTFT of the ideal frequency response. We have previously shown that the ideal lowpass filter is noncausal and cannot be physically realized. However, we can closely approach it. The ideal lowpass-filter impulse response is illustrated in Figure 11.93.

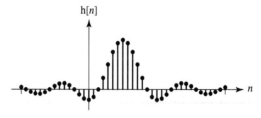

Figure 11.93
Ideal discrete-time lowpass filter impulse response

The problem in realizing this filter physically is the part of the impulse response that occurs before time $n = 0$. If we arrange to delay the impulse response by a large amount, then the signal energy of the impulse response that occurs before time $n = 0$ will become very small and we can chop it off and closely approach the ideal frequency response (Figure 11.94 and Figure 11.95).

Figure 11.94
Almost-ideal discrete-time lowpass filter impulse response

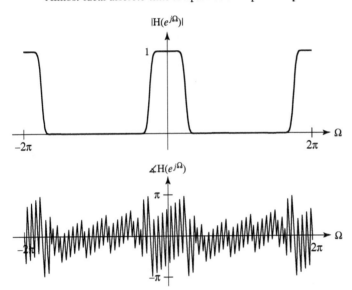

Figure 11.95
Almost-ideal discrete-time lowpass filter frequency response

The magnitude response in the stopband is so small that we cannot see its shape when it is plotted on a linear scale as in Figure 11.95. In cases like this a log-magnitude plot helps see what the real attenuation is in the stopband (Figure 11.96).

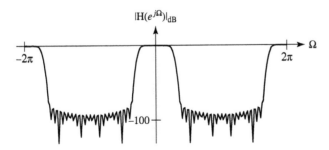

Figure 11.96
Almost-ideal discrete-time lowpass filter frequency response plotted on a dB scale

This filter has a very nice lowpass-filter magnitude response but it comes at a price. We must wait for it to respond. The closer a filter approaches the ideal, the greater time delay there is in the impulse response. This is apparent in the time delay of the impulse response and the phase shift of the frequency response. The fact that a long delay is required for filters that approach the ideal is also true of highpass, bandpass and bandstop filters and it is true for both continuous-time and discrete-time filters. It is a general principle of filter design that any filter designed to be able to discriminate between two closely spaced frequencies and pass one while stopping the other must, in a sense, "observe" them for a long time to be able to distinguish one from the other. The closer they are in frequency, the longer the filter must observe them to be able to make the distinction. That is the basic reason for the requirement for a long time delay in the response of a filter that approaches an ideal filter.

Advantages Compared to Continuous-Time Filters
One might wonder why we would want to use a discrete-time filter instead of a continuous-time filter. There are several reasons. Discrete-time filters are built with three basic elements: a delay device, a multiplier and an adder. These can be implemented with digital devices. As long as we stay within their intended ranges of operation, these devices always do exactly the same thing. That cannot be said of devices such as resistors, capacitors and operational amplifiers, which make up continuous-time filters. A resistor of a certain nominal resistance is never exactly that value, even under ideal conditions. And even if it were at some time, temperature effects or other environmental effects would change it. The same thing can be said of capacitors, inductors, transistors and so on. So discrete-time filters are more stable and reproducible than continuous-time filters.

It is often difficult to implement a continuous-time filter at very low frequencies because the component sizes become unwieldy, for example, very large capacitor values may be needed. Also, at very low frequencies thermal drift effects on components become a big problem because they are indistinguishable from signal changes in the same frequency range. Discrete-time filters do not have these problems.

Discrete-time filters are often implemented with programmable digital hardware. That means that this type of discrete-time filter can be reprogrammed to perform a different function without changing the hardware. Continuous-time filters do not have this flexibility. Also, some types of discrete-time filters are so computationally sophisticated that they would be practically impossible to implement as continuous-time filters.

Discrete-time signals can be reliably stored for very long times without any significant degradation on magnetic disk or tape or CD-ROM. Continuous-time signals can be stored on analog magnetic tape but over time the values can degrade.

By time-multiplexing discrete-time signals, one filter can accommodate multiple signals in a way that seems to be, and effectively is, simultaneous. Continuous-time filters cannot do that because to operate correctly they require that the input signal always be present.

11.5 SUMMARY OF IMPORTANT POINTS

1. Frequency response and impulse response of LTI systems are related through the Fourier transform.
2. Characterization of systems in the frequency domain allows generalized design procedures for systems that process certain types of signals.
3. An ideal filter is distortionless within its passband.
4. All ideal filters are noncausal and therefore cannot be built.
5. Filtering techniques can be applied to images as well as signals.
6. Practical discrete-time filters can be implemented as discrete-time systems using only amplifiers, summing junctions and delays.

EXERCISES WITH ANSWERS

(Answers to each exercise are in random order.)

Continuous-Time Frequency Response

1. A system has an impulse response,
$$h_{LP}(t) = 3e^{-10t}u(t),$$
and another system has an impulse response,
$$h_{HP}(t) = \delta(t) - 3e^{-10t}u(t).$$

   (a) Sketch the magnitude and phase of the frequency response of these two systems in a parallel connection.

   (b) Sketch the magnitude and phase of the frequency response of these two systems in a cascade connection.

   Answers:

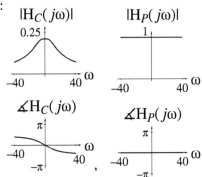

Continuous-Time Ideal Filters

2. A system has an impulse response $h(t) = 10\,\text{rect}\left(\frac{t - 0.01}{0.02}\right)$. What is its null bandwidth?

   Answer: 50 Hz

Continuous-Time Causality

3. Determine whether or not the systems with these frequency responses are causal.

   (a) $H(f) = \text{sinc}(f)$
   (b) $H(f) = \text{sinc}(f)e^{-j\pi f}$
   (c) $H(j\omega) = \text{rect}(\omega)$
   (d) $H(j\omega) = \text{rect}(\omega)e^{-j\omega}$
   (e) $H(f) = A$
   (f) $H(f) = Ae^{j2\pi f}$
   (g) $H(j\omega) = 1 - e^{-j\omega}$
   (h) $H(f) = \text{rect}(f/20)e^{-j40\pi f}$
   (i) $H(e^{j\Omega}) = \dfrac{e^{j\Omega}}{e^{j\Omega} - 0.9}$
   (j) $H(e^{j\Omega}) = \dfrac{e^{j2\Omega}}{e^{j\Omega} - 1.3}$

   Answers: 4 Causal and 6 Noncausal

Logarithmic Graphs, Bode Diagrams and Decibels

4. A system is excited by a sinusoid whose signal power is 0.01 and the response is a sinusoid of the same frequency with a signal power of 4. What is the magnitude of the transfer function of the system at the frequency of the sinusoid, expressed in decibels (dB)?

   Answer: 26 dB

5. A system is excited by a sinusoid whose amplitude is 1 µV and the response is a sinusoid of the same frequency with an amplitude of 5 V. What is the magnitude of the transfer function of the system at the frequency of the sinusoid, expressed in decibels (dB)?

   Answer: 134 dB

6. A system has a transfer function $H(s) = 10 \dfrac{s^2}{s^2 + 11s + 10}$.
   (a) Find the values of its frequency response magnitude in dB and its frequency response angle in radians at these frequencies.

   $$\omega = 0.01 \qquad \omega = 1 \qquad \omega = 10 \qquad \omega = 1000$$

   (b) Draw the overall asymptotic magnitude Bode diagram for this frequency response in the radian frequency range $10^{-2} < \omega < 10^3$.

   Answers: 20 dB and 0.011 radians, 16.9465 dB and 0.8851 radians, −80 dB and 3.1306 radians, −3.0535 dB and 2.2565 radians

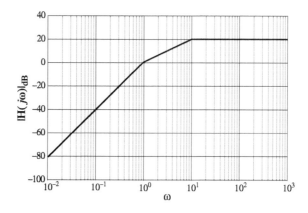

7. In an inverting op-amp amplifier, the feedback component is a 1000 Ω resistor and the component between the input voltage terminal and the operational amplifier's inverting input is a 10 µF capacitor. If the voltage transfer function is $H(f)$, what are the magnitude and phase of $H(200)$?

   Answer: 12.57 and −1.57 radians

8. An active op-amp integrator has a frequency response magnitude Bode diagram that goes through −40 dB at $\omega = 500$. At what numerical value of ω is the system frequency response magnitude 100 times smaller than it is at $\omega = 500$?

   Answer: 50,000

9. Graph the magnitude frequency responses, both on a linear-magnitude and on a log-magnitude scale, of the systems with these frequency responses, over the frequency range specified.

   (a) $H(f) = \dfrac{20}{20 - 4\pi^2 f^2 + j42\pi f}$, $-100 < f < 100$

   (b) $H(j\omega) = \dfrac{2 \times 10^5}{(100 + j\omega)(1700 - \omega^2 + j20\omega)}$, $-500 < \omega < 500$

Answers:

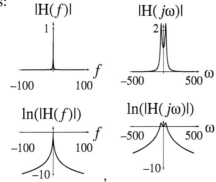

10. Draw asymptotic and exact magnitude and phase Bode diagrams for the frequency responses of the following circuits and systems.

    (a) An RC lowpass filter with $R = 1$ MΩ and $C = 0.1$ µF.

    (b) The circuit of Figure E.10b

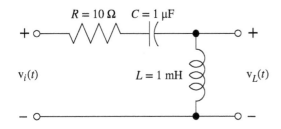

Figure E.10b

Answers:

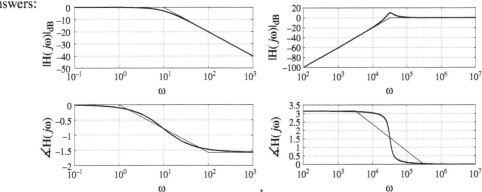

Continuous-Time Practical Passive Filters

11. Find and graph the frequency response of each of the circuits in Figure E.11 given the indicated excitation and response.

   (a) Excitation, $v_i(t)$ – Response $v_L(t)$

   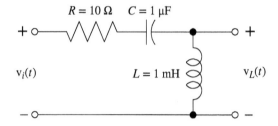

   (b) Excitation, $v_i(t)$ – Response $i_C(t)$

   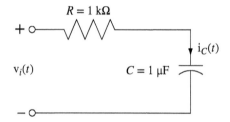

   (c) Excitation, $v_i(t)$ – Response $v_R(t)$

   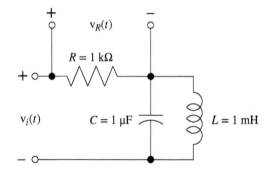

   (d) Excitation, $v_i(t)$ – Response $v_R(t)$

   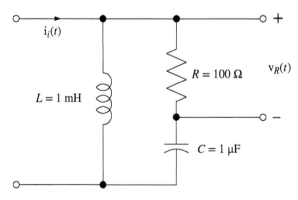

   Figure E.11

Answers:

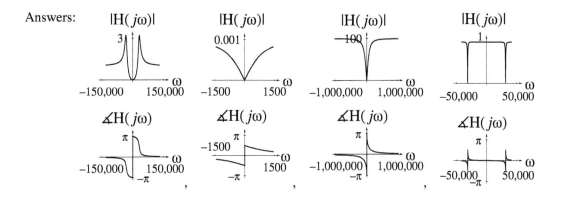

12. With reference to the circuit schematic diagram in Figure E.12

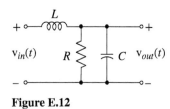

Figure E.12

(a) Find a general expression for the frequency response H($j\omega$).

(b) Is this circuit a practical lowpass, highpass, bandpass or bandstop filter?

(c) What is the slope in dB/decade of the magnitude Bode diagram for the frequency response of this filter at very low frequencies?

(d) What is the slope in dB/decade of the magnitude Bode diagram for the frequency response of this filter at very high frequencies?

Answers: Lowpass, −40 dB/decade, H($j\omega$) = $\dfrac{1}{LC} \dfrac{1}{(j\omega)^2 + j\omega/RC + 1/LC}$, 0 dB/decade

13. In Figure E.13 is a practical passive continuous-time filter. Let $C = 16$ µF and $R = 1000$ Ω.

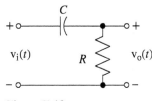

Figure E.13

(a) Find its transfer function H(f) in terms of R, C and f as variables.

(b) At what frequency f is its transfer function magnitude a minimum and what are the transfer function magnitude and phase at that frequency?

(c) At what frequency f is its transfer function magnitude a maximum and what are the transfer function magnitude and phase at that frequency?

(d) What are the magnitude and phase of the transfer function at a frequency of 10 Hz?

(e) If you keep $R = 1000\ \Omega$ and choose a new capacitor value C to make the magnitude of the transfer function at 100 Hz less than 30% of the maximum transfer function magnitude, what is the largest numerical value of C you could use?

Answers: $\{0.709, 0.7828\}$, $0.5005\ \mu F$, $H(f) = \dfrac{R}{R + 1/j2\pi fC} = \dfrac{j2\pi fRC}{j2\pi fRC + 1}$, $\{\infty, 1, 0\}$, $\{0, 0, \text{undefined}\}$

14. For the practical passive filter in Figure E.14 with transfer function $H(s) = V_{out}(s)/V_{in}(s)$,

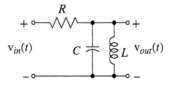

Figure E.14

(a) What is the slope, in dB per decade, of a magnitude Bode diagram of its frequency response at frequencies approaching zero and at frequencies approaching infinity?

(b) Find the frequency at which the magnitude of the transfer function is a maximum.

(c) Find the nonzero frequency at which the phase of the transfer function is zero.

(d) Find the phase shift of the transfer function just above and just below $f = 0$.

(e) Classify this filter as a practical approximation to the ideal lowpass, highpass, bandpass or bandstop filter.

Answers: $\{20 \text{ and } -20 \text{ dB/decade}\}$, 711.8, $\{\pi/2 \text{ and } -\pi/2\}$, Bandpass, 711.8

15. For each circuit in Figure E.15 the frequency response is the ratio $H(f) = \dfrac{V_o(f)}{V_i(f)}$. Which circuits have

(a) Zero frequency response at $f = 0$?

(b) Zero frequency response at $f \to +\infty$?

(c) Transfer function magnitude of one at $f = 0$?

(d) Transfer function magnitude of one at $f \to +\infty$?

(e) Transfer function magnitude nonzero and phase of zero at some frequency, $0 < f < \infty$, (at a finite, nonzero frequency)?

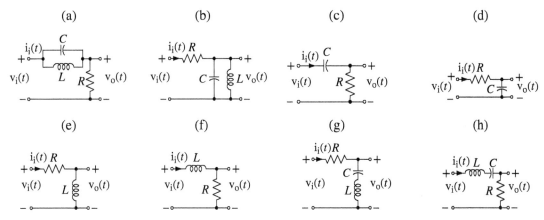

Figure E.15

Answers: {a,d,f,g}, {b,c,e,h}, {b,h}, {b,d,f,h}, {a,c,e,g}

16. The causal square wave voltage signal illustrated in Figure E.16 is the excitation for five practical passive filters numbered 1–5, also in Figure E.16. The voltage responses of the five filters are illustrated below them. Match the responses to the filters.

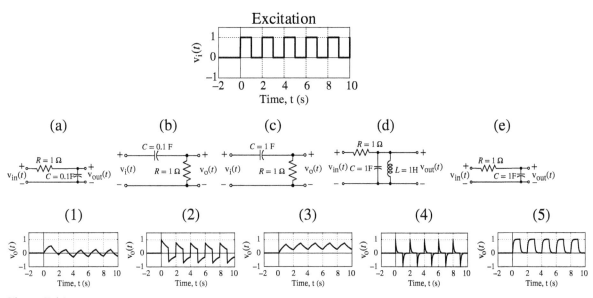

Figure E.16

Answers: 1-E, 2-D, 3-B, 4-A, 5-C

17. Classify each of these frequency responses as lowpass, highpass, bandpass or bandstop.

(a) $H(f) = \dfrac{1}{1+jf}$

(b) $H(f) = \dfrac{jf}{1+jf}$

(c) $H(j\omega) = -\dfrac{j10\omega}{100 - \omega^2 + j10\omega}$

Answers: Bandpass, Highpass, Lowpass

18. In the circuit in Figure E.18 let $R = 10\ \Omega$, $L = 10$ mH and $C = 100\ \mu$F and let $H(j\omega) = \dfrac{V_o(j\omega)}{V_i(j\omega)}$.

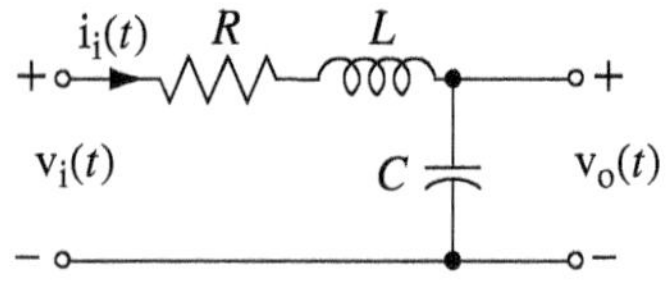

Figure E.18

(a) $H(j\omega)$ can be expressed in the form $\dfrac{A}{(j\omega)^2 + jB\omega + C}$. Find the numerical values of A, B and C.

(b) Find the numerical value of $H(0)$.

(c) Find the numerical value of $\lim_{\omega \to +\infty} H(j\omega)$.

For parts (d), (e) and (f), redefine the frequency response as $H(j\omega) = \dfrac{I_i(j\omega)}{V_i(j\omega)}$.

(d) $H(j\omega)$ can now be expressed in the form $\dfrac{j\omega A}{(j\omega)^2 + jB\omega + C}$. Find the numerical values of A, B and C.

(e) Find the numerical value of $H(0)$.

(f) Find the numerical value of $\lim_{\omega \to +\infty} H(j\omega)$.

Answers: {100, 1000, 1,000,000}, {1,000,000, 1,000, 1,000,000}, 1, 0, 0, 0

19. A passive circuit consists of a resistor R and an inductor L in parallel. Define the input signal to the system as the voltage v(t) across both the resistor and inductor and define the response signal as the total current i(t) that flows into the parallel combination of the resistor and inductor.

(a) Find an expression for the frequency response $H(f)$ as a ratio of two polynomials in f.

(b) If $R = 1\ \Omega$ and $L = 0.1$ H find the numerical value of $H(10)$.

Answers: $1.0126 e^{-j0.1578}$, $\dfrac{R + j2\pi f L}{j2\pi f RL}$

Continuous-Time Practical Active Filters

20. Match each circuit in Figure E.20 to the magnitude asymptotic Bode diagram of its frequency response, $H(j\omega) = \dfrac{V_o(j\omega)}{V_i(j\omega)}$.

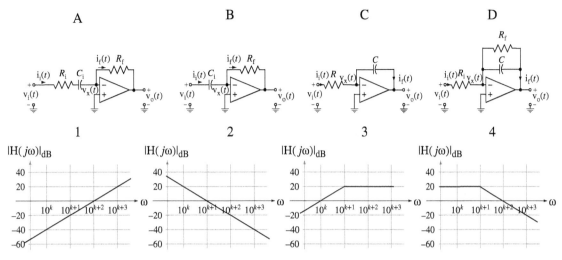

Figure E.20

Answers: A-3, B-1, C-2, D-4

21. The transfer function for the system in Figure E.20 can be written in the form

$$H(s) = \frac{b_2 s^2 + b_1 s + b_0}{s^2 + a_1 s + a_0}.$$

$$\omega_{ca} = 400,\ \omega_{cb} = 600$$

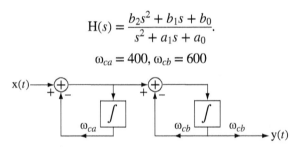

Figure E.20

(a) Find numerical values for the a's and b's.

(b) Find the system's numerical frequency response magnitude in dB and phase in radians at 150 Hz.

Answers: {−6.1193 dB, −0.60251 radians}, {0, 600, 0, 1000, 240,000}

Discrete-Time Frequency Response

22. A system has an impulse response,

$$h[n] = (7/8)^n u[n].$$

What is its half-power discrete-time-frequency bandwidth?

Answer: 0.1337 radians

23. Match each pole-zero diagram in Figure E.23 to its magnitude frequency response. (Assume that the transfer functions are of the form

$$H(s) = A\, \frac{(s - z_1)(s - z_2) \cdots (s - z_M)}{(s - p_1)(s - p_2) \cdots (s - p_N)}$$

and that $A = 1$.) In each case all finite poles and zeros are shown.

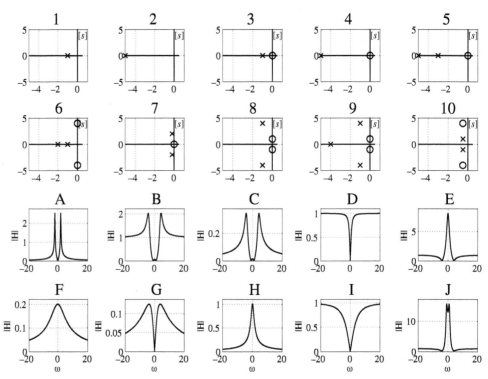

Figure E.23

Answers: 1-H, 2-F, 3-D, 4-I, 5-G, 6-E, 7-A, 8-B, 9-C, 10-J

Discrete-Time Ideal Filters

24. Classify each of the frequency responses in Figure E.24 as lowpass, highpass, bandpass or bandstop.

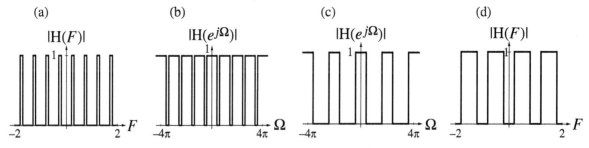

Figure E.24

Answers: Highpass, Bandpass, Lowpass, Bandstop

Discrete-Time Causality

25. For the system frequency response $H(e^{j\Omega}) = \dfrac{e^{-jA\Omega}}{1 - 0.8e^{-j\Omega}}$, what numerical range of integer values of A will produce a causal system?

Answer: $A \geq 0$

Discrete-Time Practical Filters

26. Find the frequency response, $H(e^{j\Omega}) = \dfrac{Y(e^{j\Omega})}{X(e^{j\Omega})}$, and graph the frequency response of each of the filters in Figure E.26 over the range, $-2\pi < \Omega < 2\pi$.

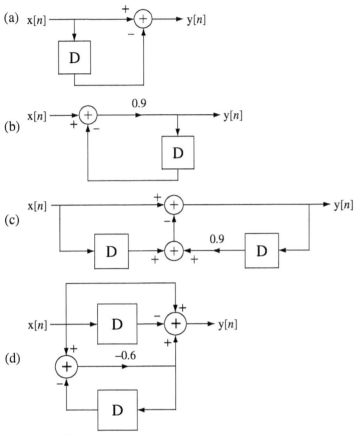

Figure E.26

Answers:

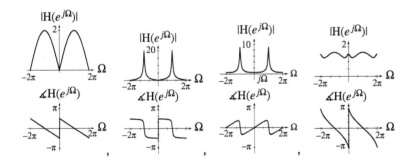

27. Find the minimum stop band attenuation of a moving-average filter with $N = 3$. Define the stop band as the frequency region, $\Omega_C < \Omega < \pi$ where Ω_c is the discrete-time frequency of the first null in the frequency response.

Answer: -9.54 dB

28. In Figure E.28 match each pole-zero diagram to its magnitude frequency response graph. (The gain constants are not all one.)

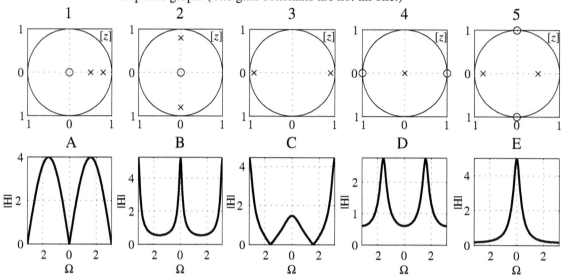

Figure E.28

Answers: 1-E, 2-D, 3-B, 4-A, 5-C

29. In Figure E.29 match each pole-zero diagram to its magnitude frequency response and unit sequence response.

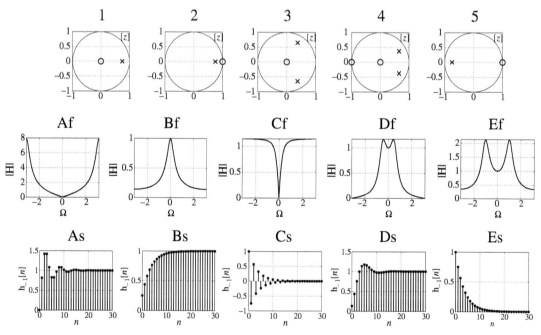

Figure E.29

Answers: 1-Bf-Bs, 2-Cf-Es, 3-Ef-As, 4-Df-Ds, 5-Af-Cs

EXERCISES WITHOUT ANSWERS

Continuous-Time Frequency Response

30. Why is it impossible to actually make a distortionless system?

31. Why is it impossible to actually make an ideal filter?

32. One problem with causal filters is that the response of the filter always lags the excitation. This problem cannot be eliminated if the filtering is done in real time but if the signal is recorded for later "off-line" filtering one simple way of eliminating the lag effect is to filter the signal, record the response and then filter that recorded response with the same filter but playing the signal back through the system *backward*. Suppose the filter is a single-pole filter with a frequency response of the form

$$H(j\omega) = \frac{1}{1 + j\omega/\omega_c}$$

where ω_c is the cutoff frequency (half-power frequency) of the filter.

(a) What is the effective frequency response of the entire process of filtering the signal forward, then backward?

(b) What is the effective impulse response?

Continuous-Time Ideal Filters

33. A signal x(t) is described by

$$x(t) = \text{rect}(1000t) * \delta_{0.002}(t).$$

(a) If x(t) is the excitation of an ideal lowpass filter with a cutoff frequency of 3 kHz, graph the excitation x(t) and the response y(t) on the same scale and compare.

(b) If x(t) is the excitation of an ideal bandpass filter with a low cutoff frequency of 1 kHz and a high cutoff frequency of 5 kHz, graph the excitation x(t) and the response y(t) on the same scale and compare.

Continuous-Time Causality

34. Determine whether or not the systems with these frequency responses are causal.

(a) $H(j\omega) = \dfrac{2}{j\omega}$

(b) $H(j\omega) = \dfrac{10}{6 + j4\omega}$

(c) $H(j\omega) = \dfrac{4}{25 - \omega^2 + j6\omega} = \dfrac{4}{(j\omega + 3)^2 + 16}$

(d) $H(j\omega) = \dfrac{4}{25 - \omega^2 + j6\omega} e^{j\omega} = \dfrac{4}{(j\omega + 3)^2 + 16} e^{j\omega}$

(e) $H(j\omega) = \dfrac{4}{25 - \omega^2 + j6\omega} e^{-j\omega} = \dfrac{4}{(j\omega + 3)^2 + 16} e^{-j\omega}$

(f) $H(j\omega) = \dfrac{j\omega + 9}{45 - \omega^2 + j6\omega} = \dfrac{j\omega + 3}{(j\omega + 3)^2 + 36} + \dfrac{6}{(j\omega + 3)^2 + 36}$

(g) $H(j\omega) = \dfrac{49}{49 + \omega^2}$

Bode Diagrams

35. Draw asymptotic and exact magnitude and phase Bode diagrams for the frequency responses of the circuits and systems in Figure E.35.

    (a)

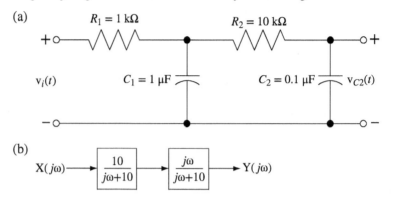

    (b)

    (c) A system whose frequency response is $H(j\omega) = \dfrac{j20\omega}{10{,}000 - \omega^2 + j20\omega}$

    Figure E.35

36. A highpass filter is made with one resistor and one practical inductor. The practical inductor can be modeled as an ideal inductor in series with a small resistance. A Bode diagram of the actual filter's frequency response deviates from the Bode diagram of the frequency response of a highpass filter made with an ideal inductor. Draw a sketch of the ideal Bode diagram and the actual Bode diagram to illustrate the difference.

37. A system has a transfer function $H(s) = 3\dfrac{s^2 + 7s}{s^2 + 8s + 4}$.

    (a) In a magnitude Bode diagram of its frequency response what are the values of all the corner frequencies (in radians/second)?

    (b) What is the slope (in dB/decade) of the magnitude Bode diagram at very low frequencies (approaching zero)?

    (c) What is the slope (in dB/decade) of the magnitude Bode diagram at very high frequencies (approaching infinity)?

Continuous-Time Practical Passive Filters

38. Find and graph the frequency response of each of the circuits in Figure E.38 given the indicated excitation and response.

    (a) Excitation $v_i(t)$ – Response $v_{C2}(t)$

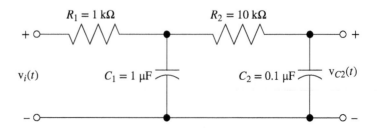

(b) Excitation $v_i(t)$ – Response $i_{C1}(t)$

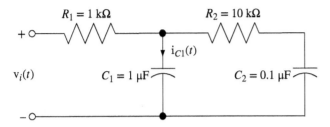

(c) Excitation $v_i(t)$ – Response $v_{R2}(t)$

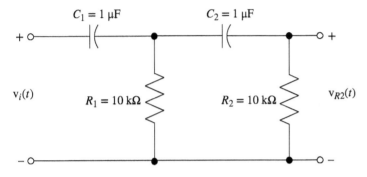

(d) Excitation $i_i(t)$ – Response $v_{R1}(t)$

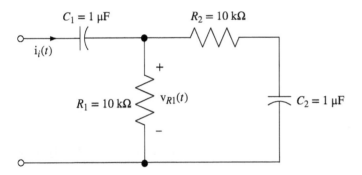

(e) Excitation $v_i(t)$ – Response $v_{RL}(t)$

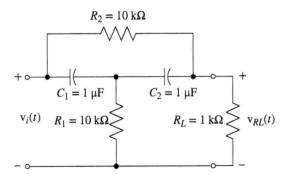

Figure E.38

39. Find, and graph versus frequency, the magnitude and phase of the input impedance $Z_{in}(j\omega) = \dfrac{V_i(j\omega)}{I_i(j\omega)}$ and frequency response $H(j\omega) = \dfrac{V_o(j\omega)}{V_i(j\omega)}$ for each of the filters in Figure E.39.

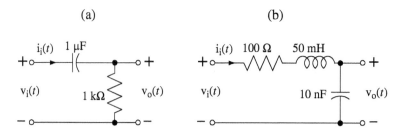

Figure E.39

40. The signal x(t) in Exercise 33 is the excitation of an *RC* lowpass filter with $R = 1\ \text{k}\Omega$ and $C = 0.3\ \mu\text{F}$. Sketch the excitation and response voltages versus time on the same scale.

Continuous-Time Filters

41. In Figure E.41 are some descriptions of filters in the form of an impulse response, a frequency response magnitude and a circuit diagram. For each of these, to the extent possible, classify the filters as ideal or practical, causal or noncausal, lowpass, highpass, bandpass or bandstop.

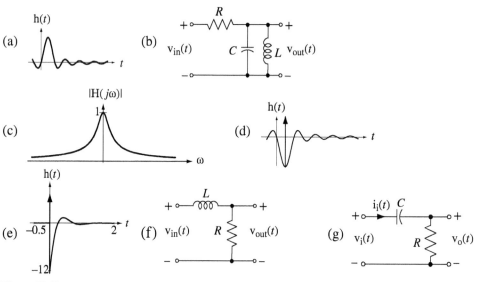

Figure E.41

Continuous-Time Practical Active Filters

42. Design an active highpass filter using an ideal operational amplifier, two resistors and one capacitor and derive its frequency response to verify that it is highpass.

43. Find the frequency response, $H(j\omega) = \dfrac{V_o(j\omega)}{V_i(j\omega)}$, of the active filter in Figure E.43 with $R_i = 1000\ \Omega$, $C_i = 1\mu\text{F}$ and $R_f = 5000\ \Omega$.

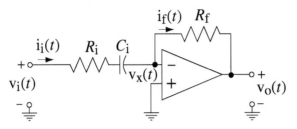

Figure E.43

(a) Find all the corner frequencies (in radians per second) in a magnitude Bode diagram of this frequency response.

(b) At very low and very high frequencies what is the slope of the magnitude Bode diagram in dB/decade?

44. In the active filters in Figure E.44 all resistors are 1 ohm and all capacitors are 1 farad. For each filter the frequency response is $H(j\omega) = \dfrac{V_o(j\omega)}{V_i(j\omega)}$. Identify the frequency response magnitude Bode diagram for each circuit.

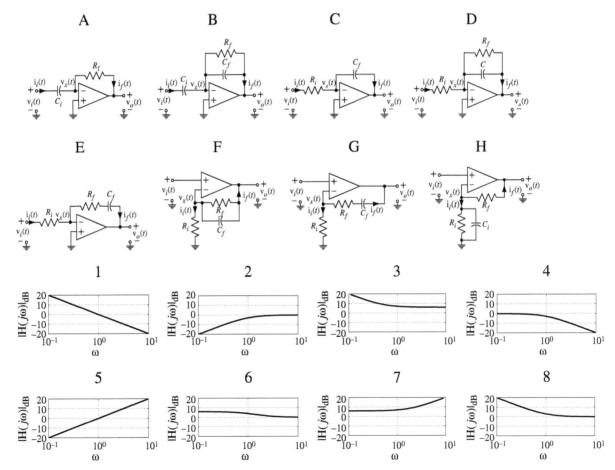

Figure E.44

45. You have available two resistors of 100 Ω and 1000 Ω, two capacitors of 10 μF and 100 μF and one ideal operational amplifier (with its noninverting input grounded) from which to design some active filters.

   (a) Draw the circuit for an active lowpass filter with a corner frequency of 159 Hz. Be sure to label the values of the resistor(s) and capacitor(s).

   (b) Draw the circuit for an active differentiator with the largest possible frequency response magnitude at 1 Hz. Be sure to label the values of the resistor(s) and capacitor(s).

   (c) Draw the circuit for an active bandpass filter. Make the low corner frequency and the high corner frequency as far apart as possible. Be sure to label the values of the resistor(s) and capacitor(s).

46. Using only resistors and capacitors put single components into the circuit diagram in Figure E.46 in the numbered positions that will make the frequency response of this filter $H(j\omega) = \dfrac{V_o(j\omega)}{V_i(j\omega)}$ bandpass in nature with two poles. You need not put values on the components, just indicate whether they are capacitors or resistors. (The triangle with a "K" in it is a voltage amplifier of gain K, not an operational amplifier. There is more than one correct answer.)

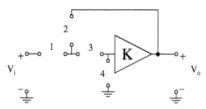

Figure E.46

47. Classify the following transfer functions of filters as lowpass, bandpass, highpass or bandstop according to these definitions:

   Lowpass: $H(0) \neq 0$, $H(j\omega)|_{\omega \to \infty} = 0$

   Highpass: $H(0) = 0$, $H(j\omega)|_{\omega \to \infty} \neq 0$

   Bandpass: $H(0) = 0$, $H(j\omega)|_{\omega \to \infty} = 0$, $|H(j\omega_0)| \neq 0$, $0 < \omega_0 < \infty$

   Bandstop: $H(0) \neq 0$, $H(j\omega)|_{\omega \to \infty} \neq 0$, $|H(j\omega_0)| = 0$ for $0 < \omega_0 < \infty$

   (a) H(s) has finite poles at $s = -2$ and $s = -7$ and a finite zero at $s = -20$.

   (b) H(s) has finite poles at $s = -2$ and $s = -7$ and a finite zero at $s = 0$.

   (c) H(s) has three finite poles in the left half-plane and no finite zeros.

   (d) H(s) has finite poles at $s = -2$ and $s = -7$ and a double finite zero at $s = 0$.

48. There are eight pole-zero graphs of system transfer functions in Figure E.48. Answer the following questions about their frequency responses, impulse responses and step responses.

   Frequency response:

   (a) Which have a phase approaching zero at very high frequencies?

(b) Which have a phase that is discontinuous at zero frequency?

(c) Which have a magnitude approaching zero at high frequencies?

Step response:

(a) Which have a step response that is nonzero in the limit $t \to \infty$?

(b) Which have a step response that is discontinuous at $t = 0$?

Impulse response:

(a) Which have an impulse response that contains an impulse?

(b) Which have an impulse response approaching zero in the limit $t \to \infty$?

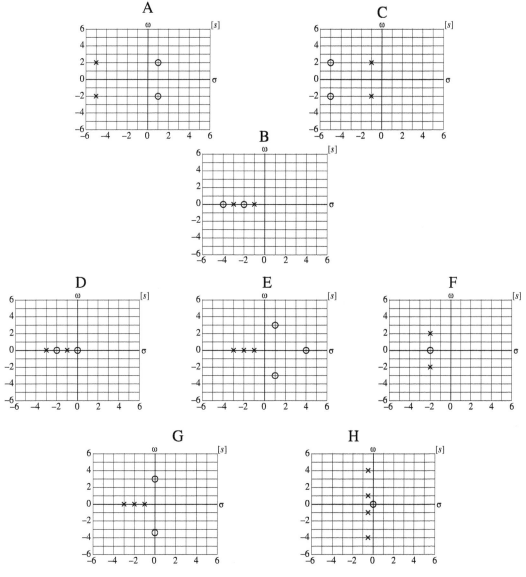

Figure E.48

49. In Figure E.49 are some pole-zero diagrams and some magnitude frequency responses. Match the frequency responses to the pole-zero diagrams.

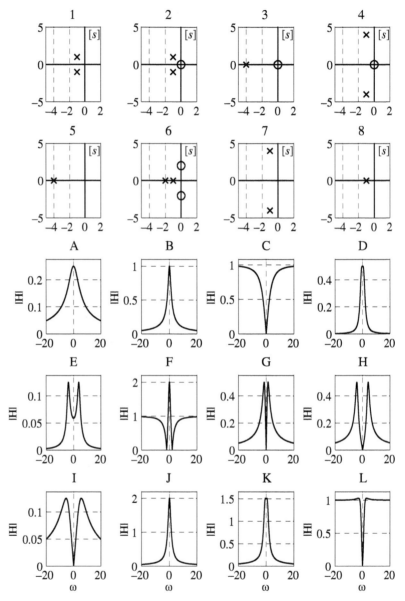

Figure E.49

Discrete-Time Causality

50. Determine whether or not the systems with these frequency responses are causal.

    (a) $H(e^{j\Omega}) = [\text{rect}(5\Omega/\pi) * \delta_{2\pi}(\Omega)]e^{-j10\Omega}$
    $h[n] = \frac{1}{10}\text{sinc}\left(\frac{n-10}{10}\right)$

(b) $H(e^{j\Omega}) = j\sin(\Omega) = \dfrac{e^{j\Omega} - e^{\Omega}}{2}$

$h[n] = (1/2)(\delta[n+1] - \delta[n-1])$

(c) $H(e^{j\Omega}) = 1 - e^{-j2\Omega}$

$h[n] = \delta[n] - \delta[n-2]$

(d) $H(e^{j\Omega}) = \dfrac{8e^{j\Omega}}{8 - 5e^{-j\Omega}} = \dfrac{e^{j\Omega}}{1 - (5/8)e^{-j\Omega}}$

$h[n] = (5/8)^{n+1}u[n+1]$

Discrete-Time Filters

51. In Figure E.51 are pairs of excitations x and responses y. For each pair, identify the type of filtering that was done, lowpass, highpass, bandpass or bandstop.

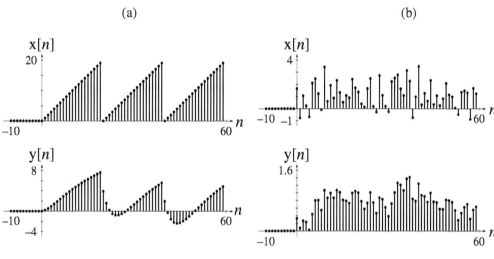

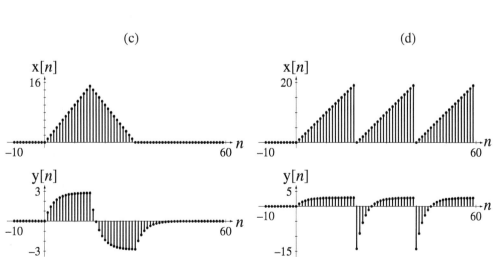

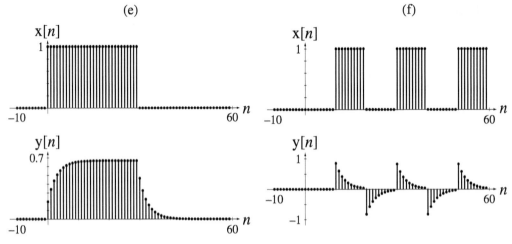

Figure E.51

52. Classify each of these frequency responses as lowpass, highpass, bandpass or bandstop.

   (a) $H(e^{j\Omega}) = \dfrac{\sin(3\Omega/2)}{\sin(\Omega/2)}$

   (b) $H(e^{j\Omega}) = j[\sin(\Omega) + \sin(2\Omega)]$

53. For each of the systems with the pole-zero diagrams in Figure E.53 find the discrete-time radian frequencies, Ω_{max} and Ω_{min}, in the range, $-\pi \leq \Omega \leq \pi$ for which the transfer function magnitude is a maximum and a minimum. If there is more than one value of Ω_{max} or Ω_{min}, find all such values.

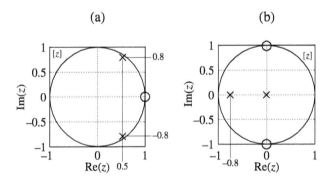

Figure E.53

54. Find the frequency response $H(e^{j\Omega}) = \dfrac{Y(e^{j\Omega})}{X(e^{j\Omega})}$ and graph it for each of the filters in Figure E.54.

(a)
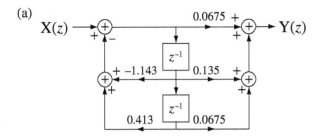

(b)
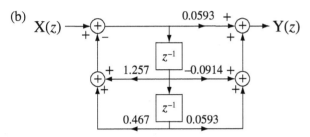

(c)
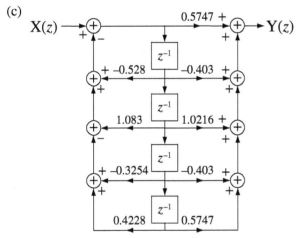

Figure E.54

55. Referring to the system block diagram in Figure E.55

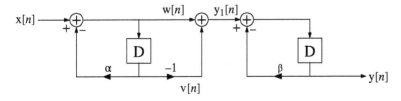

Figure E.55

(a) Write the difference equations

   i. Relating $w[n]$ to $x[n]$ without reference to $v[n]$, and

ii. Relating v[n] to x[n] without reference to w[n].

(b) Then z transform the equations from part (a) and find the transfer function

(c) Then find the transfer function $\dfrac{Y(z)}{Y_1(z)}$.

(d) The transfer function of this entire filter can be expressed in the form $H(z) = \dfrac{b_0 z^2 + b_1 z + b_2}{z^2 + a_1 z + a_2}$. If $\alpha = 0.8$ and $\beta = 0.5$, find the numerical values of the constants.

56. In Figure E.56 match each system pole-zero diagram to its step response and frequency response.

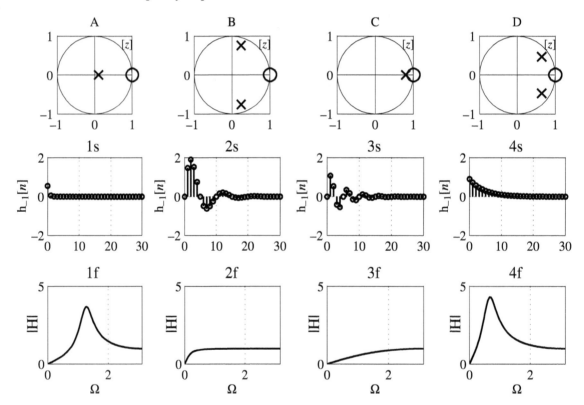

57. In Figure E.57 match the pole-zero diagrams to the corresponding magnitude frequency responses by filling in each blank with the correct letter.

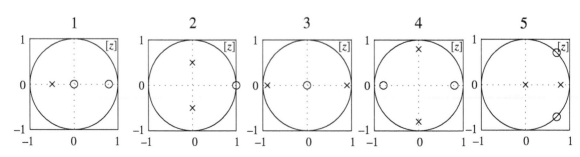

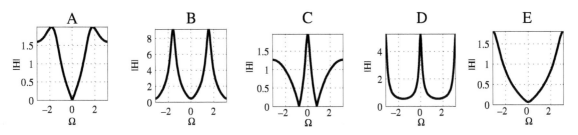

Figure E.56

58. For each system transfer function below, which type of ideal filter does it most closely approximate, lowpass, highpass, bandpass or bandstop?

(a) $H(z) = \dfrac{z-1}{z-0.9}$

(b) $H(z) = \dfrac{z^2-1}{z^2+0.8}$

12 CHAPTER

Laplace System Analysis

12.1 INTRODUCTION AND GOALS

Pierre Laplace invented the Laplace transform as a method of solving linear, constant-coefficient differential equations. Most continuous-time LTI systems are described, at least approximately, by differential equations of that type. The Laplace transform describes the impulse responses of LTI systems as linear combinations of the eigenfunctions of the differential equations that describe them. Because of this Laplace transform directly encapsulates the characteristics of a system in a powerful way. Many system analysis and design techniques are based on the use of the Laplace transform without ever directly referring to the differential equations that describe them. In this chapter we will explore some of the most common applications of the Laplace transform in system analysis.

CHAPTER GOALS

1. To apply the Laplace transform to the generalized analysis of LTI systems, including feedback systems, for stability, time-domain response to standard signals and frequency response
2. To develop techniques for realizing systems in different forms

12.2 SYSTEM REPRESENTATIONS

The discipline of system analysis includes systems of many kinds, electrical, hydraulic, pneumatic, chemical and so on. LTI systems can be described by differential equations or block diagrams. Differential equations can be transformed into algebraic equations by the Laplace transform and these transformed equations form an alternate type of system description.

Electrical systems can be described by circuit diagrams. Circuit analysis can be done in the time domain, but it is often done in the frequency domain because of the power of linear algebra in expressing system interrelationships in terms of algebraic (instead of differential) equations. Circuits are interconnections of circuit elements such as resistors, capacitors, inductors, transistors, diodes, transformers, voltage sources, current sources and so forth. To the extent that these elements can be characterized by linear frequency-domain relationships, the circuit can be analyzed

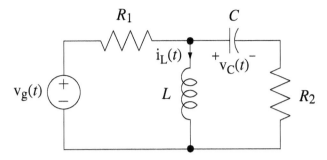

Figure 12.1
Time-domain circuit diagram of an *RLC* circuit

by frequency-domain techniques. Nonlinear elements such as transistors, diodes and transformers can often be modeled approximately over small signal ranges as linear devices. These models consist of linear resistors, capacitors and inductors plus dependent voltage and current sources, all of which can be characterized by LTI system transfer functions.

As an example of circuit analysis using Laplace methods, consider the circuit of Figure 12.1, which illustrates a circuit description in the time domain. This circuit can be described by two coupled differential equations

$$-v_g(t) + R_1\left[i_L(t) + C\frac{d}{dt}(v_C(t))\right] + L\frac{d}{dt}(i_L(t)) = 0$$

$$-L\frac{d}{dt}(i_L(t)) + v_C(t) + R_2C\frac{d}{dt}(v_C(t)) = 0.$$

If we Laplace transform both equations we get

$$-V_g(s) + R_1\{I_L(s) + C[sV_C(s) - v_c(0^+)]\} + sLI_L(s) - i_L(0^+) = 0$$

$$-[sLI_L(s) - i_L(0^+)] + V_C(s) + R_2C[sV_C(s) - v_c(0^+)] = 0.$$

If there is initially no energy stored in the circuit (it is in its zero state), these equations simplify to

$$-V_g(s) + R_1I_L(s) + sR_1CV_C(s) + sLI_L(s) = 0$$

$$-sLI_L(s) + V_C(s) + sR_2CV_C(s) = 0.$$

It is common to rewrite the equations in the form

$$\begin{bmatrix} R_1 + sL & sR_1C \\ -sL & 1 + sR_2C \end{bmatrix} \begin{bmatrix} I_L(s) \\ V_C(s) \end{bmatrix} = \begin{bmatrix} V_g(s) \\ 0 \end{bmatrix}$$

or

$$\begin{bmatrix} Z_{R_1}(s) + Z_L(s) & Z_{R_1}(s)/Z_C(s) \\ -Z_L(s) & 1 + Z_{R_2}(s)/Z_C(s) \end{bmatrix} \begin{bmatrix} I_L(s) \\ V_C(s) \end{bmatrix} = \begin{bmatrix} V_g(s) \\ 0 \end{bmatrix}$$

where

$$Z_{R_1}(s) = R_1, \quad Z_{R_2}(s) = R_2, \quad Z_L(s) = sL, \quad Z_C(s) = 1/sC.$$

The equations are written this way to emphasize the **impedance** concept of circuit analysis. The terms sL and $1/sC$ are the impedances of the inductor and capacitor,

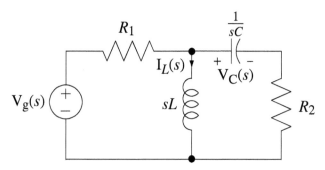

Figure 12.2
Frequency-domain circuit diagram of an RLC circuit

respectively. Impedance is a generalization of the concept of resistance. Using this concept, equations can be written directly from circuit diagrams using relations similar to Ohm's law for resistors,

$$V_R(s) = Z_R I(s) = RI(s), \quad V_L(s) = Z_L I(s) = sLI(s), \quad V_C(s) = Z_C I(s) = (1/sC)I(s).$$

Now the circuit of Figure 12.1 can be conceived as the circuit of Figure 12.2.

The circuit equations can now be written directly from Figure 12.2 as two equations in the complex frequency domain without ever writing the time-domain equations (again, if there is initially no stored energy in the circuit).

$$-V_g(s) + R_1[I_L(s) + sCV_C(s)] + sLI_L(s) = 0$$
$$-sLI_L(s) + V_C(s) + sR_2 CV_C(s) = 0$$

These circuit equations can be interpreted in a system sense as differentiation, and/or multiplication by a constant and summation of signals, in this case, $I_L(s)$ and $V_C(s)$.

$$\underbrace{R_1 I_L(s)}_{\substack{\text{multiplication} \\ \text{by a constant}}} + \underbrace{sR_1 CV_C(s)}_{\substack{\text{differentiation and} \\ \text{multiplication} \\ \text{by a constant}}} + \underbrace{sLI_L(s)}_{\substack{\text{differentiation and} \\ \text{multiplication} \\ \text{by a constant}}} = V_g(s)$$

$$\text{summation}$$

$$\underbrace{-sLI_L(s)}_{\substack{\text{differentiation and} \\ \text{multiplication} \\ \text{by a constant}}} + V_C(s) + \underbrace{sR_2 CV_C(s)}_{\substack{\text{differentiation and} \\ \text{multiplication} \\ \text{by a constant}}} = 0$$

$$\text{summation}$$

A block diagram could be drawn for this system using integrators, amplifiers and summing junctions.

Other kinds of systems can also be modeled by interconnections of integrators, amplifiers and summing junctions. These elements may represent various physical systems that have the same mathematical relationships between an excitation and a response. As a very simple example, suppose a mass m is acted upon by a force (an excitation) f(t). It responds by moving. The response could be the position p(t) of the mass in some appropriate coordinate system. According to classical Newtonian mechanics, the acceleration of a body in any coordinate direction is the force applied to the body in that direction divided by the mass of the body,

$$\frac{d^2}{dt^2}(\mathrm{p}(t)) = \frac{\mathrm{f}(t)}{m}.$$

This can be directly stated in the Laplace domain (assuming the initial position and velocity are zero) as

$$s^2 P(s) = \frac{F(s)}{m}.$$

So this very simple system could be modeled by a multiplication by a constant and two integrations (Figure 12.3).

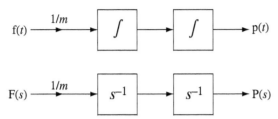

Figure 12.3
Block diagrams of $d^2 p(t)/dt = f(t)/m$ and $s^2 P(s) = F(s)/m$

We can also represent, with block diagrams, more complicated systems like Figure 12.4. In Figure 12.4, the positions $x_1(t)$ and $x_2(t)$ are the distances from the rest positions of masses m_1 and m_2, respectively. Summing forces on mass m_1,

$$f(t) - K_d x_1'(t) - K_{s1}[x_1(t) - x_2(t)] = m_1 x_1''(t).$$

Summing forces on mass m_2,

$$K_{s1}[x_1(t) - x_2(t)] - K_{s2} x_2(t) = m_2 x_2''(t).$$

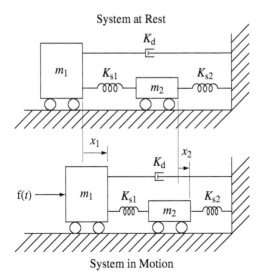

Figure 12.4
A mechanical system

Laplace transforming both equations,

$$F(s) - K_d s X_1(s) - K_{s1}[X_1(s) - X_2(s)] = m_1 s^2 X_1(s)$$
$$K_{s1}[X_1(s) - X_2(s)] - K_{s2} X_2(s) = m_2 s^2 X_2(s)$$

We can also model the mechanical system with a block diagram (Figure 12.5).

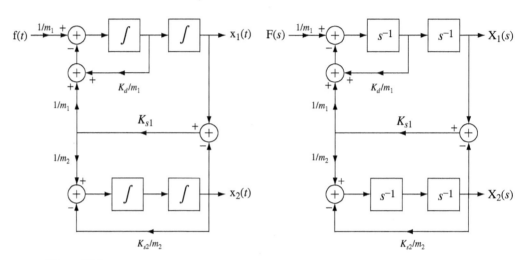

Figure 12.5
Time-domain and frequency-domain block diagrams of the mechanical system of Figure 12.4

12.3 SYSTEM STABILITY

A very important consideration in system analysis is system stability. As shown in Chapter 5, a continuous-time system is bounded-input-bounded-output (BIBO) stable if its impulse response is absolutely integrable. The Laplace transform of the impulse response is the transfer function. For systems that can be described by differential equations of the form

$$\sum_{k=0}^{N} a_k \frac{d^k}{dt^k}(y(t)) = \sum_{k=0}^{M} b_k \frac{d^k}{dt^k}(x(t))$$

where $a_N = 1$, without loss of generality, the transfer function is of the form

$$H(s) = \frac{Y(s)}{X(s)} = \frac{\sum_{k=0}^{M} b_k s^k}{\sum_{k=0}^{N} a_k s^k} = \frac{b_M s^M + b_{M-1} s^{M-1} + \cdots + b_1 s + b_0}{s^N + a_{N-1} s^{N-1} + \cdots + a_1 s + a_0}.$$

The denominator can always be factored (numerically, if necessary), so the transfer function can also be written in the form

$$H(s) = \frac{Y(s)}{X(s)} = \frac{b_M s^M + b_{M-1} s^{M-1} + \cdots + b_1 s + b_0}{(s - p_1)(s - p_2) \cdots (s - p_N)}.$$

If there are any pole-zero pairs that lie at exactly the same location in the s-plane, they cancel in the transfer function and should be removed before examining the transfer function for stability. If $M < N$, and none of the poles is repeated, then the transfer function can be expressed in partial-fraction form as

$$H(s) = \frac{K_1}{s - p_1} + \frac{K_2}{s - p_2} + \cdots + \frac{K_N}{s - p_N}.$$

and the impulse response is then of the form,

$$h(t) = (K_1 e^{p_1 t} + K_2 e^{p_2 t} + \cdots + K_N e^{p_N t})u(t)$$

where the p's are the poles of the transfer function. For $h(t)$ to be absolutely integrable, each of the terms must be individually absolutely integrable. The integral of the magnitude of a typical term is

$$I = \int_{-\infty}^{\infty} |K e^{pt} u(t)| dt = |K| \int_0^{\infty} |e^{\text{Re}(p)t} e^{j\text{Im}(p)t}| dt$$

$$I = |K| \int_0^{\infty} |e^{\text{Re}(p)t}| \underbrace{|e^{j\text{Im}(p)t}|}_{=1} dt = |K| \int_0^{\infty} |e^{\text{Re}(p)t}| dt.$$

In the last integral $e^{\text{Re}(p)t}$ is non-negative over the range of integration. Therefore

$$I = |K| \int_0^{\infty} e^{\text{Re}(p)t} dt.$$

For this integral to converge, the real part of the pole p must be negative.

> For BIBO stability, of a linear, time-invariant (LTI) system all the poles of its transfer function must lie in the *open* left half-plane (LHP).

The term *open left half-plane* means the left half-plane *not including* the ω axis. If there are simple (nonrepeated) poles on the ω axis and no poles are in the right half-plane (RHP), the system is called **marginally stable** because, even though the impulse response does not decay with time, it does not grow, either. Marginal stability is a special case of BIBO instability because in these cases it is possible to find a bounded input signal that will produce an unbounded output signal. (Even though it sounds strange, a *marginally stable* system is also a BIBO *unstable* system.)

If there is a repeated pole of order n in the transfer function, the impulse response will have terms of the general form $t^{n-1} e^{pt} u(t)$ where p is the location of the repeated pole. If the real part of p is not negative, terms of this form grow without bound in positive time, indicating there is an unbounded response to a bounded excitation and that the system is BIBO unstable. Therefore, if a system's transfer function has repeated poles, the rule is unchanged. The poles must all be in the open LHP for system stability. However, there is one small difference from the case of simple poles. If there are repeated poles on the ω axis and no poles in the right half-plane (RHP), the system is not marginally stable, it is simply unstable. These conditions are summarized in Table 12.1.

Table 12.1 Conditions for system stability, marginal stability or instability (which includes marginal stability as a special case)

Stability	Marginal Stability	Instability
All poles in the open LHP	One or more simple poles on the ω axis but no repeated poles on the ω axis and no poles in the open RHP	One or more poles in the open RHP or on the ω axis (includes marginal stability)

An analogy that is sometimes helpful in remembering the different descriptions of system stability or instability is to consider the movement of a sphere placed on different kinds of surfaces (Figure 12.6).

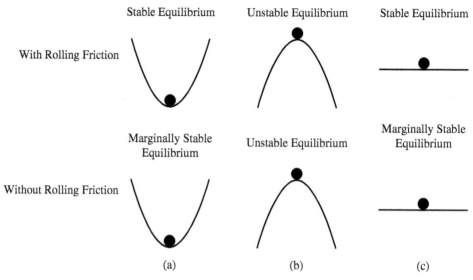

Figure 12.6
Illustrations of three types of stability

If we excite the system in Figure 12.6(a) by applying an impulse of horizontal force to the sphere, it responds by moving and then rolling back and forth. If there is even the slightest bit of rolling friction (or any other loss mechanism like air resistance), the sphere eventually returns to its initial equilibrium position. This is an example of a stable system. If there is no friction (or any other loss mechanism), the sphere will oscillate back and forth forever but will remain confined near the relative low-point of the surface. Its response does not grow with time, but it does not decay, either. In this case the system is marginally stable.

If we excite the sphere in Figure 12.6(b) even the slightest bit, the sphere rolls down the hill and never returns. If the hill is infinitely high, the sphere's speed will approach infinity, an unbounded response to a bounded excitation. This is an unstable system.

In Figure 12.6(c) if we excite the sphere with an impulse of horizontal force, it responds by rolling. If there is any loss mechanism, the sphere eventually comes to rest but not at its original point. This is a bounded response to a bounded excitation and the system is stable. If there is no loss mechanism, the sphere will roll forever without accelerating. This is marginal stability again.

EXAMPLE 12.1

Repeated pole on the ω axis

The simplest form of a system with a repeated pole on the ω axis is the double integrator with transfer function $H(s) = A/s^2$ where A is a constant. Find its impulse response.

Using $t^n u(t) \xleftrightarrow{\mathcal{L}} n!/s^{n+1}$ we find the transform pair $At\,u(t) \xleftrightarrow{\mathcal{L}} A/s^2$, a ramp function that grows without bound in positive time indicating that the system is unstable (and not marginally stable).

12.4 SYSTEM CONNECTIONS

CASCADE AND PARALLEL CONNECTIONS

Earlier we found the impulse response and frequency responses of cascade and parallel connections of systems. The results for these types of systems are the same for transfer functions as they were for frequency responses (Figure 12.7 and Figure 12.8).

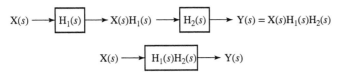

Figure 12.7
Cascade connection of systems

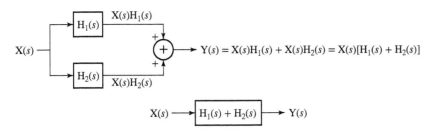

Figure 12.8
Parallel connection of systems

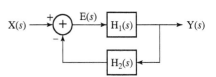

Figure 12.9
Feedback connection of systems

THE FEEDBACK CONNECTION

Terminology and Basic Relationships
Another type of connection that is very important in system analysis is the feedback connection (Figure 12.9). The transfer function $H_1(s)$ is in the **forward path** and the transfer function $H_2(s)$ is in the **feedback path**. In the control-system literature it is common to call the forward-path transfer function $H_1(s)$ the **plant** because it is usually an established system designed to produce something and the feedback-path transfer function $H_2(s)$ the **sensor** because it is usually a system added to the plant to help control it or stabilize it by sensing the plant response and feeding it back to the summing junction at the plant input. The excitation of the plant is called the **error signal** and is given by $E(s) = X(s) - H_2(s)Y(s)$ and the response of $H_1(s)$, which is $Y(s) = H_1(s)E(s)$, is the excitation for the sensor $H_2(s)$. Combining equations and solving for the overall transfer function

$$H(s) = \frac{Y(s)}{X(s)} = \frac{H_1(s)}{1 + H_1(s)H_2(s)}. \quad (12.1)$$

In the block diagram illustrating feedback in Figure 12.9 the feedback signal is subtracted from the input signal. This is a very common convention in feedback system analysis and stems from the history of feedback used as **negative feedback** to stabilize a system. The basic idea behind the term "negative" is that if the plant output signal goes too far in some direction, the sensor will feed back a signal proportional to the plant output signal, which is subtracted from the input signal and therefore tends to move the plant output signal in the opposite direction, moderating it. This, of course, assumes that the signal fed back by the sensor really does have the quality of stabilizing the system. Whether the sensor signal *actually does* stabilize the system depends on its dynamic response and the dynamic response of the plant.

It is conventional in system analysis to give the product of the forward- and feedback-path transfer functions the special name **loop transfer function** $T(s) = H_1(s) H_2(s)$ because it shows up so much in feedback system analysis. In electronic feedback amplifier design this is sometimes called the **loop transmission**. It is given the name loop transfer function or loop transmission because it represents what happens to a signal as it goes from any point in the loop, around the loop exactly one time, and back to the starting point (except for the effect of the minus sign on the summing junction). So the transfer function of the feedback system is the forward-path transfer function $H_1(s)$ divided by one plus the loop transfer function or

$$H(s) = \frac{H_1(s)}{1 + T(s)}.$$

Notice that when $H_2(s)$ goes to zero (meaning there is no feedback) that $T(s)$ does also and the system transfer function $H(s)$ becomes the same as the forward-path transfer function $H_1(s)$.

Feedback Effects on Stability

It is important to realize that feedback can have a very dramatic effect on system response, changing it from slow to fast, fast to slow, stable to unstable or unstable to stable. The simplest type of feedback is to feed back a signal directly proportional to the output signal. That means that $H_2(s) = K$, a constant. In that case the overall system transfer function becomes

$$H(s) = \frac{H_1(s)}{1 + KH_1(s)}.$$

Suppose the forward-path system is an integrator with transfer function $H_1(s) = 1/s$, which is marginally stable. Then $H(s) = \frac{1/s}{1 + K/s} = \frac{1}{s + K}$. The forward-path transfer function $H_1(s)$ has a pole at $s = 0$, but $H(s)$ has a pole at $s = -K$. If K is positive, the overall feedback system is stable, having one pole in the open LHP. If K is negative the overall feedback system is unstable with a pole in the RHP. As K is made a larger positive value the pole moves farther from the origin of the s-plane and the system responds more quickly to an input signal. This is a simple demonstration of an effect of feedback. There is much more to learn about feedback and usually a full semester of feedback control theory is needed for a real appreciation of the effects of feedback on system dynamics.

Feeding the forward-path output signal back to alter its own input signal is often called "closing the loop" for obvious reasons and if there is no feedback path the system is said to be operating "open-loop." Politicians, business executives and other would-be movers and shakers in our society want to be "in the loop." This terminology probably came from feedback loop concepts because one who is in the loop has the chance of affecting the system performance and therefore has power in the political, economic or social system.

Beneficial Effects of Feedback

Feedback is used for many different purposes. One interesting effect of feedback can be seen in a system like Figure 12.10. The overall transfer function is

$$H(s) = \frac{K}{1 + KH_2(s)}.$$

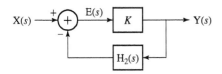

Figure 12.10
A feedback system

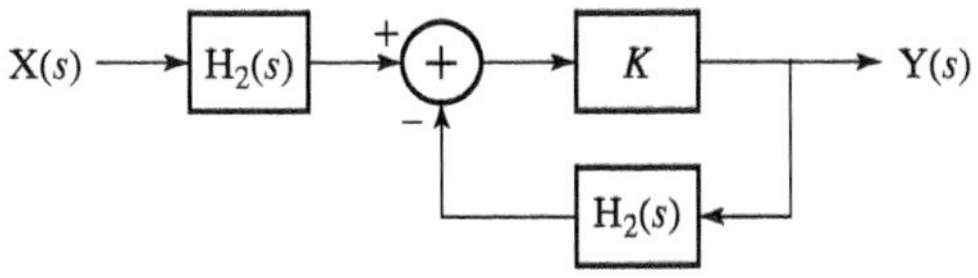

Figure 12.11
A system cascaded with another system designed to be its approximate inverse

If K is large enough, then, at least for some values of s, $KH_2(s) \gg 1$ and $H(s) \approx 1/H_2(s)$ and the overall transfer function of the feedback system performs the approximate inverse of the operation of the feedback path. That means that if we were to cascade connect a system with transfer function $H_2(s)$ to this feedback system, the overall system transfer function would be approximately one (Figure 12.11) over some range of values of s.

It is natural to wonder at this point what has been accomplished because the system of Figure 12.11 seems to have no net effect. There are real situations in which a signal has been changed by some kind of unavoidable system effect and we desire to restore the original signal. This is very common in communication systems in which a signal has been sent over a channel that ideally would not change the signal but actually does for reasons beyond the control of the designer. An **equalization filter** can be used to restore the original signal. It is designed to have the inverse of the effect of the channel on the signal as nearly as possible. Some systems designed to measure physical phenomena use sensors that have inherently lowpass transfer functions, usually because of some unavoidable mechanical or thermal inertia. The measurement system can be made to respond more quickly by cascading the sensor with an electronic signal-processing system whose transfer function is the approximate inverse of the sensor's transfer function.

Another beneficial effect of feedback is to reduce the sensitivity of a system to parameter changes. A very common example of this benefit is the use of feedback in an operational amplifier configured as in Figure 12.12.

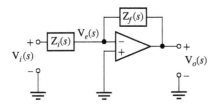

Figure 12.12
An inverting voltage amplifier using an operational amplifier with feedback

A typical approximate expression for the gain of an operational amplifier with the noninverting input grounded ($H_1(s)$ in the feedback block diagram) is

$$H_1(s) = \frac{V_o(s)}{V_e(s)} = -\frac{A_0}{1 - s/p}$$

where A_0 is the magnitude of the operational amplifier voltage gain at low frequencies and p is a single pole on the negative real axis of the s-plane. The overall transfer function can be found using standard circuit analysis techniques. But it can also be found by using feedback concepts. The error voltage $V_e(s)$ is a function of $V_i(s)$ and $V_o(s)$. Since the input impedance of the operational amplifier is typically very large compared with the two external impedances $Z_i(s)$ and $Z_f(s)$, the error voltage is

$$V_e(s) = V_o(s) + [V_i(s) - V_o(s)]\frac{Z_f(s)}{Z_i(s) + Z_f(s)}$$

or

$$V_e(s) = V_o(s)\frac{Z_i(s)}{Z_i(s) + Z_f(s)} - V_i(s)\left[-\frac{Z_f(s)}{Z_i(s) + Z_f(s)}\right].$$

So we can model the system using the block diagram in Figure 12.13.

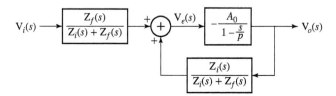

Figure 12.13
Block diagram of an inverting voltage amplifier using feedback on an operational amplifier

According to the general feedback-system transfer function

$$H(s) = \frac{Y(s)}{X(s)} = \frac{H_1(s)}{1 + H_1(s)H_2(s)}$$

the amplifier transfer function should be

$$\frac{V_o(s)}{V_i(s)\dfrac{Z_f(s)}{Z_i(s) + Z_f(s)}} = \frac{-\dfrac{A_0}{1 - s/p}}{1 + \left(-\dfrac{A_0}{1 - s/p}\right)\left(-\dfrac{Z_i(s)}{Z_i(s) + Z_f(s)}\right)}.$$

Simplifying, and forming the ratio of $V_o(s)$ to $V_i(s)$ as the desired overall transfer function,

$$\frac{V_o(s)}{V_i(s)} = \frac{-A_0 Z_f(s)}{(1 - s/p + A_0)Z_i(s) + (1 - s/p)Z_f(s)}.$$

If the low-frequency gain magnitude A_0 is very large (which it usually is), then we can approximate this transfer function at low frequencies as

$$\frac{V_o(s)}{V_i(s)} \cong -\frac{Z_f(s)}{Z_i(s)}.$$

This is the well-known ideal-operational-amplifier formula for the gain of an inverting voltage amplifier. In this case "being large" means that A_0 is large enough that the denominator of the transfer function is approximately $A_0 Z_i(s)$, which means that

$$|A_0| \gg \left|1 - \frac{s}{p}\right| \quad \text{and} \quad |A_0| \gg \left|1 - \frac{s}{p}\right| \left|\frac{Z_f(s)}{Z_i(s)}\right|.$$

Its exact value is not important as long as it is very large and that fact represents the reduction in the system's sensitivity to changes in parameter values that affect A_0 or p.

To illustrate the effects of feedback on amplifier performance let

$$A_0 = 10^7 \quad \text{and} \quad p = -100.$$

Also, let $Z_f(s)$ be a resistor of 10 kΩ and let $Z_i(s)$ be a resistor of 1 kΩ. Then the overall system transfer function is

$$\frac{V_o(s)}{V_i(s)} = \frac{-10^8}{11(1 + s/100) + 10^7}.$$

The numerical value of the transfer function at a real radian frequency of $\omega = 100$ (a cyclic frequency of $f = 100/2\pi \cong 15.9$ Hz) is

$$\frac{V_o(s)}{V_i(s)} = \frac{-10^8}{11 + j11 + 10^7} = -9.999989 + j0.000011.$$

Now let the operational amplifier's low-frequency gain be reduced by a factor of 10 to $A_0 = 10^6$. When we recalculate the transfer function at 15.9 Hz we get

$$\frac{V_o(s)}{V_i(s)} = \frac{-10^7}{11 + j11 + 10^6} = -9.99989 + j0.00011,$$

which represents a change of approximately 0.001% in the magnitude of the transfer function. So a change in the forward-path transfer function by a factor of 10 produced a change in the overall system transfer function magnitude of about 0.001%. The feedback connection made the overall transfer function very insensitive to changes in the operational amplifier gain, even very large changes. In amplifier design this is a very beneficial result because resistors, and especially resistor ratios, can be made very insensitive to environmental factors and can hold the system transfer function almost constant, even if components in the operational amplifier change by large percentages from their nominal values.

Another consequence of the relative insensitivity of the system transfer function to the gain A_0 of the operational amplifier is that if A_0 is a function of signal level, making the operational amplifier gain nonlinear, as long as A_0 is large the system transfer function is still very accurate (Figure 12.14) and practically linear.

Another beneficial effect of feedback can be seen by calculating the bandwidth of the operational amplifier itself and comparing that to the bandwidth of the inverting amplifier with feedback. The corner frequency of the operational amplifier itself in this example is 15.9 Hz. The corner frequency of the inverting amplifier with feedback is the frequency at which the real and imaginary parts of the denominator of the overall transfer function are equal in magnitude. That occurs at a cyclic frequency of $f \cong 14.5$ MHz. This is an increase in bandwidth by a factor of approximately 910,000. It is hard to overstate the importance of feedback principles in improving the performance of many systems in many ways.

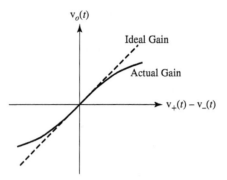

Figure 12.14
Linear and nonlinear operational amplifier gain

The transfer function of the operational amplifier is a very large number at low frequencies. So the operational amplifier has a large **voltage gain** at low frequencies. The voltage gain of the feedback amplifier is typically much smaller. So, in using feedback, we have lost voltage gain but gained gain stability and bandwidth (among other things). In effect, we have traded gain for improvements in other amplifier characteristics.

Feedback can be used to stabilize an otherwise unstable system. The F-117 stealth fighter is inherently aerodynamically unstable. It can fly under a pilot's control only with the help of a computer-controlled feedback system that senses the aircraft's position, speed and attitude and constantly compensates when it starts to go unstable. A very simple example of stabilization of an unstable system using feedback would be a system whose transfer function is

$$H_1(s) = \frac{1}{s-p}, \quad p > 0.$$

With a pole in the RHP this system is unstable. If we use a feedback-path transfer function that is a constant gain K we get the overall system transfer function,

$$H(s) = \frac{\frac{1}{s-p}}{1 + \frac{K}{s-p}} = \frac{1}{s-p+K}.$$

For any value of K satisfying $K > p$, the feedback system is stable.

Instability Caused by Feedback
Although feedback can have many very beneficial effects, there is another effect of feedback in systems that is also very important and can be a problem rather than a benefit. The addition of feedback to a stable system can cause it to become unstable. The overall feedback-system transfer function is

$$H(s) = \frac{Y(s)}{X(s)} = \frac{H_1(s)}{1 + H_1(s)H_2(s)}.$$

Even though all the poles of $H_1(s)$ and $H_2(s)$ may lie in the open LHP, the poles of $H(s)$ may not. Consider the forward and feedback transfer functions

$$H_1(s) = \frac{K}{(s+3)(s+5)} \quad \text{and} \quad H_2(s) = \frac{1}{s+4}.$$

$H_1(s)$ and $H_2(s)$ are both BIBO stable. But if we put them into a feedback system like Figure 12.9, the overall system gain is then

$$H(s) = \frac{K(s+4)}{(s+3)(s+4)(s+5)+K} = \frac{K(s+4)}{s^3 + 12s^2 + 47s + 60 + K}.$$

Whether or not this feedback system is stable now depends on the value of K. If K is 5, the poles lie at -5.904 and $-3.048 \pm j1.311$. They are all in the open LHP and the feedback system is stable. But if K is 700, the poles lie at -12.917 and $0.4583 \pm j7.657$. Two poles are in the RHP and the system is unstable.

Almost everyone has experienced a system made unstable by feedback. Often when large crowds gather to hear someone speak, a public-address system is used. The speaker speaks into a microphone. His voice is amplified and fed to one or more speakers so everyone in the audience can hear his voice. Of course, the sound emanating from the speakers is also detected and amplified by the microphone and amplifier. This is an example of feedback because the output signal of the public address system is fed back to the input of the system. Anyone who has ever heard it will never forget the sound of the public address system when it goes unstable, usually a very loud tone. And we probably know the usual solution, turn down the amplifier gain. This tone can occur even when no one is speaking into the microphone. Why does the system go unstable with no apparent input signal, and why does turning down the amplifier gain not just reduce the volume of the tone, but eliminate it entirely?

Albert Einstein was famous for the Gedankenversuch (thought experiment). We can understand the feedback phenomenon through a thought experiment. Imagine that we have a microphone, amplifier and speaker in the middle of a desert with no one around and no wind or other acoustic disturbance and that the amplifier gain is initially turned down to zero. If we tap on the microphone we hear only the direct sound of tapping and nothing from the speakers. Then we turn the amplifier gain up a little. Now when we tap on the microphone we hear the tap directly but also some sound from the speakers, slightly delayed because of the distance the sound has to travel from the speakers to our ears (assuming the speakers are farther away from our ears than the microphone). As we turn the gain up more and more, increasing the loop transfer function T, the tapping sound from the speakers rises in volume (Figure 12.15). (In Figure 12.15, p(t) is acoustic pressure as a function of time.)

As we increase the magnitude of the loop transfer function T by turning up the amplifier gain, when we tap on the microphone we gradually notice a change, not just in the volume, but also in the nature of the sound from the speakers. We hear not only the tap but also what is commonly called **reverberation**, multiple echoes of the tap. These multiple echoes are caused by the sound of the tap coming from the speaker to the microphone, being amplified and going to the speaker again and returning to the microphone again multiple times. As the gain is increased this phenomenon becomes more obvious and, at some gain level, a loud tone begins and continues, without any tapping or any other acoustic input to the microphone, until we turn the gain back down.

At some level of amplifier gain, any signal from the microphone, no matter how weak, is amplified, fed to the speaker, returns to the microphone and causes a new signal in the microphone, which is the same strength as the original signal. At this gain the signal never dies, it just keeps on circulating. If the gain is made slightly higher, the signal grows every time it makes the round trip from microphone to speaker and back. If the public address system were truly linear, that signal would increase without bound. But no real public address system is truly linear and, at some volume level, the amplifier is driving the speaker as hard as it can but the sound level does not increase any more.

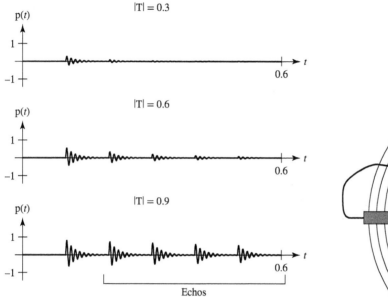

Figure 12.15
Public address system sound from tapping on the microphone for three different system loop transfer functions

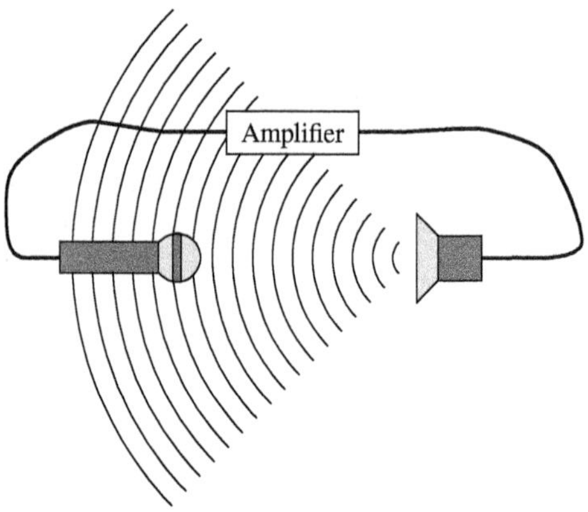

Figure 12.16
A public address system

It is natural to wonder how this process begins without any acoustic input to the microphone. First, as a practical matter, it is impossible to arrange to have absolutely no ambient sound strike the microphone. Second, even if that were possible, the amplifier has inherent random noise processes that cause an acoustic signal from the speaker and that is enough to start the feedback process.

Now carry the experiment a little further. With the amplifier gain high enough to cause the tone we move the speaker farther from the microphone. As we move the speaker away, the pitch of the loud tone changes and, at some distance, the tone stops. The pitch changes because the frequency of the tone depends on the time sound takes to propagate from the speaker to the microphone. The loud tone stops at some distance because the sound intensity from the speaker is reduced as it is moved farther away, and the return signal due to feedback is less than the original signal, and the signal dies away instead of increasing in power.

Now we will mathematically model the public address system with the tools we have learned and see exactly how feedback instability occurs (Figure 12.16). To keep the model simple, yet illustrative, we will let the transfer functions of the microphone, amplifier and speaker be the constants, K_m, K_A and K_s. Then we model the propagation of sound from the speaker to the microphone as a simple delay with a gain that is inversely proportional to the square of the distance d from the speaker to the microphone

$$p_m(t) = K \frac{p_s(t - d/v)}{d^2} \tag{12.2}$$

where $P_s(t)$ is the sound (acoustic pressure) from the speaker, $P_m(t)$ is the sound arriving at the microphone, v is the speed of sound in air and K is a constant. Laplace transforming both sides of (12.2),

$$P_m(s) = \frac{K}{d^2} P_s(s) e^{-ds/v}.$$

Then we can model the public address system as a feedback system with a forward-path transfer function

$$H_1(s) = K_m K_A K_s$$

and a feedback-path transfer function

$$H_2(s) = \frac{K}{d^2} e^{-ds/v}$$

(Figure 12.17). The overall transfer function is

$$H(s) = \frac{K_m K_A K_s}{1 - \frac{K_m K_A K_s K}{d^2} e^{-ds/v}}.$$

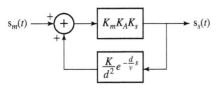

Figure 12.17
Block diagram of a public address system

The poles p of this system transfer function lie at the zeros of $1 - (K_m K_A K_s K/d^2) e^{-dp/v}$. Solving,

$$1 - \frac{K_m K_A K_s K}{d^2} e^{-dp/v} = 0 \quad (12.3)$$

or

$$e^{-dp/v} = \frac{d^2}{K_m K_A K_s K}.$$

Any value of p that solves this equation is a pole location. If we take the logarithm of both sides and solve for p we get

$$p = -\frac{v}{d} \ln\left(\frac{d^2}{K_m K_A K_s K}\right).$$

So this is *a* solution of (12.3). But it is not the only solution. It is just the only *real-valued* solution. If we add any integer multiple of $j2\pi v/d$ to p we get another solution because

$$e^{-d(p+j2n\pi v/d)/v} = e^{-dp/v} \underbrace{e^{-j2n\pi}}_{=1} = e^{-dp/v}$$

where n is any integer. That means that there are *infinitely many* poles, all with the same real part $-\frac{v}{d} \ln\left(\frac{d^2}{K_m K_A K_s K}\right)$ (Figure 12.18).

This system is a little different from the systems we have been analyzing because this system has infinitely many poles, one for each integer n. But that is not a problem in this analysis because we are only trying to establish the conditions under which the

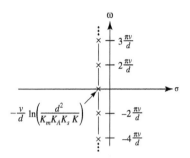

Figure 12.18
Pole-zero diagram of the public address system

system is stable. As we have already seen, stability requires that all poles lie in the open LHP. That means, in this case, that

$$-\frac{v}{d}\ln\left(\frac{d^2}{K_m K_A K_s K}\right) < 0$$

or

$$\ln\left(\frac{d^2}{K_m K_A K_s K}\right) > 0$$

or

$$\frac{K_m K_A K_s K}{d^2} < 1. \qquad (12.4)$$

In words, the product of all the transfer-function magnitudes around the feedback loop must be less than one. This makes common sense because if the product of all the transfer-function magnitudes around the loop exceeds one, that means that when a signal makes a complete round-trip through the feedback loop it is bigger when it comes back than when it left and that causes it to grow without bound. So when we turn down the amplifier gain K_A to stop the loud tone caused by feedback, we are satisfying (12.4).

Suppose we increase the loop transfer function magnitude $K_m K_A K_s K/d^2$ by turning up the amplifier gain K_A. The poles move to the right, parallel to the σ axis, and, at some gain value, they reach the ω axis. Now suppose instead we increase the loop transfer function magnitude by moving the microphone and speaker closer together. This moves the poles to the right but also away from the σ axis so that when we reach marginal stability the poles are all at higher radian frequencies.

A system that obeys this simple model can oscillate at multiple frequencies simultaneously. In reality that is unlikely. A real public address system microphone, amplifier and speaker would have transfer functions that are functions of frequency and would therefore change the pole locations so that only one pair of poles would lie on the ω axis at marginal stability. If the gain is turned up above the gain for marginal stability the system is driven into a nonlinear mode of operation and linear system analysis methods fail to predict exactly how it will oscillate. But linear system methods do predict accurately that it *will oscillate* and that is very important.

Stable Oscillation Using Feedback

The oscillation of the public address system in the last section was an undesirable system response. But some systems are designed to oscillate. Examples are laboratory function generators, computer clocks, local oscillators in radio receivers, quartz crystals in wristwatches, a pendulum on a grandfather clock and so on. Some systems are designed to oscillate in a nonlinear mode in which they simply alternate between two or more unstable states and their response signals are not necessarily sinusoidal. Free-running computer clocks are a good example. But some systems are designed to operate as an LTI system in a marginally stable mode with a true sinusoidal oscillation. Since marginal stability requires that the system have poles on the ω axis of the s-plane, this mode of operation is very exacting. The slightest movement of the system poles due to any parameter variation will cause the oscillation either to grow or decay with time. So systems that operate in this mode must have some mechanism for keeping the poles on the ω axis.

The prototype feedback diagram in Figure 12.19 has an excitation and a response. A system designed to oscillate does not have an (apparent) excitation; that is, $X(s) = 0$ (Figure 12.20). (The sign is changed on $H_2(s)$ to make the system in Figure 12.20 be just like the system in Figure 12.19 with $X(s) = 0$.) How can we have a response if we have no excitation? The short answer is, we cannot. However, it is important to realize that every system is constantly being excited whether we intend it or not. Every system has random noise processes that cause signal fluctuations. The system responds to these noise fluctuations just as it would to an intentional excitation.

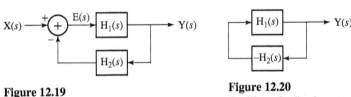

Figure 12.19
Prototype feedback system

Figure 12.20
Oscillator feedback system

The key to having a stable oscillation is having a transfer function with poles on the ω axis of the form,

$$H(s) = \frac{A}{s^2 + \omega_0^2}.$$

Then the system gain at the radian frequency ω_0 ($s = \pm j\omega_0$) is infinite, implying that the response is infinitely greater than the excitation. That could mean either that a finite excitation produces an infinite response or that a zero excitation produces a finite response. Therefore a system with poles on the ω axis can produce a stable nonzero response with no excitation.[1]

One very interesting and important example of a system designed to oscillate in a marginally stable mode is a laser. The acronym LASER stands for "Light Amplification by Stimulated Emission of Radiation." A laser is not actually a light amplifier (although, internally, light amplification does occur), it is a light oscillator. But the acronym for "Light Oscillation by Stimulated Emission of Radiation," LOSER, described itself and did not catch on.

---

[1] It is important here to distinguish between two uses of the word "stable." A BIBO stable system is one that has a bounded response to any arbitrary bounded excitation. A stable oscillation, in the context of this section, is an oscillating output signal that maintains a constant amplitude, neither growing nor decaying. If an LTI system has an impulse response that is a stable oscillation, the system is marginally stable, a special case of BIBO unstable. That is, for such a system there exists a bounded excitation that would produce an unbounded response. If we were to actually excite any real system with such an excitation, its response would grow for a while but then at some signal level would change from an LTI system to a nonlinear system (or would start to reveal that it was never actually an LTI system in the first place) and the response signal would remain bounded.

Even though the laser is an oscillator, light amplification is an inherent process in its operation. A laser is filled with a medium that has been "pumped" by an external power source in such a way that light of the right wavelength propagating through the pumped medium experiences an increase in power as it propagates (Figure 12.21).

The device illustrated in Figure 12.21 is a one-pass, **travelling-wave light amplifier**, not a laser. The oscillation of light in a laser is caused by introducing into the one-pass travelling-wave light amplifier mirrors on each end that reflect some or all of the light striking them. At each mirror some or all of the light is fed back into the pumped laser medium for further amplification (Figure 12.22).

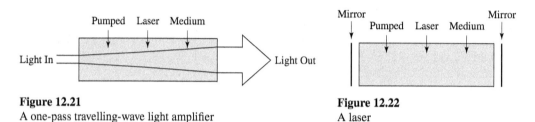

Figure 12.21
A one-pass travelling-wave light amplifier

Figure 12.22
A laser

It would be possible, in principle, to introduce light at one end of this device through a partial mirror and amplify it. Such a device is called a **regenerative travelling-wave light amplifier**. But it is much more common to make the mirror at one end as reflective as possible, essentially reflecting all the light that strikes it, and make the mirror at the other end a partial mirror, reflecting some of the light that strikes it and transmitting the rest.

A laser operates without any external light signal as an excitation. The light that it emits begins in the pumped laser medium itself. A phenomenon called **spontaneous emission** causes light to be generated at random times and in random directions in the pumped medium. Any such light that happens to propagate perpendicular to a mirror gets amplified on its way to the mirror, then reflected and further amplified as it bounces between the mirrors. The closer the propagation is to perpendicular to the mirrors, the longer the beam bounces and the more it is amplified by the multiple passes through the laser medium. In steady-state operation the light that is perpendicular to the mirrors has the highest power of all the light propagation inside the laser cavity because it has the greatest gain advantage. One mirror is always a partial mirror so some light transmits at each bounce off that mirror. This light constitutes the output light beam of the laser (Figure 12.23).

In order for light oscillation to be sustained, the loop transfer function of the system must be the real number, -1, under the assumed negative feedback sign on the prototype feedback system of Figure 12.19 or it must be the real number, $+1$, under the assumption of the oscillator system of Figure 12.20. Under either assumption, for stable oscillation, the light, as it travels from a starting point to one mirror, back to the other mirror and then back to the starting point, must experience an overall gain magnitude of one and phase shift of an integer multiple of 2π radians. This simply means that the wavelength of the light must be such that it fits into the laser cavity with exactly an integer number of waves in one round-trip path.

It is important to realize here that the wavelength of light in lasers is typically somewhere in the range from 100 nm to many microns (ultraviolet to far infrared), and typical lengths of laser cavities are in the range of a 100 μm for a laser diode to more than a meter in some cases. Therefore, as light propagates between the mirrors it may experience more than a million radians of phase shift and, even in the shortest cavities,

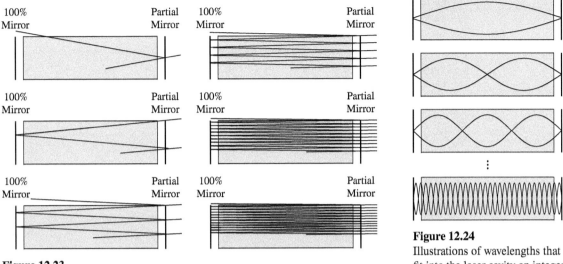

Figure 12.23
Multiple light reflections at different initial angles

Figure 12.24
Illustrations of wavelengths that fit into the laser cavity an integer number of times

the phase shift is usually a large multiple of 2π radians. So in a laser the exact wavelength of oscillation is determined by which optical wavelength fits into the round-trip path with exactly an integer number of waves. There are infinitely many wavelengths that satisfy this criterion, the wave that fits into the round trip exactly once plus all its harmonics (Figure 12.24).

Although all these wavelengths of light could theoretically oscillate, there are other mechanisms (atomic or molecular resonances, wavelength-selective mirrors, etc.) that limit the actual oscillation to a small number of these wavelengths that experience enough gain to oscillate.

A laser can be modeled by a block diagram with a forward path and a feedback path (Figure 12.25). The constants K_F and K_R represent the magnitude of the gain experienced by the electric field of the light as it propagates from one mirror to the other

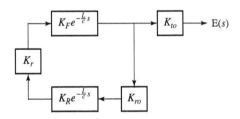

Figure 12.25
Laser block diagram

along the forward and reverse paths, respectively. The factors $e^{-(L/c)s}$ account for the phase shift due to propagation time where L is the distance between the mirrors and c is the speed of light in the laser cavity. The constant K_{to} is the electric field transmission coefficient for light exiting the laser cavity through the output partial mirror and the constant K_{ro} is the electric field reflection coefficient for light reflected at the output partial mirror back into the laser cavity. The constant K_r is the electric field reflection coefficient for light reflected at the 100% mirror back into the laser cavity. K_{to}, K_{ro} and K_r are, in general, complex, indicating that there is a phase shift of the electric field during reflection and transmission. The loop transfer function is (using the definition developed based on the sign convention in Figure 12.19)

$$\mathrm{T}(s) = -K_F K_{ro} K_R K_r e^{-(2L/c)s}.$$

Its value is -1 when

$$|K_F K_{ro} K_R K_r| = 1$$

and

$$e^{-(2L/c)s} = 1$$

or, equivalently,

$$s = -j2\pi n\left(\frac{c}{2L}\right) = -j\frac{\pi c}{L}n, \quad n \text{ any integer},$$

where the quantity $c/2L$ is the round-trip travel time for the propagating light wave. These are values of s on the ω axis at harmonics of a fundamental radian frequency $\pi c/L$. Since this is the fundamental frequency it is also the spacing between frequencies, which is conventionally called the **axial mode spacing** $\Delta\omega_{ax}$.

When a laser is first turned on the medium is pumped and a light beam starts through spontaneous emission. It grows in intensity because, at first, the magnitude of the round-trip gain is greater than one ($|K_F K_{ro} K_R K_r| > 1$). But, as it grows, it extracts energy from the pumped medium, and that reduces the gains K_F and K_R. An equilibrium is reached when the beam strength is exactly the right magnitude to keep the loop transfer function magnitude $|K_F K_{ro} K_R K_r|$ at exactly one. The pumping and light-amplification mechanisms in a laser together form a self-limiting process that stabilizes at a loop transfer function magnitude of one. So, as long as there is enough pumping power and the mirrors are reflective enough to achieve a loop transfer function magnitude of one at some very low output power, the laser will oscillate stably.

The Root-Locus Method

A very common situation in feedback system analysis is a system for which the forward-path gain $H_1(s)$ contains a "gain" constant K that can be adjusted. That is,

$$H_1(s) = K\frac{P_1(s)}{Q_1(s)}.$$

The adjustable gain parameter K (conventionally taken to be non-negative) has a strong effect on the system's dynamics. The overall system transfer function is

$$H(s) = \frac{H_1(s)}{1 + H_1(s)H_2(s)}$$

and the loop transfer function is

$$T(s) = H_1(s)H_2(s).$$

The poles of $H(s)$ are the zeros of $1 + T(s)$. The loop transfer function, can be written in the form of K times a numerator divided by a denominator

$$T(s) = K\frac{P_1(s)}{Q_1(s)}\frac{P_2(s)}{Q_2(s)} = K\frac{P(s)}{Q(s)} \qquad (12.5)$$

so the poles of H(s) occur where

$$1 + K\frac{P(s)}{Q(s)} = 0$$

which can be expressed in the two alternate forms,

$$Q(s) + KP(s) = 0 \tag{12.6}$$

and

$$\frac{Q(s)}{K} + P(s) = 0. \tag{12.7}$$

From (12.5), we see that if T(s) is proper (Q(s) is of higher order than P(s)) the zeros of Q(s) constitute all the poles of T(s) and the zeros of P(s) are all finite zeros of T(s) but, because the order of P(s) is less than the order of Q(s), there are also one or more zeros of T(s) at infinity.

The full range of possible adjustment of K is from zero to infinity. First let K approach zero. In that limit, from (12.6), the zeros of $1 + T(s)$, which are the poles of H(s), are the zeros of Q(s) and the poles of H(s) are therefore the poles of T(s) because $T(s) = KP(s)/Q(s)$. Now consider the opposite case, K approaching infinity. In that limit, from (12.7), the zeros of $1 + T(s)$ are the zeros of P(s) and the poles of H(s) are the zeros of T(s) (including any zeros at infinity). So the loop transfer function poles and zeros are very important in the analysis of the feedback system.

As the gain factor K moves from zero to infinity, the poles of the feedback system move from the poles of the loop transfer function to the zeros of the loop transfer function (some of which may be at infinity). A **root-locus** plot is a plot of the locations of the feedback-system poles as the gain factor K is varied from zero to infinity. The name "root locus" comes from the location (locus) of a root of $1 + T(s)$ as the gain factor K is varied.

We will first examine two simple examples of the root-locus method and then establish some general rules for drawing the root locus of any system. Consider first a system whose forward-path gain is

$$H_1(s) = \frac{K}{(s+1)(s+2)}$$

and whose feedback-path gain is $H_2(s) = 1$. Then

$$T(s) = \frac{K}{(s+1)(s+2)}$$

and the root-locus plot begins at $s = -1$ and $s = -2$, the poles of T(s). All the zeros of T(s) are at infinity and those are the zeros that the root locus approaches as the gain factor K is increased (Figure 12.26).

The roots of $1 + T(s)$ are the roots of

$$(s+1)(s+2) + K = s^2 + 3s + 2 + K = 0$$

and, using the quadratic formula, the roots are at $(-3 \pm \sqrt{1-4K})/2$. For $K = 0$ we get roots at $s = -1$ and $s = -2$, the poles of T(s). For $K = 1/4$ we get a repeated root at $-3/2$. For $K > 1/4$ we get two complex-conjugate roots whose imaginary parts go to plus and

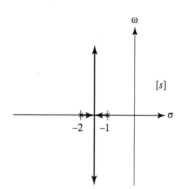

Figure 12.26
Root locus of $1 + T(s) = 1 + \dfrac{K}{(s+1)(s+2)}$

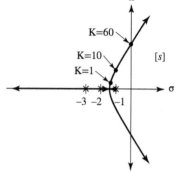

Figure 12.27
Root locus of $1 + T(s) = 1 + \dfrac{K}{(s+1)(s+2)(s+3)}$

minus infinity as K increases but whose real parts stay at $-3/2$. Since this root locus extends to infinity in the imaginary dimension with a real part that always places the roots in the LHP, this feedback system is stable for any value of K.

Now add one pole to the forward-path transfer function making it

$$H_1(s) = \frac{K}{(s+1)(s+2)(s+3)}.$$

The new root locus is the locus of solutions to the equation $s^3 + 6s^2 + 11s + 6 + K = 0$ (Figure 12.27).

At or above the value of K for which two branches of the root locus cross the ω axis, this system is unstable. So this system, which is open-loop stable, can be made unstable by using feedback. The poles are at the roots of $s^3 + 6s^2 + 11s + 6 + K = 0$. It is possible to find a general solution for a cubic equation of this type, but it is very tedious. It is much easier to generate multiple values for K and solve for the roots numerically to find the value of K that causes the poles of $H(s)$ to move into the RHP.

In Figure 12.28 we can see that $K = 60$ puts two poles exactly on the ω axis. So any value of K greater than or equal to 60 will cause this feedback system to be unstable.

K	Roots→		
0	-3	-2	-1
0.25	-3.11	-1.73	-1.16
0.5	-3.19	$-1.4 + j0.25$	$-1.4 - j0.25$
1	-3.32	$-1.34 + j0.56$	$-1.34 - j0.56$
2	-3.52	$-1.24 + j0.86$	$-1.24 - j0.86$
10	-4.31	$-0.85 + j1.73$	$-0.85 - j1.73$
30	-5.21	$-0.39 + j2.60$	$-0.39 - j2.60$
60	-6.00	$0.00 + j3.32$	$0.00 - j3.32$
100	-6.71	$0.36 + j3.96$	$0.36 - j3.96$

Figure 12.28
Roots of $s^3 + 6s^2 + 11s + 6 + K = 0$ for several values of K

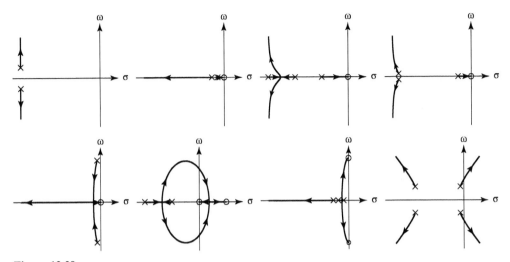

Figure 12.29
Example root-locus plots

Figure 12.29 illustrates some root-locus plots for different numbers and different locations of the poles and zeros of T(*s*). There are several rules for plotting a root locus. These rules come from rules of algebra derived by mathematicians about the locations of the roots of polynomial equations.

1. The number of branches in a root locus is the greater of the degree of the numerator polynomial and the degree of the denominator polynomial of T(*s*).
2. Each root-locus branch begins on a pole of T(*s*) and terminates on a zero of T(*s*).
3. Any region of the real axis for which the sum of the number of real poles and/or real zeros lying to its right on the real axis is odd, is a part of the root locus and all other regions of the real axis are not part of the root locus. The regions that are part of the root locus are called "allowed" regions.
4. The root locus is symmetrical about the real axis.
5. If the number of finite poles of T(*s*) exceeds the number of finite zeros of T(*s*) by an integer m, then m branches of the root locus terminate on zeros of T(*s*) that lie at infinity. Each of these branches approaches a straight-line asymptote and the angles of these asymptotes are $(2k + 1)\pi/m$, $k = 0, 1, \cdots m - 1$, with respect to the positive real axis. These asymptotes intersect on the real axis at the location

$$\sigma = \frac{1}{m}\left(\sum \text{finite poles} - \sum \text{finite zeros}\right)$$

called the **centroid** of the root locus. (These are sums of *all* finite poles and *all* finite zeros, not just the ones on the real axis.)

6. The breakaway or break-in points where the root-locus branches intersect occur where

$$\frac{d}{ds}\left(\frac{1}{\text{T}(s)}\right) = 0.$$

EXAMPLE 12.2

Root locus 1

Draw a root locus for a system whose loop transfer function is

$$T(s) = \frac{(s+4)(s+5)}{(s+1)(s+2)(s+3)}$$

The thinking steps in figuring out where the root-locus branches go are the following:

1. $T(s)$ has poles at $\sigma = -1$, $\sigma = -2$ and $\sigma = -3$ and zeros at $\sigma = -4$, $\sigma = -5$ and $|s| \to \infty$.
2. The number of root-locus branches is 3 (Rule 1).
3. The allowed regions on the real axis are in the ranges $-2 < \sigma < -1$, $-4 < \sigma < -3$ and $\sigma < -5$ (Figure 12.30) (Rule 3).

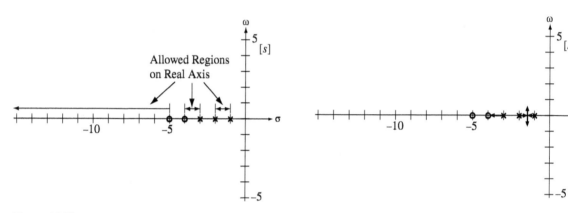

Figure 12.30
Allowed regions on the real axis

Figure 12.31
Initial stage of drawing a root locus

4. The root-locus branches must begin at $\sigma = -1$, $\sigma = -2$ and $\sigma = -3$ (Rule 2).
5. Two root-locus branches must terminate on $\sigma = -4$ and $\sigma = -5$ and the third branch must terminate on the zero at infinity (Rule 2).
6. The two root-locus branches beginning at $\sigma = -1$ and $\sigma = -2$ initially move toward each other because they must stay in an allowed region (Rule 3). When they intersect they must both become complex and must be complex conjugates of each other (Rule 4).
7. The third root-locus branch beginning at $\sigma = -3$ must move to the left toward the zero at $\sigma = -4$ (Rule 3). This branch cannot go anywhere else and, at the same time, preserve the symmetry about the real axis. So this branch simply terminates on the zero at $\sigma = -4$ (Rule 2) (Figure 12.31).
8. Now we know that the two other root-locus branches must terminate on the zero at $\sigma = -5$ and the zero at $|s| \to \infty$. They are already complex. Therefore they have to move to the left and back down to the σ axis and then one must go to the right to terminate on the zero at $\sigma = -5$ while the other one moves to the left on the real axis approaching negative infinity (Rule 2).
9. There are three finite poles and two finite zeros. That means there is only one root-locus branch going to a zero at infinity, as we have already seen. The angle that branch approaches should be π radians, the negative real axis (Rule 5). This agrees with the previous conclusion (number 8).

10. The point at which the two branches break out of the real axis and the point at which the two branches break back into the real axis must both occur where $(d/ds)(1/T(s)) = 0$ (Rule 6).

$$\frac{d}{ds}\left(\frac{1}{T(s)}\right) = \frac{d}{ds}\left[\frac{(s+1)(s+2)(s+3)}{(s+4)(s+5)}\right] = 0.$$

Differentiating and equating to zero, we get $s^4 + 18s^3 + 103s^2 + 228s + 166 = 0$. The roots are at $s = -9.47$, $s = -4.34$, $s = -2.69$ and $s = -1.50$. So the breakout point is at $\sigma = -1.50$ and the break-in point is at $\sigma = -9.47$. The root locus never moves into the RHP, so this system is stable for any non-negative value of the gain factor K (Figure 12.32).

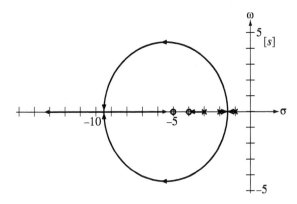

Figure 12.32
Completed root locus

(The other two solutions of $s^4 + 18s^3 + 103s^2 + 228s + 166 = 0$, $s = -4.34$ and $s = -2.69$, are the breakout and break-in points for the so-called **complementary root locus**. The complementary root locus is the locus of the poles of H(s) as K goes from zero to *negative* infinity.)

EXAMPLE 12.3

Root locus 2

Draw a root locus for a system whose forward path (plant) is the system of Figure 12.33 with $a_2 = 1$, $a_1 = -2$, $a_0 = 2$, $b_2 = 0$, $b_1 = 1$ and $b_0 = 0$, and whose feedback path (sensor) is the system of Figure 12.33 with $a_2 = 1$, $a_1 = 2$, $a_0 = 0$, $b_2 = 1$, $b_1 = 1$, $b_0 = 0$ and $K = 1$.

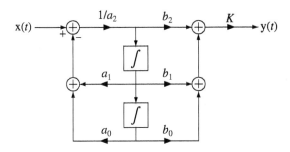

Figure 12.33
A second-order system with a gain factor K

The forward-path transfer function $H_1(s)$ and the feedback-path transfer function $H_2(s)$ are

$$H_1(s) = \frac{Ks}{s^2 - 2s + 2} \quad \text{and} \quad H_2(s) = \frac{s^2 + s}{s^2 + 2s} = \frac{s+1}{s+2}.$$

The loop transfer function is

$$T(s) = H_1(s)H_2(s) = \frac{Ks(s+1)}{(s^2 - 2s + 2)(s+2)}.$$

The poles of T are at $s = 1 \pm j$ and $s = -2$. The zeros are at $s = 0$, $s = -1$ and $|s| \to \infty$. Since $H_1(s)$ has poles in the RHP, the forward-path system is unstable.

1. The root locus has three branches (Rule 1).
2. The allowed regions on the real axis are $-1 < \sigma < 0$ and $\sigma < -2$ (Rule 3).
3. The root locus begins on the poles of T(s). So the branch that begins at $s = -2$ can only go to the left and remain in an allowed region on the real axis. It can never leave the real axis because of symmetry requirements (Rule 4). Therefore, this branch terminates on the zero at infinity.
4. The other two branches begin on complex conjugate poles at $s = 1 \pm j$. They must terminate on the remaining two zeros at $s = 0$ and $s = -2$. To reach these zeros and, at the same time preserve symmetry about the real axis (Rule 4), they must migrate to the left and down into the allowed region, $-1 < \sigma < 0$.
5. The break-in point can be found be setting $(d/ds)(1/T(s)) = 0$. The solution gives us a break-in point at $s = -0.4652$ (Figure 12.34).

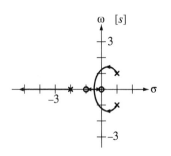

Figure 12.34
Complete root locus

In this example, the overall feedback system starts out unstable at a low K value, but as K is increased the poles that were initially in the RHP migrate into the LHP. So if K is large enough, the overall feedback system becomes stable, even though the forward-path system is unstable.

■

Tracking Errors in Unity-Gain Feedback Systems

A very common type of feedback system is one in which the purpose of the system is to make the output signal track the input signal using unity-gain feedback ($H_2(s) = 1$) (Figure 12.35).

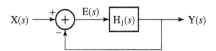

Figure 12.35
A unity-gain feedback system

This type of system is called *unity-gain* because the output signal is always compared directly with the input signal and, if there is any difference (error signal), that is amplified by the forward-path gain of the system in an attempt to bring the output signal closer to the input signal. If the forward-path gain of the system is large, that forces the error signal to be small, making the output and input signals closer together. Whether or not the error signal can be forced to zero depends on the forward-path transfer function $H_1(s)$ and the type of excitation.

It is natural to wonder at this point what the purpose is of a system whose output signal equals its input signal. What have we gained? If the system is an electronic amplifier and the signals are voltages, we have a voltage gain of one, but the input impedance could be very high and the response voltage could drive a very low impedance so that the actual power, in watts, delivered by the output signal is much greater than the actual power supplied by the input signal. In other systems the input signal could be a voltage set by a low-power amplifier or a potentiometer and the output signal could be a voltage indicating the position of some large mechanical device like a crane, an artillery piece, an astronomical telescope and so on.

Now we will mathematically determine the nature of the steady-state error. The term *steady-state* means the behavior as time approaches infinity. The error signal is

$$E(s) = X(s) - Y(s) = X(s) - H_1(s)E(s).$$

Solving for $E(s)$,

$$E(s) = \frac{X(s)}{1 + H_1(s)}.$$

We can find the steady-state value of the error signal using the final-value theorem

$$\lim_{t \to \infty} e(t) = \lim_{s \to 0} s E(s) = \lim_{s \to 0} s \frac{X(s)}{1 + H_1(s)}.$$

If the input signal is a step of the form, $x(t) = Au(t)$, then $X(s) = A/s$ and

$$\lim_{t \to \infty} e(t) = \lim_{s \to 0} \frac{A}{1 + H_1(s)}$$

and the steady-state error is zero if

$$\lim_{s \to 0} \frac{1}{1 + H_1(s)}$$

is zero. If $H_1(s)$ is in the familiar form of a ratio of polynomials in s

$$H_1(s) = \frac{b_N s^N + b_{N-1} s^{N-1} + \cdots b_2 s^2 + b_1 s + b_0}{a_D s^D + a_{D-1} s^{D-1} + \cdots a_2 s^2 + a_1 s + a_0}, \quad (12.8)$$

then

$$\lim_{t \to \infty} e(t) = \lim_{s \to 0} \frac{1}{1 + \dfrac{b_N s^N + b_{N-1} s^{N-1} + \cdots b_2 s^2 + b_1 s + b_0}{a_D s^D + a_{D-1} s^{D-1} + \cdots a_2 s^2 + a_1 s + a_0}} = \frac{a_0}{a_0 + b_0}$$

and, if $a_0 = 0$ and $b_0 \neq 0$, the steady-state error is zero. If $a_0 = 0$, then $H_1(s)$ can be expressed in the form,

$$H_1(s) = \frac{b_N s^N + b_{N-1} s^{N-1} + \cdots b_2 s^2 + b_1 s + b_0}{s(a_D s^{D-1} + a_{D-1} s^{D-2} + \cdots a_2 s + a_1)}$$

and it is immediately apparent that $H_1(s)$ has a pole at $s = 0$. So we can summarize by saying that if a stable unity-gain feedback system has a forward-path transfer function with a pole at $s = 0$, the steady-state error for a step excitation is zero. If there is no pole at $s = 0$, the steady-state error is $a_0/(a_0 + b_0)$ and the larger b_0 is in comparison with a_0, the smaller the steady-state error. This makes sense from another point of view because if the forward-path gain is of the form (12.8) the feedback-system, low-frequency gain is $b_0/(a_0 + b_0)$, which approaches one for $b_0 \gg a_0$ indicating that the input and output signals approach the same value.

A unity-gain feedback system with a forward-path transfer function $H_1(s)$ that has no poles at $s = 0$ is called a **type 0** system. If it has one pole at $s = 0$, the system is a **type 1** system. In general any unity-gain feedback system is a **type n** system where n is the number of poles at $s = 0$ in $H_1(s)$. So, summarizing using the new terminology,

1. A stable type 0 system has a finite steady-state error for step excitation.
2. A stable type n system, $n \geq 1$, has a zero steady-state error for step excitation.

Figure 12.36 illustrates typical steady-state responses to step excitation for stable type 0 and type 1 systems.

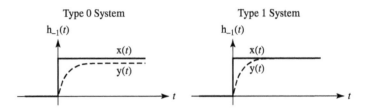

Figure 12.36
Type 0 and type 1 system responses to a step

Now we will consider a ramp excitation $x(t) = A \text{ ramp}(t) = At\, u(t)$ whose Laplace transform is $X(s) = A/s^2$. The steady-state error is

$$\lim_{t \to \infty} e(t) = \lim_{s \to 0} \frac{A}{s[1 + H_1(s)]}.$$

Again, if $H_1(s)$ is a ratio of polynomials in s,

$$\lim_{t \to \infty} e(t) = \lim_{s \to 0} \frac{1}{s} \frac{1}{1 + \dfrac{b_N s^N + b_{N-1} s^{N-1} + \cdots b_2 s^2 + b_1 s + b_0}{a_D s^D + a_{D-1} s^{D-1} + \cdots a_2 s^2 + a_1 s + a_0}}$$

or

$$\lim_{t \to \infty} e(t) = \lim_{s \to 0} \frac{a_D s^D + a_{D-1} s^{D-1} + \cdots a_2 s^2 + a_1 s + a_0}{s \begin{bmatrix} a_D s^D + a_{D-1} s^{D-1} + \cdots a_2 s^2 + a_1 s + a_0 \\ + b_N s^N + b_{N-1} s^{N-1} + \cdots b_2 s^2 + b_1 s + b_0 \end{bmatrix}}.$$

This limit depends on the values of the a's and b's. If $a_0 \neq 0$, the steady-state error is infinite. If $a_0 = 0$ and $b_0 \neq 0$, the limit is a_1/b_0 indicating that the steady-state error is a nonzero constant. If $a_0 = 0$ and $a_1 = 0$ and $b_0 \neq 0$, the steady-state error is zero. The condition, $a_0 = 0$ and $a_1 = 0$, means there is a repeated pole at $s = 0$ in the forward-path transfer function. So for a stable type 2 system, the steady-state error under ramp excitation is zero. Summarizing,

1. A stable type 0 system has an infinite steady-state error for ramp excitation.
2. A stable type 1 system has a finite steady-state error for ramp excitation.
3. A stable type n system, $n \geq 2$, has a zero steady-state error for ramp excitation.

Figure 12.37 illustrates typical steady-state responses to ramp excitation for stable type 0, type 1 and type 2 systems. These results can be extrapolated to higher-order excitations, $(At^2\, u(t), At^3\, u(t),$ etc.). When the highest power of s in the denominator of the transform

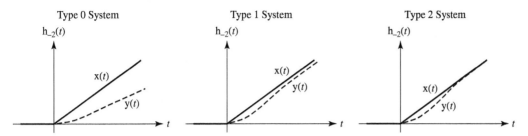

Figure 12.37
Type 0, 1 and 2 system responses to a ramp

of the excitation is the same as, or lower than, the type number (0, 1, 2, etc.) of the system, and the system is stable, the steady-state error is zero. This result was illustrated with forward-path transfer functions in the form of a ratio of polynomials but the result can be shown to be true for any form of transfer function based only on the number of poles at $s = 0$. It may seem that more poles in the forward-path transfer function at $s = 0$ are generally desirable because they reduce the steady-state error in the overall feedback system. But, generally speaking, the more poles in the forward-path transfer function, the harder it is to make a feedback system stable. So we may trade one problem for another by putting poles at $s = 0$ in the forward-path transfer function.

EXAMPLE 12.4

Instability caused by adding a pole at zero in the forward transfer function

Let the forward transfer function of a unity-gain feedback system be $H_1(s) = \dfrac{100}{s(s+4)}$. Then the overall transfer function is

$$H(s) = \frac{100}{s^2 + 4s + 100}$$

with poles at $s = -2 \pm j9.798$. Both poles are in the LHP so the system is stable. Now add a pole at zero to $H_1(s)$ and reevaluate the stability of the system.
    The new $H_1(s)$ is

$$H_1(s) = \frac{100}{s^2(s+4)}$$

and the new overall transfer function is

$$H(s) = \frac{100}{s^3 + 4s^2 + 100}$$

with poles at $s = -6.4235$ and $s = 1.212 \pm j3.755$. Two of the poles are in the RHP and the overall system is unstable.

∎

12.5 SYSTEM ANALYSIS USING MATLAB

The MATLAB **system object** was introduced in Chapter 6. The syntax for creating a system object with **tf** is

$$\text{sys = tf(num,den)}.$$

The syntax for creating a system object with **zpk** is

$$\text{sys} = \text{zpk}(z,p,k).$$

The real power of the control-system toolbox comes in interconnecting systems. Suppose we want the overall transfer function $H(s) = H_1(s)H_2(s)$ of the two systems

$$H_1(s) = \frac{s^2 + 4}{s^5 + 4s^4 + 7s^3 + 15s^2 + 31s + 75}$$

and

$$H_2(s) = 20\frac{s+4}{(s+3)(s+10)}$$

in a cascade connection. In MATLAB.

```
»num = [1 0 4];
»den = [1 4 7 15 31 75];
»H1 = tf(num,den);

»z = [-4];
»p = [-3 -10];
»k = 20
»H2 = zpk(z,p,k);

»Hc = H1*H2 ;
»Hc
Zero/pole/gain:
              20 (s+4) (s^2 + 4)
-----------------------------------------------------------------
(s+3.081) (s+3) (s+10) (s^2 + 2.901s + 5.45) (s^2 - 1.982s + 4.467)
»tf(Hc)


Transfer function:
            20 s^3 + 80 s^2 + 80 s + 320
-----------------------------------------------------------------
s^7 + 17 s^6 + 89 s^5 + 226 s^4 + 436 s^3 + 928 s^2 + 1905 s + 2250
```

If we want to know what the transfer function of these two systems in parallel would be,

```
»Hp = H1 + H2 ;
»Hp


Zero/pole/gain:
20 (s+4.023) (s+3.077) (s^2 + 2.881s + 5.486) (s^2 - 1.982s + 4.505)
-----------------------------------------------------------------
(s+3.081) (s+3) (s+10) (s^2 + 2.901s + 5.45) (s^2 - 1.982s + 4.467)
»tf(Hp)
```

Transfer function:

$$\frac{20 s^6 + 160 s^5 + 461 s^4 + 873 s^3 + 1854 s^2 + 4032 s + 6120}{s^7 + 17 s^6 + 89 s^5 + 226 s^4 + 436 s^3 + 928 s^2 + 1905 s + 2250}$$

There is also a command **feedback** for forming the overall transfer function of a feedback system.

```
>> Hf = feedback(H1,H2) ;
>> Hf
```
Zero/pole/gain:

$$\frac{(s+3)\ (s+10)\ (s^2 + 4)}{(s+9.973)\ (s^2 + 6.465s + 10.69)\ (s^2 + 2.587s + 5.163)\ (s^2 - 2.025s + 4.669)}$$

Sometimes, when manipulating system objects, the result will not be in the ideal form. It may have a pole and zero at the same location. Although there is nothing mathematically wrong with this, it is generally better to cancel that pole and zero to simplify the transfer function. This can be done using the command **minreal** (for minimum realization).

Once we have a system described, we can graph its step response with **step**, its impulse response with **impulse** and a Bode diagram of its frequency response with **bode**. We can also draw its pole-zero diagram using the MATLAB command **pzmap**. MATLAB has a function called **freqresp** that does frequency response graphs. The syntax is

$$H = freqresp(sys,w)$$

where **sys** is the MATLAB-system object, **w** is a vector of radian frequencies (ω) and **H** is the frequency response of the system at those radian frequencies. The MATLAB control toolbox also has a command for plotting the root locus of a system loop transfer function. The syntax is

$$rlocus(sys)$$

where **sys** is a MATLAB system object. There are many other useful commands in the **control** toolbox, which can be examined by typing **help control**.

12.6 SYSTEM RESPONSES TO STANDARD SIGNALS

We have seen in previous signal and system analysis that an LTI system is completely characterized by its impulse response. In testing real systems, the application of an impulse to find the system's impulse response is not practical. First, a true impulse cannot be generated and second, even if we could generate a true impulse, since it has an unbounded amplitude it would inevitably drive a real system into a nonlinear mode of operation. We could generate an approximation to the true unit impulse in the form of a very short duration and very tall pulse of unit area. Its time duration should be so small that making it any smaller would not significantly change any signals in the system. Although this type of test is possible, a very tall pulse may drive a system into nonlinearity. It is much easier to generate a good approximation to a step than to an impulse, and the step amplitude can be small enough so as to not cause the system to go nonlinear.

Sinusoids are also easy to generate and are confined to varying between finite bounds that can be small enough that the sinusoid will not overdrive the system and force it into nonlinearity. The frequency of the sinusoid can be varied to determine the frequency response of the system. Since sinusoids are very closely related to complex exponentials, this type of testing can directly yield information about the system characteristics.

UNIT-STEP RESPONSE

Let the transfer function of an LTI system be of the form

$$H(s) = \frac{N_H(s)}{D_H(s)}$$

where $N_H(s)$ is of a lower degree in s than $D_H(s)$. Then the Laplace transform of the zero-state response $Y(s)$ to $X(s)$ is

$$Y(s) = \frac{N_H(s)}{D_H(s)} X(s).$$

Let $x(t)$ be a unit step. Then the Laplace transform of the zero-state response is

$$Y(s) = H_{-1}(s) = \frac{N_H(s)}{s D_H(s)}.$$

Using the partial fraction expansion technique, this can be separated into two terms

$$Y(s) = \frac{N_{H1}(s)}{D_H(s)} + \frac{H(0)}{s}.$$

If the system is BIBO stable, the roots of $D_H(s)$ are all in the open LHP and the inverse Laplace transform of $N_{H1}(s)/D_H(s)$ is called the **natural response** or the **transient response** because it decays to zero as time t approaches infinity. The **forced response** of the system to a unit step is the inverse Laplace transform of $H(0)/s$, which is $H(0) u(t)$. The expression

$$Y(s) = \frac{N_{H1}(s)}{D_H(s)} + \frac{H(0)}{s}$$

has two terms. The first term has poles that are identical to the system poles and the second term has a pole at the same location as the Laplace transform of the unit step.

This result can be generalized to an arbitrary excitation. If the Laplace transform of the excitation is

$$X(s) = \frac{N_x(s)}{D_x(s)}$$

then the Laplace transform of the system response is

$$Y(s) = \frac{N_H(s)}{D_H(s)} X(s) = \frac{N_H(s) N_x(s)}{D_H(s) D_x(s)} = \underbrace{\frac{N_{H1}(s)}{D_H(s)}}_{\text{same poles as system}} + \underbrace{\frac{N_{x1}(s)}{D_x(s)}}_{\text{same poles as excitation}}.$$

Now let's examine the unit-step response of some simple systems. The simplest dynamic system is a first-order system whose transfer function is of the form

$$H(s) = \frac{A}{1 - s/p} = -\frac{Ap}{s - p}$$

where A is the low-frequency transfer function of the system and p is the pole location in the s-plane. The Laplace transform of the step response is

$$Y(s) = H_{-1}(s) = \frac{A}{(1-s/p)s} = \frac{A/p}{1-s/p} + \frac{A}{s} = \frac{A}{s} - \frac{A}{s-p}.$$

Inverse Laplace transforming, $y(t) = A(1 - e^{pt})u(t)$. If p is positive, the system is unstable and the magnitude of the response to a unit step increases exponentially with time (Figure 12.38).

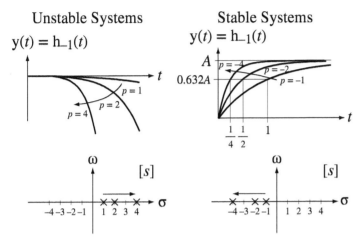

Figure 12.38
Responses of a first-order system to a unit step and the corresponding pole-zero diagrams

The speed of the exponential increase depends on the magnitude of p, being greater for a larger magnitude of p. If p is negative the system is stable and the response approaches a constant A with time. The speed of the approach to A depends on the magnitude of p, being greater for a larger magnitude of p. The negative reciprocal of p is called the **time constant** τ of the system, $\tau = -1/p$ and, for a stable system, the response to a unit step moves 63.2% of the distance to the final value in a time equal to one time constant.

Now consider a second-order system whose transfer function is of the form,

$$H(s) = \frac{A\omega_n^2}{s^2 + 2\zeta\omega_n s + \omega_n^2}, \quad \omega_n > 0.$$

This form of a second-order system transfer function has three parameters, the low-frequency gain A, the damping ratio ζ and the natural radian frequency ω_n. The form of the unit-step response depends on these parameter values. The Laplace transform of the system unit-step response is

$$H_{-1}(s) = \frac{A\omega_n^2}{s(s^2 + 2\zeta\omega_n s + \omega_n^2)} = \frac{A\omega_n^2}{s[s + \omega_n(\zeta + \sqrt{\zeta^2 - 1})][s + \omega_n(\zeta - \sqrt{\zeta^2 - 1})]}.$$

This can be expanded in partial fractions (if $\zeta \neq \pm 1$) as

$$H_{-1}(s) = A\left[\frac{1}{s} + \frac{\frac{1}{2(\zeta^2 - 1 + \zeta\sqrt{\zeta^2 - 1})}}{s + \omega_n(\zeta + \sqrt{\zeta^2 - 1})} + \frac{\frac{1}{2(\zeta^2 - 1 - \zeta\sqrt{\zeta^2 - 1})}}{s + \omega_n(\zeta - \sqrt{\zeta^2 - 1})}\right]$$

and the time-domain response is then

$$h_{-1}(t) = A\left[\frac{e^{-\omega_n(\zeta+\sqrt{\zeta^2-1})t}}{2(\zeta^2-1+\zeta\sqrt{\zeta^2-1})} + \frac{e^{-\omega_n(\zeta-\sqrt{\zeta^2-1})t}}{2(\zeta^2-1-\zeta\sqrt{\zeta^2-1})} + 1\right]u(t).$$

For the special case of $\zeta = \pm 1$ the system unit-step response is

$$H_{-1}(s) = \frac{A\omega_n^2}{(s \pm \omega_n)^2 s},$$

the two poles are identical, the partial fraction expansion is

$$H_{-1}(s) = A\left[\frac{1}{s} - \frac{\pm\omega_n}{(s \pm \omega_n)^2} - \frac{1}{s \pm \omega_n}\right]$$

and the time-domain response is

$$h_{-1}(t) = A[1 - (1 \pm \omega_n t)e^{\mp\omega_n t}]u(t) = Au(t)\begin{cases} 1 - (1 + \omega_n t)e^{-\omega_n t}, & \zeta = 1 \\ 1 - (1 - \omega_n t)e^{+\omega_n t}, & \zeta = -1 \end{cases}.$$

It is difficult, just by examining the mathematical functional form of the unit-step response, to immediately determine what it will look like for an arbitrary choice of parameters. To explore the effect of the parameters, let's first set A and ω_n constant and examine the effect of the damping ratio ζ. Let $A = 1$ and let $\omega_n = 1$. Then the unit-step response and the corresponding pole-zero diagrams are as illustrated in Figure 12.39 for six choices of ζ.

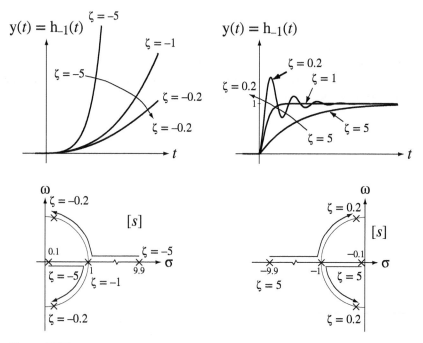

Figure 12.39
Second-order system responses to a unit step and the corresponding pole-zero diagrams

We can see why these different types of behavior occur if we examine the unit-step response

$$h_{-1}(t) = A\left[\frac{e^{-\omega_n(\zeta+\sqrt{\zeta^2-1})t}}{2(\zeta^2-1+\zeta\sqrt{\zeta^2-1})} + \frac{e^{-\omega_n(\zeta-\sqrt{\zeta^2-1})t}}{2(\zeta^2-1-\zeta\sqrt{\zeta^2-1})} + 1\right]u(t) \quad (12.9)$$

and, in particular, the exponents of e, $-\omega_n(\zeta \pm \sqrt{\zeta^2-1})t$. The signs of the real parts of these exponents determine whether the response grows or decays with time $t > 0$. For times $t < 0$ the response is zero because of the unit step $u(t)$.

Case 1: $\zeta < 0$

If $\zeta < 0$, then the exponent of e in both terms in (12.9) has a positive real part for positive time and the step response therefore grows with time and the system is unstable. The exact form of the unit-step response depends on the value of ζ. It is a simple increasing exponential for $\zeta < -1$ and an exponentially growing oscillation for $-1 < \zeta < 0$. But either way the system is unstable.

Case 2: $\zeta > 0$

If $\zeta > 0$, then the exponent of e in both terms in (12.9) has a negative real part for positive time and the step response therefore decays with time and the system is stable.

Case 2a: $\zeta > 1$

If $\zeta > 1$, then $\zeta^2 - 1 > 0$, and the coefficients of t in (12.9) $-\omega_n(\zeta \pm \sqrt{\zeta^2-1})t$ are both negative real numbers and the unit-step response is in the form of a constant plus the sum of two decaying exponentials. This case $\zeta > 1$ is called the **overdamped** case.

Case 2b: $0 < \zeta < 1$

If $0 < \zeta < 1$, then $\zeta^2 - 1 < 0$, and the coefficients of t in (12.9) $-\omega_n(\zeta \pm \sqrt{\zeta^2-1})t$ are both complex numbers in a complex-conjugate pair with negative real parts, and the unit-step response is in the form of a constant plus the sum of two sinusoids multiplied by a decaying exponential. Even though the response overshoots its final value, it still settles to a constant value and is therefore the response of a stable system. This case $0 < \zeta < 1$ is called the **underdamped** case.

The dividing line between the overdamped and underdamped cases is the case $\zeta = 1$. This condition is called **critical damping**.

Now let's examine the effect of changing ω_n while holding the other parameters constant. Let $A = 1$ and $\zeta = 0.5$. The step response is illustrated in Figure 12.40 for 3 values of ω_n.

Since ω_n is the natural radian frequency, it is logical that it would affect the ringing rate of the step response.

The response of any LTI system to a unit step can be found using the MATLAB control toolbox command step.

SINUSOID RESPONSE

Now let's examine the response of a system to a "causal" sinusoid (one applied to the system at time $t = 0$). Again let the system transfer function be of the form

$$H(s) = \frac{N_H(s)}{D_H(s)}.$$

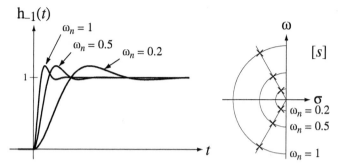

Figure 12.40
Second-order system response for three different values of ω_n and the corresponding pole-zero plots

Then the Laplace transform of the zero-state response to $x(t) = \cos(\omega_0 t)\, u(t)$, would be

$$Y(s) = \frac{N_H(s)}{D_H(s)} \frac{s}{s^2 + \omega_0^2}.$$

This can be separated into partial fractions in the form

$$Y(s) = \frac{N_{H1}(s)}{D_H(s)} + \frac{1}{2}\frac{H(-j\omega_0)}{s + j\omega_0} + \frac{1}{2}\frac{H(j\omega_0)}{s - j\omega_0} = \frac{N_{H1}(s)}{D_H(s)} + \frac{1}{2}\frac{H^*(j\omega_0)}{s + j\omega_0} + \frac{1}{2}\frac{H(j\omega_0)}{s - j\omega_0}$$

or

$$Y(s) = \frac{N_{H1}(s)}{D_H(s)} + \frac{1}{2}\frac{H^*(j\omega_0)(s - j\omega_0) + H(j\omega_0)(s + j\omega_0)}{s^2 + \omega_0^2}$$

$$Y(s) = \frac{N_{H1}(s)}{D_H(s)} + \frac{1}{2}\left\{\frac{s}{s^2 + \omega_0^2}[H(j\omega_0) + H^*(j\omega_0)] + \frac{j\omega_0}{s^2 + \omega_0^2}[H(j\omega_0) - H^*(j\omega_0)]\right\}$$

$$Y(s) = \frac{N_{H1}(s)}{D_H(s)} + \text{Re}(H(j\omega_0))\frac{s}{s^2 + \omega_0^2} - \text{Im}(H(j\omega_0))\frac{\omega_0}{s^2 + \omega_0^2}.$$

The inverse Laplace transform of the term $\text{Re}(H(j\omega_0))s/(s^2 + \omega_0^2)$ is the product of a unit step and a cosine at ω_0 with an amplitude of $\text{Re}(H(j\omega_0))$, and the inverse Laplace transform of the term $\text{Im}(H(j\omega_0))\omega_0/(s^2 + \omega_0^2)$ is the product of a unit step and a sine at ω_0 with an amplitude of $\text{Im}(H(j\omega_0))$. That is,

$$y(t) = \mathcal{L}^{-1}\left(\frac{N_{H1}(s)}{D_H(s)}\right) + [\text{Re}(H(j\omega_0))\cos(\omega_0 t) - \text{Im}(H(j\omega_0))\sin(\omega_0 t)]u(t)$$

or, using $\text{Re}(A)\cos(\omega_0 t) - \text{Im}(A)\sin(\omega_0 t) = |A|\cos(\omega_0 t + \angle A)$,

$$y(t) = \mathcal{L}^{-1}\left(\frac{N_{H1}(s)}{D_H(s)}\right) + |H(j\omega_0)|\cos(\omega_0 t + \angle H(j\omega_0))u(t).$$

If the system is stable, the roots of $D_H(s)$ are all in the open LHP and the inverse Laplace transform of $N_{H1}(s)/D_H(s)$ (the transient response) decays to zero as time t approaches infinity. Therefore the forced response that persists after the transient response has died away is a causal sinusoid of the same frequency as the excitation and with an amplitude and phase determined by the transfer function evaluated at $s = j\omega_0$. The forced response is exactly the same as the response obtained by using Fourier methods because the Fourier methods assume that the excitation is a true sinusoid (applied at time $t \to -\infty$), not a causal sinusoid and therefore there is no transient response in the solution.

EXAMPLE 12.5

Zero-state response of a system to a causal cosine

Find the total zero-state response of a system characterized by the transfer function

$$H(s) = \frac{10}{s + 10}$$

to a unit-amplitude causal cosine at a frequency of 2 Hz.

The radian frequency ω_0 of the cosine is 4π. Therefore the Laplace transform of the response is

$$Y(s) = \frac{10}{s+10} \frac{s}{s^2 + (4\pi)^2}$$

$$Y(s) = \frac{-0.388}{s+10} + \text{Re}(H(j4\pi)) \frac{s}{s^2 + (4\pi)^2} - \text{Im}(H(j4\pi)) \frac{\omega_0}{s^2 + (4\pi)^2}$$

and the time-domain response is

$$y(t) = \mathcal{L}^{-1}\left(\frac{-0.388}{s+10}\right) + |H(j4\pi)| \cos(4\pi t + \angle H(j4\pi)) u(t)$$

or

$$y(t) = \left[-0.388 e^{-10t} + \left|\frac{10}{j4\pi + 10}\right| \cos(4\pi t - \angle(j4\pi + 10))\right] u(t)$$

or

$$y(t) = [-0.388 e^{-10t} + 0.623 \cos(4\pi t - 0.899)] u(t).$$

The excitation and response are illustrated in Figure 12.41. Looking at the graph we see that the response appears to reach a stable amplitude in less than one second. This is reasonable

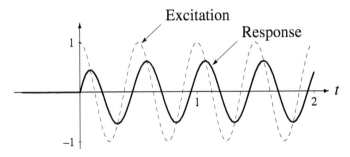

Figure 12.41
Excitation and response of a first-order system excited by a cosine applied at time $t = 0$

given that the transient response has a time constant of 1/10 of a second. After the response stabilizes, its amplitude is about 62% of the excitation amplitude and its phase is shifted so that it lags the excitation by about a 0.899 radian phase shift, which is equivalent to about a 72-ms time delay.

If we solve for the response of the system using Fourier methods, we write the frequency response from the transfer function as

$$H(j\omega) = \frac{10}{j\omega + 10}.$$

If we make the excitation of the system a true cosine, it is $x(t) = \cos(4\pi t)$ and its continuous-time Fourier transform (CTFT) is $X(j\omega) = \pi[\delta(\omega - 4\pi) + \delta(\omega + 4\pi)]$. Then the system response is

$$Y(j\omega) = \pi[\delta(\omega - 4\pi) + \delta(\omega + 4\pi)]\frac{10}{j\omega + 10} = 10\pi\left[\frac{\delta(\omega - 4\pi)}{j4\pi + 10} + \frac{\delta(\omega + 4\pi)}{-j4\pi + 10}\right]$$

or

$$Y(j\omega) = 10\pi\frac{10[\delta(\omega - 4\pi) + \delta(\omega + 4\pi)] + j4\pi[\delta(\omega + 4\pi) - \delta(\omega - 4\pi)]}{16\pi^2 + 100}.$$

Inverse Fourier transforming, $y(t) = 0.388\cos(4\pi t) + 0.487\sin(4\pi t)$ or, using

$$\text{Re}(A)\cos(\omega_0 t) - \text{Im}(A)\sin(\omega_0 t) = |A|\cos(\omega_0 t + \angle A)$$

$$y(t) = 0.623\cos(4\pi t - 0.899).$$

This is exactly the same (except for the unit step) as the forced response of the previous solution, which was found using Laplace transforms.

■

12.7 STANDARD REALIZATIONS OF SYSTEMS

The process of system design, as opposed to system analysis, is to develop a desired transfer function for a class of excitations that yields a desired response or responses. Once we have found the desired transfer function, the next logical step is to actually build or perhaps simulate the system. The usual first step in building or simulating a system is to form a block diagram that describes the interactions among all the signals in the system. This step is called **realization**, arising from the concept of making a real system instead of just a set of equations that describe its behavior. There are several standard types of system realization. We have already seen Direct Form II in Chapter 8. We will explore two more here.

CASCADE REALIZATION

The second standard system realization is the **cascade** form. The numerator and denominator of the general transfer function form

$$H(s) = \frac{Y(s)}{X(s)} = \frac{\sum_{k=0}^{M}b_k s^k}{\sum_{k=0}^{N}a_k s^k} = \frac{b_M s^M + b_{M-1}s^{M-1} + \cdots + b_1 s + b_0}{s^N + a_{N-1}s^{N-1} + \cdots + a_1 s + a_0}, \quad a_N = 1 \quad (12.10)$$

12.7 Standard Realizations of Systems

where $M \leq N$ can be factored yielding a transfer function expression of the form

$$H(s) = A \frac{s-z_1}{s-p_1} \frac{s-z_2}{s-p_2} \cdots \frac{s-z_M}{s-p_M} \frac{1}{s-p_{M+1}} \frac{1}{s-p_{M+2}} \cdots \frac{1}{s-p_N}.$$

Any of the component fractions $\frac{Y_k(s)}{X_k(s)} = \frac{s-z_k}{s-p_k}$ or $\frac{Y_k(s)}{X_k(s)} = \frac{1}{s-p_k}$ represents a subsystem that can be realized by writing the relationship as

$$H_k(s) = \underbrace{\frac{1}{s-p_k}}_{H_{k1}(s)} \underbrace{(s-z_k)}_{H_{k2}(s)} \quad \text{or} \quad H_k(s) = \frac{1}{s-p_k}$$

and realizing it as a Direct Form II system (Figure 12.42). Then the entire original system can be realized in cascade form (Figure 12.43).

$$H_k(s) = \frac{s-z_k}{s-p_k} \hspace{4cm} H_k(s) = \frac{1}{s-p_k}$$

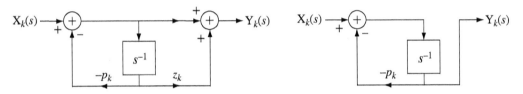

Figure 12.42
Direct Form II realization of a single subsystem in the cascade realization

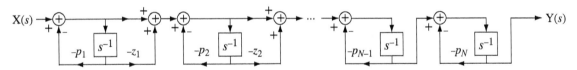

Figure 12.43
Overall cascade system realization

A problem sometimes arises with this type of cascade realization. Sometimes the first-order subsystems have complex poles. This necessitates multiplication by complex numbers and that usually cannot be done in a system realization. In such cases, two subsystems with complex conjugate poles should be combined into one second-order subsystem of the form

$$H_k(s) = \frac{s+b_0}{s^2 + a_1 s + a_0}$$

which can always be realized with real coefficients (Figure 12.44).

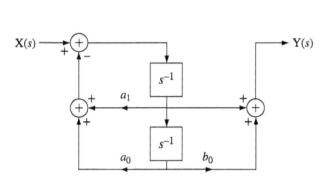

Figure 12.44
A standard-form second-order subsystem

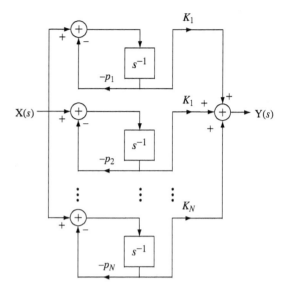

Figure 12.45
Overall parallel system realization

PARALLEL REALIZATION

The last standard realization of a system is the parallel realization. This can be accomplished by expanding the standard transfer function form (12.10) in partial fractions of the form

$$H(s) = \frac{K_1}{s - p_1} + \frac{K_2}{s - p_2} + \cdots + \frac{K_N}{s - p_N}$$

(Figure 12.45).

12.8 SUMMARY OF IMPORTANT POINTS

1. Continuous-time systems can be described by differential equations, block diagrams or circuit diagrams in the time or frequency domain.
2. A continuous-time LTI system is stable if all the poles of its transfer function lie in the open left half-plane.
3. Marginally stable systems form a subset of unstable systems.
4. The three most important types of system interconnections are the cascade connection, the parallel connection and the feedback connection.
5. The unit step and the sinusoid are important practical signals for testing system characteristics.
6. The Direct Form II, cascade and parallel realizations are important standard ways of realizing systems.

EXERCISES WITH ANSWERS

(Answers to each exercise are in random order.)

Transfer Functions

1. For each circuit in Figure E.1 write the transfer function between the indicated excitation and indicated response. Express each transfer function in the standard form

$$H(s) = A \frac{s^M + b_{N-1}s^{M-1} + \cdots + b_2 s^2 + b_1 s + b_0}{s^N + a_{D-1}s^{N-1} + \cdots + a_2 s^2 + a_1 s + a_0}.$$

(a) Excitation $v_s(t)$ Response $v_o(t)$

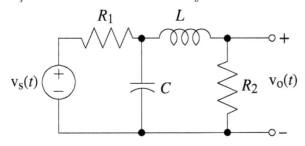

(b) Excitation $i_s(t)$ Response $v_o(t)$

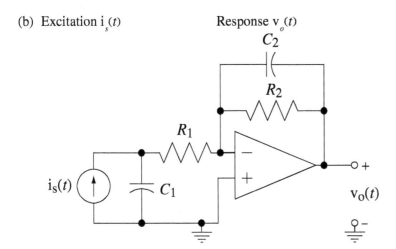

(c) Excitation $i_s(t)$ Response $i_1(t)$

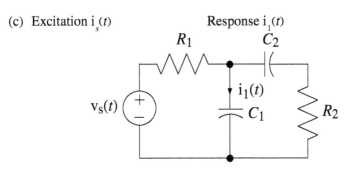

Figure E.1

Answers:

$$\frac{1}{R_1}\frac{s^2+\frac{s}{R_2C_2}}{s^2+s\left(\frac{1}{R_2C_2}+\frac{1}{R_2C_1}+\frac{1}{R_1C_1}\right)+\frac{1}{R_1R_2C_1C_2}},$$

$$-\frac{1}{R_1C_1C_2}\frac{1}{s^2+s\left(\frac{1}{R_2C_2}+\frac{1}{R_1C_1}\right)+\frac{1}{R_1R_2C_1C_2}},$$

$$\frac{R_2}{R_1LC}\frac{1}{s^2+s\left(\frac{1}{R_1C}+\frac{R_2}{L}\right)+\frac{(R_2+R_1)}{R_1LC}}$$

2. For each block diagram in Figure E.2 write the transfer function between the excitation x(t) and the response y(t).

(a)
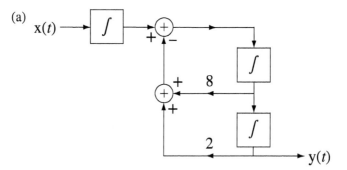

(b)
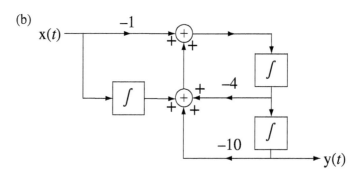

Figure E.2

Answers: $-\dfrac{s-1}{s^3+4s^2+10s}$, $\dfrac{1}{s^3+8s^2+2s}$

Stability

3. Evaluate the stability of each of these system transfer functions.

(a) $H(s)=-\dfrac{100}{s+200}$ (b) $H(s)=\dfrac{80}{s-4}$

(c) $H(s) = \dfrac{6}{s(s+1)}$ (d) $H(s) = -\dfrac{15s}{s^2+4s+4}$

(e) $H(s) = 3\dfrac{s-10}{s^2+4s+29}$ (f) $H(s) = 3\dfrac{s^2+4}{s^2-4s+29}$

(g) $H(s) = \dfrac{1}{s^2+64}$ (h) $H(s) = \dfrac{10}{s^3+4s^2+29s}$

(i) $H(s) = \dfrac{1}{s}$ (j) $H(s) = s$

(k) $H(s) = \dfrac{s^2+8}{s^2+6s+9}$ (l) $H(s) = \dfrac{s+3}{s^2+9}$

(m) $H(s) = \dfrac{s-3}{s^2+9}$ (n) $H(s) = \dfrac{s^2+8}{s^2-6s+9}$

(o) $H(s) = \dfrac{1}{s(s-3)}$ (p) $H(s) = \dfrac{1}{s^2}$

Answers: 4 Stable, 6 Marginally Stable (therefore unstable), 6 Unstable (and not marginally stable)

4. In the feedback system in Figure E.5, $H_1(s) = \dfrac{s-3}{s-2}$ and $H_2(s) = K$. What range of real values of K makes this system stable?

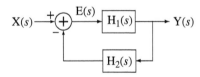

Figure E.5

Answer: $-1 < K < -2/3$

5. If $H_1(s) = \dfrac{K}{s^2-9}$ and $H_2(s) = s^2+4$, for what range of K is this system unstable?

For what range of K is this system marginally stable?

Answers: All K, $K > 9/4$

Parallel, Cascade and Feedback Connections

6. Find the overall transfer functions of the systems in Figure E.6 in the form of a single ratio of polynomials in s.

(a) $X(s) \longrightarrow \boxed{\dfrac{s^2}{s^2+3s+2}} \longrightarrow \boxed{\dfrac{10}{s^2+3s+2}} \longrightarrow Y(s)$

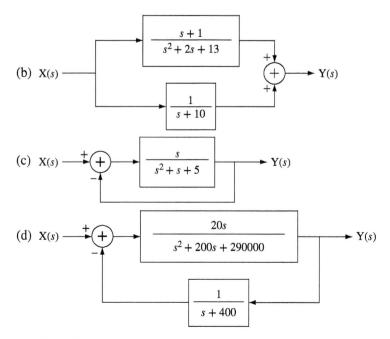

Figure E.6

Answers: $10\dfrac{s^2}{s^4 + 6s^3 + 13s^2 + 12s + 4}$, $2\dfrac{s^2 + \frac{13}{2}s + \frac{23}{2}}{s^3 + 12s^2 + 33s + 130}$,

$20\dfrac{s^2 + 400s}{s^3 + 600s^2 + 370020s + 1.16 \times 10^8}$, $\dfrac{s}{s^2 + 2s + 5}$

7. In the feedback system in Figure E.7, find the overall system transfer function for the given values of forward-path gain, K.

(a) $K = 10^6$ (b) $K = 10^5$ (c) $K = 10$
(d) $K = 1$ (e) $K = -1$ (f) $K = -10$

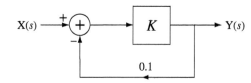

Figure E.7

Answers: ∞, 10, −1.11, 0.909, 10, 5

8. Which of the systems with these descriptions are stable?

(a) $H(s) = \dfrac{13}{s(s + 4)}$

(b) Unity-gain feedback system with $H_1(s) = \dfrac{5}{s^2 + 3}$

(c) System with impulse response $h(t) = 7\cos(22\pi t)u(t)$

(d) System with zeros at $s = 2$ and $s = -0.5$ and poles at $s = -0.8$ and $s = -1.8$

Answers: 1 Stable, 3 Unstable

9. For what range of values of K is the system in Figure E.9 stable? Graph the step responses for $K = 0$, $K = 4$ and $K = 8$.

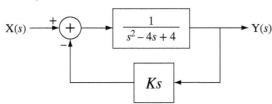

Figure E.9

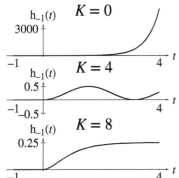

Answers: , The system is unstable if $K \leq 4$ and stable if $K > 4$

10. Graph the impulse response and the pole-zero diagram for the forward path and the overall system in Figure E.10.

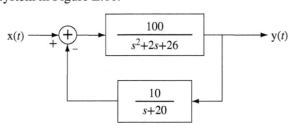

Figure E.10

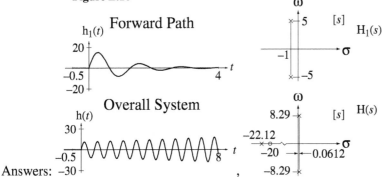

Answers:

Root Locus

11. The loop transfer function of a feedback system has 4 finite poles and 1 finite zero. What is the angle (in radians) between the asymptotes of the system root locus?

Answer: $2\pi/3$ radians

12. For each $H_1(s)$ in Figure E.12 is there a finite positive value of K for which this system is unstable?

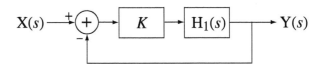

Figure E.12

(a) $H_1(s) = 1000 \dfrac{s-4}{s+10}$

(b) $H_1(s) = \dfrac{25}{s^2 + 2s + 1}$

(c) $H_1(s) = -\dfrac{10}{(s+3)(s+8)(s+22)}$

Answers: Yes, No, Yes

13. Draw the root locus for each of the systems that have these loop transfer functions and identify the transfer functions that are stable for all finite, positive real values of K.

(a) $T(s) = \dfrac{K}{(s+3)(s+8)}$

(b) $T(s) = \dfrac{Ks}{(s+3)(s+8)}$

(c) $T(s) = \dfrac{Ks^2}{(s+3)(s+8)}$

(d) $T(s) = \dfrac{K}{(s+1)(s^2 + 4s + 8)}$

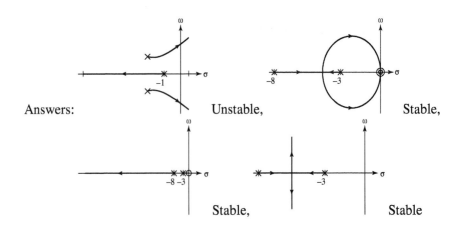

Answers: Unstable, Stable, Stable, Stable

14. Sketch a root locus for each of the diagrams in Figure E.14. The diagrams show the poles and zeros of the loop transfer function $T(s)$ of a feedback system.

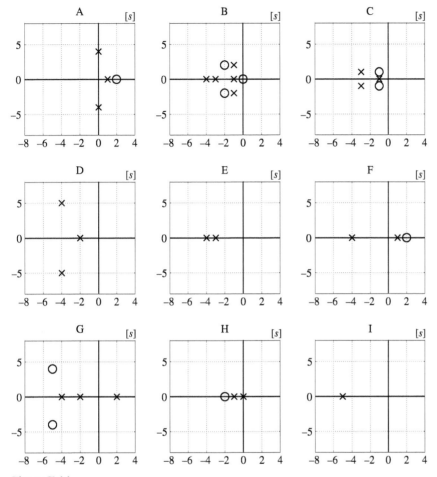

Figure E.14

For each system indicate whether it will be unstable at some finite positive value of the gain constant K.

Answers: 5 Stable and 4 Unstable

Tracking Errors in Unity-Gain Feedback Systems

15. In Figure E.15 is a block diagram of a continuous-time feedback system. Answer the questions.

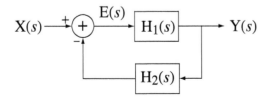

Figure E.15

(a) If $H_1(s) = \dfrac{K}{s+4}$ and $H_2(s) = \dfrac{3}{s+10}$, is this system stable for all positive values of K?

(b) If $H_1(s) = \dfrac{4}{s+5}$ and $H_2(s) = 1$, what is the steady-state error when the excitation signal is a unit step? (Steady-state error is defined as $\lim\limits_{t\to\infty} e(t)$ where $e(t) \xleftarrow{\mathcal{L}} E(s)$.)

(c) If $H_1(s) = \dfrac{4}{s(s+5)}$ and $H_2(s) = 1$, what is the steady-state error when the excitation signal is a unit step?

(d) If $H_1(s) = \dfrac{K}{s+3}$ and $H_2(s) = 1$, sketch a root locus of the system.

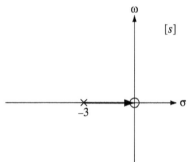

Answers: 0, 5/9, , Yes

16. A continuous-time unity-gain (tracking) feedback system has a forward-path transfer function with exactly one pole at the origin of the s-plane.

   (a) Describe in as much detail as possible the steady-state error signal in response to a step excitation.
   (b) Describe in as much detail as possible the steady-state error signal in response to a ramp excitation.

   Answer: Zero, Nonzero Finite

17. The transfer function of a continuous-time system has one finite pole in the open left half-plane at $s = s_0$ and no finite zeros. If you wanted to make its response to a step excitation approach its final value faster, how would you change s_0?

   Answer: Make s_0 more negative.

System Responses to Standard Signals

18. Using the Laplace transform, find and graph the time-domain response $y(t)$ of the systems with these transfer functions to the sinusoidal excitation, $x(t) = \cos(10\pi t)u(t)$.

   (a) $H(s) = \dfrac{1}{s+1}$
   (b) $H(s) = \dfrac{s-2}{(s-2)^2 + 16}$

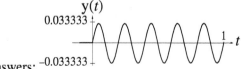

Answers: ,

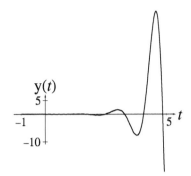

19. Find the responses of the systems with these transfer functions to a unit step and a unit amplitude, 1 Hz cosine applied at time $t = 0$. Also find the responses to a true unit amplitude, 1 Hz cosine using the CTFT and compare to the steady-state part of the total solution found using the Laplace transform.

 Use the results of the chapter in which if a cosine is applied at time $t = 0$ to a system whose transfer function is $H(s) = N(s)/D(s)$, the response is

$$y(t) = \mathcal{L}^{-1}(N_1(s)/D(s)) + |H(j\omega_0)|\cos(\omega_0 t + \angle H(j\omega_0))u(t)$$

where $N_1(s)/D(s)$ is the partial fraction involving the transfer function poles.

(a) $H(s) = \dfrac{1}{s}$
(b) $H(s) = \dfrac{s}{s+1}$
(c) $H(s) = \dfrac{s}{s^2 + 2s + 40}$
(d) $H(s) = \dfrac{s^2 + 2s + 40}{s^2}$

Answers: $\dfrac{8\pi^2\cos(2\pi t) - 3.277\sin(2\pi t)}{158.19}$, $y(t) = \dfrac{\sin(2\pi t)}{2\pi}$,

$0.3184\sin(2\pi t) - 0.0132\cos(2\pi t)$, $\dfrac{(2\pi)^2\cos(2\pi t) - 2\pi\sin(2\pi t)}{1 + (2\pi)^2}$

System Realization

20. Draw cascade system diagrams of the systems with these transfer functions.

(a) $H(s) = \dfrac{s}{s+1}$

(b) $H(s) = \dfrac{s+4}{(s+2)(s+12)}$

(c) $H(s) = \dfrac{20}{s(s^2 + 5s + 10)}$

Answers:

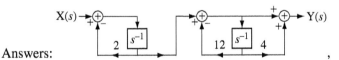

,

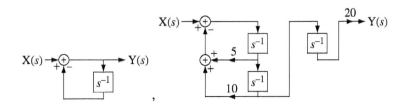

,

EXERCISES WITHOUT ANSWERS

Stability

21. A continuous-time system with a bounded impulse response is excited by $x(t) = \cos(200\pi t)u(t)$. Find and report the numerical locations for a pair of poles in the system transfer function that would cause the response to $x(t)$ to grow in positive time $(t > 0)$ without bound.

Transfer Functions

22. In the feedback system in Figure E.22, graph the response of the system to a unit step, for the time period, $0 < t < 10$, then write the expression for the overall system transfer function and draw a pole-zero diagram, for the given values of K.

    (a) $K = 20$ (b) $K = 10$ (c) $K = 1$
    (d) $K = -1$ (e) $K = -10$ (f) $K = -20$

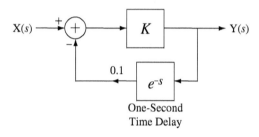

Figure E.22

23. Find the s-domain transfer functions for the circuits in Figure E.23 and then draw block diagrams for them as systems with $V_i(s)$ as the excitation and $V_o(s)$ as the response.

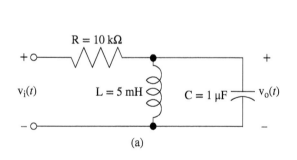

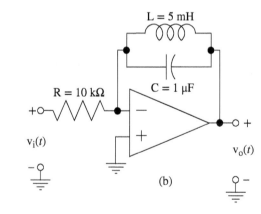

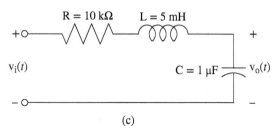

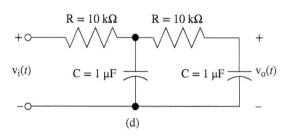

Figure E.23

Stability

24. Determine whether the systems with these transfer functions are stable, marginally stable or unstable (and not marginally stable).

    (a) $H(s) = \dfrac{s(s+2)}{s^2+8}$
    (b) $H(s) = \dfrac{s(s-2)}{s^2+8}$

    (c) $H(s) = \dfrac{s^2}{s^2+4s+8}$
    (d) $H(s) = \dfrac{s^2}{s^2-4s+8}$

    (e) $H(s) = \dfrac{s}{s^3+4s^2+8s}$

Parallel, Cascade and Feedback Connections

25. Find the expression for the overall system transfer function of the system in Figure E.25. For what positive values of K is the system stable?

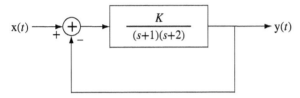

Figure E.25

26. A laser operates on the fundamental principle that a pumped medium amplifies a travelling light beam as it propagates through the medium. Without mirrors a laser becomes a single-pass travelling wave amplifier (Figure E.26.1).

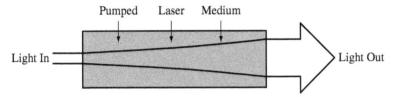

Figure E.26.1

This is a system without feedback. If we now place mirrors at each end of the pumped medium, we introduce feedback into the system.

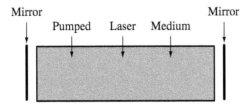

Figure E.26.2

When the gain of the medium becomes large enough the system oscillates creating a coherent output light beam. That is laser operation. If the gain of the medium is less that that required to sustain oscillation, the system is known as a regenerative travelling-wave amplifier (RTWA).

Let the electric field of a light beam incident on the RTWA from the left be the excitation of the system, $E_{inc}(s)$, and let the electric fields of the reflected light, $E_{refl}(s)$, and the transmitted light, $E_{trans}(s)$, be the responses of the system (Figure E.26.3).

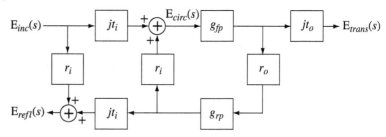

Figure E.26.3

Electric field reflectivity of the input mirror, $r_i = 0.99$

Electric field transmissivity of the input mirror, $t_i = \sqrt{1 - r_i^2}$

Electric field reflectivity of the output mirror, $r_o = 0.98$

Electric field transmissivity of the output mirror, $t_o = \sqrt{1 - r_o^2}$

Forward and reverse path electric field gains, $g_{fp}(s) = g_{rp}(s) = 1.01e^{-10^{-9}s}$

Find an expression for the frequency response $\dfrac{E_{trans}(f)}{E_{inc}(f)}$ and plot its magnitude over the frequency range $3 \times 10^{14} \pm 5 \times 10^8$ Hz.

27. A classical example of the use of feedback is the phase-locked loop used to demodulate frequency-modulated signals (Figure E.27).

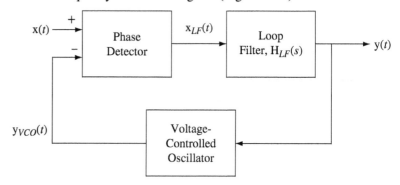

Figure E.27

The excitation, $x(t)$, is a frequency-modulated sinusoid. The phase detector detects the phase difference between the excitation and the signal produced by the voltage-controlled oscillator. The response of the phase detector is a voltage proportional to phase difference. The loop filter filters that voltage. Then the loop filter response controls the frequency of the voltage-controlled oscillator. When the excitation is at a constant frequency and the loop is "locked" the phase difference between the two phase-detector excitation signals is zero. (In an actual phase detector the phase difference is near 90° at lock. But that is not significant in this analysis since that only causes a near 90° phase shift and has no impact on system performance or stability.) As the frequency of the excitation, $x(t)$, varies, the loop detects the accompanying phase variation and tracks it. The overall response signal, y(t), is a signal proportional to the frequency of the excitation.

The actual excitation, in a system sense, of this system is not x(t), but rather *the phase of* x(t), $\phi_x(t)$, because the phase detector detects differences in phase, not voltage. Let the frequency of x(t) be $f_x(t)$. The relation between phase and frequency can be seen by examining a sinusoid. Let $x(t) = A \cos(2\pi f_0 t)$. The phase of this cosine is $2\pi f_0 t$ and, for a simple sinusoid (f_0 constant), it increases linearly with time. The frequency is f_0, the derivative of the phase. Therefore the relation between phase and frequency for a frequency-modulated signal is

$$f_x(t) = \frac{1}{2\pi} \frac{d}{dt}(\phi_x(t)).$$

Let the frequency of the excitation be 100 MHz. Let the transfer function of the voltage-controlled oscillator be $10^8 \frac{\text{Hz}}{\text{V}}$. Let the transfer function of the loop filter be

$$H_{LF}(s) = \frac{1}{s + 1.2 \times 10^5}.$$

Let the transfer function of the phase detector be $1 \frac{\text{V}}{\text{radian}}$. If the frequency of the excitation signal suddenly changes to 100.001 MHz, graph the change in the output signal $\Delta y(t)$.

28. The circuit in Figure E.28 is a simple approximate model of an operational amplifier with the inverting input grounded.

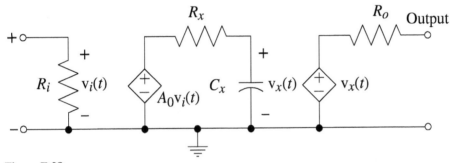

Figure E.28

$$R_i = 1\,\text{M}\Omega, \quad R_x = 1\,\text{k}\Omega, \quad C_x = 8\,\mu\text{F}, \quad R_o = 10\,\Omega, \quad A_o = 10^6$$

(a) Define the excitation of the circuit as the current of a current source applied to the noninverting input and define the response as the voltage developed between the noninverting input and ground. Find the transfer function and graph its frequency response. This transfer function is the input impedance.

(b) Define the excitation of the circuit as the current of a current source applied to the *output* and define the response as the voltage developed between the *output* and ground with the noninverting input grounded. Find the transfer function and graph its frequency response. This transfer function is the output impedance.

(c) Define the excitation of the circuit as the voltage of a voltage source applied to the noninverting input and define the response as the voltage developed between the output and ground. Find the transfer function and graph its frequency response. This transfer function is the voltage gain.

29. Change the circuit of Exercise 28 to the circuit in Figure E.29. This is a feedback circuit which establishes a positive closed-loop voltage gain of the overall amplifier. Repeat steps (a), (b) and (c) of Exercise 28 for the feedback circuit and compare the results. What are the important effects of feedback for this circuit?

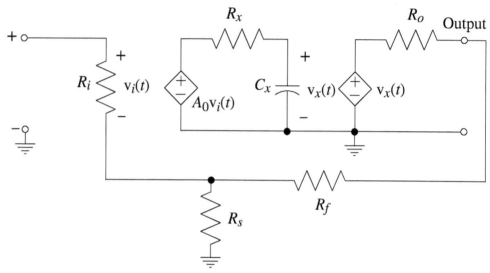

Figure E.29

$R_i = 1\,\text{M}\Omega$, $R_x = 1\,\text{k}\Omega$, $C_x = 8\,\mu\text{F}$, $R_o = 10\,\Omega$, $A_0 = 10^6$, $R_f = 10\,\text{k}\Omega$, $R_s = 5\,\text{k}\Omega$

Root Locus

30. A feedback system has a forward-path transfer function $H_1(s) = \dfrac{K(s-3)}{s+6}$ and a feedback-path transfer function $H_2(s) = \dfrac{s+10}{s^2+2s+4}$. Is there a finite, positive value of K that makes this system unstable? Explain how you know.

31. Each pole-zero diagram in Figure E.31 represents the poles and zeros of the loop transfer function ($T(s)$) of a feedback system. Sketch a root locus on each diagram. Two of these systems will become unstable at a finite, nonzero value of the gain factor "K" in the loop transfer function. Which ones? Three of these systems are marginally stable for an infinite K. Which ones?

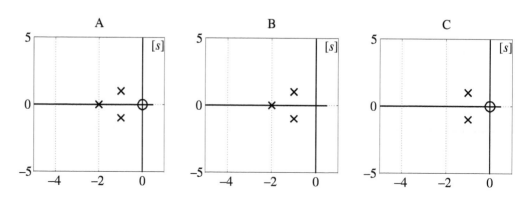

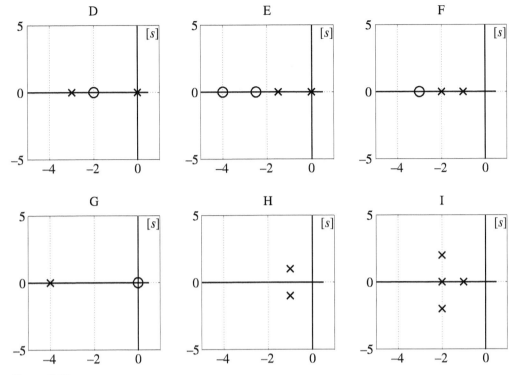

Figure E.31

Tracking Errors in Unity-Gain Feedback Systems

32. A unity-gain feedback system with overall transfer function H(s) has a forward-path transfer function $H_1(s)$. For each of the following forward-path transfer functions, assuming the overall system is stable, determine whether the responses to a unit-step and unit-ramp excitation of the overall unity-gain feedback system have zero, finite or infinite steady-state error.

   (a) $H_1(s) = \dfrac{10s}{s + 10}$

   (b) $H_1(s) = \dfrac{-7s}{(s + 4)(s + 12)}$

   (c) $H_1(s) = \dfrac{(s + 5)(s + 8)}{s^2(s + 1)(s + 25)}$

   (d) $H_1(s) = \dfrac{1}{s(s + 11)(s + 32)}$

Response to Standard Signals

33. A system with one finite pole in the left half of the s-plane and no finite zeros has a step response which approaches its final value on a 20-ms time constant. The pole is then shifted to a more negative real value on the σ axis. What happens to the time constant?

34. Given an LTI system transfer function H(s) find the time-domain response y(t) to the excitation x(t).

   (a) $x(t) = \sin(2\pi t)u(t)$, $H(s) = \dfrac{1}{s + 1}$

   (b) $x(t) = u(t)$, $H(s) = \dfrac{3}{s + 2}$

(c) $x(t) = u(t)$, $H(s) = \dfrac{3s}{s+2}$

(d) $x(t) = u(t)$, $H(s) = \dfrac{5s}{s^2 + 2s + 2}$

(e) $x(t) = \sin(2\pi t)u(t)$, $H(s) = \dfrac{5s}{s^2 + 2s + 2}$

35. Two systems A and B in Figure E.35 have the two pole-zero diagrams below. Which of them responds more quickly to a unit-step excitation (approaches the final value at a faster rate)? Explain your answer.

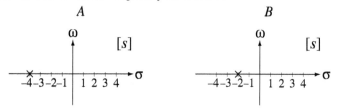

Figure E.35

36. Two systems A and B in Figure E.36 have the two pole-zero diagrams below. Which of them has a unit-step response that overshoots the final value before settling to the final value? Explain your answer.

Figure E.36

37. A second-order system is excited by a unit step and the response is as illustrated in Figure E.37. Write an expression for the transfer function of the system.

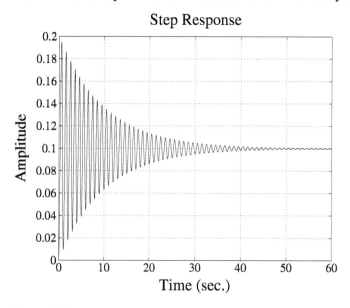

Figure E.37

38. The excitation signal, $x(t) = 20\cos(40\pi t)u(t)$, is applied to a system whose transfer function is $H(s) = \dfrac{5}{s+150}$. The response contains a transient term and a steady-state term. After the transient term has died away what is the amplitude $A_{response}$ of the response and what is the phase difference between the excitation and response signals, $\Delta\theta = \theta_{excitation} - \theta_{response}$?

System Realization

39. Draw cascade system diagrams of the systems with these transfer functions.

    (a) $H(s) = -50\dfrac{s^2}{s^3 + 8s^2 + 13s + 40}$

    (b) $H(s) = \dfrac{s^3}{s^3 + 18s^2 + 92s + 120}$

40. Draw parallel system diagrams of the systems with these transfer functions.

    (a) $H(s) = 10\dfrac{s^3}{s^3 + 4s^2 + 9s + 3}$

    (b) $H(s) = \dfrac{5}{6s^3 + 77s^2 + 228s + 189}$

13 CHAPTER

z-Transform System Analysis

13.1 INTRODUCTION AND GOALS

This chapter follows a path similar to that of Chapter 12 on system analysis using the Laplace transform, except as applied to discrete-time signals and systems instead of continuous-time signals and systems.

CHAPTER GOALS

1. To appreciate the relationship between the z and Laplace transforms
2. To apply the z transform to the generalized analysis of LTI systems, including feedback systems, for stability and time-domain response to standard signals
3. To develop techniques for realizing discrete-time systems in different forms

13.2 SYSTEM MODELS

DIFFERENCE EQUATIONS

The real power of the Laplace transform is in the analysis of the dynamic behavior of continuous-time systems. In an analogous manner, the real power of the z transform is in the analysis of the dynamic behavior of discrete-time systems. Most continuous-time systems analyzed by engineers are described by differential equations and most discrete-time systems are described by difference equations. The general form of a difference equation describing a discrete-time system with an excitation $x[n]$ and a response $y[n]$ is

$$\sum_{k=0}^{N} a_k y[n-k] = \sum_{k=0}^{M} b_k x[n-k].$$

If both $x[n]$ and $y[n]$ are causal, and we z transform both sides, we get

$$\sum_{k=0}^{N} a_k z^{-k} Y(z) = \sum_{k=0}^{M} b_k z^{-k} X(z).$$

The transfer function H(z) is the ratio of Y(z) to X(z)

$$H(z) = \frac{Y(z)}{X(z)} = \frac{\sum_{k=0}^{M} b_k z^{-k}}{\sum_{k=0}^{N} a_k z^{-k}} = \frac{b_0 + b_1 z^{-1} + \cdots + b_M z^{-M}}{a_0 + a_1 z^{-1} + \cdots + a_N z^{-N}}$$

or

$$H(z) = z^{N-M} \frac{b_0 z^M + b_1 z^{M-1} + \cdots + b_{M-1} z + b_M}{a_0 z^N + a_1 z^{N-1} + \cdots + a_{N-1} z + a_N}.$$

So the transfer function of a discrete-time system described by a difference equation is a ratio of polynomials in z just as the transfer function of a continuous-time system described by a differential equation is a ratio of polynomials in s.

BLOCK DIAGRAMS

Discrete-time systems are conveniently modeled by block diagrams just as continuous-time systems are and transfer functions can be written directly from block diagrams. Consider the system in Figure 13.1.

The describing difference equation is y[n] = 2x[n] − x[n − 1] − (1/2)y[n − 1]. We can redraw the block diagram to make it a z-domain block diagram instead of a time-domain block diagram (Figure 13.2). In the z domain the describing equation is Y(z) = 2X(z) − z^{-1}X(z) − (1/2)z^{-1}Y(z) and the transfer function is

$$H(z) = \frac{Y(z)}{X(z)} = \frac{2 - z^{-1}}{1 + (1/2)z^{-1}} = \frac{2z - 1}{z + 1/2}.$$

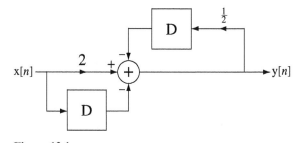

Figure 13.1
Time-domain block diagram of a system

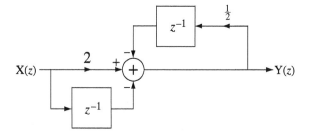

Figure 13.2
z-domain block diagram of a system

13.3 SYSTEM STABILITY

A causal discrete-time system is BIBO stable if its impulse response is absolutely summable, that is, if the sum of the magnitudes of the impulses in the impulse response is finite. For a system whose transfer function is a ratio of polynomials in z of the form

$$H(z) = \frac{b_0 z^M + b_1 z^{M-1} + \cdots + b_M}{a_0 z^N + a_1 z^{N-1} + \cdots + a_N},$$

with $M < N$ and all distinct poles, the transfer function can be written in the partial fraction form

$$H(z) = \frac{K_1}{z - p_1} + \frac{K_2}{z - p_2} + \cdots + \frac{K_N}{z - p_N}$$

and the impulse response is then of the form

$$h[n] = (K_1 p_1^{n-1} + K_2 p_2^{n-1} + \cdots + K_N p_N^{n-1})u[n-1],$$

(some of the p's may be complex). For the system to be stable each term must be absolutely summable. The summation of the absolute value of a typical term is

$$\sum_{n=-\infty}^{\infty} |Kp^{n-1}u[n-1]| = |K|\sum_{n=1}^{\infty}|p^{n-1}| = |K|\sum_{n=0}^{\infty}|p^n(e^{j\angle p})^n| = |K|\sum_{n=0}^{\infty}|p|^n \underbrace{|e^{jn\angle p}|}_{=1}$$

$$\sum_{n=-\infty}^{\infty} |Kp^{n-1}u[n-1]| = |K|\sum_{n=0}^{\infty}|p|^n.$$

Convergence of this last summation requires that $|p| < 1$. Therefore for stability, all the poles must satisfy the condition $|p_k| < 1$.

> In a discrete-time system all the poles of the transfer function must lie in the open interior of the unit circle in the z plane for system stability.

This is directly analogous to the requirement in continuous-time systems that all the poles lie in the open left half of the s-plane for system stability. This analysis was done for the most common case in which all the poles are distinct. If there are repeated poles, it can be shown that the requirement that all the poles lie in the open interior of the unit circle for system stability is unchanged.

13.4 SYSTEM CONNECTIONS

The transfer functions of components in the cascade, parallel and feedback connections of discrete-time systems combine in the same way they do in continuous-time systems (Figure 13.3 through Figure 13.5).

We find the overall transfer function of a feedback system by the same technique used for continuous-time systems and the result is

$$\boxed{H(z) = \frac{Y(z)}{X(z)} = \frac{H_1(z)}{1 + H_1(z)H_2(z)} = \frac{H_1(z)}{1 + T(z)}}, \qquad (13.1)$$

where $T(z) = H_1(z)H_2(z)$ is the loop transfer function.

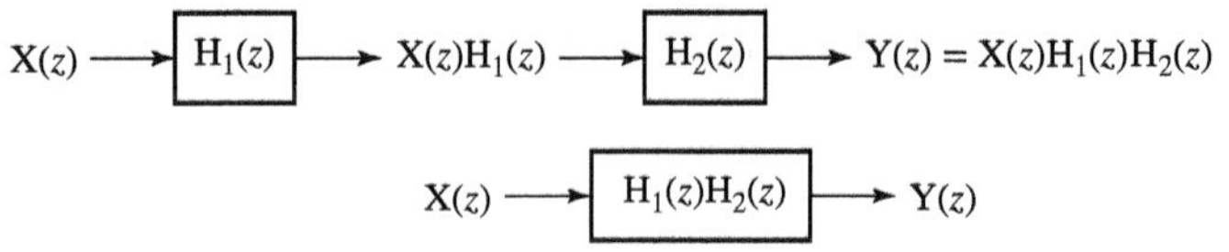

X(z) → [H₁(z)] → X(z)H₁(z) → [H₂(z)] → Y(z) = X(z)H₁(z)H₂(z)

X(z) → [H₁(z)H₂(z)] → Y(z)

Figure 13.3
Cascade connection of systems

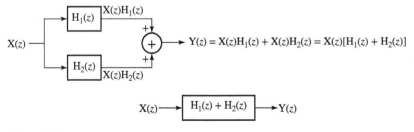

Figure 13.4
Parallel connection of systems

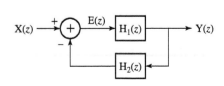

Figure 13.5
Feedback connection of systems

Just as was true for continuous-time feedback systems, a root locus can be drawn for a discrete-time feedback system for which

$$H_1(z) = K\frac{P_1(z)}{Q_1(z)} \text{ and } H_2(z) = \frac{P_2(z)}{Q_2(z)}.$$

The procedure for drawing the root locus is exactly the same as for continuous-time systems except that the loop transfer function

$$T(z) = H_1(z)H_2(z)$$

is a function of z instead of s. However, the interpretation of the root locus, after it is drawn, is a little different. For continuous-time systems, the forward-path gain K at which the root locus crosses into the right half-plane is the value at which the system becomes unstable. For discrete-time systems, the statement is the same except that "right half-plane" is replaced with "exterior of the unit circle."

EXAMPLE 13.1

Discrete-time system stability analysis using root locus

Draw a root locus for the discrete-time system whose forward-path transfer function is

$$H_1(z) = K\frac{z-1}{z+1/2}$$

and whose feedback-path transfer function is

$$H_2(z) = \frac{z-2/3}{z+1/3}.$$

The loop transfer function is

$$T(z) = K\frac{z-1}{z+1/2}\frac{z-2/3}{z+1/3}.$$

There are two zeros, at $z = 2/3$ and $z = 1$ and two poles at $z = -1/2$ and $z = -1/3$. It is apparent from the root locus (Figure 13.6) that this system is unconditionally stable for any finite positive K.

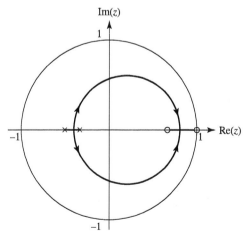

Figure 13.6
Root locus of $T(z) = K\dfrac{z-1}{z+1/2}\dfrac{z-2/3}{z+1/3}$

13.5 SYSTEM RESPONSES TO STANDARD SIGNALS

As indicated in Chapter 12, it is impractical to find the impulse response of a continuous-time system by actually applying an impulse to the system. In contrast, the discrete-time impulse is a simple well-behaved function and can be applied in a practical situation with no real problems. In addition to finding impulse response, finding the responses of systems to the unit sequence and to a sinusoid applied to the system at time $n = 0$ are also good ways of testing system dynamic performance.

UNIT-SEQUENCE RESPONSE

Let the transfer function of a system be

$$H(z) = \frac{N_H(z)}{D_H(z)}.$$

Then the unit-sequence response of the system in the z domain is

$$Y(z) = \frac{z}{z-1} \frac{N_H(z)}{D_H(z)}.$$

The unit-sequence response can be written in the partial-fraction form

$$Y(z) = z\left[\frac{N_{H1}(z)}{D_H(z)} + \frac{H(1)}{z-1}\right] = z\frac{N_{H1}(z)}{D_H(z)} + H(1)\frac{z}{z-1}.$$

If the system is stable and causal, the inverse z transform of the term $zN_{H1}(z)/D_H(z)$ is a signal that decays with time (the transient response) and the inverse z transform of the term $H(1)z/(z-1)$ is a unit sequence multiplied by the value of the transfer function at $z = 1$ (the forced response).

EXAMPLE 13.2

Unit-sequence response using the z transform

A system has a transfer function

$$H(z) = \frac{100z}{z - 1/2}.$$

Find and graph the unit-sequence response.

In the z domain the unit-sequence response is

$$H_{-1}(z) = \frac{z}{z-1}\frac{100z}{z-1/2} = z\left[\frac{-100}{z-1/2} + \frac{200}{z-1}\right] = 100\left[\frac{2z}{z-1} - \frac{z}{z-1/2}\right].$$

The time-domain, unit-sequence response is the inverse z transform which is

$$h_{-1}[n] = 100[2 - (1/2)^n]u[n]$$

(Figure 13.7).

The final value that the unit-sequence response approaches is 200, and that is the same as H(1).

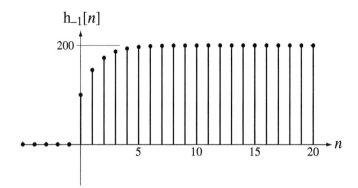

Figure 13.7
Unit-sequence response

In signal and system analysis, the two most commonly encountered systems are one-pole and two-pole systems. The typical transfer function of a one-pole system is of the form

$$H(z) = \frac{Kz}{z - p}$$

where p is the location of a real pole in the z plane. Its z-domain response to a unit-sequence is

$$H_{-1}(z) = \frac{z}{z-1}\frac{Kz}{z-p} = \frac{K}{1-p}\left(\frac{z}{z-1} - \frac{pz}{z-p}\right)$$

and its time-domain response is

$$h_{-1}[n] = \frac{K}{1-p}(1 - p^{n+1})u[n].$$

To simplify this expression and isolate effects, let the gain constant K be $1 - p$. Then

$$h_{-1}[n] = (1 - p^{n+1})u[n].$$

The forced response is $u[n]$ and the transient response is $-p^{n+1}u[n]$.

This is the discrete-time counterpart of the classic unit-step response of a one-pole continuous-time system, and the speed of the response is determined by the pole location. For $0 < p < 1$, the system is stable and the closer p is to 1, the slower the response is (Figure 13.8). For $p > 1$, the system is unstable.

A typical transfer function for a second-order system is of the form

$$H(z) = K\frac{z^2}{z^2 - 2r_0\cos(\Omega_0)z + r_0^2}.$$

The poles of $H(z)$ lie at $p_{1,2} = r_0 e^{\pm j\Omega_0}$. If $r_0 < 1$, both poles lie inside the unit circle and the system is stable. The z transform of the unit-sequence response is

$$H_{-1}(z) = K\frac{z}{z-1}\frac{z^2}{z^2 - 2r_0\cos(\Omega_0)z + r_0^2}.$$

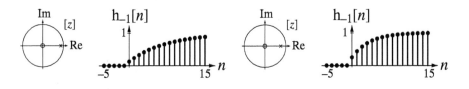
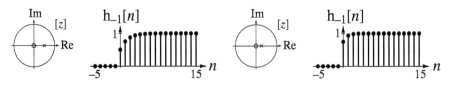

Figure 13.8
Response of a one-pole system to a unit sequence as the pole location changes

For $\Omega_0 \neq \pm m\pi$, m an integer, the partial fraction expansion of $H_{-1}(z)/Kz$ is

$$\frac{H_{-1}(z)}{Kz} = \frac{1}{1 - 2r_0\cos(\Omega_0) + r_0^2}\left[\frac{1}{z-1} + \frac{(r_0^2 - 2r_0\cos(\Omega_0))z + r_0^2}{z^2 - 2r_0\cos(\Omega_0)z + r_0^2}\right].$$

Then

$$H_{-1}(z) = \frac{Kz}{1 - 2r_0\cos(\Omega_0) + r_0^2}\left[\frac{1}{z-1} + \frac{(r_0^2 - 2r_0\cos(\Omega_0))z + r_0^2}{z^2 - 2r_0\cos(\Omega_0)z + r_0^2}\right]$$

or

$$H_{-1}(z) = H(1)\left[\frac{z}{z-1} + z\frac{(r_0^2 - 2r_0\cos(\Omega_0))z + r_0^2}{z^2 - 2r_0\cos(\Omega_0)z + r_0^2}\right]$$

which can be written as

$$H_{-1}(z) = H(1)\left(\frac{z}{z-1} + r_0\left\{[r_0 - 2\cos(\Omega_0)]\frac{z^2 - r_0\cos(\Omega_0)z}{z^2 - 2r_0\cos(\Omega_0)z + r_0^2} + \frac{1 + [r_0 - 2\cos(\Omega_0)]\cos(\Omega_0)}{\sin(\Omega_0)}\frac{zr_0\sin(\Omega_0)}{z^2 - 2r_0\cos(\Omega_0)z + r_0^2}\right\}\right).$$

The inverse z transform is

$$h_{-1}[n] = H(1)\left(1 + r_0\left\{[r_0 - 2\cos(\Omega_0)]r_0^n\cos(n\Omega_0) + \frac{1 + [r_0 - 2\cos(\Omega_0)]\cos(\Omega_0)}{\sin(\Omega_0)}r_0^n\sin(n\Omega_0)\right\}\right)u[n].$$

This is the general solution for the unit-sequence response of this kind of second-order system. If we let $K = 1 - 2r_0\cos(\Omega_0) + r_0^2$, then the system has unity gain ($H(1) = 1$).

EXAMPLE 13.3

Pole-zero diagrams and unit-sequence response using the z transform

A system has a transfer function of the form

$$H(z) = K\frac{z^2}{z^2 - 2r_0\cos(\Omega_0)z + r_0^2} \text{ with } K = 1 - 2r_0\cos(\Omega_0) + r_0^2.$$

Plot the pole-zero diagrams and graph the unit-sequence responses for

(a) $r_0 = 1/2$, $\Omega_0 = \pi/6$,   (b) $r_0 = 1/2$, $\Omega_0 = \pi/3$,
(c) $r_0 = 3/4$, $\Omega_0 = \pi/6$ and   (d) $r_0 = 3/4$, $\Omega_0 = \pi/3$.

Figure 13.9 shows the pole-zero diagrams and unit-sequence responses for the values of r_0 and Ω_0 given above.

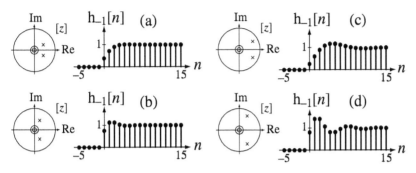

Figure 13.9
Pole-zero diagrams and unit-sequence responses of a unity-gain, second-order system for four combinations of r_0 and Ω_0

As r_0 is increased the response becomes more underdamped, ringing for a longer time. As Ω_0 is increased the speed of the ringing is increased. So we can generalize by saying that poles near the unit circle cause a more underdamped response than poles farther away from (and inside) the unit circle. We can also say that the rate of ringing of the response depends on the angle of the poles, being greater for a greater angle.

■

RESPONSE TO A CAUSAL SINUSOID

The response of a system to a unit-amplitude cosine of radian frequency Ω_0 applied to the system at time $n = 0$ is

$$Y(z) = \frac{N_H(z)}{D_H(z)} \frac{z[z - \cos(\Omega_0)]}{z^2 - 2z\cos(\Omega_0) + 1}.$$

The poles of this response are the poles of the transfer function plus the roots of $z^2 - 2z\cos(\Omega_0) + 1 = 0$, which are the complex conjugate pair $p_1 = e^{j\Omega_0}$ and $p_2 = e^{-j\Omega_0}$. Therefore $p_1 = p_2^*$, $p_1 + p_2 = 2\cos(\Omega_0)$, $p_1 - p_2 = j2\sin(\Omega_0)$ and $p_1 p_2 = 1$. Then if $\Omega_0 \neq m\pi$, m an integer and, if there is no pole-zero cancellation, these poles are distinct and the response can be written in partial-fraction form as

$$Y(z) = z\left[\frac{N_{H1}(z)}{D_H(z)} + \frac{1}{p_1 - p_2}\frac{H(p_1)(p_1 - \cos(\Omega_0))}{z - p_1} + \frac{1}{p_2 - p_1}\frac{H(p_2)(p_2 - \cos(\Omega_0))}{z - p_2}\right]$$

or, after simplification,

$$Y(z) = z\left[\left\{\frac{N_{H1}(z)}{D_H(z)} + \left[\frac{H_r(p_1)(z - p_{1r}) - H_i(p_1)p_{1i}}{z^2 - z(2p_{1r}) + 1}\right]\right\}\right]$$

where $p_1 = p_{1r} + jp_{1i}$ and $H(p_1) = H_r(p_1) + jH_i(p_1)$. This can be written in terms of the original parameters as

$$Y(z) = \left\{ z\frac{N_{H1}(z)}{D_H(z)} + \left[\begin{array}{l} \text{Re}(H(\cos(\Omega_0) + j\sin(\Omega_0)))\dfrac{z^2 - z\cos(\Omega_0)}{z^2 - z(2\cos(\Omega_0)) + 1} \\ -\text{Im}(H(\cos(\Omega_0) + j\sin(\Omega_0)))\dfrac{z\sin(\Omega_0)}{z^2 - z(2\cos(\Omega_0)) + 1} \end{array} \right] \right\}.$$

The inverse z transform is

$$y[n] = Z^{-1}\left(z\frac{N_{H1}(z)}{D_H(z)}\right) + \left[\begin{array}{l}\text{Re}(H(\cos(\Omega_0) + j\sin(\Omega_0)))\cos(\Omega_0 n) \\ -\text{Im}(H(\cos(\Omega_0) + j\sin(\Omega_0)))\sin(\Omega_0 n)\end{array}\right]u[n]$$

or, using

$$\text{Re}(A)\cos(\Omega_0 n) - \text{Im}(A)\sin(\Omega_0 n) = |A|\cos(\Omega_0 n + \measuredangle A),$$

$$y[n] = Z^{-1}\left(z\frac{N_{H1}(z)}{D_H(z)}\right) + |H(\cos(\Omega_0) + j\sin(\Omega_0))|\cos(\Omega_0 n + \measuredangle H(\cos(\Omega_0) + j\sin(\Omega_0)))u[n]$$

or

$$y[n] = Z^{-1}\left(z\frac{N_{H1}(z)}{D_H(z)}\right) + |H(p_1)|\cos(\Omega_0 n + \measuredangle H(p_1))u[n]. \quad (13.2)$$

If the system is stable, the term

$$Z^{-1}\left(z\frac{N_{H1}(z)}{D_H(z)}\right)$$

(the transient response) decays to zero with discrete time, and the term $|H(p_1)|\cos(\Omega_0 n + \measuredangle H(p_1))u[n]$ (the forced response) is equal to a sinusoid after discrete time $n = 0$ and persists forever.

EXAMPLE 13.4

System response to a causal cosine using the z transform

The system of Example 13.2 has a transfer function

$$H(z) = \frac{100z}{z - 1/2}.$$

Find and graph the response to $x[n] = \cos(\Omega_0 n)u[n]$ with $\Omega_0 = \pi/4$.

In the z domain the response is of the form

$$Y(z) = \frac{Kz}{z - p}\frac{z[z - \cos(\Omega_0)]}{z^2 - 2z\cos(\Omega_0) + 1} = \frac{Kz}{z - p}\frac{z[z - \cos(\Omega_0)]}{(z - e^{j\Omega_0})(z - e^{-j\Omega_0})}$$

where $K = 100$, $p = 1/2$ and $\Omega_0 = \pi/4$. This response can be written in the partial-fraction form,

$$Y(z) = Kz\left[\underbrace{\frac{\frac{p[p - \cos(\Omega_0)]}{(p - e^{j\Omega_0})(p - e^{-j\Omega_0})}}{z - p}}_{\text{transient response}} + \underbrace{\frac{Az + B}{z^2 - 2z\cos(\Omega_0) + 1}}_{\text{forced response}}\right]$$

Using (13.2),

$$y[n] = Z^{-1}\left(100z \frac{(1/2)[1/2 - \cos(\pi/4)]}{(1/2 - e^{j\pi/4})(1/2 - e^{-j\pi/4})}\right) + \left|\frac{100e^{j\pi/4}}{e^{j\pi/4} - 1/2}\right|\cos\left(\Omega_0 n + \angle\frac{100e^{j\pi/4}}{e^{j\pi/4} - 1/2}\right)u[n]$$

$$y[n] = [-19.07(1/2)^n + 135.72\cos(\pi n/4 - 0.5)]u[n] \qquad (13.3)$$

(Figure 13.10).

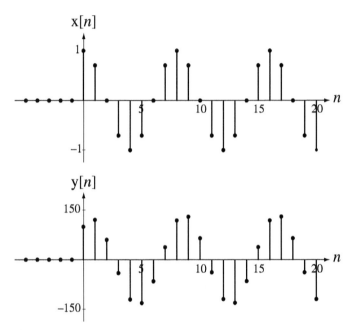

Figure 13.10
Causal cosine and system response

For comparison let's find the system response to a true cosine (applied at time $n \to -\infty$) using the discrete-time Fourier transform (DTFT). The transfer function, expressed as a function of radian frequency Ω using the relationship $z = e^{j\Omega}$, is

$$H(e^{j\Omega}) = \frac{100\,e^{j\Omega}}{e^{j\Omega} - 1/2}.$$

The DTFT of $x[n]$ is

$$X(e^{j\Omega}) = \pi[\delta_{2\pi}(\Omega - \Omega_0) + \delta_{2\pi}(\Omega + \Omega_0)].$$

Therefore the response is

$$Y(e^{j\Omega}) = \pi[\delta_{2\pi}(\Omega - \Omega_0) + \delta_{2\pi}(\Omega + \Omega_0)]\frac{100\,e^{j\Omega}}{e^{j\Omega} - 1/2}$$

or

$$Y(e^{j\Omega}) = 100\pi\left[\sum_{k=-\infty}^{\infty}\frac{e^{j\Omega}}{e^{j\Omega} - 1/2}\delta(\Omega - \Omega_0 - 2\pi k) + \sum_{k=-\infty}^{\infty}\frac{e^{j\Omega}}{e^{j\Omega} - 1/2}\delta(\Omega + \Omega_0 - 2\pi k)\right].$$

Using the equivalence property of the impulse,

$$Y(e^{j\Omega}) = 100\pi \sum_{k=-\infty}^{\infty}\left[\frac{e^{j(\Omega_0+2\pi k)}}{e^{j(\Omega_0+2\pi k)}-1/2}\delta(\Omega-\Omega_0-2\pi k) + \frac{e^{j(-\Omega_0+2\pi k)}}{e^{j(-\Omega_0+2\pi k)}-1/2}\delta(\Omega+\Omega_0-2\pi k)\right].$$

Since $e^{j(\Omega_0+2\pi k)} = e^{j\Omega_0}$ and $e^{j(-\Omega_0+2\pi k)} = e^{-j\Omega_0}$ for integer values of k,

$$Y(e^{j\Omega}) = 100\pi \sum_{k=-\infty}^{\infty}\left[\frac{e^{j\Omega_0}\delta(\Omega-\Omega_0-2\pi k)}{e^{j\Omega_0}-1/2} + \frac{e^{-j\Omega_0}\delta(\Omega+\Omega_0-2\pi k)}{e^{-j\Omega_0}-1/2}\right]$$

or

$$Y(e^{j\Omega}) = 100\pi \left[\frac{e^{j\Omega_0}\delta_{2\pi}(\Omega-\Omega_0)}{e^{j\Omega_0}-1/2} + \frac{e^{-j\Omega_0}\delta_{2\pi}(\Omega+\Omega_0)}{e^{-j\Omega_0}-1/2}\right].$$

Finding a common denominator, applying Euler's identity and simplifying,

$$Y(e^{j\Omega}) = \frac{100\pi}{5/4-\cos(\Omega_0)}\left\{\begin{array}{l}(1-(1/2)\cos(\Omega_0))[\delta_{2\pi}(\Omega-\Omega_0)+\delta_{2\pi}(\Omega+\Omega_0)]\\ +(j/2)\sin(\Omega_0)[\delta_{2\pi}(\Omega+\Omega_0)-\delta_{2\pi}(\Omega-\Omega_0)]\end{array}\right\}.$$

Finding the inverse DTFT,

$$y[n] = \frac{50}{5/4-\cos(\Omega_0)}\{[1-(1/2)\cos(\Omega_0)]2\cos(\Omega_0 n) + \sin(\Omega_0)\sin(\Omega_0 n)\}$$

or, since $\Omega_0 = \pi/4$,

$$y[n] = 119.06\cos(\pi n/4) + 65.113\sin(\pi n/4) = 135.72\cos(\pi n/4 - 0.5).$$

This is exactly the same (except for the unit sequence u[n]) as the forced response in (13.3). ∎

13.6 SIMULATING CONTINUOUS-TIME SYSTEMS WITH DISCRETE-TIME SYSTEMS

z-TRANSFORM-LAPLACE-TRANSFORM RELATIONSHIPS

We explored in earlier chapters important relationships between Fourier transform methods. In particular we showed that there is an information equivalence between a discrete-time signal $x[n] = x(nT_s)$ formed by sampling a continuous-time signal and a continuous-time impulse signal $x_\delta(t) = x(t)\delta_{T_s}(t)$ formed by impulse sampling the same continuous-time signal, with $f_s = 1/T_s$. We also derived the relationships between the DTFT of $x[n]$ and the continuous-time Fourier transform (CTFT) of $x_\delta(t)$ in Chapter 10. Since the z transform applies to a discrete-time signal and is a generalization of the DTFT and a Laplace transform applies to a continuous-time signal and is a generalization of the CTFT, we should expect a close relationship between them also.

Consider two systems, a discrete-time system with impulse response $h[n]$ and a continuous-time system with impulse response $h_\delta(t)$ and let them be related by

$$h_\delta(t) = \sum_{n=-\infty}^{\infty} h[n]\delta(t-nT_s). \quad (13.4)$$

This equivalence indicates that everything that happens to $x[n]$ in the discrete-time system, happens in a corresponding way to $x_\delta(t)$ in the continuous-time system

13.6 Simulating Continuous-Time Systems with Discrete-Time Systems

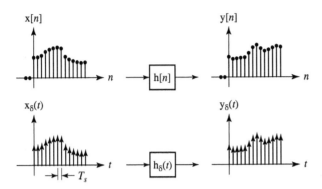

Figure 13.11
Equivalence of a discrete-time and a continuous-time system

(Figure 13.11). Therefore it is possible to analyze discrete-time systems using the Laplace transform with the strengths of continuous-time impulses representing the values of the discrete-time signals at equally spaced points in time. But it is notationally more convenient to use the z transform instead.

The transfer function of the discrete-time system is

$$H(z) = \sum_{n=0}^{\infty} h[n]z^{-n}$$

and the transfer function of the continuous-time system is

$$H_\delta(s) = \sum_{n=0}^{\infty} h[n]e^{-nT_s s}.$$

If the impulse responses are equivalent in the sense of (13.4), then the transfer functions must also be equivalent. The equivalence is seen in the relationship,

$$H_\delta(s) = H(z)|_{z \to e^{sT_s}}.$$

It is important at this point to consider some of the implications of the transformation $z \to e^{sT_s}$. One good way of seeing the relationship between the s and z complex planes is to map a contour or region in the s-plane into a corresponding contour or region in the z plane. Consider first a very simple contour in the s-plane, the contour $s = j\omega = j2\pi f$ with ω and f representing real radian and cyclic frequency, respectively. This contour is the ω axis of the s-plane. The corresponding contour in the z plane is $e^{j\omega T_s}$ or $e^{j2\pi f T_s}$ and, for any real value of ω and f, must lie on the unit circle. However the mapping is not as simple as the last statement makes it sound.

To illustrate the complication, map the segment of the imaginary axis in the s-plane $-\pi/T_s < \omega < \pi/T_s$ that corresponds to $-f_s/2 < f < f_s/2$ into the corresponding contour in the z plane. As ω traverses the contour $-\pi/T_s \to \omega \to \pi/T_s$, z traverses the unit circle from $e^{-j\pi}$ to $e^{+j\pi}$ in the counterclockwise direction, making one complete traversal of the unit circle. Now if we let ω traverse the contour $\pi/T_s \to \omega \to 3\pi/T_s$, z traverses the unit circle from $e^{j\pi}$ to $e^{+j3\pi}$, which is exactly the same contour again because $e^{-j\pi} = e^{j\pi} = e^{j3\pi} = e^{j(2n+1)\pi}$, n any integer. Therefore it is apparent that the transformation $z \to e^{sT_s}$ maps the ω axis of the s-plane into the unit circle of the z plane, *infinitely many times* (Figure 13.12).

This is another way of looking at the phenomenon of aliasing. All those segments of the imaginary axis of the s-plane of length $2\pi/T_s$ look exactly the same when translated into the z plane because of the effects of sampling. So, for every point on the

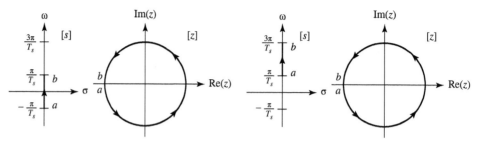

Figure 13.12
Mapping the ω axis of the s-plane into the unit circle of the z plane

imaginary axis of the s-plane there is a corresponding unique point on the unit circle in the z plane. But this unique correspondence does not work the other way. For every point on the unit circle in the z plane there are infinitely many corresponding points on the imaginary axis of the s-plane.

Carrying the mapping idea one step farther, the left half of the s-plane maps into the interior of the unit circle in the z plane and the right half of the s-plane maps into the exterior of the unit circle in the z plane (infinitely many times in both cases). The corresponding ideas about stability and pole locations translate in the same way. A stable continuous-time system has a transfer function with all its poles in the open left half of the s-plane and a stable discrete-time system has a transfer function with all its poles in the open interior of the unit circle in the z plane (Figure 13.13).

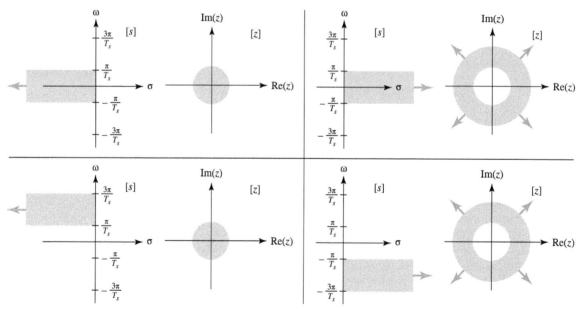

Figure 13.13
Mapping of the regions of the s-plane into regions in the z plane

IMPULSE INVARIANCE

In Chapter 10 we examined how continuous-time signals are converted to discrete-time signals by sampling. We found that, under certain conditions, the discrete-time signal was a good representation of the continuous-time signal in the sense that it preserved

13.6 Simulating Continuous-Time Systems with Discrete-Time Systems

all or practically all of its information. A discrete-time signal formed by properly sampling a continuous-time signal in a sense *simulates* the continuous-time signal. In this chapter we examined the equivalence between a discrete-time system with impulse response h[n] and a continuous-time system with impulse response

$$h_\delta(t) = \sum_{n=-\infty}^{\infty} h[n]\delta(t - nT_s).$$

The system whose impulse response is $h_\delta(t)$ is a very special type of system because its impulse response consists only of impulses. As a practical matter, this is impossible to achieve because the transfer function of such a system, being periodic, has a nonzero response at frequencies approaching infinity. No real continuous-time system can have an impulse response that contains actual impulses, although in some cases that might be a good approximation for analysis purposes.

To simulate a continuous-time system with a discrete-time system we must first address the problem of forming a useful equivalence between a discrete-time system, whose impulse response must be discrete, and a continuous-time system, whose impulse response must be continuous. The most obvious and direct equivalence between a discrete-time signal and a continuous-time signal is to have the values of the continuous-time signal at the sampling instants be the same as the values of the discrete-time signal at the corresponding discrete times $x[n] = x(nT_s)$. So if the excitation of a discrete-time system is a sampled version of a excitation of a continuous-time system, we want the response of the discrete-time system to be a sampled version of the response of the continuous-time system (Figure 13.14).

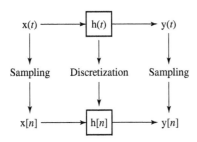

Figure 13.14
Signal sampling and system discretization

The most natural choice for h[n] would be $h[n] = h(nT_s)$. Since h[n] is not actually a signal occurring in this system, but rather a function that characterizes the system, we cannot accurately say that Figure 13.14 indicates a sampling process. We are not sampling a signal. Instead we are **discretizing** a system. The choice of impulse response for the discrete-time system $h[n] = h(nT_s)$ establishes an equivalence between the impulse responses of the two systems. With this choice of impulse response, if a unit continuous-time impulse excites the continuous-time system and a unit discrete-time impulse of the same strength excites the discrete-time system, the response y[n] is a sampled version of the response y(t) and $y[n] = y(nT_s)$. But even though the two systems have equivalent impulse responses in the sense of $h[n] = h(nT_s)$ and $y[n] = y(nT_s)$, that does not mean that the system responses to other excitations will be equivalent in the same sense. A system design for which $h[n] = h(nT_s)$ is called an **impulse invariant** design because of the equivalence of the system responses to unit impulses.

It is important to point out here that if we choose to make $h[n] = h(nT_s)$, and we excite both systems with unit impulses, the responses are related by $y[n] = y(nT_s)$, but we cannot say that $x[n] = x(nT_s)$ as in Figure 13.14. Figure 13.14 indicates that the discrete-time excitation is formed by sampling the continuous-time excitation. But if the continuous-time excitation is an impulse we cannot sample it. Try to imagine sampling a continuous-time impulse. First, if we are sampling at points in time at some finite rate to try to "catch" the impulse when it occurs, the probability of actually seeing the impulse in the samples is zero because it has zero width. Even if we could sample exactly when the impulse occurs we would have to say that $\delta[n] = \delta(nT_s)$ but this makes no sense because the amplitude of a continuous-time impulse at its time of occurrence is not defined (because it is not an ordinary function), so we cannot establish the corresponding strength of the discrete-time impulse $\delta[n]$.

SAMPLED-DATA SYSTEMS

Because of the great increases in microprocessor speed and memory and large reductions in the cost of microprocessors, modern system design often uses discrete-time subsystems to replace subsystems that were traditionally continuous-time subsystems to save money or space or power consumption and to increase the flexibility or reliability of the system. Aircraft autopilots, industrial chemical process control, manufacturing processes, automobile ignition and fuel systems are examples. Systems that contain both discrete-time subsystems and continuous-time subsystems and mechanisms for converting between discrete-time and continuous-time signals are called **sampled-data** systems.

The first type of sampled-data system used to replace a continuous-time system, and still the most prevalent type, comes from a natural idea. We convert a continuous-time signal to a discrete-time signal with an analog-to-digital converter (ADC). We process the samples from the ADC in a discrete-time system. Then we convert the discrete-time response back to continuous-time form using a digital-to-analog converter (DAC) (Figure 13.15).

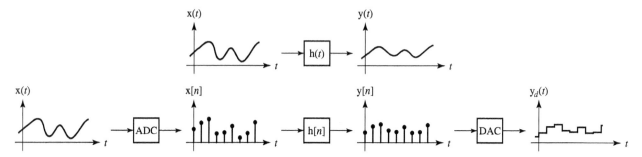

Figure 13.15
A common type of sampled-data simulation of a continuous-time system

The desired design would have the response of the sampled-data system be very close to the response that would have come from the continuous-time system. To do that we must choose h[n] properly and, in order to do that, we must also understand the actions of the ADC and DAC.

It is straightforward to model the action of the ADC. It simply acquires the value of its input signal at the sampling time and responds with a number proportional to that signal value. (It also quantizes the signal, but we will ignore that effect as negligible in this analysis.) The subsystem with impulse response h[n] is then designed to make the sampled-data system emulate the action of the continuous-time system whose impulse response is h(t).

The action of the DAC is a little more complicated to model mathematically than the ADC. It is excited by a number from the discrete-time subsystem, the strength of an impulse, and responds with a continuous-time signal proportional to that number,

13.6 Simulating Continuous-Time Systems with Discrete-Time Systems

which stays constant until the number changes to a new value. This can be modeled by thinking of the process as two steps. First let the discrete-time impulse be converted to a continuous-time impulse of the same strength. Then let the continuous-time impulse excite a zero-order hold (first introduced in Chapter 10) with an impulse response that is rectangular with height one and width T_s beginning at time $t = 0$

$$h_{zoh}(t) = \begin{cases} 0, & t < 0 \\ 1, & 0 < t < T_s \\ 0, & t > T_s \end{cases} = \text{rect}\left(\frac{t - T_s/2}{T_s}\right)$$

(Figure 13.16).

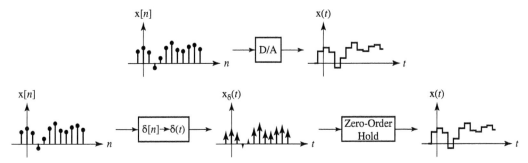

Figure 13.16
Equivalence of a DAC and a discrete-time-to-continuous-time impulse conversion followed by a zero-order hold

The transfer function of the zero-order hold is the Laplace transform of its impulse response $h_{zoh}(t)$, which is

$$H_{zoh}(s) = \int_{0^-}^{\infty} h_{zoh}(t)e^{-st}\,dt = \int_{0^-}^{T_s} e^{-st}\,dt = \left[\frac{e^{-st}}{-s}\right]_{0^-}^{T_s} = \frac{1 - e^{-sT_s}}{s}.$$

The next design task is to make $h[n]$ emulate the action of $h(t)$ in the sense that the overall system responses will be as close as possible. The continuous-time system is excited by a signal $x(t)$ and produces a response $y_c(t)$. We would like to design the corresponding sampled-data system such that if we convert $x(t)$ to a discrete-time signal $x[n] = x(nT_s)$ with an ADC, process that with a system to produce the response $y[n]$, then convert that to a response $y_d(t)$ with a DAC, then $y_d(t) = y_c(t)$ (Figure 13.17).

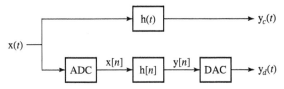

Figure 13.17
Desired equivalence of continuous-time and sampled-data systems

This cannot be accomplished exactly (except in the theoretical limit in which the sampling rate approaches infinity). But we can establish conditions under which a good approximation can be made, one that gets better as the sampling rate is increased.

As a step toward determining the impulse response h[n] of the subsystem, first consider the response of the continuous-time system, *not to* x(t), *but rather to* $x_\delta(t)$ defined by

$$x_\delta(t) = \sum_{n=-\infty}^{\infty} x(nT_s)\delta(t - nT_s) = x(t)\delta_{T_s}(t).$$

The response to $x_\delta(t)$ is

$$y(t) = h(t) * x_\delta(t) = h(t) * \sum_{m=-\infty}^{\infty} x(nT_s)\delta(t - mT_s) = \sum_{m=-\infty}^{\infty} x[m]h(t - mT_s)$$

where x[n] is the sampled version of x(t), $x(nT_s)$. The response at the nth multiple of T_s is

$$y(nT_s) = \sum_{m=-\infty}^{\infty} x[m]h((n - m)T_s). \tag{13.5}$$

Compare this to the response of a discrete-time system with impulse response $h[n] = h(nT_s)$ to $x[n] = x(nT_s)$ which is

$$y[n] = x[n] * h[n] = \sum_{m=-\infty}^{\infty} x[m]h[n - m]. \tag{13.6}$$

By comparing (13.5) and (13.6) it is apparent that the response y(t) of a continuous-time system with impulse response h(t) at the sampling instants nT_s to a continuous-time impulse-sampled signal

$$x_\delta(t) = \sum_{n=-\infty}^{\infty} x(nT_s)\delta(t - nT_s)$$

can be found by finding the response of a system with impulse response $h[n] = h(nT_s)$ to $x[n] = x(nT_s)$ and making the equivalence $y(nT_s) = y[n]$ (Figure 13.18).

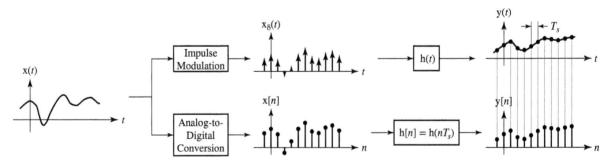

Figure 13.18
Equivalence, at continuous times nT_s and corresponding discrete times n of the responses of continuous-time and discrete-time systems excited by continuous-time and discrete-time signals derived from the same continuous-time signal

Now, returning to our original continuous-time and sampled-data systems, modify the continuous-time system as illustrated in Figure 13.19. Using the equivalence in Figure 13.18, $y[n] = y(nT_s)$.

13.6 Simulating Continuous-Time Systems with Discrete-Time Systems

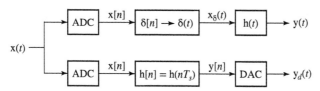

Figure 13.19
Continuous-time and sampled-data systems when the continuous-time system is excited by $x_\delta(t)$ instead of $x(t)$

Now change both the continuous-time system and discrete-time system impulse responses by multiplying them by the time between samples T_s (Figure 13.20). In this modified system we can still say that $y[n] = y(nT_s)$ where now

$$y(t) = x_\delta(t) * T_s h(t) = \left[\sum_{n=-\infty}^{\infty} x(nT_s)\delta(t - nT_s)\right] * h(t)T_s = \sum_{n=-\infty}^{\infty} x(nT_s)h(t - nT_s)T_s, \quad (13.7)$$

$$y[n] = \sum_{m=-\infty}^{\infty} x[m] h[n - m] = \sum_{m=-\infty}^{\infty} x[m] T_s h((n - m)T_s).$$

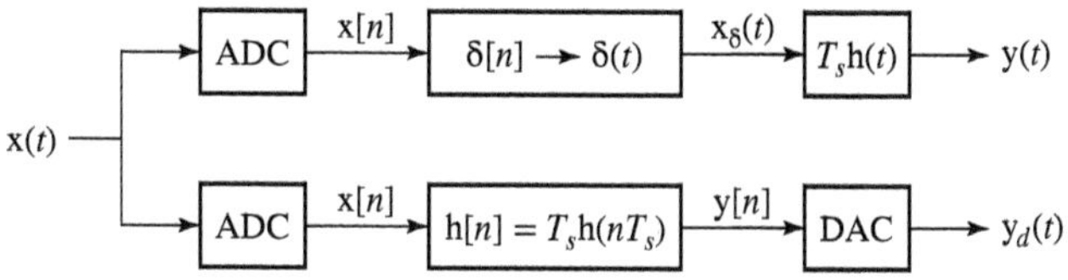

Figure 13.20
Continuous-time and sampled-data systems when their impulse responses are multiplied by the time between samples T_s

The new subsystem impulse response is $h[n] = T_s h(nT_s)$ and $h(t)$ still represents the impulse response of the original continuous-time system. Now in (13.7) let T_s approach zero. In that limit, the summation on the right-hand side becomes the convolution integral first developed in the derivation of convolution in Chapter 5,

$$\lim_{T_s \to 0} y(t) = \lim_{T_s \to 0} \sum_{n=-\infty}^{\infty} x(nT_s)h(t - nT_s) T_s = \int_{-\infty}^{\infty} x(\tau)h(t - \tau) d\tau,$$

which is the signal $y_c(t)$, the response of the original continuous-time system in Figure 13.17 to the signal $x(t)$. Also, in that limit, $y[n] = y_c(nT_s)$. So, in the limit, the spacing between points T_s approaches zero, the sampling instants nT_s merge into a continuum t, and there is a one-to-one correspondence between the signal values $y[n]$ and the signal values $y_c(t)$. The response of the sampled-data system $y_d(t)$ will be indistinguishable from the response $y_c(t)$ of the original system to the signal $x(t)$. Of course, in practice we can never sample at an infinite rate, so the correspondence $y[n] = y_c(nT_s)$ can never be exact, but it does establish an approximate equivalence between a continuous-time and a sampled-data system.

There is another conceptual route to arriving at the same conclusion for the discrete-time-system impulse response $h[n] = T_s h(nT_s)$. In the development above we formed a continuous-time impulse signal

$$x_\delta(t) = \sum_{n=-\infty}^{\infty} x(nT_s)\delta(t - nT_s)$$

whose impulse strengths were equal to samples of the signal x(t). Now, instead, form a modified version of this impulse signal. Let the new correspondence between x(t) and $x_\delta(t)$ be that the strength of an impulse at nT_s is approximately the *area under* x(t) in the sampling interval $nT_s \leq t < (n+1)T_s$ not the *value at* nT_s. The equivalence between x(t) and $x_\delta(t)$ is now based on (approximately) equal areas (Figure 13.21). (The approximation gets better as the sampling rate is increased.)

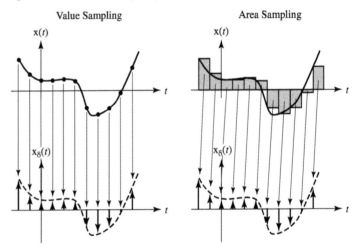

Figure 13.21
A comparison of value sampling and area sampling

The area under x(t) is approximately $T_s x(nT_s)$ in each sampling interval. Therefore the new continuous-time impulse signal would be

$$x_\delta(t) = T_s \sum_{n=-\infty}^{\infty} x(nT_s)\delta(t - nT_s).$$

If we now apply this impulse signal to a system with impulse response h(t) we get exactly the same response as in (13.7)

$$y(t) = \sum_{n=-\infty}^{\infty} x(nT_s)h(t - nT_s) T_s$$

and, of course, the same result that $y[n] = y_c(nT_s)$ in the limit as the sampling rate approaches infinity. All we have done in this development is associate the factor T_s with the excitation instead of with the impulse response. When the two are convolved the result is the same. If we sampled signals setting impulse strengths equal to signal areas over a sampling interval, instead of setting them equal to signal values at sampling instants, then the correspondence $h[n] = h(nT_s)$ would be the design correspondence between a continuous-time system and a sampled-data system that simulates it. But, since we don't sample that way (because most ADC's do not work that way) we instead associate the factor T_s with the impulse response and form the correspondence $h[n] = T_s h(nT_s)$.

EXAMPLE 13.5

Design of a sampled-data system to simulate a continuous-time system

A continuous-time system is characterized by a transfer function

$$H_c(s) = \frac{1}{s^2 + 40s + 300}.$$

Design a sampled-data system of the form of Figure 13.15 to simulate this system. Do the design for two sampling rates $f_s = 10$ and $f_s = 100$ and compare step responses.

The impulse response of the continuous-time system is

$$h_c(t) = (1/20)(e^{-10t} - e^{-30t})u(t).$$

The discrete-time-subsystem impulse response is then

$$h_d[n] = (T_s/20)(e^{-10nT_s} - e^{-30nT_s})u[n]$$

and the corresponding z-domain transfer function is

$$H_d(z) = \frac{T_s}{20}\left(\frac{z}{z - e^{-10T_s}} - \frac{z}{z - e^{-30T_s}}\right).$$

The step response of the continuous-time system is

$$h_{-1c}(t) = \frac{2 - 3e^{-10t} + e^{-30t}}{600} u(t).$$

The response of the subsystem to a unit sequence is

$$h_{-1d}[n] = \frac{T_s}{20}\left[\frac{e^{-10T_s} - e^{-30T_s}}{(1 - e^{-10T_s})(1 - e^{-30T_s})} + \frac{e^{-10T_s}}{e^{-10T_s} - 1}e^{-10nT_s} - \frac{e^{-30T_s}}{e^{-30T_s} - 1}e^{-30nT_s}\right]u[n]$$

and the response of the D/A converter is

$$h_{-1d}(t) = \sum_{n=0}^{\infty} y[n] \text{rect}\left(\frac{t - T_s(n + 1/2)}{T_s}\right)$$

(Figure 13.22).

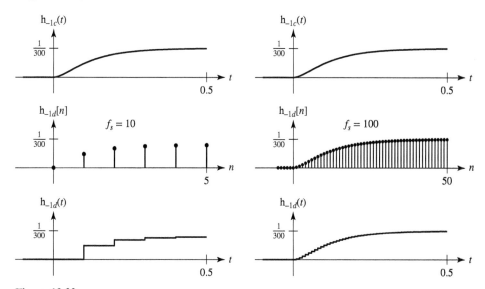

Figure 13.22
Comparison of the step responses of a continuous-time system and two sampled-data systems that simulate it with different sampling rates

For the lower sampling rate the sampled-data system simulation is very poor. It approaches a forced response value that is about 78 percent of the forced response of the continuous-time system. At the higher sampling rate the simulation is much better with a forced response approaching a value that is about 99 percent of the forced response of the continuous-time system. Also, at the higher sampling rate, the difference between the continuous-time response and the sampled-data-system response is much smaller than at the lower sampling rate.

We can see why the disparity between forced values exists by examining the expression,

$$y[n] = \frac{T_s}{20}\left[\frac{e^{-10T_s} - e^{-30T_s}}{(1 - e^{-10T_s})(1 - e^{-30T_s})} + \frac{e^{-10T_s}}{e^{-10T_s} - 1}e^{-10nT_s} - \frac{e^{-30T_s}}{e^{-30T_s} - 1}e^{-30nT_s}\right]u[n].$$

The forced response is

$$y_{forced} = \frac{T_s}{20}\frac{e^{-10T_s} - e^{-30T_s}}{(1 - e^{-10T_s})(1 - e^{-30T_s})}.$$

If we approximate the exponential functions by the first two terms in their series expansions, as $e^{-10T_s} \approx 1 - 10T_s$ and $e^{-30T_s} \approx 1 - 30T_s$ we get $y_{forced} = 1/300$, which is the correct forced response. However, if T_s is not small enough, the approximation of the exponential function by the first two terms of its series expansion is not very good and actual and ideal forced values are significantly different. When $f_s = 10$, we get $e^{-10T_s} = 0.368$ and $1 - 10T_s = 0$ and $e^{-30T_s} = 0.0498$ and $1 - 30T_s = -2$, which are terrible approximations. But when $f_s = 100$ we get $e^{-10T_s} = 0.905$ and $1 - 10T_s = 0.9$ and $e^{-30T_s} = 0.741$ and $1 - 30T_s = 0.7$, which are much better approximations.

■

13.7 STANDARD REALIZATIONS OF SYSTEMS

The realization of discrete-time systems very closely parallels the realization of continuous-time systems. The same general techniques apply and the same types of realizations result.

CASCADE REALIZATION

We can realize a system in cascade form from the factored form of the transfer function

$$H(z) = A\frac{z - z_1}{z - p_1}\frac{z - z_2}{z - p_2} \cdots \frac{z - z_M}{z - p_M}\frac{1}{z - p_{M+1}}\frac{1}{z - p_{M+2}} \cdots \frac{1}{z - p_N}$$

where the numerator order is $M \leq N$ (Figure 13.23).

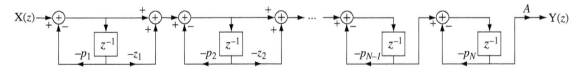

Figure 13.23
Overall cascade system realization

PARALLEL REALIZATION

We can express the transfer function as the sum of partial fractions

$$H(z) = \frac{K_1}{z - p_1} + \frac{K_2}{z - p_2} + \cdots + \frac{K_N}{z - p_N}$$

and realize the system in parallel form (Figure 13.24).

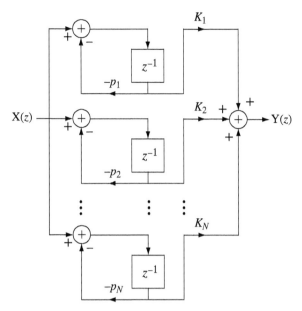

Figure 13.24
Overall parallel system realization

Discrete-time systems are actually built using digital hardware. In these systems the signals are all in the form of binary numbers with a finite number of bits. The operations are usually performed in fixed-point arithmetic. That means all the signals are quantized to a finite number of possible values and therefore are not exact representations of the ideal signals. This type of design usually leads to the fastest and most efficient system, but the round-off error between the ideal signals and the actual signals is an error that must be managed to avoid noisy, or in some cases even unstable, system operation. The analysis of such errors is beyond the scope of this text but, generally speaking, the cascade and parallel realizations are more tolerant and forgiving of such errors than the Direct Form II canonical realization.

13.8 SUMMARY OF IMPORTANT POINTS

1. It is possible to do analysis of discrete-time systems with the Laplace transform through the use of continuous-time impulses to simulate discrete time. But the z transform is notationally more convenient.
2. Discrete-time systems can be modeled by difference equations or block diagrams in the time or frequency domain.
3. A discrete-time LTI system is stable if all the poles of its transfer function lie in the open interior of the unit circle.
4. The three most important types of system interconnections are the cascade connection, the parallel connection and the feedback connection.
5. The unit sequence and sinusoid are important practical signals for testing system characteristics.
6. Discrete-time systems can closely approximate the actions of continuous-time systems and the approximation improves as the sampling rate is increased.
7. The Direct Form II, cascade and parallel realizations are important standard ways of realizing systems.

EXERCISES WITH ANSWERS

(Answers to each exercise are in random order.)

Stability

1. Evaluate the stability of the systems with each of these transfer functions.

   (a) $H(z) = \dfrac{z}{z-2}$

   (b) $H(z) = \dfrac{z}{z^2 - 7/8}$

   (c) $H(z) = \dfrac{z}{z^2 - 3z/2 + 9/8}$

   (d) $H(z) = \dfrac{z^2 - 1}{z^3 - 2z^2 + 3.75z - 0.5625}$

   (e) $H(z) = \dfrac{z-1}{(z^2 + \sqrt{2}z + 1)^2}$

   (f) $H(z) = \dfrac{z+2}{\left(z+\dfrac{1}{2}\right)\left(z - e^{j\frac{\pi}{3}}\right)\left(z - e^{-j\frac{\pi}{3}}\right)}$

   (g) $H(z) = \dfrac{z}{(z - 0.8 + j0.8)(z - 0.8 - j0.8)}$

   Answers: 1 Stable and 6 Unstable

2. Determine whether these systems are stable or unstable and explain how you know.

   (a) A system described by $2y[n] + 3y[n-1] = x[n]$.

   (b) A feedback system with $H_1(z) = \dfrac{0.7z}{z^2 + 0.6z + 0.5}$ and $H_2(z) = z^{-1}$.

   Answers: Both Unstable

3. A discrete-time feedback system has a forward path transfer function $H_1(z) = \dfrac{Kz}{z - 0.5}$ and a feedback path transfer function $H_2(z) = 4z^{-1}$. For what range of values of K is the system stable?

   Answer: $-1/8 < K < 3/8$

Parallel, Cascade and Feedback Connections

4. A feedback discrete-time system has a transfer function,

   $$H(z) = \dfrac{K}{1 + K\dfrac{z}{z - 0.9}}.$$

   For what range of K's is this system stable?

   Answer: $K > -0.1$ or $K < -1.9$

5. Find the overall transfer functions of the systems in Figure E.5 in the form of a single ratio of polynomials in z.

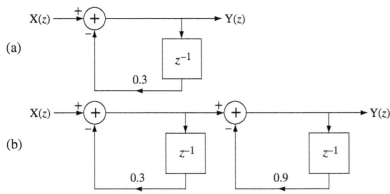

Figure E.5

Answers: $\dfrac{z}{z+0.3}$, $\dfrac{z^2}{z^2+1.2z+0.27}$

Response to Standard Signals

6. Find the time-domain responses y[n] of the systems with these transfer functions to the unit-sequence excitation $x[n] = u[n]$.

   (a) $H(z) = \dfrac{z}{z-1}$

   (b) $H(z) = \dfrac{z-1}{z-1/2}$

   Answers: $(1/2)^n u[n]$, $y[n] = \text{ramp}[n+1]$

7. A discrete-time system has a transfer function, $H(z) = \dfrac{z(z-0.5)}{(z-0.2)(z+0.8)}$.

   (a) What is the final value of its response to a unit-sequence excitation?

   (b) If a discrete-time impulse of strength 8 is applied at time $n = 7$, and that is the only excitation of the system for all time, find the value of the response at time, $n = 12$.

   Answers: $-3.4086, 0.3472$

8. A discrete-time system has a transfer function,

   $$H(z) = \dfrac{0.5z^2}{z^2+1.2z+0.27}.$$

   If a unit sequence u[n] is applied as an excitation to this system, what are the numerical values of the responses y[0], y[1] and y[2]?

   Answers: $0.485, -0.1, 0.5$

9. Find the responses y[n] of the systems with these transfer functions to the excitation $x[n] = \cos(2\pi n/8)\, u[n]$. Then show that the steady-state response is the same as would have been obtained by using DTFT analysis with an excitation $x[n] = \cos(2\pi n/8)$.

   (a) $H(z) = \dfrac{z}{z-0.9}$

   (b) $H(z) = \dfrac{z^2}{z^2-1.6z+0.63}$

   Answers: $1.3644\cos(2\pi n/8 - 1.0518)$, $1.9293\cos(2\pi n/8 - 1.3145)$

Root Locus

10. Draw a root locus for each system with the given forward and feedback path transfer functions.

    (a) $H_1(z) = K \dfrac{z-1}{z+\dfrac{1}{2}}$, $H_2(z) = \dfrac{4z}{z-0.8}$

    (b) $H_1(z) = K \dfrac{z-1}{z+\dfrac{1}{2}}$, $H_2(z) = \dfrac{4}{z-0.8}$

    (c) $H_1(z) = K \dfrac{z}{z-\dfrac{1}{4}}$, $H_2(z) = \dfrac{z+\dfrac{1}{5}}{z-\dfrac{3}{4}}$

    (d) $H_1(z) = K \dfrac{z}{z-\dfrac{1}{4}}$, $H_2(z) = \dfrac{z+2}{z-\dfrac{3}{4}}$

    (e) $H_1(z) = K \dfrac{1}{z^2 - \dfrac{1}{3}z - \dfrac{2}{9}}$, $H_2(z) = 1$

Answers:

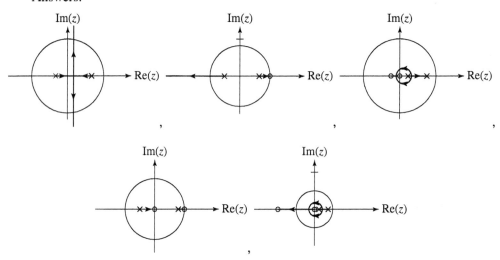

11. Draw a root locus for each of the pole-zero diagrams of loop transfer functions of feedback systems in Figure E.11.

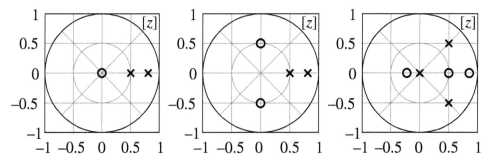

Figure E.11

Answers:

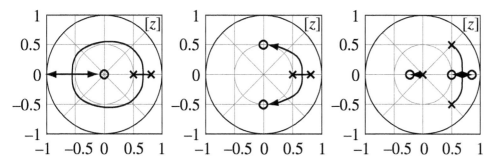

12. Sketch a root locus for each pole-zero map of a loop transfer function in Figure E.12. Then, for each one, indicate whether the system is unstable at a finite, positive value of the gain constant K.

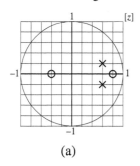 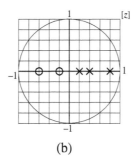

(a)   (b)

Figure E.12

Answers: One Unstable and One Stable

Laplace-Transform-z-Transform Relationship

13. Sketch regions in the z plane corresponding to these regions in the s-plane.

    (a) $0 < \sigma < \frac{1}{T_s}$, $0 < \omega < \frac{2\pi}{T_s}$

    (b) $-\frac{1}{T_s} < \sigma < 0$, $-\frac{\pi}{T_s} < \omega < 0$

    (c) $-\infty < \sigma < \infty$, $0 < \omega < \frac{2\pi}{T_s}$

    Answers: The entire z plane,

Sampled-Data Systems

14. Using the impulse-invariant design method, design a discrete-time system to approximate the continuous-time systems with these transfer functions at the

sampling rates specified. Compare the impulse and unit-step (or sequence) responses of the continuous-time and discrete-time systems.

(a) $H(s) = \dfrac{6}{s+6}$, $f_s = 4$ Hz

(b) $H(s) = \dfrac{6}{s+6}$, $f_s = 20$ Hz

Answers:

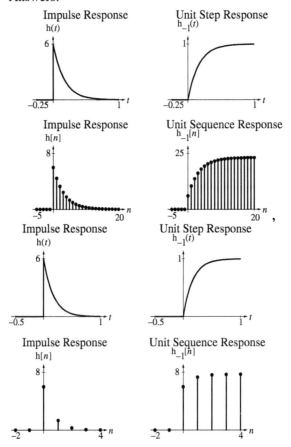

System Realization

15. Draw a cascade-form block diagram for each of these system transfer functions.

(a) $H(z) = \dfrac{z}{(z + 1/3)(z - 3/4)}$

(b) $H(z) = \dfrac{z - 1}{4z^3 + 2z^2 + 2z + 3}$

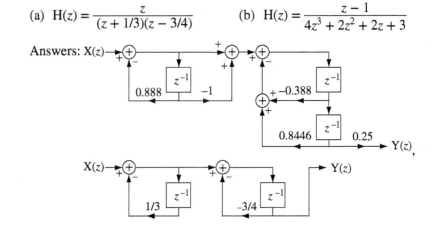

16. Draw a parallel-form block diagram for each of these system transfer functions.

    (a) $H(z) = \dfrac{z}{(z+1/3)(z-3/4)}$

    (b) $H(z) = \dfrac{8z^3 - 4z^2 + 5z + 9}{7z^3 + 4z^2 + z + 2}$

    Answers:

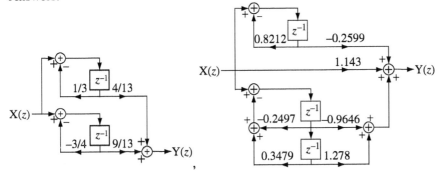

EXERCISES WITHOUT ANSWERS

Stability

17. A discrete-time feedback system has a forward-path transfer function $H_1(z) = K$ and a feedback-path transfer function $H_2(z) = 3(1 + 2z^{-1})$. For what range of real values of K is this system stable?

18. If $(1.1)^n \cos(2\pi n/16)\,u[n] \xleftrightarrow{\mathcal{Z}} H_1(z)$, and $H_2(z) = H_1(az)$ and $H_1(z)$, and $H_2(z)$ are transfer functions of systems #1 and #2, respectively, what range of values of a will make system #2 stable and physically realizable?

Root Locus

19. The loop transfer function of a discrete-time feedback system is
$$T(z) = \frac{z(z+0.2)}{(z-0.1)(z-0.8)(z^2+0.6)}.$$
What regions of the real axis in the z plane are part of the root locus?

Parallel, Cascade and Feedback Connections

20. A feedback system has a forward path transfer function $H_1(z) = \dfrac{Kz}{z - 0.5}$ and a feedback path transfer function $H_2(z) = 4z^{-1}$. For what range of values of K is the system stable?

Response to Standard Signals

21. If the system below is excited by a unit impulse, what are the values of $y[0]$, $y[1]$, $y[2]$ and $y[7]$?

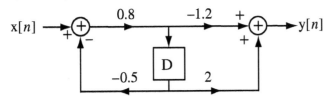

22. A system has a transfer function
$$H(z) = \frac{z}{z^2 + z + 0.24}.$$

If a unit sequence u[n] is applied to this system, what are the values of the responses y[0], y[1] and y[2]?

23. Find the responses y[n] of the systems with these transfer functions to the unit-sequence excitation $x[n] = u[n]$.

   (a) $H(z) = \dfrac{z}{z^2 - 1.8z + 0.82}$

   (b) $H(z) = \dfrac{z^2 - 1.932z + 1}{z(z - 0.95)}$

24. In Figure E.24 are six pole-zero diagrams for six discrete-time system transfer functions. Answer the following questions about them.

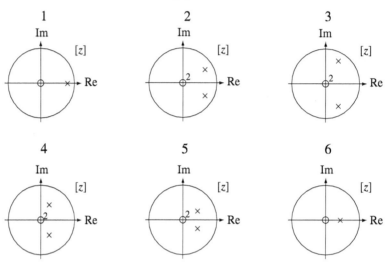

Figure E.24

(a) Which of these systems have an impulse response that is monotonic? (Monotonic means always moving in the same direction, always rising or always falling, not oscillating or ringing.)

(b) Of those systems which have a monotonic impulse response which one has the fastest response to a unit sequence? (Fastest means approaching its final value more quickly.)

(c) Of those systems which have an oscillatory or ringing impulse response, which one system rings at the fastest rate and has the largest overshoot in its response?

25. Answer the following questions.

   (a) A digital filter has an impulse response, $h[n] = 0.6^n \, u[n]$. If it is excited by a unit sequence, what is the final value of the response?

   (b) A digital filter has a transfer function $H(z) = \dfrac{10z}{z - 0.5}$. At what discrete-time radian frequency, Ω, is its magnitude response a minimum? ($z = e^{j\Omega}$)

   (c) A digital filter has a transfer function $H(z) = \dfrac{10(z - 1)}{z - 0.3}$. At what radian frequency, Ω, is its magnitude response a minimum? ($z = e^{j\Omega}$)

   (d) A digital filter has a transfer function $H(z) = \dfrac{2z}{z - 0.7}$. What is the magnitude of its response at a discrete-time radian frequency of $\Omega = \pi/2$? ($z = e^{j\Omega}$)

Laplace-Transform-z-Transform Relationship

26. For any given sampling rate f_s the relationship between the s- and z planes is given by $z = e^{sT_s}$ where $T_s = 1/f_s$. Let $f_s = 100$.

    (a) Describe the contour in the z plane that corresponds to the entire negative σ axis in the s-plane.

    (b) What is the minimum length of a line segment along the ω axis in the s-plane that corresponds to the entire unit circle in the z plane?

    (c) Find the numerical values of two different points in the s-plane s_1 and s_2 that correspond to the point $z = 1$ in the z plane.

Sampled-Data Systems

27. Using the impulse-invariant design method, design a discrete-time system to approximate the continuous-time systems with these transfer functions at the sampling rates specified. Compare the impulse and unit-step (or sequence) responses of the continuous-time and discrete-time systems.

    (a) $H(s) = \dfrac{712s}{s^2 + 46s + 240}$, $f_s = 20$ Hz

    (b) $H(s) = \dfrac{712s}{s^2 + 46s + 240}$, $f_s = 200$ Hz

System Realization

28. Draw a parallel-form block diagram for each of these system transfer functions.

    (a) $H(z) = (1 + z^{-1}) \dfrac{18}{(z - 0.1)(z + 0.7)}$

    (b) $H(z) = \dfrac{\dfrac{z}{z-1}}{1 + \dfrac{z}{z-1} \dfrac{z^2}{z^2 - 1/2}}$

General

29. In Figure E.29 are some descriptions of discrete-time systems in different forms. Answer the following questions about them.

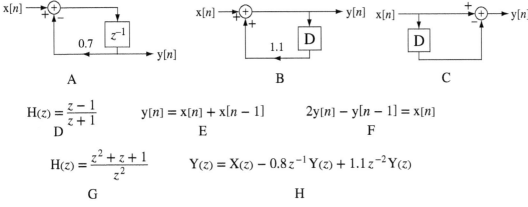

Figure E.29

(a) Which of these systems are unstable (including marginally stable)?

(b) Which of these systems have one or more zeros on the unit circle?

14 CHAPTER

Filter Analysis and Design

14.1 INTRODUCTION AND GOALS

One of the most important practical systems is the filter. Every system is, in one sense, a filter because every system has a frequency response that attenuates some frequencies more than others. Filters are used to tailor the sound of music according to personal tastes, to smooth and eliminate trends from signals, to stabilize otherwise unstable systems, to remove undesirable noise from a received signal, and so on. The study of the analysis and design of filters is a very good example of the use of transform methods.

CHAPTER GOALS

1. To become familiar with the most common types of optimized continuous-time filters, to understand in what sense they are optimal and to be able to design them to meet specifications

2. To become familiar with the filter design and analysis tools in MATLAB

3. To understand how to convert one type of filter to another through a change of variable

4. To learn methods of simulating optimized continuous-time filters with discrete-time filters and to understand the relative advantages and disadvantages of each method

5. To explore both infinite-duration-impulse-response and finite-duration-impulse-response discrete-time filter designs and to understand the relative advantages and disadvantages of each method

14.2 ANALOG FILTERS

In this chapter continuous-time filters will be referred to as **analog** filters and discrete-time filters will be referred to as **digital** filters. Also, when discussing both analog and digital filters the subscript a will be used to indicate functions or parameters applying to analog filters and the subscript d will be used similarly for functions or parameters applying to digital filters.

BUTTERWORTH FILTERS

Normalized Butterworth Filters

A very popular type of analog filter is the **Butterworth** filter, named after British applied physicist Stephen Butterworth, who invented it. An nth order lowpass Butterworth filter has a frequency response whose squared magnitude is

$$|H_a(j\omega)|^2 = \frac{1}{1 + (\omega/\omega_c)^{2n}}.$$

The lowpass Butterworth filter is designed to be **maximally flat** for frequencies in its passband $\omega < \omega_c$, meaning its variation with frequency in the passband is monotonic and approaches a zero derivative as the frequency approaches zero. Figure 14.1 illustrates the frequency response of a Butterworth filter with a corner frequency of $\omega_c = 1$ for four different orders n. As the order is increased the filter's magnitude frequency response approaches that of an ideal lowpass filter.

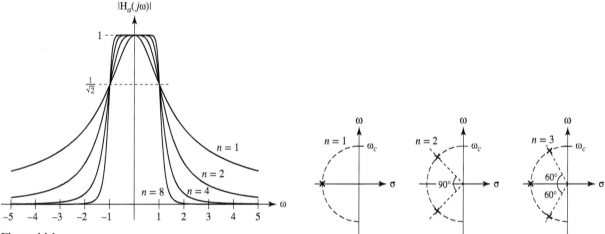

Figure 14.1
Butterworth filter magnitude frequency responses for a corner frequency, $\omega_c = 1$, and four different orders

Figure 14.2
Butterworth filter pole locations

The poles of a lowpass Butterworth filter lie on a semicircle of radius ω_c in the open left half-plane (Figure 14.2). The number of poles is n and the angular spacing between poles (for $n > 1$) is always π/n. If n is odd, there is a pole on the negative real axis and all the other poles occur in complex conjugate pairs. If n is even, all the poles occur in complex conjugate pairs. Using these properties, the transfer function of a unity-gain lowpass Butterworth filter can always be found and is of the form

$$H_a(s) = \frac{1}{(1 - s/p_1)(1 - s/p_2)\cdots(1 - s/p_n)} = \prod_{k=1}^{n} \frac{1}{1 - s/p_k} = \prod_{k=1}^{n} -\frac{p_k}{s - p_k}$$

where the p_k's are the pole locations.

The MATLAB signal toolbox has functions for designing analog Butterworth filters. The MATLAB function call,

$$[\text{za,pa,ka}] = \text{buttap(N)};$$

returns the finite zeros in the vector **za**, the finite poles in the vector **pa** and the gain coefficient in the scalar **ka**, for an **N**th order, unity-gain, Butterworth lowpass filter with a corner frequency, $\omega_c = 1$. (There are no finite zeros in a lowpass Butterworth filter transfer function so **za** is always an empty vector and, since the filter is unity-gain, **ka** is always one. The zeros and gain are included in the returned data because this form of returned data is used for other types of filters, for which there may be finite zeros and the gain may not be one.)

```
>> [za,pa,ka] = buttap(4) ;
>> za
za =
   []
>> pa
pa =
  -0.3827 + 0.9239i
  -0.3827 - 0.9239i
  -0.9239 + 0.3827i
  -0.9239 - 0.3827i
>> ka
ka =
    1
```

Filter Transformations

Once a design has been done for a lowpass Butterworth filter of a given order with a corner frequency $\omega_c = 1$, the conversion of that filter to a different corner frequency and/or to a highpass, bandpass or bandstop filter can be done with a change of the frequency variable. MATLAB allows the designer to quickly and easily design an nth order lowpass Butterworth filter with unity gain and a corner frequency $\omega_c = 1$. Denormalizing the gain to a nonunity gain is trivial since it simply involves changing the gain coefficient. Changing the corner frequency or the filter type is a little more involved.

To change the frequency response from a corner frequency $\omega_c = 1$ to a general corner frequency $\omega_c \neq 1$, make the independent-variable change $s \to s/\omega_c$ in the transfer function. For example, a first-order, unity-gain, normalized Butterworth filter has a transfer function

$$H_{norm}(s) = \frac{1}{s+1}.$$

If we want to move the corner frequency to $\omega_c = 10$, the new transfer function is

$$H_{10}(s) = H_{norm}(s/10) = \frac{1}{s/10 + 1} = \frac{10}{s + 10}.$$

This is the transfer function of a unity-gain lowpass filter with a corner frequency $\omega_c = 10$.

The real power of the filter transformation process is seen in converting a lowpass filter to a highpass filter. If we make the change of variable $s \to 1/s$ then

$$H_{HP}(s) = H_{norm}(1/s) = \frac{1}{1/s + 1} = \frac{s}{s+1}$$

and $H_{HP}(s)$ is the transfer function of a first-order, unity-gain, highpass Butterworth filter with a corner frequency $\omega_c = 1$. We can also simultaneously change the corner frequency by making the change of variable $s \to \omega_c/s$. We now have a transfer function with one finite pole and one finite zero at $s = 0$. In the general form of the transfer function of a normalized lowpass Butterworth filter

$$H_{norm}(s) = \prod_{k=1}^{n} \frac{-p_k}{s - p_k}$$

when we make the change of variable $s \to 1/s$ we get

$$H_{HP}(s) = \left[\prod_{k=1}^{n} \frac{-p_k}{s - p_k}\right]_{s \to 1/s} = \prod_{k=1}^{n} \frac{-p_k}{1/s - p_k} = \prod_{k=1}^{n} \frac{p_k s}{p_k s - 1} \prod_{k=1}^{n} \frac{s}{s - 1/p_k}.$$

The poles are at $s = 1/p_k$. They are the reciprocals of the normalized lowpass filter poles, all of which have a magnitude of one. The reciprocal of any complex number is at an angle that is the negative of the angle of the complex number. In this case, since the magnitudes of the poles have not changed, the poles move to their complex conjugates and the overall constellation of poles is unchanged. Also, there are now n zeros at $s = 0$. If we make the change of variable $s \to \omega_c/s$, the poles have the same angles but their magnitudes are now all ω_c instead of one.

Transforming a lowpass filter into a bandpass filter is a little more complicated. We can do it by using the change of variable

$$s \to \frac{s^2 + \omega_L \omega_H}{s(\omega_H - \omega_L)}$$

where ω_L is the lower positive corner frequency of the bandpass filter and ω_H is the higher positive corner frequency. For example, let's design a first-order, unity-gain, bandpass filter with a passband from $\omega = 100$ to $\omega = 200$ (Figure 14.3).

$$H_{BP}(s) = H_{norm}\left(\frac{s^2 + \omega_L \omega_H}{s(\omega_H - \omega_L)}\right) = \frac{1}{\frac{s^2 + \omega_L \omega_H}{s(\omega_H - \omega_L)} + 1} = \frac{s(\omega_H - \omega_L)}{s^2 + s(\omega_H - \omega_L) + \omega_L \omega_H}$$

$$H_{BP}(j\omega) = \frac{j\omega(\omega_H - \omega_L)}{-\omega^2 + j\omega(\omega_H - \omega_L) + \omega_L \omega_H}$$

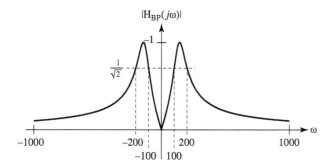

Figure 14.3
Magnitude frequency response of a unity-gain, first-order bandpass Butterworth filter

Simplifying and inserting numerical values,

$$H_{BP}(j\omega) = \frac{j100\omega}{-\omega^2 + j100\omega + 20,000} = \frac{j100\omega}{(j\omega + 50 + j132.2)(j\omega + 50 - j132.2)}.$$

The peak of the bandpass response occurs where the derivative of the frequency response with respect to ω is zero.

$$\frac{d}{d\omega}H_{BP}(j\omega)$$
$$= \frac{(-\omega^2 + j\omega(\omega_H - \omega_L) + \omega_L\omega_H)j(\omega_H - \omega_L) - j\omega(\omega_H - \omega_L)(-2\omega + j(\omega_H - \omega_L))}{[-\omega^2 + j\omega(\omega_H - \omega_L) + \omega_L\omega_H]^2} = 0$$

$$(-\omega^2 + j\omega(\omega_H - \omega_L) + \omega_L\omega_H) + 2\omega^2 - j\omega(\omega_H - \omega_L) = 0$$

$$\Rightarrow \omega^2 + \omega_L\omega_H = 0 \Rightarrow \omega = \pm\sqrt{\omega_L\omega_H}.$$

So the natural radian frequency is $\omega_n = \pm\sqrt{\omega_L\omega_H}$. Also, to conform to the standard second-order system transfer function form,

$$j2\zeta\omega_n\omega = j\omega(\omega_H - \omega_L) \Rightarrow \zeta = \frac{\omega_H - \omega_L}{2\sqrt{\omega_L\omega_H}}.$$

So the damping ratio is $\zeta = \frac{\omega_H - \omega_L}{2\sqrt{\omega_H\omega_L}}$.

Finally, we can transform a lowpass filter into a bandstop filter with the transformation

$$s \to \frac{s(\omega_H - \omega_L)}{s^2 + \omega_L\omega_H}.$$

Notice that for a lowpass filter of order n the degree of the denominator of the transfer function is n, but for a bandpass of order n the degree of the denominator of the transfer function is $2n$. Similarly, for a highpass filter the denominator degree is n and for a bandstop filter the degree of the denominator is $2n$.

MATLAB Design Tools

MATLAB has commands for the transformation of normalized filters. They are

    lp2bp Lowpass to bandpass analog filter transformation
    lp2bs Lowpass to bandstop analog filter transformation
    lp2hp Lowpass to highpass analog filter transformation
    lp2lp Lowpass to lowpass analog filter transformation

The syntax for lp2bp is

$$[\text{numt},\text{dent}] = \text{lp2bp}(\text{num},\text{den},\text{w0},\text{bw})$$

where num and den are vectors of coefficients of s in the numerator and denominator of the normalized lowpass filter transfer function, respectively, w0 is the center frequency of the bandpass filter, and bw is the bandwidth of the bandpass filter (both in rad/s), and numt and dent are vectors of coefficients of s in the numerator and denominator of the bandpass filter transfer function. The syntax of each of the other commands is similar.

As an example, we can design a normalized lowpass Butterworth filter with buttap.

```
»[z,p,k] = buttap(3) ;
»z
z =
   []
```

```
»p
p =
 -0.5000 + 0.8660i
 -1.0000
 -0.5000 - 0.8660i
»k
k =
 1
```

This result indicates that a third-order normalized lowpass Butterworth filter has the frequency response

$$H_{LP}(s) = \frac{1}{(s+1)(s+0.5+j0.866)(s+0.5-j0.866)}.$$

We can convert this to a ratio of polynomials using MATLAB system-object commands.

```
»[num,den] = tfdata(zpk(z,p,k),'v') ;
»num

num =

 0 0 0 1

»den

den =

 1.0000  2.0000 + 0.0000i  2.0000 + 0.0000i  1.0000 + 0.0000i
```

This result indicates that the normalized lowpass frequency response can be written more compactly as

$$H_{LP}(s) = \frac{1}{s^3 + 2s^2 + 2s + 1}.$$

Using this result we can transform the normalized lowpass filter to a denormalized bandpass filter with center frequency $\omega = 8$ and bandwidth $\Delta\omega = 2$.

```
»[numt,dent] = lp2bp(num,den,8,2) ;
»numt
numt =
 Columns 1 through 4
    0  0.0000 - 0.0000i  0.0000 - 0.0000i  8.0000 - 0.0000i
 Columns 5 through 7
  0.0000 - 0.0000i  0.0000 - 0.0000i  0.0000 - 0.0000i
»dent
dent =
 1.0e+05 *
 Columns 1 through 4
  0.0000  0.0000 + 0.0000i  0.0020 + 0.0000i  0.0052 + 0.0000i
 Columns 5 through 7
  0.1280 + 0.0000i  0.1638 + 0.0000i  2.6214 - 0.0000i
»bpf = tf(numt,dent) ;
»bpf
```

```
Transfer function:
1.542e-14 s^5 + 2.32e-13 s^4 + 8 s^3 + 3.644e-11 s^2 + 9.789e-11 s +
9.952e-10
-----------------------------------------------------------------
s^6 + 4 s^5 + 200 s^4 + 520 s^3 + 1.28e04 s^2 + 1.638e04 s + 2.621e05
»
```

This result indicates that the bandpass-filter transfer function can be written as

$$H_{BP}(s) = \frac{8s^3}{s^6 + 4s^5 + 200s^4 + 520s^3 + 12{,}800s^2 + 16{,}380s + 262{,}100}.$$

(The extremely small nonzero coefficients in the numerator of the transfer function reported by MATLAB are the result of round-off errors in the MATLAB calculations and have been neglected. Notice they were zero in **numt**.)

CHEBYSHEV, ELLIPTIC AND BESSEL FILTERS

We have just seen how the MATLAB command **buttap** can be used to design a normalized Butterworth filter and how to denormalize it to other Butterworth filters. There are several other MATLAB commands that are useful in analog filter design. There are four other "...ap" commands, **cheb1ap**, **cheb2ap**, **ellipap** and **besselap** that design normalized analog filters of optimal types other than the Butterworth filter. The other optimal analog filter types are the Chebyshev (sometimes spelled Tchebysheff or Tchebischeff) filter, the Elliptic filter (sometimes called the Cauer filter) and the Bessel filter. Each of these filter types optimizes the performance of the filter according to a different criterion.

The Chebyshev filter is similar to the Butterworth filter but has an extra degree of design freedom (Figure 14.4).

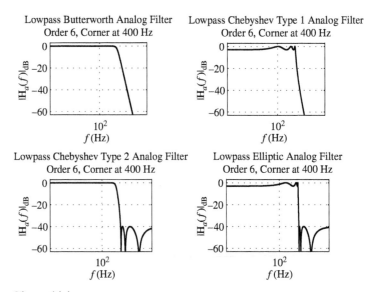

Figure 14.4
Typical magnitude frequency responses of Butterworth, Chebyshev and Elliptic filters

14.2 Analog Filters

The Butterworth filter is called *maximally flat* because it is monotonic in the pass and stopbands and approaches a flat response in the passband as the order is increased. There are two types of Chebyshev filter, types one and two. The type-one Chebyshev has a frequency response that is not monotonic in the passband but is monotonic in the stopband. Its frequency response ripples (varies up and down with frequency) in the passband. The presence of ripple in the passband is usually not in itself desirable but it allows the transition from the passband to the stopband to be faster than a Butterworth filter of the same order. In other words, we trade passband monotonicity for a narrower transition band. The more ripple we allow in the passband, the narrower the transition band can be. The type-two Chebyshev filter is just the opposite. It has a monotonic passband and ripple in the stopband and, for the same filter order, also allows for a narrower transition band than a Butterworth filter.

The Elliptic filter has ripple in both the passband and stopband and, for the same filter order, it has an even narrower transition band than either of the two types of Chebyshev filter. The Bessel filter is optimized on a different basis. The Bessel filter is optimized for linearity of the phase in the passband rather than flat magnitude response in the passband and/or stopband, or narrow transition band.

The syntax for each of these normalized analog filter designs is given below.

```
[z,p,k] = cheb1ap(N,Rp) ;
[z,p,k] = cheb2ap(N,Rs) ;
[z,p,k] = ellipap(N,Rp,Rs) ;
[z,p,k] = besselap(N) ;-
```

where N is the order of the filter, Rp is allowable ripple in the passband in dB and Rs is the minimum attenuation in the stopband in dB.

Once a filter has been designed, its frequency response can be found using either bode, which was introduced earlier, or freqs. The function freqs has the syntax

$$H = freqs(num,den,w) ;$$

where H is a vector of responses at the real radian-frequency points in the vector w, and num and den are vectors containing the coefficients of s in the numerator and denominator of the filter transfer function.

EXAMPLE 14.1

Comparison of fourth-order bandstop Butterworth and Chebyshev filters using MATLAB

Using MATLAB, design a normalized fourth-order lowpass Butterworth filter, transform it into a denormalized bandstop filter with a center frequency of 60 Hz and a bandwidth of 10 Hz, then compare its frequency response with a type-one Chebyshev bandstop filter of the same order and corner frequencies and an allowable ripple in the pass band of 0.3 dB.

```
%   Butterworth design

%   Design a normalized fourth-order Butterworth lowpass filter
%   and put the zeros, poles and gain in zb, pb and kb

[zb,pb,kb] = buttap(4) ;
```

```
%   Use MATLAB system tools to obtain the numerator and
%   denominator coefficient vectors, numb and denb

[numb,denb] = tfdata(zpk(zb,pb,kb),'v') ;

%   Set the cyclic center frequency and bandwidth and then set
%   the corresponding radian center frequency and bandwidth

f0 = 60 ; fbw = 10 ; w0 = 2*pi*f0 ; wbw = 2*pi*fbw ;

%   Denormalize the lowpass Butterworth to a bandstop Butterworth

[numbsb,denbsb] = lp2bs(numb,denb,w0,wbw) ;

%   Create a vector of cyclic frequencies to use in plotting the
%   frequency response of the filter. Then create a corresponding
%   radian-frequency vector and compute the frequency response.

wbsb = 2*pi*[40:0.2:80]' ; Hbsb = freqs(numbsb,denbsb,wbsb) ;

%   Chebyshev design

%   Design a normalized fourth-order type-one Chebyshev lowpass
%   filter and put the zeros, poles and gain in zc, pc and kc

[zc,pc,kc] = cheb1ap(4,0.3) ; wc = wb ;

%   Use MATLAB system tools to obtain the numerator and
%   denominator coefficient vectors, numc and denc

[numc,denc] = tfdata(zpk(zc,pc,kc),'v') ;

%   Denormalize the lowpass Chebyshev to a bandstop Chebyshev

[numbsc,denbsc] = lp2bs(numc,denc,w0,wbw) ;

%   Use the same radian-frequency vector used in the Butterworth
%   design and compute the frequency response of the Chebyshev
%   bandstop filter.

wbsc = wbsb ; Hbsc = freqs(numbsc,denbsc,wbsc) ;
```

The magnitude frequency responses are compared in Figure 14.5. Notice that the Butterworth filter is monotonic in the passbands while the Chebyshev filter is not, but that the Chebyshev filter has a steeper slope in the transition between pass and stopbands and slightly better stopband attenuation.

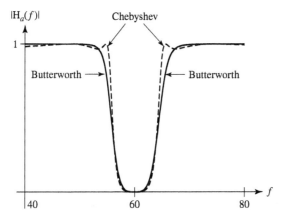

Figure 14.5
Comparison of the Butterworth and Chebyshev magnitude frequency responses

14.3 DIGITAL FILTERS

The analysis and design of analog filters is a large and important topic. An equally large and important topic (maybe even more important) is the design of digital filters that simulate some of the popular kinds of standard analog filters. Nearly all discrete-time systems are filters in a sense because they have frequency responses that are not constant with frequency.

SIMULATION OF ANALOG FILTERS

There are many optimized standard filter design techniques for analog filters. One very popular way of designing digital filters is to simulate a proven analog filter design. All the commonly used standard analog filters have *s*-domain transfer functions that are ratios of polynomials in *s* and therefore have impulse responses that endure for an infinite time. This type of impulse response is called an **infinite-duration impulse response (IIR)**. Many of the techniques that simulate the analog filter with a digital filter create a digital filter that also has an IIR infinite-duration impulse response, and these types of digital filters are called **IIR filters**. Another popular design method for digital filters produces filters with a **finite-duration impulse response** and these filters are called **FIR filters**.

In the following discussion of simulation of analog filters with digital filters the analog filter's impulse response will be $h_a(t)$, its transfer function will be $H_a(s)$, the digital filter's impulse response will be $h_d[n]$ and its transfer function will be $H_d(z)$.

FILTER DESIGN TECHNIQUES

IIR Filter Design

Time-Domain Methods
Impulse-Invariant Design One approach to digital-filter design is to try to make the digital filter response to a standard digital excitation a sampled version of the analog filter response to the corresponding standard continuous-time excitation. This idea leads to the **impulse-invariant** and **step-invariant** design procedures. Impulse-invariant design makes the response of the digital filter to a discrete-time unit impulse a

sampled version of the response of the analog filter to a continuous-time unit impulse. Step-invariant design makes the response of the digital filter to a unit sequence a sampled version of the response of the analog filter to a unit step. Each of these design processes produces an IIR filter (Figure 14.6).

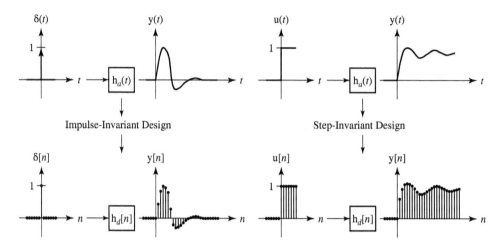

Figure 14.6
The impulse-invariant and step-invariant digital-filter design techniques

We know from sampling theory that we can *impulse sample* the analog filter impulse response $h_a(t)$ to form $h_\delta(t)$ whose Laplace transform is $H_\delta(s)$ and whose continuous-time Fourier transform (CTFT) is

$$H_\delta(j\omega) = f_s \sum_{k=-\infty}^{\infty} H_a(j(\omega - k\omega_s)).$$

where $H_a(s)$ is the analog filter's transfer function and $\omega_s = 2\pi f_s$. We also know that we can *sample* $h_a(t)$ to form $h_d[n]$ whose z transform is $H_d(z)$ and whose discrete-time Fourier transform (DTFT) is

$$H_d(e^{j\Omega}) = f_s \sum_{k=-\infty}^{\infty} H_a(jf_s(\Omega - 2\pi k)). \tag{14.1}$$

So it is apparent that the digital-filter's frequency response is the sum of scaled aliases of the analog filter's frequency response and, to the extent that the aliases overlap, the two frequency responses must differ. As an example of impulse-invariant design, let $H_a(s)$ be the transfer function of a second-order, Butterworth lowpass filter with low-frequency gain of A and cutoff frequency of ω_c radians per second.

$$H_a(s) = \frac{A\omega_c^2}{s^2 + \sqrt{2}\omega_c s + \omega_c^2}$$

Then, inverse Laplace transforming,

$$h_a(t) = \sqrt{2}A\omega_c e^{-\omega_c t/\sqrt{2}} \sin(\omega_c t/\sqrt{2})u(t).$$

Now sample at the rate f_s to form $h_d[n] = \sqrt{2}A\omega_c e^{-\omega_c n T_s/\sqrt{2}} \sin(\omega_c n T_s/\sqrt{2})u[n]$ (Figure 14.7) and

$$H_d(z) = \sqrt{2}A\omega_c \frac{ze^{-\omega_c T_s/\sqrt{2}} \sin(\omega_c T_s/\sqrt{2})}{z^2 - 2e^{-\omega_c T_s/\sqrt{2}} \cos(\omega_c T_s/\sqrt{2})z + e^{-2\omega_c T_s/\sqrt{2}}}$$

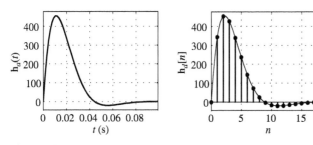

Figure 14.7
Analog and digital filter impulse responses

or

$$H_d(e^{j\Omega}) = \sqrt{2}A\omega_c \frac{e^{j\Omega}e^{-\omega_c T_s/\sqrt{2}}\sin(\omega_c T_s/\sqrt{2})}{e^{j2\Omega} - 2e^{-\omega_c T_s/\sqrt{2}}\cos(\omega_c T_s/\sqrt{2})e^{j\Omega} + e^{-2\omega_c T_s/\sqrt{2}}}. \quad (14.2)$$

Equating the two forms (14.1) and (14.2),

$$H_d(e^{j\Omega}) = f_s \sum_{k=-\infty}^{\infty} \frac{A\omega_c^2}{[jf_s(\Omega - 2\pi k)]^2 + j\sqrt{2}\omega_c f_s(\Omega - 2\pi k) + \omega_c^2}$$

$$= \sqrt{2}A\omega_c \frac{e^{j\Omega}e^{-\omega_c T_s/\sqrt{2}}\sin(\omega_c T_s/\sqrt{2})}{e^{j2\Omega} - 2e^{-\omega_c T_s/\sqrt{2}}\cos(\omega_c T_s/\sqrt{2})e^{j\Omega} + e^{-2\omega_c T_s/\sqrt{2}}}.$$

If we let $A = 10$ and $\omega_c = 100$ and sample at a rate of 200 samples/second, then

$$H_d(e^{j\Omega}) = 2000 \sum_{k=-\infty}^{\infty} \frac{1}{[j2(\Omega - 2\pi k)]^2 + j2\sqrt{2}(\Omega - 2\pi k) + 1}$$

or

$$H_d(e^{j\Omega}) = 1000\sqrt{2}\frac{e^{j\Omega}e^{-1/2\sqrt{2}}\sin(1/2\sqrt{2})}{e^{j2\Omega} - 2e^{-1/2\sqrt{2}}\cos(1/2\sqrt{2})e^{j\Omega} + e^{-1/\sqrt{2}}}$$

$$= \frac{343.825e^{j\Omega}}{e^{j2\Omega} - 1.31751e^{j\Omega} + 0.49306}.$$

As a check compare the two forms at $\Omega = 0$.

The complete digital-filter frequency response is shown in Figure 14.8. The heavy line is the actual frequency response and the light lines are the individual scaled aliases of the analog filter's frequency response. The difference between the analog filter's response at zero frequency and the digital filter's response at zero frequency is about -2% due to the effects of aliasing.

This filter can be realized directly from its transfer function in Direct Form II.

$$H_d(z) = \frac{Y_d(z)}{X_d(z)} = \frac{343.825z}{z^2 - 1.31751z + 0.49306}$$

or

$$z^2 Y_d(z) - 1.31751z Y_d(z) + 0.49306 Y_d(z) = 343.825z X_d(z)$$

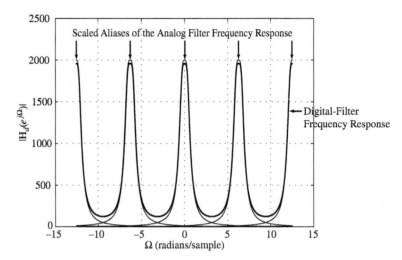

Figure 14.8
Digital-filter frequency response showing the effects of aliasing

Rearranging and solving for $Y_d(z)$

$$Y_d(z) = 343.825 z^{-1} X_d(z) + 1.31751 z^{-1} Y_d(z) - 0.49306 z^{-2} Y_d(z).$$

Then, inverse z transforming

$$y_d[n] = 343.825\, x_d[n-1] + 1.31751\, y_d[n-1] - 0.49306\, y_d[n-2] \text{ (Figure 14.9)}.$$

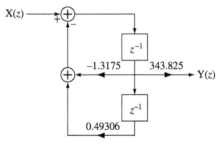

Figure 14.9
Block diagram of a lowpass filter designed using the impulse-invariant method

To illustrate a subtlety in this design method, consider a first-order lowpass analog filter whose transfer function is

$$H_a(s) = \frac{A\omega_c}{s + \omega_c} \Rightarrow H_a(j\omega) = \frac{A\omega_c}{j\omega + \omega_c}$$

with impulse response

$$h_a(t) = A\omega_c e^{-\omega_c t} u(t).$$

Sample at the rate f_s to form $h_d[n] = A\omega_c e^{-\omega_c n T_s} u[n]$ and

$$H_d(z) = A\omega_c \frac{z}{z - e^{-\omega_c T_s}} \Rightarrow H_d(e^{j\Omega}) = A\omega_c \frac{e^{j\Omega}}{e^{j\Omega} - e^{-\omega_c T_s}} \qquad (14.3)$$

and the frequency response can be written in the two equivalent forms

$$H_d(e^{j\Omega}) = f_s \sum_{k=-\infty}^{\infty} \frac{A\omega_c}{jf_s(\Omega - 2\pi k) + \omega_c} = A\omega_c \frac{e^{j\Omega}}{e^{j\Omega} - e^{-\omega_c T_s}}.$$

Let $a = 10$, $\omega_c = 50$ and $f_s = 100$ and again check the equality at $\Omega = 0$.

$$f_s \sum_{k=-\infty}^{\infty} \frac{A\omega_c}{jf_s(\Omega - 2\pi k) + \omega_c} = \sum_{k=-\infty}^{\infty} \frac{50{,}000}{-j200\pi k + 50} = 1020.7$$

$$A\omega_c \frac{e^{j\Omega}}{e^{j\Omega} - e^{-\omega_c T_s}} = 500 \frac{1}{1 - e^{-1/2}} = 1270.7$$

These two results, which should ideally be equal, differ by almost 25% at $\Omega = 0$. The two frequency responses are illustrated in Figure 14.10.

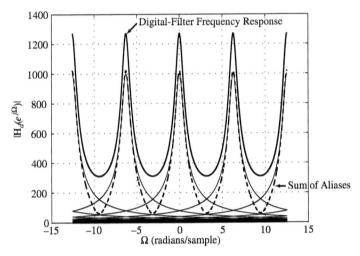

Figure 14.10
Digital-filter frequency response showing an apparent error between two frequency responses that should be equal

The question, of course, is why are they different? The error arises from the statement above that the digital filter impulse response found by sampling the analog filter impulse response is $h_d[n] = A\omega_c e^{-\omega_c n T_s} u[n]$. The analog impulse response has a discontinuity at $t = 0$. So what should the sample value be at that point? The impulse response $h_d[n] = A\omega_c e^{-\omega_c n T_s} u[n]$ implies that the sample value at $t = 0$ is $A\omega_c$. But why isn't a sample value of zero just as valid since the discontinuity extends from zero to $A\omega_c$? If we replace the first sample value of $A\omega_c$ with $A\omega_c/2$, the average of the two limits from above and below of the analog filter's impulse response at $t = 0$, then the two formulas for the digital-filter frequency response agree exactly. So it would seem that when sampling at a discontinuity the best value to take is the average value of the two limits from above and below. This is in accord with Fourier transform theory for which the Fourier transform representation of a discontinuous signal always goes through the midpoint of a discontinuity. This problem did not arise in the previous analysis of the second-order Butterworth lowpass filter because its impulse response is continuous.

Given the error in the first-order lowpass digital-filter design due to sampling at a discontinuity, one might suggest that, to avoid the problem, we could simply delay the analog filter's impulse response by some small amount (less than the time between samples) and avoid sampling at a discontinuity. That can be done and the two forms of the digital-filter's frequency response again agree exactly.

The MATLAB signal toolbox has a command `impinvar` that does impulse-invariant digital-filter design. The syntax is

$$[\text{bd,ad}] = \text{impinvar(ba,aa,fs)}$$

where `ba` is a vector of coefficients of s in the numerator of the analog filter transfer function, `aa` is a vector of coefficients of s in the denominator of the analog filter transfer function, `fs` is the sampling rate in samples/second, `bd` is a vector of coefficients of z in the numerator of the digital-filter transfer function and `ad` is a vector of coefficients of z in the denominator of the digital-filter transfer function. Its transfer function is not identical to the impulse-invariant design result given here. It has a different gain constant and is shifted in time, but the impulse response shape is the same (see Example 14.2).

EXAMPLE 14.2

Digital bandpass filter design using the impulse-invariant method

Using the impulse-invariant design method, design a digital filter to simulate a unity-gain, second-order, bandpass, Butterworth analog filter with corner frequencies 150 Hz and 200 Hz and a sampling rate of 1000 samples/second. The transfer function is

$$H_a(s) = \frac{9.87 \times 10^4 s^2}{s^4 + 444.3s^3 + 2.467 \times 10^6 s^2 + 5.262 \times 10^8 s + 1.403 \times 10^{12}}$$

and the impulse response is

$$h_a(t) = [246.07 e^{-122.41t} \cos(1199.4t - 1.48) + 200.5 e^{-99.74t} \cos(977.27t + 1.683)] u(t)$$

Compare the frequency responses of the analog and digital filters.

This impulse response is the sum of two exponentially damped sinusoids with time constants of about 8.2 ms and 10 ms, and sinusoidal frequencies of $1199.4/2\pi \approx 190.9$ and $977.27/2\pi \approx 155.54$ Hz. For a reasonably accurate simulation we should choose a sampling rate such that the sinusoid is oversampled and there are several samples of the exponential decay per time constant. Let the sampling rate f_s be 1000 samples/second. Then the discrete-time impulse response would be

$$h_d[n] = [246.07 e^{-0.12241n} \cos(1.1994n - 1.48) + 200.5 e^{-0.09974n} \cos(0.97727n + 1.683)] u[n].$$

The z transform of this discrete-time impulse response is the transfer function,

$$H_d(z) = \frac{48.4z^3 - 107.7z^2 + 51.46z}{z^4 - 1.655z^3 + 2.252z^2 - 1.319z + 0.6413}.$$

The analog and digital filters' impulse responses are illustrated in Figure 14.11.

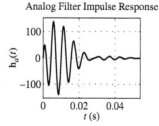
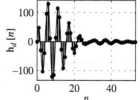

Figure 14.11
Analog and digital filter impulse responses

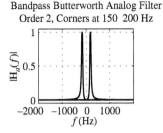
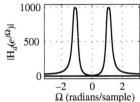

Figure 14.12
Magnitude frequency responses of the analog filter and its digital simulation by the impulse-invariant method

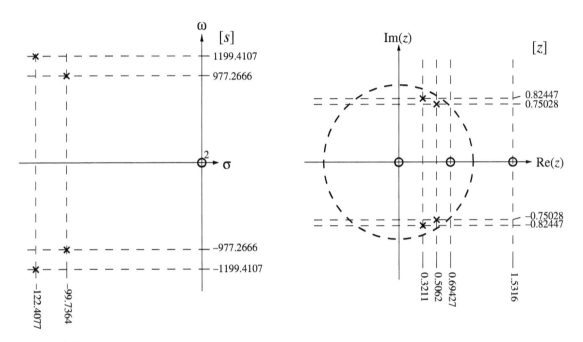

Figure 14.13
Pole-zero diagrams of the analog filter and its digital simulation by the impulse-invariant method

The magnitude frequency responses of the analog and digital filters are illustrated in Figure 14.12 and their pole-zero diagrams are in Figure 14.13.

Two things immediately stand out about this design. First, the analog filter has a response of zero at $f = 0$ and the digital filter does not. The digital-filter's frequency response at $\Omega = 0$ is about 0.85% of its peak frequency response. Since this filter is intended as a bandpass filter, this is an undesirable design result. The gain of the digital filter is much greater than the gain of the analog filter. The gain could be made the same as the analog filter by a simple adjustment of the multiplication factor in the expression for $H_d(z)$. Also, although the frequency response does peak at the right frequency, the attenuation of the digital filter in the stopband is not as good as the analog filter's attenuation. If we had used a higher sampling rate the attenuation would have been better.

Doing this design with MATLAB's `impinvar` command,

```
>> [bd,ad] = impinvar([9.87e4 0 0],[1 444.3 2.467e6 5.262e8 1.403e12],1000)
bd =
```

                 -0.0000 0.0484 -0.1077 0.0515
ad =
                 1.0000 -1.6547 2.2527 -1.3188 0.6413

The resulting transfer function is

$$H_M(z) = \frac{Y(z)}{X(z)} = \frac{0.0484z^2 - 0.1077z + 0.0515}{z^4 - 1.6547z^3 + 2.2527z^2 - 1.3188z + 0.6413}.$$

Compare this to the result above

$$H_d(z) = \frac{48.4z^3 - 107.7z^2 + 51.46z}{z^4 - 1.655z^3 + 2.252z^2 - 1.319z + 0.6413}.$$

The relation between them is

$$H_M(z) = (z^{-1}/f_s)H_d(z).$$

So MATLAB's version of impulse-invariant design divides the transfer function by the sampling rate, changing the filter's gain constant and multiplies the transfer function by z^{-1}, delaying the impulse response by one unit in discrete time. Multiplication by a constant and a time shift are the two things we can do to a signal without distorting it. Therefore the two impulse responses, although not identical, have the same shape.

■

Step-Invariant Design A closely related design method for digital filters is the step-invariant method. In this method the unit-sequence response of the digital filter is designed to match the unit-step response of the analog filter at the sampling instants. If an analog filter has a transfer function $H_a(s)$, the Laplace transform of its unit-step response is $H_a(s)/s$. The unit-step response is the inverse Laplace transform

$$h_{-1a}(t) = \mathcal{L}^{-1}\left(\frac{H_a(s)}{s}\right).$$

The corresponding discrete-time unit-sequence response is then

$$h_{-1d}[n] = h_{-1a}(nT_s).$$

Its z transform is the product of the z-domain transfer function and the z transform of a unit sequence,

$$\mathcal{Z}(h_{-1d}[n]) = \frac{z}{z-1}H_d(z).$$

We can summarize by saying that, given an s-domain transfer function $H_a(s)$, we can find the corresponding z-domain transfer function $H_d(z)$ as

$$H_d(z) = \frac{z-1}{z}\mathcal{Z}\left(\mathcal{L}^{-1}\left(\frac{H_a(s)}{s}\right)_{(t)\to(nT_s)\to[n]}\right).$$

In this method we sample the analog unit-step response to get the digital unit-sequence response. If we *impulse sample* the analog filter unit-step response $h_{-1a}(t)$ we form $h_{-1\delta}(t)$ whose Laplace transform is $H_{-1\delta}(s)$ and whose CTFT is

$$H_{-1\delta}(j\omega) = f_s\sum_{k=-\infty}^{\infty}H_{-1a}(j(\omega - k\omega_s)),$$

where $H_{-1a}(s)$ is the Laplace transform of the analog filter's unit-step response and $\omega_s = 2\pi f_s$. We also know that we can *sample* $h_{-1a}(t)$ to form $h_{-1d}[n]$ whose z transform is $H_{-1d}(z)$ and whose DTFT is

$$H_{-1d}(e^{j\Omega}) = f_s \sum_{k=-\infty}^{\infty} H_{-1a}(jf_s(\Omega - 2\pi k)) \quad (14.4)$$

Relating this result to the analog and digital transfer functions

$$H_{-1d}(e^{j\Omega}) = \frac{e^{j\Omega}}{e^{j\Omega} - 1} H_d(e^{j\Omega})$$

and

$$H_{-1a}(j\omega) = H_a(j\omega)/j\omega$$

$$H_d(e^{j\Omega}) = \frac{e^{j\Omega} - 1}{e^{j\Omega}} H_{-1d}(e^{j\Omega}) = \frac{e^{j\Omega} - 1}{e^{j\Omega}} f_s \sum_{k=-\infty}^{\infty} \frac{H_a(jf_s(\Omega - 2\pi k))}{jf_s(\Omega - 2\pi k)}.$$

EXAMPLE 14.3

Digital bandpass filter design using the step-invariant method

Using the step-invariant method, design a digital filter to approximate the analog filter whose transfer function is the same as in Example 14.2

$$H_a(s) = \frac{9.87 \times 10^4 s^2}{s^4 + 444.3 s^3 + 2.467 \times 10^6 s^2 + 5.262 \times 10^8 s + 1.403 \times 10^{12}}$$

with the same sampling rate $f_s = 1000$ samples/second.

The unit-step response is

$$h_{-1a}(t) = [0.2041 e^{-122.408 t} \cos(1199.4 t + 3.1312) + 0.2041 e^{-99.74 t} \cos(977.27 t + 0.01042)] u(t).$$

The unit-sequence response is

$$h_{-1d}[n] = [0.2041(0.8847)^n \cos(1.1994 n + 3.1312) + 0.2041(0.9051)^n \cos(0.97727 n + 0.0102)] u[n].$$

The digital-filter transfer function is

$$H_d(z) = \frac{0.03443 z^3 - 0.03905 z^2 - 0.02527 z + 0.02988}{z^4 - 1.655 z^3 + 2.252 z^2 - 1.319 z + 0.6413}.$$

The step responses, magnitude frequency responses and pole-zero diagrams of the analog and digital filters are compared in Figure 14.14, Figure 14.15 and Figure 14.16.

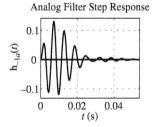

Figure 14.14
Step responses of the analog filter and its digital simulation by the step-invariant method

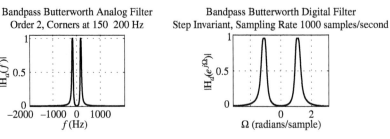

Figure 14.15
Magnitude frequency responses of the analog filter and its digital simulation by the step-invariant method

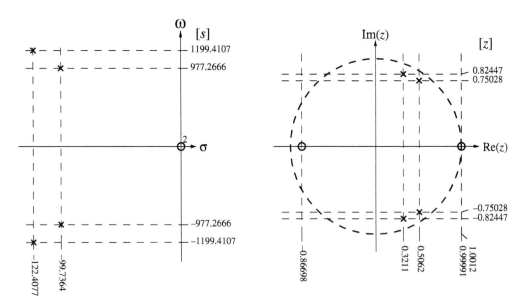

Figure 14.16
Pole-zero diagrams of the analog filter and its digital simulation by the step-invariant method

In contrast with the impulse invariant design, this digital filter has a response of zero at $\Omega = 0$. Also, the digital filter peak passband frequency response and the analog filter peak passband frequency response differ by less than 0.1%.

Finite-Difference Design Another method for designing digital filters to simulate analog filters is to approximate the differential equation describing the linear system with a difference equation. The basic idea in this method is to start with a desired transfer function of the analog filter $H_a(s)$ and find the differential equation corresponding to it in the time domain. Then continuous-time derivatives are approximated by finite differences in discrete time and the resulting expression is a digital-filter transfer function approximating the original analog filter transfer function. For example, suppose that

$$H_a(s) = \frac{1}{s+a}.$$

Since this is a transfer function it is the ratio of the response $Y_a(s)$ to the excitation $X_a(s)$.

$$\frac{Y_a(s)}{X_a(s)} = \frac{1}{s+a}$$

Then

$$Y_a(s)(s+a) = X_a(s).$$

Taking the inverse Laplace transform of both sides,

$$\frac{d}{dt}(y_a(t)) + a\,y_a(t) = x_a(t).$$

A derivative can be approximated by various finite-difference expressions and each choice has a slightly different effect on the approximation of the digital filter to the analog filter. Let the derivative in this case be approximated by the forward difference

$$\frac{d}{dt}(y_a(t)) \cong \frac{y_d[n+1] - y_d[n]}{T_s}.$$

Then the difference-equation approximation to the differential equation is

$$\frac{y_d[n+1] - y_d[n]}{T_s} + a\,y_d[n] = x_d[n]$$

and the corresponding recursion relation is

$$y_d[n+1] = x_d[n]T_s + (1 - aT_s)y_d[n].$$

The digital-filter transfer function can be found by z transforming the equation into

$$z(Y_d(z) - y_d[0]) = T_s X_d(z) + (1 - aT_s)Y_d(z).$$

Transfer functions are computed based on the assumption that the system is initially in its zero state. Therefore $y_d[0] = 0$ and

$$H_d(z) = \frac{Y_d(z)}{X_d(z)} = \frac{T_s}{z - (1 - aT_s)}. \tag{14.5}$$

A block diagram realization of this filter is illustrated in Figure 14.17.

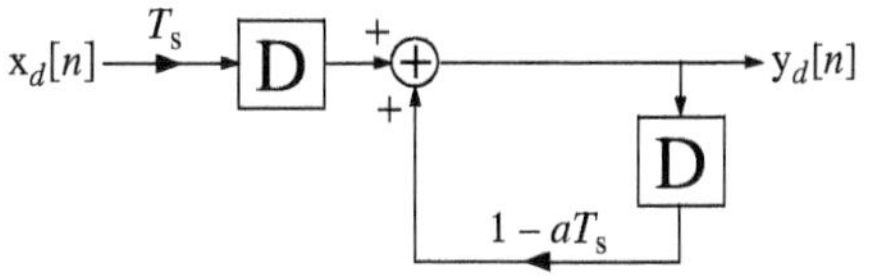

Figure 14.17
Block diagram of a digital-filter designed by approximating a differential equation with a difference equation using forward differences

The digital filter could also have been based on a backward-difference approximation to the derivative,

$$\frac{d}{dt}(y_a(t)) \cong \frac{y_d[n] - y_d[n-1]}{T_s}$$

or a central difference approximation to the derivative,

$$\frac{d}{dt}(y_a(t)) \cong \frac{y_d[n+1] - y_d[n-1]}{2T_s}.$$

We can systematize this method by realizing that every s in an s-domain expression represents a corresponding differentiation in the time domain,

$$\frac{d}{dt}(x_a(t)) \xleftrightarrow{\mathcal{L}} sX_a(s)$$

(again with the filter initially in its zero state). We can approximate derivatives with forward, backward or central differences,

$$\frac{d}{dt}(x_a(t)) \cong \frac{x_a(t+T_s) - x_a(t)}{T_s} = \frac{x_d[n+1] - x_d[n]}{T_s},$$

$$\frac{d}{dt}(x_a(t)) \cong \frac{x_a(t) - x_a(t-T_s)}{T_s} = \frac{x_d[n] - x_d[n-1]}{T_s}$$

or

$$\frac{d}{dt}(x_a(t)) \cong \frac{x_a(t+T_s) - x_a(t-T_s)}{2T_s} = \frac{x_d[n+1] - x_d[n-1]}{2T_s}.$$

The z transforms of these differences are

$$\frac{x_d[n+1] - x_d[n]}{T_s} \xleftrightarrow{\mathcal{Z}} \frac{z-1}{T_s} X_d(z),$$

$$\frac{x_d[n] - x_d[n-1]}{T_s} \xleftrightarrow{\mathcal{Z}} \frac{1-z^{-1}}{T_s} X_d(z) = \frac{z-1}{zT_s} X_d(z)$$

or

$$\frac{x_d[n+1] - x_d[n-1]}{2T_s} \xleftrightarrow{\mathcal{Z}} \frac{z-z^{-1}}{2T_s} X_d(z) = \frac{z^2-1}{2zT_s} X_d(z).$$

Now we can replace every s in an s-domain expression with the corresponding z-domain expression. Then we can approximate the s-domain transfer function,

$$H_a(s) = \frac{1}{s+a}$$

with a forward-difference approximation to a derivative,

$$H_d(z) = \left(\frac{1}{s+a}\right)_{s \to \frac{z-1}{T_s}} = \frac{1}{\frac{z-1}{T_s} + a} = \frac{T_s}{z - 1 + aT_s}, \tag{14.6}$$

which is exactly the same as (14.5). This avoids the process of actually writing the differential equation and substituting a finite difference for each derivative.

There is one aspect of finite-difference digital-filter design that must always be kept in mind. It is possible to approximate a stable analog filter and create an unstable digital filter using this method. Take the transfer function in (14.5) as an example. It has a pole at $z = 1 - aT_s$. The analog filter's pole is at $s = -a$. If the analog filter is stable $a > 0$ and $1 - aT_s$ is at a location $z = \text{Re}(z) < 1$ on the real axis of the z plane. If aT_s is greater than or equal to two, the z-plane pole is outside the unit circle and the digital filter is unstable.

A digital filter's transfer function can be expressed in partial fractions, one for each pole, and some poles may be complex. A pole at location $s = s_0$ in the s plane maps into a pole at $z = 1 + s_0 T_s$ in the z plane. So the mapping $s_0 \to 1 + s_0 T_s$ maps the ω axis of the s plane into the line $z = 1$ and the left half of the s plane into the region of the z plane to the left of $z = 1$. For stability the poles in the z plane should be inside the unit circle. Therefore this mapping does not guarantee a stable digital-filter design. The s_0's are determined by the analog filter so we cannot change them. Therefore, to solve the instability problem we could reduce T_s which means we would increase the sampling rate.

If, instead of using a forward difference, we had used a backward difference in (14.6) we would have gotten the digital-filter transfer function

$$H_d(z) = \left(\frac{1}{s+a}\right)_{s \to \frac{z-1}{zT_s}} = \frac{1}{\frac{z-1}{zT_s} + a} = \frac{zT_s}{z - 1 + azT_s} = \frac{1}{1 + aT_s} \frac{zT_s}{z - 1/(1 + aT_s)}.$$

Now the pole is at $z = 1/(1 + aT_s)$. The mapping $a \to 1/(1 + aT_s)$ maps positive values of a (for stable analog filters) into the real axis of the z plane between $z = 0$ and $z = 1$. The pole is inside the unit circle and the system is stable, regardless of the values of a and T_s. More generally, if the analog filter has a pole at $s = s_0$, the digital filter has a pole at $z = 1/(1 - s_0 T_s)$. This maps the ω axis in the s plane into a circle in the z plane of radius $1/2$ centered at $z = 1/2$ and maps the entire left-half of the s plane into the interior of that circle (Figure 14.18).

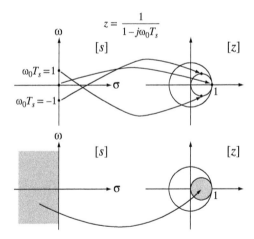

Figure 14.18
Mapping $z = 1/(1 - s_0 T_s)$

Although this mapping of poles guarantees a stable digital filter from a stable analog filter, it also restricts the type of digital filter that can be effectively designed using this method. Lowpass analog filters with poles on the negative real axis of the s plane become lowpass digital filters with poles on the real axis of the z plane in the interval $0 < z < 1$. If the analog filter has poles at $\sigma_0 \pm j\omega_0$ with $\omega_0 \gg \sigma_0$, meaning the analog filter is tuned to strongly respond at frequencies near ω_0, and if $\omega_0 T_s > 1$, the z-plane poles will not lie close to the unit circle and its response will not be nearly as strong near the equivalent discrete-time frequency.

EXAMPLE 14.4

Bandpass filter design using the finite-difference method

Using the difference-equation design method with a backward difference, design a digital filter to simulate the analog filter of Example 14.2 whose transfer function is

$$H_a(s) = \frac{9.87 \times 10^4 s^2}{s^4 + 444.3 s^3 + 2.467 \times 10^6 s^2 + 5.262 \times 10^8 s + 1.403 \times 10^{12}}$$

using the same sampling rate, $f_s = 1000$ samples/second. Compare the frequency responses of the two filters.

If we choose the same sampling rate as in Example 14.2, $f_s = 1000$, the z-domain transfer function is

$$H_d(z) = \frac{0.169 z^2 (z-1)^2}{z^4 - 1.848 z^3 + 1.678 z^2 - 0.7609 z + 0.1712}.$$

The impulse responses, magnitude frequency responses and pole-zero diagrams of the analog and digital filters are compared in Figure 14.19, Figure 14.20 and Figure 14.21.

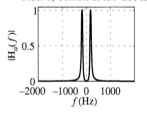

Figure 14.19
Impulse responses of the analog filter and its digital simulation using the finite-difference method

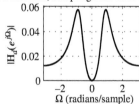

Figure 14.20
Magnitude frequency responses of the analog filter and its digital simulation using the finite-difference method

The digital filter impulse response does not look much like a sampling of the analog filter impulse response and the width of the digital filter passband is much too large. Also, the attenuation at higher frequencies is very poor. This result is much worse than the previous two designs.

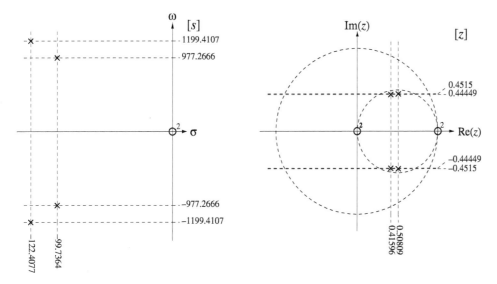

Figure 14.21
Pole-zero diagrams of the analog filter and its digital simulation using the finite-difference method

EXAMPLE 14.5

Lowpass filter design using the finite-difference method

Using the difference-equation design method with a forward difference, design a digital filter to simulate the analog filter whose transfer function is

$$H_a(s) = \frac{1}{s^2 + 600s + 4 \times 10^5}$$

using a sampling rate, $f_s = 500$ samples/second.
    The z-domain transfer function is

$$H_d(z) = \frac{1}{\left(\frac{z-1}{T_s}\right)^2 + 600\frac{z-1}{T_s} + 4 \times 10^5}$$

or

$$H_d(z) = \frac{T_s^2}{z^2 + (600T_s - 2)z + (1 - 600T_s + 4 \times 10^5 T_s^2)}$$

or

$$H_d(z) = \frac{4 \times 10^{-6}}{z^2 - 0.8z + 1.4}.$$

This result looks quite simple and straightforward but the poles of this z-domain transfer function are outside the unit circle and the filter is unstable, even though the s-domain transfer function is stable. Stability can be restored by increasing the sampling rate or by using a backward difference.

Frequency-Domain Methods

Direct Substitution and the Matched z-Transform A different approach to the design of digital filters is to find a direct change of variable from s to z that maps the s plane into the z plane, converts the poles and zeros of the s-domain transfer function into appropriate corresponding locations in the z plane and converts stable analog filters into stable digital filters. The most common techniques that use this idea are the **matched-z transform, direct substitution** and the **bilinear transformation**. This type of design process produces an IIR filter (Figure 14.22).

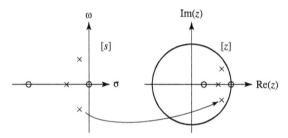

Figure 14.22
Mapping of poles and zeros from the s plane to the z plane

The direct substitution and matched z-transform methods are very similar. These methods are based on the idea of simply mapping the poles and zeros of the s-domain transfer function into the z domain through the relationship $z = e^{sT_s}$.

For example, to transform the analog filter frequency response,

$$H_d(s) = \frac{1}{s+a}$$

which has a pole at $s = -a$, we simply map the pole at $-a$ to the corresponding location in the z plane. Then the digital filter pole location is e^{-aT_s}. The direct substitution method implements the transformation $s - a \to z - e^{aT_s}$ while the matched z-transform method implements the transformation $s - a \to 1 - e^{aT_s}z^{-1}$. The z-domain transfer functions that result (in this case) are

Direct Substitution:

$$H_d(z) = \frac{1}{z - e^{-aT_s}} = \frac{z^{-1}}{1 - e^{-aT_s}z^{-1}}, \text{ with a pole at } z = e^{-aT_s} \text{ and no finite zeros}$$

Matched z-Transform:

$$H_d(z) = \frac{1}{1 - e^{-aT_s}z^{-1}} = \frac{z}{z - e^{-aT_s}}, \text{ with a pole at } z = e^{-aT_s} \text{ and a zero at } z = 0$$

Notice that the matched z-transform result is exactly the same result as was obtained using the impulse-invariant method and the direct substitution result is the same except for a single sample delay due to the z^{-1} factor. For more complicated s-domain transfer functions the results of these methods are not so similar. These methods do not directly involve any time-domain analysis. The design is done entirely in the s and z domains. The transformations $s - a \to z - e^{aT}$ and $s - a \to 1 - e^{aT}z^{-1}$ both map a pole in the open left-half of the s plane into a pole in the open interior of the unit circle in the z plane. Therefore, stable analog filters are transformed into stable digital filters.

EXAMPLE 14.6

Digital bandpass filter design using the matched-z transform

Using the matched-z transform design method, design a digital filter to simulate the analog filter of Example 14.2 whose transfer function is

$$H_a(s) = \frac{9.87 \times 10^4 s^2}{s^4 + 444.3s^3 + 2.467 \times 10^6 s^2 + 5.262 \times 10^8 s + 1.403 \times 10^{12}}$$

using the same sampling rate, $f_s = 1000$ samples/second. Compare the frequency responses of the two filters.

This transfer function has a double zero at $s = 0$ and poles at $s = -99.7 \pm j978$ and at $s = -122.4 \pm j1198.6$. Using the mapping,

$$s - a \rightarrow 1 - e^{aT} z^{-1},$$

we get a z-domain double zero at $z = 1$, a double zero at $z = 0$ and poles at

$$z = 0.5056 \pm j0.7506 \quad \text{and} \quad 0.3217 \pm j0.8242$$

and a z-domain transfer function,

$$H_d(z) = \frac{z^2(98700z^2 - 197400z + 98700)}{z^4 - 1.655z^3 + 2.252z^2 - 1.319z + 0.6413}$$

or

$$H_d(z) = 98700 \frac{z^2(z-1)^2}{z^4 - 1.655z^3 + 2.252z^2 - 1.319z + 0.6413}.$$

The impulse responses, magnitude frequency responses and pole-zero diagrams of the analog and digital filters are compared in Figure 14.23, Figure 14.24 and Figure 14.25.

If this design had been done using the direct substitution method, the only differences would be that the zeros at $z = 0$ would be removed, the impulse response would be the same except delayed two units in discrete time, the magnitude frequency response would be exactly the same and the phase of the frequency response would have a negative slope with a greater magnitude.

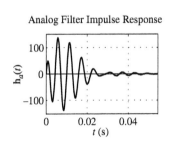
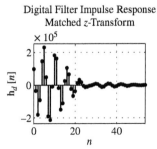

Figure 14.23
Impulse responses of the analog filter and its digital simulation by the matched-z transform method

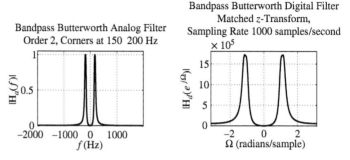

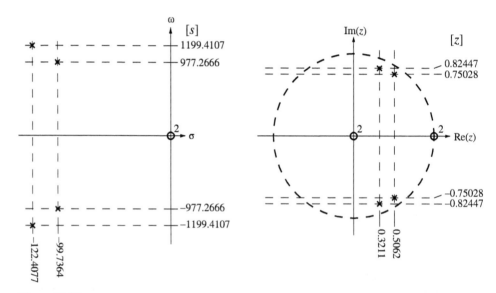

Figure 14.24
Frequency responses of the analog filter and its digital simulation by the matched-z transform method

Figure 14.25
Pole-zero diagrams of the analog filter and its digital simulation by the matched-z transform method

The Bilinear Method The impulse-invariant and step-invariant design techniques try to make the digital filter's discrete-time-domain response match the corresponding analog filter's continuous-time-domain response to a corresponding standard excitation. Another way to approach digital-filter design is to try to make the frequency response of the digital filter match the frequency response of the analog filter. But, just as a discrete-time-domain response can never exactly match a continuous-time-domain response, the frequency response of a digital filter cannot exactly match the frequency response of an analog filter. One reason, mentioned earlier, that the frequency responses cannot exactly match is that the frequency response of a digital filter is inherently periodic. When a sinusoidal continuous-time signal is sampled to create a sinusoidal discrete-time excitation, if the frequency of the continuous-time signal is changed by an integer multiple of the sampling rate, the discrete-time signal does not change at all. The digital filter cannot tell the difference and responds the same way as it would respond to the original signal (Figure 14.26).

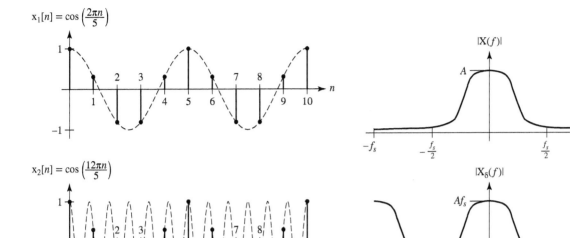

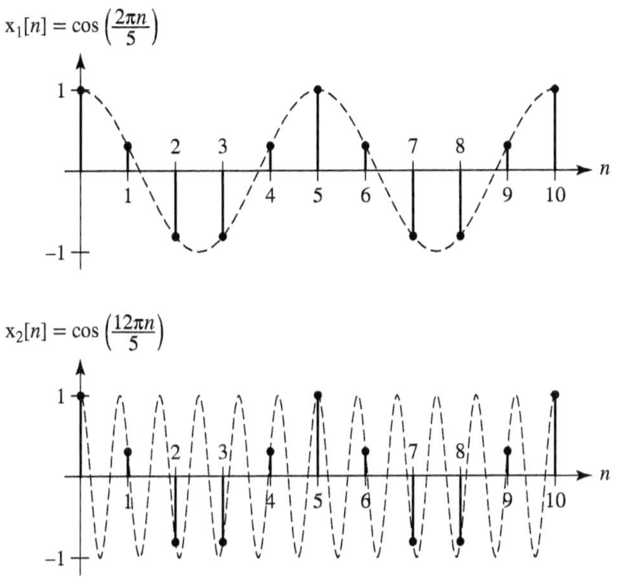

Figure 14.26
Two identical discrete-time signals formed by sampling two different sinusoids

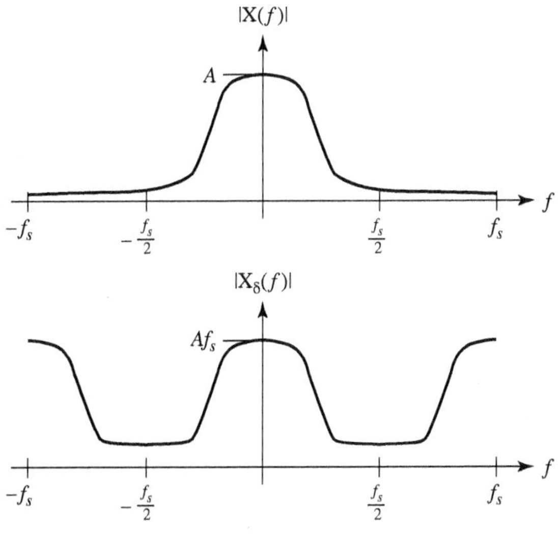

Figure 14.27
Magnitude spectrum of a continuous-time signal and a discrete-time signal formed by impulse sampling it

According to the sampling theorem, if a continuous-time signal can be guaranteed never to have any frequency components outside the range $|f| < f_s/2$, then when it is sampled at the rate f_s the discrete-time signal created contains all the information in the continuous-time signal. Then, when the discrete-time signal excites a digital filter, the response contains all the information in a corresponding continuous-time signal. So the design process becomes a matter of making the digital-filter frequency response match the analog filter frequency response only in the frequency range $|f| < f_s/2$, not outside. In general this still cannot be done exactly but it is often possible to make a good approximation. Of course, no signal is truly bandlimited. Therefore in practice we must arrange to have very little signal power beyond half the sampling rate instead of no signal power (Figure 14.27).

If a continuous-time excitation does not have any frequency components outside the range $|f| < f_s/2$, any nonzero response of an analog filter outside that range would have no effect because it has nothing to filter. Therefore, in the design of a digital filter to simulate an analog filter, the sampling rate should be chosen such that the response of the analog filter at frequencies $|f| > f_s/2$, is approximately zero. Then all the filtering action will occur at frequencies in the range $|f| < f_s/2$. So the starting point for a frequency-domain design process is to specify the sampling rate such that

$$X(f) \cong 0 \quad \text{and} \quad H_a(f) \cong 0, \quad |f| > f_s/2$$

or

$$X(j\omega) \cong 0 \quad \text{and} \quad H_a(j\omega) \cong 0, \quad |\omega| > \pi f_s = \omega_s/2.$$

Now the problem is to find a digital-filter transfer function that has approximately the same shape as the analog filter transfer function we are trying to simulate in the frequency range $|f| < f_s/2$. As discussed earlier, the straightforward method to accomplish this goal would be to use the transformation $e^{sT_s} \to z$ to convert a desired transfer function $H_a(s)$ into the corresponding $H_d(z)$. The transformation,

$e^{sT_s} \to z$, can be turned around into the form $s \to \ln(z)/T_s$. Then the design process would be

$$H_d(z) = H_a(s)\big|_{s \to \frac{1}{T_s}\ln(z)}.$$

Although this development of the transformation method is satisfying from a theoretical point of view, the functional transformation $s \to \ln(z)/T_s$ transforms an analog filter transfer function in the common form of the ratio of two polynomials into a digital-filter transfer function, which involves a ratio of polynomials, not in z but rather in $\ln(z)$, making the function transcendental with infinitely many poles and zeros. So, although this idea is appealing, it does not lead to a practical digital-filter design.

At this point it is common to make an approximation in an attempt to simplify the form of the digital-filter transfer function. One such transformation arises from the series expression for the exponential function

$$e^x = 1 + x + \frac{x^2}{2!} + \frac{x^3}{3!} + \cdots = \sum_{k=0}^{\infty} \frac{x^k}{k!}.$$

We can apply that to the transformation $e^{sT_s} \to z$ yielding

$$1 + sT_s + \frac{(sT_s)^2}{2!} + \frac{(sT_s)^3}{3!} + \cdots \to z.$$

If we approximate this series by the first two terms, we get

$$1 + sT_s \to z$$

or

$$s \to \frac{z-1}{T_s}.$$

The approximation $e^{sT_s} \cong 1 + sT_s$ is a good approximation if T_s is small and gets better as T_s gets smaller and, of course, f_s gets larger. That is, this approximation becomes very good at high sampling rates. Examine the transformation $s \to (z-1)/T_s$. A multiplication by s in the s domain corresponds to a differentiation with respect to t of the corresponding function in the continuous-time domain. A multiplication by $(z-1)/T_s$ in the z domain corresponds to a forward difference divided by the sampling time T_s of the corresponding function in the discrete-time domain. This is a forward-difference approximation to a derivative. As mentioned in the finite-difference method, the two operations, multiplication by s and by $(z-1)/T_s$ are analogous. So this method has the same problem as the finite-difference method using forward differences. A stable analog filter can become an unstable digital filter.

A very clever modification of this transformation solves the problem of creating an unstable digital filter from a stable analog filter and at the same time has other advantages. We can write the transformation from the s domain to the z domain as

$$e^{sT_s} = \frac{e^{sT_s/2}}{e^{-sT_s/2}} \to z,$$

approximate both exponentials with an infinite series

$$\frac{1 + \frac{sT_s}{2} + \frac{(sT_s/2)^2}{2!} + \frac{(sT_s/2)^3}{3!} + \cdots}{1 - \frac{sT_s}{2} + \frac{(sT_s/2)^2}{2!} - \frac{(sT_s/2)^3}{3!} + \cdots} \to z$$

and then truncate both series to two terms

$$\frac{1 + sT_s/2}{1 - sT_s/2} \to z$$

yielding

$$s \to \frac{2}{T_s}\frac{z-1}{z+1} \quad \text{or} \quad z \to \frac{2 + sT_s}{2 - sT_s}.$$

This mapping from s to z is called the **bilinear** z transform because the numerator and denominator are both linear functions of s or z. (Don't get the terms *bilinear* and *bilateral* z transform confused.) The bilinear z transform transforms any stable analog filter into a stable digital filter because it maps the entire open left half of the s plane into the open interior of the unit circle in the z plane. This was also true of matched-z transform and direct substitution but the correspondences are different. The mapping $z = e^{sT_s}$ maps any strip $\omega_0/T_s < \omega < (\omega_0 + 2\pi)/T_s$ of the s plane into the entire z plane. The mapping from s to z is unique but the mapping from z to s is not unique. The bilinear mapping $s \to (2/T_s)(z-1)/(z+1)$ maps each point in the s plane into a unique point in the z plane and the inverse mapping $z \to (2 + sT_s)/(2 - sT_s)$ maps each point in the z plane into a unique point in the s plane. To see how the mapping works consider the contour $s = j\omega$ in the s plane. Setting $z = (2 + sT_s)/(2 - sT_s)$ we get

$$z = \frac{2 + j\omega T_s}{2 - j\omega T_s} = 1\angle 2\tan^{-1}\left(\frac{\omega T_s}{2}\right) = e^{j2\tan^{-1}\left(\frac{\omega T_s}{2}\right)}$$

which lies entirely on the unit circle in the z plane. Also the contour in the z plane is traversed exactly once for $-\infty < \omega < \infty$. For the more general contour $s = \sigma_0 + j\omega$, σ_0 a constant, the corresponding contour is also a circle but with a different radius and centered on the Re(z) axis such that as ω approaches $\pm\infty$, z approaches -1 (Figure 14.28).

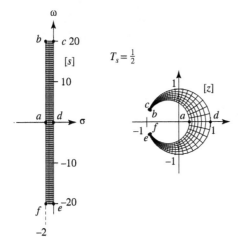

Figure 14.28
Mapping of an s-plane region into a corresponding z-plane region through the bilinear z transform

As the contours in the *s* plane move to the left, the contours in the *z* plane become smaller circles whose centers move closer to the $z = -1$ point. The mapping from *s* to *z* is a one-to-one mapping but the distortion of regions becomes more and more severe as *s* moves away from the origin. A higher sampling rate brings all poles and zeros in the *s* plane nearer to the $z = 1$ point in the *z* plane where the distortion is minimal. That can be seen by taking the limit as T_s approaches zero. In that limit, *z* approaches +1.

The important difference between the bilinear *z* transform method and the impulse invariant or matched *z*-transform methods is that there is no aliasing using the bilinear *z* transform because of the unique mapping between the *s* and *z* planes. However, there is a warping that occurs because of the way the $s = j\omega$ axis is mapped into the unit circle $|z| = 1$ and vice versa. Letting $z = e^{j\Omega}$, Ω real, determines the unit circle in the *z* plane. The corresponding contour in the *s* plane is

$$s = \frac{2}{T_s} \frac{e^{j\Omega} - 1}{e^{j\Omega} + 1} = j\frac{2}{T_s} \tan\left(\frac{\Omega}{2}\right)$$

and, since $s = \sigma + j\omega$, $\sigma = 0$ and $\omega = (2/T_s) \tan(\Omega/2)$ or, inverting the function, $\Omega = 2 \tan^{-1}(\omega T_s/2)$ (Figure 14.29).

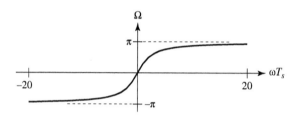

Figure 14.29
Frequency warping caused by the bilinear transformation

For low frequencies, the mapping is almost linear but the distortion gets progressively worse as we increase frequency because we are forcing high frequencies ω in the *s* domain to fit inside the range $-\pi < \Omega < \pi$ in the *z* domain. This means that the asymptotic behavior of an analog filter as *f* or ω approaches positive infinity occurs in the *z* domain at $\Omega = \pi$, which, through $\Omega = \omega T_s = 2\pi f T_s$, is at $f = f_s/2$, the Nyquist frequency. Therefore, the warping forces the full infinite range of continuous-time frequencies into the discrete-time frequency range $-\pi < \Omega < \pi$ with a nonlinear invertible function, thereby avoiding aliasing.

The MATLAB signal toolbox has a command **bilinear** for designing a digital filter using the bilinear transformation. The syntax is

[bd,ad] = bilinear(ba,aa,fs)

or

[zd,pd,kd] = bilinear(za,pa,ka,fs)

where **ba** is a vector of numerator coefficients in the analog filter transfer function, **aa** is a vector of denominator coefficients in the analog filter transfer function, **bd** is a vector of numerator coefficients in the digital-filter transfer function, **ad** is a vector of denominator coefficients in the digital-filter transfer function, **za** is a vector of analog filter zero locations, **pa** is a vector of analog filter pole locations, **ka** is the gain factor of the analog filter, `fs` is the sampling rate in samples/second, `zd` is a vector of digital filter zero locations, `pd` is a vector of digital filter pole locations and `kd` is the gain factor of the digital filter. For example,

```
»za = [] ; pa = -10 ; ka = 1 ; fs = 4 ;
»[zd,pd,kd] = bilinear(za,pa,ka,fs) ;
»zd
zd =
 -1
»pd
pd =
 -0.1111
»kd
kd =
 0.0556
```

EXAMPLE 14.7

Comparison of digital lowpass filter designs using the bilinear transformation with different sampling rates

Using the bilinear transformation, design a digital filter to approximate the analog filter whose transfer function is

$$H_a(s) = \frac{1}{s+10}$$

and compare the frequency responses of the analog and digital filters for sampling rates of 4 Hz, 20 Hz and 100 Hz.

Using the transformation $s \to \frac{2}{T_s}\frac{z-1}{z+1}$,

$$H_d(z) = \frac{1}{\frac{2}{T_s}\frac{z-1}{z+1}+10} = \left(\frac{T_s}{2+10T_s}\right)\frac{z+1}{z - \frac{2-10T_s}{2+10T_s}}.$$

For a 4 samples/second sampling rate,

$$H_d(z) = \frac{1}{18}\frac{z+1}{z+\frac{1}{9}}.$$

For a 20 samples/second sampling rate,

$$H_d(z) = \frac{1}{50}\frac{z+1}{z-\frac{3}{5}}.$$

For a 100 samples/second sampling rate,

$$H_d(z) = \frac{1}{210}\frac{z+1}{z-\frac{19}{21}}.$$

(Figure 14.30).

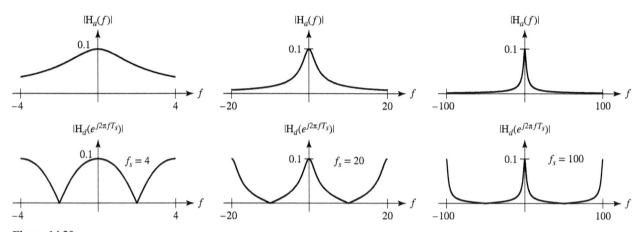

Figure 14.30
Magnitude frequency responses of the analog filter and three digital-filters designed using the bilinear transform and three different sampling rates

EXAMPLE 14.8

Digital bandpass filter design using the bilinear transformation

Using the bilinear-z transform design method, design a digital filter to simulate the analog filter of Example 14.2 whose transfer function is

$$H_a(s) = \frac{9.87 \times 10^4 s^2}{s^4 + 444.3s^3 + 2.467 \times 10^6 s^2 + 5.262 \times 10^8 s + 1.403 \times 10^{12}}$$

using the same sampling rate $f_s = 1000$ samples/second. Compare the frequency responses of the two filters.

Using the transformation $s \rightarrow (2/T_s)(z-1)/(z+1)$ and simplifying,

$$H_d(z) = \frac{12.38z^4 - 24.77z^2 + 12.38}{z^4 - 1.989z^3 + 2.656z^2 - 1.675z + 0.711}$$

or

$$H_d(z) = 12.38 \frac{(z+1)^2(z-1)^2}{z^4 - 1.989z^3 + 2.656z^2 - 1.675z + 0.711}.$$

The impulse responses, magnitude frequency responses and pole-zero diagrams of the analog and digital filters are compared in Figure 14.31, Figure 14.32 and Figure 14.33.

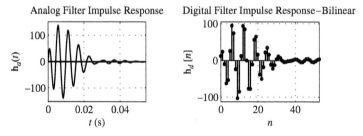

Figure 14.31
Impulse responses of the analog filter and its digital simulation by the bilinear-z transform method

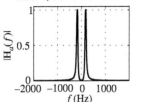
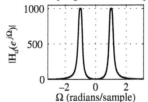

Figure 14.32
Magnitude frequency responses of the analog filter and its digital simulation by the bilinear method

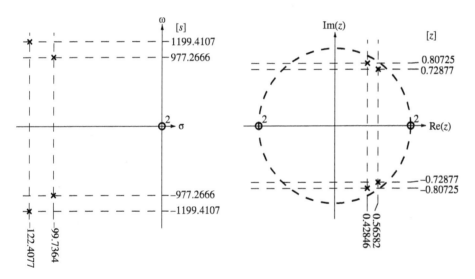

Figure 14.33
Pole-zero diagrams of the analog filter and its digital simulation by the bilinear-z transform method

---

FIR Filter Design

Truncated Ideal Impulse Response Even though the commonly used analog filters have infinite-duration impulse responses, because they are stable systems their impulse responses approach zero as time t approaches positive infinity. Therefore, another way of simulating an analog filter is to sample the impulse response, as in the impulse-invariant design method, but then truncate the impulse response beginning at discrete time $n = N$ where it has fallen to some low level, creating a finite-duration impulse response (Figure 14.34). Digital filters that have finite-duration impulse responses are called FIR filters.

The technique of truncating an impulse response can also be extended to approximating noncausal filters. If the part of an ideal filter's impulse response that occurs before time $t = 0$ is insignificant in comparison with the part that occurs after time $t = 0$, then it can be truncated, forming a causal impulse response. It can also be truncated after some later time when the impulse response has fallen to a low value, as previously described (Figure 14.35).

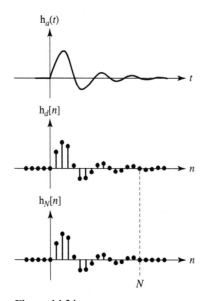

Figure 14.34
Truncation of an IIR impulse response to an FIR impulse response

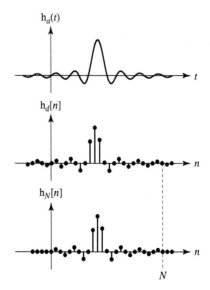

Figure 14.35
Truncation of a noncausal impulse response to a causal FIR impulse response

Of course, the truncation of an IIR response to an FIR response causes some difference between the impulse and frequency responses of the ideal analog and actual digital filters, but that is inherent in digital-filter design. So the problem of digital-filter design is still an approximation problem. The approximation is just done in a different way in this design method.

Once the impulse response has been truncated and sampled, the design of an FIR filter is quite straightforward. The discrete-time impulse response is in the form of a finite summation of discrete-time impulses

$$h_N[n] = \sum_{m=0}^{N-1} a_m \delta[n-m]$$

and can be realized by a digital filter of the form illustrated in Figure 14.36.

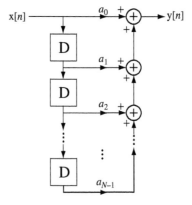

Figure 14.36
Prototypical FIR filter

One essential difference between this type of filter design and all the others presented so far is that there is no feedback of the response to combine with the excitation to produce the next response. This type of filter has only **feedforward** paths. Its transfer function is

$$H_d(z) = \sum_{m=0}^{N-1} a_m z^{-m}.$$

This transfer function has $N-1$ poles, all located at $z = 0$, and is absolutely stable regardless of the choice of the coefficients a.

This type of digital filter is an approximation to an analog filter. It is obvious what the difference between the two impulse responses is, but what are the differences in the frequency domain? The truncated impulse response is

$$h_N[n] = \begin{cases} h_d[n], & 0 \le n < N \\ 0, & \text{otherwise} \end{cases} = h_d[n]w[n]$$

and the DTFT is

$$H_N(e^{j\Omega}) = H_d(e^{j\Omega}) \circledast W(e^{j\Omega})$$

(Figure 14.37).

As the nonzero length of the truncated impulse increases, the frequency response approaches the ideal rectangular shape. The similarity in appearance to the convergence of a CTFS is not accidental. A truncated CTFS exhibits the Gibb's phenomenon in the reconstructed signal. In this case, the truncation occurs in the continuous-time domain and the ripple, which is the equivalent of the Gibb's phenomenon, occurs in the frequency domain. This phenomenon causes the effects marked as **passband ripple** and as **side lobes** in Figure 14.37. The peak amplitude of the passband ripple does not diminish as the truncation time increases but it is more and more confined to the region near the cutoff frequency.

We can reduce the ripple effect in the frequency domain, without using a longer truncation time, by using a "softer" truncation in the time domain. Instead of windowing the original impulse response with a rectangular function we could use a differently shaped window function that does not cause such large discontinuities in the truncated impulse response. There are many window shapes whose Fourier transforms have less ripple than a rectangular window's Fourier transform. Some of the most popular are the following:

1. von Hann or Hanning

$$w[n] = \frac{1}{2}\left[1 - \cos\left(\frac{2\pi n}{N-1}\right)\right], \quad 0 \le n < N$$

2. Bartlett

$$w[n] = \begin{cases} \dfrac{2n}{N-1}, & 0 \le n \le \dfrac{N-1}{2} \\ 2 - \dfrac{2n}{N-1}, & \dfrac{N-1}{2} \le n < N \end{cases}$$

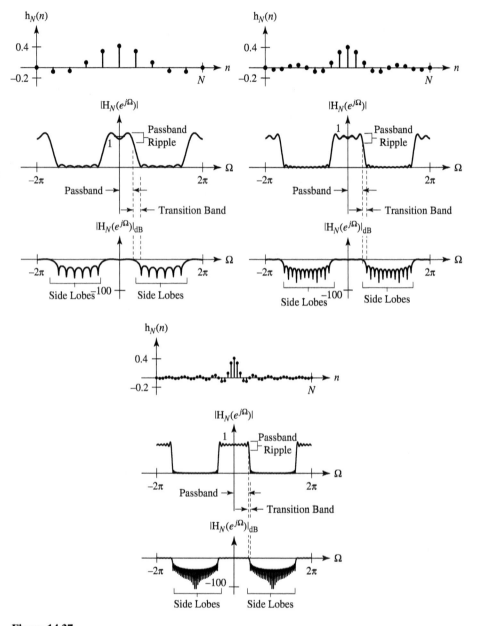

Figure 14.37
Three truncated ideal-lowpass-filter discrete-time impulse responses and their associated magnitude frequency responses

3. Hamming

$$w[n] = 0.54 - 0.46 \cos\left(\frac{2\pi n}{N-1}\right), \quad 0 \leq n < N$$

4. Blackman

$$w[n] = 0.42 - 0.5 \cos\left(\frac{2\pi n}{N-1}\right) + 0.08 \cos\left(\frac{4\pi n}{N-1}\right), \quad 0 \leq n < N$$

5. Kaiser

$$w[n] = \frac{I_0\left(\omega_a\sqrt{\left(\frac{N-1}{2}\right)^2 - \left(n - \frac{N-1}{2}\right)^2}\right)}{I_0\left(\omega_a \frac{N-1}{2}\right)}$$

where I_0 is the modified zeroth order Bessel function of the first kind and ω_a is a parameter that can be adjusted to trade off between transition-band width and side-lobe amplitude (Figure 14.38).

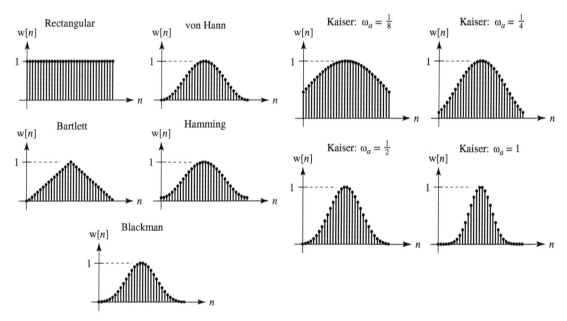

Figure 14.38
Window functions ($N = 32$)

The transforms of these window functions determine how the frequency response will be affected. The magnitudes of the transforms of these common window functions are illustrated in Figure 14.39.

Looking at the magnitudes of the transforms of the window functions, it is apparent that, for a fixed N, two design goals are in conflict. When approximating ideal filters with FIR filters we want a very narrow transition band and very high attenuation in the stopband. The transfer function of the FIR filter is the convolution of the ideal filter's transfer function with the transform of the window function. So the ideal window function would have a transform that is an impulse, and the corresponding window function would be an infinite-width rectangle. That is impossible, so we must compromise. If we use a finite-width rectangle, the transform is the Dirichlet function and we get the transform illustrated in Figure 14.39 for a rectangle, which makes a relatively fast transition from the peak of its central lobe to its first null, but then the sinc function rises again to a peak that is only about 13 dB below the maximum. When we convolve it with an ideal lowpass filter frequency response the transition band

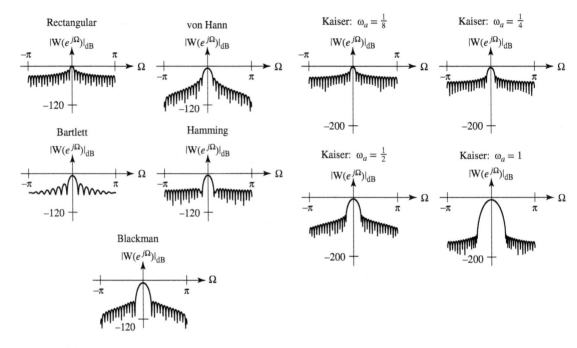

Figure 14.39
Magnitudes of z transforms of window functions ($N = 32$)

is narrow (compared to the other windows) but the stopband attenuation is not very good. Contrast that with the Blackman window. The central lobe width of its transform magnitude is more than twice that of the rectangle so the transition band will not be as narrow. But once the magnitude goes down, it stays more than 60 dB down. So its stopband attenuation is much better.

Another feature of an FIR filter that makes it attractive is that it can be designed to have a linear phase response. The general form of the FIR impulse response is

$$h_d[n] = h_d[0]\delta[n] + h_d[1]\delta[n-1] + \cdots + h_d[N-1]\delta[n-(N-1)],$$

its z transform is

$$H_d(z) = h_d[0] + h_d[1]z^{-1} + \cdots + h_d[N-1]z^{-(N-1)}$$

and the corresponding frequency response is

$$H_d(e^{j\Omega}) = h_d[0] + h_d[1]e^{-j\Omega} + \cdots + h_d[N-1]e^{-j(N-1)\Omega}.$$

The length N can be even or odd. First, let N be even and let the coefficients be chosen such that

$$h_d[0] = h_d[N-1], h_d[1] = h_d[N-2], \cdots, h_d[N/2-1] = h_d[N/2]$$

(Figure 14.40).

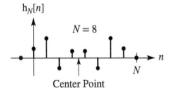

Figure 14.40
Example of a symmetric discrete-time impulse response for $N = 8$

This type of impulse response is **symmetric** about its center point. Then we can write the frequency response as

$$H_d(e^{j\Omega}) = \begin{cases} h_d[0] + h_d[0]e^{-j(N-1)\Omega} + h_d[1]e^{-j\Omega} + h_d[1]e^{-j(N-2)\Omega} + \cdots \\ + h_d[N/2-1]e^{-j(N/2-1)\Omega} + h_d[N/2-1]e^{-jN\Omega/2} \end{cases}$$

or

$$H_d(e^{j\Omega}) = e^{-j\left(\frac{N-1}{2}\right)\Omega} \begin{cases} h_d[0]\left(e^{j\left(\frac{N-1}{2}\right)\Omega} + e^{-j\left(\frac{N-1}{2}\right)\Omega}\right) \\ + h_d[1]\left(e^{j\left(\frac{N-3}{2}\right)\Omega} + e^{-j\left(\frac{N-3}{2}\right)\Omega}\right) + \cdots \\ + h_d[N/2-1](e^{-j\Omega} + e^{j\Omega}) \end{cases}$$

or

$$H_d(e^{j\Omega}) = 2e^{-j\left(\frac{N-1}{2}\right)\Omega} \begin{cases} h_d[0]\cos\left(\left(\frac{N-1}{2}\right)\Omega\right) + h_d[1]\cos\left(\left(\frac{N-3}{2}\right)\Omega\right) + \cdots \\ + h_d[N/2-1]\cos(\Omega) \end{cases}.$$

This frequency response consists of the product of a factor $e^{-j((N-1)/2)\Omega}$ that has a linear phase shift with frequency and some other factors, which have real values for all Ω. Therefore, the overall frequency response is linear with frequency (except for jumps of π radians at frequencies at which the sign of the real part changes). In a similar manner it can be shown that if the filter coefficients are **antisymmetric**, meaning

$$h_d[0] = -h_d[N-1], h_d[1] = -h_d[N-2], \quad \cdots, \quad h_d[N/2-1] = -h_d[N/2]$$

then the phase shift is also linear with frequency. For N odd the results are similar. If the coefficients are symmetric,

$$h_d[0] = h_d[N-1], \quad h_d[1] = h_d[N-2], \quad \cdots, \quad h_d\left[\frac{N-3}{2}\right] = h_d\left[\frac{N+1}{2}\right]$$

or antisymmetric,

$$h_d[0] = -h_d[N-1], h_d[1] = -h_d[N-2], \quad \cdots, \quad h_d\left[\frac{N-3}{2}\right]$$

$$= -h_d\left[\frac{N+1}{2}\right], h_d\left[\frac{N-1}{2}\right] = 0$$

the phase frequency response is linear. Notice that in the case of N odd there is a center point and, if the coefficients are antisymmetric, the center coefficient $h_d[(N-1)/2]$ must be zero (Figure 14.41).

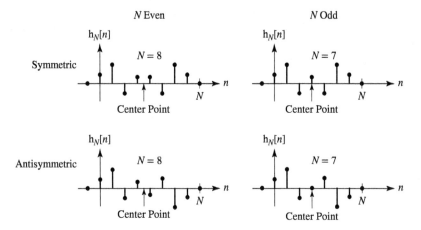

Figure 14.41
Examples of symmetric and antisymmetric discrete-time impulse responses for N even and N odd

EXAMPLE 14.9

Digital lowpass FIR filter design by truncating the ideal impulse response

Using the FIR method, design a digital filter to approximate a single-pole lowpass analog filter whose transfer function is

$$H_a(s) = \frac{a}{s+a}.$$

Truncate the analog filter impulse response at three time constants and sample the truncated impulse response with a time between samples that is one-fourth of the time constant, forming a discrete-time function. Then divide that discrete-time function by a to form the discrete-time impulse response of the digital filter.

(a) Find and graph the magnitude frequency response of the digital filter versus discrete-time radian frequency Ω.
(b) Repeat part (a) for a truncation time of five time constants and a sampling rate of 10 samples per time constant.

The impulse response is

$$h_a(t) = ae^{-at}\,u(t).$$

The time constant is $1/a$. Therefore the truncation time is $3/a$, the time between samples is $1/4a$, and the samples are taken at discrete times $0 \le n \le 12$. The FIR impulse response is then

$$h_d[n] = ae^{-n/4}(u[n] - u[n-12]) = a\sum_{m=0}^{11} e^{-m/4}\delta[n-m].$$

The z-domain transfer function is

$$H_d(z) = a\sum_{m=0}^{11} e^{-m/4}z^{-m}$$

and the frequency response is

$$H_d(e^{j\Omega}) = a\sum_{m=0}^{11} e^{-m/4}(e^{j\Omega})^{-m} = a\sum_{m=0}^{11} e^{-m(1/4+j\Omega)}.$$

For the second sampling rate in part (b), the truncation time is $5/a$, the time between samples is $1/10a$, and the samples are taken at discrete times, $0 \leq n \leq 50$. The FIR impulse response is then

$$h_d[n] = ae^{-n/10}(u[n] - u[n-50]) = a \sum_{m=0}^{49} e^{-m/4} \delta[n-m].$$

The z-domain transfer function is

$$H_d(z) = a \sum_{m=0}^{49} e^{-m/10} z^{-m}$$

and the frequency response is

$$H_d(e^{j\Omega}) = a \sum_{m=0}^{49} e^{-m/10}(e^{j\Omega})^{-m} = a \sum_{m=0}^{49} e^{-m(1/10+j\Omega)}$$

(Figure 14.42).

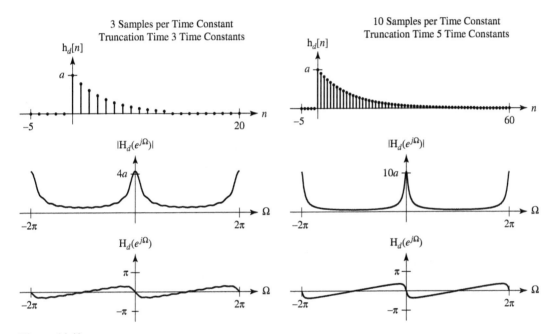

Figure 14.42
Impulse responses and frequency responses for the two FIR designs

The effects of truncation of the impulse response are visible as the ripple in the frequency response of the first FIR design with the lower sampling rate and shorter truncation time.

∎

EXAMPLE 14.10

Communication-channel digital-filter design

A range of frequencies between 900 MHz and 905 MHz is divided into 20 equal-width channels in which wireless signals may be transmitted. To transmit in any of the channels, a transmitter

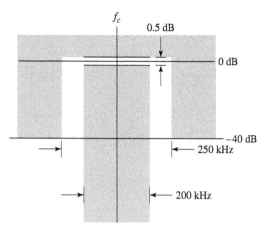

Figure 14.43
Specification for spectrum of the transmitted signal

must send a signal whose amplitude spectrum fits within the constraints of Figure 14.43. The transmitter operates by modulating a sinusoidal carrier whose frequency is the center frequency of one of the channels, with the baseband signal. Before modulating the carrier, the baseband signal, which has an approximately flat spectrum, is prefiltered by an FIR filter, which ensures that the transmitted signal meets the constraints of Figure 14.43. Assuming a sampling rate of 2 million samples/second, design the filter.

We know the shape of the ideal baseband analog lowpass filter's impulse response

$$h_a(t) = 2Af_m \operatorname{sinc}(2f_m(t - t_0))$$

where f_m is the corner frequency. The sampled impulse response is

$$h_d[n] = 2Af_m \operatorname{sinc}(2f_m(nT_s - t_0)).$$

We can set the corner frequency of the ideal lowpass filter to about halfway between 100 kHz and 125 kHz, say 115 kHz or 5.75% of the sampling rate. Let the gain constant A be one. The time between samples is 0.5 μs. The filter will approach the ideal as its length approaches infinity. As a first try set the mean-squared difference between the filter's impulse response and the ideal filter's impulse response to be less than 1% and use a rectangular window. We can iteratively determine how long the filter must be by computing the mean-squared difference between the filter and a very long filter. Enforcing a mean-squared error of less than 1% sets a filter length of 108 or more. This design yields the frequency responses of Figure 14.44.

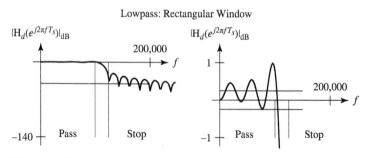

Figure 14.44
Frequency response of an FIR filter with a rectangular window and less than 1% error in impulse response

This design is not good enough. The passband ripple is too large and the stopband attenuation is not great enough. We can reduce the ripple by using a different window. Let's try a Blackman window with every other parameter remaining the same (Figure 14.45).

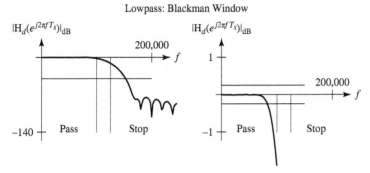

Figure 14.45
Frequency response of an FIR filter with a Blackman window and less than 1% error in impulse response

This design is also inadequate. We need to make the mean-squared error smaller. Making the mean-squared error less than 0.25% sets a filter length of 210 and yields the magnitude frequency response in Figure 14.46.

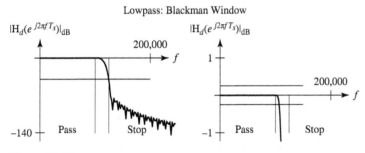

Figure 14.46
Frequency response of an FIR filter with a Blackman window and less than 0.25% error in impulse response

This filter meets specifications. The stopband attenuation just barely meets the specification and the passband ripple easily meets specification. This design is by no means unique. Many other designs with slightly different corner frequencies, mean-squared errors or windows could also meet the specification. ■

Optimal FIR Filter Design   There is a technique for designing filters without windowing impulse responses or approximating standard analog filter designs. It is called Parks-McClellan optimal equiripple design and was developed by Thomas W. Parks and James H. McClellan in the early 1970s. It uses an algorithm developed in 1934 by Evgeny Yakovlevich Remez called the Remez exchange algorithm. An explanation of the method is beyond the scope of this text but it is important enough that students should be aware of it and able to use it to design digital filters.

The Parks-McClellan digital-filter design is implemented in MATLAB through the command `firpm` with the syntax

$$B = firpm(N,F,A)$$

where B is a vector of N+1 real symmetric coefficients in the impulse response of the FIR filter, which has the best approximation to the desired frequency response described by F and A. F is a vector of frequency band edges in pairs, in ascending order between 0 and 1 with 1 corresponding to the Nyquist frequency or half the sampling frequency. At least one frequency band must have a nonzero width. A is a real vector the same size as F, which specifies the desired amplitude of the frequency response of the resultant filter B. The desired response is the line connecting the points (F(k),A(k)) and (F(k+1),A(k+1)) for odd k. `firpm` treats the bands between F(k+1) and F(k+2) for odd k as transition bands. Thus the desired amplitude is piecewise linear with transition bands.

This description serves only as an introduction to the method. More detail can be found in MATLAB's help description.

EXAMPLE 14.11

Parks-McClellan design of a digital bandpass filter

Design an optimal equiripple FIR filter to meet the magnitude frequency response specification in Figure 14.47.

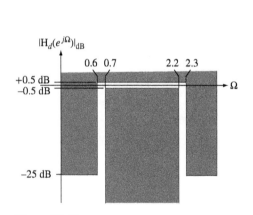

Figure 14.47
Bandpass filter specification

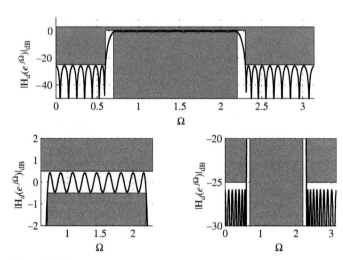

Figure 14.48
Frequency response of an optimal equiripple FIR bandpass filter with $N = 70$

The band edges are at $\Omega = \{0, 0.6, 0.7, 2.2, 2.3, \pi\}$ and the desired amplitude responses at those band edges are $A = \{0, 0, 1, 1, 0, 0\}$. The vector F should therefore be

$$F = \Omega/\pi = \{0, 0.191, 0.2228, 0.7003, 0.7321, 1\}$$

After a few choices of N, it was found that a filter with $N = 70$ met the specification (Figure 14.48).

MATLAB Design Tools

In addition to the MATLAB features already mentioned in earlier chapters and in earlier sections of this chapter, there are many other commands and functions in MATLAB that can help in the design of digital filters.

Probably the most generally useful function is the function `filter`. This is a function that actually digitally filters a vector of data representing a finite-time piece of a discrete-time signal. The syntax is `y=filter(bd,ad,x)` where `x` is the vector of data to be filtered and `bd` and `ad` are vectors of coefficients in the recursion relation for the filter. The recursion relation is of the form,

$$ad(1)*y(n) = bd(1)*x(n) + bd(2)*x(n-1) + \ldots + bd(nb+1)*x(n-nb)$$
$$- ad(2)*y(n-1) - \ldots - ad(na+1)*y(n-na).$$

(written in MATLAB syntax, which uses (·) for arguments of all functions without making a distinction between continuous-time and discrete-time functions). A related function is `filtfilt`. It operates exactly like `filter` except that it filters the data vector in the normal sense and then filters the resulting data vector backward. This makes the phase shift of the overall filtering operation identically zero at all frequencies and doubles the magnitude effect (in dB) of the filtering operation.

There are four related functions, each of which designs a digital filter. The function `butter` designs an *N*th order lowpass Butterworth digital filter through the syntax `[bd,ad]=butter[N,wn]` where `N` is the filter order and `wn` is the corner frequency expressed as a fraction of *half the sampling rate* (not the sampling rate itself). The function returns filter coefficients `bd` and `ad`, which can be used directly with `filter` or `filtfilt` to filter a vector of data. This function can also design a bandpass Butterworth filter simply by making `wn` a row vector of two corner frequencies of the form `[w1,w2]`. The passband of the filter is then $w1 < w < w2$ in the same sense of being fractions of *half the sampling rate*. By adding a string `'high'` or `'stop'` this function can also design highpass and bandstop digital filters.

Examples:

`[bd,ad] = butter[3,0.1]`	lowpass third-order Butterworth filter, corner frequency $0.5f_s$
`[bd,ad] = butter[4,[0.1 0.2]]`	bandpass fourth-order Butterworth filter, corner frequencies of $0.05f_s$ and $0.1f_s$
`[bd,ad] = butter[4,0.02,'high']`	highpass fourth-order Butterworth filter, corner frequency $0.1f_s$
`[bd,ad] = butter[2,[0.32 0.34],'stop']`	bandstop second-order Butterworth filter, corner frequencies $0.16f_s$ and $0.17f_s$

(There are also alternate syntaxes for `butter`. Type `help butter` for details. It can also be used to do analog filter design.)

The other three related digital-filter design functions are `cheby1`, `cheby2` and `ellip`. They design Chebyshev and Elliptical filters. Chebyshev and Elliptical filters have a narrower transition region for the same filter order than Butterworth filters but do so at the expense of passband and/or stopband ripple. Their syntax is similar except that maximum allowable ripple in dB must also be specified in the passband and the minimum attenuation in dB must be specified in the stop band.

Several standard window functions are available for use with FIR filters. They are `bartlett`, `blackman`, `boxcar` (rectangular), `chebwin` (Chebyshev), `hamming`, `hanning` (von Hann), `kaiser` and `triang` (similar to, but not identical to, `bartlett`).

The function **freqz** finds the frequency response of a digital filter in a manner similar to the operation of the function **freqs** for analog filters. The syntax of **freqz** is

$$[H,W] = \text{freqz}(bd,ad,N) ;$$

where **H** is the complex frequency response of the filter, **W** is a vector of discrete-time frequencies in radians (not radians per second because it is a discrete-time frequency) at which **H** is computed, **bd** and **ad** are vectors of coefficients of the numerator and denominator of the digital-filter transfer function and **N** is the number of points.

The function **upfirdn** changes the sampling rate of a signal by upsampling, FIR filtering and down sampling. Its syntax is

$$y = \text{upfirdn}(x,h,p,q) ;$$

where **y** is the signal resulting from the change of sampling rate, **x** is the signal whose sampling rate is to be changed, **h** is the impulse response of the FIR filter, **p** is the factor by which the signal is upsampled by zero insertion before filtering and **q** is the factor by which the signal is downsampled (decimated) after filtering.

These are by no means all of the digital signal processing capabilities of MATLAB. Type **help signal** for other functions.

Example 14.12

Filtering a discrete-time pulse with a highpass Butterworth filter using MATLAB

Digitally filter the discrete-time signal

$$x[n] = u[n] - u[n - 10]$$

with a third-order highpass digital Butterworth filter whose discrete-time radian corner frequency is $\pi/6$ radians.

```
%    Use 30 points to represent the excitation, x, and the response, y

N = 30 ;

%    Generate the excitation signal

n = 0:N-1 ; x = uDT(n) - uDT(n-10) ;

%    Design a third-order highpass digital Butterworth filter

[bd,ad] = butter(3,1/6,'high') ;

%    Filter the signal

y = filter(bd,ad,x) ;
```

The excitation and response are illustrated in Figure 14.49.

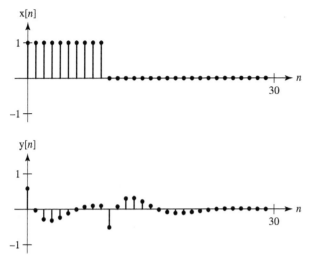

Figure 14.49
Excitation and response of a third-order highpass digital Butterworth filter

14.4 SUMMARY OF IMPORTANT POINTS

1. The Butterworth filter is maximally flat in both the pass and stopbands and all its poles lie on a semicircle in the left half of the s plane.
2. A lowpass Butterworth filter can be transformed into a highpass, bandpass or bandstop filter by appropriate variable changes.
3. Chebyshev, Elliptic and Bessel filters are filters optimized on a different basis than Butterworth filters. They can also be designed as lowpass filters and then transformed into highpass, bandpass or bandstop filters.
4. One popular design technique for digital filters is to simulate a proven analog filter design.
5. Two broad classifications of digital filters are inifinite-duration impulse response (IIR) and finite-duration impulse response (FIR).
6. The most popular types of IIR digital-filter design are the impulse invariant, step invariant, finite difference, direct substitution, matched z and bilinear methods.
7. FIR filters can be designed by windowing ideal impulse responses or by the Parks-McClellan equiripple algorithm.

EXERCISES WITH ANSWERS

(Answers to each exercise are in random order.)

Continuous-Time Filters

1. Using only a calculator, find the transfer function of a third-order ($n = 3$) lowpass Butterworth filter with cutoff frequency, $\omega_c = 1$, and unity gain at zero frequency.

   Answer: $\dfrac{1}{s^3 + 2s^2 + 2s + 1}$

2. Using MATLAB, find the transfer function of an eighth-order lowpass Butterworth filter with cutoff frequency, $\omega_c = 1$, and unity gain at zero frequency.

   Answer: $\dfrac{1}{\begin{bmatrix} s^8 + 5.126s^7 + 13.1371s^6 + 21.8462s^5 + 25.6884s^4 \\ +21.8462s^3 + 13.1371s^2 + 5.126s + 1 \end{bmatrix}}$

3. What are the numerical s-plane finite pole and finite zero locations for Butterworth filters of order N with corner frequencies f_c in Hz or ω_c in radians/second. (For repeated poles or zeros list them multiple times.)

   (a) Lowpass, $N = 2$, $\omega_c = 25$

   (b) Highpass, $N = 2$, $f_c = 5$

   Answers: {Poles at $10\pi/e^{\pm j3\pi/4} = 22.2(1 \pm j)$ and Double Zero at zero.},
   {Poles at $25e^{\pm j3\pi/4} = 17.68(-1 \pm j)$, No finite zeros.}

4. Using MATLAB design Chebyshev Type 1 and Elliptic fourth-order analog highpass filters with a cutoff frequency of 1 kHz. Let the allowed ripple in the passband be 2 dB and let the minimum stopband attenuation be 60 dB. Graph the magnitude Bode diagram of their frequency responses on the same scale for comparison. How wide is the transition band for each filter?

   Answers: The transition band for the Elliptic filter goes from 445 Hz to 1000 Hz for a width of 555 Hz. The transition band for the Chebyshev Type 1 filter goes from 274 Hz to 1000 Hz for a width of 726 Hz.

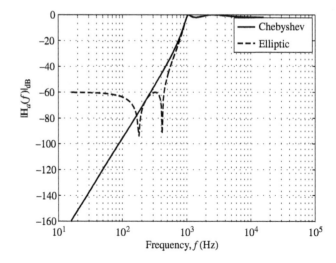

Finite-Difference Filter Design

5. What is the transition from s to z that is used in the difference-equation design technique to approximate a backward difference?

   Answer: $s \to \dfrac{1 - z^{-1}}{T_s}$

6. Using the finite-difference technique with backward differences, design a digital filter that approximates the lowpass filter whose transfer function is $H(s) = \frac{1}{s+1}$. Is there a finite sampling rate for which the digital filter is unstable? If so, provide one.

   Answer: $\frac{zT_s}{z(1+T_s)-1}$, Filter is absolutely stable.

7. Using the finite-difference method and all backward differences, design digital filters to approximate analog filters with these transfer functions. In each case, if a sampling frequency is not specified choose a sampling frequency that is 10 times the magnitude of the distance of the farthest pole or zero from the origin of the "s" plane. Graphically compare the step responses of the digital and analog filters.

   (a) $H_a(s) = s$, $f_s = 1$ MHz

   (b) $H_a(s) = 1/s$, $f_s = 1$ kHz

   (c) $H_a(s) = \frac{2}{s^2 + 3s + 2}$

Answers:

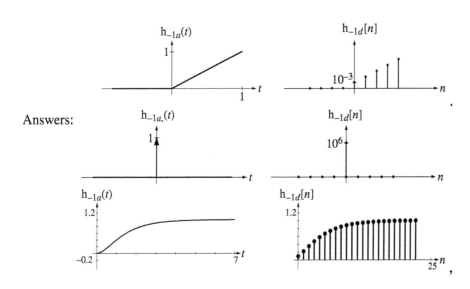

Matched z-Transform and Direct Substitution Filter Design

8. A continuous-time filter with transfer function $H(s) = \frac{10(s-8)}{s^2 + 7s + 12}$ is approximated by a digital filter using the matched z-transform technique. It is desired that all the zeros and poles of the digital-filter's transfer function $H(z)$ be located within a distance of 0.2 of the $z = 1$ point in the z plane. What is the minimum numerical sampling rate required?

   Answer: 43.88

9. A digital-filter designed by the matched z-transform technique using a sampling rate of 10 samples/second has a pole at $z = 0.5$. The sampling rate is changed to 50 samples/second. What is the new numerical location of the pole?

Answer: 0.8706

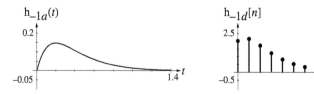

Bilinear z-Transform Filter Design

10. The transfer function, $H(s) = \dfrac{s-2}{s(s+4)}$, is approximated by a digital-filter designed using the bilinear transformation with a sampling rate of 10 samples/second. The digital filter can be realized by a block diagram of the form shown in Figure E.10. Enter into the block diagram the numbers in the empty rectangles.

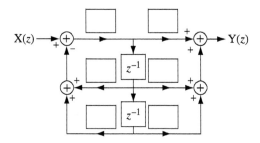

Figure E.10

Answers: 1, −0.1667, 0.667, 0.0375, −0.4125, −0.08333

FIR Filter Design

11. Using a rectangular window of width 50 and a sampling rate of 10,000 samples/second, design an FIR digital filter to approximate the analog filter whose transfer function is

$$H_a(s) = \dfrac{2000s}{s^2 + 2000s + 2 \times 10^6}$$

Compare the frequency responses of the analog and digital filters.

Answers: $h_d[n] = 2000\sqrt{2}(0.9048)^n \cos(0.1n + 0.7854)(u[n] - u[n-50])$,

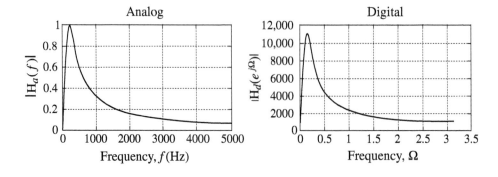

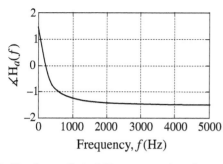
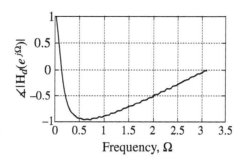

12. Design a digital filter approximation to each of these ideal analog filters by sampling a truncated version of the impulse response and using the specified window. In each case choose a sampling frequency which is 10 times the highest frequency passed by the analog filter. Choose the delays and truncation times such that no more than 1% of the signal energy of the impulse response is truncated. Graphically compare the magnitude frequency responses of the digital and ideal analog filters using a dB magnitude scale versus linear frequency.

Answers:

(a) Lowpass–Rectangular Window

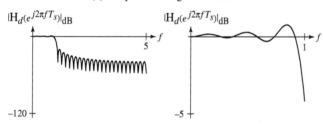

(b) Lowpass–Von Hann Window

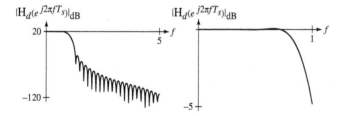

Digital-Filter Design Method Comparison

13. A continuous-time filter has a transfer function $H(s) = 4\dfrac{s-2}{s(s+2)}$. It is approximated by three digital-filter design methods, matched z-transform, direct substitution and bilinear z transform using a sampling rate $f_s = 2$. What are the numerical pole and zero locations of these digital filters?

Answers: {2.718, 1, 0.368}, {3, −1, 1, 0.333}, {0, 2.718, 1, 0.368}

EXERCISES WITHOUT ANSWERS

Analog Filter Design

14. Among Butterworth, Chebyshev Type 1, Chebyshev Type 2 and Elliptic filters, identify which type or types of lowpass filter have the characteristic described.

    (a) Magnitude frequency response in the passband monotonically approaches a slope of zero at zero frequency. (Monotonic means "always moving in the same direction." That is, always moving up or always moving down or "not rippling.")

(b) Magnitude frequency response in the stopband monotonically approaches zero as frequency approaches infinity.

(c) Fastest transition from passband to stopband for a given filter order.

(d) Slowest transition from passband to stopband for a given filter order.

15. Thermocouples are used to measure temperature in many industrial processes. A thermocouple is usually mechanically mounted inside a "thermowell," a metal sheath which protects it from damage by vibration, bending stress or other forces. One effect of the thermowell is that its thermal mass slows the effective time response of the thermocouple/thermowell combination compared with the inherent time response of the thermocouple alone. Let the actual temperature on the outer surface of the thermowell in Kelvins be $T_s(t)$ and let the voltage developed by the thermocouple in response to temperature be $v_t(t)$. The response of the thermocouple to a one-Kelvin step change in the thermowell outer-surface temperature from T_1 to $T_1 + 1$ is

$$v_t(t) = K\left[T_1 + \left(1 - e^{-\frac{t}{0.2}}\right)u t\right]$$

where K is the thermocouple temperature-to-voltage conversion constant.

(a) Let the conversion constant be $K = 40$ µV/K. Design an active filter that processes the thermocouple voltage and compensates for its time lag making the overall system have a response to a one-Kelvin step thermowell-surface temperature change that is itself a step of voltage of 1 mV.

(b) Suppose that the thermocouple also is subject to electromagnetic interference (EMI) from nearby high-power electrical equipment. Let the EMI be modeled as a sinusoid with an amplitude of 20 µV at the thermocouple terminals. Calculate the response of the thermocouple-filter combination to EMI frequencies of 1 Hz, 10 Hz and 60 Hz. How big is the apparent temperature fluctuation caused by the EMI in each case?

16. Design a Chebyshev Type 2 bandpass filter of minimum order to meet these specifications.

Passband: 4 kHz to 6 kHz, Gain between 0 dB and −2 dB
Stopband: <3 kHz and >8 kHz, Attenuation >60 dB

What is the minimum order?
Make a Bode diagram of its magnitude and phase frequency response and check it to be sure the pass and stopband specifications are met.
Make a pole-zero diagram.
What is the time of occurrence of the peak of its impulse response?

Impulse-Invariant and Step-Invariant Filter Design

17. Using the impulse-invariant design method, design a discrete-time system to approximate the continuous-time systems with these transfer functions at the

sampling rates specified. Compare the impulse and unit-step (or sequence) responses of the continuous-time and discrete-time systems.

(a) $H_a(s) = \dfrac{712s}{s^2 + 46s + 240}$, $f_s = 20$ Hz

(b) $H_a(s) = \dfrac{712s}{s^2 + 46s + 240}$, $f_s = 200$ Hz

Finite-Difference Filter Design

18. Using the difference-equation method and all backward differences, design digital filters to approximate analog filters with these transfer functions. In each case choose a sampling frequency which is 10 times the magnitude of the distance of the farthest pole or zero from the origin of the "s" plane. Graphically compare the step responses of the digital and analog filters.

(a) $H_a(s) = \dfrac{s^2}{s^2 + 3s + 2}$ (b) $H_a(s) = \dfrac{s + 60}{s^2 + 120s + 2000}$

(c) $H_a(s) = \dfrac{16s}{s^2 + 10s + 250}$

Matched z-Transform and Direct Substitution Filter Design

19. A continuous-time filter with a transfer function $H(s) = \dfrac{K}{s + 10}$ is approximated by a digital-filter designed using both the direct substitution and matched z-transform methods. Then the digital filters are used as the forward-path transfer functions $H_1(z)$ in two discrete-time feedback systems both with feedback transfer functions $H_2(z) = 1$. The gain K is gradually increased from 0 upward. Which system will go unstable at a finite value of K? Explain your answer.

Bilinear z-Transform Filter Design

FIR Filter Design

20. Graph the frequency response of an FIR filter designed using the Parks-McClellan algorithm that meets the specification in Figure E.20 with the shortest possible impulse response.

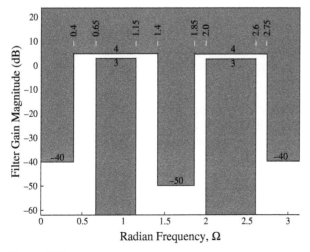

Figure E.20

Digital-Filter Design Method Comparison

21. A lowpass digital filter is designed first using the impulse-invariant technique and then using the step-invariant technique. Which one is guaranteed to have the correct response magnitude at zero frequency and why?

22. In designing a digital filter to approximate an analog lowpass filter, which design technique guarantees that the digital filter's magnitude response is zero at $\Omega = \pm\pi$?

23. A digital filter has a transfer function, $H_1(z) = \dfrac{z^2}{z^2 + az + b}$. If it is excited by a unit sequence, the first three values of its response, $y_1[n]$, are

$$y_1[0] = 3, \quad y_1[1] = -4, \quad y_1[2] = 2.$$

    If a digital filter with a transfer function, $H_2(z) = z^{-1}H_1(z)$, is excited by a unit sequence, what are the first three numerical values of its response, $y_2[n]$?

24. Of the three window types, Rectangular, von Hann and Blackman, if they are all used to design the same type of digital FIR filter with the same number of samples in the impulse response

    (a) Which one yields the narrowest transition from passband to stopband?

    (b) Which one yields the greatest stopband attenuation?

25. What is a disadvantage of finite-difference digital-filter design using forward differences?

26. Generally, in order to design a digital filter with a certain passband gain, transition bandwidth and stopband attenuation, which type of digital filter accomplishes the design goal with fewer multiplications and additions, IIR or FIR?

27. We have studied six methods for approximating an analog filter by an IIR digital filter.

    Impulse Invariant, Step Invariant, Finite Difference, Matched-z, Direct Substitution and Bilinear

    (a) Which method(s) can be done without using any time-domain functions in the process?

    (b) Which method(s) squeeze(s) the entire continuous-time frequency from zero to infinity into the discrete-time radian-frequency range zero to π.

    (c) Which two methods are almost the same, differing only by a time-domain delay?

    (d) Which method can design an unstable digital filter while trying to approximate a stable analog filter and, if constrained to only stable designs, restricts the flexibility of the design to only certain types of filters?

APPENDIX I

Useful Mathematical Relations

$$e^x = 1 + x + \frac{x^2}{2!} + \frac{x^3}{3!} + \frac{x^4}{4!} + \cdots$$

$$\sin(x) = x - \frac{x^3}{3!} + \frac{x^5}{5!} - \frac{x^7}{7!} + \cdots$$

$$\cos(x) = 1 - \frac{x^2}{2!} + \frac{x^4}{4!} - \frac{x^6}{6!} + \cdots$$

$$\cos(x) = \cos(-x) \quad \text{and} \quad \sin(x) = -\sin(-x)$$

$$e^{jx} = \cos(x) + j\sin(x)$$

$$\sin^2(x) + \cos^2(x) = 1$$

$$\cos(x)\cos(y) = \frac{1}{2}[\cos(x-y) + \cos(x+y)]$$

$$\sin(x)\sin(y) = \frac{1}{2}[\cos(x-y) - \cos(x+y)]$$

$$\sin(x)\cos(y) = \frac{1}{2}[\sin(x-y) + \sin(x+y)]$$

$$\cos(x+y) = \cos(x)\cos(y) - \sin(x)\sin(y)$$

$$\sin(x+y) = \sin(x)\cos(y) + \cos(x)\sin(y)$$

$$A\cos(x) + B\sin(x) = \sqrt{A^2 + B^2}\cos(x - \tan^{-1}(B/A))$$

$$\frac{d}{dx}[\tan^{-1}(x)] = \frac{1}{1+x^2}$$

$$\int u\,dv = uv - \int v\,du$$

$$\int x^n \sin(x)\, dx = -x^n \cos(x) + n \int x^{n-1} \cos(x)\, dx$$

$$\int x^n \cos(x)\, dx = x^n \sin(x) - n \int x^{n-1} \sin(x)\, dx$$

$$\int x^n e^{ax}\, dx = \frac{e^{ax}}{a^{n+1}}[(ax)^n - n(ax)^{n-1} + n(n-1)(ax)^{n-2} + \ldots + (-1)^{n-1} n!(ax) + (-1)^n n!], \quad n \geq 0$$

$$\int e^{ax} \sin(bx)\, dx = \frac{e^{ax}}{a^2 + b^2}[a \sin(bx) - b \cos(bx)]$$

$$\int e^{ax} \cos(bx)\, dx = \frac{e^{ax}}{a^2 + b^2}[a \cos(bx) + b \sin(bx)]$$

$$\int \frac{dx}{a^2 + (bx)^2} = \frac{1}{ab} \tan^{-1}\left(\frac{bx}{a}\right)$$

$$\int \frac{dx}{(x^2 \pm a^2)^{\frac{1}{2}}} = \ln\left|x + (x^2 \pm a^2)^{\frac{1}{2}}\right|$$

$$\int_0^\infty \frac{\sin(mx)}{x}\, dx = \begin{cases} \pi/2, & m > 0 \\ 0, & m = 0 \\ -\pi/2, & m < 0 \end{cases} = \frac{\pi}{2}\operatorname{sgn}(m)$$

$$|Z|^2 = ZZ^*$$

$$\sum_{n=0}^{N-1} r^n = \begin{cases} \frac{1-r^N}{1-r}, & r \neq 1 \\ N, & r = 1 \end{cases}$$

$$\sum_{n=0}^{\infty} r^n = \frac{1}{1-r}, \quad |r| < 1$$

$$\sum_{n=k}^{\infty} r^n = \frac{r^k}{1-r}, \quad |r| < 1$$

$$\sum_{n=0}^{\infty} n r^n = \frac{r}{(1-r)^2}, \quad |r| < 1$$

$$\frac{e^{j\pi n}}{e^{j\pi n/N_0}} \operatorname{drcl}\left(\frac{n}{N_0}, N_0\right) = \delta_{N_0}[n], \quad n \text{ and } N_0 \text{ integers}$$

$$\operatorname{drcl}\left(\frac{n}{2m+1}, 2m+1\right) = \delta_{2m+1}[n], \quad n \text{ and } m \text{ integers}$$

APPENDIX II

Continuous-Time Fourier Series Pairs

Continuous-time Fourier series (CTFS) for a periodic function with fundamental period $T_0 = 1/f_0 = 2\pi/\omega_0$ represented over the period T.

$$x(t) = \sum_{k=-\infty}^{\infty} c_x[k] e^{j2\pi kt/T} \xleftrightarrow[T]{\mathcal{FS}} c_x[k] = \frac{1}{T} \int_T x(t) e^{-j2\pi kt/T} dt$$

In these pairs k, n, and m are integers.

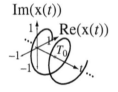

$$e^{j2\pi kt/T_0} \xleftrightarrow[mT_0]{\mathcal{FS}} \delta[k-m]$$

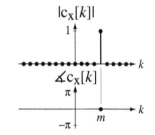

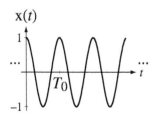

$$\cos(2\pi t/T_0) \xleftrightarrow[mT_0]{\mathcal{FS}} (1/2)(\delta[k-m] + \delta[k+m])$$

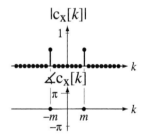

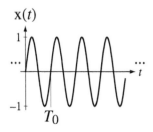

$$\sin(2\pi t/T_0) \xleftrightarrow[mT_0]{\mathcal{FS}} (j/2)(\delta[k+m] - \delta[k-m])$$

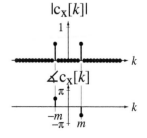

Appendix II Continuous-Time Fourier Series Pairs

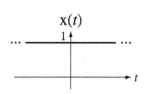

$$1 \xleftrightarrow[T]{\mathscr{FS}} \delta[k]$$

T is arbitrary

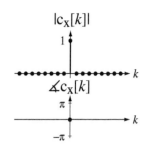

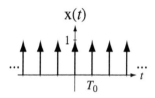

$$\delta_{T_0}(t) \xleftrightarrow[mT_0]{\mathscr{FS}} f_0 \delta_m[k]$$

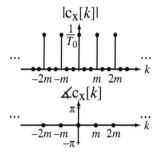

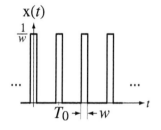

$$(1/w)\,\mathrm{rect}(t/w) * \delta_{T_0}(t) \xleftrightarrow[T_0]{\mathscr{FS}} f_0\,\mathrm{sinc}(wkf_0)$$

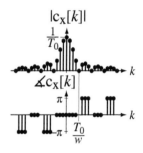

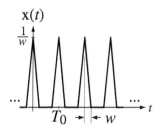

$$(1/w)\,\mathrm{tri}(t/w) * \delta_{T_0}(t) \xleftrightarrow[T_0]{\mathscr{FS}} f_0\,\mathrm{sinc}^2(wkf_0)$$

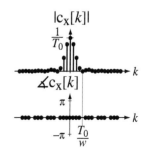

Appendix II Continuous-Time Fourier Series Pairs

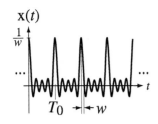

$$(1/w)\,\text{sinc}(t/w) * \delta_{T_0}(t) \xleftrightarrow[T_0]{\mathscr{FS}} f_0\,\text{rect}(wkf_0)$$

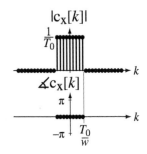

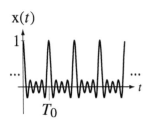

$$\text{drcl}(f_0 t, 2M+1) \xleftrightarrow[T_0]{\mathscr{FS}} \frac{u[n+M] - u[n-M-1]}{2M+1}$$

M an integer

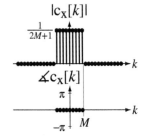

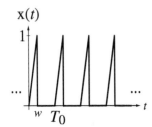

$$\frac{t}{w}[u(t) - u(t-w)] * \delta_{T_0}(t) \xleftrightarrow[T_0]{\mathscr{FS}}$$

$$\frac{1}{wT_0}\frac{[j(2\pi kw)/T_0 + 1]e^{-j(2\pi kw/T_0)} - 1}{(2\pi kw/T_0)^2}$$

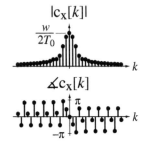

APPENDIX III

Discrete Fourier Transform Pairs

Discrete Fourier Transform (DFT) for a periodic discrete-time function with fundamental period N_0 represented over the period N.

$$x[n] = \frac{1}{N}\sum_{k=\langle N\rangle} X[k]e^{j2\pi kn/N} \xleftrightarrow[N]{\mathcal{DFT}} X[k] = \sum_{n=\langle N\rangle} x[n]e^{-j2\pi kn/N}$$

In all these pairs k, n, m, q, N_w, N_0, N, n_0, and n_1 are integers.

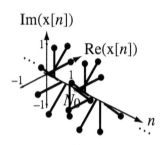

$$e^{j2\pi n/N_0} \xleftrightarrow[mN_0]{\mathcal{DFT}} mN_0 \delta_{mN_0}[k-m]$$

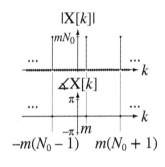

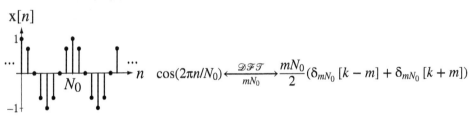

$$\cos(2\pi n/N_0) \xleftrightarrow[mN_0]{\mathcal{DFT}} \frac{mN_0}{2}(\delta_{mN_0}[k-m] + \delta_{mN_0}[k+m])$$

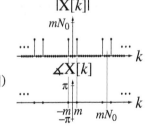

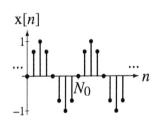

$$\sin(2\pi n/N_0) \xleftrightarrow[mN_0]{\mathcal{DFT}} \frac{jmN_0}{2}(\delta_{mN_0}[k+m] - \delta_{mN_0}[k-m])$$

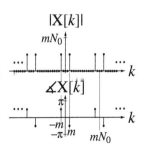

Appendix III Discrete Fourier Transform Pairs

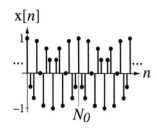

$$\cos(2\pi q n/N_0) \underset{mN_0}{\overset{\mathcal{DFT}}{\longleftrightarrow}} \frac{mN_0}{2}(\delta_{mN_0}[k - mq] + \delta_{mN_0}[k + mq])$$

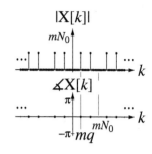

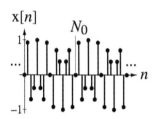

$$\sin(2\pi q n/N_0) \underset{mN_0}{\overset{\mathcal{DFT}}{\longleftrightarrow}} \frac{jmN_0}{2}(\delta_{mN_0}[k + mq] - \delta_{mN_0}[k - mq])$$

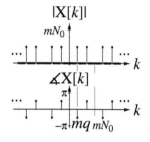

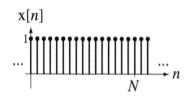

$$1 \underset{N}{\overset{\mathcal{DFT}}{\longleftrightarrow}} N\delta_N[k]$$

N is arbitrary

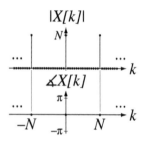

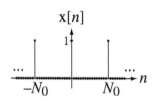

$$\delta_{N_0}[n] \underset{mN_0}{\overset{\mathcal{DFT}}{\longleftrightarrow}} m\delta_{mN_0}[k]$$

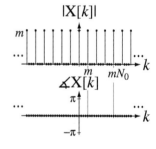

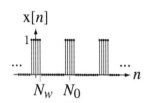

$$(u[n + N_w] - u[n - N_w - 1]) * \delta_{N_0}[n] \underset{N_0}{\overset{\mathcal{DFT}}{\longleftrightarrow}} (2N_w + 1)\,\mathrm{drcl}(k/N_0, 2N_w + 1)$$

N_w an integer

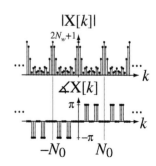

Appendix III Discrete Fourier Transform Pairs **A-9**

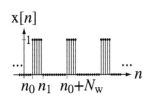

$$(\mathrm{u}[n - n_0] - \mathrm{u}[n - n_1]) * \delta_{N_0}[n] \xleftrightarrow[N_0]{\mathcal{DFT}}$$

$$\frac{e^{-j\pi k(n_1+n_0)/N_0}}{e^{-j\pi k/N_0}}(n_1 - n_0)\,\mathrm{drcl}(k/N_0, n_1 - n_0)$$

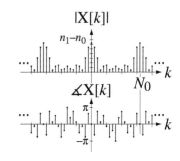

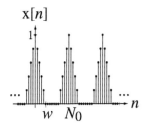

$$\mathrm{tri}(n/w) * \delta_{N_0}[n] \xleftrightarrow[N_0]{\mathcal{DFT}} w\,\mathrm{sinc}^2(wk/N_0) * \delta_{N_0}[k]$$

$$\mathrm{tri}(n/N_w) * \delta_{N_0}[n] \xleftrightarrow[N_0]{\mathcal{DFT}} N_w\,\mathrm{drcl}^2(k/N_0, N_w)$$

N_w an integer

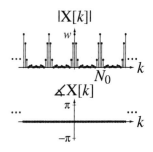

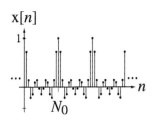

$$\mathrm{sinc}(n/w) * \delta_{N_0}[n] \xleftrightarrow[N_0]{\mathcal{DFT}} w\,\mathrm{rect}(wk/N_0) * \delta_{N_0}[k]$$

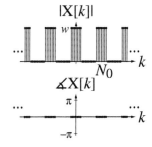

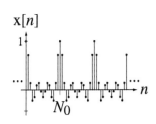

$$\mathrm{drcl}(n/N_0, 2M+1) \xleftrightarrow[N_0]{\mathcal{DFT}}$$

$$\frac{\mathrm{u}[n+M] - \mathrm{u}[n-M-1]}{2M+1} * N_0 \delta_{N_0}[k]$$

M an integer

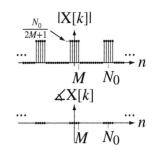

IV APPENDIX

Continuous-Time Fourier Transform Pairs

$$x(t) = \int_{-\infty}^{\infty} X(f) e^{+j2\pi ft} df \xleftrightarrow{\mathcal{F}} X(f) = \int_{-\infty}^{\infty} x(t) e^{-j2\pi ft} dt$$

$$x(t) = \frac{1}{2\pi} \int_{-\infty}^{\infty} X(j\omega) e^{+j\omega t} d\omega \xleftrightarrow{\mathcal{F}} X(j\omega) = \int_{-\infty}^{\infty} x(t) e^{-j\omega t} dt$$

For all the periodic time functions, the fundamental period is $T_0 = 1/f_0 = 2\pi/\omega_0$.

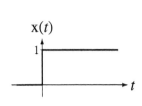

$$u(t) \xleftrightarrow{\mathcal{F}} (1/2)\delta(f) + 1/j2\pi f$$

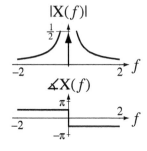

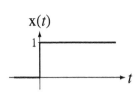

$$u(t) \xleftrightarrow{\mathcal{F}} \pi\delta(\omega) + 1/j\omega$$

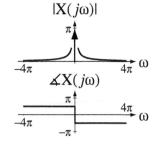

Appendix IV Continuous-Time Fourier Transform Pairs

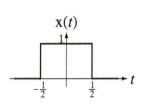

$$\text{rect}(t) \xleftrightarrow{\mathcal{F}} \text{sinc}(f)$$
$$\text{rect}(t) \xleftrightarrow{\mathcal{F}} \text{sinc}(\omega/2\pi)$$

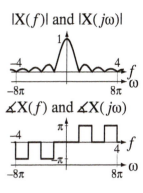

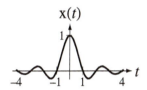

$$\text{sinc}(t) \xleftrightarrow{\mathcal{F}} \text{rect}(f)$$
$$\text{sinc}(t) \xleftrightarrow{\mathcal{F}} \text{rect}(\omega/2\pi)$$

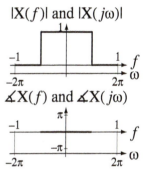

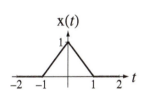

$$\text{tri}(t) \xleftrightarrow{\mathcal{F}} \text{sinc}^2(f)$$
$$\text{tri}(t) \xleftrightarrow{\mathcal{F}} \text{sinc}^2(\omega/2\pi)$$

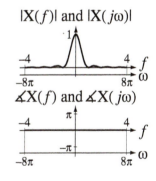

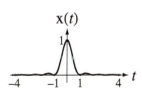

$$\text{sinc}^2(t) \xleftrightarrow{\mathcal{F}} \text{tri}(f)$$
$$\text{sinc}^2(t) \xleftrightarrow{\mathcal{F}} \text{tri}(\omega/2\pi)$$

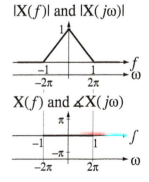

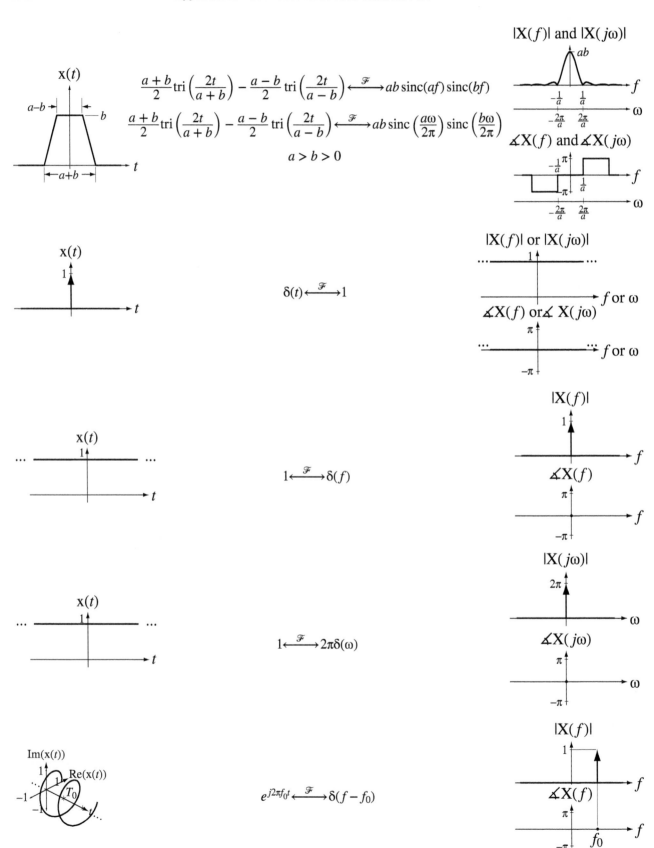

Appendix IV Continuous-Time Fourier Transform Pairs

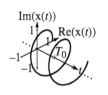

$$e^{j\omega_0 t} \xleftrightarrow{\mathscr{F}} 2\pi\delta(\omega - \omega_0)$$

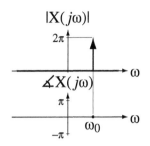

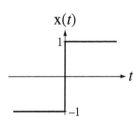

$$\mathrm{sgn}(t) \xleftrightarrow{\mathscr{F}} 1/j\pi f$$

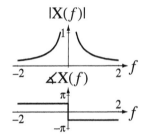

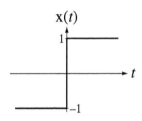

$$\mathrm{sgn}(t) \xleftrightarrow{\mathscr{F}} 2/j\omega$$

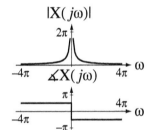

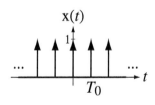

$$\delta_{T_0}(t) \xleftrightarrow{\mathscr{F}} f_0 \delta_{f_0}(f)$$

$$f_0 = 1/T_0$$

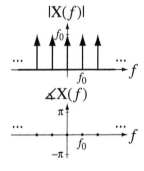

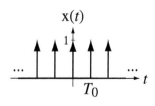

$$\delta_{T_0}(t) \xleftrightarrow{\mathscr{F}} \omega_0 \delta_{\omega_0}(\omega)$$

$$\omega_0 = 2\pi/T_0$$

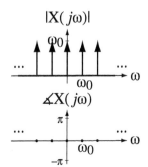

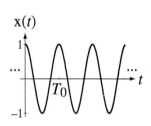

$$\cos(2\pi f_0 t) \xleftrightarrow{\mathscr{F}} \frac{1}{2}[\delta(f-f_0)+\delta(f+f_0)]$$

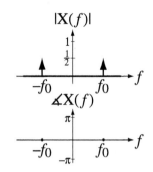

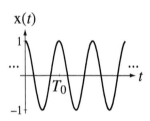

$$\cos(\omega_0 t) \xleftrightarrow{\mathscr{F}} \pi[\delta(\omega-\omega_0)+\delta(\omega+\omega_0)]$$

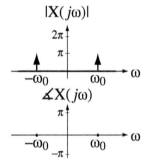

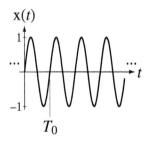

$$\sin(2\pi f_0 t) \xleftrightarrow{\mathscr{F}} \frac{j}{2}[\delta(f+f_0)-\delta(f-f_0)]$$

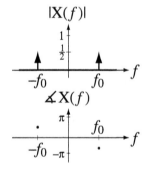

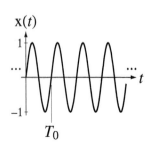

$$\sin(\omega_0 t) \xleftrightarrow{\mathscr{F}} j\pi[\delta(\omega+\omega_0)-\delta(\omega-\omega_0)]$$

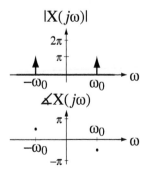

Appendix IV   Continuous-Time Fourier Transform Pairs

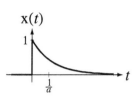

$$e^{-at}\,u(t) \overset{\mathcal{F}}{\longleftrightarrow} \frac{1}{j\omega + a},\ \mathrm{Re}(a) > 0$$

$$e^{-at}\,u(t) \overset{\mathcal{F}}{\longleftrightarrow} \frac{1}{j2\pi f + a},\ \mathrm{Re}(a) > 0$$

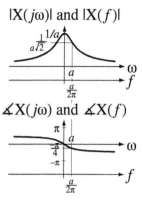

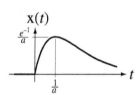

$$t e^{-at}\,u(t) \overset{\mathcal{F}}{\longleftrightarrow} \frac{1}{(j\omega + a)^2},\ \mathrm{Re}(a) > 0$$

$$t e^{-at}\,u(t) \overset{\mathcal{F}}{\longleftrightarrow} \frac{1}{(j2\pi f + a)^2},\ \mathrm{Re}(a) > 0$$

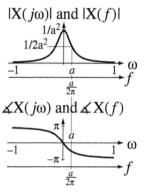

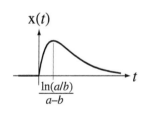

$$\frac{e^{-at} - e^{-bt}}{b - a}\,u(t) \overset{\mathcal{F}}{\longleftrightarrow} \frac{1}{(j\omega + a)(j\omega + b)},\ \begin{array}{l}\mathrm{Re}(a) > 0 \\ \mathrm{Re}(b) > 0 \\ a \neq b\end{array}$$

$$\frac{e^{-at} - e^{-bt}}{b - a}\,u(t) \overset{\mathcal{F}}{\longleftrightarrow} \frac{1}{(j2\pi f + a)(j2\pi f + b)},\ \begin{array}{l}\mathrm{Re}(a) > 0 \\ \mathrm{Re}(b) > 0 \\ a \neq b\end{array}$$

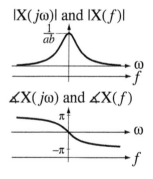

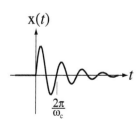

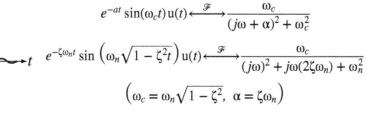

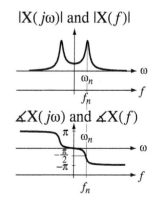

Appendix IV Continuous-Time Fourier Transform Pairs

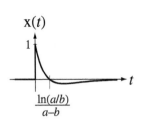

$$\frac{ae^{-at} - be^{-bt}}{a - b}u(t) \xleftrightarrow{\mathcal{F}} \frac{j\omega}{(j\omega + a)(j\omega + b)}, \quad \begin{array}{l} \text{Re}(a) > 0 \\ \text{Re}(b) > 0 \\ a \neq b \end{array}$$

$$\frac{ae^{-at} - be^{-bt}}{a - b}u(t) \xleftrightarrow{\mathcal{F}} \frac{j2\pi f}{(j2\pi f + a)(j2\pi f + b)}, \quad \begin{array}{l} \text{Re}(a) > 0 \\ \text{Re}(b) > 0 \\ a \neq b \end{array}$$

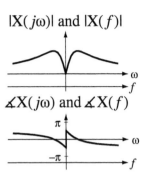

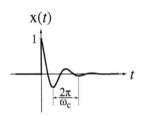

$$e^{-\alpha t}\cos(\omega_c t)\,u(t) \xleftrightarrow{\mathcal{F}} \frac{j\omega + \alpha}{(j\omega + \alpha)^2 + \omega_c^2}$$

$$e^{-\zeta\omega_n t}\cos\left(\omega_n\sqrt{1-\zeta^2}\,t\right)u(t) \xleftrightarrow{\mathcal{F}} \frac{j\omega + \zeta\omega_n}{(j\omega)^2 + j\omega(2\zeta\omega_n) + \omega_n^2}$$

$$\left(\omega_c = \omega_n\sqrt{1-\zeta^2},\ \alpha = \zeta\omega_n\right)$$

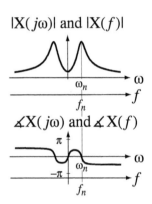

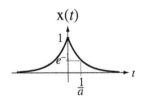

$$e^{-a|t|} \xleftrightarrow{\mathcal{F}} \frac{2a}{\omega^2 + a^2},\ \text{Re}(a) > 0$$

$$e^{-a|t|} \xleftrightarrow{\mathcal{F}} \frac{2a}{(2\pi f)^2 + a^2},\ \text{Re}(a) > 0$$

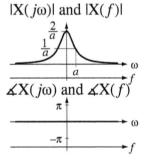

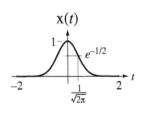

$$e^{-\pi t^2} \xleftrightarrow{\mathcal{F}} e^{-\pi f^2}$$

$$e^{-\pi t^2} \xleftrightarrow{\mathcal{F}} e^{-\omega^2/4\pi}$$

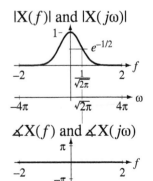

APPENDIX V

Discrete-Time Fourier Transform Pairs

$$x[n] = \int_1 X(F) \, e^{j2\pi Fn} dF \xleftrightarrow{\mathcal{F}} X(F) = \sum_{n=-\infty}^{\infty} x[n] e^{-j2\pi Fn}$$

$$x[n] = \frac{1}{2\pi} \int_{2\pi} X(e^{j\Omega}) \, e^{j\Omega n} d\Omega \xleftrightarrow{\mathcal{F}} X(e^{j\Omega}) = \sum_{n=-\infty}^{\infty} x[n] e^{-j\Omega n}$$

For all the periodic time functions, the fundamental period is $N_0 = 1/F_0 = 2\pi/\Omega_0$. In all these pairs, n, N_W, N_0, n_0, and n_1 are integers.

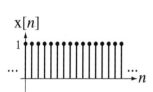

$1 \xleftrightarrow{\mathcal{F}} \delta_1(F)$

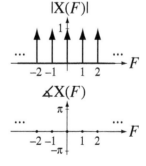

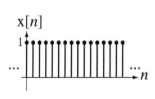

$1 \xleftrightarrow{\mathcal{F}} 2\pi\delta_{2\pi}(\Omega)$

A-17

Appendix V   Discrete-Time Fourier Transform Pairs

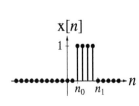

$$u[n - n_0] - u[n - n_1] \xleftrightarrow{\mathscr{F}}$$
$$\frac{e^{-j\pi F(n_1+n_0)}}{e^{-j\pi F}}(n_1 - n_0)\,\mathrm{drcl}(F, n_1 - n_0)$$
$$u[n - n_0] - u[n - n_1] \xleftrightarrow{\mathscr{F}}$$
$$\frac{e^{-j\Omega(n_1+n_0)/2}}{e^{-j\Omega/2}}(n_1 - n_0)\,\mathrm{drcl}\!\left(\frac{\Omega}{2\pi}, n_1 - n_0\right)$$

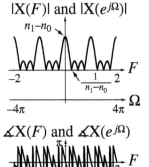

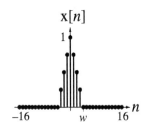

$$\mathrm{tri}(n/w) \xleftrightarrow{\mathscr{F}} w\,\mathrm{drcl}^2(F, w)$$
$$\mathrm{tri}(n/w) \xleftrightarrow{\mathscr{F}} w\,\mathrm{drcl}^2(\Omega/2\pi, w)$$

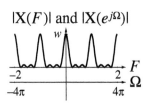

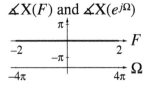

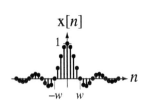

$$\mathrm{sinc}(n/w) \xleftrightarrow{\mathscr{F}} w\,\mathrm{rect}(wF) * \delta_1(F)$$
$$\mathrm{sinc}(n/w) \xleftrightarrow{\mathscr{F}} w\,\mathrm{rect}(w\Omega/2\pi) * \delta_{2\pi}(\Omega)$$

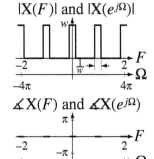

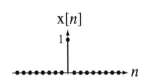

$$\delta[n] \xleftrightarrow{\mathscr{F}} 1$$

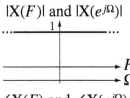

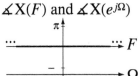

Appendix V Discrete-Time Fourier Transform Pairs

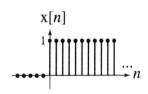

$$u[n] \xleftrightarrow{\mathscr{F}} \frac{1}{1 - e^{-j2\pi F}} + \frac{1}{2}\delta_1(F)$$

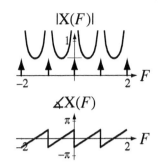

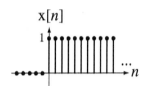

$$u[n] \xleftrightarrow{\mathscr{F}} \frac{1}{1 - e^{-j\Omega}} + \pi\delta_{2\pi}(\Omega)$$

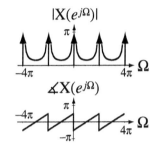

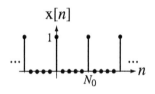

$$\delta_{N_0}[n] \xleftrightarrow{\mathscr{F}} (1/N_0)\delta_{1/N_0}(F) = F_0\delta_{F_0}(F)$$

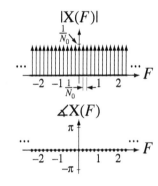

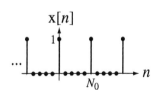

$$\delta_{N_0}[n] \xleftrightarrow{\mathscr{F}} (2\pi/N_0)\delta_{2\pi/N_0}(\Omega) = \Omega_0\delta_{\Omega_0}(\Omega)$$

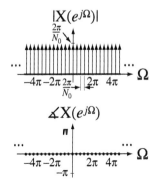

Appendix V Discrete-Time Fourier Transform Pairs

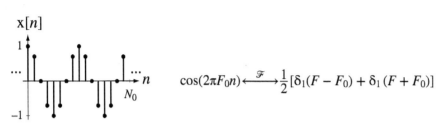
$$\cos(2\pi F_0 n) \xleftrightarrow{\mathscr{F}} \tfrac{1}{2}[\delta_1(F-F_0)+\delta_1(F+F_0)]$$
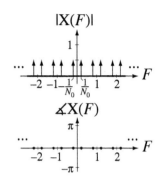

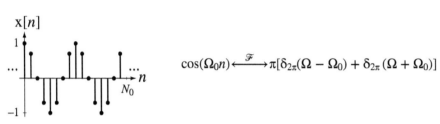
$$\cos(\Omega_0 n) \xleftrightarrow{\mathscr{F}} \pi[\delta_{2\pi}(\Omega-\Omega_0)+\delta_{2\pi}(\Omega+\Omega_0)]$$
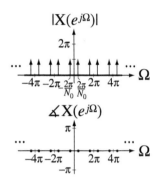

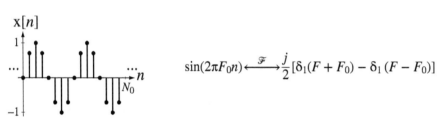
$$\sin(2\pi F_0 n) \xleftrightarrow{\mathscr{F}} \tfrac{j}{2}[\delta_1(F+F_0)-\delta_1(F-F_0)]$$
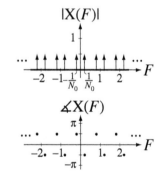

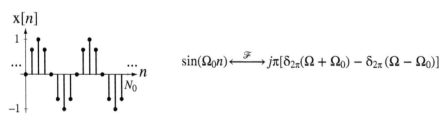
$$\sin(\Omega_0 n) \xleftrightarrow{\mathscr{F}} j\pi[\delta_{2\pi}(\Omega+\Omega_0)-\delta_{2\pi}(\Omega-\Omega_0)]$$
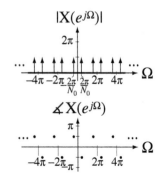

Appendix V Discrete-Time Fourier Transform Pairs

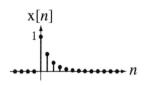

$$\alpha^n u[n] \xleftrightarrow{\mathscr{F}} \frac{1}{1 - \alpha e^{-j\Omega}}$$

$$\alpha^n u[n] \xleftrightarrow{\mathscr{F}} \frac{1}{1 - \alpha e^{-j2\pi F}}, \quad |\alpha| < 1$$

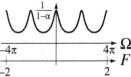

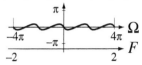

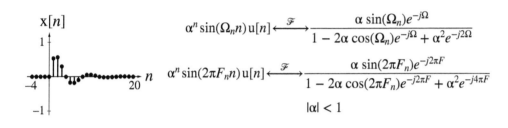

$$\alpha^n \sin(\Omega_n n) u[n] \xleftrightarrow{\mathscr{F}} \frac{\alpha \sin(\Omega_n) e^{-j\Omega}}{1 - 2\alpha \cos(\Omega_n) e^{-j\Omega} + \alpha^2 e^{-j2\Omega}}$$

$$\alpha^n \sin(2\pi F_n n) u[n] \xleftrightarrow{\mathscr{F}} \frac{\alpha \sin(2\pi F_n) e^{-j2\pi F}}{1 - 2\alpha \cos(2\pi F_n) e^{-j2\pi F} + \alpha^2 e^{-j4\pi F}}$$

$$|\alpha| < 1$$

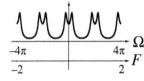

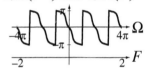

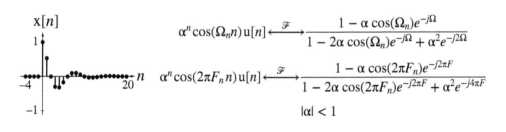

$$\alpha^n \cos(\Omega_n n) u[n] \xleftrightarrow{\mathscr{F}} \frac{1 - \alpha \cos(\Omega_n) e^{-j\Omega}}{1 - 2\alpha \cos(\Omega_n) e^{-j\Omega} + \alpha^2 e^{-j2\Omega}}$$

$$\alpha^n \cos(2\pi F_n n) u[n] \xleftrightarrow{\mathscr{F}} \frac{1 - \alpha \cos(2\pi F_n) e^{-j2\pi F}}{1 - 2\alpha \cos(2\pi F_n) e^{-j2\pi F} + \alpha^2 e^{-j4\pi F}}$$

$$|\alpha| < 1$$

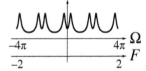

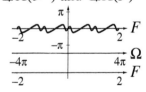

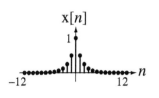

$$\alpha^{|n|} \xleftrightarrow{\mathscr{F}} \frac{1 - \alpha^2}{1 - 2\alpha \cos(2\pi F) + \alpha^2}$$

$$\alpha^{|n|} \xleftrightarrow{\mathscr{F}} \frac{1 - \alpha^2}{1 - 2\alpha \cos(\Omega) + \alpha^2}, \quad |\alpha| < 1$$

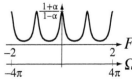

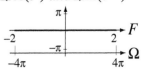

VI APPENDIX

Tables of Laplace Transform Pairs

CAUSAL FUNCTIONS

$$\delta(t) \xleftrightarrow{\mathcal{L}} 1, \quad \text{All } s$$

$$u(t) \xleftrightarrow{\mathcal{L}} \frac{1}{s}, \quad \text{Re}(s) > 0$$

$$u_{-n}(t) = \underbrace{u(t) * \cdots * u(t)}_{(n-1)\,\text{convolutions}} \xleftrightarrow{\mathcal{L}} \frac{1}{s^n}, \quad \text{Re}(s) > 0$$

$$t\,u(t) \xleftrightarrow{\mathcal{L}} \frac{1}{s^2}, \quad \text{Re}(s) > 0$$

$$e^{-\alpha t} u(t) \xleftrightarrow{\mathcal{L}} \frac{1}{s + \alpha}, \quad \text{Re}(s) > -\alpha$$

$$t^n u(t) \xleftrightarrow{\mathcal{L}} \frac{n!}{s^{n+1}}, \quad \text{Re}(s) > 0$$

$$t e^{-\alpha t} u(t) \xleftrightarrow{\mathcal{L}} \frac{1}{(s + \alpha)^2}, \quad \text{Re}(s) > -\alpha$$

$$t^n e^{-\alpha t} u(t) \xleftrightarrow{\mathcal{L}} \frac{n!}{(s + \alpha)^{n+1}}, \quad \text{Re}(s) > -\alpha$$

$$\sin(\omega_0 t) u(t) \xleftrightarrow{\mathcal{L}} \frac{\omega_0}{s^2 + \omega_0^2}, \quad \text{Re}(s) > 0$$

$$\cos(\omega_0 t) u(t) \xleftrightarrow{\mathcal{L}} \frac{s}{s^2 + \omega_0^2}, \quad \text{Re}(s) > 0$$

$$e^{-\alpha t} \sin(\omega_c t) u(t) \xleftrightarrow{\mathcal{L}} \frac{\omega_c}{(s + \alpha)^2 + \omega_c^2}, \quad \text{Re}(s) > -\alpha$$

$$e^{-\alpha t}\cos(\omega_c t)u(t) \xleftrightarrow{\mathcal{L}} \frac{s+\alpha}{(s+\alpha)^2+\omega_c^2}, \quad \operatorname{Re}(s) > -\alpha$$

$$e^{-\alpha t}\left[A\cos(\omega_c t) + \left(\frac{B-A\alpha}{\beta}\right)\sin(\omega_c t)\right]u(t) \xleftrightarrow{\mathcal{L}} \frac{As+B}{(s+\alpha)^2+\omega_c^2}$$

$$e^{-\alpha t}\left[\sqrt{A^2+\left(\frac{B-A\alpha}{\omega_c}\right)^2}\cos\left(\omega_c t - \tan^{-1}\left(\frac{B-A\alpha}{A\omega_c}\right)\right)\right]u(t) \xleftrightarrow{\mathcal{L}} \frac{As+B}{(s+\alpha)^2+\omega_c^2}$$

$$e^{-\frac{C}{2}t}\left[A\cos\left(\sqrt{D-\left(\frac{C}{2}\right)^2}\,t\right) + \frac{2B-AC}{\sqrt{4D-C^2}}\sin\left(\sqrt{D-\left(\frac{C}{2}\right)^2}\,t\right)\right]u(t) \xleftrightarrow{\mathcal{L}} \frac{As+B}{s^2+Cs+D}$$

$$e^{-\frac{C}{2}t}\left[\sqrt{A^2+\left(\frac{2B-AC}{\sqrt{4D-C^2}}\right)^2}\cos\left(\sqrt{D-\left(\frac{C}{2}\right)^2}\,t - \tan^{-1}\left(\frac{2B-AC}{A\sqrt{4D-C^2}}\right)\right)\right]u(t) \xleftrightarrow{\mathcal{L}} \frac{As+B}{s^2+Cs+D}$$

ANTICAUSAL FUNCTIONS

$$-u(-t) \xleftrightarrow{\mathcal{L}} \frac{1}{s}, \quad \operatorname{Re}(s) < 0$$

$$-e^{-\alpha t}u(-t) \xleftrightarrow{\mathcal{L}} \frac{1}{s+\alpha}, \quad \operatorname{Re}(s) < -\alpha$$

$$-t^n u(-t) \xleftrightarrow{\mathcal{L}} \frac{n!}{s^{n+1}}, \quad \operatorname{Re}(s) < 0$$

NONCAUSAL FUNCTIONS

$$e^{-\alpha|t|} \xleftrightarrow{\mathcal{L}} \frac{1}{s+\alpha} - \frac{1}{s-\alpha}, \quad -\alpha < \operatorname{Re}(s) < \alpha$$

$$\operatorname{rect}(t) \xleftrightarrow{\mathcal{L}} \frac{e^{s/2}-e^{-s/2}}{s}, \quad \text{All } s$$

$$\operatorname{tri}(t) \xleftrightarrow{\mathcal{L}} \left(\frac{e^{s/2}-e^{-s/2}}{s}\right)^2, \quad \text{All } s$$

VII APPENDIX

z-Transform Pairs

CAUSAL FUNCTIONS

$$\delta[n] \xleftrightarrow{\mathcal{Z}} 1, \quad \text{All } z$$

$$u[n] \xleftrightarrow{\mathcal{Z}} \frac{z}{z-1} = \frac{1}{1-z^{-1}}, \quad |z|>1$$

$$\alpha^n u[n] \xleftrightarrow{\mathcal{Z}} \frac{z}{z-\alpha} = \frac{1}{1-\alpha z^{-1}}, \quad |z|>|\alpha|$$

$$n u[n] \xleftrightarrow{\mathcal{Z}} \frac{z}{(z-1)^2} = \frac{z^{-1}}{(1-z^{-1})^2}, \quad |z|>1$$

$$n^2 u[n] \xleftrightarrow{\mathcal{Z}} \frac{z(z+1)}{(z-1)^3} = \frac{1+z^{-1}}{z(1-z^{-1})}, \quad |z|>1$$

$$n \alpha^n u[n] \xleftrightarrow{\mathcal{Z}} \frac{z\alpha}{(z-\alpha)^2} = \frac{\alpha z^{-1}}{(1-\alpha z^{-1})^2}, \quad |z|>|\alpha|$$

$$n^m \alpha^n u[n] \xleftrightarrow{\mathcal{Z}} (-z)^m \frac{d^m}{dz^m}\left(\frac{z}{z-\alpha}\right), \quad |z|>|\alpha|$$

$$\frac{n(n-1)(n-2)\cdots(n-m+1)}{m!} \alpha^{n-m} u[n] \xleftrightarrow{\mathcal{Z}} \frac{z}{(z-\alpha)^{m+1}}, \quad |z|>|\alpha|$$

$$\sin(\Omega_0 n) u[n] \xleftrightarrow{\mathcal{Z}} \frac{z \sin(\Omega_0)}{z^2 - 2z \cos(\Omega_0) + 1} = \frac{\sin(\Omega_0) z^{-1}}{1 - 2\cos(\Omega_0) z^{-1} + z^{-2}}, \quad |z|>1$$

$$\cos(\Omega_0 n) u[n] \xleftrightarrow{\mathcal{Z}} \frac{z[z - \cos(\Omega_0)]}{z^2 - 2z \cos(\Omega_0) + 1} = \frac{1 - \cos(\Omega_0) z^{-1}}{1 - 2\cos(\Omega_0) z^{-1} + z^{-2}}, \quad |z|>1$$

$$\alpha^n \sin(\Omega_0 n) u[n] \xleftrightarrow{\mathcal{Z}} \frac{z\alpha \sin(\Omega_0)}{z^2 - 2\alpha z \cos(\Omega_0) + \alpha^2} = \frac{\alpha \sin(\Omega_0) z^{-1}}{1 - 2\alpha \cos(\Omega_0) z^{-1} + \alpha^2 z^{-2}}, \quad |z| > |\alpha|$$

$$\alpha^n \cos(\Omega_0 n) u[n] \xleftrightarrow{\mathcal{Z}} \frac{z[z - \alpha \cos(\Omega_0)]}{z^2 - 2\alpha z \cos(\Omega_0) + \alpha^2} = \frac{1 - \alpha \cos(\Omega_0) z^{-1}}{1 - 2\alpha \cos(\Omega_0) z^{-1} + \alpha^2 z^{-2}}, \quad |z| > |\alpha|$$

ANTICAUSAL FUNCTIONS

$$-u[-n-1] \xleftrightarrow{\mathcal{Z}} \frac{z}{z-1}, \quad |z| < 1$$

$$-\alpha^n u[-n-1] \xleftrightarrow{\mathcal{Z}} \frac{z}{z-\alpha}, \quad |z| > |\alpha|$$

$$-n\alpha^n u[-n-1] \xleftrightarrow{\mathcal{Z}} \frac{\alpha z}{(z-\alpha)^2}, \quad |z| > |\alpha|$$

NONCAUSAL FUNCTIONS

$$\alpha^{|n|} \xleftrightarrow{\mathcal{Z}} \frac{z}{z-\alpha} - \frac{z}{z-1/\alpha}, \quad |\alpha| < |z| < |1/\alpha|$$

BIBLIOGRAPHY

Analog Filters

Huelsman, L. and Allen, P., *Introduction to the Theory and Design of Active Filters*, New York, NY, McGraw-Hill, 1980

Van Valkenburg, M., *Analog Filter Design*, New York, NY, Holt, Rinehart and Winston, 1982

Basic Linear Signals and Systems

Brown, R. and Nilsson, J., *Introduction to Linear Systems Analysis*, New York, NY, John Wiley and Sons, 1966

Chen, C., *Linear System Theory and Design*, New York, NY, Holt, Rinehart and Winston, 1984

Cheng, D., *Analysis of Linear Systems*, Reading, MA, Addison-Wesley, 1961

ElAli, T, and Karim, M., *Continuous Signals and Systems with MATLAB*, Boca Raton, FL, CRC Press, 2001

Gajic, Z., *Linear Dynamic Systems and Signals*, Upper Saddle River, NJ, Prentice Hall, 2003

Gardner, M. and Barnes, J., *Transients in Linear Systems*, New York, NY, John Wiley and Sons, 1947

Gaskill, J., *Linear Systems, Fourier Transforms and Optics*, New York, NY, John Wiley and Sons, 1978

Haykin, S. and VanVeen, B., *Signals and Systems*, New York, NY, John Wiley & Sons, 2003

Jackson, L., *Signals, Systems and Transforms*, Reading, MA, Addison-Wesley, 1991

Kamen, E. and Heck, B., *Fundamentals of Signals and Systems*, Upper Saddle River, NJ, Prentice Hall, 2007

Lathi, B., *Signal Processing and Linear Systems*, Carmichael, CA, Berkeley-Cambridge, 1998

Lathi, B., *Linear Systems and Signals*, New York, NY, Oxford University Press, 2005

Lindner, D., *Introduction to Signals and Systems*, New York, NY, McGraw-Hill, 1999

Neff, H., *Continuous and Discrete Linear Systems*, New York, NY, Harper & Row, 1984

Oppenheim, A. and Willsky, A., *Signals and Systems*, Upper Saddle River, NJ, Prentice Hall, 1997

Phillips, C. and Parr, J., *Signals, Systems, and Transforms*, Upper Saddle River, NJ, Prentice Hall, 2003

Schwartz, R. and Friedland, B., *Linear Systems*, New York, McGraw-Hill, 1965

Sherrick, J., *Concepts in System and Signals*, Upper Saddle River, NJ, Prentice Hall, 2001

Soliman, S. and Srinath, M., *Continuous and Discrete Signals and Systems*, Englewood Cliffs, NJ, Prentice Hall, 1990

Varaiya, L., *Structure and Implementation of Signals and Systems*, Boston, MA, Addison-Wesley, 2003

Ziemer, R., Tranter, W, and Fannin, D., *Signals and Systems Continuous and Discrete*, Upper Saddle River, NJ, Prentice Hall, 1998

Circuit Analysis

Dorf, R. and Svoboda, J., *Introduction to Electric Circuits*, New York, NY, John Wiley and Sons, 2001

Hayt, W., Kemmerly, J. and Durbin, S., *Engineering Circuit Analysis*, New York, NY, McGraw-Hill, 2002

Irwin, D., *Basic Engineering Circuit Analysis*, New York, NY, John Wiley and Sons, 2002

Nilsson, J. and Riedel, S., *Electric Circuits*, Upper Saddle River, NJ, Prentice Hall, 2000

Paul, C., *Fundamentals of Electric Circuit Analysis*, New York, NY, John Wiley and Sons, 2001

Thomas, R. and Rosa, A., *The Analysis and Design of Linear Circuits*, New York, NY, John Wiley and Sons, 2001

Communication Systems

Couch, L., *Digital and Analog Communication Systems*, Upper Saddle River, NJ, Prentice Hall, 2007

Lathi, B., *Modern Digital and Analog Communication Systems*, New York, NY, Holt, Rinehart and Winston, 1998

Roden, M., *Analog and Digital Communication Systems*, Upper Saddle River, NJ, Prentice Hall, 1996

Shenoi, K., *Digital Signal Processing in Telecommunications*, Upper Saddle River, NJ, Prentice Hall, 1995

Stremler, F., *Introduction to Communication Systems*, Reading, MA, Addison-Wesley, 1982

Thomas, J., *Statistical Communication Theory*, New York, NY, John Wiley and Sons, 1969

Ziemer, R. and Tranter, W, *Principles of Communications*, New York, NY, John Wiley and Sons, 1988

Discrete-Time Signals and Systems and Digital Filters

Bose, N., *Digital Filters: Theory and Applications*, New York, NY, North-Holland, 1985

Cadzow, J., *Discrete-Time Systems*, Englewood Cliffs, NJ, Prentice Hall, 1973

Childers, D. and Durling, A., *Digital Filtering and Signal Processing*, St. Paul, MN, West, 1975

DeFatta, D., Lucas, J. and Hodgkiss, W., *Digital Signal Processing: A System Design Approach*, New York, NY, John Wiley and Sons, 1988

Gold, B. and Rader, C., *Digital Processing of Signals*, New York, NY, McGraw-Hill, 1969

Hamming, R., *Digital Filters*, Englewood Cliffs, NJ, Prentice Hall, 1989

Ifeachor, E. and Jervis, B., *Digital Signal Processing*, Harlow, England, Prentice Hall, 2002

Ingle, V. and Proakis, J., *Digital Signal Processing Using MATLAB*, Thomson-Engineering, 2007

Kuc, R., *Introduction to Digital Signal Processing*, New York, NY, McGraw-Hill, 1988

Kuo, B., *Analysis and Synthesis of Sampled-Data Control Systems*, Englewood Cliffs, NJ, Prentice Hall, 1963

Ludeman, L., *Fundamentals of Digital Signal Processing*, New York, NY, John Wiley and Sons, 1987

Oppenheim, A., *Applications of Digital Signal Processing*, Englewood Cliffs, NJ, Prentice Hall, 1978

Oppenheim, A. and Shafer, R., *Digital Signal Processing*, Englewood Cliffs, NJ, Prentice Hall, 1975

Peled, A. and Liu, B., *Digital Signal Processing: Theory Design and Implementation*, New York, NY, John Wiley and Sons, 1976

Proakis, J. and Manolakis, D., *Digital Signal Processing: Principles, Algorithms and Applications*, Upper Saddle River, NJ, Prentice Hall, 1995

Rabiner, L. and Gold, B., *Theory and Application of Digital Signal Processing*, Englewood Cliffs, NJ, Prentice Hall, 1975

Roberts, R. and Mullis, C., *Digital Signal Processing*, Reading, MA, Addison-Wesley, 1987

Shenoi, K., *Digital Signal Processing in Telecommunications*, Upper Saddle River, NJ, Prentice Hall, 1995

Stanley, W., *Digital Signal Processing*, Reston, VA, Reston Publishing, 1975

Strum, R. and Kirk, D., *Discrete Systems and Digital Signal Processing*, Reading, MA, Addison-Wesley, 1988

Young, T., *Linear Systems and Digital Signal Processing*, Englewood Cliffs, NJ, Prentice Hall, 1985

The Fast Fourier Transform

Brigham, E., *The Fast Fourier Transform*, Englewood Cliffs, NJ, Prentice Hall, 1974

Cooley, J. and Tukey, J., "An Algorithm for the Machine Computation of the Complex Fourier Series," *Mathematics of Computation*, Vol. 19, pp. 297–301, April 1965

Fourier Optics

Gaskill, J., *Linear Systems, Fourier Transforms and Optics*, New York, NY, John Wiley and Sons, 1978

Goodman, J., *Introduction to Fourier Optics*, New York, NY, McGraw-Hill, 1968

Related Mathematics

Abramowitz, M. and Stegun, I., *Handbook of Mathematical Functions*, New York, NY, Dover, 1970

Churchill, R., *Operational Mathematics*, New York, NY, McGraw-Hill, 1958

Churchill, R., Brown, J., and Pearson, C., *Complex Variables and Applications*, New York, NY, McGraw-Hill, 1990

Craig, E., *Laplace and Fourier Transforms for Electrical Engineers*, New York, NY, Holt, Rinehart and Winston, 1964

Goldman, S., *Laplace Transform Theory and Electrical Transients*, New York, NY, Dover, 1966

Jury, E., *Theory and Application of the z-Transform Method*, Malabar, FL, R. E. Krieger, 1982

Kreyszig, E., *Advanced Engineering Mathematics*, New York, NY, John Wiley and Sons, 1998

Matthews, J. and Walker, R., *Mathematical Methods of Physics*, New York, NY, W. A. Benjamin, 1970

Noble, B., *Applied Linear Algebra*, Englewood Cliffs, NJ, Prentice Hall, 1969

Scheid, F., *Numerical Analysis*, New York, NY, McGraw-Hill, 1968

Sokolnikoff, I. and Redheffer, R., *Mathematics of Physics and Modern Engineering*, New York, NY, McGraw-Hill, 1966

Spiegel, M., *Complex Variables*, New York, NY, McGraw-Hill, 1968

Strang, G., *Introduction to Linear Algebra*, Wellesley, MA, Wellesley-Cambridge Press, 1993

Random Signals and Statistics

Bendat, J. and Piersol, A., *Random Data: Analysis and Measurement Procedures*, New York, NY, John Wiley and Sons, 1986

Cooper, G. and McGillem, C., *Probabilistic Methods of Signal and System Analysis*, New York, NY, Oxford University Press, 1999

Davenport, W. and Root, W., *Introduction to the Theory of Random Signals and Noise*, New York, NY, John Wiley and Sons, 1987

Fante, R., *Signal Analysis and Estimation*, New York, John Wiley and Sons, 1988

Leon-Garcia, A., *Probability and Random Processes for Electrical Engineering*, Reading, MA, Addison-Wesley, 1994

Mix, D., *Random Signal Processing*, Englewood Cliffs, NJ, Prentice Hall, 1995

Papoulis, A., and Pillai, S. *Probability, Random Variables and Stochastic Processes*, New York, NY, McGraw-Hill, 2002

Thomas, J., *Statistical Communication Theory*, New York, NY, Wiley-IEEE Press, 1996

Specialized Related Topics

DeRusso, P., Roy, R. and Close, C., *State Variables for Engineers*, New York, NY, John Wiley and Sons, 1998

INDEX

A

A matrix, diagonalizing, 780–781
absolute bandwidth, 517
accumulation (or summation), 94–96
accumulation property, 416, 422–423, 424
acoustic energy, 736, 737
acquisition, of signals, 446
active filters, 536–545
active highpass filter, design of, 540–542
active integrator, 538
active RLC realization, of a biquadratic filter, 545
ADC response, 447
additive system, 133–134
air pressure variations, 14
aliases, 452
aliasing, 454–457, 462–463, 495, 661
almost-ideal discrete-time lowpass filter, 565
alternate state-variable choices, 777
ambiguity problem, 94
American Standard Code for Information Interchange (ASCII), 5
amplifier, 124, 145
amplifier transfer function, 602
amplitude modulation, 41, 738–753, 755–757
amplitude scaling, 36, 37, 43–44, 89
analog and digital filter impulse responses, 694
analog filters, 680–689
analog modulation and demodulation, 738–753
analog multiplier, 141, 142
analog recording device, 446
analog signals, 3
analog voltage, converting to a binary bit pattern, 448
analog-to-digital converter (ADC), 4, 447, 664
angle modulation, 744–752
  exercises, 757–758
anti-aliasing filter, RC filter as, 456–457
anticausal signal, 139, 423
antiderivative, of a function of time, 48
antisymmetric filter coefficients, 719–720
aperiodic convolution, 244, 479–480
aperiodic function, 54, 56
aperiodic signals, 255, 323–324
aperture time, 447
area property, of the convolution integral, 179
area sampling, compared to value sampling, 668

arguments
  of functions, 20, 34
  in MATLAB, 25
artificial systems, 118
associativity property, of convolution, 179, 181, 196, 200
asymptotes, 525
asynchronous demodulation, 743
asynchronous transmission, 5
attenuated signal, 547
attenuation, 540
audio amplifier, 253, 511
audio compact disk (CD), 461–462
audio range, 509
audio-amplifier controls, 510
automobile suspension system, model of, 119
axial mode spacing, 612

B

backward difference
  approximation, 699
  of a discrete-time function, 94, 96
band-limited periodic signals, 467–470
  exercise, 495
bandlimited signals, 242, 453, 457–458, 517
  exercises, 492
bandpass Butterworth analog filter, 694
bandpass Butterworth digital filter, 694
bandpass discrete-time filter, 556
bandpass filter design, 702–703
bandpass filter(s), 128, 390, 392, 511, 512, 519, 535, 683. *See also* causal bandpass filter
bandpass signals, sampling, 461–463
bandpass-filter transfer function, 686
bandstop discrete-time filter, 556
bandstop filter(s), 128, 511, 512–513, 519, 684. *See also* causal bandstop filter
bandwidth, 517
Bartlett window function, 715, 717
`bartlett` window function, in MATLAB, 725
baseband signal
  relation with modulated carrier, 743
  transmission, 739
basis vectors, 313
beat frequency, 741
bel(B), 521
Bell, Alexander Graham, 521
Bessel filter, 686–689
Bessel function, of the first kind, 751
`besselap` command, 687

best possible approximation, 242
BIBO stable system, 138, 200
BIBO unstable system, 152, 154
bilateral Laplace transform, 380
`bilinear` command, in MATLAB, 710–711
bilinear method, 706–711
bilinear transformation, 704, 711–713
bilinear z transform, 709
  exercise, 730
binary numbers, 671
biological cell, as a system, 120
biquadratic RLC active filter, 544–546
biquadratic transfer function, 513
Blackman window, 718
Blackman window function, 716, 717
`blackman` window function, in MATLAB, 725
block diagrams, 145–146
  of convolution, 173
  of discrete-time systems, 651
  representing systems, 124–126
Bode, Hendrik, 523
`bode` command, in MATLAB, 389
Bode diagrams, 521–531, 623
  exercises, 568–569
Bode plot, 523
bounded excitation, producing an unbounded response, 138–139, 154
bounded-input-bounded-output (BIBO) stable system. *See* BIBO stable system
`boxcar` (rectangular) window function, in MATLAB, 725
bridged-T network, response of, 384–385
brightness, of top row of pixels, 551
`buttap` command, in MATLAB, 684–685
`butter` function, in MATLAB, 725
Butterworth filters, 681–686, 687, 688
Butterworth lowpass filter, 457

C

capacitor values, 541
capacitor voltage, 763
capacitors, 128, 533, 540
carrier, modulating, 738
cascade and parallel connections, 599
cascade connection
  exercises, 672
  of system, 330–331, 358, 409
  of two systems, 181, 200
cascade realization, 630–631, 670
causal bandpass filter, 519, 550
causal bandstop filter, 519, 550

causal discrete-time system, as BIBO stable, 651
causal energy signal, sampling, 478
causal exponential, z transform of, 420–421
causal exponentially damped sinusoid, z transform of, 420–421
causal highpass filter, 519, 550
causal lowpass filter, 519, 550
causal signal, 139
causal sinusoid, 431, 657–658
causal system, 139
causality, 139, 140, 548
causally-filtered brightness, 551
central difference approximation, 700
centroid of the root locus, 615
change of period property, 245, 316
change-of-scale-in-z property, 416, 419
change-of-scale property, 421
channel, 1
`cheb1ap` command, 687
`cheb2ap` command, 687
`chebwin` (Chebyshev) window function, in MATLAB, 725
`cheby1` function, in MATLAB, 725
`cheby2` function, in MATLAB, 725
Chebyshev (Tchebysheff or Tchebischeff) filter, 686–689, 725
checkerboard pattern, filtered, 553, 554
chopper-stablilized amplifier, 761–762
circuits, 127
clipped signal, 515
clock, driving a computer, 81
closed-loop system, 125
code, 447
combinations, of even and odd signals, 100–101
comment lines, in MATLAB, 25
communication
 between people, 16
 time delay, 39
communication system analysis, 735–755
communication systems, 1, 735–737
communication-channel digital filter design, 721–722
commutativity property, 179, 196
compact trigonometric Fourier series, 237–239
complementary root locus, 617
complex conjugate pair of poles, 530–532
complex CTFS, 234
complex exponential excitation, 182–183, 203–204
complex exponential excitation and response, 357, 408
complex exponentials, 22, 144
complex sinusoids, 22, 55, 145, 184, 230
components, system as an assembly of, 124

compound interest, accruing, 154
computers, as discrete-time systems, 81
conjugation property, 245, 316, 331, 416
constant, as special case of sinusoid, 231–232, 309
constant-K bandpass filter, 542–543
continuous independent variables, signals as functions of, 17
continuous signals, 239
continuous time, 164–186, 207–208, 329
contiguous-pulse approximation, 169
continuous-space function, of spatial coordinates, 551
continuous-time Butterworth filters, exercises, 727, 728
continuous-time causality, exercise, 567
continuous-time communication systems, 735–752
continuous-time convolution, 169–187, 479–480
continuous-time derivatives, approximating, 698
continuous-time exponential, 85
continuous-time feedback systems, 12–13, 126–127
continuous-time filters, 510–542, 554, 566
continuous-time Fourier methods, 229–274
continuous-time Fourier series. See CTFS
continuous-time Fourier transform. See CTFT
continuous-time frequency response exercise, 567–568
continuous-time functions, 20–21
continuous-time ideal filters, exercises, 567
continuous-time impulse function, 471
continuous-time LTI system, as BIBO stable, 138
continuous-time numerical convolution, 198
continuous-time practical active filters, exercises, 574–575
continuous-time practical passive filters, exercises, 570–574
continuous-time pressure signal, 14–15
continuous-time problem, solving, 147
continuous-time sampling, 447–478
continuous-time signal functions, summary of, 34
continuous-time signals, 3–4, 5, 7
 compared to discrete-time, 80
 estimating CTFT of, 480
 graphing convolution of, 198
 mathematical description of, 19–56
 sampling, 79–80, 451
continuous-time sinusoids, 82–83
continuous-time state equations, exercises, 791–793
continuous-time system response, exercise, 793

continuous-time systems, 119–142, 763–781
 approximate modeling of, 146
 as BIBO stable, 182
 feedback in, 12
 frequency response of, 185–186
 interpretation of the root locus, 653
 response to periodic excitation, 246–248
 simulating with discrete-time systems, 660–669
continuous-value signal, 4
continuums, 4
`control` toolbox, in MATLAB, 393–395
`conv` command, in MATLAB, 196
`conv` function, in MATLAB, 198, 198–199
`convD` function, in MATLAB, 562–564
convergence, 239–241, 327
convergence factor, 260, 356
convolution, 6, 164, 229
 in discrete time, 196
 exercises, 209–213, 215–219
 finding response of a system using, 201–203
 graphical and analytical examples of, 173–177, 192–194
 in time property, 377
 as two general procedures, 174
 of two unit rectangles, 180
convolution integral, 32, 173, 179, 667
convolution method, 186–188
convolution operator, 173
convolution properties, 178–180, 196–197, 333, 416
convolution result, graphing, 193
convolution sum
 computing with MATLAB, 197–198
 for system response, 191
Cooley, James, 322
coordinated notation, for singularity functions, 33
corner frequency, 525
cosine accumulation, graphing, 98
cosine-wave frequency modulation, 752
cosine(s), 52
 carriers modulated by, 749
 sampled, 464
Cramer's rule, 385
critical damping, 627
critical radian frequency, 144
CTFS (continuous-time Fourier series), 230–255
 DFT approximating, 250–252
 properties, 244, 245
 relation to CTFT exercises, 293
CTFS harmonic function, 340, 750
 computing with DFT, 478
 from a DFT harmonic function, 468
 estimating, 248
 exercises, 281–284

CTFS harmonic function (*Continued*)
    of a periodic signal using CTFT, 266
    of a rectangular wave, 238
CTFS pairs, 236, 244
CTFS representation, of a continuous periodic signal, 239
CTFT (continuous-time Fourier transform), 6, 255–280
    approximating with DFT, 478
    of convolution of signals, 272
    DFT approximating, 274–276
    of an impulse-sampled signal, 454
    limitations of, 354
    of a modulated sinusoid, 266
    of scaled and shifted rectangle, 272
    of the signum and unit-step functions, 262–263
    of a single continuous-time rectangle, 340
    system analysis using, 277–281
    of time-scaled and time-shifted sines, 271
    total area under a function using, 271
    of the unit-rectangle function, 264
    using differentiation property, 270–271
CTFT pairs, 259
CTFT-CTFS-DFT relationships, exercises, 495–496
CTFT-DFT relationship, 470–471
CTFT-DTFT relationship, 471–474
cumsum function, in MATLAB, 49, 96
cumulative integral, 48
cup anemometer, 120

D

damped sinusoid, 145
damping factor, 144
damping ratio, 144
decaying exponential shape, signal with, 44
deci prefix, 522
decibel (dB), 521–523
decimation, 90–91, 481, 482
definite integral, 48
delay, 145
demodulation, 739–740, 744
derivation, 169–174, 189–192
derivative, generalized
    exercises, 67
derivative of the phase, controlling, 746
derivatives
    of even and odd functions, 53
derivatives of functions
    exercises, 66
deterministic signal, 4
DFT (discrete Fourier transform), 249, 310–311
    approximating CTFS, 250–252
    approximating CTFS harmonic function, 273
    approximating CTFT, 274–276
    defined, 334
    exercises, 342–344, 497–500
    of a periodically repeated rectangular pulse, 316–317
    properties, 315–321
    signal processing using, 470–480
    using to find a system response, 338–340
DFT harmonic function, 310
    based on one fundamental period, 323
    of a discrete-time function, 468
    period of, 323
DFT pairs, 320
DFT transform pair, 315
diagonalization, 779–782
    exercise, 794
diff function, in MATLAB, 96
difference, 94
difference equations
    describing discrete-time systems, 650
    for a discrete-time system, 203
    exercises solving, 438–439
    with initial conditions, 424
    modeling discrete-time systems, 146–151
    solution of, 424–425
difference-equation description, exercise, 794
differencing and accumulation, 94
    exercises, 109–110
differencing property, of the convolution sum, 196
differential equations
    approximating difference equation, 699
    exercises solving, 399
    with initial conditions, 383–385
    modeling systems using, 120–127
    solution of, 121
differential-equation description, exercise, 794
differentiation, 47–50
differentiation property
    of the convolution integral, 179
    of the CTFT, 269–270
    z transform using, 422
differentiators, 527
digital bandpass filter design
    bilinear transformation, 712–713
    impulse-invariant method, 694–696
    matched-z transform, 705–706
    Parks-McClellan, 724
    step-invariant method, 697–699
digital filters, 446, 680, 689–722
    creating unstable, 700–701
    frequency response as periodic, 706
    frequency response matching analog filter, 706–707
    functions designing, 725
digital hardware, 671
digital image processing, on computers, 8
digital lowpass filter designs, 711–712, 720–721
digital signal processing (DSP), 446
digital signals, 4, 5–6
digital simulation, by impulse-invariant method, 695
digital-filter frequency response, 691–692
digital-to-analog converter (DAC), 448, 664–665
diode, as statically nonlinear component, 140, 141
Direct Form II, 359
    realization, 359–360, 419
    system, 511
    system realization, 409–410, 410
    system realization exercise, 396, 435
direct substitution, 704
direct substitution method, 704–705
direct terms, vector of, 374
diric function, in MATLAB, 320–321
Dirichlet conditions, 237
Dirichlet function, 319, 717
discontinuities, functions with, 23–32
discontinuous function, 21
discontinuous signals, 240–241
discrete Fourier transform. *See* DFT
discrete independent variable, signals as functions of, 18
discrete time, 186–203, 329
    exercises, 214
discrete-space function, 551
discrete-time causality, exercise, 576
discrete-time convolution, 189–204, 479–480
discrete-time delay, 11
discrete-time DSBSC modulation, 754
discrete-time exponentials, 85–86
discrete-time feedback system, 151
discrete-time filters, 546–564
discrete-time Fourier methods, 307–338
discrete-time Fourier series. *See* DTFS
discrete-time Fourier transform. *See* DTFT
discrete-time frequency response, 555
    exercises, 575–576
discrete-time functions, 80
    continuous-time singularity functions and, 86–89
    domain of, 90
    examples, 81
    graphing, 81, 92–94
    summations of, 101
discrete-time ideal filters, exercises, 576
discrete-time impulses, MATLAB function for, 86
discrete-time numerical convolution, 196
discrete-time practical filters, exercises, 577–578

discrete-time pulse, filtering, 726
discrete-time radian frequency, representing, 425
discrete-time sampling, 481–485
discrete-time signal functions, summary of, 89
discrete-time signals, 3, 6
   from continuous-time signals, 454
   examples, 80
   sampling, 481
   simulating continuous-time signals, 663
discrete-time sinusoidal-carrier amplitude modulation, 753–754
discrete-time sinusoids, 82–84
discrete-time state equations, exercises, 794–795
discrete-time system objects, 428–429
discrete-time system response, exercise, 795–796
discrete-time system stability, analysis, 653
discrete-time systems, 11–12, 145–154, 330, 331
   equivalence with continuous-time systems, 660
   feedback in, 12
   frequency response of, 425
   modeled by block diagrams, 651
   periodic frequency response, 426, 548
   properties of, 152
   realization of, 670–671
   simulating continuous-time systems, 660–669
   state-space analysis of, 782–790
discrete-time time scaling, 482
discrete-time unit ramp, 96
discrete-value signals, 4
discretizing, a system, 663
distortion, 515–516, 547
distortionless system, 516, 547
distributivity property, of convolution, 181, 201
"divide-and-conquer" approach, to solving linear-system problems, 134
domain, of a function, 20
double-sideband suppressed-carrier (DSBSC)
   modulation, 738–741
   signal sampling, 485
double-sideband suppressed carrier modulation, 378
double-sideband transmitted carrier (DSBTC), 741–744
downsampling, 483
DTFS (discrete-time Fourier series), 307–310, 341
DTFS harmonic function, 310
DTFT (discrete-time Fourier transform), 323–338
   of any discrete-time signal, 454
   approximating with DFT, 478
   compared to other Fourier methods, 340
   convergence, 327
   of a decimated signal, 483
   defined, 334
   derivation and definition, 324–325
   derived from the z transform, 546
   of a discrete-time function, 547
   of a discrete-time signal, 481, 482
   exercises, 344
   generalized, 326–327
   generalizing, 407–408
   of modulation, carrier and modulated carrier, 754
   numerical computation of, 334–338
   of a periodic impulse, 330
   properties, 327–333
   of a system response, 430–431
   of a window function, 473
DTFT pairs, 325, 326–327
dynamic system, 140

E

ear-brain system, 229
`eig` command in MATLAB, 781–782
eigenfunctions, 22, 121
electromagnetic energy propagation, 736
electromechanical feedback system, 772
`ellip` function, in MATLAB, 725
`ellipap` command, 687
Elliptic filter (Cauer filter), 686–689, 725
encoded response, 448
encoding, 448
encoding signals, 4–5
energy signals, 59, 60
energy spectral density, 270
envelope detector, 742
equalization filter, 601
equalization system, 762
equation of motion, 9
equivalence, of continuous-time and discrete-time systems, 666
equivalence property, of the impulse, 30, 279, 331
error signal, 599
Euler's identity, 22, 184, 230
even and odd functions
   combinations of, 52–53
   exercises, 110–111
even and odd parts, of a function, 50–51
even and odd signals, 98–101
   exercises, 67–69
even function, 49
excitation harmonic function, 248
excitations, 1, 176
exercises, 157–158
existence of z transform, exercise, 435

exponentials, 82, 85–86
exponentials (`exp`), 21

F

F-117 stealth fighter, 12, 604
fast Fourier transform (FFT), 250, 321–322
feedback, 12
feedback connection, 599
   beneficial effects of feedback, 601–604
   exercises, 672
   instability caused by feedback, 604–608
   root-locus method, 612–615
   stability, feedback effects on, 600
   stable oscillation using feedback, 608–612
   of systems, 652
   terminology and basic relationships, 599–600
   tracking errors in unity-gain feedback systems, 618–621
feedback systems, 12–14, 149–152
feedback-path transfer function, 599, 618
feedback-system transfer function, 602, 604
feedforward paths, 715
fft algorithm, implementing DFT on computers, 338–340
`fft` function, in MATLAB, 250, 318, 322
`fftshift` function, in MATLAB, 237, 275
filter classifications, 516, 548–554
`filter` function, in MATLAB, 725
filter transformations, 682–684
filtering, images, 549–552
filters, 509, 510
   continuous-time, 510–542
   design techniques, 689–721
   effects on signals, 558–560
   processing signals, 6
   uses of, 680
`filtfilt` function, in MATLAB, 725
final value theorem, 416
finite difference design, 698–700, 729
finite-difference method, 702–703, 703
finite-duration impulse response, 689, 713
FIR filter design, 713–722
   exercises, 730–731
FIR filters, 689
`firpm` command, in MATLAB, 724–725
first backward difference property, 416
first time derivative property, 381
first-order hold, 460–461
first-order systems, 143
fixed-point arithmetic, 671
fluid system, 9–10
fluid-mechanical system, modeling, 122–123
FM (frequency modulation), 746, 747
forced response, 139, 145, 432
forced response of the system, 624

forced response values, 669–670
forcing function, 121–122, 165
forward and inverse discrete-time Fourier transforms, exercises, 345–348
forward and inverse Laplace transforms, exercises, 286–293, 396–397
forward and inverse z transforms
    examples of, 418–422
    exercises, 435–438
forward CTFT, 478
forward DFT, 311, 311–314, 321
forward difference, of a discrete-time function, 94, 95
forward Fourier transform, 355
forward Laplace transform, 355
forward transfer function, 621
forward z transform, defined, 407
forward-path output signal, 600
forward-path transfer function, 599, 600, 618
Fourier, Jean Baptiste, 230
Fourier method comparisons, 340
Fourier methods matrix, 340
Fourier series, 230
    of even and odd periodic functions, 243
    exercises, 281–282
    extending to aperiodic signals, 255–260
    numerical computation of, 248–255
Fourier transform, 255
    alternate definitions of, 355–356
    generalized, 260–264
    generalizing, 355–357
    as not a function of time, 258
    numerical computation of, 273–280
Fourier transform pairs, 258, 264
Fourier transform properties, 265–270
Fourier transform representation, of a discontinuous signal, 693
Fourier-series tables and properties, 244–248
freqresp function, in MATLAB, 623
freqs function, 687
frequency, 15, 453
frequency compression, 268
frequency differentiation property, 265
frequency domain, 6, 229
frequency modulation (FM), 746, 747
frequency multiplexing, 737–738
frequency response(es), 184–185, 204–205
    of a bandpass filter, 562
    of discrete-time and continuous-time lowpass filters, 555
    of discrete-time systems, 425
    in everyday life, 509
    of a filter, 512
    of ideal filters, 517, 518, 548
    of a lowpass filter, 554–555
    phase of, 386
    from pole-zero diagram, 387–389
    shaping, 510

    of a system, 270, 331–332
    from a transfer function, 427–428
frequency scaling property, 265, 268, 330, 382
frequency shifting, 38
frequency shifting property, 245, 265, 266, 268, 316, 329
frequency warping, 710
frequency-domain methods, 704–711
frequency-domain resolution, 334
frequency-independent gain, 389, 527–528
frequency-scaling property, 419
freqz function, in MATLAB, 726
full-wave rectifier, as not invertible, 143–144
functions
    combinations of, 34–36
    with discontinuities, 23–32
    even and odd parts, 98
    exercises, 105–107
    fundamental period of, 101
    graphing accumulation of in MATLAB, 97
    graphing combinations, 35–36
    with integrals, 48
    sums, products and quotients of, 35
    types of, 20
fundamental cyclic frequency, 53
fundamental period, 53, 247
    of CTFS representation, 236
    of a function, 101
    of a signal, 55–56
fundamental radian frequency, 53

G

gain, as opposite of attenuation, 540
"gate" function, unit rectangle function as, 33
gcd function, in MATLAB, 56
generalized CTFT, 326
generalized derivative, 29
generalized DTFT, 326–327
generalized Fourier transform, 260–264, 356
generalized Fourier-transform pair, 261
Gibbs, Josiah Willard, 240
Gibbs phenomenon, 240
Gibb's phenomenon, 715
graphic equalizer, 513–514, 546
graphing function, scaling and shifting with MATLAB, 45–46
greatest common divisor (GCD), 55

H

half-power bandwidth, 517
Hamming window function, 716, 717
hamming window function, in MATLAB, 725
hanning (von Hann) window function, in MATLAB, 725

harmonic function, 233
harmonic number, 233, 310
harmonic response, 246–248
Heaviside, Oliver, 25
highpass active filters, cascade of two inverting, 541
highpass discrete-time filter, 556
highpass filters, 128, 387, 392, 511, 512, 519, 535. *See also* causal highpass filter
    design of active, 540–542
    frequency response of, 390
    response to sinusoids, 557–558
high-spatial-frequency information, in an image, 553
highway bridge, as a system, 120
home-entertainment audio system, 509
homogeneity, 131–132
homogeneous solution, 121, 164
homogeneous system, 131
human body, as a system, 120
human ear, response to sounds, 509–510

I

ideal bandpass filter, 516, 548
ideal discrete-time filters, 548
ideal filters, 509, 515–520
    discrete-time, 547–553
    frequency responses, 516–517
    impulse and frequency responses of, 548
    as noncausal, 518
ideal highpass filter, 516, 548
ideal interpolation, 458–459
ideal lowpass filter, 515, 516, 548
ideal operational amplifier, 537
ideal-lowpass-filter impulse response, 564
IIR filter design, 689–710
IIR filters, 689
image-processing techniques, application of, 8
images, 7–8, 549–552
impedance, 533–534
impedance concept of circuit analysis, 593
impinvar command, in MATLAB, 694–695, 695
impulse invariance, 662–664
impulse invariant design, 663
impulse modulation, 452
impulse response(s), 164–168, 173, 181, 182, 186–188
    of any discrete-time system, 554
    of continuous-time systems, 165–168
    of discrete-time and RC lowpass filters, 555
    of a distortionless system, 516, 547
    exercises, 208, 215
    of a filter, 547
    of ideal filters, 517, 518, 549–550

of an LTI system, 173
for the moving-average filter, 562
of an *RC* lowpass filter, 176, 526
of the *RLC* bandpass filter, 536–537
of a system, 188–189, 201
at three outputs, 558
time delay in, 566
truncating ideal, 720–721
of a zero-order hold, 460
impulse sample, 690
impulse sampling, 452
exercises, 489–491
interpolation and, 458
impulse train, 32
impulse-invariant design, 690–694
exercise, 489–491
MATLAB's version of, 696
impulse-invariant method, digital bandpass filter design, 694–696
impulses, graphical representations of, 30
indefinite integral, 48
independent variable, 34
inductor current, 764
inductors, equations for, 533
infinite energy, 58, 59
infinite-duration impulse response (IIR), 689. *See also* IIR filters
infinitely many samples, availability of, 459
information, 15
inhomogeneous system, 131
initial value theorem, 416
inner product, of complex sinusoids, 235
in-phase part, 465
input signals, 1, 119
inputs, 1
instantaneous frequency, 745
instrumentation system, in an industrial process, 514
integer multiple, of the fundamental frequency, 467
integrals
of even and odd functions, 53
exercises, 66
of functions, 48
integration, 47–50
integration property, 277, 748
integrators, 124, 125, 527, 538
interference, 16
interpolation, 90, 458–461, 481, 483–485
exercises, 493–494
intrinsic functions, in MATLAB, 21
invariant functions, 49
inverse CTFT, 263–264, 479
inverse DFT, 310–311
approximating the inverse DTFT, 336
defined, 334, 335
of a periodic function, 474
inverse DTFT

exact and approximate, 336
MATLAB program finding, 337–338
of a periodically repeated rectangle, 333–334
of two periodic shifted rectangles, 328–329
using the DFT, 334–335
inverse Fourier transform, 357
inverse Fourier transform integral, 263–264
inverse Laplace transform, 360, 365–366
using partial-fraction expansion, 367–368, 368, 371–372
inverse unilateral Laplace transform, 380
inverse *z* transform, 410, 415–416, 432, 433, 654, 658
inverse *z*-transform methods, 417–422
invertibility, 142–143
invertible system, 142
inverting amplifier, 537

K

Kaiser window function, 717, 718
`kaiser` window function, in MATLAB, 725
Kirchhoff's voltage law, 128
Kronecker delta function, 86

L

Laplace, Pierre Simon, 355
Laplace system analysis, 592
goals, 592
standard realizations of systems, 630–632
system analysis using MATLAB, 621–623
system connections, 599–621
system representations, 592–596
system responses to standard signals, 623–629
system stability, 596–598
Laplace transform, 184
analysis of dynamic behavior of continuous-time systems, 650
counterpart to, 406
development of, 355–358
exercises, 395
existence of, 360–362
generalizing CTFT, 407
making Fourier transform more directly compatible with, 258
of a noncausal exponential signal, 365–366
properties, 377–379
of the system response, 624
of time-scaled rectangular pulses, 378
Laplace transform pairs, 357, 362–366, 379
Laplace-transform-*z*-transform relationship, exercise, 675
`lcm` function, in MATLAB, 55

leakage
minimizing, 473
reducing, 474
least common multiple (LCM), 54
left-sided signal, 362, 411, 412
Leibniz's formula, 138
L'Hôpital's rule, 167, 239, 320
light waves, Doppler shift with, 41
linear, time-invariant system, 134–135
linear algebra theory, 778
linear system, 134
linear system dynamics, 383
linearity, 134, 171
linearity property, 245, 265, 271, 316, 332, 377, 416, 418, 423
linearizing, a system, 137
local oscillator, 741
log-amplified signal, 547
logarithmic graphs, 523
exercises, 568
logarithmic scale, uniform spacing on, 514
log-magnitude graph, 523
loop transfer function, 600, 612
loop transmission, 600
lowpass Butterworth filter
converting to a highpass, 682
maximally flat, 681
transforming into a bandpass filter, 683
transforming into a bandstop filter, 684
lowpass discrete-time filter, 754
lowpass filter, 128, 390, 511, 519, 520, 532–534, 538–539. *See also* causal lowpass filter
lowpass filter design, 703
LTI discrete-time system, 153
LTI systems, 134
excited by sinusoids, 230
frequency response of a cascade of, 270
impulse responses of, 354
response of, 375–377
response to a complex-exponential excitation, 357
system and output equations of, 766
testing for causality, 139

M

magnitude Bode diagrams, 523, 526, 530
magnitude spectrum, of a general bandpass signal, 462
magnitude-frequency-response Bode diagram, 523
marginally stable system, 597
matched-*z* transform, 704, 705–706
matched-*z* transform and direct substitution filter design, exercise, 729
mathematical functions, describing signals, 19, 35
mathematical model, 9

mathematical voltage-current relations, 128
MATLAB
  arguments, 26
  `bartlett` window function, 725
  `bilinear` command, 710–711
  `blackman` window function, 725
  `bode` command, 389
  `boxcar` (rectangular) window function, 725
  `buttap` command, 684–685
  `butter` function, 725
  `chebwin` (Chebyshev) window function, 725
  `cheby1` function, 725
  `cheby2` function, 725
  comment lines, 25
  computing convolution sum, 197–198
  `control` toolbox, 393–395
  `conv` command, 197
  `conv` function, 198–199
  `convD` function, 562–564
  creating functions in, 25
  `cumsum` function, 49, 96
  design tools, 684–685, 725–726
  designing analog Butterworth filters, 681
  `diff` function, 41–42, 87
  `dirac` function, 31
  `diric` function, 320
  `eig` command, 781–782
  `ellip` function, 725
  exponentials and sinusoids in, 21
  `fft` function, 250, 322
  `fftshift` command, 275
  `fftshift` function, 252
  `filter` function, 725
  `filtfilt` function, 725
  finding inverse DTFT, 337–338
  `firpm` command, 724–725
  `freqresp` function, 623
  `freqz` function, 726
  function for discrete-time impulses, 86
  `gcd` function, 56
  graphic function scaling and shifting, 45–46
  graphing function combinations, 35–36
  `hamming` window function, 725
  `hanning` (von Hann) window function, 725
  `heaviside` intrinsic function, 25
  `impinvar` command, 694–695, 695
  `int` function, 48
  intrinsic functions, 21
  invoking a function, 34
  `kaiser` window function, 725
  `lcm` function, 56
  m file for the ramp function, 27
  `minreal` command, 623
  name, 25

NaN constant, 25
numerical integration functions in, 49
`pzmap` command, 389
`residue` function, 374–375
`rlocus` command, 623
`sign` function, 24
simulating a discrete-time system, 11
`stem` command, 81
system-object commands, 685–686
system objects, 393–395, 428–429
`tf` (transfer function) command, 393–394
`tfdata` command, 394
tools for state-space analysis, 790
transformation of normalized filters, 684
`triang` window function, 725
`upfirdn` function, 726
use of, 18
`zpk` command, 393
`zpkdata` command, 394
matrix transfer function, 775
maximally flat Butterworth filter, 681
McClellan, James H., 723
McLaurin series, 28
measurement instruments, 120
mechanical systems, 9
  modeling, 120–122
  state-space analysis of, 772–775
memory, 139–140
minimum error, of Fourier-series partial sums, 242–244
minimum sampling rate, reducing, 461
`minreal` command, in MATLAB, 623
modified CTFS harmonic functions, 256
  for rectangular-wave signals, 257
modulated carrier, 739
modulation, 738, 753
modulation index, 741
moving-average digital filter, 194–195
moving-average filter, 196, 560–563
multipath distortion, 762
multiple bandstop filter, 561
multiplication-convolution duality, 330
multiplication-convolution duality property, 244, 245, 265, 266, 316, 475

N

name, in MATLAB, 25
narrow-bandpass-signal spectrum, 461
narrowband FM, 747
narrowband PM, 747
natural radian frequency, 144
natural response, 624
natural systems, 118
negative amplitude-scaling factor, 37
negative feedback, 599
negative sine function, signal shape of, 44
noise, 1, 4, 16, 17, 520–521
noise removal, 520–521

nonadditive system, 133–134
noncausal filter, 155
noncausal lowpass filter, 553
noncausal signal-processing systems, 155
noninverting amplifier, 537
noninverting amplifier transfer function, 537
nonlinear systems, 137, 140
normalized analog filter designs, 687
normalized Butterworth filters, 681–683
normalized filters, MATLAB commands for transformation of, 684
null bandwidth, 517
numerical computation
  of discrete-time Fourier transform, 334–338
  of Fourier series, 248–255
  of Fourier transform, 273–280
numerical convolution, 196
numerical CTFT, exercise, 294
numerical integration, cumsum function, 49
numerical integration functions, in MATLAB, 49
Nyquist, Harry, 453
Nyquist frequency, 454
Nyquist rates, 453
  exercise, 491
  of signals, 456–457
  sinusoids sampled above, below and at, 464–466

O

octave intervals, filters spaced at, 514
odd functions, 49, 53
Ohm's law, 140
one-finite-pole, one-finite-zero highpass filter, 390
one-finite-pole lowpass filter, 389
one-pole system, unit-sequence response, 655
one-real-pole system, 525
one-real-zero system, 526
one-sided Laplace transform, 380
open left half-plane, 597
open loop system, 125
operational amplifiers, 537–538
  saturation in real, 141
optimal FIR filter design, 723–724
order, of a system, 764
orthogonal basis vectors, 312–313
orthogonal complex sinusoids, 234
orthogonality
  exercises, 282, 342
  harmonic function and, 234–236
oscillator feedback system, 609
output equations, 764, 765, 766
output signals, 1
outputs, 1

overdamped case, 627
overmodulation, 743
oversampled signal, 453

P

parallel, cascade and feedback connections, exercises, 635–637, 672
parallel connections
　of systems, 652
　of two systems, 181, 200–201
parallel realization, 632, 670
parallel response, ADC, 447
parallel RLC circuit, 764
parentheses, indicating a continuous-time function, 81
Parks, Thomas W., 723
Parks-McClellan design, of a digital bandpass filter, 724
Parks-McClellan optimal equiripple design, 723
Parseval des Chênes, Marc-Antoine, 270
Parseval's theorem, 245, 265, 270, 288, 316, 332
partial-fraction expansion, 367–377, 418
passband, 510
　filter distortionless within, 516
　ripple, 687, 715, 716
　signal transmission, 739
passive filters, 532–535
pendulum, analyzing, 137–138
period
　of a function, 53
　in a periodic signal, 255
periodic convolution, 244, 479
periodic even signal, 243
periodic excitation, response of a continuous-time system and, 246–248
periodic functions, 54, 101–102
　exercises, 111–112
periodic impulse, 32
periodic odd function, 244
periodic signals, 53–55, 101–102, 139
　average signal power calculation, 58
　with discontinuities, 240–241
　exercises, 69–70
　as power signals, 58
periodically repeated sinc function, 318
periodic-impulse sampling, 481–483
periodicity, of the DTFT, 331
periodic-repetition relationship, sampling and, 474–478
phase, 267, 269
phase Bode diagram, 524, 526, 530
phase modulation (PM), 745
phase-locked loop, 741
photographs, 551
physical systems, as filters, 536
picket fencing, 476

pitch, 15
pixels, 551
PM (phase modulation), 747
point spread function, 554
pole, of a Laplace transform, 364
　of an analog filter, 695
　exercises, 439–440
　frequency response and, 385–393, 425–427
　pole-zero diagrams, 364
　of system transfer functions, 426
　using the z transform, 656–657
pole-zero plots, 421, 427–428
power of signals, finding, 59
power signals, 59, 60
power spectral density, 15–16
power spectrum, 514, 520
practical filters, 532–544, 554–565
practical interpolation, 459
propagation delay, in ordinary conversation, 38
prototype feedback system, 609
public address system, 606, 607
pulse amplitude modulation, exercises, 487
pure sinusoids, 431
`pzmap` command, in MATLAB, 389, 623

Q

quadrature part, 465
qualitative concepts, 449–450
quantization, 448
quantized response, 448
quantizing signals, 4–5

R

radian frequency, 547
ramp function, 26
random signals, 4, 6, 558–560
range, of a function, 20
rate, 453
rational function, 183
RC circuit, frequency response of, 528–529
RC filter, as an anti-aliasing filter, 456–457
RC lowpass filter, 135, 170, 526, 532
real exponential functions, 21
real systems, eigenfunctions of, 135
real-time filtering, of time signals, 552–553
real-valued sines and cosines, replacing, 234–235
real-valued sinusoids, 21
receiver, 1, 739
rectangular pulses, convolution of, 177
rectangular-rule integration, 171
rectangular wave, CTFS harmonic function of, 238
recursion, 419, 784–785
red shift, 41

regenerative travelling-wave light amplifier, 610
region of convergence (ROC), 361, 362, 364–365, 380, 412
Remez, Evgeny Yakovlevich, 723
Remez exchange algorithm, 723
`residue` function, of MATLAB, 374–375
residues, vector of, 374
resistive voltage divider, 140
resistors, 128, 533, 540
resonant frequency, 535
response harmonic function, 247
responses, 1
result, in MATLAB, 25
reverberation, 605
RF signal transmission, 739
right-sided signal, 361–362, 411
ripple effect, reducing in the frequency domain, 715
RLC circuit, 143
`rlocus` command, in MATLAB, 623
ROC (region of convergence), 361, 362, 364–365, 380, 412
root locus
　for discrete-time feedback system, 653
　exercises, 637–639, 674
root-locus method, 612–615
root-locus plot, 613
running integral, 48–49

S

Sa function, 239
Sallen-Key bandpass filter, 542–543
sample-and-hold (S/H), 447
sampled-data systems, 664–670
　designing, 668–669
　exercise, 675
sampled sinc function, 264
sampling, 79–80, 446
　at a discontinuity, 693
　exercises, 487–488
　a signal, 3
sampling methods, 447–449
sampling period or interval, 80
sampling property, of the impulse, 31–32
sampling rate, 449–450, 461, 463–464, 706–707
sampling signals, 4–5
sampling theorem, 449–453
satellite communication system, propagation delay, 38
scaled aliases, of an analog filter's frequency response, 690, 691, 692
scaling, 36–45, 89
　exercises, 62–65, 107–109
scaling property, 31, 179, 272
script file, 51
s-domain differentiation, 377–378, 378–379

s-domain shifting property, 378
second-order complex pole pair, 531
second-order complex zero pair, 532
second-order system, 143–144, 655
second-order system transfer function, 625
sequential-state machines, 81
serial response, ADC, 447
Shannon, Claude, 450
shifting, 37–46, 89
    exercises, 62–65, 107–109
shifting property, 31
side lobes, 715, 717
sidebands, 738
signal energy, 56–57
    exercises, 70–71
    finding signal power using MATLAB, 103–105
    finding using MATLAB, 59–60
    per unit cyclic frequency, 271
    of a signal, 102–103
    of a sinc signal, 332–333
signal energy and power
    exercises, 112–113
signal functions, exercises, 61
signal power, 58, 103
signal processing, using the DFT, 470–480
signal reconstruction, 460, 461
signal transmission, types of, 739
signals, 1
    approximated by constants, 231
    approximated by periodic functions, 54
    examples of, 19
    finding Nyquist rates of, 456–457
    spatially separating, 737
    switching on or off, 23
    system responses to standard, 654–660
    types of, 3–8
signal-to-noise ratio (SNR), 16, 17, 521
signum function, 24–25, 87
simultaneous shifting and scaling, 43–44
sinc function
    carriers modulated by, 749
    definition of, 239
    similarity to Dirichlet function, 320
sinc signal, signal energy of, 332–333
sines, 52, 465
sine-wave phase, of a carrier, 746
single-input, single-output system, 1, 124
single-negative-real-zero subsystem, 526
single-sideband suppressed-carrier (SSBSC) modulation, 743–744
singularity functions, 23, 33, 86–89
sinusoid response, 627–629
sinusoidal signal, signal power of, 58
sinusoids, 22, 82–84
    adding to constants, 231
    in discrete-time signal and system analysis, 82

    multiplied by unit sequences, 432
    real and complex, 230
    responses to, 229
    sampling, 464–466
    signal as burst of, 44
    system responses to, 431–432
smoothing filter, 561
sound, 14, 229
space, functions of, 7
space shifting, 38
spatial dimension, independent variable as, 38
spatial variables, 8
spectra, of PM and FM signals, 747–748
spectrum analyzer, 520
s-plane region, mapping, 708, 709
spontaneous emission, 610
square brackets [ ]
    indicating a discrete-time function, 81
    in MATLAB, 87
square wave, representing, 135
square-wave phase, of a carrier, 746
ss function, 782, 790
SSBSC (single-sideband suppressed-carrier) modulation, 743–744
ssdata function, 790
ss2ss function, 782, 790
stability, 138, 181
    exercises, 213–214, 219–220, 635, 672
    impulse response and, 200
    types of, 598
stable analog filter, becoming unstable digital filter, 708
standard realizations of systems, 630
    cascade realization, 630–631
    parallel realization, 632
standard signals, response to exercises, 673
standard signals, system responses to, 623
    exercises, 640–641
    sinusoid response, 627–629
    unit-step response, 624–627
start bit, 5
state equations, diagonalizing using MATLAB, 781
state space, 764
state transition matrix, 767, 785
state variables, 763, 777, 778
state vector, 764
state-space analysis
    characteristics of, 764
    MATLAB tools for, 782
    of a mechanical system, 772–775
    of a two-input, two-output system, 769–772
    using state variables, 763
static nonlinear components, 140
static nonlinearity, 140–141
static system, 140
statically nonlinear system, 154–155

stem command, in MATLAB, 81
step response, 181, 182
step-invariant design, 689, 696
step-invariant method, 696–700
stop bits, 5
stopbands, 510
straight-line signal reconstruction, 460
strength, of an impulse, 30
strictly bandlimited signals, 453, 517
subfunctions, 51
sum property, of the convolution sum, 196
summing junction, 11, 124–125, 145
superposition
    applying to find approximate system response, 171
    applying to linear systems, 134
    finding response of a linear system, 136
    finding response to a square wave, 135
    for LTI systems, 145
suppressed carrier, 739
symbolic integration, int function, 48
symmetric impulse response, 719
synchronous demodulation, 741
synthetic division, 417
system analysis
    using CTFT, 277–281
system and output equations, 764–775, 783–787
system connections, 181–182, 200–201, 652–653
system discretization, signal sampling and, 663
system equations, 765–766
system modeling, 119–121, 122, 145–154
system models
    exercises, 156–157
system objects, in MATLAB, 393–395, 428–429
system properties, 127–140, 152–155
    exercises, 158–160
system realization, 359
    exercises, 641, 676
system response
    exercises, 294
    to standard signals, 654–660
    to system excitation, 191
    using DTFT and DFT, 336–338
system stability, 651–652
system-object commands, in MATLAB, 685–686
systems, 1, 3
    defining, 118
    examples of, 8–14

T

tf (transfer function) command, in MATLAB, 393–394
tfdata command, in MATLAB, 394

thermocouples, 732
thermostat, 12, 119
thermowell, 732
time compression, for discrete-time
    functions, 90–91
time constant, 625
time derivative properties, 381
time differentiation property, 245, 265, 377
time expansion, 90–91
time expansion property, 416
time expression, 268
time index, 83
time integration property, 245, 265, 377,
    379, 381
time invariance, 132–133
time invariant system, 132
time limited signals, 58, 457–458
time multiplexing, 737
time reversal property, 245, 316, 416
time reversed function, 39
time scaling, 39–43, 89–93, 329
time scaling property, 245, 265, 272, 316,
    377, 381, 382
time shifting property, 245, 316, 416
time shifting, 37–39, 42–43, 89
time signals, 7
time translation, 37
time variant system, 132, 153
time-domain block diagram, of a system, 651
time-domain methods, 689–693
time-domain response, of a one-pole
    system, 655
time-domain system analysis, 164–203
time-limited signals, 361, 410–411
    exercises, 492
time-scaling property, 268, 330
time-shifted signal, 515
time-shifted unit-step function, 38
time-shifting property, 265, 267–268, 272,
    273, 277, 317, 378, 381, 418, 419, 422
tonal sound, 15
tone, 15
Toricelli's equation, 123, 146, 147, 148
total area property, 265
total harmonic distortion (THD), 253–255
total system response, 430–431
tracking errors in unity-gain feedback systems
    exercises, 639–640
trajectory, 764
transfer function, 182–183, 358, 775–777, 787
    common kind of, 385
    for discrete-time systems, 204, 408
    exercises, 633–634
    frequency response and, 184, 205–207
    using time-shifting property, 418–419
transform method comparisons, 430–434
transformation, 6
transformations, 778, 787

transient response, 624
transmitted carrier, 741
transmitter, 1
travelling-wave light amplifier, 610
triang window function, in MATLAB, 725
triangular pulses, convolution of, 178
trigonometric form, of the CTFS, 233
trigonometric Fourier series, 237
truncated ideal impulse response, 713–718
Tukey, John, 322
tuning, a radio receiver, 741
two-dimensional signal, images as, 549
two-finite-pole lowpass filter, 391
two-finite pole system, 390
two-input, two-output system, state-space
    analysis of, 769–772
two-input OR gate, in a digital logic system,
    154–155
two-pole highpass filter, 540
two-pole system. *See* second-order system
two-sided Laplace transform, 380
two-stage active filter, frequency response
    of, 538–539
type-one Chebyshev bandstop filter, 687–688
type-one Chebyshev filter, 687
type-two Chebyshev filter, 687

U

unbounded response, 138–139
unbounded zero-state response, 152
uncertainty principle, of Fourier
    analysis, 269
undamped resonance, 393
underdamped case, 627
underdamped highpass filter, 391, 392
underdamped low pass filter, 391
underdamped system, 144
undersampled signal, 453
undersampling, ambiguity caused by, 465
uniform sampling, 80
unilateral Laplace transform, 379–385
unilateral Laplace transform integral,
    exercise, 399
unilateral Laplace-transform pairs, 382
unilateral z transform, 423–424
unilateral z-transform properties, exercises,
    438
unit discrete-time periodic impulse or
    impulse train, 88
unit doublet, 33
unit function, 239
unit impulse, 29–30
unit pulse response, of an RC lowpass filter,
    170
unit ramp function, 26–27
unit rectangle function, 33, 264, 272
unit rectangles, convolution of, 180
unit sequence, defined, 96

unit step, integral relationship with unit
    ramp, 27
unit triangle function, 180
unit triplet, 33
unit-area rectangular pulse, 28
unit-area triangular pulse, 29
unit-impulse function, 86–87
unit-pulse response, 170
unit-ramp function, 88–89
unit-sample function. *See* unit-impulse
    function
unit-sequence function, 87–88
unit-sequence response
    as accumulation of unit-impulse
        response, 201
    impulse response and, 201
    at three outputs, 559
    using the z transform, 654–655, 656–657
    in the z domain, 654
unit-sinc function, 238
unit-step function, 24–25, 29
unit-step response, 624–627
    of an RC lowpass filter, 176
    of a one-pole continuous-time system, 655
unity-gain feedback systems, tracking errors
    in, 618–621
    exercises, 639–640
unity-gain system, 618
unstable digital filter, avoiding, 708
upfirdn function, in MATLAB, 726
upsampling, 483

V

value, returned by a function, 20
value sampling, compared to area
    sampling, 668
vector of state variables, 766
voiced sound, 15
voltage divider, RC lowpass filter as, 533
voltage gain, 604
voltage response, determinants of, 129
voltage signal, ASCII-encoded, 5
voltage-current relationships, for resistors,
    capacitors and inductors, 533
von Hann or Hanning window function,
    715, 717

W

water level
    differential equations for, 123
    versus time for volumetric inflows, 10
weight, of an impulse, 30
wideband FM spectrum, with cosine
    modulation, 752–753
window function, 472–473
window shapes, 715
windowing, 473
windows, exercise, 497

Z

z transform, 406–432
    analysis of dynamic behavior of discrete-time systems, 650
    existence of, 410–413
    as a generalization of DTFT, 407
    of a noncausal signal, 413–414
    of the state transition matrix, 786
    of a unit-sequence response, 655–656
z transform pairs, 413–416
z-domain block diagram, of a system, 651
z-domain differentiation property, 416
z-domain response, to a unit-sequence, 655
z-transform pair, 407
z-transform properties, 416
z-transform-Laplace-transform relationships, 660–662
zero, of a Laplace transform, 364
zero padding, 334
zero-input response, 122, 149–150
zero-order hold, 460
zero-state response, 122, 130, 562
    of a discrete-time system, 787–790
    of a system, 629
    to a unit-sequence excitation, 153
zpk command, in MATLAB, 393
zpkdata command, in MATLAB, 394

$$\delta(t) = 0 \;,\; t \neq 0$$
$$\int_{t_1}^{t_2} \delta(t)\, dt = \begin{cases} 1 \;,\; t_1 < 0 < t_2 \\ 0 \;,\; \text{otherwise} \end{cases}$$

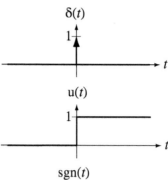

$$u(t) = \begin{cases} 1 \;,\; t > 0 \\ 1/2 \;,\; t = 0 \\ 0 \;,\; t < 0 \end{cases}$$

$$\mathrm{sgn}(t) = \begin{cases} 1 \;,\; t > 0 \\ 0 \;,\; t = 0 \\ -1 \;,\; t < 0 \end{cases}$$

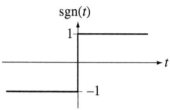

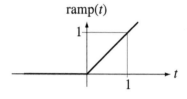

$$\mathrm{ramp}(t) = \begin{cases} t \;,\; t \geq 0 \\ 0 \;,\; t < 0 \end{cases}$$

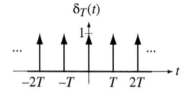

$$\delta_T(t) = \sum_{n=-\infty}^{\infty} \delta(t - nT)$$

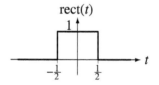

$$\mathrm{rect}(t) = \begin{cases} 1 \;,\; |t| < 1/2 \\ 1/2 \;,\; |t| = 1/2 \\ 0 \;,\; |t| > 1/2 \end{cases}$$

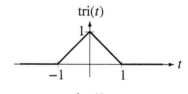

$$\mathrm{tri}(t) = \begin{cases} 1 - |t| \;,\; |t| < 1 \\ 0 \;,\; |t| \geq 1 \end{cases}$$

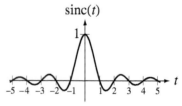

$$\mathrm{sinc}(t) = \frac{\sin(\pi t)}{\pi t}$$

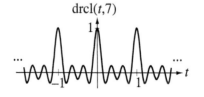

$$\mathrm{drcl}(t, N) = \frac{\sin(\pi N t)}{N \sin(\pi t)}$$

$$\delta[n] = \begin{cases} 1, & n=0 \\ 0, & n \neq 0 \end{cases}$$

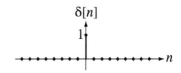

$$u[n] = \begin{cases} 1, & n \geq 0 \\ 0, & n < 0 \end{cases}$$

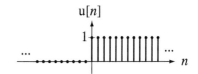

$$\operatorname{sgn}[n] = \begin{cases} 1, & n > 0 \\ 0, & n = 0 \\ -1, & n < 0 \end{cases}$$

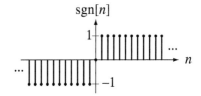

$$\operatorname{ramp}[n] = \begin{cases} n, & n \geq 0 \\ 0, & n < 0 \end{cases} = n u[n]$$

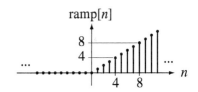

$$\delta_N[n] = \sum_{m=-\infty}^{\infty} \delta[n - mN]$$

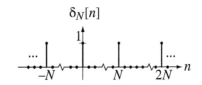

CTFT

$$\delta(t) \xleftrightarrow{\mathscr{F}} 1 \ , \ 1 \xleftrightarrow{\mathscr{F}} \delta(f) \ , \ \delta_{T_0}(t) \xleftrightarrow{\mathscr{F}} f_0 \delta_{f_0}(f)$$

$$u(t) \xleftrightarrow{\mathscr{F}} \frac{1}{2}\delta(f) + \frac{1}{j2\pi f}$$

$$\sin(2\pi f_0 t) \xleftrightarrow{\mathscr{F}} \frac{j}{2}[\delta(f+f_0) - \delta(f-f_0)]$$

$$\cos(2\pi f_0 t) \xleftrightarrow{\mathscr{F}} \frac{1}{2}[\delta(f-f_0) + \delta(f+f_0)]$$

$$\text{rect}(t) \xleftrightarrow{\mathscr{F}} \text{sinc}(f) \ , \ \text{sinc}(t) \xleftrightarrow{\mathscr{F}} \text{rect}(f)$$

$$\text{tri}(t) \xleftrightarrow{\mathscr{F}} \text{sinc}^2(f) \ , \ \text{sinc}^2(t) \xleftrightarrow{\mathscr{F}} \text{tri}(f)$$

$$e^{-at} u(t) \xleftrightarrow{\mathscr{F}} \frac{1}{j\omega + a} \ , \ \text{Re}(a) > 0$$

$$\omega = 2\pi f$$

DTFT

$$1 \xleftrightarrow{\mathscr{F}} \delta_1(F) \ , \ \delta[n] \xleftrightarrow{\mathscr{F}} 1 \ , \ \delta_{N_0}[n] \xleftrightarrow{\mathscr{F}} (1/N_0)\delta_{1/N_0}(F)$$

$$u[n] \xleftrightarrow{\mathscr{F}} \frac{1}{1-e^{-j2\pi F}} + \frac{1}{2}\delta_1(F)$$

$$\sin(2\pi F_0 n) \xleftrightarrow{\mathscr{F}} \frac{j}{2}[\delta_1(F+F_0) - \delta_1(F-F_0)]$$

$$\cos(2\pi F_0 n) \xleftrightarrow{\mathscr{F}} \frac{1}{2}[\delta_1(F-F_0) + \delta_1(F+F_0)]$$

$$u[n-n_0] - u[n-n_1] \xleftrightarrow{\mathscr{F}} \frac{e^{-j\pi F(n_0+n_1)}}{e^{-j\pi F}}(n_1-n_0)\,\text{drcl}(F, n_1-n_0)$$

$$\text{tri}(n/w) \xleftrightarrow{\mathscr{F}} w\,\text{drcl}^2(F, w)$$

$$\text{sinc}(n/w) \xleftrightarrow{\mathscr{F}} w\text{rect}(wF) * \delta_1(F)$$

$$\alpha^n u[n] \xleftrightarrow{\mathscr{F}} \frac{1}{1-\alpha e^{-j\Omega}} \ , \ |\alpha| < 1$$

$$\Omega = 2\pi F$$

Laplace Transform

$$\delta(t) \xleftrightarrow{\mathcal{L}} 1 \quad, \quad \text{All } s$$

$$u(t) \xleftrightarrow{\mathcal{L}} \frac{1}{s} \quad, \quad \text{Re}(s) > 0$$

$$t\,u(t) \xleftrightarrow{\mathcal{L}} \frac{1}{s^2} \quad, \quad \text{Re}(s) > 0$$

$$t^n e^{-\alpha t}\,u(t) \xleftrightarrow{\mathcal{L}} \frac{n!}{(s+\alpha)^{n+1}} \quad, \quad \text{Re}(s) > -\alpha$$

$$e^{-\alpha t} \sin(\omega_n t)\,u(t) \xleftrightarrow{\mathcal{L}} \frac{\omega_n}{(s+\alpha)^2 + \omega_n^2} \quad, \quad \text{Re}(s) > -\alpha$$

$$e^{-\alpha t} \cos(\omega_n t)\,u(t) \xleftrightarrow{\mathcal{L}} \frac{s+\alpha}{(s+\alpha)^2 + \omega_n^2} \quad, \quad \text{Re}(s) > -\alpha$$

z Transform

$$\delta[n] \xleftrightarrow{\mathcal{Z}} 1 \quad, \quad \text{All } z$$

$$u[n] \xleftrightarrow{\mathcal{Z}} \frac{z}{z-1} = \frac{1}{1-z^{-1}} \quad, \quad |z| > 1$$

$$n\,u[n] \xleftrightarrow{\mathcal{Z}} \frac{z}{(z-1)^2} = \frac{z^{-1}}{(1-z^{-1})^2} \quad, \quad |z| > 1$$

$$n^m \alpha^n\,u[n] \xleftrightarrow{\mathcal{Z}} (-z)^m \frac{d^m}{dz^m}\left(\frac{z}{z-\alpha}\right) \quad, \quad |z| > |\alpha|$$

$$\alpha^n \sin(\Omega_0 n)\,u[n] \xleftrightarrow{\mathcal{Z}} \frac{z\alpha \sin(\Omega_0)}{z^2 - 2\alpha z \cos(\Omega_0) + \alpha^2} \quad, \quad |z| > |\alpha|$$

$$\alpha^n \cos(\Omega_0 n)\,u[n] \xleftrightarrow{\mathcal{Z}} \frac{z[z - \alpha \cos(\Omega_0)]}{z^2 - 2\alpha z \cos(\Omega_0) + \alpha^2} \quad, \quad |z| > |\alpha|$$